Urban Energy Transition

Urban Energy Transition

From Fossil Fuels to Renewable Power

Edited by

Peter Droege

University of Newcastle, NSW, Australia
World Council for Renewable Energy

ELSEVIER

AMSTERDAM • BOSTON • HEIDELBERG • LONDON • NEW YORK • OXFORD
PARIS • SAN DIEGO • SAN FRANCISCO • SINGAPORE • SYDNEY • TOKYO

Elsevier
Linacre House, Jordan Hill, Oxford OX2 8DP, UK
Radarweg 29, PO Box 211, 1000 AE Amsterdam, The Netherlands

First edition 2008

British Library Cataloguing in Publication Data
Droege, Peter
Urban energy transition: form fossil fuels to renewable power
1. Cities and towns – Energy consumption 2. Greenhouse gas mitigation 3. Greenhouse gas mitigation –
Government policy 4. City and town life – Environmental aspects 5. City planning – Environmental aspects
I. Title
333.7'9'091732

Library of Congress Catalog Number: 2007941706

ISBN: 978-0-08-045341-5

For information on all Elsevier publications
visit our website at books.elsevier.com

Typeset by Charon Tec Ltd (A Macmillan Company), Chennai, India

Printed and bound in Hungary

08 09 10 11   11 10 9 8 7 6 5 4 3 2 1

Cover design by Peter Droege
About the cover image: overall household energy requirement in Greater Sydney and surrounding local
government areas, scaled from red, at 270 Gigajoules per person a year, to green, at 150 GJ. The map represents
energy used in households directly as well as energy embodied in goods and services consumed. It roughly cor-
relates with income and lifestyle, and socio-demographic makeup of urban, suburban and rural communities. It
corrects the popular assumption that the denser, central city areas are more energy-wise. As long as commercial
energy is derived primarily from fossil-fuel combustion big city life also represents the highest source of greenhouse gas emissions in geographic terms. *Map generated by Chris Dey for the chapter 'Direct versus embodied energy –
the need for urban lifestyle transitions' by Manfred Lenzen, Richard Wood and Barney Foran.*

With gratitude to the Renewable Energy & Energy Efficiency Partnership (REEEP) for its assistance in
developing this volume.

Contents

Urban Energy Transition: An Introduction

PETER DROEGE
University of Newcastle, Australia

Around the world, cities and urban communities plant the seeds to a great transformation, unprecedented in history in its reach and magnitude. The growing footprint of contemporary and especially wealthy cities is well documented. Their carbon belching thirst for fossil fuels, their demand for an ever rising stream of global resources, their contribution to land clearing, second to fossil fuel combustion in concentrating greenhouse gases in the atmosphere – all these conditions are not only painfully understood, but have begun to drive important shifts in urban energy and environmental policy making. Urban communities in developing countries confront energy transition challenges that are only superficially different from those of the more industrialized world. Their challenge is to stabilize a growing hunger for secure energy supplies, avoid polluting and wasteful industries and power systems, and – not unlike their more developed sisters – shun development directions that hardwire costly and inefficient mobility patterns for generations to come. Enlightened community leaders and governments are sharply attuned to the need to enhance human health and urban livelihood, and construct bridges of access, equity and empowerment. They seek to nurture more vital and autonomous rural and peri-urban regions, to help reduce and even deflect migrational pressures, but they also hope to craft new development directions that radically depart from the congested coal-and-petroleum path that has been blazed by the economically dominant world, ever since the industrial revolution as referred to by both Lenzen *et al.* and Kenworthy in this book.

Energy, Cities, Evolution and Innovation

Fundamental features in the relation between urban life and energy use are common to Southern and Northern cities. While urban settlements are often more transport-energy efficient – i.e. when looking at per capita petroleum energy expended – than suburban or peri-urban areas, given that nearly all motorized movements are oil based – their overall energy requirements soar high above those of less urbanized communities as referred to by both Lenzen *et al.* and Kenworthy in this book. Only in this transport sense can cities rightfully be seen to be more 'energy-wise', by affording greater functional densities and hence more compact, concentrated or combined land-use patterns. Here lies a civilization challenge that can only be satisfactorily resolved in a combination of efficiency, cultural innovations and the shift to renewable power. The vitality of cities – their very power as a cultural concentrator, market, and production-consumption engine – is based on the very

1

densities, resulting synergies and serendipitous encounters they engender. Since their rise in the Bronze Age cities have been central to advanced civilization's conception of cultural and economic accomplishment and supremacy. As a corollary to their exalted status in the architecture of post-agrarian societies they also engender far greater levels of energy consumption when looking at the entire energy requirement spectrum – as described by Manfred Lenzen and his colleagues in their chapter on urban energy embodiment. This elevated energy and general resource intensity of urban economies and lifestyles is hence likely always to have been a basic feature of cities – long resulting in deleterious impacts on surrounding countryside and forest areas. Yet the great difference today is not merely the very proliferation of cities against the background of the fossil-fuel charged population explosion, and not alone the conspicuous and inconspicuous levels in globalized forms of consumption – but also the basic fact that their overwhelming commercial energy input is fossil, with all of its devastating consequences.

Positive signals arise from the current global energy conundrum, common to cities in more and less wealthy regions of the world. The present energy transition triggers a technological and logistical innovation wave, affecting areas as disparate as personal and public transport systems; efficiency in computing, industrial processes and building design; innovations in facility construction and use; fiscal, funding and investment models; and in renewable energy generation, storage and management itself. This wave has reinforced many governments in their nascent forms of action, and mobilized business leaders in venture capital finance, equity funds and infrastructure investment across Europe, India, China or the United States – yielding many new companies and a net growth in jobs and flow-on in economic benefits. Yet despite numerous and clear signals of progress and advance into new and sustainable directions of development, the folklore of traditional international energy policy has it that this type of innovation is too expensive or otherwise beyond reach of the developing world. Such statements risk being read as promoting outmoded infrastructures, rather than as genuine concern for improving the livelihood of the poor. Why should the developing world not avail itself of locally sourced, unlimited, non-polluting and income generating means of indigenous renewable energy generation – rather than fall into the trap of antiquated models of electricity supply that are cheap only because their external costs have been discounted, they are heavily subsidised, or both?

A fresh generation of urban community leaders embraces new – and newly rediscovered – approaches to city planning and design, with local energy liberation in both developing and developed parts of the world in mind. Supported by experts, leading businesses and international networks they move to bring about a range of related community development, production and consumption level and other economic changes of particular significance to life in cities, city regions and states. To be sure: earlier urban energy technology transitions have been dramatic, too – and none more so than the rapid, epic spread of coal and oil combustion that underpinned urban development as we know it, from the late eighteenth, during the nineteenth and especially throughout the twentieth centuries, yielding our present resource and climate predicaments. Over a short time span this massive, complex energy revolution spawned global electrification, the meteoric rise of motors and machines and the very age of mechanization and automation they represent, of telecommunications and petroleum, coal and gas powered transport on the ground, across the seas and through the air. All these innovations helped boost the primacy and spread of cities, and of urban economies and their inordinately energy-draining lifestyles.

The currently commencing energy transition is different. For one, a far greater level of collective consciousness underpins it: today's future choices seem clearer to us than what they were to observers and decision-makers one or two centuries ago. The need to change direction is more widely appreciated today than during earlier times, driven by manifest constraints as much as new opportunities. Our recent, fossil-fuel charged past was accompanied by extraordinary futurist visions – the delirious genre of the cities-of-tomorrow, particularly powerful during the first half of the twentieth century. Machine inspired mirages of modern things-to-come projected grand futures and dazzled champions of urban change. Yet in this very pursuit advanced civilizations stumbled into a new global reality few dared to fathom or reflect on. The implications of change did not reach public policy discourse until at least half a century after the great fossil-fuelled growth visions began to enthral, inspiring aspirations of ever-rising prosperity around the world. It was only in the 1970s when the global urbanization wave became too massive to ignore, and the unpleasant prospect of a sprawling, crumbling civilization started to seriously spoil the view of future horizons – the very time when the first oil supply shock hit, more than a decade after the inevitable oil decline had first been mooted by an industry insider (Droege 2006). Fears over the looming climate catastrophe had been publicly expressed for a full decade longer: for instance, in the proto-Gore educational film produced by Frank Capra for Bell Labs, *The Unchained Goddess* (Capra and Hurtz 1958). Yet, even in these very moments of lucidity the link between the worldwide energy revolution driving progress throughout the twentieth century, the triggering of an unprecedented population explosion, the unfolding global urban reality and the stability of the planet's ecosystem had not been appreciated. Since that time, future urban development scenarios have become much more tangible, and unnervingly so.

New Perspectives: Lifting the Gaze from the Ground Below

This transition also differs in its philosophical and scientific outlooks, and in its very paradigms of progress. The all-consuming preoccupation with fossil and nuclear energy is rooted in the spirit of the late eighteenth and nineteenth centuries, and the popular obsession with the rising science of geology, the great new frontier of that era, caricatured so succinctly by Bill Bryson in his *Short History of Nearly Everything* (Bryson 2003). The intensive study of underground rock formations triggered a massive, mole-like pursuit of mineral mining. The nineteenth century still rules: mining very much colours mainstream discourse on energy today: oil wells, uranium deposits, coal seams describe sources of 'our energy'. And, within this family of historical and philosophical affinities, methane hydrate, oil shale and tar sands are frequently presented as 'new' energy resources – a sad irony since they only represent the fading days of the golden age of energy mining, and the inexor-able end of the fossil incineration age. In anticipation of the inevitable, calls for clamping down on the global resource conflagration have grown more agitated. Advanced renewable energy systems are finally beginning to be pursued on a global scale. Their deployment is vastly different from the mining and processing of combustible carbonic carcasses. In almost all its guises renewable power is stellar in nature: derived from our own star, the Sun. At the closest distance of 146 million kilometres it projects enough power to heat the entire surface and atmosphere of our planet by 300 degrees Celsius (C), that is, from 0 degree kelvin (K) to some 18 degrees C – and rising. It gives rise to plant and animal life. It drives the worldwide water cycle and causes atmospheric air masses to shift at a global scale. It has been calculated that wind power alone – gauged at a height of some 80 metres and counting only those 13% of calculation points yielding good wind speeds – is capable of generating some five to eight times the electricity now generated by all commercial sources combined (Archer and Jacobson 2005).

Structural Differences Between Non-Renewable and Renewable Power Systems

Direct renewable energy sources – such as sunshine, wind, water and wave flow, or the cycles of ebb and tide – are freely available and hence not tradable commodities until converted into electricity, hydrogen or heat. Adding to this unique strength – seen as weakness in a primary resource and commodity-geared world – are continually lowered conversion or realization costs. End-user prices drop dramatically with expanding market application and uptake. By contrast, conventional and geological, i.e. fossil and nuclear, sources such as oil, gas or radioactive ore are not only inevitably rising in extraction cost and market price but also trapped in their dual character as mineral commodities and energy sources, greatly complicating and even polluting policy making. Paradoxically and tragically, the very shortcoming of rising scarcity and price exerts enormous temptations yet we press on with their promotion and use. The very innovation of carbon trading confirms this: it bestows the mythical status of a precious and valued commodity upon a set of toxic waste products: the greenhouse gas emerging from smoke stacks, tail pipes and ravaged forest soils.

In a theoretical and macro-economic sense, overall hardware or plant costs may be comparable between non-renewable and renewable systems – but these are subject to such greatly different investment and depreciation cycles, as well as distribution and ownership patterns that it is impossible to draw such comparisons with any practical purpose. Renewable and efficiency technology engenders and thrives on locally and community based arrangements, and is most readily spawned, deployed, owned, and managed in distributed ways. Herein lies a great challenge to a smooth energy transition: so many established ownership patterns, institutional arrangements and vested interests feel at risk, and need to be considered and engaged in effecting change.

Short-term and narrow individual, corporate or public revenue interests aside, the incentives to change are overwhelming across both net fossil fuel producing and importing countries. National economies from Australia to Malaysia, Russia and Venezuela are at great risk through their overdependence on oil, coal gas and uranium revenue streams – while some 40 less and least developed nations have to outlay more in oil imports each year than their national economies yield in available revenue (Scheer 2006). Yet there are those who still advocate economically and environmentally costly fossil and nuclear solutions for the 'Third World', whether based on endogenously or exogenously extracted sources, as a path from poverty over locally and community controlled, autonomous, sustainable solutions. Indeed, these would be unattractive to the purveyors of large-scale, conventional power systems: they are locally decentralized and hence outside central supply government control, or unprofitable to transnational operations.

A Revolution in Local Practice: Efficiency and Renewable Generation

For close to half a century humanity has marched into its urban destiny with eyes at least half open. The energy connection has become popularly understood, and entered the mainstream policy discourse. Yet while the transformation to renewable technology practice has long been possible and also widely envisioned, the political restrictions on energy change have also become clear, and many painfully so. Still, many recent successes, most of them local or state-level innovations based on renewable systems and efficiency mandates, open up great perspectives of optimism, scope for hope and the opportunity for urban renewal, genuine prosperity and equity in livelihood generation – and even the reversal of some mistakes of the past.

Many communities and local leaders sense a mounting degree of anxiety – and some even despair – about a planet trapped in a vicious vortex of manmade, increasingly chaotic climate destabilization dynamics, a fatal affliction primarily induced by the widespread burning of coal, oil and natural gas. The pronounced and seemingly sudden arrival, accelerating speed and increasingly devastating reach of this cycle begins to unnerve urban communities. To many, cities are the home, hope and expression of their highest aspirations – and yet each of the fossil- or uranium-powered, high consumption cities is a massive atmospheric destruction agent or contributes to radiation risks. Each delightful and proud city serves its local communities – and also acts globally in a process of pollution distribution, through the embodied and direct resource and waste streams it engenders. Efforts to resolve such conflicts and contradictions abound, and the vectors of an urban evolution become apparent. Countless signs of hope and practical action appear around the globe. While many national and international institutions are either still oblivious to the problem or agonisingly slow to act because of fear of offending some powerful force, cities and towns far more readily express the profound local concern manifested in a multitude of community-based and municipally supported commitments and programs.

There is rising optimism that the once seemingly limit-free, now increasingly risky and costly fossil fuel regime can be replaced entirely by renewable systems. As a sign of this growing confidence, and going beyond older, hypothetical studies to prove this potential, three German renewable systems producers have recently demonstrated its practical realization, at least in a technological sense. *Enercon, Solar World* and *Schmack Biogas* – with support from Germany's Federal Ministry for Economic Development – linked up 25 decentralized wind, solar, biomass and hydropower facilities in a controlled, small-scale simulation of a conventional power supply system. The experiment demonstrated that this hybrid source approach can replace the capacity of large generators, meeting demand without the fluctuations often feared with renewable systems relying on intermittent sources such as wind or sunshine (Informationszentrum 2007, Eurosolar 2007). This initiative is not an isolated innovation but serves to exemplify the advanced state of evolution, and the trajectory of a rapid urban evolution already in train.

In Austria, villages and towns have long pursued increasing degrees of urban renewable energy autonomy. Solar hamlets and bio-fuelled regions have a well-established tradition in the country's rural regions. Across energy-conscious Switzerland, citywide efficiency drives also have been long maintained with sophistication and success. Davos, host of the World Economic Forum and international luxury ski destination, ranks among the more visible urban communities that aspire to reduce local carbon footprints. Germany's extraordinarily successful renewable energy feed-in legislation introduced in the mid-1990s has been applied to 47 other countries. It helped bring about local and building-based integration of solar power generation capacity, and helped revitalize agriculture through the introduction of regional energy farming (Alt 2007). During the late 1990s it was supported with a 100 000 Solar Roof program, implemented nation-wide well before its completion date. Beyond building integration and roof installation, clusters of small towns in Bavaria have recently formed cooperative ventures to host some of the world's largest photovoltaic farms. Also in Germany, the small farming town of Jühnde has become famous for becoming entirely fossil-energy independent – powered through locally sourced and produced bio-fuel for both stationary and transport use.

After significant improvements in the years leading up to and continuing during the 1980s, efficiency gains had nearly stalled across many industrialized countries, and in some instances even reversed during the 1990s. But now, this appears to change again. Numerous cities have begun to discuss and some even set greenhouse-gas emissions targets, preparing staged-reduction frameworks. Veteran leaders in municipal innovation such as Sacramento, California, have acquired – *municipalized* – state or private energy suppliers to control their own destinies. Here, community efforts in the 1980s led to the early retirement of a relatively new nuclear power plant and the seizing of greater consumer control over the countywide Municipal Utility District – the sixth largest customer-owned electric utility in the United States, since then investing in cogeneration, natural gas, solar PV and wind power production. State governments in the United States from California to Massachusetts have developed a range of efficiency and renewable energy initiatives. To support such measures New Jersey even legislated, in mid-2007, an 80% emissions reduction by 2050. And also in mid-2007, the Delaware Sustainable Energy Utility (SEU) has been established as a bond issue supported public policy institution driving local innovation in efficiency and renewable generation (SEU 2007).

State and local utilities and distributed energy service companies are being introduced at a record rate around the world. In Denmark, the best example for successfully established, locally owned renewable energy assets is the city of Copenhagen. It is the new home to one of the world's largest cooperatives, with thousands of individual and corporate shareholders holding personal financial stakes in the successful planning, design and building of the city's proud new offshore wind farm, *Middelgrunden*. In the regional and local regulatory sense, dramatic innovations take place, too. The city of Barcelona introduced a Solar Ordinance in the 1990s, requiring that all new and refurbished apartments and homes derive 60% of their hot water, and some other energy use, from solar sources. The model, originally developed for the city of Berlin, proved so successful that it has been adopted first by dozens of cities across Spain and is now a broadly applied principle. Chicago, one of the renewable city leaders, legislates for 'cool roofs' while over the past decade New York has become home to some of the most advanced renewable energy buildings found in dense inner cities today. Partial roof greening has become mandatory in Tokyo in the city's battle against the heat island syndrome.

In China some real property tycoons compete with commercial towers that are energy independent or even net energy producers. The Chinese industrial new town initiative of Dongtan, announced by United Kingdom engineers Arup in 2006 for an area north of Shanghai, is said to become the world's first renewably powered industrial city, while Abu Dhabi's renewable energy company Masdar unveiled in May 2007 plans for a six square kilometre, 'zero carbon/zero waste' new city project to be built by 2009, designed by London's Foster and Partners. Elsewhere across the developing world a growing number of initiatives of empowering cities, informal settlements of the poor, and rural communities with renewable energy capabilities can be found. Many are linked to bilateral aid initiatives or multilateral vehicles, from non-profit or quasi-non-governmental networks such as the Global Village Energy Partnership (GVEP), the Renewable Energy and Energy Efficiency Partnership (REEEP), the United Nations Development and Environment Programmes (UNDP, UNEP), and even the World Bank. The most promising urban efforts are locally or regionally conceived and applied – yet like the internationally supported programs, these still remain exceptional. The rural sector, in many ways virtuously deprived of early, now increasingly unsustainable electrification schemes, fares better than urban communities in benefiting from renewable energy innovations. Bangladesh is home to a world's largest micro-lender, Grameen Bank, whose founder Muhamad Yunis received the recent Nobel Peace Prize. Grameen Shakti is its solar power lender and provider, empowering tens of thousands of women as heads of households and leading rural community members across several developing countries.

Premise and Scope of this Book

This book is a selective compilation of working concepts, technological directions and country-specific organizational perspectives – aspects that promise to yield a better systems-based understanding of policy frameworks and action agendas. The book features basic and advanced, practical and principled thought, across technology, carbon emissions methods, community engagement strategies and various urban renewable energy and efficiency implementation techniques. The aim is to show everyday energy transitional practice to be distinct from, and yet intimately linked to, the various established realms of the conventionally powered urban development culture. The book also focuses on urban aspects of efficiency gains in embodied, supplied and end-use energy – not only because of the near-term scarcity of renewable power, but also in order to help make its widespread introduction both affordable and achievable. And finally, the book attempts a global – or at least cross-cultural – perspective, examining dimensions of universal interest.

The unfolding transition spawns a new field of discourse. Informed by the dynamic forces of a global urban energy shifts, it is beginning to take shape across loosely shared clusters of discourse, policy patterns and related subgenres of technological innovation, social action research and critical writing. Many disciplines are linked – as diverse as economics, community development, architecture and urban design, transport planning, energy policy, renewable and efficiency technology – to name but a few. The transformation of the very energy base of cities is the focus of a growing research, practice and policy activities – in transport, better building design, higher levels of employment, more meaningful cultural development and other means of strengthening local livelihood.

The techniques and dynamics that characterize this change have moved onto the stages of public policy and civic action. They articulate a nexus of not always resolved or even clearly-defined issues, yet usefully linking concerns that for too long have been thriving in splendid sectoral isolation. This segregation has been nurtured during a hundred years of growing separation between energy issues and virtually all other urban service dimensions: energy systems and their supply industries, support businesses, finance arrangements and regulatory controls on the one hand and the worlds of urban planning and design, transport infrastructure, building construction, property development and civic life in general on the other. In urban-energy-based efforts across the world, community development, health and prosperity objectives loom large. In the not-too-distant past urban communities and civil society were insulated from energy questions, sheltered by the now failing model. Periodic crisis-triggered moments of alertness aside: to the majority of its citizens cities appeared as supplied by a technologically obscure, remote, seemingly limitless and efficient, and low risk set of sources – when ignoring petrol price hikes, air smog alerts and the occasional nuclear plant scare. At the same time, beneath the perfect façade, powerful industrial and political forces engineered the transition from a low emission to a high pollution society, in the name of progress for all. And this transition was packaged in enticing visions of an improbably yet sincerely promised future, a future based on limitless growth, infinite fossil and mineral resource streams and an endless capacity to pollute with impunity. The great, much-heralded globalization project of the late twentieth century is founded on this very illusion.

The book sets out to explore urban agenda horizons at this point in time, in the years of the Peak, identified by critical expert organizations as 2006 and 2010, respectively, as the apex of the global oil production and consumption wave (Campbell 2005, Schindler and Zittel 2007). The aim of this exploration is to identify key domains and features of the policy and practice landscape that serves as the very setting for this historical phenomenon. The background to this book is not only on the transformation of the well-established energy base of contemporary urban systems, but also on the impending end of the most fundamental drivers of modern cities: the great fossil fuel revolution of the past century. The book affords glimpses into key areas of this transformation, and gives voice to an eclectic chorus, composed of veteran policy experts in this transformation; visionary innovators; and a large cohort of other new and devoted actors and agents of change, working on the ground, in local governments, laboratories, consultancies, across countries and world regions at various levels of development.

Recapitulating the Features of an Urban World at Risk

The first decade of the twenty-first century awakens to an urban world at risk – to a higher degree than ever before, despite widespread euphoria about urban prosperity across the industrialized countries. Climate hazards, and, given current depletion trends, inevitably arising fossil fuel and uranium supply uncertainties, combine with widespread and systemic environmental damage, pollution and mounting health costs, to undermine urban civilization. Cities in developing countries are in an as precarious condition as urban agglomerations in advanced industrial states are: for different reasons and to varying degrees – but all due to their overwhelming dependence on a narrow set of declining and damaging set of energy systems.

The world economy is decidedly urban. Financial flows are organized in city networks: here are the centres of global and local economic transactions. Cities are central settings for social control, political discourse and cultural exchange. Here leading social images and ideals are produced and packaged, realities assembled and aspirations articulated. Throughout history the major centres dominated the political life of their respective eras. All globally dominant political systems have been engendered in urban societies and represented in their collective spaces and public institutions – shaping these in turn, to paraphrase Winston Churchill's 'We shape our buildings, and afterwards our buildings shape us.'

But cities, their form, economies and growth dynamics are also defined by the energy system dominating their era. And increasingly, the manner in which both interact defines the risk profile of our age, any given city, nation or that of the global system of political, cultural and trade relations. International exchange, spreading urban networks or eras of large and dominant cities are not new – primary cities from eighteenth century BC Babylon to eighteenth century AD London are testimony to this (Droege 2006). But the sheer explosion of urban development and the emergence of super-sized cities as a worldwide phenomenon are unprecedented. The growth of the world population's urban share in the second half of the twentieth century was rehearsed in the earlier expansion of a number of cities across the United Kingdom, the United States and elsewhere. From the 1970s on, a burgeoning research literature genre formed to give meaning to these and related phenomena (Hall 1977; Doxiades and Papaioannou 1974; Friedman and Wulff 1976), described as *world cities, global cities, mega-cities* or referred to as the *global urban system*, all signifying the rule of markets, corporate control, seats of power, home to security agencies and, above all, innovation in surface, air and sea transport, defence and the rule of advanced telecommunications.

Cheap and abundant coal and petroleum power have driven urban expansion, in a process that saw the structural transformation of many national economies, lengthening and securing fuel supply chains – and increasingly disconnecting urban areas from agrarian regions. Global city formation is largely a fossil major phenomenon. Startlingly ignored in both urban or energy engineering literature, the primary danger to worldwide prosperity and security is not excessive urban expansion or the population explosion, but the most plausible driver of this expansion: excessive, growing fossil fuel use and its damaging impact on the global biosphere – yet also its unavoidable, possibly sharp and uncontrolled decline (Droege 2006). The fossil revolution has delivered great prosperity to industrialized states, but the finite and localized nature of oil resources, coal mines and uranium deposits threatens it if no action is taken. Only forty oil fields supply 60% of worldwide oil consumption, and of these many lie in contested, even war ravaged areas of the world (Scheer 2006). High grade uranium is a particularly limited resource: were it technically even possible for it to substitute for oil and gas production it would be depleted within a decade or so.

Fossil Cities, the Contemporary Conundrum

The world relies on fossil sources for its energy and survival: the share is over 80% (followed by renewable energy – hydropower, solar, wind and combustibles, at least 13% – and nuclear energy, between 2.3 and 7%: this variance is due to inconsistencies in the definition of primary energy flows: see Lenzen *et al.* in this book, citing International Energy Agency data; other sources have named 85% fossil share in commercial energy supply – see USGS 2007). Ten years ago, across the cities of the Organisation of Economic Co-operation and Development member states, three-quarters of this fossil energy flow was accounted for by cities, for stationary and transport use (OECD 1995). Fossil fuels are the basic engine of the world economy, and cities are its physical articulation. Across the globe, transportation is essentially fossil fuel based, with virtually all motorized movements in the air, on sea, road or rail, oil or gas driven (Lenzen *et al.* 2003).

The mounting dependence on the advanced – and by definition fragile – artefact *city* is a major risk in itself, but the present energy crisis threatens the very foundation of modern urbanity. This risk reaches deeper than mere infrastructure dependencies. Hermann Scheer enumerates the multiple energy crises engulfing the world today: global poverty levels are tied to the fossil fuel economy – some 40 of the poorer countries spend more on petroleum imports than their export earnings; nuclear crises loom due to misguided hankerings for atomic power on the part of a number of developing countries; water depletion crises brought about by pandemic pollution, abuse and global-warming induced precipitation changes – and glacial melting. These are magnified by the enormous thirst of old-fashioned thermal power stations still prevailing today; health crises are endemic to most industrialized and technologically emerging societies through fossil-based air, water and soil pollution; and agricultural crises are looming because of the worldwide reliance on petrochemical pesticides, fertilizers, harvesting, distribution and processing systems. (Scheer 2006)

Depleting fuels

The literature alerting us to the threat of an imminent global natural gas and oil supply peak is sizeable, with central implications for cities (Campbell and Laherre 1995; Klare 2002; Heinberg 2003; Goodstein 2004; Campbell 1998, 2003, 2005). All actively consumed finite resources follow the bell curve of depletion; hence the neat in-tandem production peaks predicted to roll across the countries and regions of Europe, Asia Pacific and the Middle East. Liquid fuel supply reserves may stretch to the middle of this century; but the statistical plateau of global oil production – the combined peak of all wells – may already be behind us. Price increases make the mining of so-called unconventional and speculative resources – steam or CO_2 injection of declining oil wells, methane ice – *clathrate* – exploration, tar sand mining and oil shale production financially feasible. In addition to the mounting financial burden they bring, they are also environmentally costly, inherently water and energy intensive practices yielding diminishing return in the larger economical sense. While delivering profits for some they merely delay the inevitable. Projections of global oil production plateaus have not differed fundamentally since Marion King Hubbert presented his incisive observation half a century ago, accurately timing the US oil production peak for some

15 years later, 1970. Later, he predicted the global peak to occur by 2000 while a more recent estimate has been adjusted to 2010 (Hubbert 1956, 1971; Campbell 2005). The more recent Energy Watch Group commissioned study by Schindler and Zittel has not only placed the global oil peak at the year 2006, but also projected the total 'super-peak' of all fossil and uranium supplies combined to take place prior to 2015 (Schindler and Zittel 2007).

In historical terms, the global supply peak is with us, as even acknowledged by both OPEC and the International Energy Agency (European Tribune 2006/2007) – at a time when the global fossil fuel demand continues to rise, and a burgeoning global popula-tion has grown addicted to uninterrupted supply. Unless fundamental changes occur, and a transition to new and sustainable energy practices takes place, a rapidly widening supply gap is unavoidable, bringing about price surges and mounting military missions around the globe. The vast bulk of oil resources are limited to a shrinking number of brit-tle regions: the Middle East, Africa, and the Caspian Sea. And like natural gas or oil, coal is geographically limited: only six countries produce 90% of global coal reserves. The present drive into increasingly costly and damaging recovery methods only confirms that we have entered the dusk of the fossil fuel era, while continued and overwhelming fossil depend-ency increases the risk of a cataclysmic supply disruption to cities.

Urban Risks from Climate Change

When trends are projected to 2030, global energy demand increases by 60% – and 85% of the increase would be supplied by oil, gas and coal. Given the impending supply peak, this is a most unlikely scenario – but for a very long time such myths have been nurtured to keep consumption habits and policy frames unchanged; see, for example, the relatively recent, now superseded, positions of the International Energy Agency documented in its *2004 World Energy Outlook* (IEA 2004). Even if fossil fuel supplies did turn out to be magic-ally unlimited their continued and unfettered use has become increasingly unpalatable, due to the need to swiftly respond to anthropogenic global warming. If living standards and livelihoods are to be maintained and even improved neither 'clean-coal' nor new nuclear power can deliver sufficient action in the short time span required.

The fossil fuel era has spawned modern cities – and brought them to the brink of a dual calamity. As shown above, one risk is posed by finite supplies; the other by its combus-tion. Fossil fuel burning causes three quarters of the human-triggered warming trend of the earth's biosphere (IPCC 2007, page 5), and 25% is due to land clearing and soil disturbance. Tragically, it took a good century to arrive at the present level of commonly shared cer-tainty since Nobel Prize winning physicist Svante Arrhenius began to link the greenhouse effect first postulated by Joseph Fourier in 1827 to atmospheric carbon dioxide concentra-tions (Arrhenius 1896), and later, with colleague Arvid Högom, to the release of 'carbonic' waste gases through rampant coal combustion. This relative certainty is still giving rise to contestation and all kinds of attempts at obfuscating causal relations and delaying action.

Yet the most tangible results of this effect mask even more pernicious risks, such as a steep dive of the global economy, or the catastrophic impact on the oceanic thermodynamic and chemical regime, threatening abrupt climate change or the biological collapse of large stretches of oceans. Urban fresh water sources are being depleted at alarming rates, exacer-bated by rapidly retreating glaciers and snow pack, and may soon endanger urban centres, regions and settlements along mountainous regions across the Andes, Alps, Rockies, Himalayas and other frozen water cradles from Afghanistan to Peru, from India to Canada, and from Switzerland to the United States. The livelihood of hundreds of urban dwellers is eroded. Elsewhere, too, urban areas face an uncertain future where subterranean and surface water resources fail because shifting precipitation patterns overstretch frail fresh water resources, already stressed by generations of inefficiency, pollution and abuse. Early victims include cities in regions of the world as diverse as Australia, China or the United States. Unsustainable modes of water consumption in agriculture, industry, mining and households have only compounded the levels of freshwater depletion accounted for by the fossil fuel and nuclear power regime. The spectacular thirst of thermal electricity-generating plants – coal, oil and nuclear – epitomize the structural wastefulness of our inherited energy regime. The water uptake of a standard 500 megawatt coal fired power plant can rival that of a small city of 250 000.

Inundation, flooding, storm surge and hurricane damage are other dramatic and tangibly felt features of fossil emissions-induced urban costs. More frequent and increasingly severe weather events are not a distant, possible greenhouse hazard but have long been tracked as a statistical feature of doing business in the insurance industry, recorded in the books of all major reinsurers. Such global economic damage is said to have mounted ten-fold since 1950, with a good deal of the loss incurred in urban regions. Insurance payouts rose 60-fold in the United States (Mills 2005). Islands like Tuvalu have begun to disappear, while sea-level rise compensating measures focused on urban assets, from China to the Maldives to Italy, already absorb significant capital planning and construction budgets. By contrast, poorer exposed island nations or low lying coastal bound countries – and their frequently crowded cities and towns – are fiscally overtaxed when faced with any physical adaptation measures. Those least able to adapt are also often those apt to suffer most – and also least to be blamed for the climate calamity: their greenhouse gas equivalent emissions are lowest of all.

These changes are the most popularly reported today, but do not represent the most pernicious climate hazard to the fragile balance of urban life by far. While many climate change adaptation programs focus only on the most obvious emergency response techniques called for, the risks and already incurred costs of social and economic impacts are at least as profound. Economic risks include potentially massive and negative shifts in agricultural, trade and industrial productivity. Possible social costs encompass health damage such as rising annual and regional spikes in heat stress and weather related disease vectors; rising psychosocial stress – and massive demographic shifts such as climate migration: an impending global diaspora of weather shift refugees. The precise nature of change is subject to as much speculation as the strategic art of scenario building – already informing planning teams in offices from the Pentagon to Chinese, Russian and European security agencies. Much will depend on the fate of the Gulf Stream and other warm ocean flows given the massive amount of cold fresh melt water released into the Atlantic Ocean. The abrupt arrival of near-Arctic conditions in the Northern hemisphere, and subsequent parching of equator rainforests may well be upon us in the next decades (Schwartz and Randall 2003). The gravest and ultimate threat is the move beyond a climate tipping point, and the onset of unstoppable climate chaos fed by terrestrial feedback mechanisms, and the massive release of greenhouse gases from warming oceans, drying land and dying forests.

The Nature of the Challenge

Likely urban costs and planning responses have been at the horizon to anyone who cared to look and think, at least since the mid-1980s – yet, as a whole, the urban management and planning community is yet to rise from its long, petroleum induced slumber. Much lip service is paid to broad 'ecologically sustainable development' and 'smart growth' notions; and more recently adaptation efforts now entered the urban mind. Yet most still don't quite manage to focus on the transformation of the energy regime as *the* adaptation mandate of this time – professional frames constructed in the nineteenth and early twentieth century makes this switch institutionally and psychologically challenging. Instead, 'the market' is once again called to the rescue. Considerable faith is placed in carbon emissions trading, in the hope that the economy can trade itself out of trouble. Yet so far emissions counting, carbon trading and clean development mechanism-based (CDM) programs, a potential source of funding to important projects, are too frequently used to leverage imaginary, dubious and even destructive initiatives especially across the developing world. Its focus on narrowly defined projects – rather than more effective but hard-to-budget programmatic efforts, institution building or transport initiatives – is increasingly decried; see also Shobhakar Dhakal's contribution in this book. As currently practiced, carbon trade mechanisms cannot by themselves be reliably expected to help lower emissions and greenhouse gas concentration levels, certainly not at the speed and breadth required to make an impact on present petroleum supply and climate risk profiles (Lohmann 2006; Byrne and Glover 2000). They must be accompanied by structural changes in the energy supply regime; government managed renewable energy feed-in legislation (Mendonça 2007); dramatic innovations in all forms of energy efficiency – including in the transport-land use nexus; a shift to local sufficiency and regionally autonomous structures and an unwavering focus on the construction of regional renewable energy infrastructures – including intelligent grid networks and distributed, ubiquitous generation and storage systems. Unfortunately, there is no consensus across the carbon trading community that these important goals constitute shared aspirations. What makes matters even more troubling is that its very tenets are questionable, to say the least. The thresholds today promoted as relatively 'safe' by the Kyoto-associated frameworks and groups are a far cry from that. The atmospheric greenhouse gas (GHG) concentration level of 450 parts per mission (ppm) promoted to the level suitable to keep global warming at 2 degrees celsius over pre-industrial levels carries, by IPCC calculations, a 50% risk of missing that aim. A safer level would be 350 ppm – but the atmosphere today has already a GHG concentration of 385 ppm. Even more unnervingly, the 2 degree warming increase portrayed as 'safe' would almost certainly be catastrophic, given that ice melts have proceeded much faster than anyone anticipated, and world-wide signs of stress and crisis abound – already at the 0.7 degrees the world has warmed during the twentieth century. The feedback tipping point may already be behind us – hence 100% renewable energy scenarios founded on drastic efficiency measures should be discussed as central aims, rather than the various 'economically practical energy mix' cocktails presently promoted.

Repairing the Damage Wreaked by Fossil Fuel

Fossil fuels will grow more expensive due to supply difficulties and growing demand; prospecting, production and processing cost increases as the global hunt for combustibles enters increasingly marginal regions; but also due to growing carbon penalties. The internalization of external costs brought about by increased incidences of cardiovascular disease, cancer, or oil spills, also awaits implementation: China's air pollution levels cause an annual US$50 billion in health damage; and US$70 billion are expended in such costs each year across Europe (Geller 2002). The lowering of conventional energy subsidies, too, can help build local renewable energy markets and be a boost to the profitable world of efficiency and conservation. Such energy transitions boost employment in renewable, more labour productive industries (Kammen 2004). In spatial planning, a number of measures can be taken to lower the risks of climate change. These involve transport policy and technology changes, regional planning and agricultural reform, and institutional change. In transport alone, change, while slow in the past, promises dramatic improvements: in improved planning and urban design to boost pedestrian movements and bicycle use, public transport, lighter, more efficient vehicles, sustainable bio-energy engines – and a move to renewably powered electric cars that in many ways are more efficient, cost-effective, widely practical and affordable than hydrogen solutions. All of these measures help to cut reliance on high risk global fossil fuel supply lines and begin to foster local and regional systems of resource autonomy instead (Droege 2006; Scheer 2002, 2006).

Urban Energy Transition as Community Challenge

The United States Federal Emergency Management Agency in 1980 prepared a report that argued for an end to fossil-fuel domination in the USA, as a matter of national security (FEMA 1980). The report was shelved since the actions recommended would have threatened oil and coal profits. Yet the underlying notion is today again popularly regarded as sound. Only energy transition strategies aimed at efficiency boosts and the rapid proliferation of renewable energy supply systems can combat fuel supply risks and successfully address fundamental global poverty issues. The highly misleading and inappropriate technological answers currently being offered, such as carbon capture and sequestration (CCS) or greater reliance on nuclear generation, offer no solution. Instead, urban energy policy and practice agendas are now being strengthened, building on a great store of strength in technology development and research. Renewable energy sources and their support systems should be pursued in the search for a more equitable, sustainable and prosperous urban world. Many city leaders correctly perceive this not primarily as an engineering or urban planning challenge, but as a social equity, political and economic development task. No technological or logistical barrier hinders a switch away from the present, dangerous levels of nuclear and fossil energy reliance. The urban energy transition is a challenge of culture, community and civilization, and of bold institutional, political and policy changes.

The most promising response to the twin threat of oil depletion and climate chaos is to engage in tightly legislated, community supported, infrastructure change, to help overcome anachronistic urban power regimes and regional development dynamics, and construct the market frameworks in which distributed renewable power can thrive. The experience of cities around the world shows that sun, wind, water and sustainable forms of bio-energy can be introduced readily and effectively, while rapidly increasing efficiency and conservation. The institutional and technological tools deployed are as diverse as the communities that are developing them.

References

Alt, F. (2007). Deutsches EEG hat weltweit 47 Nachahmer. Retrieved on 7 July 2007 from http://sonnenseite.kjm4.de/ref.php?id=2c85c951843ms54

Archer, C.L. and Jacobson, M.Z. (2005). Evaluation of global wind power. *Journal of Geophysical Research,* **110**. http://www.stanford.edu/group/efmh/winds/global_winds.html

Arrhenius, S. (1896). On the influence of carbonic acid in the air upon the temperature of the ground. *Philosophical Magazine,* **41**, 237–276.

ASPO – The Association for the Study of Peak Oil. February 2007. *Newsletter,* **74**. http://www.peakoil.ie/newsletter/en/htm/Newsletter74.htm#794

Bryson, B. (2003). *A Short History of Nearly Everything*. Black Swan.

Byrne, J. and Glover, L. (2000). Climate shopping: putting the atmosphere up for sale. *Environment, Economy and Society Series*, Australian Conservation Foundation, TELA.

Campbell, C.J. and Laherrere, J.H. (1995). *The World's Oil Supply 1930–2050*. PetroConsultants SA.

Campbell, C.J. (1998). *The End of Cheap Oil*. PetroConsultants SA.

Campbell, C.J. (2003). *The Essence of Oil and Gas Depletion*. Multi-Science Publishing Company and PetroConsultants SA.

Campbell, C.J. (2005). Revision of the depletion model. The Association for the Study of Peak Oil and Gas (ASPO) – Article 624, *Newsletter* No. 58.

Capra, F. and Hurtz, W.T. (1958). *Unchained Goddess*. Bell Telephone Science Series (VHS).

Doxiades, K.A. and Papaioannou, J.G. (1974). *Ecumenopolis: The Inevitable City of the Future*. Athens Center of Ekistics.

Droege, P. (2006). *The Renewable City-comprehensive guide to an urban revolution*. Wiley.

European Tribune (2006/2007). IEA predicts shortages within five years, http://www.eurotrib.com/story/2006/12/17/155626/19 and OPEC says peak oil (and $100 oil) is near, http://www.eurotrib.com/story/2006/12/17/155626/19. Retrieved on 12 July 2007.

Eurosolar (2007). Energiegipfel sind ungeeignet für Umbau der Energiewirtschaft: Zuverlässige Vollversorgung mit 100% Erneuerbaren Energien schon jetzt machbar. Press release 29 June 2007. Berlin and Bonn offices.

FEMA – Federal Emergency Management Agency (1980). *Dispersed, Decentralised and Renewable Energy Sources: Alternatives to National Vulnerability and War*. FEMA.

Friedman, J. and Wulff, R. (1976). *The Urban Transition: Comparative Studies of Newly Industrializing Societies*. Arnold Press.

Geller, H. (2002). *Energy Revolution: Policies for a Sustainable Future*. Island Press.

Goodstein, D. (2004). *Out of Gas: The End of the Age of Oil*. W.W. Norton & Company.

Heinberg, R. (2003). *The Party's Over: Oil, War and the Fate of Industrial Societies*. New Society Publishers.

Hubbert, M.K. (1956). Nuclear energy and the fossil fuels. Publication No. 95, June 1956. Shell Development Company.

Hubbert, M.K. (1971). *Energy and Power*. The Scientific American.

IEA – International Energy Agency (2004). *World Energy Outlook*. IEA.

Informationszentrum für Erneuerbare Energien (2007). EE100 – das Regenerative Kombikraftwer (Deutschland hat unendlich viel Energie). http://www.unendlich-viel-energie.de/uploads/media/Pr_sentation_EE_100_Regeneratives_Kombikraftwerk.pdf

IPCC – Intergovernmental Panel on Climate Change (2007). *Climate Change 2007: The Physical Science Basis – Summary for Policymakers*. IPCC.

Kammen, D.M., Kapadia, K. and Fripp, M. (2004). *Putting Renewables to Work: How Many Jobs Can the Clean Energy Industry Generate?* Renewable and Appropriate Energy Laboratory (RAEL) Report. University of California at Berkeley.

Klare, M.T. (2002). *Resource Wars: The New Landscape of Global Conflict*. Owl Books.

Lenzen, M., Dey, C. and Hamilton, C. (2003). Climate change, in *Handbook of Transport and the Environment* (eds D.A. Hensher and K.J. Button). Elsevier.

Lohmann, L. (2006). Carbon trading – a critical conversation on climate change, privatization and power. *Development Dialogue No. 48, September 2006*. The Dag Hammarskjöld Centre.

Mendonça, M. (2007). Feed-in Tariffs-Accelerating the Deployment of Renewable Energy. Earthscan.

Mills, E. (2005). Insurance in a climate of change. *Science* 309(5737), 1040–1044, 12 August 2005.

OECD – Organisation for Economic Co-operation and Development (1995). *Urban Energy Handbook*. OECD.

Scheer, H. (2002) (Original in German, 1999). *The Solar Economy – Renewable Energy for a Sustainable Global Future*. Earthscan Publications Ltd.

Scheer, H. (2006). *Energy Autonomy. The Economic, Social and Technological Case for Renewable Energy*. Earthscan/James & James.

Schindler, J. and Werner Zittel. Crude Oil – The Supply Outlook. Energy Watch Group.

Schwartz, P. and Randall, D. (2003) An abrupt climate change scenario and its implication for United States security, October. http://www.environmentaldefense.org/documents/3566_AbruptClimateChange.pdf

SEU 2007 (2007). Sustainable Energy Utility. Retrieved on 12 July 2007 from http://www.seu-de.org

PART I
Principles and Drivers

Chapter 1

Solar City: Reconnecting Energy Generation and Use to the Technical and Social Logic of Solar Energy

HERMANN SCHEER

Increasingly as each year passes, a long overlooked and forgotten fact will move more sharply into focus: without resources there can be no economic activity. I call this the physical conception of economics: all economic activity consists in essence of converting materials in their raw state into finished products using converted energy; this includes the energy for communication, consumption, transport and distribution. In essence the economic process is first and foremost a process of converting resources. But it matters very much – not solely for ecological reasons – what resources we use and how we use them.

1.1 No Possible Change within the Conventional Energy System

The determining factor of an energy system is the energy sources it makes available. All subsequent steps follow on from this choice: the energy conversion technologies and infrastructure required for its optimal use, and the energy habits, social opportunities and types of business, as well as the settlement and transport structures which result from it. We need a sociological perspective taking in energy-related, technological and economic aspects in order to be able to understand the energy system of today and identify the consequences which will result from the impending historic changeover from fossil and atomic energies to renewables.

The impact that the existing atomic/fossil energy system has had on shaping society is particularly apparent in the world's two antithetical socio-cultural poles: the big cities of the industrialized societies on the one hand, and the rural areas of the Third World on the other. It was the fossil energy system which developed with the industrial revolution that paved the way for the growing megacities of the industrial modern age. Now that fossil energies are nearing exhaustion, the megacities are threatened with collapse, taking with them the cultural forms of civil society which have developed since the beginning of the modern era in the wake of enlightenment and secularization, the industrial society and democratization.

The widespread notion that the energy supply introduced in the industrialized countries represents the optimum in terms of energy management and hence should also serve as a role model for developing countries is sociologically unsound. It begins with the third step in the development of the energy system – that is to say after the choice of energy source and the technologies applied to use the energy – and regards this system's functional model as a constant and indispensable factor. New energy options are measured against this to see if they fit this well-tuned system. It is therefore seen as an incontrovertible truth, permitting no serious doubt as to its necessity and seeming to warrant no further justification.

Where future energy supply is concerned, however, everything necessarily and urgently points to the use of renewable energies. There is no getting away from the fact that fossil energy reserves will run out, as will uranium reserves. For the future, therefore, there are only two energy options: renewable energies or nuclear energy from fast breeder reactors (because the reprocessing of spent fuel elements and their use in such reactors can make the fissile material go further by around a factor of 60) as well as, in the future, from nuclear fusion reactors. The past has clearly shown, however, that the nuclear option is more a theoretical than a practicable solution. There is still no fast breeder reactor fit for purpose and in stable permanent operation today, despite the many billions spent on research and development. And whether or not a nuclear fusion reactor will ever become a reality is in the lap of the gods. In any event the costs are likely to be extremely high compared with present-day nuclear fission reactors, quite apart from the considerable additional safety and security risks they would pose both politically and ecologically.

1.2 Renewable Energies as an Energetic Imperative

Even in terms of energy costs alone, it is already extremely unlikely that they could ever be lower than those of renewable energies, the cost of which will undoubtedly continue to fall in the wake of further technical improvements, new technologies, new forms of use and, above all, an increase in plant production and the development of production technologies. Renewable energies truly represent the only realistic option. The question is not whether a system with renewable energies comes to pass but simply how long it will take civilization and its responsible actors to recognize and realize this. And this, it must be said, is a question of central importance. If it takes too long, that is to say, if the changeover continues to be blocked or even just delayed, the disruption to civilization will be huge. The ethnosociologist Hans-Peter Duerr (2002) describes the 10 000 years of the history of civilization as a history from 'nomads to the monads', forced to eke out their existence in conditions in which society is no longer viable. This is a feasible scenario above all because of the worldwide dominance of the atomic/fossil energy system, which, if we continue to cling on to it, must lead to the collaboration of the two ecosystems, the ecological and the economic. In *Zorn und Zeit* (*Anger and Time*), Peter Sloterdijk (2006) talks of the death of two entropies: energy related and moral.

Implementing the renewable energies' option rapidly is an 'energetic imperative', as defined back at the beginning of the twentieth century by Nobel Prize winning chemist Wilhelm Ostwald. He argued that humankind was living off the depleting capital of fossil fuel reserves and that economic activity could continue only by harnessing the constant supply of energy provided by the sun (Ostwald 1912). But making the changeover is more than a matter of simply switching energy source. It will lead inevitably – if it is done cost-effectively and productively to take advantage of the particular characteristics of renewable energies – to a structural revolution in energy supply as a whole. This inevitability results from the fact that there is an optimum productivity specific to each resource. It is seldom known from the outset what this is, with what technologies it can be accomplished and under what external conditions (from political to economic). In general the road taken to reach that point is likely to be circuitous and bumpy.

There is, however, an inherent productivity logic which is indeed discernible at the outset – if instead of considering individual technologies in isolation one thinks in terms of the systemic context which results from their use. In order to identify this, it is necessary to look at the flow of a resource from the place it is tapped or collected through to its final active use. The analytical guideline for this is: the greater the number of technical transformations that are needed between energy generation and the end users and the more infrastructure is needed to convey the energy flow, the more ineffective, complex and ultimately costly the system is. What is particularly striking in this respect is the key difference between non-renewable and renewable energies: while the end use of energy is always decentralized – where people work and live – the first step in the active process of making the energy available happens

- in the case of fossil energies and nuclear power in relatively few places in the world: in coal mines, oil and gas fields and uranium mines. From there the energy chain to the energy consumers is formed everywhere along the energy flows organized for it. Between the beginning and the end of this chain the infrastructure to deliver the energy is tailored to these energy flows: initially in increasingly concentrated form upstream to the point the primary energy is transformed into useful energy and subsequently downstream when the flows branch out widely and become progressively de-concentrated until they reach the energy users;
- in the case of renewable energies, by contrast, in a potentially infinite number of places where they have to be harvested by technical means (wind power plants, PV modules, solar collectors, energy crops, wave energy plants, hydroelectric plants), since this is natural environmental energy which occurs everywhere. The sole exceptions are large-scale hydroelectric plants which require a dam or a large fall of water constructed for the purpose, or also large-area solar power stations in (semi-) arid regions of the world and offshore wind parks. Whether these actually provide more economical, i.e. cheaper, energy depends in turn on the overall cost of the required system including installation, transport and distribution costs. But the vast number of large areas with particularly intense solar radiation or wind frequency, areas with particularly fertile soil or highly wooded regions with many existing bodies of flowing water, provide huge potential which is relatively easy to harness. This accounts for the differences in harvest potential and determines the preferred choice of renewable energy in each case. What is available everywhere is solar radiation. As the natural energy supply has a lower energy density than fossil energy reserves, there is always a need for plants to harvest the energy to be widely dispersed.

1.3 Energy Generation and Energy Use: from Disconnection to Reconnection

In the past the production of fossil and/or atomic energy led to a progressive disconnection of the areas of primary energy production from those where the useful energy was consumed, creating a growing need for infrastructure. With the evolution of very densely populated urban economic areas and, at the same time, the progressive depletion of domestic sources of primary energy and a growing demand for energy, the conventional energy system has inevitably become ever more concentrated as it has had to internationalize and ultimately globalize.

For as long as there were no efficient energy technologies in the history of human settlement, no rapid and large-volume transport capacities and no transport infrastructure, the settlement areas had to remain directly connected to the areas of energy generation and food production. The fields and forests which in pre-industrial times provided the energy resources for the towns and cities had to be some 40 times larger – depending on soil quality and climatic growing conditions – than the actual settlement areas. For as long as this natural limit to growth was respected, supply crises were unlikely to occur. Economic conditions seldom allowed excessive wealth; at the same time their stability was threatened only by the effects of war. While there were certainly differences in living conditions, caused either by nature or different levels of technical and cultural development, in pre-industrial times these were less pronounced than they became globally in the fossil fuel-based industrial age. It was the industrial revolution, with its possibilities of concentrated energy supply and improved transport, which first created the basis for the growth in high density urban areas and hence the shift of human settlements from the countryside to the towns and cities. This development depended on the ability to transport food and energy from increasingly remote areas to the urban settlements ever faster and cheaper. Initially the industrial urban centres mushroomed in the areas with large coal deposits, later spreading out along the main lines and major flows of the concentrating energy supply. The more manifold these methods of energy transport and use became, the more the energy supply and transport infrastructure was expanded and the more and the faster the energy flows poured into the industrial metropolises. In this way they also became centres of energy-intensive services. The faster the fossil megacities grew, the more inorganically they sprawled. The fastest and greatest growth surges paved the way for the supra-regional electricity grid and the mass production of the car, which increasingly shaped urban development.

As a result of the ongoing concentration of the conventional energy system and the separation of energy production and consumption, infrastructure costs grew steadily and have for a long time been higher than extraction costs or the costs of operating the power stations. The more the energy flow increased, however, the less these costs were a factor for consideration. In this way pro rata plant costs were kept relatively low. They become more noticeable, however, as smaller volumes of energy are conveyed through the system, which must inevitably lead overall to a considerable rise in final energy costs. For this reason, quite apart from any others, there is resistance to the substitution of non-renewable by renewable energies which do not have to make use of this system – or attempts are made to compel them to use the system.

Even many of the advocates of renewable energies do not think beyond this system because they, too, see its infrastructure – at least the downstream part of the chain – as an independent and enduring constant. They think in terms of maintaining the core of the existing system and simply swapping energy sources. In this case renewable energies would simply be integrated in the system or the outdated system copied using renewables. This is the point of view which gives rise to the proposals to focus future energy supply on solar power stations in the sun-rich Sahara or on major offshore wind parks at sea and then, as before, to transport the power in high, medium and low voltage networks and to distribute it in distant urban metropolises. Or – in the case of fuel – to establish a centralized system of hydrogen production and then convey the fuel long distance to the fuel stations. In terms of the infrastructure needed, the big transnational energy companies supporting it once again seem indispensable.

Yet this model for renewable energies represents a shortsighted and structurally conservative viewpoint. Exploiting the possibility of sourcing natural primary energy over a wide area in the towns and cities means connecting the areas of active energy use with those of the energy harvest – that is to say with less and in some cases, in the urban centres in particular, with no infrastructure requirement of any kind. It is no coincidence that it is mostly towns and cities which have autonomously launched the first groundbreaking initiatives because they are not so bound up in the 'politico-energy industry complex' as governments. The large number of small towns or counties – in Germany or Austria, for example – which are striving to achieve or even have already achieved 100% energy autonomy at municipal level is striking. Without the scheme providing cost-covering compensation for solar power fed into the grid in German towns, there would have been no industrial basis for the Federal Government's '100 000 roofs' program to promote solar power in 1999 and the subsequent massive increase in statutory compensation for solar electricity introduced by the EEG in 2000. Local authority initiatives, therefore, can pave the way for general legislation. If those towns which have led the way had turned for advice before they started to scientific experts who would have compared the costs of the project with theoretical projects elsewhere, none of these initiatives would probably have got off the ground. It is possible that these municipal projects were not always 'cost-efficient'; they have, nevertheless, improved the quality of life and the social atmosphere of their local communities and created local and/or regional jobs.

In terms of the technology required for electricity supply, in addition to the facilities to harvest and convert renewable energies, it is also necessary to mobilize storage facilities – from battery systems to fly wheels, from compressed air to super condensers, from hydrogen to thermo-chemical storage systems – and last but not least, hybrid systems.

To highlight a lack of storage possibilities as one of the principal drawbacks of renewable energies, however, is to take a superficial viewpoint. No modern energy supply system can exist without storage, because there is no simultaneity between primary energy generation or harvesting, and final energy use. In the case of fossil energy, not only do the coal heap, oil tank or gasometer (large urban natural gas pressure vessel) fulfil a storage function; so, too, do the transport and distribution systems – that is to say the infrastructure, which acts as an interim store. The same is true for renewable energies from biomass, except that in this case it is possible to have a regional rather than an international supply chain. The only forms of energy it is not possible to store before they are converted to electricity are solar and wind power. The grid can fulfil this role in part as long as the amounts of electricity generated in this way are relatively small. But specific decentralized technologies which store the energy after electricity generation offer the optimum solution. If the grid alone were to be used, the outdated dependency on infrastructure would persist. This dependency can be reduced, however, if wind and solar power facilities are spread over a wide area and little or no use is made of the transmission grid, making use instead of regional and local networks. In this case medium and low voltage electricity is fed in and taken out there.

If complementary storage techniques come into the frame, the picture changes completely. In this case if respective system costs are compared:

- the costs of conventional energy supply (comprising primary energy, conversion, transport and distribution infrastructure costs, as well as environmental impact costs)
- are set against those of energy from renewable energies with plant and storage technology costs, but – with the exception of bio energy – without current fuel costs and with little or even no infrastructure requirements.

Where the latter is concerned, it is safe to assume that, based on series production of plant with the associated economies of scale, this will be not only more cost-effective than the outdated system but in many respects more culturally attractive. The advantages, therefore, are not just ecological. The system can be operated autonomously and provides guaranteed security of supply and certainty of costing since it is not affected by fluctuations in energy prices. Moreover, it is faster to import and, because it operates on a modular basis, avoids both undersupply and overcapacity: investments can be made on the basis of demand, largely avoiding bad investments.

True deregulation of the electricity markets will aid this trend: liberalization, as recent experiences from California to southern Europe have shown, jeopardizes security of supply in large-scale systems and makes investors less willing to invest to maintain grid quality. It also makes them less willing to invest in large new power stations since their capacity utilization over decades can no longer be guaranteed.

1.4 Looking Back to Look Forward

This brings us back to the original vision of electricity pioneer Thomas Edison. At that time there were two different basic concepts regarding electricity supply which caused a conflict of ideas (and business conflict) between the two pioneers in the USA, Edison and Westinghouse.

Centralized electricity supply structures arose not because they were more economical per se but essentially because electricity can be transported significantly faster and more cleanly than fuel. This is a particularly important factor for energy supply for cities. In order to understand this, one must first know something about the history of electricity supply.

The conflict between Edison and Westinghouse marks the beginning of the story. Edison's vision was for all buildings to generate their own electricity, while Westinghouse's vision was for electricity to be supplied to houses from mains. Since electricity was generated from fossil fuels and water power, Westinghouse had the better 'systemic' view. Since it is impossible to produce electricity from water power in the cities, its use necessitates mains which deliver the electricity on a carriage paid basis rapidly and cleanly. Edison's concept, by contrast, required fossil fuels to be delivered to every building, resulting in numerous individual fireplaces, an idea to which people were resistant since they were already fed up with coal-fired heating. While his concept gave more freedom, for city dwellers it had more of a direct environmental impact and was less convenient. For similar reasons the over five million wind power plants that existed in the rural Midwest of America in the 1930s were abandoned in a flash once the electricity cables were laid in the wake of the Tennessee Valley Program under President Roosevelt's New Deal. The Program created large hydro-electric power stations and coal-fired power stations which required the construction of overland cables to the farms in order to fully utilize their capacity. This provided the farmers with electricity around the clock in contrast to the electricity generated by the windmills which they had little or no possibility of storing.

The conditions that led to the model of a universally networked electricity supply with large power stations as production centres tend not to apply, however, in the case of renewable energies. The use of solar power in towns and cities, integrated in roofs, facades and windows and the use of wind power even in high-rise buildings in big cities (as in the Bahrain World Trade Center), cuts out the need to transport fuel. Solar radiation is 'supplied' automatically and costs nothing. There is also the possibility of turning biogas from the large volumes of food waste in the cities into electricity and of generating electricity from wind power and biomass in the hinterland of cities and in rural areas. There are plenty of examples where this is happening: cities and municipal utility companies which run their entire public bus service on biofuel, supply their citizens with electricity and heat from biomass power stations and heat entire housing estates with solar heat which they store in the ground for winter.

The use of renewable energies in buildings and in urban planning, and the adaptation of buildings and orientation to the existing natural environment, are critical elements in transforming energy supply at municipal level. At the same time urban planning – in terms of distances between residential areas, work, recreation and shopping areas – is a key factor in determining transport behaviour and hence energy consumption. There is a need to substantially cut local energy consumption by ensuring that urban planning is oriented to the need to eliminate traffic. The local authorities can use their competences to set the framework for using renewable energies in buildings and in urban planning, taking into account bioclimatic conditions. Making the changeover to solar power in building and urban planning represents a historic turning point in building culture and urban development: the criteria for solar architecture and bioclimatic urban planning are necessarily to determine the entire future. The sooner a start is made, the more future-proof urban development will be.

As a consequence of this development, long-distance transmission networks will in the future become redundant as electricity supply becomes fragmented. Emergency power supply units, tens of thousands of which are located in large utility buildings such as hospitals, will become main units. In addition to mains electricity supply, there will be decentralized production locations for the producer's own and neighbourhood use, area supply for residential areas and island systems for small towns, either fully self-sufficient or supplemented by spur lines. Electricity supply is returning – not completely and immediately but increasingly – to Edison's vision, and this is only practicable on the basis of renewable energies. In what forms this process takes place depends on external factors (legislation), sociological factors (information, level of education, cultural awareness and values) and technological factors.

As this development continues, utilization of power station capacity will decrease, making electricity more expensive. As the number of autonomous producers grows, the grid system will have to pass on its costs to fewer customers. The old system will grow obsolete and lose its economic advantages. Long-distance line electricity supply will have to concentrate on key areas in order to remain commercial. The transmission network will become less densely meshed: the more large power stations are taken out of commission, the more high voltage lines will become redundant. Voltage compensation will become more difficult and regular energy more complicated and expensive to supply. The tendency among large consumers will be to produce their peak load themselves. The regional medium voltage sector will grow in importance in comparison with the local distribution networks, but capacity utilization will also fall abruptly. And with each of these stages, decentralized solutions will look increasingly attractive and technological creativity will focus on this area. This will be a process stretching over a considerable time – just as today's electricity supply situation developed over decades.

A similar development is in the offing in the heat energy and fuel market in relation to both grid-bound and non grind-bound energy. Solar heating in a house starts to pay for itself through savings in fuel costs once the price of oil hits around 80 dollars a barrel, and well before that figure in the case of new buildings which have optimized integral solar systems. The numbers of customers for natural gas and crude oil will fall, and the same applies to electricity supplies – as the demand for heating and hot water which currently accounts for 30% of total energy demand in Germany will fall due to replacement either by PV-operated ground heat pumps or by solar thermal power systems. Mains gas supply is also experiencing losses in terms of market share and function, while the importance of local island gas networks to transport biogas is increasing.

The oil industry's highly branched marketing system which supplies the oil merchants and fuel stations from a few central refineries is also becoming more regionalized. Regional biofuel production centres are being established, processing raw biomass and setting up regional marketing structures. New production cooperatives for biofuels are being set up or municipal utility companies are taking on this function as biomass customers. At the same time they are selling secondary usage products, fertilizers or feedstuffs made from the material left over from biofuel production.

Power and heat cogeneration, the growth of which is often restricted by the lack of a grid-bound heat market, can, with the aid of a Stirling engine, for example, become a three-step power/heat/power generation system which enables more electricity to be made from the surplus heat for the producer's own requirements or to feed into the local networks, thereby creating new local and regional synergies.

1.5 From Global Energy Supply to the City as Power Station

In my book *The Solar Economy* I compare three different prototype energy supply systems: the current system comprising in each case separately operating providers and consumers of electricity, fuel and directly used thermal energy; the system the energy industry wants to see which seeks to integrate these functions in a highly concentrated manner; and a system consisting of an integrated decentralized supply structure, the optimized function mechanisms of which are described above, with numerous other possible variations and possibilities to provide greater flexibility not touched on here. In each case a distinction is made between the energy sources (renewable source, RS; conventional fossil source, CS), the producers and/or providers of energy technologies (G), the providers of useful energy (S) and the energy consumers and/or users (C) as well as those producing for their own consumption (OS). It is important to note that the first and second prototypes consist of long energy flows, and the third prototype short flows. The latter is the less complex system and hence ultimately has the greater potential productivity and reliability at potentially lower costs with assured price stability (see Fig. 1.1).

Several increasingly interlinked factors are driving developments in this direction: the approaching exhaustion of fossil reserves, the manifest ecological compulsion to act, the increasing technological and industrial profiling of renewable energy, energy efficiency and generally decentralized energy conversion and storage plants, and the development of stand-alone and standby current consuming devices which produce their electricity fully or partially themselves with PV modules and micro fuel cells.

If in the future, as centralized providers start to be replaced by decentralized providers and the numbers of those producing for their own consumption grow, less and less use is made of the transmission and also distribution networks, all calculations will be thrown into confusion. As this process progresses – in a way which cannot be predicted – it will, however, be impossible to abandon mains services because many people will still be reliant on them, some of them permanently. The alternative to mains supply would be to make self-supply obligatory, a move which would be senseless and unworkable for functional and social reasons since it would create widespread structural distortions and extremely unequal production conditions and prices. In order to safeguard general energy supply, therefore, there will necessarily be a trend to take the energy supply infrastructure back into the public sector and guarantee it under local authority, regional or national responsibility.

An additional problem is that private network operators cannot forever rely on being able to use the countryside free of charge. This creates a new political problem. If there are

Hierarchical divisions within the conventional energy system between source and consumer

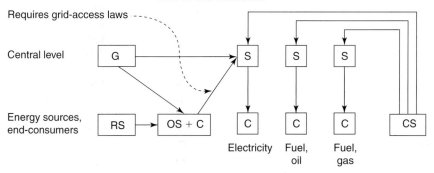

Integrated hierarchical energy supplies – the industry model

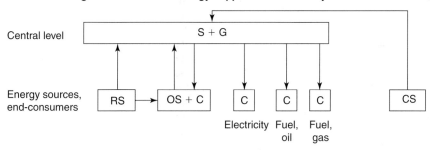

Integrated distributed energy supplies using renewable energy

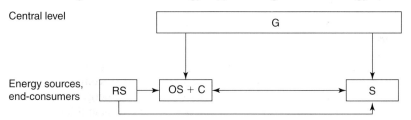

Note: RS = renewable source; OS + C = own supply and consumer; C = consumer; CS = conventional source; S = supplier; G = generation plant

Fig. 1.1 The advantages of an integrated, distributed and renewable energy based supply model (bottom) when compared with that of the conventional, segregated and hierarchical system as it historically developed through the age of industrialisation to become currently prevalent (top), and the more integrated and highly hierachical model presently advanced by the conventional energy industry (center).

fewer and fewer people using the infrastructure, those who remain connected will either have to pay ever rising connection costs or the state/public at large will have to subsidize the mains supply to ensure that costs for users are not too high. It is likely therefore that in order to compensate the latter, the state will resort to obligatory connection to the mains even for those who do not use the infrastructure – or alternatively those who are not connected will be taxed. This would signify a trend to return to de-liberalization, although it would also lead to intensified efforts to make mains operation more effective. This, however, will have to be based on the criterion of supply security at the same price for those energy users who are dependent on the infrastructure and not according to the criterion of the highest possible capital returns. It is therefore a public and not a private sector criterion – a criterion of the modern commons.

The three prototypes described signify three phases of resource use and/or technology development: looking back to the history of modern energy supply it is apparent that this could not have been created without the public sector providing the infrastructure. In this first phase there were still too few users to have made it economical for the private sector. In order to make the advantages of the system available to all, the public authorities had to provide an outlay in advance. Where, as in the developing countries, only the towns and cities have benefited because the economy lacked the strength to do more, a seemingly hopeless gulf was created between the cities and rural areas. The cities of the Third World have grown rapidly as a result of the exodus from the land, only incomparably faster and more comprehensively than happened in the industrialized countries. The process has impoverished both the cities and the rural areas. The second phase is where we are at present: where infrastructure, as in the successful industrialized countries, was built and expanded to provide blanket coverage and everybody became a user, it no longer seemed necessary to keep infrastructure provision under public control. Increasingly it seemed desirable to privatize the infrastructure, the users of which, it seemed, would always be dependent on it, creating a relatively risk-free natural monopoly.

The more this trend progresses, however, and the users are progressively less able to pay for the infrastructure, the whole business starts to run into trouble. Even collecting the charges for the infrastructure becomes impracticable if the number of self-suppliers grows too large or, as in the Third World, the number of people who can no longer pay their bill simply tap into the grid where they can. If they are cut off from using the infrastructure, however, mass protest is the inevitable consequence. This is one of the reasons why privatizing infrastructure – electricity and water supply – in the mass cities in the Third World failed because what was intended to be a high volume business became instead a subsidising business.

Today we are at the beginning of the third phase when the system, after a period of convergence, is diversifying again, leading on the one hand to a gradually declining number of energy users who remain dependent on the broad networked structures of centralized electricity and gas supply, and on the other hand a gradually increasing number of players who are producing and supplying energy to the grid and are at the same time customers of the grid. Some are in this way developing into full individual energy producers; others are creating their own energy patches and districts. The decentralized structures – through to the Solar City – are becoming increasingly attractive, offering diversity and flexibility. They are turning the city into a solar power station. While conventional energy supply is a permanent and rising cost factor, solar energy is becoming an autonomous value added factor for the city and its inhabitants. The city is becoming more prosperous and a better place to live. The clean air is good for public health, making outdoor life more enjoyable and promoting communication. The first step is to realize that it is finally time to make full use of the most important infrastructure in the city: the Sun.

References

Duerr, H.-P. (2002). *Der Mythos vom Zivilisationsprozess*. Suhrkamp Verlag.
Ostwald, W. (1912). *Der energetische Imperativ*. Akademie Verlagsgesellschaft, pp. 81ff.
Sloterdijk, P. (2006). *Zorn und Zeit*. Suhrkamp Verlag.

Chapter 2

Undoing Atmospheric Harm: Civil Action to Shrink the Carbon Footprint

JOHN BYRNE, LADO KURDGELASHVILI AND KRISTEN HUGHES

Center for Energy and Environmental Policy, 278 Graham Hall, University of Delaware, Newark, DE 19716

Global climate change represents the major environmental challenge of the modern era. An imposing body of scientific evidence links climate change to anthropogenic greenhouse gas (GHG) emissions (IPCC 2007; The Royal Society 2005; AMS 2003; NRC 2001). Carbon dioxide (CO_2) contributes more than three quarters of GHG emissions from human activity. In turn, more than 95% of global CO_2 emissions are due to fossil fuel burning and land use change (WRI 2006). Historical data show that carbon concentrations have increased 35% from pre-industrial levels of 280 parts per million (ppm), to reach the current 380 ppm (CDIAC 2006). This change in atmospheric chemistry coincides with a temperature increase of 0.6 ± 0.2°C in the twentieth century (IPCC 2007, 2001).

The Intergovernmental Panel on Climate Change (IPCC) projects that, due to cumulative GHG emissions from human activity (especially over the past century), average global temperatures are likely to increase between 2 and 4.5°C by the year 2100 compared to 1990 (IPCC 2007; 2001), unless major efforts are made quickly to reduce them. Such temperature increases correspond to carbon concentration levels of 541 to 963 ppm, respectively.[1] Warming at the high end of this range could have widespread catastrophic consequences (Schneider and Lane, 2006). The widely proposed protective threshold is carbon concentrations of no more than 450 ppm (Oppenheimer and Petsonk, 2005; Hansen, 2004; Parry *et al.* 2001).

With the international scientific community – led by the IPCC – largely in agreement regarding the role of anthropogenic behaviour in forcing a new climate, calls for immediate and sizeable policy action to arrest the problem are mounting. Recently, the Stern Review on the Economics of Climate Change, prepared for the Office of the Prime Minister and the Chancellor of the Exchequer of the United Kingdom, has announced the need for cuts of 80% or more in anthropocentric emissions involving all nations (Stern 2006). In June 2005, the national science academies of the G8 nations[2] (including the US) and those

---

[1] These values are mid-points for IPCC scenarios, and reported ranges for concentration and temperature are higher (1.4 to 5.8 °C for temperature and 490 to 1260 ppm for CO_2 concentration).

[2] The G8 nations include the United Kingdom, France, Russia, Germany, the US, Japan, Italy, and Canada.

of India, Brazil and China released a joint statement citing the urgency for significant and immediate responses to the emerging climate crisis (The Royal Society 2005). These proposals share a common perspective: namely, the need *now* for global policy innovation in changing the direction of present development paths which are wedded to conventional energy systems.

The political challenge of responding to climate change is daunting. Liberal democracies have shown particular difficulty in abandoning their commitment to a cornucopian politics of ceaseless economic growth in order to launch the transformative social changes needed to avert a significantly warmer world (Byrne and Yun 1999). The problem is especially evident in the case of the US, whose national government has refused to accept even modest reductions in its GHG emissions.

This chapter explores the politics of transformation needed to end the dangerous experiment in climate change now under way due to the failure of industrialized societies to limit their GHG footprint. First, international political negotiations are analysed and the need for an explicit commitment to carbon equity is shown as essential to the realization of climate sustainability. The Gini coefficient, a traditional economic measure of income distribution equity, is employed to explore how unequal carbon distributions – now and in the future – can prevent the achievement of climate mitigation, even if many nations act responsibly and ambitiously to alter their present energy pathways.

The chapter then examines political strategy in the face of US national governmental intransigence. Through an inventory of American civil society initiatives, a case is built for understanding the grassroots revolt under way in the US to challenge its national political posture. An alliance with this locality-focused revolt is recommended strategically as a means to undermine the American national government's support for climate inaction. But it is further recommended on the ground that eventually the politics of climate *action* must extend beyond the rhetorical level, where nations establish and enforce GHG reduction targets, to the level of practice, wherein social transformation is actually undertaken. An era of sustainability and equity in practice is ultimately the province of civil society and, specifically, its urban industrial communities in this instance because of their dominant role in GHG emissions. Urban action and innovation is essential if climate sustainability and carbon equity (defined below) are to be realized.[3]

2.1 Changing the Sky

Certainly for children, and for many (perhaps most) in their adulthood, the notion of human alteration of the atmosphere is baffling. The blue canopy is rarely conceived in human scale. Rather, it most often inspires the poesy of the human mind and spirit, in which we celebrate and bow in deference to the heavens and their celestial rulers. A distinct trait of modernity is that we now engage the heavens as a project of science and economics. We may have little choice, as we have changed its chemistry and the mechanics

---

[3] Data from UN and US sources indicate that urban communities account for more than three quarters of global CO_2 emissions, despite containing less than one half of the human population. The wealthy urban communities of North America, Europe, Japan, Australia and New Zealand release over 40% of the world's annual CO_2 emissions, although they are home to only 13% of the human community. By contrast, most of rural life is sustenance based and releases carbon at very modest rates – its population share of 54% accounts for less than 30% of yearly anthropocentric emissions (UN 2004; EIA 2004).

Table 2.1.   Summary of IPCC scenarios.

Family		A1		A2	B1	B2
Scenario group	A1B	A1T	A1FI	A2	B1	B2
Population in 2050 and 2100 (Billion)	8.7	8.7	8.7	11.3	8.7	9.3
	7.1	7.0	7.1	15.1	7.0	10.4
World GDP in 2050 and 2100 (Trillion 1990 US$)	181	187	164	82	136	110
	529	550	525	243	328	235
Share of zero carbon in primary energy in 2100 (%)	65	85	31	28	52	49
Cumulative carbon emissions from fossil fuels (1990–2100, GtC)	1437	1038	2128	1773	989	1160
Cumulative carbon emissions from land use change (1990–2100, GtC)	62	31	61	89	−6	4
Carbon concentration in 2100 (ppm)	710	580	960	850	540	620
Mean temperature increase in 2100 compared to 1990 level (°C)	3.0	2.5	4.5	3.8	2	2.7

Source: IPCC 2001.

of climate associated with the molecular composition of the sky (for a discussion of this paradox, see Byrne and Glover 2005).

Future CO_2 concentrations in the atmosphere are partly a function of the amount of CO_2 released from fossil fuels and land use change over time. The US Energy Information Administration (EIA 2006, 2007) provides projections of world, regional and national CO_2 emissions from the use of fossil fuels for the next 25 years (2006–2030). The IPCC has produced 40 pathways, grouped within four sets of scenarios referred to as 'families', to capture the range of physical impacts associated with alternative assumptions about global social and economic trends. These families in turn have been grouped under the headings A1, A2, B1 and B2 (see Table 2.1).

The A1 scenario family describes a future with rapid economic growth and global population that peaks in mid-century and declines thereafter. Its subgroups map changes in the world energy system – a fossil intensive path (A1FI), a path with increasing reliance on non-fossil energy sources (A1T), or a path with a mix of fossil and non-fossil sources (A1B). Under the A2 scenario family, economic growth is regionally concentrated among Southern[4] nations and population is continuously increasing. The B1 scenario family describes a world with rapid change in economic structures, toward a service and information economy and with the same population trends as in the A1 scenario. The B2 scenario family emphasizes local solutions for economic, social and environmental sustainability, where global population continually increases but at a rate lower than trends in A2. In B2, intermediate levels of economic development coincide with less rapid and more diverse technological change than in the A1 and B1 storylines (IPCC 2001).

---

[4] 'South' and 'Southern' are used here to refer to the countries of Latin America, Africa, and Asia (excepting Japan, Australia and New Zealand), which are characterized by comparatively low annual CO_2 emissions per capita.

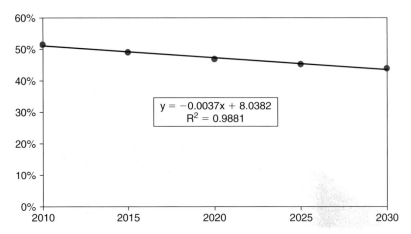

Fig. 2.1. Projected global share of industrialized countries' CO_2 emissions from fossil fuel burning under a BAU scenario (%).

Among these scenarios, A1B represents a mid-range path which can be considered a 'business as usual' (BAU) scenario for future CO_2 emissions.[5] After combining the EIA's reference case projections of fossil fuel-based CO_2 emissions for 2006–2030 and the IPCC's A1B projections for 2031–2100, BAU projections are made for future world CO_2 emissions from fossil fuels. These results were then combined with the IPCC's projections for emissions related to land use change for 2006–2100, under the A1B scenario.

An important aspect of future CO_2 emissions involves the location of these emissions. Currently, most fossil fuel-based CO_2 emissions originate in industrialized countries. However, rapid economic development and high population growth are expected to significantly increase the share of such emissions attributed to Southern countries. Nevertheless, on a per capita basis, emissions from industrialized countries are likely to remain higher for some time. Projections by the EIA provide the total fossil fuel-based CO_2 emissions of the industrialized (Annex 1) and industrializing (non-Annex 1) countries for 2006–2030 (EIA, 2006). A statistical curve fitting of EIA projections for the two groups indicates industrialized countries will account for a slowly declining global share of fossil fuel emissions (see Fig. 2.1). For the period 2031–2100, CO_2 emissions among industrialized and industrializing countries are derived by assuming the same trend for change in the national shares of emissions as we found during 2006–2030. This assumption means that the projected share of fossil fuel-based CO_2 emissions from Annex 1 will decrease from 51% in 2010 to 17% in 2100. The resulting BAU emission scenario (expressed in total CO_2 emissions and per capita emissions) is presented in Figs. 2.2 and 2.3, respectively.

Researchers at the Center for Energy and Environmental Policy (CEEP) have investigated scenarios for large CO_2 emission reductions since the early 1990s. A 1998 publication sought to fix scenario parameters in a manner that would satisfy: (1) a *sustainability* condition based on IPCC assessments of needed global CO_2 emission cuts to halt warming risk at

---

[5] For example, if we average carbon concentration levels and mean temperature increases under the scenarios in Table 2.1, the average value will be 710 ppm and the temperature increase will be 3°C, which are the concentration level and temperature increase for the A1B scenario.

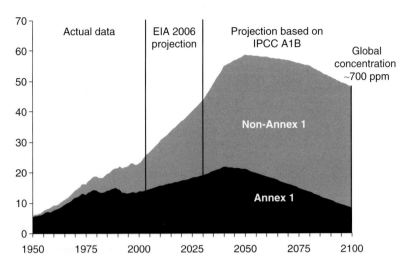

Fig. 2.2. CO_2 Emissions from fossil fuel burning under a BAU scenario (in $GtCO_2$).

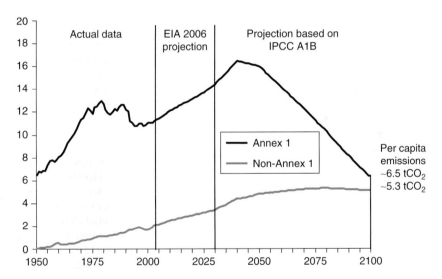

Fig. 2.3. Per capita CO_2 emissions from fossil fuel burning under a BAU scenario (tCO_2 per capita).

current levels; and (2) an *equity* principle in which the biospheric carbon store is equally shared (Byrne *et al.* 1998). The sustainability condition used in the modelling reported here is a level of emissions that will not increase carbon concentrations above 450 ppm by 2050.[6] This agrees with the most recent IPCC assessment (2007). To determine this pathway, the interrelation of carbon concentration levels and emissions are first established, and then a corresponding emissions path is derived. For this purpose, CEEP researchers relied upon

---

[6] The recent report of the UNDP (2007) on human development uses a 450 ppm concentration as well.

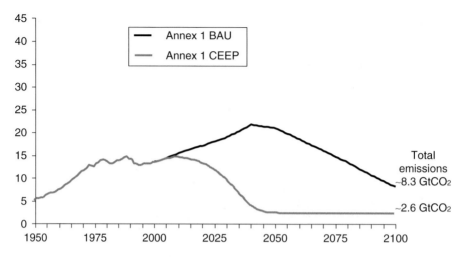

Fig. 2.4. CO_2 emissions from fossil fuel burning in Annex 1 countries under the BAU and CEEP scenarios (in $GtCO_2$).

a mixed-layer, pulse response function model (known as the Bern model).[7] In addition, assumptions were made regarding carbon uptake from the land biosphere and GHG emissions due to land use change based on IPCC (2001) and Joos *et al.* (2001).[8]

To satisfy the model's equity principle, a projection of future population growth is built on United Nations population projections through 2050 (UN, 2005); afterward, it is assumed that human population would stabilize at the 2050 level. Finally, an emissions path must stipulate reductions in carbon emissions by industrialized and industrializing countries. Starting in 2010, Annex 1 emissions are modelled to decline rapidly in order to reach a 2050 rate consistent with the specified sustainability and equity requirements. For Southern countries, carbon emissions are modelled to increase through 2040 and then decline, but at rates which are always slower than those for Annex 1.

Specifically, for Annex 1 countries, emissions would follow the BAU scenario until 2010 and subsequently decline to a level of 2 tCO_2 per person by 2050 (see Fig. 2.4). Total emissions from Annex 1 countries in 2100 would equal 2.6 $GtCO_2$. For non-Annex 1 countries, emissions would follow the BAU scenario until 2025, after which they grow slower than the BAU case. After 2040, non-Annex 1 emissions begin to decline, reaching the level of 2 tCO_2 per person in 2060 (see Fig. 2.5).[9] This scenario (hereinafter called the CEEP scenario)

---

[7]The model was developed by the Climate and Environmental Physics Institute at the University of Bern, Switzerland. Detailed discussion of this model, due to its technical complexity, is beyond the scope of this chapter. Readers can consult the IPCC's discussion of the model (1997); see also Joos and Bruno (1996) and Joos *et al.* (1996).

[8]Specifically, we used the assumption in the A1B scenario of emissions from land use change equal to 0.4 PgC per year, which implies that excess carbon released from the land biosphere is 2.7 PgC (based on Joos *et al.* 2001).

[9]The level of 2 tCO_2 per person in 2100 is based on the sustainable and equity rate of 3.3 tCO_2 per person at the 1990 world population level developed by Byrne *et al.* (1998). However, the 1998 Byrne *et al.* paper included all greenhouse gases that are not available for many countries, the analysis reported here addresses only CO_2. As a result, estimates of atmospheric stability (i.e. 450 ppm of CO_2) under the different scenarios presented in this chapter may understate the needed level of reductions.

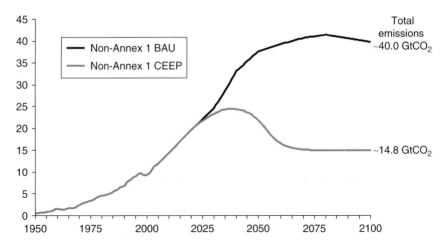

Fig. 2.5. CO_2 emissions from fossil fuel burning in non-Annex 1 countries under the BAU and CEEP scenarios (in $GtCO_2$).

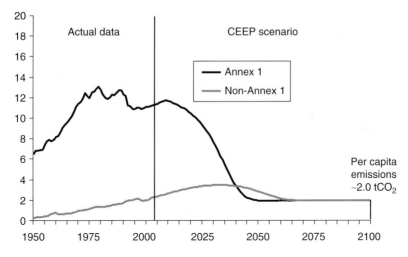

Fig. 2.6. CO_2 emissions from fossil fuel burning under the CEEP scenario (in $GtCO_2$).

expects Southern nations will take some time to adjust emission trends, while industrialized nations are subjected to an obligation of rapid emission reductions in order for the human community to meet the sustainability target.

In Figs. 2.6 and 2.7, resulting projections under the CEEP scenario are presented. Figure 2.6 shows that, for Annex 1 countries, per capita emissions should rapidly decline after 2010 to 2 tCO_2 per person in 2050. For the non-Annex 1 countries, per capita emissions can grow until 2040, reaching 3.5 tCO_2 per person and then declining to a global average of 2 tCO_2 per person after 2060. Figure 2.7 indicates that, under the CEEP scenario, the achievement of a sustainable level of carbon concentration (i.e. 450 ppm by 2050) requires emissions to be nearly one third of those forecasted in the BAU scenario.

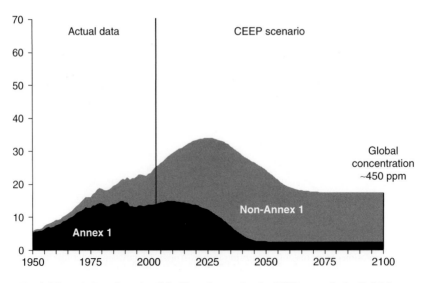

Fig. 2.7. Total CO$_2$ emissions from fossil fuel burning under the CEEP scenario (in GtCO$_2$).

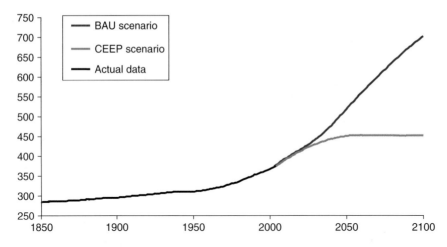

Fig. 2.8. Projections of atmospheric carbon concentration under the BAU and CEEP scenarios (in ppm).

For comparison, the carbon concentration paths for the BAU and CEEP scenarios are presented in Fig. 2.8. Under the BAU scenario, carbon concentrations will gradually increase and reach approximately 520 ppm in 2050 and 700 ppm in 2100. By contrast, under the CEEP scenario, concentrations stabilize around 450 ppm by 2050. Obviously, the BAU scenario violates the specified sustainability condition and equity principle.

2.2 Carbon Emission Allocations Under an Equity Consideration

Equity has both theoretical and practical groundings in concerns for sustainability. As previously explored in the literature, equity can be used to consider how different populations,

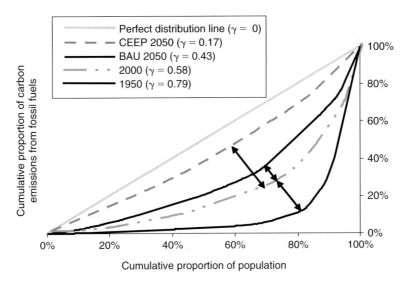

Fig. 2.9. Lorenz curves for global carbon emissions.

regions, and life forms – now and across generations – will be impacted by various environmental harms (Haughton 1999). Applied to the challenge of climate change, equity has been defined in terms of per capita GHG emissions that may safely and fairly be released by the global human population. To derive this amount, CEEP researchers first looked to the GHG emissions levels for 2050 identified in the 1990s by the IPCC as compatible with scenarios for a stabilized climate based on global carbon sink capacity (Byrne *et al.* 1998, p. 337). Then, this target volume for GHG emissions was divided by global population, approximately 5.2 billion people, in 1989. This calculation produced a yearly target of some 3.3 tonnes CO_2 equivalent released per person (Byrne *et al.* 1998, p. 337), apportioned equally among the residents of all nations regardless of their economic or political power. The per person target upholds a 'democratic principle that no human being is entitled to greater access to our atmospheric commons' (Byrne *et al.* 2001, p. 451). However, one should note here that increases in world population since 1989 have necessitated a further reduction in the allowable amount of GHG emissions per person, causing the original target of 3.3 tonnes CO_2 equivalent to fall to 2.0 tonnes by 2050.

With an equitable and sustainable emission rate in hand, the next challenge is to compare national and regional efforts to meet this rate over time. One notable method for measuring equity is the Gini coefficient (Stiglitz 1993), which is from Lorenz curves developed more than 100 years ago (Lorenz 1905) for the purpose of characterizing the extent of inequality in a community's income distribution. While the Gini coefficient has largely been used to gauge income inequality, it can be applied to measuring the inequality of carbon emissions among different nations or regions.

To plot a Lorenz curve for carbon emissions, per capita emissions were at first sorted from low to high values. Then, the percentage of total cumulative carbon emissions corresponding to the cumulative percentage of population is plotted. The derived Lorenz curves for the actual years of 1950 and 2000, as well as projected emissions under the BAU and CEEP scenarios for the year 2050, are presented in Fig. 2.9. As can be seen from this analysis, historical inequality decreased from 1950 to 2000, because non-Annex 1 emissions increased. If carbon emissions continue under the BAU scenario, inequality will continue to decline but will nonetheless be extensive in 2050 (see the discussion of Gini coefficients below). By contrast, under the CEEP scenario significant improvement will occur in the equitable distribution of carbon emissions.

After plotting the Lorenz curves, corresponding Gini (γ) coefficients are derived.[10] In 1950, γ was 0.79, which indicates that vast inequality in per capita carbon emissions existed at a global scale. In 2000, γ was reduced to 0.58, still indicating significant inequality. Under the BAU scenario, the Gini coefficient by 2050 will reach 0.43, indicating some improvement yet still failing to reach a more equitable allocation of emissions among nations. If the CEEP scenario is followed, by 2050 γ will have significantly decreased, reaching 0.17.

Thus, three distinct pathways exist toward equity, albeit 'equity' of very different types. The shift in the Lorenz curve from 1950 to 2000 depicts an increase in equity resulting from higher overall GHG emissions throughout the world, where developing nations have begun to increase their emissions toward the significant scale of releases demonstrated by industrialized countries. The Lorenz curve shift from 2000 to the BAU 2050 scenario portrays a world where equity increases because of higher non-Annex 1 emissions, while world emissions overall reach a plateau. In both instances, equity improvement occurs *without* corresponding progress in sustainability. In fact, both the historical and BAU cases increase warming risk, with the BAU path doing so at alarming levels (leading to CO_2 concentrations of more than 710 ppm by 2100). By contrast, the shift from 2000 to the CEEP 2050 scenario represents a type of equity under which non-Annex 1 nations at first emit increasing amounts of GHGs, while Annex 1 countries simultaneously achieve significant cuts. Importantly, aggregated global emissions – to include non-Annex 1 actors – fall to sustainable levels by 2050 and avert warming risk by the end of the century.

In this regard, the CEEP scenario achieves equity of national effort to realize climate sustainability, while the BAU case achieves neither. This underscores a key finding: it is not possible to avert warming risk if unequal efforts to reduce CO_2 emissions are maintained. The implication of this finding is next explored for the case of the largest CO_2 emitter in the world – the US – and the decision to date of its national government to refuse participation in global accord to cut emissions.

2.3 Impact of US (In)Action on Climate Sustainability and Carbon Equity

On sustainability and equity grounds, the CEEP scenario is far superior to the BAU path. Implementing the CEEP scenario requires both groups of nations (i.e. Annex 1 and non-Annex 1) to make commitments for reducing and curbing their emissions. This is particularly urgent in the case of the US, which accounts for more than 40% of emissions by Annex 1 countries (CDIAC 2006).

To reach sustainable emission levels not exceeding 450 ppm in 2050, all industrialized nations need to begin reducing their emissions no later than 2010. To demonstrate this point, an analysis is presented in which US CO_2 emissions are initially projected to climb according to the EIA (2006) forecast, and later emissions are reduced to the 2 tonnes CO_2 per capita level. Figure 2.10 displays the path of emissions projections for this scenario. The rest of the world is assumed to follow the CEEP scenario. The derived results show that even if other industrialized and non-Annex 1 nations meet their proportionate obligations under a 450 ppm policy regime, global concentrations increase to a probable 480 ppm by the conclusion of the century. Even an extreme case in which both the EU and Japan lower their emissions by 2030 to zero will result in a probable global concentration of 465 ppm (see Fig. 2.11).

---

[10] For a mathematically evolved discussion of the Gini coefficient and Lorenz curve, see Gastwirth (1972).

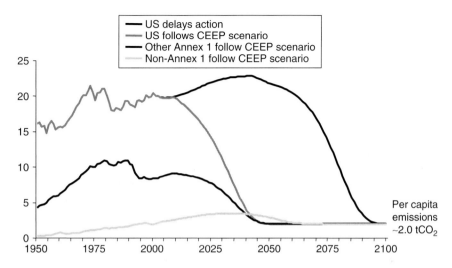

Fig. 2.10. Per capita CO_2 emissions from fossil fuel burning, in the US and other countries under different scenarios (tCO_2 per person).

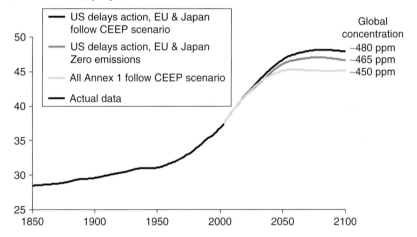

Fig. 2.11. CO_2 concentrations under different scenarios (ppm).

The zero emissions case is not only unlikely but will also increase the inequity of carbon emissions among nations, as demonstrated in Fig. 2.12. Under Scenario 1, where all Annex 1 countries follow the CEEP scenario, the Gini coefficient falls to 0.17. For Scenario 2 in which the US delays action, γ almost doubles to 0.33. If the EU and Japan try to compensate for US inaction, γ rises to 0.39 because inequality persists between Annex 1 and non-Annex 1 countries *and* inequality within the Annex 1 group is exacerbated by delayed US action (see Scenario 3 in Fig. 2.12).

2.4 American Civil Society in Revolt: Breaking Ranks with the National Government

As illustrated above, climate sustainability *cannot* succeed without robust participation by the US. Yet, US national policy is built on inaction and delay of the type modelled above.

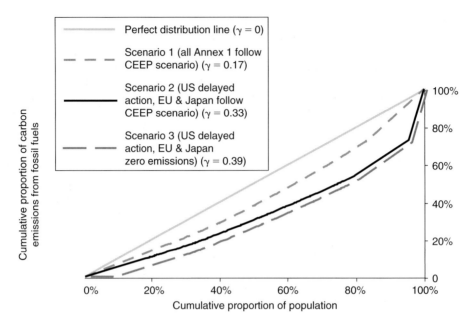

Fig. 2.12. Lorenz curves under different emission scenarios in 2050.

This raises a fundamental political problem: how shall the world community interact with American society to address the need for significant and rapid action.

Understandably, attention has been focused on US intransigence in UNFCCC treaty negotiations. Our argument here should not be construed as, in any sense, a call for diminished pressure on US national policy and its leadership. As we discuss below, however, there is evidence of a sizeable and growing divide between American national policy and civil society. This divide offers a second response to the political problem: engagement of American communities prepared to participate in the repair of the atmospheric commons. The politics of this strategy are merited not only by the possibility of overcoming US national governmental inaction, but it may also more properly locate the ground and momentum of the social change needed to halt the warming risk. As evident in the discussion below, major reductions in CO_2 emissions require *community transformation*.

National and international reduction targets and corresponding commitments of funds to support social action are essential components of greenhouse politics, but these agendas can neither embody the diversity of strategic actions needed, nor can they stand for community will and action – the crucible of transformative change. Indeed, what we describe here as a civil revolt against national policy underscores the incompleteness of nationally and internationally organized politics, even when the challenge is surely global in character.

An extraordinarily diverse collection of American states and cities are now working to fill the void left by the US national government in taking real action to sustain life in the greenhouse (Byrne *et al.* 2007, 2006a). Through their efforts, a politics of climate protection is forming which is intimately linked with goals of greater economic security, better public health, and improved quality of life. In this vein, ecological and community political agendas are merging while recognizing the differences of locale as a source of political innovation in responding to warming risk.

Municipal governments in the US represent an increasingly influential force for climate protection. Despite their increasing location as the headquarters of global economic activity amid a more general urbanizing trend, cities in many cases are simultaneously displaying an interest in addressing a global ecological agenda. The phenomenon of cities acting as forces for climate protection can be seen in the proliferation of individual policies alongside wider American city participation in climate-conscious policy networks. For example, the Cities for Climate Protection (CCP) campaign, established by ICLEI (International Council for Local Environmental Initiatives (ICLEI), 2007) has linked 650 local governments throughout the globe to cut CO_2 emissions 20% from 2000 levels by 2010. One hundred and seventy one US cities have adopted the CCP target, complemented by local action plans and programs which monitor and report progress (ICLEI 2007). At present, these US municipalities represent nearly one fifth of the country's population.

The US Mayors Climate Protection Agreement was adopted in February 2005. Its 435 participating cities have pledged to meet or go beyond targets originally established for the US under the Kyoto Protocol and also to push officials at state and federal levels to act accordingly. Yet another framework, the International Solar Cities Initiative (ISCI), calls on cities throughout the world – to include those in the US – to act more aggressively to reduce carbon emissions to levels consistent with that necessary to achieve climate stabilization by 2050. Its target is 2.0–3.3 tonnes CO_2-equivalent released per person annually, apportioned equally among the nations (and cities) of the world (Byrne *et al.* 2006a).

For cities that have undertaken aggressive efforts to control their greenhouse gas emissions, chosen pathways vary widely, reflecting the goals and needs of diverse communities. In Austin, Texas, a city of 650000 residents, local leaders have linked greenhouse gas reduction to two major platforms pursued via the city's municipal utility, Austin Energy (AE). The utility is working to source some 20% and 15% of energy demand in 2020 from renewable energy and energy efficiency, respectively (City of Austin 2003). The renewable energy goal includes a 100 MW solar commitment. These efforts aim to meet 'realistically achievable' goals set in 1997 for lowering CO_2 emissions by 4.5 million tonnes, a 25% cut compared to business-as-usual projections of 16.7 million tonnes in 2010 (City of Austin 1997, 2001).

Austin and its municipal utility have sought these measures for a number of reasons. First, should the city ever be forced to open its service area to competition, AE's provision of green power products was deemed capable of helping the utility retain its popularity among customers (Sustainable Energy Task Force 1998). This is an important consideration, as AE traditionally has provided a major source of revenue for the city of Austin. Second, as a municipal utility, AE has long encouraged conservation and energy efficiency in the community as a means of avoiding expensive additions to its generation capacity and grid infrastructure. Third, as the city over time has grown more reliant on natural gas for electricity generation, this fuel has become more expensive and its price trends are punctuated by volatile swings in amount. The city therefore seeks to proactively manage its vulnerability to reliance on spiking fossil fuel costs by utilizing wind and increasingly solar energy to meet local energy demand. Additionally, by carefully cultivating its reliance on alternative energy, Austin aims to stake a lucrative claim within the growing international market for advanced energy and related technology.

While all these factors have proven significant in helping the community to devise its ambitious programs for a more sustainable energy future, a major push for such initiatives can also be linked to the conscious efforts of local citizens and community groups who have worked since the 1970s to keep environmental issues on the energy agenda in Austin. In turn, the community has benefited from their foresight. Under AE's GreenChoice program to buy electricity from renewable sources, households and businesses pay an alternative fuel charge, which is currently higher than conventional charges, but it remains fixed throughout the term of the agreement, so that participants are protected from volatile conventional fuel prices. The use of solar PV and solar thermal installations in this program is supported by city-funded rebates, indicative of a community politics of sustainability (City of Austin 2003; Austin Energy 2004a).

For energy efficiency and conservation, AE is distinguished as a global leader in green building. Utility representatives work with the construction industry and end users to market innovative technologies and processes, offering training in related fields and rating both new and refurbished structures according to their energy needs. Rebates are provided to customers for the purchase of energy efficient products and appliances, alongside free or for-payment audits to help customers identify cost-effective improvements. Renters and homeowners are eligible for free services such as caulking, weather stripping, solar screen shading, ductwork sealing, and attic insulation (Austin Energy 2004b). To address transport, utility and city officials have pushed for a national commitment among automakers for the development and dissemination of the 'Plug-In Hybrid Vehicle' (Austin Energy 2005), which could be charged from the grid on electricity generated at night from wind farms. If wind were to play a substantial role in such electric generation, the relative cost of each 'electric' gallon of fuel to power the vehicles could prove vastly cheaper compared to current prices of gasoline – even as the cars themselves generate substantially less greenhouse gases (Austin Energy 2005).

While various AE programs are funded by customer rates and city grants and low and zero interest loans, the utility has simultaneously succeeded in achieving lower rates through avoiding construction of new electric plants (Regelson 2005). Aided by approximately $5 million in rebates for 40 000 area apartment units, utility expenses have decreased up to 40% for some Austin residents. Approximately 1100 homes recently achieved a 'star' rating or better under AE's Green Building programs, and 19 commercial buildings have received LEED (US Green Building Council's Leadership in Energy and Efficiency Design) silver certification (Magnusson 2005). Additionally, GreenChoice has achieved the greatest US sales of utility-sponsored green power (NREL 2004), generating 340 GWh of electricity from renewable sources and lowering CO_2 emissions by nearly 255 000 tons (ICLEI, n.d.). Collectively, Austin's efforts have allowed the utility to forego electricity generation equivalent to that of a 500 MW electric plant (Austin Energy 2003). In light of the city's successes to date in transforming its energy system, the Austin city council in February 2007 approved an advanced Climate Protection Plan. Looking ahead to 2020, targets entail increasing the role of renewable energy to meet 30% of overall energy demand, while offsetting demand for an additional 700 MW by enhanced energy efficiency and conservation. By 2015, new single-family homes should be 'zero net-energy capable', with other new buildings achieving at least a 75% increase in their energy efficiency (City of Austin 2007, p. 2).

The city of Chicago (population 2.8 million) has similarly acted to achieve a more sustainable future as 'the most environmentally friendly city in America', a goal extensively promoted by Chicago's mayor. Its efforts have centred, primarily, on lowering local energy demand and increasing the energy efficiency of both public and private sectors. Chicago has taken such action in part as a means to improve regional air quality, which has negatively affected human health and increased costs for industrial and other economic actors seeking to meet federal environmental standards (City of Chicago 2001, 2004). Moreover, heatwaves in 1995 and 1999 placed serious strains on the local electric grid in Chicago, leading to power outages and contributing to the deaths of hundreds of area residents (Regelson 2005). The city has subsequently sought to apply funds from a $100 million settlement negotiated with the local private energy utility, Commonwealth Edison (ComEd), to help mitigate energy demand and improve electric reliability within Chicago.

To reach these goals, the city has put forward a number of initiatives. Under its Rebuild Chicago program, the municipal government has supported partnerships by which zero or low interest loans are offered to manufacturers and other firms in helping them lower the energy intensity of their industrial processes. Meanwhile, private buildings meeting green standards benefit from an expedited process in receiving needed permits (City of Chicago Department of Environment 2006). As a result of improvements being made at present in Chicago's public buildings – representing approximately 15 million ft^2 – the city should avoid energy and related expenses of some $6 million each year. Further economic and environmental gains are expected as new municipal buildings in Chicago are now required to achieve at least LEED silver certification (Widholm 2006).

Chicago also has sought to dramatically enhance the natural beauty and greenery of the local environment. Particular efforts here have involved support for green roofs, which can now be spotted growing on 2.5 million ft^2 of both commercial and residential buildings, as well as City Hall (McCarthy 2006). Some 30 000 trees are planted annually (Schneider 2006; Johnston 2006). These efforts simultaneously aim to improve air quality within Chicago while also helping to lower the urban heat island effect.

Finally, in addition to the city's targeted push for greater energy efficiency, the municipal government of Chicago set a goal in 2001 for 20% of its electricity to come from renewable resources by 2006 (City of Chicago 2001). Although the goal was later pushed back to 2010, the city is working with public and private partners, including ComEd, in placing solar energy installations throughout Chicago. Noteworthy examples of solar power can now be found atop the Chicago Center for Green Technology, as well as local schools and museums (Chicago Solar Partnership 2006). The emphasis on distributed clean electricity generation serves not only to enhance the reliability of electric service, but also helps reduce pollution and encourages local consumer interest and awareness regarding solar energy and alternative resources.

Through these initiatives, Chicago aims to improve environmental conditions in the city, lower energy costs for industrial, commercial, and residential energy users, and facilitate the emergence of a new industrial sector – one specifically linked to clean energy technology and services. More broadly, the city can lower its greenhouse gas emissions some 7% beneath its 1990 levels by 2012, a policy target established as part of Chicago's participation in the US Mayors Climate Protection Agreement (City of Chicago Department of Environment 2006; City of Seattle Mayor's Office 2006). Environmental *and* economic sustainability thus figure prominently in Chicago's political agenda for the twenty-first century.

As with Chicago, the city of San Francisco is taking major steps to alter its energy future by promoting energy efficiency and utilizing the savings to invest in renewable energy development. San Francisco's effort in this regard has been motivated by several factors. First, the community – like much of California – was subjected to the 2000–2001 energy crisis in the state, following the implementation of electricity sector restructuring (Beck 2002). At that time, service interruptions and escalating electric prices had major impacts on the residents and economy of the city. In more technical realms, San Francisco's location on a peninsula means that the city must rely significantly on electric imports into the community. With mounting demands on its limited energy infrastructure due to population and economic growth, local officials have expressed concern that existing transmission capacity may not prove sufficient to area needs, with the potential for both reliability and price impacts (Smeloff *et al.* 2002). Moreover, as a means to take pressure off of existing transmission lines serving San Francisco, older power stations reliant on fossil fuels have long been required to operate within the city. Within the neighbourhoods where these plants have been located, local residents have experienced pollution and high rates of breast cancer and asthma (Greenaction 2007). Furthermore, San Francisco's bay location means that the city is vulnerable to rising sea levels, which could threaten much of its existing urban infrastructure (SFE and SFPUC 2004).

To lessen its reliance on fossil fuel electric plants and regional transmission systems, while improving local environmental health, San Francisco has adopted a multi-pronged approach. San Francisco voters passed a $100 million bond initiative as the funding mechanism for sustainability investments. San Francisco (which is both a city and a county) is seeking to alter the method by which it receives electric service in the community. At present, the city receives electric service from a private utility – Pacific Gas & Electric (PG&E). However, through a new plan for community choice aggregation, the local government of San Francisco is seeking to devise a framework by which it may offer blocs of residents and businesses alternative service bids from entities other than PG&E. Within the terms of service, the alternative electric service provider would be compelled to meet targets set by San Francisco for renewable energy, energy efficiency, conservation, distributed generation, and related measures to meet 360 MW of community energy demand, compared to typical daytime demand of 850 MW (SFE 2004). The new energy target could assist the city in meeting its goal to reduce community greenhouse gas emissions by 2012 from a business as usual projection of 10.8 million tonnes released annually, down to 7.2 million tonnes by that year – representing a 20% decrease (SFE and SFPUC 2004).

San Francisco promotes the installation of solar photovoltaic technologies in both public and private structures, to include the largest municipal solar installation in the US (atop the Moscone Convention Center). Together with energy efficiency measures, Moscone's PV installation is offsetting the demand for approximately 4 million kWh each year and avoiding the release of 35 000 tonnes of CO_2 that would otherwise be emitted from the use of fossil fuels (Moscone Center 2005). Efforts for alternative energy development are complemented by a number of initiatives to promote more sustainable construction. To lessen energy demand in buildings, San Francisco has passed a Green Building Ordinance under which new and renovated municipal structures larger than 5000 square feet are required to achieve LEED silver standards (SFE 2006). Through programs such as the Mayor's Energy Conservation Account, the Energy Watch Program, the Power Savers Program, and the Peak Energy Program, the city is working to reduce local energy demand by some 55 MW by 2008 and 107 MW by 2012 (SFE 2007; Smeloff *et al.* 2002).

US states are also taking sizable steps to reduce their impacts on the world's climate. Some 28 US states and Puerto Rico have adopted Climate Action Plans (CAPs) establishing goals for lowering greenhouse gas emissions through a number of diverse activities (EPA 2007). The range of such activities span energy efficiency, renewable energy, waste management and recycling, public transportation, the use of alternative fuels in fleets, and land use. Specific types of policies, and examples of state action in such areas, are briefly reviewed below.

Oregon mandates that new power plants offset 17% of their expected CO_2 emissions (Oregon Department of Energy 2004) through direct reductions or through contributions made to a fund managed by the state's Climate Trust. Similar power plant regulations have been adopted by Washington State (Washington Department of Ecology 2004). In California, Assembly Bill 32 mandates that the California Air Resource Board (CARB) put forward regulations by which to lower the state's greenhouse gas emissions by 2020 to 1990 baselines (CARB 2006). CARB is also to set GHG emissions standards for light trucks and cars by 2009. While the national government is challenging these standards, California's actions will and are already affecting the US, considering that states on both the east and west coasts have enacted compatible rules following California's example (Council of State Governments Eastern Regional Conference 2006; Freeman 2006).

The state of New York has established targets to lower its greenhouse gas emissions by 5% against 1990 levels in 2010, and 10% against 1990 levels by 2020 (New York State Energy Research and Development Authority (NYSERDA), 2002). Relevant strategies to pursuing such goals span energy efficiency and energy demand reduction, to greater reliance on renewable energy and the use of distributed generation (Center for Clean Air Policy 2003).

Delaware has set goals to lower its greenhouse gas emissions 30% below BAU by 2019, and has approved the formation of a 'sustainable energy utility' (SEU) to undertake energy policies supportive of climate protection (Delaware SEU (Sustainable Energy Utility) Task Force, 2007). The initiative is funded by a Sustainable Energy Bond and a charge on electricity sales which, together, will transform energy capital investment from its supply bias to a demand-side focus.

New Jersey's commitment to lower greenhouse emissions has grown considerably from its original target of 3.5% below 1990 levels by 2005 (New Jersey Climate Change Workgroup 1999). While modest by today's standards, this commitment included a signed Letter of Intent with the Netherlands for shared action in establishing a system for emissions banking. With private and public actors partnering under the program, New Jersey pursued 'covenants' for greater reliance on efficient technologies, reduction of waste, and conservation of both energy and open space (New Jersey Sustainable State Institute 2004). Success over the seven years of the initiative led to a dramatic step in 2007 – the governor of New Jersey issued an executive order under which the state must decrease its emissions 20% below 1990 levels by 2020, and by 80% by 2050 (nj.com, 2007). Representing the most aggressive climate planning commitment in the US, New Jersey – the country's most urbanized state – demonstrates how deeply climate action is embedded in the fabric of American civil society.

As well, states have joined together to pursue regional initiatives for climate protection. In the northeast states, the relative scarcity of native fossil energy supplies – resulting in often high energy prices – has contributed to an interest in alternative energy development *and* controlling carbon emissions (US Department of Energy (DOE) 2003). In 2003, 11 northeast states collectively established a regional cap-and-trade program to control carbon emissions from power plants (Union of Concerned Scientists 2003), known as the Regional Greenhouse Gas Initiative (RGGI). Under this framework, participating states have pledged to stabilize power plant emissions at 137 million metric tons from 2009 to 2015. Then, from 2015 to 2020, emissions must fall 10% beneath the cap (RGGI 2005). Present participation under RGGI includes ten states (Connecticut, Delaware, Maine, Maryland, Massachusetts, New Hampshire, New Jersey, New York, Rhode Island, and Vermont) that have adopted a timeline for emissions reduction (RGGI 2007). Other states may join RGGI, whose partnership structure is open ended and welcomes additional participants.

State agencies in the northeast that seek to protect air quality are additionally working to devise cooperative actions for greenhouse gas reduction, as part of the Northeast States for Coordinated Air Use Management (NESCAUM). Beyond the northeast region, NESCAUM has worked 'to promote harmonized GHG accounting and reporting standards' with the California Climate Action Registry (California Climate Action Registry, 2005).

On the US west coast, the governors of Oregon, California, and Washington in 2003 established the West Coast Governor's Global Warming Initiative (WCGGWI (2004)) to examine the likely impacts of emerging climate and alternative energy policies. The WCGGWI (2004) additionally signalled its support for the creation of a regional system for emissions trading compatible with that enacted by RGGI, and the harmonization of GHG vehicle emissions standards. The governors of New York and California even put forward plans in October 2006 to link California's GHG reduction program with RGGI (Young 2006), allowing an opportunity for wider shared action among west and east coast states.

Together, the RGGI and WCGGWI states release approximately 1000 MMT CO_2 each year, some 20% of US CO_2 emissions (Fontaine 2005; EIA 2003). Representing approximately 30% of the US population, these states as members of a coordinated plan for greenhouse gas reduction should increase pressure on the national government to harmonize their efforts in ways compatible with the smooth operation of interstate commerce.

In assessing the impact of WCGGWI and RGGI efforts to lower emissions from power plants, such actions over the coming decade could result in an emissions decrease of 21% against present forecasts (based on data from the US Energy Information Administration (see Byrne *et al.* 2007).

Beyond dedicated policies to directly address climate protection, US states are also acting in diverse ways to support renewable energy and energy efficiency. Several policies are promoted not simply for their carbon impacts but, equally important, as tools to lower pollution, reinforce energy security, lessen volatility in energy prices, and create opportunities for new markets and jobs related to cleaner technologies. For example, the US is now home to the largest market in the world for customer-driven electricity from renewable sources (Bird *et al.* 2002). Its 'citizens' market is comprised of green pricing, competitive green power products, and 'green tags' markets in which individuals, communities and organizations can invest directly in renewable energy plant through the purchase of shares or 'tags'. In 36 states, approximately 600 utilities now offer green pricing options, helping to foster the development of 800 MW of renewable capacity (Bird and Swezey 2006). Commercial and industrial customers are also boosting demand for green power purchases as a means to improve their community standing, fulfil in-house environmental targets, and lower regulatory risks (Hanson and Van Son 2003; Holt *et al.* 2001). These initiatives have led to the development of 1710 MW of renewable capacity (Bird and Swezey 2006).

Many states have also required utilities to procure a minimum amount of their electricity from renewable sources, under Renewable Portfolio Standard (RPS) policies. By summer 2007, 24 states and the District of Columbia had passed RPS policies, with an additional 14 states examining such legislation. While RPS policies differ in scope and level of success (van der Linden *et al.* 2005), many states have appeared to support more far-reaching policies over time. Certain states have acted to reinforce existing laws, to enlarge targets, or to hasten compliance timelines, where their RPS policies have been active for at least three years (Rickerson 2005). Examples include New Jersey's enlarged target for 23% by 2021, with a 2% solar 'carveout' (DSIRE 2007), New York's enhanced target of 24% by 2013 (DSIRE 2007), and California's accelerated RPS schedule from 20% by 2017 to 20% by 2010 (Doughman *et al.* 2004; California Public Utilities Commission 2006). According to the Union of Concerned Scientists (2006), 44 900 MW of new renewable capacity will come online by 2020 due to current state RPS policies.

Markets for tradable renewable energy credits (RECs) have also begun to proliferate, as most US states with RPS programs allow utilities to obtain their mandated supply beyond state boundaries. Such markets are found in Connecticut, Delaware, Maine, Maryland, Massachusetts, New Jersey, Texas and Washington, DC, and regional authorities have designed systems to track credit trades in Texas, the Northeast, and the Mid-Atlantic, with other efforts targeted for the West and the upper Midwest (Porter and Chen 2004; Wingate and Lehman 2003). The goal of such systems as part of RPS initiatives lies in cultivating wider cooperation among many states in meeting RPS targets and developing renewable energy markets.

By summer 2007, 21 state public benefit funds (PBFs) could be found in the US, with 15 dedicated to renewable energy development (DSIRE 2007). These PBFs are made possible by charges placed on electricity sales within individual states, usually at the level of $0.001 to $0.003 per kWh (Kushler *et al.* 2004). In turn, they generate yearly deposits into state renewable energy accounts approximating \$500 million, so that by 2017 \$4 billion will have been earmarked for renewable energy projects (Union of Concerned Scientists 2004). While wind projects in recent years have received the majority of these funds (Bolinger and Wiser 2006), PBFs are also strong forces for the installation of PV technology. For example, under the California Solar Initiative (CSI) of 2006, \$2.35 billion in PBF funding will go toward the development of 3000 MW of solar electricity by 2017 (Go Solar California! 2006).

Accrued monies from PBFs may also be spent on energy efficiency, related research and development, and household weatherization for low and moderate income families, often in the form of rebates or production credits. Energy efficiency is the largest area of investment, where states often apply their PBFs in ways that enhance commitments made by utilities and other actors. An annual investment of \$1.2 billion is expected from PBFs for energy efficiency in 21 US states until 2015 (DSIRE 2007; American Council for an Energy-Efficient Economy 2005). Even larger investments are likely, since the PBF monies do not include programs funded by cities, municipal utilities, rural electric cooperatives, and investor owned utilities, or state policies aiming to add to public spending with tax incentives for energy-efficiency investments and minimum efficiency standards for appliances (Alliance to Save Energy 2005).

In the transport sector, states are working innovatively to reduce greenhouse gas emissions from vehicles. California's zero emission vehicle standard has received much attention and is being imitated in New York and other states. Other state policies promote vehicle labels, tax incentives, and feebates as mechanisms to enhance the attractiveness of high efficiency and low emission vehicle models. Still other state initiatives advance the use of biofuels or the market competitiveness of hybrid-electric and fuel cell vehicles (Curtin and Gangi 2006). As well, 'smart growth' programs seek to foster urban land use planning where individuals can rely on walking, biking, or public transit rather than individual use of automobiles (Prindle *et al.* 2003).

A recently published estimate by this chapter's authors of the impacts of state and regional policies supporting energy efficiency, renewable energy development (including RGGI and WCGGWI) indicates that scale effects of grassroots action are substantial (Byrne *et al.* 2007). Transport policies were not included, as comprehensive figures on impacts from state policies are not yet available. A decrease in emissions totalling 1663 million tonnes CO_2 by 2020 is attributable to state energy efficiency policies. State RPS policies are forecasted to cause CO_2 emissions reductions of an additional 111 million tonnes by 2020. Projected decreases from the RGGI and WCGGWI programs by 2020 will provide an additional 48 million tonnes. Collectively, the three commitments represent an emissions reduction of 1,822 million tonnes of CO_2 by 2020, compared with the BAU case of 2812 million tons of CO_2 (US Energy Information Administration (EIA) 2007), or a *65% decrease in emissions* (see Byrne *et al.* 2007 for a detailed discussion of the methodology used in this assessment).

2.5 Toward a Grassroots Politics of Climate Sustainability

The political momentum built in US cities, states and regions to initiate climate mitigation and related efforts is to be contrasted with inaction by the US national government in addressing the climate challenge. Support for climate protection can be found in polling of Americans which points to 83% support among the country's citizens for greater national leadership in addressing climate change, and even deeper support for state and community action to address climate concerns (Opinion Research Corporation 2006). If the

American people appear to support such initiatives, the question becomes why are states, cities and regions leading the way, rather than the national government?

US national politics has for decades exhibited a troubling amenability to the interests of fossil fuel and automaker lobbies (Leggett 2001; Public Citizen 2005; NRDC 2001). A recent example of this influence can be found in the history of the National Energy Policy Development Group, which took input 'principally' from actors associated with such interests (US General Accounting Office (GAO) 2003). At the same time, the national administration has been noted for the presence of individuals with backgrounds in the auto, mining, natural gas, electric, and oil industries, in positions at the White House, the Environmental Protection Agency, and the Departments, respectively, of Energy, Commerce, and the Interior (Bogardus, 2004; Drew and Oppel Jr 2004; NRDC, 2001).

State-level politics may be able to obviate this influence through their efforts to allow a more direct citizen influence upon decision making. For example, 23 states permit citizens to petition for a direct vote (Initiative and Referendum Institute 2007), a strategy that has helped ensure the advancement of environmentally minded initiatives within states in recent years, such as the State of Washington's enactment by ballot of an RPS proposal in 2006 (Initiative and Referendum Institute 2007).

An additional reason for the ability of states and cities to push forward green-minded energy and climate policies may also be found in their traditional jurisdictional authority over many spheres of activity with links to such concerns. Examples here include the regulation of energy utilities, as well as states' and cities' responsibilities to address public health, local economic development, land use, and urban planning. A noteworthy rationale can also be found in the objective among many states and cities of avoiding the most damaging scenarios of climate change, which hold the potential to destroy local geography and natural resources and to threaten longstanding economic sectors. Rising oceans from melting ice caps portend saltwater contamination of drinking water as well as crop losses in Rhode Island (Rhode Island Greenhouse Gas Action Plan Stakeholder Process 2002). In New Jersey, hurricanes and droughts stand to have a larger impact on the state's many seashore developments (New Jersey Office of the Governor 2005; New Jersey Climate Change Workgroup 1999). Throughout the coming century, Connecticut's coastline could witness sea level rises of 56 cm, making necessary some $500 million to $3 billion in protective measures and negatively impacting forests and fisheries (Dutzik *et al.* 2004).

Another major driver of locale-based action for climate protection can be found in the relative diversity of the industrial makeup of many cities and states, where fossil fuel and automobile companies seldom dominate local economies. Moreover, policies to promote renewable energy and reduce overall energy demand can enhance demand for new technologies and industries that create local jobs (CEEP 2005; Hoerner and Barrett 2004; Union of Concerned Scientists 2005; Sterzinger and Svrcek 2004a, b). Also, community-based renewable energy initiatives can enhance landowner revenue and the municipal tax base, with such investments showing the potential for a greater multiplier effect on communities' economies than that associated with fossil fuels (CEEP 2005; Hopkins 2003).

The beneficial economic impacts of renewable energy development can be linked in part to its reliance on 'free' fuel inputs, particularly as greater reliance on natural gas-sourced electricity in the US has resulted in growing power price volatility (Henning *et al.* 2003; Klass 2003; Zarnikau 2005). Greater reliance on renewable sources of electricity, energy efficiency and conservation as tools to lessen energy demand can work as a hedge against overreliance on natural gas (Delaware SEU (Sustainable Energy Utility) Task Force 2007; Rickerson *et al.* 2005; Biewald *et al.* 2003). For example, the Lawrence Berkeley National Laboratory has found that, for every 1% of demand replaced by green energy development, a price decline of as much as 2% takes place for boiler fuel natural gas (Wiser *et al.* 2005). This consideration has been cited in examinations of RPS proposals undertaken in Colorado, Delaware, Maryland, New York, Pennsylvania, and Texas, as a potential beneficial impact on electricity prices (CEEP 2005; Deyette and Clemmer 2005; Binz 2004; New York State Department of Public Service *et al.* 2004; Pletka *et al.* 2004; Chen *et al.* 2003).

In these and myriad other ways, grassroots politics in the US[11] is adopting a climate sustainability agenda in defiance of a national politics of inaction. The aphorism 'all politics are local' is mistaken if, at least in this case, it means that global scale problems command little local interest. Instead, it would appear more appropriate to interpret its meaning as a recognition that enduring political commitments, and especially those commitments seeking transformative change, are very often built from the bottom up, not the other way around.

2.6 Civil Strategy to Decarbonize the Human Footprint

The above analysis demonstrates the need to combine global agendas of carbon equity and climate sustainability if we are to successfully address the problem of climate change. Halting the current experiment in warming risk requires *all* industrialized societies to transform their social and economic structures in a manner that is consistent with the carbon cycle *and* social justice. Without transformation along both dimensions, it is unlikely that a global commitment to significantly and rapidly reduce greenhouse gas emissions can be mounted.

Additionally, it is now clear that irresponsible climate action by just one nation – the US – can result in the loss of sustainability for the entire world. If the US delays action, even a commitment by Europe and Japan to *zero* emissions cannot avert warming risk. This vital point cannot be stressed too strongly.

To resolve the current stalemate, the global political and scientific communities need to confront the necessity of an equitable foundation to strategies and policies for GHG reduction. This challenge involves recognizing equity not simply as a target for sustainable rates of energy use, but as a political basis for motivating all nations to adopt and act upon climate protective measures. In particular, if nations with burgeoning populations and economies such as China and India are to be expected to act in concert with international accords, the US – the world's largest violator of the atmospheric commons – must commit to a transformation of its social and economic structures. One political strategy is to sanction the US for its delayed action in the hope that compliance can be forced.

But another option – which does not preclude the sanction strategy – is to forge alliances with American civil society through, for example, locality focused partnerships with US states and cities. The methods to be pursued here remain subject to the democratic negotiations of the global political community, yet the need for such intervention cannot be forestalled. American civil society has undertaken precisely the changes in social and economic structure that its national leadership refuses to consider. Ultimately, compacts with civil society across the world will produce the transformation to renewable cities (Droege, 2007) and a global solar economy (Scheer 2004; see also Byrne *et al.* 2006b). International treaties among governments can call for action, but these agreements cannot produce results.[12] Our political challenge, therefore, is not only the creation of a rhetoric of climate justice, but its practice and this, after all, is the province of civil society.

---

[11]There is fast-growing literature of local policy leadership which includes initiatives in Europe and Asia that cannot be discussed here (please see, for example, Kim 2006).

[12] This fundamental point is sometimes overlooked. Recent interest in restoring US governmental leadership in the climate arena seems to make this mistake. See Claussen and Diringer (2007).

References

Alliance to Save Energy (2005). State Energy Efficiency Index. Available on the Internet: http://www.ase.org/content/article/detail/2356

American Council for an Energy-Efficient Economy (2005). Summary Table of Public Benefit Programs and Electric Utility Restructuring. Available on the Internet: http://www.aceee.org/briefs/mktabl.htm

American Meteorological Society(AMS) (2003). Climate change research: issues for the atmospheric and related sciences. *Bull. AMS*, **84**, 508–515.

Austin Energy (2003). City of Austin, Austin Energy to receive awards for energy efficiency, green power. Press Release (November 4).

Austin Energy (2004a). Austin Energy launches solar program. Press Release (May 27).

Austin Energy (2004b). Free weatherization program helps elderly, disabled, low-income save on energy bills. Press Release (February 5).

Austin Energy (2005). Austin Kicks off Plug-in Hybrid Campaign. Press Release (August 22). Available on the Internet: http://www.austinenergy.com/

Beck, R.W. (2002). *Energy services study summary*. San Francisco Local Agency Formation Commission, San Francisco, CA.

Biewald, B., Chen, C., Sommer, A., *et al.* (2003). *Comments on the RPS Cost Analyses of the Joint Utilities and the DPS Staff*. Synapse Energy Economics, Cambridge, MA.

Binz, R.J. (2004). *The Impact of a Renewable Portfolio Standard on Retail Electricity Rates in Colorado*. Public Policy Consulting, Denver, CO.

Bird, L. and Swezey, B. (2006). *Green Power Marketing in the United States: A Status Report* (9th ed. NREL/TP-640-40904). National Renewable Energy Laboratory, Golden, CO.

Bird, L., Wustenhagen, R. and Aabakken, J. (2002). A review of international green power markets: recent experience, trends, and market drivers. *Renewable and Sustainable Energy Reviews*, **6**(6), 513–536.

Bogardus, K. (2004). A Pipeline of Influence. The Center for Public Integrity. Available on the Internet: http://www.publicintegrity.org/oil/report.aspx?aid=348

Bolinger, M. and Wiser, R. (2006). *The Impact of State Clean Energy Fund Support for Utility-Scale Renewable Energy Projects*. Lawrence Berkeley National Laboratory and Clean Energy States Alliance, Berkeley, CA.

Byrne, J. *et al.* (2001). The Postmodern Greenhouse: Creating Virtual Carbon Reductions from Business-as-Usual Energy Politics. *Bulletin of Science, Technology & Society*, **21**(6), 443–455.

Byrne, J. and Glover, L. (2005). Ellul and the weather. *Bulletin of Science, Technology & Society*, **25**(1), 4–16. Special issue 'Celebrating the Intellectual Gifts and Insights of Jacques Ellul'.

Byrne, J., Hughes, K., Rickerson, W. and Kurdgelashvili, L. (2007). American policy conflict in the greenhouse: divergent trends in federal, regional, state, and local green energy and climate change policy. *Energy Policy*, **35**(9), 4555–4573.

Byrne, J., Hughes, K., Toly, N. and Wang, Y.-D. (2006a). Can cities sustain life in the greenhouse? *Bulletin of Science, Technology & Society*, **26**(2), 84–95.

Byrne, J., Toly, N. and Glover, L. (2006b). *Transforming Power*. Transaction Publishers, London and New Brunswick, New Jersey.

Byrne, J., Wang, Y.-D., Lee, H. and Kim, J.-D. (1998). An equity- and sustainability-based policy response to global climate change. *Energy Policy*, **26**, 335–343.

Byrne, J. and Yun, S.J. (1999). Efficient global warming: contraindications in liberal democratic responses to global environmental problems. *Bulletin of Science, Technology & Society*, **19**(6), 493–500.

California Air Resources Board (CARB) (2006). Timeline – California Global Warming Solutions Act of 2006. Available on the Internet: http://www.arb.ca.gov/cc/factsheets/ab32timeline.pdf

California Climate Action Registry (2005). Partners. Available on the Internet: http://www.climateregistry.org/Default.aspx?TabID=3333&refreshed=true

California Public Utilities Commission (2006). PUC Striving Towards RPS Goal of 20% by 2010. Available on the Internet: http://www.cpuc.ca.gov/static/energy/electric/renewableenergy/060224_rpssummary.htm

Carbon Dioxide Information Analysis Center (CDIAC) (2006). Online trends: a compendium of data on global change. Available on the Internet: http://cdiac.esd.ornl.gov/trends/co2/contents.htm

Center for Clean Air Policy (2003). *Recommendations to Governor Pataki for Reducing New York State Greenhouse Gas Emissions*. Center for Clean Air Policy in collaboration with the New York Greenhouse Gas Task Force, Washington, DC.

Center for Energy and Environmental Policy (CEEP) (2005). *The Potential Economic Impacts of a Renewable Portfolio Standard in Delaware*. University of Delaware, Center for Energy and Environmental Policy, Newark, DE. Prepared for the Delaware General Assembly. Available on the Internet: http://ceep.udel.edu/publications/2005_es_Delaware%20Senate_RPS%20briefing%20paper.pdf

Chen, C., White, D., Woolf, T. and Johnston, L. (2003). *The Maryland Renewable Portfolio Standard: An Assessment of Potential Cost Impacts*. Synapse Energy Economics, Inc., Cambridge, MA.

Chicago Solar Partnership (2006). Photovoltaic Installations around Chicago. Available on the Internet: http://www.chicagosolarpartnership.org/

City ofAustin (1997). *City of Austin: Carbon Dioxide Reduction Strategy*. City of Austin, Austin, TX.

City of Austin (2001). *The City of Austin Sustainable Communities Initiative*. Available on the Internet: http://www.ci.austin.tx.us/sustainable/

City of Austin (2003). *Austin Energy Strategic Plan – Public Document*. City of Austin, Austin, TX.

City of Austin (2007). Austin Climate Protection Plan. Available on the Internet: http://www.ci.austin.tx.us/council/downloads/mw_acpp_points.pdf

City of Chicago (2001). *Energy Plan*. Department of Environment, Chicago, IL.

City of Chicago (2004). Daley urges Chicagoans: do your part to protect environment. Press Release (April 22).

City of Chicago (2006). *Environmental Action Agenda: Executive Summary*. Department of Environment, Chicago, IL.

City of Seattle Mayor's Office (2006). U.S. Mayors Climate Protection Agreement. Available on the Internet: http://www.ci.seattle.wa.us/mayor/climate/

Claussen, E. and Diringer, E. (2007). A new climate treaty: US leadership after Kyoto. *Harvard International Review*, **29**(1). Available on the Internet: http://hir.harvard.edu/articles/1594/

Council of State Governments Eastern Regional Conference (2006). States ring in the New Year with tougher greenhouse gas rules. *Weekly Bulletin* (January 17).

Curtin, S. and Gangi, J. (2006). *State Activities that Promote Fuel Cells and Hydrogen Infrastructure Development*. Breakthrough Technologies Institute, Inc., Washington, DC.

Delaware SEU (Sustainable Energy Utility) Task Force (2007). Homepage. Available on the Internet: http://www.seu-de.org

Deyette, J. and Clemmer, S. (2005). *Increasing the Texas Renewable Energy Standard: Economic and Employment Benefits*. Union of Concerned Scientists, Cambridge, MA.

Doughman, P., Johnson, D., Lieberg, *et al.* (2004). *Accelerated Renewable Energy Development*. California Energy Commission, Sacramento, CA.

Drew, C. and Oppel, R. Jr. (2004). Friends in the White House come to coal's aid. *New York Times* (August 9).

Droege, P. (2007). *The Renewable City*. Wiley and Sons, Hoboken, New Jersey.

DSIRE (2007). Database of State Incentives for Renewable Energy. Available on the Internet: www.dsireusa.org

Dutzik, T., Ridlington, E. and Suter, B. (2004). *Connecticut Responds to Global Warming: An Analysis of Connecticut's Emission Reduction Goals, Current Strategies, and Opportunities for Progress*. Clean Water Fund and the State PIRGs, Hartford, CT.

U.S. Environmental Protection Agency (EPA). (2007). Climate Change-State and Local Governments: State Action Plans. Available on the Internet: http://www.rggi.org.

Fontaine, P. (2005). A new world order. *Public Utilities Fortnightly*, **143**(2), 26–28.

Freeman, S. (2006). States adopt California's greenhouse gas limits. *Washington Post* (January 3).

Gastwirth, J.L. (1972). The estimation of the Lorenz curve and Gini index. *The Review of Economics and Statistics*, **54**, 306–316.

Go Solar California! (2006). Homepage. Available on the Internet: http://www.gosolarcalifornia.ca.gov/index.html

Greenaction (2007). Fact Sheet. Available on the Internet: http://www.greenaction.org/hunterspoint/factsheet.html

Hansen, J. (2004). Defusing the Global warming time bomb. *Scientific American*, **290**, 68–77.

Hanson, C. and Van Son, V. (2003). *Renewable Energy Certificates: An Attractive Means for Corporate Customers to Purchase Renewable Energy (Corporate Guide to Green Power Markets Installment 5)*. World Resources Institute, Washington, DC.

Haughton, G. (1999). Environmental justice and the sustainable city. *Journal of Planning Education and Research*, **18**(3), 233–243.

Henning, B., Sloan, M. and de Leon, M. (2003). *Natural Gas and Energy Price Volatility*. American Gas Foundation, Arlington, VA. (Prepared for the Oak Ridge National Laboratory).

Hoerner, J.A. and Barrett, J. (2004). *Smarter, Cleaner, Stronger: Secure Jobs, a Clean Environment and Less Foreign Oil*. Redefining Progress, Oakland, CA.

Holt, E., Wiser, R., Fowlie, M., *et al.* (2001). *Understanding Non-residential Demand for Green Power*. National Wind Coordinating Committee, Washington, DC.

Hopkins, B. (2003). *Renewable Energy and State Economies*. The Council of State Governments.

Intergovernmental Panel on Climate Change (IPCC) (1997). Implications of Proposed CO_2 Emissions Limitations. IPCC Technical Paper 4. Available on the Internet: http://www.ipcc.ch/pub/IPCCTP. IV(E).pdf

Intergovernmental Panel on Climate Change (IPCC) (2001). Climate Change 2001: Synthesis Report (ed. R.T. Watson). Cambridge University Press, Cambridge, England. Available on the Internet: http://www.grida.no/climate/ipcc_tar/vol4/english/index.htm

Intergovernmental Panel on Climate Change (IPCC) (2007). Climate Change 2007: Physical Science Basis. Cambridge University Press, Cambridge, England. Also available on the Internet: http://ipcc-wg1.ucar.edu/wg1/wg1-report.html

International Council for Local Environmental Initiatives (ICLEI) (2007). Climate Protection. Available on the Internet: http://www.iclei.org/index.php?id=800

International Council for Local Environmental Initiatives (ICLEI) (n.d.). Case Study: Austin, Texas (Local Government Renewables Portfolio Standard). Available on the Internet: www.greenpowergovs. org/wind/Austin%20case%20study.html

Johnston, S. (2006). The Green Development: making Chicago an environmentally friendly city. *Greater Philadelphia Regional Review* (Winter).

Johnston, S. (2007). Chicago's green business strategy. *Conscious Choice* (March).

Joos, F., Bruno, M., Fink, R., *et al.* (1996). An efficient and accurate representation of complex oceanic and biospheric models of anthropogenic carbon uptake. *Tellus*, **48B**, 397–417.

Joos, F., Prentice, C., Stitch, S., *et al.* (2001). Global warming feedbacks on terrestrial carbon uptake under the Intergovernmental Panel on Climate Change (IPCC) emission scenarios. *Global Biogeochemical Cycles*, **15**, 891–907.

Kim, J. (ed.) (2006). Solar cities: linking climate change and human settlement. *Bulletin of Science, Technology & Society*, **34**(2). Special issue.

Klass, D.L. (2003). A critical assessment of renewable energy usage in the USA. *Energy Policy*, **31**(4), 353–367.

Kushler, M., York, D. and Witte, P. (2004). Five years, in *An Examination of the First Half-Decade of Public Benefits Energy Efficiency Policies (Report No. U041)*. American Council for an Energy-Efficient Economy, Washington, DC.

Leggett, J.K. (2001). *The Carbon War: Global Warming and the End of the Oil Era*. Routledge, London.

Lorenz, M.O. (1905). Methods of measuring the concentration of wealth. *Publications of the American Statistical Association*, **9**, 209–219.

Magnusson, J. (2005). America's Top 10 Green Cities. *The Green Guide* (April 19). Available on the Internet: www.thegreenguide.com

McCarthy, K. (2006). Chicago Approves Big Grants for Green Roof Retrofits. Construction.com (July 19).

Moscone Center (2005). Moscone Solar Featured at World Environment Day. Press Release (July 1).

National Renewable Energy Laboratory (NREL) (2004). NREL Highlights Leading Utility Green Power Programs. Press Release (March 4).

Natural Resources Defense Council (NRDC) (2001). The Bush Cheney Energy Plan: Players, Profits and Paybacks. Available on the Internet: http://www.nrdc.org/bushrecord/airenergy_policy.asp

New Jersey Climate Change Workgroup (1999). *New Jersey Sustainability Greenhouse Gas Action Plan*. New Jersey Department of Environmental Protection Division of Science, Research and Technology, Trenton, NJ.

New Jersey Office of the Governor (2005). Corzine Announces Landmark Regional Agreement to Combat Global Warming. Available on the Internet: http://www.nj.gov/cgi-bin/governor/njnewsline/view_article.pl?id=2851

New Jersey Sustainable State Institute (2004). *Living with the Future in Mind: Goals and Indicators for New Jersey's Quality of Life*. (3rd ed.) New Jersey Sustainable State Institute, New Brunswick, NJ.

New York State Department of Public Service, NYSERDA, Sustainable Energy Advantage, and La Capra Associates (2004). *New York Renewable Portfolio Standard Cost Study Report II, Vol. A*. State of New York Public Service Commission, Albany, NY.

New York State Energy Research and Development Authority (NYSERDA) (2002). *2002 State Energy Plan and Final Environmental Impact Statement*. New York State Energy Research and Development Authority, Albany, NY.

Nj.com (2007). Warming to the Task. Available on the Internet: http://www.nj.com/timesoftrenton/stories/index.ssf?/base/news-0/1171516157208510.xml&coll=5

Opinion Research Corporation (2006). *Global Warming & Alternative Energy: A Leadership Survey*. Prepared for the Civil Society Institute and 40MPG.org, Princeton, NJ.

Oppenheimer, M. and Petsonk, A. (2005). Article 2 of UNFCCC: Historical Origins, Recent Interpretations. *Climatic Change*, **73**, 195–226.

Oregon Department of Energy (2004). Oregon Carbon Dioxide Emission Standards for New Energy Facilities. Available on the Internet: http://www.energy.state.or.us/climate/ccnewst.pdf

Parry, M., Arnell, N., McMichael, T., *et al.* (2001). Millions at risk: defining critical climate change threats and targets. *Global Environmental Change*, **11**, 181–183.

Porter, K. and Chen, D. (2004). *Results of a Survey Regarding a Potential Midwest Renewable Energy Tracking System*. National Council on Electricity Policy and Exeter Associates, Columbia, MD.

Prindle, W.R., Dietsch, N., Elliott, R.N., *et al.* (2003). *Energy Efficiency's Next Generation: Innovation at the State Level (E031)*. American Council for an Energy-Efficient Economy, Washington, DC.

Public Citizen (2005). The Best Energy Bill Corporations Could Buy: Summary of Industry Giveaways in the 2005 Energy Bill. Available on the Internet: http://www.citizen.org/cmep/energy_enviro_nuclear/electricity/energybill/2005/articles.cfm?ID=13980

Regelson, K. (2005). *Sustainable Cities: Best Practices for Renewable Energy & Energy Efficiency*. Sierra Club – Rocky Mountain Chapter, Denver, CO.

Regional Greenhouse Gas Initiative (RGGI) (2005). Memorandum of Understanding. Available on the Internet: http://www.rggi.org/docs/mou_12_20_05.pdf

Rhode Island Greenhouse Gas Action Plan Stakeholder Process (2002). *Rhode Island Greenhouse Gas*. Rhode Island Department of Environmental Management, Rhode Island State Energy Office, Providence, RI.

Rickerson, W. (2005). End users as economic agents of energy sustainability: distributed energy policy and structural change in the electricity industry. Master's Analytical Paper. University of Delaware, Center for Energy and Environmental Policy, Newark, DE.

Rickerson, W., Wong, H., Byrne, J., *et al.* (2005). Bracing for an uncertain energy future: renewable energy and the US electricity industry. *Risk Management Matters*, **3**(1), 46–61. Available on the Internet: http://ceep.udel.edu/publications/energysustainability/2005_es_renewables&risk.pdf

Royal Society (The) (2005). Joint Science Academies' Statement: Global Response to Climate Change. Available on the Internet: http://www.royalsoc.ac.uk/document.asp?latest=1&id=3222

San Francisco Department of the Environment (SFE) (2004). San Francisco calls for 360 MW renewables. Press Release (May 13).

San Francisco Department of the Environment (SFE) (2006). Mayor Newsom announces expedited permit processing for green buildings. Press Release (September 28).

San Francisco Department of the Environment (SFE) (2007). Energy Efficiency. Available on the Internet: http://www.sfenvironment.org/our_programs/topics.html?ssi=6&ti=14

San Francisco Department of the Environment and San Francisco Public Utilities Commission (SFE and SFPUC) (2004). Executive summary, in *Climate Action Plan for San Francisco: Local Actions to Reduce Greenhouse Gas Emissions*. SFE and SFPUC, San Francisco, CA.

Scheer, H. (2004). *The Solar Economy*. Earthscan, London.

Schneider, K. (2006). To Revitalize a City, Try Spreading Some Mulch. *New York Times* (May 17).

Schneider, S. H. and Lane, J. (2006). An overview of 'dangerous' climate change, in *Avoiding Dangerous Climate Change. Scientific Symposium on Stabilisation of Greenhouse Gases February 1st to 3rd, 2005*. Met Office, Exeter, United Kingdom. Published by the Department for Environment, Food and Rural Affairs.

Smeloff, E., Broomhead, C., Dowers, D. and Kelly, A. (2002). *The Electricity Resource Plan: Choosing San Francisco's Energy Future*. San Francisco Public Utilities Commission and San Francisco Department of Environment, San Francisco, CA.

Stern, N. (2006). *The Economics of Climate Change – The Stern Review*. Cambridge University Press, Cambridge, England.

Sterzinger, G. and Svrcek, M. (2004a). *Solar PV Development: Location of Economic Activity*. Renewable Energy Policy Project, Washington, DC.

Sterzinger, G. and Svrcek, M. (2004b). *Wind Turbine Development: Location of Manufacturing Activity (Technical Report)*. Renewable Energy Policy Project, Washington, DC.

Stiglitz, J.E. (1993). *Economics*. W.W. Norton & Co., New York.

Sustainable Energy Task Force (1998). *Choose Clean Energy: Establish Austin as a Leader in Sustainable Energy*. City of Austin, Austin, TX.

United Nations (UN) (2004). *Urban and Rural Areas 2003*. Population Division, New York.

United Nations (UN) (2005). World Population Prospects: The 2004 Revision Population Database. Available on the Internet: http://esa.un.org/unpp/

United Nations Development Programme (UNDP) (2007). *Human Development 2007/2008: Fighting climate change: Human solidarity in a divided world*. UNDP, New York.

Union of Concerned Scientists (2003). *Year-End Review of State Level Clean Energy Campaigns*. EnergyNet Policy Update (December 30).

Union of Concerned Scientists (2004). *State Public Benefits Funding for Energy Efficiency, Renewables, and R&D (as of December 2004) (Table D-1)*. Author, Cambridge, MA.

Union of Concerned Scientists (2005). Renewing America's Economy: A 20% National Renewable Electricity Standard Will Create Jobs and Save Consumers Money. Available on the Internet: http://www.ucsusa.org/clean_energy/renewable_energy/page.cfm?pageID=1505

Union of Concerned Scientists (2006). *Renewable Electricity Standards at Work in the States (Fact Sheet)*. Union of Concerned Scientists, Cambridge, MA.

US Department of Energy (DOE) (2003). Regional Energy Profile: Northeast Data Abstract. Available on the Internet: http://www.eia.doe.gov/emeu/reps/abstracts/northeast.html

US Energy Information Administration (EIA) (2003). Emissions of Greenhouse Gases: Executive Summary. Available on the Internet: http://www.eia.doe.gov/oiaf/1605/gg03rpt/executive_summary.html

US Energy Information Administration (EIA) (2004). *International Energy Annual 2003*. Washington, DC: US Department of Energy.

US Energy Information Administration (EIA) (2006). Projected International Carbon Dioxide Emissions from Energy Use to 2030 (Reference Case). Available on the Internet: http://www.eia.doe.gov/emeu/international/carbondioxide.html

US Energy Information Administration (EIA) (2007). Annual Energy Outlook 2007. Available on the Internet: http://www.eia.doe.gov/oiaf/aeo/index.html

US General Accounting Office (GAO) (2003). *Energy Task Force: Process used to Develop the National Energy Policy (GAO-03-894)*. US General Accounting Office, Washington, DC.

van der Linden, N.H., Uyterlinde, M.A., Vrolijk, C., *et al.* (2005). *Review of International Experience with Renewable Energy Obligation Support Mechanisms*. Energy research Centre of the Netherlands, Petten, Netherlands.

Washington Department of Ecology (2004). Ch. 173-407 WAC: Carbon Dioxide Mitigation Requirements for Electrical Generating Facilities. Available on the Internet: http://www.ecy. wa.gov/laws-rules/archive/wac173407.html

West Coast Governor's Global Warming Initiative (WCGGWI) (2004). Staff Recommendations to the Governors. Available on the Internet: http://www.ef.org/westcoastclimate/WCGGWI_Nov_ 04%20Report.pdf

Widholm, P. (2006). *Sustainable Buildings: Permit Program Speeds Greening of Chicago.* Midwest Construction (November).

Wingate, M. and Lehman, M. (2003). *The Current Status of Renewable Energy Certificate Tracking Systems in North America.* Center for Resource Solutions, San Francisco, CA. (Prepared for the Commission for Environmental Cooperation)

Wiser, R., Bolinger, M. and St Clair, M. (2005). *Easing the Natural Gas Crisis: Reducing Natural Gas Prices through Increased Deployment of Renewable Energy and Energy Efficiency (LBNL-56756).* Lawrence Berkeley National Laboratory, Berkeley, CA.

World Resource Institute (WRI) (2006). World's GHG Emissions Flow Chart. Available on the Internet: http://cait.wri.org/figures/World-FlowChart.pdf

Young, S. (2006). Schwarzenegger Expands Green Initiatives; Gov. Schwarzenegger to Link California to Northeast's Program of Reducing Greenhouse Gases. ABC News. Available on the Internet: http://www.abcnews.go.com/US/print?id=2571862/www.abcnews.go.com/US/print?id=2571862

Zarnikau, J. (2005). A review of efforts to restructure Texas' electricity market. *Energy Policy*, **33**(1), 15–25.

Chapter 3

Urbanization, Increasing Wealth and Energy Transitions: Comparing Experiences between the USA, Japan and Rapidly Developing Asia-Pacific Economies

PETER J. MARCOTULLIO[1] AND NIELS B. SCHULZ[2]

[1]*Distinguished Lecturer, Hunter College, City University of New York, 695 Park Ave, New York, NY and Adjunct Senior Fellow, United Nations University, Institute of Advanced Studies, UNU Office at the UN, 2 UN Plaza, DC2-2060, New York, NY 10017;* [2]*Research Associate, Imperial College London, Energy Futures Lab, Urban Energy Systems, 101, Skempton Building, South Kensington Campus, SW7 2AZ, UK*

3.1 Introduction

The world's human population is undergoing a transition from being largely rural to urban. By 2008, the global urban population will be, for the first time in history, greater than 50% (United Nations Population Fund 2007). As such, urban growth and accompanying changes with urbanization are increasingly being recognized as one of the critical development issues of the twenty-first century.[1]

Energy use and related issues of poverty, health, carbon emissions, etc. are high on popular, international and academic agendas (Arrow 2007; Gore 2006; The Stern Review 2006; United Nations Development Programme 2007). Energy supply and consumption have been linked to a large spectrum of development concerns including sustainable development, industrial development, air and atmosphere pollution and climate change (United Nations 2006a).

---

[1]Due to this recognition, several international research and academic institutions have begun urbanization research programs and projects including, *inter alia*, the International Human Dimension Programme's (IHDP) Urbanization and Global Environmental Change (UGEC) project, the Global Carbon Project's (GCP) Urban and Regional Carbon Management (URCM) project, the Convention on Biological Diversity's (CBD) Cities and Biodiversity: Achieving the 2010 Biodiversity Target, the United Nations University's (UNU) Sustainable Urban Futures (SUF) Programme, the Alliance for Global Sustainability's (AGS) new forum on 'New Thinking on Urban Futures', and the Third World Academy of Sciences' (TWAS) Cities, Science and Sustainability project.

With its large population and rapidly rising economic wealth, the Asia-Pacific region has become an important environmental focal point for both consumption of resources and generation of emissions at the local, regional and global scales. This region includes several large and growing economies of differing per capita incomes. The region's population is expected to grow 25% over the next 25 years to reach almost 2.5 billion (United Nations 2006b). Predicted economic growth (year 2006 through year 2010) for the Association of Southeast Asian Nations (ASEAN) (5.6%–6.5% annually) is higher than that for the world (3.1–3.5%), USA (2.5–3.5%) and Japan (1.2–2.8%). China's forecasted annual growth (6.6–8.6%) is more than double the world average during the same period (The Economist Intelligence Unit 2006). Due largely to rapid economic growth, under business as usual scenarios, energy use in ASEAN and East Asia will at least double over the next 20–30 years (Aldhous 2005; ASEAN 2002; International Energy Agency (IEA) 2006).

Urbanization itself accounts for a vast amount of energy resources. First, buildings can account for 40–60% of total urban energy consumption. Second, cities are centres of resource consumption including food, etc. Third, transporting goods and services typically account for about 25% of energy consumption and may increase during the shift from rural to urban lifestyles (Schurr *et al.* 1979). Finally, with lower percentages of the population engaged in agricultural activities and the need to supply food to larger non-agricultural populations, primary sector activities become more resource and energy intensive (Jones 1991).

Interestingly, the relationship between urbanization and energy supply and consumption has been less studied (for exceptions see Dhakal 2004; Jones 1991; Shen *et al.* 2005). Importantly, there has been a lack of comparative studies of urbanization, energy supply and consumption and rising incomes.[2]

One pathway for examining the relationships between urbanization, rising incomes and energy supply and consumption is to focus on energy transitions. Energy transitions are a change from one state of an energy system to another one, for example from comparatively low levels of energy use relying on non-commercial, traditional, renewable fuels to high levels of energy use relying on commercial, modern, fossil-based fuels (Gruebler 2004). Energy transitions have been historically documented for the USA (Marchetti 1988; Melosi 1985), but these transitions have been examined over either time or wealth. Comparative studies of this type have demonstrated the differences between energy transitions among economies developing at different points in time (Marcotullio and Schulz 2007). We now need to understand better the role of urbanization in energy transitions.

What is the relationship between urbanization and energy transitions? Do the relationships that held through the history of the developed world still hold for currently developing nations? This chapter focuses on these questions by comparing energy transitions among a select set of nations (USA, Japan, China, Hong Kong, Indonesia, Malaysia, the Philippines, Singapore, South Korea, Thailand and Vietnam) in an attempt to outline energy transitions and identify differences in transition experiences.

The second section of the chapter presents the perspectives used in the analyses. Thereafter, the third section presents the data sources, the analyses performed and the claims that guide the research. The fourth section compares the urbanization trends between

---

[2] Dhakal (2004) provides one of the first comparative studies of urban energy use focusing on Tokyo, Seoul, Beijing and Shanghai. As excellent as this study is, its focus is on current differences and not historical trajectories.

the sample economies. Then in the fifth section, the chapter shifts to focus on the results of comparisons of energy transitions over income, time and urbanization levels. In the sixth section we discuss the results and in the seventh section we provide limits and caveats to the findings. In the eighth and last section we conclude with policy recommendations.

3.2 Background: Linkage between Development, Urbanization and Energy Transitions

The basis for this study relies on theories related to three important development trends. First, there are theories related to environmental and energy transitions. We apply urban environmental transition theory to energy transitions. Second, we briefly explore the drivers of transitions. We argue that transitions are affected by a number of different factors, both directly and indirectly. Third, we focus on how these drivers have changed over the past century and most intensely over the past few decades. We claim that changes in drivers have time- and space-related effects. These effects shift human/natural interactions and therefore have significant impacts on environmental and energy transitions. We describe the effects and various time/space concepts that provide ways in which to understand the unique circumstances experienced by developing economies today.

3.2.1 Urban environmental transition theory

What are the environmental challenges that cities undergo as they develop? Urban environmental transition theory provides a powerful tool for addressing this question (see, for example, McGranahan *et al.* 2001; McGranahan and Songsore 1994). Rather than the simple notion described by the environmental Kuznet's curve (EKC) of an 'inverted U'-shaped function relating environmental degradation to rising incomes (see, for example, Grossman and Krueger 1995), the urban environmental transition identifies layers of changes in the relationship between affluence and urban environmental burdens. The claims are based upon an empirical tendency for urban environmental burdens to be more dispersed and delayed in higher income cities than in lower income cities. In summary, the theory suggests that in poor cities, environmental challenges are localized, immediate and health threatening. In middle-income, rapidly developing cities, environmental burdens are citywide or regional, somewhat more delayed in their impacts and a threat to both health and ecological sustainability. In affluent or high income cities, environmental burdens are global, intergenerational and primarily a threat to sustainability.

The theory includes the addition of geographic and temporal scale to notions of urban sustainability and as such questions whether urban sustainability has been achieved by any city in the world. Essentially, by including the differences in environmental burdens and the scale at which the impacts are felt, the theory predicts that at different levels of income different problems dominate, but that no city has addressed all issues.

Historical urban research that associates urban growth and environmental impacts suggests that in the past, urban environmental burdens were addressed by simply dispersing the associated harms to greater scales. Urban environmental historians in the USA have also noticed the change in environmental burdens over time. Melosi (2000) identifies how environmental challenges associated with water supply, sanitation and solid waste management have undergone a series of changes over time and have increasingly spread to wider geographical spaces. Tarr (1996) suggests that urban environmental history can be fundamentally characterized as the search for larger and larger sinks we used to unload waste streams. Both these historians have identified changes in type and geographic and temporal aspects of environmental burdens that are comparable to urban environmental transition theory.

Those working in the area of energy have also identified historical transitions at both the macro-level for developed countries (Elias and Victor 2005; Gruebler 1998; Gruebler 2004; Marchetti 1988; Nakicenovic 1988; Smil 1994) and for developing countries at the household level (Elias and Victor 2005; Smith 1987; Victor and Victor 2002). The first transition was associated with the Neolithic revolution and energy technologies that are associated with the shift from hunting and gathering to agriculture (Diamond 1997). It was not until the industrial revolution that societies turned from biomass and animate power supplemented with wind and water power as supplies to fossil fuels (Gruebler 1998). The industrial revolution that signed the start of a trend of sequential changes in primary energy supplies with increasingly higher energy densities; from coal to petroleum to natural gas and nuclear power. With each change in energy source came a reorganization of economic activity and new environmental consequences.

Urban environmental transition theory has been applied to different contexts, including economies in rapidly developing Asia (Bai and Imura 2000; Webster 1995). These applications have either simply described differences between cities of different income levels or included changes in the speed of transitions. Bai and Imura (2000), for example, insist that environmental transitions can be observed within Asian cities and that they have occurred in sequence, from traditional to industrial, to modern environmental challenges, albeit in a faster manner than previously experienced. These previous studies have missed important aspects of the current development context. That is, how environmental challenges within rapidly developing countries are emerging at lower levels of income and in a more overlapping or simultaneous fashion (Marcotullio 2005b; Marcotullio and Lee 2003). At the household level, recent studies identify few societies experiencing a regular and consistent path from traditional fuel consumption to the use of electricity and other modern fuels (Barnes *et al.* 2005).

What might be the influences affecting transitions and why would they change over time? Many emphasize the importance of income in influencing transitions (Leach 1992; Pachauri 2004). The authors of the urban environmental transition theory argue that affluence is only one of many factors explaining these shifts.[3] Elias and Victor (2005) suggest that climate, resource endowments, and distance to markets are non-income related aspects that force energy transitions. We argue that there are many more influences on transitions at the macro-level, as will be described in the next section. Unfortunately, as many of these influences were overshadowed by a focus on economic growth, their affects have been largely ignored.

3.2.2 Drivers of urban environmental transition change

There are potentially many drivers of change that help to produce the patterns identified in urban environmental transition theory. 'Drivers' can include any natural or human-induced

---

[3]McGranahan *et al.* (2001) stress that transitions do not reflect human preferences at different levels of economic development. Rather, they suggest that transitions reflect social inequities and the failure to accommodate human preferences, preferences that are not easily represented and negotiated within current socio-economic and political systems.

factors that cause a change in the environment. Given the large number of factors in urban environmental change, a useful distinction between different types include those that have 'direct' (where the impact between the drivers and the impact can be measured) and those that have 'indirect' (where the impact between the drivers and the change cannot be measured) influence.

We can see the effect of these different drivers on transitions within cities in the developed world in urban environmental history. For example, the rapid population, economic and physical expansion of New York City after the completion of the Erie Canal, low levels of technology adoption in water supply, miasma disease theory, health and fires crises, public interest in wresting control of water supply from the private sector and elite interest in making the city one of the leading cities within the then growing USA combined to promote development of the Croton Water Supply system cities in the nation. Given the intensity of these drivers, New York City was one of the first, located several miles outside the city, around the 1830s–1840s to implement a comprehensive water supply system (and even went out of the city to get water) (Burrows and Wallace 1999). The centralization of water and the importance of large infrastructure projects associated with water supply became a 'path dependent' outcome of development in this arena for many US cities (Melosi 2000). Other cities in the nation soon followed suit, but to their later dismay made decisions to extract water from nearby surface water bodies (see, for example, Philadelphia and Chicago).

This important transition ushered in a new era of environmental change and facilitated the growth of cities in the USA. It was the beginning of the industrial revolution in the country, and rapid economic, social and political changes occurred. The rapid increases in water use, the failure of the contemporary drainage system, the threat of disease within an increasingly concentrated population, the ability of the political machine to use public works for rent seeking purposes, among other factors, prompted the development of the sanitation system in New York City around the 1870s–1880s. Importantly, the 'unintended consequences' of developing a water supply system were an important force in the development of a sewage system. That is, once large amounts of water were brought into the city, a new crisis arose as to how to get the increasingly larger volume of used water out of the city. Hence, across the USA, cities' sanitation systems were developed after water supply systems (Tarr 1999).

It was not until the turn of the century (a full 50 years after water supply systems were successfully implemented), however, and in adoption of chlorine powder to urban water supplies around the country, that the link between density and disease was broken (Melosi 2000). At that point, urban typhoid fever levels plunged dramatically. That is, the development of water supply and sanitation systems along with advances in germ theory and water treatment technologies helped cities in the country largely overcome traditional health burdens (Melosi 2000).

Overcoming health burdens allowed cities to further increase in density, which also created massive markets for products. Commercial districts separated from residential areas within cities and the beginnings of a mass market developed, particularly within the metropolitan areas of the country. Industrial production on a large scale developed within a national economy and with this came the beginnings of chemical pollution.

At about this time, another transition was occurring within cities of the developed world. Changes from horse and other animate powered modes of transportation to motor vehicles depended upon the development and the mass production of the technology, increases in average incomes, shifts in housing arrangements and the structure of cities, improvements in street paving, the rise in importance of engineers in city planning, the health impacts of horses and changes in the perceived use of streets, among other influences.

Interestingly, health advocates at that time promoted the automobile as an answer to horse pollution and hence as an environmental solution (McShane 1994).[4]

In this perspective, development proceeded through waves that followed linked changes in the economy. This relationship between these drivers and development patterns (whether contingently or structurally linked), created sequential shifts in many transitions (urbanization, demographic, health, nutrition, environmental, energy, etc.) in the Western experience. The long waves of economic growth (Kondratieff 1979) focused on economic and technical change, but included a number of other shifts over a 50–60-year period (Berry 1997). Work associating these shifts to patterns of historic growth in European and North American societies concluded that waves of development did exist, but too much emphasis was placed on the potential structural linkages with the economy (Maddison 1991).

We argue that the strength of the structural linkages is as important to current policy thinking as how conditions were addressed. Historically, in the USA addressing urban environmental issues took on a 'first-things-first' methodology (Warner 1955), which had (and continues to have) a strong (and inappropriate) presence in development thinking today. We argue that even if long waves of development did exist and were related to economic structural adjustments, they are very different if not absent today. Indeed, the emergence, timing and speed of environmental conditions have altered. These changes have been due to a shift in the drivers of the past, which together have changed time/space dynamics.

3.2.3 Time- and space-related effects and changes in the drivers of change

There has been a large body of literature that suggests the contemporary development context is significantly different from previous eras (see, for example, Held *et al.* 1999). Importantly, globalization, defined by the stretching of a number of human relationships over space, is altering the way human activities and perceptions unfold. Globalization and domestic influences, over the past 30–40 years, have had particularly strong time- and space- related effects on the human-environment relation.

Time-related effects are changes in development patterns as a result of changing speed and efficiency of human socio-economic activities. Time-related effects draw places closer together and create urban dynamics across the globe forcing convergence among urban areas. That is, they create similar conditions across cities of different social, cultural and political histories and economic levels. Space-related effects concentrate increasingly diverse phenomena unevenly in spatial nodes (i.e. within and among cities) and create urban dynamics across the globe forcing divergence among urban areas. Space-related effects increase differences among cities, concentrating what was once unique across an

---

[4] The horse, in the late nineteenth century city, rivalled humans in creating waste. In the USA, at the turn of the century, there were 3 million to 3.5 million horses in use. Engineers estimated that a city horse produced more than 20 pounds of manure and several gallons of urine daily, most of which ended up in the streets. Cumulative totals of manure produced by urban horses were staggering. For example, 26 000 horses used in Brooklyn and 12 500 horses in Milwaukee yielded about 200 and 133 tons of manure daily, respectively. In the mid-1880s, the discharges of 100 000 horses and mules, pulling 18 000 horse cars over 3500 miles of track nationwide, cluttered the nation's streets, corroded the metal streetcar tracks and also threatened the health of city dwellers. Moreover, since the life expectancy of a city horse was only about two years, carcasses were plentiful and difficult to move. New York City scavengers removed 15 000 dead horses in 1880. Often dead horses lay in the streets for days before they were carted away. It is not surprising therefore that the automobile was received as an environmental benefit (Melosi 2001; Tarr 1993).

entire nation, into its cities. Massey (1996), for example, has pointed out how different and diverse phenomena are increasingly concentrated in cities.

There is a significant history of studies of time/space effects. Within the literature, there are three interlinked ways of thinking about how these effects relate including: *time/space convergence*, *time/space distanciation* and *time/space compression*. *Time/space convergence* refers to the decrease in the friction of distance between places. It refers to the apparent convergence of settlements linked by transport technology. As transport evolved, travel time would be reduced between them, giving the sensation that they had moved closer together. The velocity at which settlements are moving together may be called the time/space convergence rate (Janelle 1968, 1969). This notion is often expressed as the 'world is an increasingly smaller place'.

Time/space distanciation refers to the stretching of social systems and relationships across space and time. The argument is that people interact in two ways: face to face, and remotely through transport and communications technologies. The first way of interaction occurs more often between people living in different nations, more frequently due to air travel. The second modality has become increasingly important with globalization, 'distanciating' social relationships. Together, during the contemporary period, it is not necessary for people to be physically present at a particular location to be important social actors, as these relationships have been stretched over space (Giddens 1990).

Time/space compression refers to 'the annihilation of space through time' that lies at the core of the capitalist dynamic (Harvey 1989). While the concepts of *time/space convergence* and *time/space distanciation* do not offer an explanation for why social relations and development patterns have been stretched across space and subsequently dramatically changed the development context, *time/space compression* does. The argument is that this is one of the central processes of capitalist development. As 'time is money' the tendency for relations under this mode of production is to find ways to speed up the 'circuits of capital' so as to reduce the 'turnover time of capital' (i.e. the amount of time it takes to convert investment into a profit). As a result, technologies and policies to facilitate these processes facilitate *time/space compression*. The effect of time/space compression is disorienting and disruptive on both the balance of class power and social and cultural life.

This concept encompasses the descriptive accounts of *time/space convergence* and *distanciation*, making them a result of *time/space compression*. Ultimately the argument places an economic rationale at the core of change and not surprisingly this has been criticized by cultural scholars (Murray 2006).

To these three concepts of time- and space-related effects, we add a fourth, *time/space telescoping*. *Time/space telescoping* is also a descriptive narrative similar to *time/space convergence* and *time/space distanciation*. It is evident in the shifting patterns associated with development, such as environmental transitions, such that contemporary conditions and transitions occur *sooner* (at lower levels of income) change *faster* (over time) and emerge *more simultaneously* (as sets of challenges) than had previously been experienced by the now developed world (Marcotullio 2005b). Moreover, there are a number of different direct and indirect influences, including global economic, demographic and institutional shifts as well as local land use and policy influences that have helped to create these trends (Marcotullio 2005a), so it is theoretically different from *time/space compression*.

The notion of *time/space telescoping* stresses that the result of these changes in drivers is more than the speeding up development. China, for example, is not simply undergoing a quicker version of what the UK or the USA had experienced during the late nineteenth and early twentieth centuries. Rather, while speed is important, the addition of conditions and challenges appearing at lower levels of income and the layering of previous sequential development patterns, make the current context much more complex and bewildering.

Surprisingly, despite the diversity, complexity and rapidity of change, some of the conditions in the now developing world (i.e. those related to energy consumption) are more efficient or less environmentally harmful than those experienced by the developed world, as measured in a number of ways (i.e. supply and consumption of energy per capita).[5] That is not to say that these conditions are *good* or *good enough*, but rather they are indeed significantly better than experienced in the past. Moreover, when examining the environmental impact of urban activities, at least in terms of energy consumption and related CO_2 emissions, for many developing countries, the rapidly developing world is growing in a much less environmentally harmful manner. These are the notions that this study examines.

3.3 Data and Analysis

Our analyses incorporate several types of data from different sources including: (1) historic energy supplies in the USA (1850–2001) and historic energy consumption data in the USA (1900–2001); (2) recent (1960–2000) energy supply and consumption data in developing and other countries; (3) per capita income (Geary-Khamis international dollars) for all countries analysed; (4) historic per cent GDP created by the industrial sector for the USA (1900–1997); (5) recent (1960–2000) per cent GDP created by the industrial sector for developing and other countries; (6) historic urbanization levels for all economies (from 1850 to 2001 for the USA and 1950 to 2000 for all other economies); and (7) historical data on CO_2 emissions due to technical energy consumption and concrete production for the periods mentioned above. A detailed description of these data, the sources and limitations can be found in a previous study conducted by the authors (Marcotullio and Schulz 2007).

The analysis requires connecting energy supply and consumption estimates from time series (i.e. indexed by calendar year) to economic growth (i.e. indexed by constant-dollar per capita GDP) and urbanization levels (as per cent of total). Maddison (2001) provides gross domestic product (GDP) over time, at purchasing power parity (PPP). The UN (1999; 2006b) provides the urbanization levels for most countries.[6] The International Energy Agency provides data on energy supply and consumption (International Energy Agency (IEA) 2002a, 2002b). US historical census data provide urbanization levels for the USA, which were also matched to UN data, after 1950 (US Bureau of the Census 1975). Table 3.1 presents the ranges for the different variables for each country in the analysis.

Using the Geary-Khamis dataset, the USA per capita income was ~$1800 in 1850, ~$4100 in 1900, ~$10000 in 1950 and ~$28000 in 2000. The analyses for hypothesis 2, *faster*, was restricted to countries with a minimum current income of more than $1800 for energy supply and $4100 for energy consumption, because of the necessity of making valid comparisons with the USA's experiences. Table 3.2a presents the comparative income ranges for each country with the USA and related total primary energy supply (TPES) and total final consumption (TFC) range figures.

Similarly, we match urbanization levels between countries. In 1850, the USA was ~15% urban. In 1900, the country was ~40% urban. In 1950, the nation was ~64% urban and in

---

[5] In terms of health issues, during the 1990s there was discussion of the double burden of disease, as developing country residents were often exposed to both traditional and modern risks. That is, the concern was that given new combined risks, health in developing countries would decline. Despite the emergence of these new risks, however, longevity and other health indicators have continued to improve.
[6] The UN provides historical urbanization level data for most countries starting from 1950 for five-year intervals. Annual levels were calculated by estimating five-year annual average increases.

Table 3.1. Comparative descriptive statistics: GDP per capita, urbanization, TPES per capita and TFC per capita.

	Period of study	Range of GDP per capita (G–K$)	Range in urbanization level (%)	Range of TPES (toe/capita)	Range of TFC (toe/capita)
South Korea	1971–2000	2522–12343	40.7–81.8	0.52–4.04	0.41–2.67
Singapore	1971–2000	4904–22207	100–100	1.40–7.36	0.60–2.48
China	1971–2000	799–3425	17.4–32.1	0.47–0.92	0.22–0.68
Thailand	1971–2000	1725–6877	13.3–21.6	0.37–1.18	0.26–0.85
Malaysia	1971–2000	2180–7872	33.5–57.3	0.52–2.27	0.39–1.41
Hong Kong	1971–2000	5968–21503	87.7–100	0.86–2.53	0.62–1.68
Indonesia	1971–2000	1235–3655	17.1–40.9	0.31–0.69	0.29–0.54
Vietnam	1971–2000	710–1790	18.3–19.7	0.35–0.47	0.05–0.43
Philippines	1971–2000	1808–2425	33.0–58.6	0.36–0.54	0.27–0.34
Japan	1960–2000	3986–21069	62.5–78.7	0.86–4.12	0.60–2.81
USA (TPES)	1850–2001	1806–28129	15.4–77.2	1.97–8.45	
USA (TFC)	1900–2001	4091–28129	39.6–77.2		2.54–6.24

Table 3.2a. Date and GDP per capita ranges for supply and consumption comparisons over time.

	TPES Period of comparison	USA Period of comparison	GDP/capita range (G–K$)	TFC Period of comparison	USA Period of comparison	GDP/capita range (G–K$)
South Korea	1971–2000	1872–1967	2522–14343	1979–2000	1900–1967	4091–14343
Singapore	1971–2000	1906–1987	4904–22207	1971–2000	1906–1987	4904–22207
China	1988–2000	1850–1891	1806–3425	–	–	–
Thailand	1973–2000	1850–1938	1806–6877	1989–2000	1900–1938	4091–6877
Malaysia	1971–2000	1861–1940	2180–7872	1983–2000	1900–1940	4091–7872
Hong Kong	1971–2000	1923–1986	5968–21503	1971–2000	1923–1986	5968–21503
Indonesia	1980–2000	1850–1892	1806–3655	–	–	–
Philippines	1971–2000	1850–1867	1806–2358	–	–	–
Japan	1960–2000	1898–1985	3986–21069	1961–2000	1900–1985	4091–21069

Table 3.2b. Date and urbanization level ranges for supply and consumption comparisons over urbanization ranges.

	TPES Period of comparison	USA Period of comparison	Urbanization range (%)	TFC Period of comparison	USA Period of comparison	Urbanization range (%)
South Korea	1971–1994	1902–2000	40.7–77.2	1971–1994	1902–2000	40.7–77.2
China	1971–2000	1855–1886	17.4–32.1	–	–	–
Thailand	1976–2000	1850–1864	15.4–21.6	–	–	–
Malaysia	1971–2000	1888–1942	33.5–57.3	1978–2000	1900–1942	39.6–57.3
Indonesia	1971–2000	1855–1903	17.1–40.9	–	–	–
Vietnam	1971–2000	1857–1860	18.3–19.7	–	–	–
Philippines	1971–2000	1888–1943	33.0–58.6	1982–2000	1900–1943	39.6–58.6
Japan	1960–1989	1949–2000	62.5–77.2	1960–1989	1949–2000	62.5–77.2

2000 it was ~72% urban. Using urbanization ranges allows for comparisons between different countries and the USA experiences than at similar income ranges. For example, while the Philippines' GDP per capita data didn't allow for a comparison with the USA, similar urbanization levels between economies did. Moreover, given that two economies were 'city-states' (Singapore and Hong Kong) for much of their history, we excluded them from comparisons over similar urbanization levels. Table 3.2b shows the comparative income ranges for each country with the USA, over similar urbanization levels, and the associated TPES and TFC range figures.

We use a variety of different, but straightforward analyses to examine the differences in trends between economies, both in terms of GDP per capita and urbanization levels. We examine the sooner hypothesis by identifying whether the nations in our database experience supplies in the more advanced carriers or the consumption of more advanced energy technologies at lower economic growth and urbanization levels than those of the USA. More advanced carriers include all those in the database except biomass and coal. We use a binary test, recording whether or not there were significant levels of supply or consumption of these flows at income levels under those of the USA. The significant level was arbitrarily identified as 0.01 tonnes oil equivalent per capita (10 kg oil equivalent per capita) per year.

To examine whether changes occurred faster for those economies developing under intensive globalization forces, we compare rates of change (over time) in energy supply and consumption using the beta values of the ordinary least squares analysis. These slopes provide an indicator of rate of change. We make all comparisons at similar income ranges between individual countries and the USA in order to adequately compare levels of economic income. We also examine rates of change under similar urbanization ranges and over changes in per cent urbanization level.

To examine whether the transition sequence is experienced in a similar manner across economies, we first identify the transitions that the USA underwent during its development. We equate energy transitions with those periods when one carrier's numerical share of total energy supply surpasses another. We then identify the income and urbanization levels, at which these transitions occurred and the amount of time and percentage urban share change between transitions.

USA transitions in the supply of energy (Fig. 3.1) over time demonstrate the sequential nature of the experience. Before 1850, wood (biomass) held the largest share of energy use among carriers, accounting for approximately 82% of all energy consumption. In the early 1880s, coal took the lead in total share. Use of coal reached its relative peak around 1910 when it absorbed approximately 80% of total share of energy supply. Oil and gas reached a 1% share of the market around the 1860s and overcame coal around 1946. According to these data, oil and gas reached a peak market share around 1978, when together they accounted for 78% of the total world energy use. Subsequently oil and gas energy use dropped, but only slightly (to around a 64% share in 2000). This slight drop is due to the relative increase in natural gas use while oil supplies fell. In 1973, nuclear power came on the scene with over a 1% share. Nuclear surpassed biomass in 1974. By 2001, nuclear power made up less than 9% and modern renewable sources made up less than 0.5% of the nation's total energy supplies. This view of energy transitions provides the common understanding of how energy transitions evolved over time.

To examine and compare the total amount of energy consumed during similar levels of economic growth and urbanization, we simply summed energy consumed by product. We also matched the energy consumed within the industrial sectors of each economy and calculated an intensity figure (energy consumed per $) and compared these figures over similar

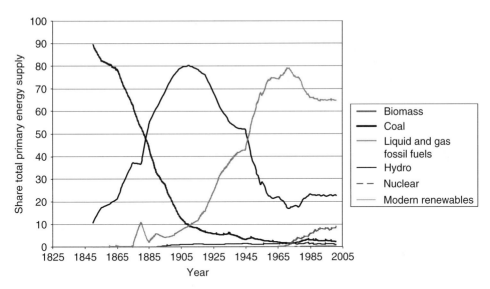

Fig. 3.1. USA energy transitions.

GDP per capita levels. We deem the more efficient industries as those with the lower ratio at a given GDP level (i.e. lower numerical value for industrial energy intensity).

Finally, we compare the production of CO_2 emissions, as tonnes of carbon, over similar economic growth periods and over similar urbanization levels. As in the previous analysis, those economies that had overall lower levels of CO_2 emissions are considered to have lower global systemic environmental impact.

3.4 Comparison of Urbanization Trends: USA, Japan and Rapidly Developing Asia-Pacific Economies

Typically, when discussing urbanization in the Asia-Pacific, demographers and urban geographers emphasize the current scale of urbanization and growth of large cities in the region (see, for example, Douglass 1998; Douglass 2000; Lo and Marcotullio 2000; Lo and Yeung 1996; United Nations Population Fund 2007). During the first half of the twentieth century, when the now developed world was rapidly urbanizing, populations increased from 300 to 400 million in all of Europe (a 0.7% growth rate)[7] and from 90 to 170 million in the USA (a 1.2% annual average growth rate). Compare these population sizes to those of contemporary developing Asia-Pacific, with China in the lead (approximately 1.3 billion), followed by Indonesia (approximately 215 million), Philippines (approximately 85 million), Vietnam (approximately 82 million) and Thailand (approximately 65 million). Each of these economy's populations, between 1970 and 2000, has grown at over 1.4% annually, and some have experienced population growth exceeding 2.0% annually (Indonesia and Vietnam).

Within the region, since the 1980s, massive populations have moved into cities. From 1980 to 2005 approximately 335 million people were added to Chinese cities and in Indonesia,

---

[7] Europe's average annual rates of population increase were highest between 1800 and 1900, as many countries in this part of the world were the first industrializers. During this period growth rates reached 1.0%.

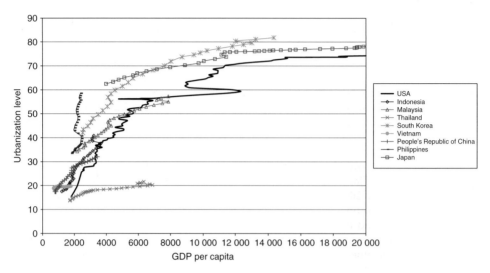

Fig. 3.2. Comparative urbanization levels by GDP per capita, USA and selected Asia-Pacific economies. [Plate 1]

during this same 25-year period, 74 million additional people were added to the nation's urban areas. Indeed, Eastern and Southeast Asia experienced a growth of 375 and 152 million people, in their respective region's cities during this period (United Nations 2006b).

The swelling of the urban population has resulted in the rise of large and megacities. In 1980, in China, for example, there were approximately 42 cities of larger than 1 million and no city in the country was larger than 10 million. By 2005, there were approximately 95 cities larger than 1 million and two were in the megacity category (Shanghai and Beijing). Within the region, in 1980, there were approximately 67 cities of larger than 1 million and one (Tokyo) larger than 10 million. By 2005, there were 131 cities of larger than 1 million and six (Tokyo, Shanghai, Jakarta, Osaka-Kobe, Beijing and Manila) larger than 10 million (United Nations 2006b).

McGee (2007) has suggested that these large urban areas continue to grow and are the force behind the growth of many small and medium sized cities that are located close by or sometimes within the urban field of the megacities.[8] These mega-urban regions are new and are now and will continue to be part of the urban landscape in the region.

Certainly, the scale of urbanization and the size of urban centres are important considerations in explaining differences between the Western experience and what those of the developing world are currently undergoing. At the same time, however, there are indications of other differences, not as often discussed, which are nevertheless significant. These include the timing and speed of urbanization.

By timing, we refer to the economic income level at which urbanization levels change. What is often missed in the contemporary literature is that urbanization in many parts of the world is occurring at lower levels of economic income than in the past (Fig. 3.2 [Plate 1]). That is to say, that at any particular GDP per capita level, most countries within the Asia-Pacific region are at higher levels of urbanization than was the USA. One important

---

[8]An opposite view is that megacities are not growing, but that the medium and smaller sized cities in the world are the faster urban growth zones (United Nations Population Fund 2007). McGee's argument suggests that we need to look beyond political boundaries, as most of the so-called rapidly developing cities are in the economic and social orbit of the megacities.

exception is Thailand, whose urbanization and economic development patterns are particularly unique in that the country has increased its wealth, but not urbanized in proportion. This may be due to the unique primacy of Bangkok within the urban system of the country and the lower appeal of other major urban centres in the country (Muscat 1994).

The other factor of importance is the speed with which urbanization is occurring. The differences in speed can be seen at the national level and in terms of individual city growth rates. Table 3.3 compares urbanization rates, measured in terms of increases in per cent urban levels over time, of the USA and several Asia-Pacific economies, at similar levels of economic development. In each case, except for Thailand, urbanization levels increased at faster rates than they did for the USA.

We can see further evidence of the rapid speed of urbanization in the region by comparing the experiences of Japan, South Korea and the USA. The USA was approximately 37% urban in 1895 and by 2000 it reached 77% urban. This means the nation experienced an increase in its urbanization level by 40% in more than 100 years. Japan was approximately 38% urban in 1940 and by 2000 it reached 78% urban. This country increased the urban share of the population also by 40%, but experienced this change over a 60-year period. South Korea, on the other hand, was approximately 42% urban in 1950 and by 2000 it was 81% urban. This economy experienced an increase in its urbanization level of approximately 40%, but did so within 50 years or half the time experienced by those in the USA.

Moreover, differences in speed of urban change can also been seen in terms of individual city growth rates. Within the USA, New York City, one of the fastest growing cities during the nation's industrial development, grew from 200000 residents in 1830 to more than 1 million in 1860, reaching almost 7 million in the late 1920s when immigration constraints came into effect. During one day at the height of an immigrant wave in 1907, approximately 12000 people queued up on Ellis Island for entry to the US and during that year 1.2 million people were received in New York (Muller 1993). Manhattan Island reached 2.3 million people by 1910 and according to Ken Jackson, noted New York City historian, by that time had obtained residential densities higher than any city in the world to that point, and possibly since then. Urban growth in parts of the Asia-Pacific has been even more spectacular. For example, around 1980, Shenzhen, China, had a population of approximately 350000. Today, the city's population has reached 8 million, translating into a 12.3% annual population growth rate for 27 years.

Table 3.3. Comparative change in urbanization level at similar income ranges (per cent/year).

		USA
South Korea	1.46	0.49
China	0.51	0.47
Thailand	0.24	0.51
Malaysia	0.79	0.50
Indonesia	0.94	0.47
Philippines	0.92	0.49
Japan[*]	0.87	0.47

[*] For this analysis, Japanese data includes the range 1920–2000.
In 1960, the year that the energy data began, Japan was approximately 63% urbanized.

These differences in timing and speeds of urbanization associate with significant differences in the urban energy transitions experienced by Asia-Pacific economies and those of the USA. We review the comparisons of aspects of these transitions in the next section.

3.5 Comparisons of the Energy Transitions: USA, Japan and Rapidly Developing Asian Economies

In this section, we present the results of analyses between urbanization, energy supply and consumption and income trends. We find support for the expected differences in relationships between these variables for rapidly developing economies and those of the USA. Most of the economics in our sample from the developing world demonstrate sooner urbanization (see above) and sooner use of energy carriers and consumption trends. Moreover, when comparing changes in energy consumption over time, most experience faster growth rates than those of the USA, when compared over similar ranges of income. Furthermore, the linear and sequential transitions experienced by the USA are not evident in rapidly developing Asia-Pacific economies. At the same time, it also appears that not all economies experience sooner development of energy carriers by urbanization level. In terms of speed of change over urbanization share, supply and consumption levels are typically lower in Asia-Pacific economies than over similar urbanization ranges of the USA. Despite all these seemingly more chaotic circumstances, developing economies are growing in wealth and urbanization with more efficient energy supply and consumption patterns than those of the USA. These more efficient patterns have led to less global systemic environmental impact. The general trends are explored in more detail below.

3.5.1 Sooner

The evidence for the sooner trend can be seen in both the analysis for energy supply and consumption. We find that for Japan and the rapidly developing economies in the Asia-Pacific, many of the carriers appear sooner on the income scale than that of the USA (Table 3.4).

For crude oil and petrol, seven of the ten nations in Asia experienced significant supply levels at income levels below that of the USA. Those that did not (including Singapore, Hong Kong and Japan), may have, but the data do not go back far enough to identify the point of emergence. The patterns for natural gas are slightly different. In terms of the emergence of this carrier, five of the nine economies from the region experienced significant supply levels at an income below that of the USA. In this case, those that experienced the emergence of this carrier at equal or higher levels of income include South Korea, Singapore and Hong Kong. The comparison for Japan and the USA is inconclusive.

For hydro power, the results suggest that two of the six economies that have experienced the emergence of this carrier did so at lower levels of income than that of the USA. Those that did not follow this pattern include South Korea, Thailand and Malaysia. Data for Japan do not go back far enough to draw conclusions.

For nuclear and modern renewable supplies, most of the economies that are using these technologies do so at sometimes much lower levels of income than the USA. For example, South Korea experienced the emergence of nuclear power at approximately $4000 GDP per capita while the technology emerged in the USA at approximately $14300. Vietnam has been deploying modern renewable energy supplies at approximately $2300 per capita while they first emerged in the USA at approximately $18500 GDP per capita. The exception

Table 3.4. Comparison of emergence of various energy carriers, dates, GDP per capita and urbanization levels.

	Emergence of crude oil and petrol as primary energy carriers			Emergence of natural gas as primary energy carrier			Emergence of hydro as primary energy carrier			Emergence of nuclear as primary energy carrier			Emergence of other renewables as primary energy carrier		
	Year	GDP per capita	Urbanization level	Year	GDP per capita	Urbanization level	Year	GDP per capita	Urbanization level	Year	GDP per capita	Urbanization level	Year	GDP per capita	Urbanization level
South Korea	na	before 2522	before 40.7	1987	8704	68.3	1990	8704	73.8	1978	4064	53.0	–	–	–
Singapore	na	before 4904	before 100.0	1992	15537	100.0	–	–	–	–	–	–	1996	19963	100.0
China	na	before 799	before 17.4	1977	895	18.3	1993	2277	28.7	–	–	–	–	–	–
Thailand	na	before 1725	before 13.3	1982	2744	17.3	1995	6573	19.9	–	–	–	–	–	–
Malaysia	na	before 2180	before 33.5	1974	2688	36.8	1983	4096	44.2	–	–	–	–	–	–
Hong Kong	na	before 5968	before 88.1	1996	21075	100.0	–	–	–	–	–	–	–	–	–
Indonesia	na	before 1235	before 17.1	1994	3146	34.5	–	–	–	–	–	–	1995	3348	35.6
Vietnam	na	before 754	before 18.3	1998	1672	19.6	1993	1214	19.6	–	–	–	–	–	–
Philippines	na	before 1808	before 33.0	–	–	–	na	–	–	–	–	–	1979	2323	37.1
Japan	na	before 3986	before 62.5	na	before 3986	before 62.5	na	before 3986	before 62.5	1970	9714	71.2	1983	14307	76.5
USA	1876	2570	27.2	1885	3270	31.5	1885	3270	31.5	1967	14330	72.5	1980	18577	73.7

Notes: Emergence occurs when value >0.01 toe/capita. na = data not available, emergence occurred prior to dates in database. "–" not yet emerged.

to this rule is renewable energy supplies in Singapore, where they have appeared at approximately $20 000 per capita.

The comparisons of the emergence of carriers by levels of urbanization reveal different patterns. Here there are, at least, three categories of differences. First, there are economies, which consistently applied energy carriers at lower levels of urbanization than the USA. For example, China, Thailand and Vietnam all experienced the emergence of crude oil and petrol, natural gas, and hydro power at lower levels of urbanization than that of the USA. Indonesia also experienced the emergence of crude oil and petrol and modern renewable sources at lower levels of urbanization than the USA, but had urbanized to a slightly higher level when natural gas emerged (34.5% in Indonesia compared to 31.5% in the USA). The second group includes those that experienced the emergence of these carriers at higher levels of urbanization. For example, South Korea and Malaysia experienced natural gas at higher levels of urbanization and South Korea experienced the emergence of hydro power at higher levels of urbanization than that of the USA. Japan experienced the emergence of modern renewable sources at higher levels of urbanization, but experienced the emergence of nuclear power at approximately the same level of urbanization (71.2% for Japan, as compared to 72.5% for the USA). The third group includes the 'city-states' (Singapore and Hong Kong) which are mostly urbanized and therefore experienced the emergence of carriers at higher levels of urbanization.

In terms of consumption, we compared the emergence of electricity, using the same 0.01 toe per capita level for significance (Table 3.5). In this case, eight of the ten economies experience significant consumption levels at income levels lower than those of the USA. The two economies that did not, Singapore and Hong Kong, were both cities. Japan also experienced sooner consumption of significant electricity energy consumption than did the USA.

The comparison of electricity consumption by urbanization level also suggests three different categories of differences. China, Thailand, Malaysia, Indonesia, Vietnam and the Philippines experienced electricity consumption at lower levels of urbanization than the USA. South Korea and Japan had equal or higher levels of urban population shares when they experienced the emergence of electricity at an important energy carrier. Hong Kong and Singapore both experienced higher levels of urbanization for the same comparison.

Table 3.5.   Comparison of emergence of electricity, dates, GDP per capita and urbanization levels.

	Emergence of electricity			
	Year	Level (toe/capita)	GDP per capita	Urbanization level
South Korea	Before 1971	0.024	Before 2522	Before 42.0
Singapore	Before 1971	0.085	Before 4904	Before 100.0
China	Before 1971	0.012	Before 799	17.4
Thailand	1972	0.011	1748	14.0
Malaysia	Before 1971	0.025	Before 2180	Before 34.3
Hong Kong	Before 1971	0.112	Before 5968	Before 88.1
Indonesia	1989	0.011	2352	29.5
Vietnam	1984	0.010	895	19.5
Philippines	Before 1971	0.018	Before 1807	Before 33.5
Japan	Before 1960	0.087	Before 3986	Before 62.5
USA	**1905**	**0.011**	**4642**	**42.5**

Notes: Emergence occurs when value > 0.01toe/capita.

3.5.2 *Faster*

There are several ways to compare urbanization, income and changes in energy supply and consumption. We choose two representative comparisons. First, we compare the change in energy supplies over time at similar levels of income. Second, we compare the change in energy supply over urbanization levels at similar levels of urbanization. The first comparison tells us something about how quickly energy supplies and consumption changed over time during similar economic growth ranges. The second comparison tells us something about how quickly energy supplies and consumption changed over urbanization levels during similar urbanization ranges.

The results for these comparisons for total primary energy supply and total final consumption are presented in Table 3.6a and b and Table 3.7a and b. In terms of changes in total supply over time at similar GDP per capita income ranges, eight of the ten Asia-Pacific economies experienced faster rates of change than that of the USA. Exceptions include Thailand (which underwent nearly the same TPES increases as that of the USA) and Hong Kong (which underwent slower TPES increases than that of the USA). In terms of comparison of changes in total final consumption, all six comparisons yielded faster growth in Asia-Pacific economies than that of the USA.

In comparison, however, the reverse is typically true for changes in TPES and TFC over similar urbanization levels. For example, for energy supplies, two of the four comparable economies grew faster than that of the USA (China and Thailand). China's supplies grew at slightly faster rates (27.3 koe/capita for every per cent increase in urban population)

Table 3.6. Comparative change in total primary energy supply (TPES) under similar GDP per capita income and urbanization ranges.

a. Changes in supply over similar income ranges (koe/capita/year)

	Change in TPES	USA Change in TPES
South Korea	124.27	43.59
Singapore	187.21	58.55
China	17.12	11.08
Thailand	32.07	33.18
Malaysia	61.24	36.90
Hong Kong	59.45	75.94
Indonesia	15.53	11.28
Philippines	5.10	−5.49
Japan	76.29	57.67

b. Changes in supply over similar urbanization levels (koe/capita/% urban)

	Change in TPES	USA Change in TPES
South Korea	60.04	142.13
China	27.27	25.59
Thailand	116.99	−10.36
Malaysia	76.77	83.22
Indonesia	16.32	38.56
Vietnam	−16.08	−14.47
Philippines	5.53	83.99
Japan	161.99	286.82

Table 3.7.   Comparative change in total final consumption (TFC) under similar GDP per capita income and urbanization ranges.

a. Changes in consumption over similar income ranges (koe/capita/year)

	Change in total final consumption	USA Change in total final consumption
South Korea	105.29	29.38
Singapore	74.79	34.77
Thailand	31.69	13.64
Malaysia	53.09	13.82
Hong Kong	43.26	42.78
Japan	45.4	36.0

b. Changes in consumption over urbanization levels (koe/capita/% urban)

	Change in total final consumption	USA Change in total final consumption
South Korea	40.70	76.88
Malaysia	57.47	40.16
Philippines	2.44	44.64
Japan	107.14	105.34

than that of the USA (25.6 koe/capita for every per cent increase in urban population). For Thailand, similar urbanization ranges for the USA produced negative change in supply of energy. Upon closer examination, the USA comparative period with the Philippines includes a period of turmoil leading to and including the US civil war, during which time total energy supply decreased.

For consumption comparisons over urbanization levels, changes in Malaysia and Japan were faster than those of the USA. In the other two cases, changes in consumption levels in the USA outpaced those of South Korea and the Philippines.

To examine these differences in more detail we break down supply and consumption changes into carriers, products and sectors. For example, to explore differences in total energy supply we first look at changes by carrier (Table 3.8a). Within those that experienced faster increases in energy supplies, typically crude oil, petrol and natural gas were the fastest growing carriers. For these economies, the rapid expansion of petrol, oil and natural gas outstripped growth in the USA. For South Korea, Hong Kong and Japan the increases in coal supplies were also important. Moreover, for South Korea and Japan increases in nuclear power energy sources were also faster than those of the USA. For the Philippines similar economic growth periods for the USA produced negative change in supply of energy. Like comparisons at similar urbanization with Thailand, the USA period for this comparison includes the US civil war. This may be regarded as part of historical contingency rather than a structural pattern.

When we compare the details of changes in energy supply carriers over changes in urbanization levels we find interesting differences between Asia-Pacific economies and the USA experience. For example, petrol, oil and natural gas increases in Asia economies were lower per increase in urbanization level than that of the USA, except for Thailand, Malaysia and Indonesia, meaning that these latter economies during similar urbanization levels, experienced faster growth in petrol, oil and natural gas supplies when compared to the USA. For coal, South Korean growth was faster per level of urbanization than that of the USA.

Table 3.8a. Comparative change in primary energy supply over similar GDP ranges, by carrier (koe/capita/year).

	Coal	Biomass	Petrol, oil and NG	Hydro	Nuclear	Modern Renewables	Total primary energy supply	Coal	Biomass	Petrol, oil and NG	Hydro	Nuclear	Modern Renewables	Total primary energy supply
								USA						
South Korea	19.71	1.20	81.85	0.14	21.36	0.03	124.27	5.68	−12.15	32.39	16.73	0.03	0.00	43.59
Singapore	0.04	−0.14	186.85	0.00	0.00	0.44	187.21	−27.15	−2.47	52.07	31.14	3.73	0.15	58.55
China	9.92	−0.78	7.10	0.58	0.36	0.00	17.12	31.11	−24.40	3.00	1.35	0.00	–	11.08
Thailand	5.63	0.84	25.43	0.12	0.00	0.00	32.07	37.40	−21.19	12.02	4.17	0.00	0.00	33.18
Malaysia	2.57	0.08	57.79	0.81	0.00	0.00	61.24	35.73	−20.00	15.12	5.17	0.00	0.00	36.90
Hong Kong	36.98	−0.10	19.65	0.00	0.00	0.00	59.45	−23.98	−1.11	56.49	37.63	5.27	0.21	75.94
Indonesia	2.96	−0.19	11.97	0.16	0.00	0.63	15.53	31.77	−24.81	2.83	1.46	0.00	0.00	11.28
Philippines	1.57	−1.05	0.39	0.13	0.00	4.05	5.10	12.34	−18.29	0.45	0.00	0.00	0.00	−5.49
Japan	4.55	1.78	49.74	−0.07	19.39	0.90	76.29	−22.24	−3.61	50.29	29.37	2.70	0.10	57.67

Table 3.8b. Comparative change in primary energy supply over similar GDP ranges, by carrier (koe/capita/percent urban).

	Coal	Biomass	Petrol, oil and NG	Hydro	Nuclear	Modern Renewables	Total primary energy supply	Coal	Biomass	Petrol, oil and NG	Hydro	Nuclear	Modern Renewables	Total primary energy supply
								USA						
South Korea	12.51	0.25	35.80	0.15	11.33	0.01	60.04	−52.06	−5.89	183.60	2.11	13.43	0.64	142.13
China	21.12	−1.02	6.27	0.72	0.19	0.00	27.27	64.90	−50.30	10.98	0.00	0.00	0.00	25.59
Thailand	20.05	4.25	91.92	0.51	0.00	0.00	116.99	25.40	−36.59	0.83	0.00	0.00	0.00	−10.36
Malaysia	3.23	0.11	72.43	1.02	0.00	0.00	76.77	34.68	−23.44	70.00	1.93	0.00	0.00	83.22
Indonesia	2.45	−0.31	13.47	0.16	0.00	0.55	16.32	87.90	−58.08	7.82	0.93	0.00	0.00	38.56
Vietnam	3.37	0.89	−27.94	7.61	0.00	0.00	−16.08	10.13	−25.95	1.35	0.00	0.00	0.00	−14.47
Philippines	1.70	−1.20	0.68	0.13	0.00	4.22	5.53	33.37	−23.26	71.87	1.95	0.00	0.00	83.99
Japan	1.01	1.96	136.59	−0.07	21.82	0.67	161.99	10.46	3.74	208.07	84.55	56.36	2.77	286.82

Table 3.9a. Comparative change in energy consumption, by product, over similar GDP ranges (koe/capita/year).

	Electricity	Total final consumption	Electricity	Total final consumption
South Korea	17.69	105.29	5.84	29.38
Singapore	17.63	74.79	10.61	34.77
Thailand	6.59	31.69	2.49	13.64
Malaysia	11.56	53.09	2.59	13.82
Hong Kong	13.45	43.26	13.47	42.78
Japan	14.19	45.44	9.54	36.01

Table 3.9b. Comparative change in energy consumption, by product, over similar urbanization ranges (koe/capita/% urban).

	Electricity	Total final consumption	Electricity	Total final consumption
South Korea	6.13	40.70	27.45	76.88
Malaysia	11.75	57.47	5.88	40.16
Philippines	0.71	2.44	6.26	44.64
Japan	86.32	107.14	72.49	105.34

For all other categories in these comparisons, except for the growth of modern renewable sources in the Philippines, changes occurred faster in the USA.

Increases in electricity consumption followed similar patterns as total energy consumption (Table 3.9a). In all cases, growth in electricity consumption was equal to or greater than that of the USA under similar GDP per capita income ranges. For electricity consumption per per cent urbanization level, growth was faster in Malaysia and Japan than that of the USA, but slower in South Korea and the Philippines.

Exploring changes in energy consumption by sector provides further insights into differences (Table 3.10a). Over time, during similar GDP per capita ranges, changes in energy consumption in the industrial and transport sectors in Asia-Pacific economies were typically faster than that of the USA. For speed of change in industrial sector consumption, Hong Kong and Japan experienced slower rates of increase than that of the USA. The commercial sector's energy consumption in South Korea, Thailand, Malaysia and Japan grew at faster rates than that of the USA. In no case did an Asia-Pacific economy experience faster energy growth in the residential sector when compared to the experiences of the USA.

Over urbanization ranges, we find that growth in industrial energy consumption is greater per level of urbanization than that of the USA for three of the four Asia-Pacific economies (South Korea, Malaysia and Japan) that could be compared. Interestingly, however, energy consumption is greater per level of urbanization in the commercial sector for Malaysia and the Philippines than that of the USA. Furthermore, in Malaysia, transport energy consumption grew at a greater rate per level of urbanization than that of the USA. The largest differences in energy consumption levels were in the residential sector where in all cases increases in the USA were much higher than for those in the Asia-Pacific economies.

Table 3.10a. Comparative change in energy consumption over similar GDP ranges, by sector (koe/capita/year).

	Industrial	Commercial	Residential	Transport	Total consumption	USA Industrial	USA Commercial	USA Residential	USA Transport	USA Total consumption
South Korea	47.63	21.25	−1.19	29.31	105.29	7.77	7.90	16.61	9.82	29.06
Singapore	36.91	5.62	1.95	29.84	74.79	6.85	10.78	17.15	13.60	34.71
Thailand	16.17	2.25	−0.99	13.15	31.69	−0.66	0.26	2.99	5.13	13.69
Malaysia	23.65	4.03	2.97	20.44	53.09	−1.81	0.28	4.13	5.78	13.86
Hong Kong	−0.55	10.52	3.47	29.56	43.26	13.63	13.87	20.18	15.98	42.81
Japan	8.13	11.35	8.40	15.84	45.44	8.83	9.95	16.45	13.31	35.93

Table 3.10b. Comparative change in energy consumption over similar urbanization levels, by sector (koe/capita/% urban).

	Industrial	Commercial	Residential	Transport	Total consumption	USA Industrial	USA Commercial	USA Residential	USA Transport	USA Total consumption
South Korea	16.02	7.70	2.73	11.06	40.70	11.71	25.65	38.51	34.10	76.90
Malaysia	25.02	4.60	2.86	22.75	57.47	−24.16	0.73	11.34	10.34	17.19
Philippines	1.89	1.44	−4.04	4.67	2.44	2.26	0.68	9.66	17.14	44.74
Japan	37.38	19.17	16.09	28.00	107.14	−0.77	32.83	28.06	54.18	105.34

Table 3.11. Comparison of timing of transitions and level of economic development.

	Biomass to coal	Biomass to liquid and gas fossil fuels	Biomass to hydro	Coal to liquid and gas fossil fuels	Coal to hydro	Biomass to advanced technologies	Coal to advanced technologies
South Korea	na	na	na	na	–	na	–
Singapore	na	na	na	na	–	–	–
China	na	1997	–	–	–	–	–
Thailand	na	1980	–	na	–	–	–
Malaysia	–	na	–	na	–	–	–
Hong Kong	na	na	–	na	–	–	–
Indonesia	–	1989	–	na	–	–	–
Vietnam	–	–	–	na	–	–	–
Philippines	na	na	–	na	–	1983	2000
Japan	na	na	na	1963	–	na	–
	1st set of transitions			2nd set of transitions		3rd set of transitions	
USA	1883	1916	1926	1950	1958	1974	–

Transitions occur when the share of one carrier passes that of the other. na: data not available, occurrence prior to record. '–': transition has yet to occur.

3.5.3 More simultaneously

The more simultaneous trend can be partially demonstrated through a comparison of the timing of transitions (Table 3.11). Importantly, the table demonstrates that the USA sequential pattern of transition from one energy form to another did not occur in the other economies. That is, for none of the economies studied was there evident a similar pattern as the sequencing experienced by the USA (i.e. many transitions occurred out of order for developing economies when compared against the USA). For example, Indonesia has yet to experience a biomass to coal transition, but has already experienced a biomass to oil and natural gas transition. Also many countries had experienced some transitions before the database started, but have a history of transitions that have yet to occur falling out of timing with those of the USA.

When transitions did occur, they often were at lower income and urbanization levels than that of the USA. For example, the transition from biomass to liquid and gaseous fossil fuels took place in China in 1997 when the nation stood at 30.6% urban and had a GDP per capita of $2973. The same transition occurred in Thailand in 1980 when the nation was at 17% urban and had a GDP per capita of $2554. In Indonesia, this transition occurred in 1989, when the economy stood at 29.5% urban and had a GDP per capita of $2352. In the USA, this transition occurred in 1916, when the nation was at 48.9% urban and had a GDP per capita of $5459.

3.5.4 More efficiently

We compared the total consumption of energy across sectors for similar ranges of income and urbanization levels as well as the measured differences in industrial output per energy input and found in both cases that most economies were able to develop in a more efficient energetic manner when compared to that of the USA.

Table 3.12 presents the comparative total energy consumption by carrier for the developing countries and the USA and Japan over similar economic growth and urbanization ranges. In all cases, the supply of energy was much (sometime over ten times) greater per capita in the USA than in the developing country. For example, during the economic growth of South Korea, the average South Korean was supplied with 34 tonnes of oil equivalent worth of energy. For the same economic growth range the average American was supplied 267 tonnes of oil equivalent worth of energy.

We find the same differences when comparing total energy consumption over similar urbanization levels. For example, to move the country from and to similar levels of urbanization, the average South Korean consumed 24.1 tonnes of oil equivalent worth of energy, while the average American consumed 446.1 tonnes of oil equivalents.

The comparison of energy consumption in the industrial sector per GDP generated in the industrial sector (Fig. 3.3) obviates the fact that developing countries have used less energy per industrial GDP value than that of the USA at any level of comparable economic growth. As these figures are aggregates of all industries, however, care must be taken in making specific evaluations.

3.5.5 With lower systemic global environmental impact

Finally, we compared the total carbon dioxide emissions from each economy to those of the USA over similar levels of economic growth and urbanization levels (Figs 3.4 [Plate 2] and 3.5 [Plate 5]). These data were not calculated using the energy data appearing in our charts and tables, but rather were collected from a different source. They include total carbon dioxide emissions from technical fuel use and cement production, but not all societal activities.[9] The data represent a proxy for the total carbon emissions and therefore systemic global envir-onmental impact.

The graphs demonstrate that for all levels of income and for almost all levels of urbanization, the USA had produced more carbon emissions per capita than any other economy or group of economies in the database. Singapore also comes close but consistently emits lower levels of carbon than those of the USA.[10] In terms of urbanization, it is interesting to note that Thailand is producing more emissions per level of urbanization than that of the USA.

3.6 Discussion

Why are some economies experiencing the emergence of energy carriers and consumption patterns at lower levels of income than the developed world and specifically the USA? There are, of course, many different reasons for this outcome. Certainly, technological advances are important. Simply, when now developed world economies were in their developing phase, many new technologies (such as cell phones or automobiles) were not yet innovated.

---

[9] These data do not include, for example, carbon dioxide released due to landcover change and other agricultural activities.

[10] The CO_2 emission data from Marland *et al.* include those emissions of bunker fuel for international freight shipping and aviation, which contribute to more than 50% of Singapore's emissions in 2000 (Schulz 2007). While this procedure is justified from the view of total emission accounting it differs from the national responsibility according to IPCC guidelines, which excludes such emissions from national liability. Moreover, Singapore hosts one of the world largest industrial petroleum refining complexes. Much of its industrial sector emissions are waste products due to the refining of crude oil for export. Those products will be finally consumed for energetic use in other national economies, raising questions of liability for life cycle emissions upstream during fuel production.

Table 3.12a. Comparative energy consumption during similar GDP ranges, by major product (toe/capita).

	Coal	Biomass	Petrol, oil and NG	Electricity	TFC	USA Coal	USA Biomass	USA Petrol, oil and NG	USA Electricity	USA TFC
South Korea	4.92	0.02	25.10	4.53	34.81	121.83	21.97	114.04	9.00	266.84
Singapore	0.00	0.04	41.81	8.91	50.76	115.65	22.68	203.76	23.13	365.26
Thailand	0.57	1.83	4.98	1.11	8.49	89.77	15.82	26.83	1.74	134.16
Malaysia	0.53	1.29	13.02	2.41	17.25	93.07	16.34	30.27	1.97	141.64
Hong Kong	0.03	0.29	24.55	8.32	33.19	69.87	14.96	192.77	21.80	299.43
Japan	7.44	0.38	55.71	15.40	79.07	127.56	25.73	196.59	21.53	371.43

Table 3.12b. Comparative energy consumption during similar urbanization ranges, by major product (toe/capita).

	Coal	Biomass	Petrol, oil and NG	Electricity	TFC	USA Coal	USA Biomass	USA Petrol, oil and NG	USA Electricity	USA TFC
South Korea	5.67	0.00	15.76	2.65	24.12	126.21	26.72	257.06	35.73	446.07
Malaysia	0.55	1.64	15.03	2.68	19.91	97.06	16.83	34.04	2.25	150.18
Philippines	0.19	2.79	2.55	0.59	6.11	99.32	17.07	36.00	2.42	154.80
Japan	6.13	0.18	37.67	9.44	53.46	21.02	9.86	209.61	32.44	273.28

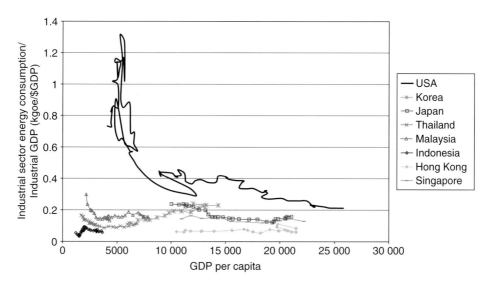

Fig. 3.3. Comparative change in industrial efficiency over GDP per capita, USA and selected Asia-Pacific economies.

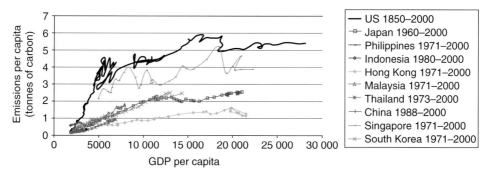

Fig. 3.4. Comparative changes in carbon emissions by GDP per capita, USA and selected Asia-Pacific economies. [Plate 2]

Hence, the developing world is benefiting from earlier technological developments and improvements, for example, from electricity and modern energy supplies.

Many of these technologies, however, were developed in different parts of the world and certainly trade of technology and information exchange also play a role in the diffusion of usage. We expect that those economies that are more open are experiencing a sooner emergence of some carriers than those that are not.

There are positive aspects of sooner adoption of energy technologies. Adopting certain carriers can lead to efficiency gains on the part of the economy as well as social benefits. In some cases, scholars have claimed that by using modern technologies at lower levels of income, the developing world can bypass problems experienced by the now developed world (i.e. 'leap-frog' over challenges). This notion suggests that developing countries have the opportunity to 'do it right the first time' by installing clean efficient technology, among other changes (Goldemberg 1998; Ho 2005). This is confirmed by studies that have identified trends of increasing energy efficiency experienced by developing 'late-comers' (Smith 1993).

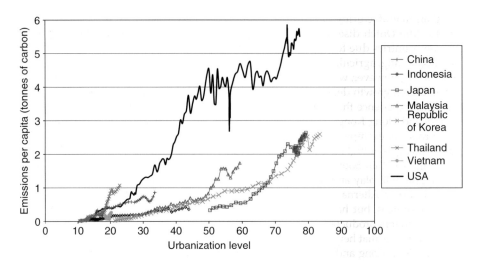

Fig. 3.5. Comparative carbon emissions per level of urbanization, USA and selected Asia-Pacific economies. [Plate 3]

The advantages of some of these technologies, the importance of electricity, for example, in terms of both health and social advances cannot be underestimated (Nye 2001).

In other studies, there is evidence that globalization and foreign direct investment (drivers associated with globalization) have facilitated some 'leaf-frogging' in the energy area (Melnik and Goldemberg 2002).[11] In the developing Asia-Pacific, there is also some evidence that industries are implementing cleaner production technologies and processes in industry (see, for example, Angel and Rock 2000). In terms of our analyses we catch some of these differences when comparing the experiences of Malaysia, Thailand and China, which are open economies, to those of Japan and South Korea, which have kept trade and foreign direct investment inward flows lower. The former typically experienced sooner development of carriers, while the later had differential results.

Others suggest leap-frogging is dependent upon a host of legal, political and institutional frameworks (Ho 2005). Therefore establishing appropriate conditions for leap-frogging requires a host of abilities including institutional capacity. Certainly, governments in the region have been eager to develop their energy supplies and spread them beyond urban areas, so national policy plays a role also.

Moreover, rapid social acceptance of the technology, in terms of the use of various technologies is also an important factor, among many others, that plays a role in the sooner aspect of current development patterns.

Why then would some economies not experience the sooner aspect of time/space telescoping? Besides the openness to globalization, in the energy arena, there is also the importance of natural endowments in energy transitions. Sachs and Warner (1995) provide an analysis of how during the post-war era, the economic performance of resource-rich countries was weaker than resource-poor economies. Following Matsuyama (1992) they argue that land-intensive economies with open economies will promote agriculture

---

[11] The rapid diffusion of information technologies such as mobile phones is an example of how current technology can be provided by private companies, often under competitive conditions. Mobile phones bypass the large investments needed by traditional copper wire telephone networks.

rather than manufacturing. The move away from manufacturing results in shrinkage in the sector (the Dutch disease) and slower growth. This situation is currently happening in African countries due to increasingly heavy inflows of Chinese investments in primary industries; mining, agricultural and oil (The Economist 2006).

In our study, however, we also found this to be true. The question is does this rapid manu-facturing-driven growth depend upon domestic energy resource supplies? Those economies that did not experience the sooner development of energy carriers were city-states (such as Singapore and Hong Kong), which by definition have low resources, or have low levels of the resources. For example, in terms of natural gas and hydro power, South Korea developed these technologies at higher levels of income as compared to the USA (both of which are in low quantities within South Korea). This suggests that the availability, domestically, of the energy supply will play an important role when the economy can begin to develop it as an import-ant carrier. Alternatively, in the case of hydro development in Thailand, the country has large resources, but has met with political opposition against new large hydro power plans and therefore production has been slow (Todoc *et al.* 2007). In this case, it may have been political influences that helped to bring out results. At the same time, however, in the cases of Singapore, Hong Kong and South Korea, these economies focused on developing industrial power rapidly and concentrated on the most modern carriers despite the fact that they have low domestic resources (fossil fuels and in the case of South Korea, nuclear power). Hence, the (often poor resource) rapid developers did not necessarily develop domestic energy supplies, even if they had them, particularly if they were not fossil fuels or modern carriers. Rather, it has been the most energy dense sources that nations focus on developing first.

In terms of the timing of significant electricity consumption, only Singapore and Hong Kong were economies that may have developed supplies after those of the USA (and in each case data are not available to fully compare differences). At the same time, no other economy experienced significant consumption levels at income levels higher than those of the USA during its initial electricity consumption periods. This makes sense as increasing electricity and modern energy consumption is part of current development planning.

What does it mean when economies develop energy carriers and consume modern sup-plies at lower levels of urbanization than that of the USA? There are, at least, two possible explanations for this finding. First, the use of these carriers or consumption patterns could be due to more intensive per capita use and consumption in Asia-Pacific cities than those of the USA or that the use of the carriers and consumption patterns is more widespread (including rural areas) than experienced by the USA. Both notions suggest that urbaniza-tion is occurring under different energy conditions. In the past, the developed world's urban growth was significantly altered with changes in and quality of energy sources and consumption patterns. If energy carriers are appearing at lower levels of urbanization, these will arguably place different pressures on urban growth and urban patterns. If what is occurring is the spread of energy supply and use beyond urban areas and therefore lower-ing the urbanization levels at which significance supply and consumption appears, then it translates into significant advances in rural use. This is seen in some of the countries within the region. For example, Vietnam reported that by the end of 2004, the national power grid reached 900 poor communes. All districts and 90% of communes through the country have electricity (United Nations Development Programme 2007).[12] At the same level of income and urbanization, however, the USA could not boast this claim.

---

[12] This is certainly not true for all countries in the region. In Cambodia, for example, over 92% of the population is dependent on fuel wood as its primary energy source. Rural areas rely almost exclu-sively on wood for their energy needs (United Nations Development Programme 2007).

Those economies that did not experience this phenomenon include city-states (for obvious reasons) and South Korea and Malaysia for some of the energy carriers. In the latter cases, the results suggest that urbanization may not be proceeding in a more efficient manner than that of the USA, but as economic growth is so rapid, it is overwhelming other patterns. That is, the use of higher quality energy supplies has come later in the urbanization process, but earlier in the wealth generation process. As such, given that different energy sources create different pressures for urban development, we would expect that these economies will not only face different pressures than did the USA, but also face different circumstances than their neighbours.

The results of the faster increases in supply and consumption, particularly for the rapid developers, are not surprising. Energy use and economic development are linked (at least for the initial periods of growth). As countries grow, they use more energy to help organize more complex activities. Faster uptake of energy, however, also comes with increased complexity of management. Rapidly developing countries therefore need to build and manage complex energy systems faster than previously demanded.

Given the rapidity of change, the question whether these economies are keeping up with energy infrastructure is questionable and what impacts this has on energy supply and consumption is not well understood. Studies in rapidly developing Asia suggest that infrastructure development is not keeping up with the demands created through economic growth (Brockman and Williams 1996). Furthermore, given rapid economic growth, trends in environmental impact from lack of infrastructure may lag. For example, a recent study suggests that there is a different relationship between provision of road infrastructure and road transportation fuel consumption and consequent carbon dioxide emissions in developing Pacific Asian economies than in the USA. Essentially, adding kilometres of paved road in rapidly developing Asia countries results in much greater increases in road transportation fuel consumption than it did in the USA. As Asia-Pacific economic growth slows and infrastructure catches up with demand (i.e. overcomes the so-called, 'infrastructure bottleneck'), levels of global emissions from rapidly developing countries may approach patterns set by developed countries including those of the USA (Marcotullio and Williams 2007).

When comparing transition experiences the findings suggest that energy transitions no longer exist for developing countries. That is, the sequential patterns of development, in terms of energy use, are not observable in our sample. Those in the USA have long believed that changes made in supply, from one major source to another, are an obvious improvement – more, better and cheaper energy. Thus, it has been generally accepted that the United States witnessed major shifts in energy sources: from wood and waterpower to coal (or from renewable to non-renewable sources) in the mid-nineteenth century; and from coal to petroleum and natural gas in the early twentieth century. Because of these perceptions, scholars have argued that a 'single-source mentality' developed (Melosi 1985, p. 9). Hence, energy sources have been regarded as competitive rather than complementary meaning that there was 'one best way' that prevailed over each energy era.

For the rapidly developing world, complementarity is more often experienced than competitiveness among supplies, as the findings suggest several different carriers are used simultaneously and that transitions do not follow sequential patterns. Certainly, the slow transition from one energy source to another is not distinguishable. Energy mixes may not be due to energy source scarcity or quality, but due to price, technology, transportation, accessibility to sources, environmental impact, consumer preference and several other economic and non-economic influences. Given that each of the influences has a trajectory of its own, we may no longer expect to see energy transitions as they have occurred in the West. This finding may be the consequence of a change from long waves of development or specific to energy transitions. In either case, the findings question whether long waves of development and historical environmental transitions still exist.

Notwithstanding all these differences, it is exciting to note the lower levels of energy use, intensity and subsequent carbon emissions per capita from the developing world. The reason why this is occurring, however, is not entirely clear. It could be that these economies are simply more efficient and are developing under new technologies, and social systems that facilitate economic growth with lower energy demand. On the other hand, it could be, as explained earlier, due to infrastructure bottlenecks, which once resolved will lead to massive increases in consumption. At the very least, it suggests a reconceptualization of the relationships between energy, urbanization and increasing wealth. A fuller understanding will demand focused comparative studies.

There are planning implications for future planning strategies in these findings. For developing countries, a diverse portfolio of energy sources is not only a better strategy than concentrating on one source, it seems the logical outcome of current conditions. In doing so, economies lower the risks related to price hikes in one area (such as those experienced during the 'oil shocks' of the 1970s) and create more resilient energy systems.

At the same time, as planning an energy system includes strategic decisions, it requires addressing a number of trade-offs associated with choosing dominance in one path over others. In some cases, what the rapid developing world is facing are new choices. For example, when choosing a mix of energy carriers for an economy's growth, relying extensively on liquid fossil fuel sources, because they are currently economical, can 'lock-in' that economy more quickly than in the past. Civil uses of nuclear technologies for electricity generation are both expensive and complicated and also 'lock-in' an economy to a long-term commitment to this source. In order to balance future requirements to flexible solutions, integrated environmental-energy policies are even more necessary in the developing context than in the developed world.

Planning strategies are based upon predictions of future trends. The idea underlying some energy future predictions is that the previously experienced transitions are stable and long-term trends. For example, the post-war natural gas trends were predicted to rise and overtake those of petroleum, nuclear power is predicted to rise thereafter and this would be followed by fusion, which is predicted to emerge as an important force at the end of the twenty-first century (Marchetti 1988). If developing countries are not using fuels as predicted by the substitution model, these patterns would be less useful in prediction. The study also suggests that time- and space-related effects are important drivers of these emissions. Understanding the development of these drivers and how they change greenhouse gas emissions is crucial for scenario development. Therefore in answer to the call from the Intergovernmental Panel on Climate Change, these effects should be subject to further study (Nakicenovic and Swart 2000).

Finally, this study implicates the changing susceptibility of energy transitions to planning in general. As mentioned, in the past energy transitions were stable. This may partly be because of the lack of competition between sources. Before 1820 the major fuel source was biomass and up to the 1880s, it was between coal and wood. Now, there are a number of different sources available making the market more complex and the use of fuels more relatively price dependent. As oil prices reach US$100 a barrel or more, or as more roads are built in countries developing their private transportation sector, the use of liquid fossil fuels will be impacted. On the one hand, with increasing prices in one fuel, there are a number of other sources from which to choose (including making liquid fuels from biomass), hence energy use and/or carbon emissions will change. On the other hand, building more roads helps to lock-in carbon intensive practices at lower levels of national income. From this analysis energy planning is more important today and energy use and consumption may show greater response than in the past when traditional structural shift in technologies were more important. Using planning to create sustainable energy transitions is a potential trajectory to the post-fossil fuel urban era (Droege 2004).

3.7 Qualifications

There are several qualifications and caveats to this study. First, generalizing these patterns beg the question of how 'development' can be measured and meaningfully compared across space and time. Is it fair to compare turn of the twentieth century USA with 1980s Thailand, just because they may have similar GDP levels per capita (PPP)? Using purchasing power parity (PPP) per capita indicators refines international comparisons by allowing income standardization across price differences in goods and services between countries. That is, a PPP value for income in one country will match the ability of citizens to purchase the same amount of an exact set of goods and services in another country as well as their own. While using PPP per capita values is more appropriate than simply comparing GDP or GDP per capita figures, it still leaves much of what development encompasses out of the picture.

The level of GDP per capita or economic growth does not speak to changes in social or political structures, for example. Those attempting to address these problems often supplement economic data with social indicators including material possession acquisition such as telephones, televisions, radios and the use of banks, schools, cinemas and provision of housing, medical or educational services (see, for example, any *Human Development Report* of the United Nations Development Programme). To further refine studies of development, others focus on the reduction or elimination of poverty, inequality and unemployment in the context of economic growth. Development, in this view concerns equity and distributive justice at all scales. Recently, economists have turned toward definitions that include improving the quality of life for citizens, broadly defined, and especially the poor (World Bank 1991). These various definitions suggest that 'Development must therefore be conceived by a multidimensional process involving major changes in social structure, popular attitudes and national institutions, as well as the acceleration of economic growth, the reduction of inequality and the eradication of poverty' (Todaro 1997, p. 16).

The notion of *time/space telescoping* fundamentally questions the underlying understanding of development portrayed only by simple economic growth. We have been careful in separating the economic growth from development and do not intend to conflate them. Rather our point is to demonstrate that various differences in development experiences (in this case energy development) exist at similar levels of economic income and during similar economic growth periods. These differences suggest that using economic income is not sufficient to understand how nations are developing.

Using urbanization level as another variable helps to demonstrate the complexity of development. In some cases economic growth outstripped that of urbanization and in some of our case economies the reverse was true. Including more variables, together with the price of fuel, will tell us more as to why some economies experienced their carriers and consumption patterns.

Second, there are data quality concerns. At the general level the data used are not comprehensive and leave out many details. Some examples have already been noted and include the use of animate power for energy and the carbon dioxide emissions related to land cover change. One important point is that many economies in the developing world are built upon large informal markets, meaning that the values of goods and services bought and sold does not appear in national accounts. This lack of information deflects the level of GDP per capita in these countries downward. Hence, the GDP per capita estimates provided are conservative. Differences between levels in the USA and developing economies may not be as great as mentioned. Moreover, the data quality in the IEA category for some categories is questionable including, for example, the category 'Combustible, renewables and waste' (biomass).[13] For this category, we chose nations where the data quality was evaluated as 'High to Medium' (non-OECD Europe), 'High' (Latin America) or 'High to Low' (Asia) (International Energy Agency (IEA) 2002a). This may not have been as effective as we had wished. Moreover, aggregating data threatens to produce the ecological fallacy. Trends identified at the national level cannot be assumed to occur at other scales. In this case, our finding that industrial intensity within the secondary economic sector is more efficient in developing countries than in the USA, for example, needs further exploration. One exciting opportunity is to compare the historical trends within a set of industries to further identify exactly which industries are more efficient than others (for single year detailed comparisons see, for example, the work of Ernst Worrell).[14]

Third, our tests could not in and of themselves, provide evidence to reject the null hypothesis in each case. Rather they provide an indicative comparative synthesis at the broadest level. Certainly, this study is not the final word on this subject, but rather the opening proposal.

Finally, using only the USA as a comparative example of the experience of the developed world is unfortunate. The USA's history has proven to be extreme in terms of materials consumption (Wernick 1996) and energy use (Ayres *et al.* 2003; Cleveland *et al.* 1984). Further comparisons are necessary including those with Western European countries, Australia and New Zealand. At the same time, anecdotal information suggests that the USA's more unique trends, even among developed countries, lessens between 1820 and 1913. For example, Madisson (2003, 2005) points out that a comparison of primary energy consumption between the US and the UK converged during this period. That is, in 1820 the differences were significant. The US was consuming 2.45 tonnes of oil equivalent energy (toe) per capita while the UK was consuming only 0.61 toe per capita. In 1870 the differences were significantly reduced with the US still at 2.45 toe per capita, but the UK energy use expanded to 2.21 toe per capita. In 1913, the differences were 4.47 for the USA and 3.24 for the UK. By 1950, the differences grew again. The US consumed 5.68 toe per capita while the UK consumed only 3.14 toe per capita. Thereafter this difference grew. This convergence is important and suggests that the transitions within the USA during this period were not that different from what was happening in other developed countries (i.e. the UK). This was also within the period which many of the comparisons with developing economies in the article are made. Nevertheless, leaving out other developed countries limits our ability to make final conclusions concerning the comparative experiences between the generalized categories of developed and developing countries.

3.8 Conclusions

We conclude with some general policy recommendations for the Asia Pacific and for the developing world. We suggest that economies in the region will be well served by long-term

---

[13]This category also includes industrial and municipal waste which is defined as that waste produced by industry, commercial, residential and public services collected by local authorities for disposal in a central location for the production of heat and/or power.

[14]See for example, http://ies.lbl.gov/, http://ies.lbl.gov/staff/worrellieua.html

energy policies. Given the lack of clear transition trends and the rapid increases in supply and consumption, nations should consider ways in which (including the diversity of energy sources) rapid urbanization and economic growth can be most efficiently achieved. That is, despite the higher efficiency achieved by all these economies compared to the USA, the scale of energy needs for urbanization and the large populations within the region still threatens the local, regional and global environment. Furthermore, the reliance of many economies on liquid fossil fuels in a post-peak era is not sustainable. Given the study's results, particular attention might be paid to the transport and industrial sectors. It is these sectors where petrol and oil consumption is increasing most rapidly.

Urban centres typically do not have energy policies, but given the size of the agglomerations in the region, there may be a call to rationalize energy use at this scale. Energy policies for urban centres in developing countries, however, should be considered carefully. Compact city policy, for example, may not be appropriate for many locations, as the cities are already compact and further compaction in combination with current urban industrial economic structures may exacerbate exposure to air pollutants. Rather, ways to expand energy supply, through renewable sources, promote the use of efficient fuels and technologies, reduce demand and control motor vehicle use are just as if not more appropriate. Certainly national and urban energy policies must be more 'home grown' than taken off the shelf from the developed world.

Finally, it is from the Asia-Pacific that some of the most exciting advances in energy efficiency are already being applied. These include Singapore's electric area pricing scheme and the bus rapid transit systems that are promoted throughout the region. Indeed, there is much for the developed world to learn from the Asia-Pacific experience, not only in terms of the energy histories, but also in terms of current policies.

Acknowledgements

Versions of this study were presented at an IHDP workshop at Arizona State University and the UNU workshop on 'Sustainable Urban Futures'. The authors would like to thank those at these meetings for their feedback. We'd also like to thank Peter Droege for this encouragement and critical comments during the writing of this manuscript.

References

Aldhous, P. (2005). China's burning ambition. *Nature*, **435**, 1152–1154.

Angel, D.P. and Rock, M.T. (eds) (2000). *Asia's Clean Revolution: Industry, Growth and the Environment.* Greenleaf, Sheffield.

Arrow, K.J. (2007). Global climate change: a challenge to policy. *The Economists' Voice*, **4**. Article 2.

ASEAN (2002). GHG emissions – a serious environment concern in energy. *ASEAN Energy Bulletin*, **6**, 18.

Ayres, R.U., Ayres, L.W. and Warr, B. (2003). Energy, power and work in the US economy, 1900–1998. *Energy* 28, 219–273.

Bai, X. and Imura, H. (2000). A comparative study of urban environment in East Asia: stage model of urban environmental evolution. *International Review for Environmental Strategies*, **1**, 135–158.

Barnes, D.F., Krutilla, K. and Hyde, W.F. (2005). *The Urban Household Energy Transition, Social and Environmental Impacts in the Developing World.* Resources for the Future, Washington, DC.

Berry, B.J.L. (1997). Long waves and geography in the twenty-first century. *Futures*, **29**, 301–310.

Brockman, R.A.C. and Williams, A. (eds) (1996). *Urban Infrastructure Finance.* Asian Development Bank, Manila.

Burrows, E.G. and Wallace, M. (1999). *Gotham, A History of New York City to 1989*. Oxford University Press, New York.

Cleveland, C.J., Costanza, R., Hall, C.A.S. and Kaufmann, R. (1984). Energy and the US economy: a biophysical perspective. *Science* 225, 890–897.

Dhakal, S. (2004). *Urban Energy Use and Greenhouse Gas Emissions in Asian Mega-cities, Policies for a Sustainable Future*. IGES, Kanagawa.

Diamond, J. (1997). *Guns, Germs, and Steel: The Fates of Human Societies*. W.W. Norton and Company, Inc., New York.

Douglass, M. (1998). East Asian urbanization: patterns, problems and prospects, in *Walter H. Shorenstein Distinguished Lecture Series: Cities and the Regional Dynamics of East Asia*, Stanford University, Asia/Pacific Research Center.

Douglass, M. (2000). Mega-urban regions and world city formation: globalization, the economic crisis and urban policy issues in Pacific Asia. *Urban Studies*, **37**, 2315–2335.

Droege, P. (2004). Renewable energy and the city. *Encyclopedia of Energy*, **5**, 301–311.

Elias, R.J. and Victor, D.G. (2005). *Energy transitions in developing countries: a review of concepts and literature*. Program on Energy and Sustainable Development, Stanford University, Stanford. 33 pp.

Giddens, A. (1990). *The Consequences of Modernity*. Stanford University Press, Stanford.

Goldemberg, J. (1998). Leapfrog energy technologies. *Energy Policy*, **26**, 729–741.

Gore, A. (2006). *An Inconvenient Truth, The Crisis of Global Warming*. Rosedale Inc., New York.

Grossman, G.M. and Krueger, A.B. (1995). Economic growth and the environment. *Quarterly Journal of Economics*, **110**, 353–377.

Gruebler, A. (1998). *Technology and Global Change*. Cambridge University Press, Cambridge.

Gruebler, A. (2004). Transitions in energy use, in *Encyclopedia of Energy*, Volume 6, pp. 163–177.

Harvey, D. (1989). *The Condition of Postmodernity: An Enquiry into the Origins of Cultural Change*. Blackwell, Oxford.

Held, D., McGrew, A., Goldblatt, D. and Perraton, J. (1999). *Global Transformations, Politics, Economics and Culture*. Polity Press, Cambridge.

Ho, P. (2005). Green industries in newly industrializing countries: Asian-style leapfrogging? *International Journal of Environment and Sustainable Development*, **4**, 209–226.

International Energy Agency (IEA) (2002a). Energy Balances – non-OECD Member Countries. OECD Organisation for Economic Co-operation and Development and IEA World Energy Statistics and Balances.

International Energy Agency (IEA) (2002b). Energy Balances – OECD Member Countries. OECD Organisation for Economic Co-operation and Development and IEA World Energy Statistics and Balances.

International Energy Agency (IEA) (2006). *Forecasts from Energy Policies of IEA Countries – Energy Balances*. OECD, Paris.

Janelle, D.G. (1968). Central place development in a time/space framework. *Professional Geographer*, **20**, 5–10.

Janelle, D.G. (1969). Spatial reorganization: a model and concept. *Ann. Ass. Am. Geogr.* **59**, 348–364.

Jones, D.W. (1991). How urbanization affects energy-use in developing countries. *Energy Policy*.

Kondratieff, N.D. (1979). The long waves of economic life. *Review*, **II**, 519–562.

Leach, G. (1992). The energy transition. *Energy Policy*, **20**, 116–123.

Lo, F.-C. and Marcotullio, P.J. (2000). Globalization and urban transformations in the Asia Pacific region: a review. *Urban Studies*, **37**, 77–111.

Lo, F.-C. and Yeung, Y.-M. (eds) (1996). *Emerging World Cities in Pacific Asia*. UNU Press, Tokyo.

Maddison, A. (1991). *Dynamic Forces in Capitalist Development: A Long-Run Comparative View*. Oxford University Press, Oxford.

Maddison, A. (2001). *The World Economy, A Millennial Perspective*. OECD, Paris.

Maddison, A. (2003). Growth accounts, technological change, and the role of energy in Western growth. *Economica e Energia* XIII.

Maddison, A. (2005). *Evidence submitted to the Select Committee on Economic Affairs, House of Lords, London, for the inquiry into 'Aspects of the Economics of Climate Change'*. House of Lords, London. 4 pp.

Marchetti, C. (1988). Infrastructures for movement: past and future, in *Cities and their Vital Systems, Infrastructure Past, Present and Future* (eds J.H. Ausubel and R. Herman), pp. 146–174. National Academy Press, Washington DC.

Marcotullio, P.J. (2005a). Shifting drivers of change, time/space telescoping and urban environmental transitions in the Asia Pacific, in *Managing Urban Futures, Sustainability and Urban Growth in Developing Countries* (eds M. Keiner, M. Koll-Schretzenmayr and W.A. Schmid), pp. 103–124. Ashgate Publishing Ltd, Aldershot.

Marcotullio, P.J. (2005b). *Time/space telescoping and urban environmental transitions in the Asia Pacific*. United Nations University Institute of Advanced Studies, Yokohama.

Marcotullio, P.J. and Lee, Y.-S.F. (2003). Urban environmental transitions and urban transportation systems: a comparison of the North American and Asian experiences. *International Development Planning Review*, **25**, 325–354.

Marcotullio, P.J. and Schulz, N. (2007). Comparison of energy transitions between the USA and developing and industrializing economies. *World Development*, **35**, (10): 1650–1683.

Marcotullio, P.J. and Williams, E. (2007). Exploring effects of an 'infrastructure bottleneck' on road transportation CO_2 emission in Asia Pacific countries. *International Journal of Environment and Pollution*, **30**, 27–44.

Massey, D.S. (1996). The age of extremes: Concentrated affluence and poverty in the twenty-first century. *Demography*, **33**, 395–412.

Matsuyama, K. (1992). Agricultural productivity, comparative advantage, and economic growth. *Journal of Economic Theory*, **58**, 317–334.

McGee, T. (2007). Urban space in the twenty first century, thoughts on the shaping of the global *urban* future, in *United Nations University Sustainable Urban Futures Workshop*, New York.

McGranahan, G., Jacobi, P., Songsore, J. *et al.* (2001). *The Citizens at Risk, from Urban Sanitation to Sustainable Cities*. Earthscan, London.

McGranahan, G. and Songsore, J. (1994). Wealth, health, and the urban household: weighing environmental burden in Jakarta, Accra, Sao Paulo. *Environment* 36, 4–11, 40–45.

McShane, C. (1994). *Down the Asphalt Path, the Automobile and the American City*. Columbia University Press, New York.

Melnik, O. and Goldemberg, J. (2002). Foreign direct investment and decoupling between energy and gross domestic product in developing countries. *Energy Policy*, **30**, 87–89.

Melosi, M. (1985). *Coping with Abundance, Energy and Environment in Industrial America*. Temple University, Philadelphia.

Melosi, M.V. (2000). *The Sanitary City: Urban Infrastructure in America from Colonial Times to the Present*. Johns Hopkins Press, Baltimore.

Melosi, M.V. (2001). *Effluent America, Cities, Industry, Energy and the Environment*. Pittsburgh University Press, Pittsburgh.

Muller, T. (1993). *Immigrants and the American City*. New York University Press, New York.

Murray, W.E. (2006). *Geographies of Globalization*. Routledge, London.

Muscat, R.J. (1994). *The Fifth Tiger, a Study of Thai Development Policy*. UNU Press and M.E. Sharpe, Tokyo and Armonk.

Nakicenovic, N. (1988). Dynamics and replacement of US transport, in *Cities and their Vital Systems, Infrastructure Past, Present and Future* (eds J.H. Ausubel and R. Herman), pp. 175–221. National Academy Press, Washington, DC.

Nakicenovic, N. and Swart, R. (2000). *Special Report on Emissions Scenarios*. Cambridge University Press, Cambridge.

Nye, D.E. (2001). *Electrifying America, Social Meanings of a New Technology*. MIT Press, Cambridge, MA.

Pachauri, S. (2004). An analysis of cross-sectional variations in total household energy requirements in India using micro survey data. *Energy Policy*, **32**, 1723–1735.

Sachs, J.D. and Warner, A.M. (1995). *Natural Resource Abundance and Economic Growth*. National Bureau of Economic Research, Cambridge, MA. 54 pp.

Schulz, N.B (2007). The direct material inputs into Singapore's development . Journal of Industry Ecology 11(2): 117–131.

Schurr, S.H., Darmstadter, J., Perry, H., *et al.* (1979). *Energy in America's Future, the Choices Before Us.* The Johns Hopkins University Press for Resources for the Future, Baltimore.

Shen, L., Cheng, S., Gunson, A.J. and Wan, H. (2005). Urbanization, sustainability and utilization of energy and mineral resources in China. *Cities*, **22**, 287–302.

Smil, V. (1994). *Energy in World History.* Westview Press, Boulder.

Smith, K. (1987). The biofuel transition. *Pacific and Asian Journal of Energy*, **1**, 13–32.

Smith, K. (1993). The most important chart in the world, in *UN University Lectures*, 28 pp. UNU, Tokyo.

Tarr, J. (1999). Decisions about wastewater technology, 1850–1932, in *The American Cities and Technology Readers* (ed. G.R. Roberts), pp. 154–162. Routledge and the Open University, London.

Tarr, J.A. (1993). Urban pollution – many long years ago, in *The Great Metropolis, Poverty and Progress in New York City* (ed. K.T. Jackson), pp. 163–170. American Heritage Custom Publishing Group, New York.

Tarr, J.A. (1996). *The Search for the Ultimate Sink, Urban Pollution in Historical Perspective.* University of Akron Press, Akron.

The Economist (2006). China in Africa. *The Economist*, **381**, 53–54.

The Economist Intelligence Unit (2006). *Country forecast, Asia and Australasia, regional overview.* The Economist Intelligence Unit, London.

The Stern Review (2006). The economics of climate change. Office of the Prime Minister, London.

Todaro, M.P. (1997). *Economic Development.* Longman, London and New York.

Todoc, J., Todoc, M. and Lefevre, T. (2007). Thailand, in *Energy Indicators for Sustainable Development: Country Studies on Brazil, Cuba, Lithuania, Mexico, Russian Federation, Slovakia and Thailand* (eds United Nations and International Atomic Energy Association), pp. 409–463. UN, New York.

United Nations (1999). *World Urbanization Prospects: The 1999 Revision.* UN Population Division, Department of Economic and Social Affairs, New York.

United Nations (2006a). Policy options and possible actions to expedite implementation: energy for sustainable development, Report of the Secretary-General, 16 pp. Commission on Sustainable Development, New York.

United Nations (2006b). *World Urbanization Prospects: The 2006 Revision.* UN Population Division, Department of Economic and Social Affairs, New York.

United Nations Development Programme (2007). *A Review of Energy in National MDG Reports.* United Nations Development Programme, New York.

United Nations Population Fund (2007). *State of the World Population 2007.* Unleashing the Potential of Urban Growth, UNFPA, New York.

US Bureau of the Census (1975). *Historical Statistics of the United States, Colonial Times to 1970, Part 1.* US Bureau of the Census, Washington, DC.

Victor, N.M. and Victor, D.G. (2002). *Macro patterns in the use of traditional biomass fuels.* Stanford University, Stanford. 31 pp.

Warner, S.B. (1955). Public health reform and the depression of 1873–1878. *Bulletin of the History of Medicine*, **29**, 503–516.

Webster, D. (1995). The urban environment in Southeast Asia: challenges and opportunities, in *Southeast Asian Affairs* (ed. Institute of Southeast Asian Studies), pp. 89–107. ISAS, Singapore.

Wernick, I.K. (1996). Consuming materials: the American way. *Technological Forecasting and Social Change*, **53**, 111–122.

World Bank (1991). *World Development Report.* World Bank and Oxford University Press, New York.

Chapter 4
Direct versus Embodied Energy – The Need for Urban Lifestyle Transitions

MANFRED LENZEN, RICHARD WOOD AND BARNEY FORAN

Centre for Integrated Sustainability Analysis, A28, The University of Sydney NSW 2006, Australia

4.1 Introduction: What is Embodied Energy?

Living means consuming, and consuming means needing energy. Most of the energy resources we consume today are non-renewable – hence there is the obvious problem of their depletion, among other problems of environmental pollution and climate change. We as householders are familiar with using energy in our homes – *residential energy* – mainly in the form of electricity for our appliances, as gas for cooking, or as firewood for space heating. We also use *transport energy* in the form of petrol in our private cars, and possibly in boats and other vehicles. The consumption of residential and transport energy is called *direct energy consumption* (Bin and Dowlatabadi 2005, p. 199). We experience direct energy daily, as heat, cooling, or motion. More than that, we experience that heat dissipates, and motion ceases when the energy source is switched off. We therefore have quite an intuitive feeling for the fact that consuming energy involves the irreversible use of a finite resource.

Consuming also means needing goods and services, ranging from 'material' items such as food, clothing, water, appliances, vehicles and equipment to 'immaterial' amenities such as entertainment, public transport, insurance and personal care. These commodities come to us ready to use, and they usually do not evoke any association with energy. However, producing goods and services requires substantial amounts of energy to be used – on farms, in factories, in power plants, and by corporate and public vehicles. The energy that is needed throughout the entire life cycle of a final consumer item – good or service – starting with the transformation of raw materials and ending with its final disposal, is often called the energy *embodied* in the consumer item.[1] We have often no idea how much energy is needed to produce the things we buy, and whether all of the embodied energy is higher than the direct energy we experience using in our daily lives.

These questions (and a fair few suggested answers for that matter) are not new; in fact, some of the 1970s references cited in this chapter read as if they were written in the twenty-first century. What have improved though are the analytical tools applied to these questions and the quantitative sophistication of their answers. The next section explores embodied

---

[1] Alternative terms are 'embedded energy', 'indirect energy', 'life-cycle energy', 'cradle-to-grave energy', or 'supply-chain energy'.

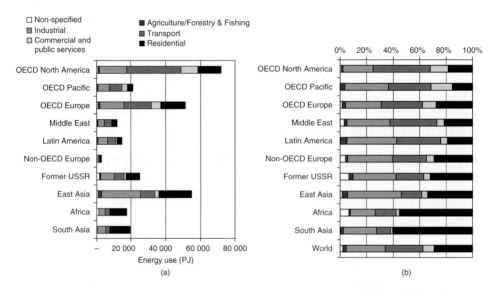

Fig. 4.1. World energy consumption by region and usage (International Energy Agency 2007). (a) Absolute values in PJ (10^{15} joules), (b) Percentage distribution across usage categories.

energy research and its findings about consumer items, trends over time, and the role of international trade. In particular, we investigate whether there are certain influential socio-economic-demographic traits, whether embodied energy exceeds direct energy or not, and whether international trade displaces the environmental and resource impacts of wealthy countries' energy-hungry consumption into low income countries. Following is an empirical case study of households in Australia's largest city, Sydney. The chapter concludes with an outline of the implications of embodied energy for urban lifestyles, both present and future.

It should also be noted that what can be said of energy can also be said of other resources, and also pollutants. Consuming also means using water, or emitting greenhouse gases. Once again, we can experience direct water use as it comes out of our taps, and direct greenhouse gas emissions from our cars and fireplaces. Rarely though do we think about irrigation water embodied in our food,[2] or greenhouse gas emissions embodied in our aluminium window frames. The indicator dealt with in this chapter – energy – is a good proxy for other environmental impacts, especially emissions from fuel combustion, such as CO_2, NO_2, and SO_2, for example (Schipper 1998).

4.2 Embodied Energy – An International Perspective

4.2.1 A comparison of direct and industrial energy use

Most of the world's energy is consumed in OECD North America, followed by East Asia and OECD Europe (Fig. 4.1a). Most of this energy is fossil energy (80%), followed by renewables (hydro, solar, and combustibles, 13%) and nuclear energy (7%) (International Energy Agency 2007). Residential energy constitutes between 20% and 60% of total energy

---

[2] In this case, the term 'virtual water' is often used.

Table 4.1. Per capita world energy consumption (GJ) by region (International Energy Agency 2007).

Region	World	South Asia	Africa	East Asia	Former USSR	Non-OECD Europe	Latin America	Middle East	OECD Europe	OECD Pacific	OECD North America
Energy use (GJ/cap)	47	15	20	30	88	49	35	56	98	107	171

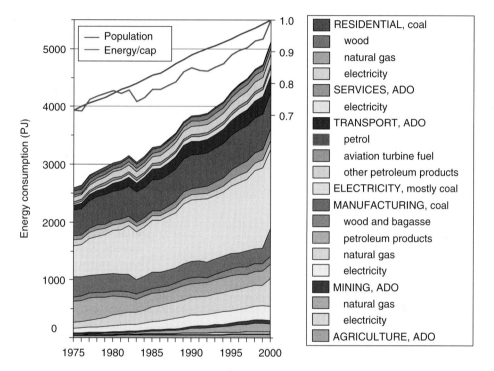

Fig. 4.2. Energy consumption in Australia 1975–2000 (Australian Bureau of Agricultural and Resource Economics 2006). Areas represent energy consumed minus derived fuels produced. Brown curve: Australian population; red curve: per capita energy consumption (both indexed to 2000 = 1). ADO = Automotive Diesel Oil. [Plate 4]

consumption (Fig. 4.1b), but this portion is mainly correlated with the level of per capita income. In Asia and Africa, residential energy is mainly supplied by combustibles such as wood, and animal and crop waste, and represents around half of national energy consumption. In contrast, residential energy is mainly supplied by commercial fuels such as gas or fossil-fuelled electricity in higher income regions, and its portion has shrunk to about 20% of national energy consumption (International Energy Agency 2007; Leach 1998).

In per capita terms, OECD countries are the top energy consumers, followed by the former Soviet Union and non-OECD Europe as well as the Middle East. Below world average are Latin America, Asia and Africa (Table 4.1).

The transport sector in Fig. 4.1 contains an unknown amount of petrol used for private vehicles. Examining more detailed Australian data (Fig. 4.2 [Plate 4]) reveals that in

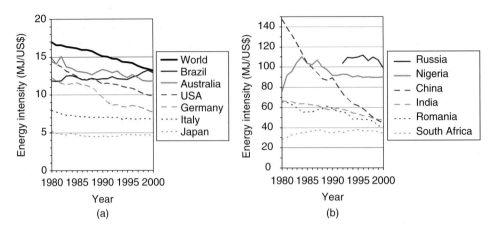

Fig. 4.3. World energy intensity trends (MJ/US$) for selected countries (World Resources Institute 2006): (a) energy-efficient economies, (b) energy-intensive economies. [Plate 5]

Australia petrol consumption (about 600 PJ/year) actually exceeds energy use in the house (about 400 PJ/year),[3] and that together they represent about 20% of Australian energy consumption. Urban consumers consume more energy per capita than rural consumers (Lenzen 1998).

Similarly, in China in 2002, direct energy amounted to about 30% of total energy consumption (Wei *et al.* 2007). Even though rural residents still outnumbered city dwellers by a factor of 1.7, the latter collectively used 2–2.5 times more energy than the former.

Non-residential energy is traditionally reported as energy used in industry (Australian Bureau of Agricultural and Resource Economics 2006). However, if we follow Adam Smith (1776) in arguing that 'the sole end and purpose of production is consumption', industrial energy is ultimately expended for the sake of producing commodities that someone will finally consume. In fact, the entire philosophy of *life-cycle assessment* builds on the notion that energy (and other resources and pollutants) is passed on by being embodied in the intermediate products and materials that are then passed on between producers, until they reach the final consumer.

Accordingly, looking at overall energy use from a consumer's perspective, it becomes clear that in high income countries, the energy embodied in consumer items significantly exceeds direct energy. Hence, if the debate concluded that attention must be devoted to those aspects of our lives that consume most of our energy, then these aspects are clearly related to embodied and not direct energy (Bin and Dowlatabadi 2005).

This does not mean that we should ignore direct energy; there are certainly savings that are easily implemented without undue loss of comfort. Especially in low income countries, where the proportion of residential energy is high, energy intensities are generally high as well, leaving ample scope for reductions (Fig. 4.3b [Plate 5]). However in high income countries with a highly urbanized population, where direct energy makes up only about 25%, many technologies have been efficient for some time, and further reductions

---

[3] With respect to electricity use we follow accounting conventions and allocate electricity to residential energy, while the primary fuels (coal etc) used in the power plant *net of* electricity produced are counted as embodied energy.

are likely to be less effective (Fig. 4.3a [Plate 5]). This means that rather than focusing on reductions of energy use in urban households of high income countries, through technological options that may be costly and/or difficult to implement, there may be 'low hanging fruit' elsewhere.[4] There are options for changes of consumption habits involving substantial reductions in the form of embodied energy (Lenzen and Murray 2001; Melasniemi-Uutela 1999; Schipper 1993), for example simply the reduction of food intakes to recommended dietary levels, or switching to alternative items fulfilling the same need (Lenzen and Dey 2002). In the literature, they are generally referred to as *lifestyle changes*.

4.2.2 Lifestyles and energy

In the 1970s researchers started to appraise the connection of lifestyles and energy mostly because of concerns about the stability of oil supply (Mazur and Rose 1974). Two decades later the concerns started to be about climate change (Wolvén 1991). Many of the more quantitative investigations exploit data on household expenditure in order to characterize different lifestyles, as well as input/output analysis in order to calculate their *embodied energy requirement* (Bin and Dowlatabadi 2005; Weber and Perrels 2000). The latter is achieved by multiplying every expenditure item of the household by an *energy intensity* (Bullard and Herendeen 1975; Herendeen 1974). The most comprehensive survey of such studies to date is contained in a five-country analysis by Lenzen and co-workers (Lenzen *et al.* 2006).

The overarching finding of this kind of research is that energy requirements increase with overall household expenditure, which in turn depends on household income. The relationship is, however, not a proportional one, but per capita energy requirements show some saturation towards higher per capita expenditures (Fig. 4.4). This is because as societies become more affluent, their consumer baskets change to incorporate a higher proportion of services, which require less energy compared to food and other manufactured items.

In fact, the saturating expenditure/energy relationship can be explained by an *expenditure elasticity of energy*, which in the case of Fig. 4.4 is about 0.9. This means that if expenditure increases by 10%, energy requirements will increase by only 9%. Interestingly, this elasticity is higher for low income economies than for high income economies (Lenzen *et al.* 2006). In transitional economies with a large part of the population in the process of rapid building-up of appliance and car stocks, this elasticity is even higher than 1, such as measured for transport energy in Brazil (Cohen *et al.* 2005) and all direct energy in 1970s Hong Kong (Newcombe 1979). We will return to this topic in section 4.3.1 below.

While in low income and transiting economies, direct energy consumption is increasing with urbanization, electrification, and increasing work opportunities and incomes (Leach 1998; Cohen *et al.* 2005; Newcombe 1979; Kulkarni *et al.* 1994; Qiu *et al.* 1994), in wealthy economies such as Australia, direct energy is practically independent of income, which – together with the consumer basket effect – determines the 'saturating' shape of the overall household energy requirement (Fig. 4.5). This effect was measured as a time trend in Hong Kong: while in the 1970s direct energy use was heavily skewed towards rich households (Newcombe 1979), by the end of the 1980s this effect had largely levelled out (Hills 1994). At high incomes, direct energy is an economic necessity, and not consumed in

---

[4] Many decision-makers have taken advantage of this basic idea, for example in emissions trading, or the Clean Development Mechanism and Activities Implemented Jointly initiatives under the Kyoto Protocol: if a company or a country found it too costly and/or difficult to reduce emissions on-site/domestically, it should be encouraged to reduce emissions elsewhere if those reductions are more cost-effective.

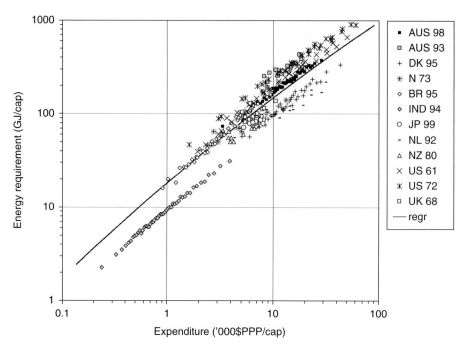

Fig. 4.4. Country comparison of energy requirements as a function of household expenditure ('000$PPP/cap).[5] Results from (Lenzen *et al.* 2006) (large symbols): Australia (AUS 98), Brazil (BR 95), Denmark (DK 95), India (IND 94), and Japan (JP 99). Results from other studies (grey small symbols): US 1961 (Herendeen and Tanaka 1976), UK 1968 (Roberts 1975), US 1972 (Herendeen *et al.* 1981), Norway 1973 (Herendeen 1978), New Zealand 1980 (Peet *et al.* 1985), Netherlands 1992 (Vringer and Blok 1995), Australia 1993 (Lenzen 1998) 'regr' denotes a curve fit. Data points lying below the fitted curve indicate national fuel mixes with subaverage energy intensity, and vice versa.

larger amounts when incomes increase. Figure 4.5 confirms the conclusions of section 4.0 in that embodied energy significantly exceeds direct energy.[6]

It has been proposed that as societies develop and become wealthier, energy use and associated environmental impact may initially increase, but then decrease again once these

---

[5] The World Bank (http://www.worldbank.org/depweb/english/modules/glossary.htm#ppp) defines purchasing power parities (PPP) as 'a method of measuring the relative purchasing power of different countries' currencies over the same types of goods and services. Because goods and services may cost more in one country than in another, PPP allows us to make more accurate comparisons of standards of living across countries.'

[6] One note of caution is due here: one and the same country-specific energy multiplier is applied to certain commodities, no matter whether they are bought by low or high income households. It is, however, likely that when high income households buy the same commodity as a low income household, they choose one of higher quality at a higher price. It may hence be that, even though the more pricey item in reality embodies the same amount of energy (or maybe even less, because it was made using more hand-labour), it will be charged with more embodied energy than the cheaper item, because input/output analysis assumes a proportionality between money and energy flows. What this means for Figs 4.4 and 4.5 is that the 'dips' in the curves towards higher incomes should probably be more pronounced than shown.

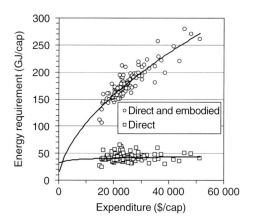

Fig. 4.5. Direct and embodied energy of Australian households. Compiled using household expenditure data (Australian Bureau of Statistics 2000), energy statistics (Australian Bureau of Agricultural and Resource Economics 2006), and input/output tables (Australian Bureau of Statistics 2004) for 1999.

societies have attained a level of prosperity that allows them to improve resource efficiency and environmental conditions (the so-called 'environmental Kuznets hypothesis').[7] This is true, for example, for SO_2 emissions because there are clean technologies, which producers in wealthy countries can afford to install, and hence governments can afford to legislate. However, without exception, energy requirements analyses find that while a weak saturation exists, there appears to be no wealth threshold above which energy requirements will actually start to decrease. This is because there exists today no renewable energy technology that, once economical and affordable, provides for unrestrained energy needs.[8]

4.2.3 Driving forces of energy consumption over time

Energy requirement studies provide insights about cross-sections of countries at a particular point in time, but do not necessarily allow extrapolating these trends over time. A number of factors have influenced energy consumption in the past, and will continue to do so in the future. An obvious 'upwards' driving force is population – the more people there are the more energy is needed. Similarly, affluence potentially drives up energy use, since wealthier people demand more commodities. On the other hand, these upwards trends can in principle be offset by improvements in energy efficiency, structural changes in the economy, and compositional changes in final demand, for example by shifting consumptive emphasis from goods to services. In the following we examine trends in both direct and embodied energy consumption. While direct energy is often investigated using detailed bottom-up analyses (Schipper and Ketoff 1985) or index decomposition approaches (Ang 2000), Structural Decomposition Analysis (SDA (Dietzenbacher and Los 1997;

---

[7] For further reading see (Pearson 1994; Selden and Song 1994; Shafik 1994; Grossman and Krueger 1995; Stern *et al.* 1996; Cole *et al.* 1997; Ekins 1997; Ehrhardt-Martinez *et al.* 2002; Stern 2003).

[8] The same holds for CO_2 and other emissions for which there is no cost-effective end-of-pipe retention technology.

Hoekstra and van den Bergh 2002)) is often applied to unravel embodied energy trends over time.[9]

4.2.3.1 Residential energy

In high income countries, the per capita requirement of residential energy has changed remarkably little during the past 25 years. Unander *et al.* (2004) show that in the USA, Canada, France, the UK, Denmark and Sweden, people used as much energy in their homes in 1973 as they did in 1998.[10] This is due to equal but opposing trends, such as the simultaneous increase in floor space of houses and efficiency of space heating devices and other building and equipment stock (Schipper and Ketoff 1985).[11] Between 1973 and 1990, space heating was the energy demand category growing fastest (Schipper and Ketoff 1985); however, between 1990 and 1998 it was overtaken by appliance energy (Unander *et al.* 2004). It seems that while the need for space heat was able to saturate at some stage, rising incomes provided time-poor consumers with continuing opportunities to purchase new types of appliances that did automatically what previously had to be done manually (Schipper and Ketoff 1985, pp. 393ff).[12] Over time, these appliances penetrate successive household cohorts as they climb up the income ladder (Newcombe 1979).

Between 1976 and 1995 US energy consumption for miscellaneous items such as micro-wave ovens, bed heaters, swimming pool pumps, air cleaners, video equipment, coffee makers and computers has grown more than 4% per year, which is twice as fast as the growth of energy used in traditional appliances (Sanchez *et al.* 1998). In 1995, miscellane-ous electricity uses by the top 50 products amounted to 235 TWh of electricity per year, which converts to roughly 25 GJ of primary fossil energy per household per year.[13] The energy embodied in those products is probably in the order of 250 GJ per household.[14] In Japan, a similar long-term trend is observed, involving increasing penetration of rice cookers, electric blankets (*kotatsu*) and carpets, bath heaters, air conditioners, and micro-wave ovens (Nakagami 1996). Hong Kong underwent a similar trend for air conditioners, heaters, stoves and washing machines (Newcombe 1979; Hills 1994), and India for refrig-erators, fans, television sets (Kulkarni *et al.* 1994). While energy-intensive uses may vary among cultures (for example, lighting and heating in Norway versus bathing in Japan (Wilhite *et al.* 1996)), the growth of appliance stock and/or living space is common to households of all provenances.

While in most OECD countries, the growth in appliance ownership and residential energy use has been relatively slow, Germany, Italy and Japan have experienced higher growth rates because of their post-war recovery (Schipper and Ketoff 1985; Nakagami

---

[9] Many decomposition case studies are couched in CO_2 rather than energy terms; however, since the majority of CO_2 emissions stem from energy use, we regard CO_2 decompositions as relevant for the purpose of this chapter.

[10] Exceptions are Norway, where high incomes combined with cheap and abundant hydro-electricity have led to about 65% of homes being electrically heated (Unander *et al.* 2004).

[11] Such *rebound* effects are well known in energy research: money saved on energy bills through energy-efficient devices is spent on new appliances and other goods, the embodied energy of which often exceeds the previous energy savings (Lenzen and Dey 2002).

[12] Newcombe (1979) refers to these appliances as 'energy slaves' employed by upper income groups.

[13] Assuming 100 million households, and a power generation efficiency of 33%.

[14] Assuming each household buys one product of 50 types each at $500, and that the energy intensity of equipment manufacturing is about 10 MJ/$.

1996; Schipper *et al.* 1997; 1989). Even higher growth rates are perhaps experienced in economies in transition, where traditional biofuels are rapidly replaced by commercial fuels[15] as newly urbanized aspiring households acquire appliances for convenience, comfort and status (Leach 1998; Garcia *et al.* 1994; Pongsapich and Wongsekiarttirat 1994), often leading to shortages and blackouts (Tyler 1994).

4.2.3.2 Transport energy

In absolute terms, passenger transport by private car represents the majority of transport energy use, which in turn has been one of the fastest growing sectors of OECD economies (Lenzen *et al.* 2003; Schipper and Fulton 2003). Linked principally to income, both car ownership and mileage per vehicle have increased, although the latter only slightly (Schipper *et al.* 1997; Scholl *et al.* 1996).

The energy efficiency of private mobility (MJ/passenger-km) has stayed about constant (Schipper *et al.* 1997), which is due to the combination of more fuel-efficient engines on the one hand, but larger engines and increases in travel activity on the other (Scholl *et al.* 1996). Whether this trend can be reversed in the future depends on many factors. Further future reductions of fuel intake per vehicle-km of 25% by 2020 are technically possible (Schipper and Fulton 2003). The effect of fuel taxes may be diminished by relatively price-inelastic demand for mobility.[16] Alternative renewable transport fuels are still fraught with problems such as efficient storage of hydrogen (Johansson 2003), or the negative impact of large-scale biofuel cropping on biodiversity (UN-Energy 2007). While these factors can all contribute to reducing the effect of personal transport on energy resource depletion, they may have to be complemented with more long-term structural measures such as public transport systems and urban planning (Smith and Raemaekers 1998) which are aimed at the main driving force – demand for travel.

Especially in Asia's rapidly growing large metropolises, increased mobility is expressed by income-driven increased car ownership. By 2020, car ownership in China could increase by a factor of 20 compared to 2000 levels (Schipper *et al.* 2001). Further growth may only be curbed by untenable congestion and/or air pollution (Sathaye *et al.* 1994).

4.2.3.3 Embodied energy

There is one outstanding phenomenon that can explain why efficiency improvements have not led to a proportional improvement in overall environmental and resource pressure. Better technology will often save the consumer not only time, effort and energy, but also money. These savings are invariably spent on other (new) purposes, leading to what is called a *rebound effect*. At the very least, the impacts of the rebound consumption will partly cancel the efficiency improvements achieved, such as when people will drive more and heat more after switching to more energy-efficient vehicles and appliances. But often rebounds occur in patterns that actually undo and overturn any efficiency gains (Lenzen and Dey 2002; Heyes and Liston-Heyes 1993).

---

[15] The 'energy ladder'(Smith *et al.* 1994), or 'transition ladder' (Leach 1998). Since this fuel transition is driven by urbanization, Leach (1998) refers to this process as the 'urban energy transition', which interestingly coincides with the title of this book.

[16] Scholl *et al.* (1996) find that despite significant differences in petrol prices, Europeans and Americans spend roughly the same percentage of their incomes on car travel.

Decomposing time series spanning three decades, Wood and Wachsmann (2003; 2005) both come to the conclusion that the growth of final consumption, of which households form the largest component, represents the strongest upward driving force for overall national energy consumption in economies as different as Australia and Brazil. These recent findings confirm results from previous studies on energy and CO_2 emissions by Wier (1998) for Denmark, Proops and co-workers (1993) for Germany and the UK, and Common (1992) for Australia. Mélanie and co-workers (1994) decompose overall final consumption into population growth and per capita affluence components. They present graphs for Australia, Canada, France, the UK, the USA, and China, showing that in all of these countries, personal affluence growth outstrips all other positive and negative drivers of CO_2 emissions, including population growth and improvements in energy intensity. Similarly, Hamilton and Turton (1999) show convincingly that without exception affluence is the dominant driving force of greenhouse gas emissions in the USA, Japan, the EC, Australia, New Zealand, and Canada.

In the world's most populous nations, India (Mukhopadhyay and Chakraborty 1999) and China (Lin and Polenske), increases in final demand levels in the 1980s outstripped technological improvements by a factor of seven and three, respectively. In the Chinese case, capital investment proved to be the strongest driving force, with household consumption and exports following close behind. Similarly, in Chung's analysis of Korea (Chung and Rhee 2001), the effect of the growth of the economy in accelerating CO_2 emissions is four times stronger than all retarding impacts combined. In Taiwan (Chen and Rose 1990), only strong exports orientation and rapid increase of material inputs between 1971 and 1984 exceeded final demand as positive drivers for energy use. A decade later (Chang and Lin 1998), this trend was continuing unabated, and domestic final demand had overtaken exports as the main driving force.

Interestingly, in Brazil, Germany and the UK, overall energy consumption increased because of final consumption even though direct energy use decreased over some subperiods (Wachsmann 2005; Proops *et al.* 1993). This clearly demonstrates how the desire for increased material wealth can negate reductions achieved for the direct energy needs more obvious to the householder.

If the past and present are in any way indicative for the foreseeable future, energy requirements and resource depletion will increase as people strive towards the affluent lifestyles of the high income world. In order for energy transitions to be truly effective, they will have to address embodied energy before direct energy. In the absence of readily available technological fixes, this means that energy transitions must involve radical lifestyle transitions.

4.2.4 *Embodied energy trade*

In our globalized world, there are very few people left whose consumption habits have no or very little impact on the rest of the world. Life in modern urban centres of the affluent part of the world is underpinned by a complex trade network that funnels resources in to cities and wastes out (Folke *et al.* 1997; Rees and Wackernagel 1996). While the notion of a 'resource hinterland' seems to suggest that the area supplying these resources borders the outskirts of the city, in reality this hinterland is scattered all over the world.

As the commodities consumed by city dwellers are imported from foreign countries, so is the energy embodied in them 'foreign energy'. Translating financial trade balances of countries into embodied energy trade balances allows identifying net suppliers and net demanders of embodied energy. In analogy to the National Accounting Identity

$$\text{Gross Domestic Product (GDP)} + \text{Imports} - \text{Exports}$$
$$= \text{Gross National Expenditure (GNE)},$$

where GDP in an embodied energy account represents the energy that a national production system embodies into commodities, no matter where these are consumed, while GNE represents the energy that is embodied in what is consumed nationally, no matter where it is produced (compare with (Bourque 1981)). The trade balance is the difference between imports and exports. There exists a large number of studies that present energy and other resource and environmental balances for single countries; they are probably most comprehensively reviewed by Wiedmann *et al.* (2007).

It appears that resource-rich energy-intensive economies top the list of net energy exporters, while population-dense, service-oriented economies top the list of net energy importers (Table 4.2). Through occupying markets for value-added commodities, the latter countries have successfully displaced energy- and resource-intensive production processes abroad, along with the environmental pressure these processes involved (Muradian *et al.* 2002). This phenomenon is widely known as pollution leakage. The net import of

Table 4.2. National Embodied Energy Accounts (PJ, 10^{15} joules) of the world's top ten producers, importers, exporters and consumers of embodied energy. Compiled using a multi-region input/output model of the world economy (World Resources Institute 2006; United Nations Statistics Division 2007).

Rank	GDP		Imports		Exports	
1	United States of America	96197	United States of America	13732	United States of America	6319
2	China	51600	Germany	6003	China	5738
3	Russian Federation	25949	China	5760	Russian Federation	5144
4	India	22609	Japan	4549	Germany	4673
5	Japan	21711	United Kingdom	4064	Saudi Arabia	4127
6	Germany	14547	France	3789	Japan	3240
7	France	11167	Italy	3197	Canada	2812
8	Canada	10501	Netherlands	2856	Korea, Republic of	2203
9	United Kingdom	9513	Canada	2512	France	2194
10	Korea, Republic of	8547	Belgium	2231	United Kingdom	1551

Rank	GNE		Trade balance, top 10		Trade balance, bottom 10	
1	United States of America	103610	Russian Federation	4069	United States of America	−7413
2	China	51623	Saudi Arabia	3675	United Kingdom	−2512
3	India	23543	Venezuela	899	Hong Kong	−2201
4	Japan	23020	Nigeria	841	Italy	−1872
5	Russian Federation	21880	Qatar	608	France	−1596
6	Germany	15877	Kuwait	443	Spain	−1450
7	France	12763	Iran	417	Netherlands	−1367
8	United Kingdom	12026	Canada	299	Germany	−1331
9	Canada	10202	Libya	279	Japan	−1309
10	Italy	9127	Kazakhstan	271	Singapore	−1179

Table 4.3. Structural Path Analysis of the world's embodied energy trade (PJ, 10^{15} joules), calculated using a multi-region input/output model of the world economy (World Resources Institute 2006; United Nations Statistics Division 2007). Paths are interpreted, for example, as 'the commodity trade from Japan to China embodies 427 PJ of energy'. 'nes' = not elsewhere specified.

Rank	Energy (PJ)	Path
1	2123	Canada > United States of America
2	1286	China > United States of America
3	993	United States of America > Canada
4	865	Mexico > United States of America
5	635	United States of America > Mexico
6	635	Japan > United States of America
7	491	Nigeria > United States of America
8	489	China > Japan
9	471	Korea, Republic of > China
10	467	Venezuela > United States of America
11	427	Japan > China
12	401	Saudi Arabia > Other Asia, nes > China
13	380	Germany > United States of America
14	370	Germany > France
15	357	United States of America > Japan
16	335	Saudi Arabia > United States of America
17	318	Russian Federation > Areas, nes > India
18	315	China > Hong Kong
19	302	Saudi Arabia > Japan
20	300	Russian Federation > Germany

embodied energy into the USA is of the order of the total energy consumption of an entire medium-sized economy such as Australia.

At one end on a continuum, countries with either resource abundance or cheap labour will deliver raw energy or embodied energy in basic goods at historically low prices. At the other end of the continuum, countries that import low cost direct or embodied energy generally exchange in return sophisticated goods and services costing much more than their physical inputs. Historically, exactly this exchange of embodied energy has provided cities their success, and with the ability to retain and cement power, affluence and influence for its citizens.

Each of the countries in Table 4.2 will in general have a large number of trading partners that either supply or receive the energy embodied in the traded commodities. Structural Path Analysis (Treloar 1997) is a method for extracting single paths that link the original source of energy use with the location of final consumption. Based on an input/output analysis of a one-sector world economy, the most important structural paths stretch only one node, that is they involve two countries (Table 4.3). The USA dominates the list of embodied energy 'sinks', which is fed by Canada and China, among others. Ranks 12 and 17 show paths stretching two nodes, where China and India are the recipients of embodied energy originating in Saudi Arabia and Russia, respectively, but traded via other countries.[17]

[17] If only two-node and higher order paths were analysed, city-territories such as Hong Kong are often situated as intermediaries (compare Newcombe 1975). Prominent paths are China > Hong Kong > China (158 PJ), China > Hong Kong > USA (59 PJ), Japan > Hong Kong > China (38 PJ), Korea > Hong Kong > China (24 PJ), USA > Hong Kong > China (23 PJ), China > Hong Kong > Japan (18 PJ), and Singapore > Hong Kong > China (17 PJ).

Taking into account that global trade exacerbates even further the discrepancy between direct and embodied energy use, residences in modern societies such as the USA, Europe and Japan may be equipped with efficient technology, but in addition to domestic embodied energy, their inhabitants consume a substantial amount of energy embodied in goods produced at high energy intensities, often in low income countries such as Mexico, China, Nigeria, and Venezuela. This once more underlines the need for taking embodied energy and lifestyle changes into account.

To account for the energy realities of globalized trade would require revamping the System of National Accounts in that energy and greenhouse gas accounts be attributed to the country of consumption rather than the country of production (Munksgaard and Pedersen 2001; Lenzen *et al.* 2004). Attributing responsibility for energy and emissions alongside trade flows would overturn many core assumptions in the economic and political settings of the globe. Implemented, it would lead to substantial transformation, because any country seeking to reduce its national energy consumption or greenhouse gas emissions would need to limit imports of goods from production chains with high energy/greenhouse gas content, and perhaps advantage local production chains.

4.3 Sydney – A Case Study

Cities are almost entirely dependent on the influx of resources from outside,[18] and urbanization is proceeding rapidly around the world. Australia is one of the most urbanized countries in the world, and Sydney is its largest city, with roughly 4.5 million residents. In addition to growing personal incomes and expenditures – as described in the previous sections – we can expect the energy needs of Sydney residents to be influenced over time by a host of other societal factors: examples are the tendency for young people to move into single-person households, to marry late, and to have fewer children; more extensive child-care facilities and generous maternity-leave policies allowing women to enter the workforce; longer opening hours and more consumption possibilities; longer hours spent commuting to work and family social events;[19] improved health and longevity, and increased leisure activity of the elderly; legislation such as tax rules, vehicle road worthiness criteria and building construction standards; regional population density and climate; and cultural traditions such as holiday cottages, expectations of comfort, etc. (Schipper 1998; Schipper *et al.* 1989). Unfortunately, data were not available on all of the above factors, but only on a set of 11 socio-economic-demographic variables (Table 4.4).[20]

The results presented in the following were obtained by (1) applying generalized input/output analysis (Lenzen 2001) to the most recent Australian Household Expenditure Survey (Australian Bureau of Statistics 2000),[21] (2) extracting the set of socio-economic-demographic data from the survey, and using these data as explanatory variables in a multiple-regression

---

[18] This circumstance becomes drastically clear for residents of a city-territory, such as Hong Kong (Newcombe 1975; Part I and Part II).

[19] Per unit of time, travel exhibits by far the highest energy intensity of all household activities (Schipper *et al.* 1989).

[20] Especially climate has been shown to significantly influence residential energy use in Australia (Newman 1982), so future studies could add data on this variable. However, for a comparison of Sydney SLAs the omission of climate data is inconsequential, since the mean annual temperature is similar across all SLAs.

[21] At the spatial level of Statistical Divisions (SDs) and Statistical Sub-Divisions (SSDs).

Table 4.4.   Explanatory variables for energy consumption.

Explanatory variable	Symbol	Comments
Per capita household income	Inc	Weekly
Household size	Size	Number of people in the household
Average age	Age	Calculated from age bracket occupancies; in units of 100 years
Average qualification	Qual	Dummy variable: Postgraduate Degree = 5, Graduate Diploma and Graduate Certificate = 4, Bachelor Degree = 3, Advanced Diploma and Diploma = 2, Certificate = 1
Population density	Dens	Calculated as a weighted sum with population weights; in units of '000 people per km^2
Tenure type	Ten	Dummy variable: Owners without a mortgage = 5, Owners with a mortgage = 4, Renters from state or territory housing authority = 3, Other renters = 2, Other = 1
Employment status	Empl	% of employed persons in household aged 18–64
Provenance	Prov	% of householders born overseas
Car ownership	Car	Number of cars per person
Travel to work by car	Trav	% of householders
Dwelling type	Dwel	Dummy variable: Separate house = 4, Semi-detached House = 3, Flat = 2, Other = 1

analysis (Lenzen *et al.* 2004), and (3) applying the regression formula to the same set of variables, but extracted for smaller spatial units from the Australian National Census (Australian Bureau of Statistics 2007).[22] While task 1 is documented elsewhere (Lenzen 1998; Lenzen *et al.* 2006), tasks 2 and 3 are described in the following two subsections.

4.3.1 Explaining energy requirements

Before regressing energy data against potential explanatory variables, the latter have to be tested for pairwise correlation. Take, for example, the variables 'travel by car', 'dwelling type', and 'population density' which, in Australia, seem to be highly correlated (Table 4.5). This is probably because where population density is low (rural areas and city fringes), the percentage of people driving to work is high (correlation coefficient of −0.71), and the dwelling type index is high (−0.78), indicating a high percentage of separate houses instead of flats. In principle, any of these variables could explain part of the energy requirement, but not all of them simultaneously, since they really measure one and the same thing. Analysts often isolate the original causal variable; in this case it is likely that the availability of space enables the abundance of separate houses, but the sparse occupation of space also brings with it a lack of public transit. For this reason we have omitted the variables 'travel by car' and 'dwelling type' from our regression.[23]

---

[22]At the spatial level of Statistical Local Areas (SLAs).

[23]All other variables were considered in a step-up-type iterative selection process: first, a regression with one variable – the one showing the highest correlation with energy consumption – was carried out, then the next most-correlated variable added, and so on, until the adjusted R^2 value of the regression did not increase anymore.

Table 4.5. Pearson product moment correlation coefficients between explanatory variables, extracted from about 6900 household samples.

	Size	Age	Qual	Dens	Ten	Empl	Prov	Car	Trav	Dwel
Inc	−0.16	0.25	*0.59*	0.36	−0.05	*0.56*	0.33	−0.22	−0.46	*−0.50*
Size		*−0.58*	−0.09	−0.17	0.18	0.14	0.10	−0.10	0.10	0.16
Age			0.23	0.23	0.29	−0.11	0.07	0.05	−0.13	−0.21
Qual				*0.54*	−0.12	0.11	*0.52*	−0.18	*−0.60*	*−0.64*
Dens					−0.19	0.08	0.39	−0.40	**−0.71**	**−0.78**
Ten						−0.02	0.00	0.18	0.36	0.34
Empl							−0.08	0.01	−0.09	−0.07
Prov								−0.41	−0.35	−0.41
Car									0.62	*0.51*
Trav										**0.84**

Table 4.6. Results of a multiple regression of the per capita energy requirement against variables in Table 4.4. Rows marked 'Residential' etc. list regression coefficients.[24] Rows marked t list results from a Students-t test for significance. Coefficients in bold font are significant at the 99% significance level, regular font – 95%, italic – 90%, underlined – not significant.

	Inc	Size	Age	Qual	Dens	Ten	Empl	Prov	Car
Residential	*0.07*	<u>−0.10</u>	*1.41*	*−0.16*	*0.00*	*0.02*	*0.18*	*−0.29*	*0.92*
t	*0.48*	<u>2.00</u>	*1.98*	*1.29*	*0.08*	*0.17*	*0.51*	*0.89*	*1.87*
Transport	**0.32**	*0.02*	<u>1.03</u>	*−0.07*	<u>−0.07</u>	*0.05*	*−0.01*	*0.01*	*0.33*
t	**3.61**	*0.54*	<u>2.33</u>	*0.95*	<u>2.54</u>	*0.78*	*0.05*	*0.06*	*1.09*
Embodied	**0.40**	*−0.06*	<u>0.66</u>	**0.20**	*0.00*	*−0.01*	*0.04*	*−0.10*	*−0.34*
t	**6.48**	**2.84**	<u>2.19</u>	**3.65**	*0.03*	*0.20*	*0.27*	*0.76*	*1.63*
Total	**0.34**	**−0.06**	*0.83*	<u>0.12</u>	*0.00*	*0.00*	*0.07*	*−0.13*	*−0.09*
t	**6.41**	**2.96**	**3.15**	<u>2.55</u>	*0.23*	*0.02*	*0.51*	*1.13*	*0.47*

Other relatively highly correlated pairs are car ownership and travel to work by car (+), employment status and income (+), and qualification and income (+), all for obvious reasons. People with high qualifications tend to live in densely populated areas, and hence in flats rather than separate houses. Interestingly, people born overseas tend to have higher qualifications.

A multiple regression yields that (Table 4.6)

- residential energy is best explained by household size: the more people occupy the household the lower the per capita residential energy consumption;
- increasing age also explains increasing per capita residential energy, independent of higher income or lower household sizes older households may have;

---

[24] We chose the functional form $\ln(E) = \alpha + \beta \ln(Inc) + \sum_i \gamma_i x_i$, where the x_i are *Size, Age, Qual*, etc., and α, β and γ_i are regression coefficients (Wier *et al.* 2001). Using $\partial \ln(E)/LE = E^{-1}$, it can easily be shown that $\beta = (\partial E/E)/(\partial Inc/Inc)$ is the income elasticity of the energy requirement, which describes the percentage change in the energy requirement as a result of a percentage change in per capita income. Similarly, $\gamma_{Size} = (\partial E/E)/(\partial Size)$ describes the percentage change in the energy requirement as a result of the addition of one family member, $\gamma_{Age} = (\partial E/E)/(\partial Age)$ the percentage change in the energy requirement as a result of an increase in the average age by one year, and so on.

- car ownership appears to explain residential energy, although the connection is unclear;
- per capita transport energy tends to be high when incomes are high, for households in less densely populated areas, and for older households;[25]
- per capita embodied energy is clearly and most strongly driven by income, followed by qualification (independent of income);
- per capita embodied energy decreases with increasing household size, which is probably due to people sharing things such as appliances;
- per capita embodied energy increases with age, independent of higher income or lower household sizes older households may have;
- tenure type, employment status and provenance do not appear to influence energy consumption;
- the variability of the total energy requirement is dominated by embodied energy.

We measure the income elasticity of the total energy requirement at 0.34 (Table 4.6), which means that for a 10% increase in expenditure, the energy requirement increases only by 3.4%. This is due to the fact that when income increases, the proportion of services compared to goods in the consumer basket increases, so that each additionally purchased commodity tends to be less energy intensive.[26] The size and age elasticities of the energy requirement are −6% per additional household member, and +0.8% per year, respectively.

Applying the regression coefficients in Table 4.6 to Australian average socio-economic-demographic variables yields a per capita energy requirement of 180 GJ, which is in excellent agreement with the energy requirement of 176 GJ/cap calculated directly from household expenditure data (Australian Bureau of Statistics 2000).

4.3.2 A spatial view of direct and embodied energy consumption

There are a number of studies that attribute energy use or greenhouse gas emissions to residents of a city, and depict the result as shaded maps (Newcombe 1976; Kalma and Newcombe 1976; VandeWeghe and Kennedy 2007). This has even been done for Sydney, our own case study (Kalma *et al.* 1972). However, to our knowledge, these studies only deal with residential and not with embodied energy. The first publication to present embodied energy in a spatial context is Lenzen *et al.*'s analysis of Sydney (2004). The work presented in the following case study replicates this previous study's methodology, but updates the results, and significantly increases the spatial resolution by applying the regression coefficients in Table 4.6 to information on lifestyles contained in the Australian Census, which holds data on all of the variables in Table 4.6 at a detailed spatial level.

---

[25] Interestingly, Schipper *et al.* (1989) finds that in the 1980s US, elderly people spent more time at home and drove less, so that their residential and transport energy requirements were above and below average, respectively. Already in 1989, Schipper conjectures that 'tomorrow's energetic retirees . . . could carry with them their mobility patterns of younger years, while they continue to live in homes originally built to house families with two or three children'. In our regression we see exactly this phenomenon: Both residential and transport energy use are accelerated by average household age.

[26] The income elasticity of 0.34 reported here is lower than the expenditure elasticities in Fig. 4.5, which range between 0.7 and 1.0. This is because (a) a multivariate regression as in Table 4.6 involves explanatory variables in addition to income, and as such the income elasticity of 0.34 is lower than would have resulted from a univariate regression with only income as a variable; and (b) Fig. 4.5 depicts energy versus expenditure, while Table 4.6 analyses energy versus income. Expenditure elasticities are always higher than income elasticities because expenditure is a better proxy for the energy requirement than income (Newcombe 1979; Wier *et al.* 2001).

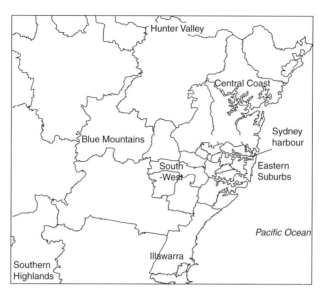

Fig. 4.6. Regions of Greater Sydney.

The maps shown in the following sections contain the city of Greater Sydney, which includes the Blue Mountains and the Lower Central Coast. Bordering are the Illawarra and Southern Highland regions, the Hunter Valley, and the Pacific Ocean (Fig. 4.6). These maps may suggest viewing the embodied energy issue as an issue of geographic location; however, we ask the reader to interpret spatial units as entities of certain socio-economic-demographic characteristics. Where the maps show 'hot spots' of high embodied energy, this is where the *drivers* of embodied energy – for example, affluence – are strong. Hence, we will investigate why certain areas feature certain characteristics, which then lead to consumption habits, and in turn to the embodied energy pattern.[27]

4.3.2.1 Residential energy
The majority of residential energy in Australia is consumed as electricity, natural gas, and firewood (Table 4.7). In our analysis of Sydney, firewood is likely to play a minor role compared to the whole nation, so that our multiple regression is most sensitive to households' electricity and gas bills.

Residential energy appears to be high in the SLAs around Sydney Harbour, and in rural areas to the north-west of the city (Fig. 4.7 [Plate 6]). Our regression (Table 4.6) shows that per capita residential energy is high for small and old households, so one would expect to see a spatial correlation of residential energy with these factors.

Household size exerts the strongest influence (Table 4.6) and is therefore shown for comparison in Fig. 4.8 [Plate 7]. Indeed, for the city itself, we find that residential energy is low where many people share a household, and vice versa. This could simply be because

---

[27] Maps are available also for greenhouse gas emissions, water use and the Ecological Footprint for SLAs in all Australian States and Territories (http://www.acfonline.org.au/consumptionatlas).

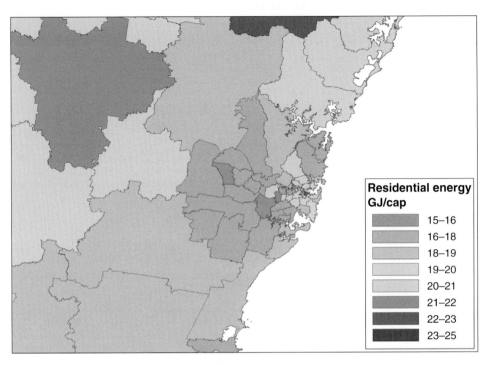

**Residential energy
GJ/cap**

	15–16
	16–18
	18–19
	19–20
	20–21
	21–22
	22–23
	23–25

Fig. 4.7. Per capita residential energy in Greater Sydney SLAs. [Plate 6]

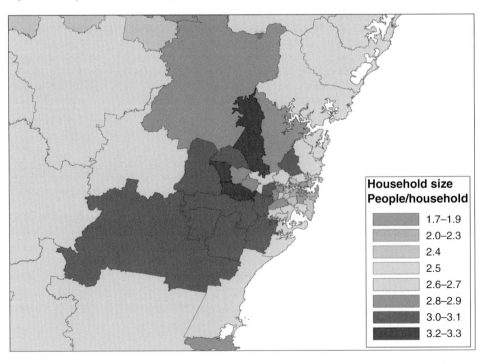

**Household size
People/household**

	1.7–1.9
	2.0–2.3
	2.4
	2.5
	2.6–2.7
	2.8–2.9
	3.0–3.1
	3.2–3.3

Fig. 4.8. Household size in Greater Sydney SLAs (Australian Bureau of Statistics 2007). [Plate 7]

Table 4.7. Breakdown of Australian residential energy into fuel types (Australian Bureau of Agricultural and Resource Economics 2006).

Energy carrier	% of total
Electricity	44.1
Natural gas	30.1
Firewood	21.1
LPG, LNG	2.6
Solar energy	1.0
Heating oil	0.6
Diesel oil	0.4
Kerosene	0.1
Black and brown coal	0.04

these people share energy services such as light and heat that are largely independent of how many people live in a flat or house.

Sydney's age structure (older households in the seaside suburbs, younger households towards the south-west) reinforces the spatial distribution in Fig. 4.7 [Plate 6]. The two outer high energy SLAs (Lithgow and Cessnock/Hunter Valley) are caused by the two remaining explanatory factors (high car ownership and low percentage of people born overseas), and are perhaps an artefact of the regression analysis.

4.3.2.2 *Transport energy*
Energy used for private passenger transport (almost all petrol) is clearly related to geographical location, with energy use increasing with increasing distance from the city centre (Fig. 4.9 [Plate 8]). People living in the centre and along major bus and train arteries have at their disposal various transport options, leading to below-average petrol consumption.

Population density is the strongest explanatory variable for transport energy (Table 4.6), and this relationship is well depicted by contrasting the maps for both factors (Figs 4.9 [Plate 8] and 4.10 [Plate 9]). Major train lines only reach west into the Blue Mountains, north toward the Central NSW Coast, and south to the Illawarra and Southern Highlands regions. Compared to urban rail transport in other megacities, these services are infrequent and slow. Frequent buses run mostly in SLAs denser than $500\,km^{-2}$, so that most people living well outside of the city proper rely entirely on the private car.

4.3.2.3 *Embodied energy*
By far the most significant relationship in our entire regression ties embodied energy to income (Table 4.6). This is why perhaps the most striking resemblance can be found between the maps depicting these two variables (Figs 4.11 [Plate 10] and 4.12 [Plate 11]).

First, note that the magnitude of the scale in Fig. 4.11 [Plate 10] is five to ten times higher than those in Figs 4.6 and 4.8 [Plate 7], once more emphasizing the dominance of embodied energy. Any visitor to Sydney can easily tell from the size and style of houses, and the water views, that the most affluent people congregate around Sydney Harbour, along the

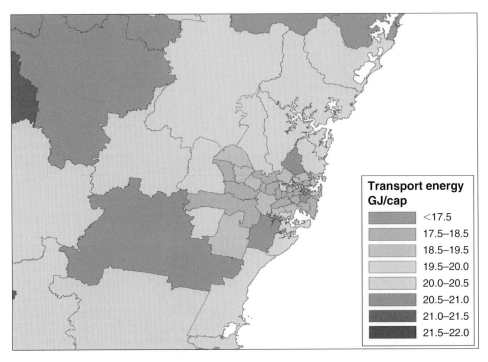

Transport energy
GJ/cap

	<17.5
	17.5–18.5
	18.5–19.5
	19.5–20.0
	20.0–20.5
	20.5–21.0
	21.0–21.5
	21.5–22.0

Fig. 4.9. Per capita transport energy in Greater Sydney SLAs. [Plate 8]

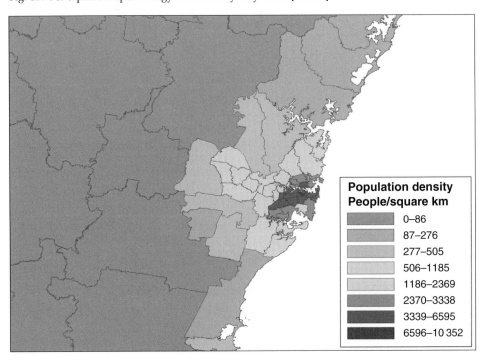

Population density
People/square km

	0–86
	87–276
	277–505
	506–1185
	1186–2369
	2370–3338
	3339–6595
	6596–10 352

Fig. 4.10. Population density in Greater Sydney SLAs (Australian Bureau of Statistics 2007). [Plate 9]

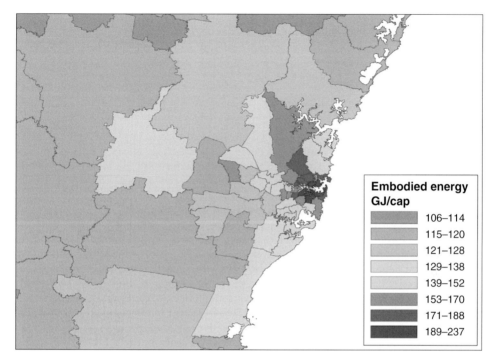

Fig. 4.11. Per capita embodied energy in Greater Sydney SLAs. [Plate 10]

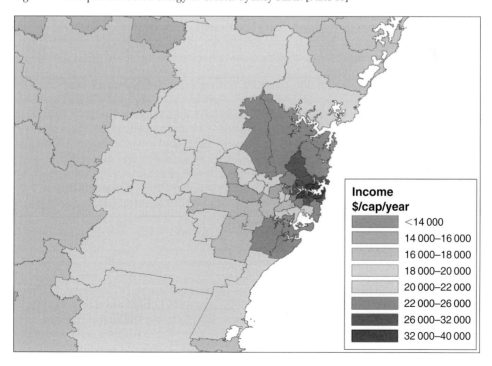

Fig. 4.12. Per capita annual income in Greater Sydney SLAs (Australian Bureau of Statistics 2007). [Plate 11]

Eastern Suburbs beaches, and on the North Shore. This is where some of the nation's most wealthy and vigorous consumers reside, and where embodied energy accumulates.

In contrast, the south-west of Sydney and the rural areas to the west (except the Blue Mountains) are characterized by young, larger families with mostly low incomes and, as a consequence, low embodied energy budgets.

4.3.2.4 Energy – a way of life

Sydney adheres strongly to the conventional measures of success implied by the last three centuries of industrialization and globalization. Its metabolism is based on the manner in which it attracts global citizens seeking modern lifestyles with the best-quality goods and services, and is facilitated by a well-connected port and airport. The city 'believes' in economic growth, affluence, population growth and a continuing real-estate boom. Occupancy rates for domestic and commercial building stocks remain high and reap good financial returns for their owners.

In terms of direct energy, Sydney faces many structural impediments caused by its settlement geography and history, which surface as a range of engineering difficulties, particularly for better urban transit. In energy terms it is mostly dependent on black-coal electricity generated in the Hunter Valley to the north of the city, thus locking in a high carbon content for its electricity. The strategic encouragement of car transport by continual freeway construction along with a massive underinvestment in efficient urban transit further imposes a higher carbon content of mobility than technology suggests is available. The relative age (in Australian terms) of the inner city means that retrofitting of building stock can be physically difficult and expensive. Finally, the marginalized and fragmented nature of local, state and federal representation means that energy politics has never mustered sufficient medium-term support and financing to achieve a 'great leap forward' in the city's energy metabolism. There are few politicians prepared to put their votes at risk for difficult transitions that might take two decades to return real energy savings, particularly given the dominant conventional view that equates embodied energy consumption with success.

In embodied energy terms, the residents of Sydney are generally aspirational of a lifestyle that implies style, fashion and a diversity of possessions. They have higher skills on average than the rest of Australia, get higher pay and thus consume more in embodied energy terms. In the wealth and consumption stakes Sydney is Australia's showcase city in world and national terms. Within its city boundaries it displays a strong gradient in per capita embodied energy requirements from the longer settled and more affluent areas around its harbour and coastlines, to the less affluent suburbs at the western and southern extremities. These embodied energy gradients reflect social equity barriers that have possibly locked in less affluent residents to energy dependence as a way of life. Since social disadvantage usually means poorer health and educational outcomes, any hypothetical energy transition will be doubly difficult because of lower levels of human capital and understanding of why these transitions should take place.

Given this hierarchy of global and functional influences, it seems unlikely that our city of interest, Sydney, or for that matter any first-world city, will be able to effect an energy transition with sufficient leverage to reformat its metabolism and thus meet its global environmental goals. Notwithstanding this dour conclusion, there are five policy strategies which Sydney's managers must work towards as follows. The first is to create new attributes of wealth, human development and status where social richness and community vitality replace volumetric consumption as the key driving forces and measures of success of advanced societies. The second is to cap absolute population numbers over the next two to three human generations, while achieving a population structure that is reasonably balanced by age (no booms and busts) and age by spatial location (no very old and very young suburbs). The third is to introduce carbon rationing and trading for consumers on a full production chain basis so that the full embodied greenhouse content is accounted for, and priced into consumer transactions at the point of sale. The fourth is to form a multimodal city well linked by low carbon transport options and to constrain single-person vehicle transport. The fifth is to mandate leading-edge energy regulations for all new buildings, and, in parallel, begin retrofitting suburb by suburb the existing built infrastructure to constrain energy use, while meeting reasonable economic and social requirements.

4.3.3 Forecasting energy requirements

Just before the manuscript for this book went to print, the 2006 Australian Census was released by the Australian Bureau of Statistics, prompting the main Sydney daily newspaper to describe Australians as 'richer, older, and lonelier than before'(Wade *et al.* 2007). These comments refer to increasing real incomes, an ageing society, and the trend towards single or couple households. Such developments have become daily topics of debate among politicians and the general public alike, and therefore it is intriguing to examine the energy implications of a few future scenarios.

The business-as-usual (BAU) scenario assumes that the historical 2% real economic growth (World Resources Institute 2006) continues until 2050. By that time the average Australian is assumed to live in suburbs that are as dense as city fringes are today ($800 \, \text{km}^{-2}$). Immigration and procreation are assumed to balance the age pyramid, and household sizes will stay the same (two generous assumptions). According to our lifestyle regression, under this scenario, even though affluence rises, the per capita energy requirement will stay about constant at 180 GJ/cap, thanks to (a) energy efficiency improvements of 0.8%/year or an overall 40%, (b) higher population densities that make better use of transport infrastructure, and (c) the 'consumer basket effect' described at the end of the previous subsection. Overall (national) energy consumption will be 70% higher than in 1999, because Australia's population will have grown by 70% from 20 million to just above 30 million.

The 'Fast' scenario assumes an accelerated real growth of 3%/year, which among other influences will drive the trend towards smaller households. An aged Australian population is assumed to live in families typical for today's North Shore (see Figs 4.8 [Plate 7 and 4.12

Table 4.8. Results for three simple future scenarios for 2050. All scenarios assume the continuation of historical trends for the overall energy intensity (-0.8%/year, technology driven) and population ($+1\%$/year, immigration plus procreation) (Foran and Poldy 2002).

Scenario	Australia, average	BAU	Fast	Slow
Income growth		2%	3%	1%
Income	$337	$924	$1520	$559
Household size	2.8	2.8	2.2	3.6
Age	35	35	38	33
Population density	554	800	554	1000
Per capita energy requirement	180	180	225	141
Total energy requirement (1999 = 1)	1	1.7	2.1	1.3

Plate 11]; 2.2 people/household, 38 years), and new settlements are assumed to sprawl by copying the existing spatial structure (554 km^{-2}). Under this scenario, socio-economic-demographic trends outrun efficiency gains, and per capita energy requirements will exceed 220 GJ. Australia's total energy use will more than double until 2050.

The 'Slow' scenario features only 1%/year real growth, and consolidated urban and family structures. A younger Australian population is assumed to live in households typical for today's Fairfield-Liverpool (a low income Sydney suburb, 3.6 people/household, 33 years), but at densities of 1000 km^{-2}.[28] Under this scenario, per capita energy requirements will have significantly decreased to about 140 GJ. Australia's total energy use will still increase by 30%.

Under every scenario, the growth of Australia's energy metabolism is underpinned by more people wanting a more comfortable and convenient life, travelling more often and further, and enjoying more material wealth. This shows once again, and drastically, that an urban energy transition cannot be viewed without considering embodied energy.

4.4 Conclusions: Technological vs Lifestyle Transition

As people flock to the cities in search of opportunities, and societies become more urbanized, they also become more affluent, and their energy systems more efficient. At the same time the demands of people for material wealth, comfort and convenience increase rapidly, causing the increase of industrial energy demand, which in turn often outstrips all energy efficiency gains. Generally the cities are the places where money is made, and which become the home of the wealthy, while the countryside is the place where resources are taken. Urban centres thus become sinks of rural resources and energy (Various authors 2007).

Many growing urban metropolises are reaching their limits of domestic resource availability or environmental pollution, while at the same time further efficiency improvements become costly. By increasingly drawing on cheap resources and energy beyond their borders, facilitated by an increasingly globalized international trade, they export their environmental pressure into resource-rich, but often technology-poor, low wage regions. The domestic resource hinterland turns into the global resource hinterland.

Human civilization is thus a story of urban civilization which, over the past 300 years of industrialization, has become rich and successful on the back of energy exploitation, almost exclusively by combusting fossil fuels that accumulate in form of embodied energy and atmospheric greenhouse gas concentrations. This development has now led to the prospects of diminishing oil supply and runaway global climate change. Thus the affluent citizens of the world's cities are responsible historically for this situation, and are now central to any future prospect of retreating to safer conditions.

In these affluent, urbanized societies, direct energy is less important than embodied energy. More so, the latter has been increasing so strongly that it appears that measures aimed at improving technology on their own have not been able to counteract trends of looming resource depletion and climate change (Trainer 1997).[29] Combining this with the fact that the overarching driving force for the magnitude and the rate of increase of energy

---

[28] For comparison, Australian inner city centres have about 2000–4000 km^{-2}.

[29] Caldwell (1976) has argued that, given the structure of social institutions of modern industrial societies, a cheap and inexhaustible energy supply may pose far greater problems than energy shortages: 'There is no quicker way to destroy a society than to confer upon it power that it lacks wisdom, consensus, restraint, and institutional means to handle.'

consumption is personal affluence, we conclude that the urban energy transition can only be effective if technology transitions are complemented by far-reaching lifestyle transitions (Trainer 1995).

It is possible that such lifestyle transitions are doomed to failure. Cities seem to have escaped, for the time being at least, the physical dependence on their immediate hinterland, which may have led to a feeling of invincibility where people's aspirations have transgressed natural limits and are now driven only by human ingenuity. Thus, having formally escaped physical realities, city dwellers fail to read the danger signs of their existence and may even deny those signs totally. This manifests itself in the missing knowledge-concern-action link: By and large, people become concerned once they know about an environmental problem, but there appear to be no signs whatsoever that concerned people cause less environmental impact than people who are not concerned, or do not even know about environmental issues (Kempton 1993; Stokes *et al.* 1994; Vringer *et al.*; Hastings 2007; Wilby 2007; McKibben 2003).

To consider transforming the escalator of aspirations, from one that is forever outrunning unhappiness, to one that allows fulfilment, beggars belief as to what drives a modern city resident in a time-poor and globally connected world. Changing the storyline of the century just gone will require focused social engineering for many human generations.

Like most of the general public, public policy in developed countries has also failed to appreciate, or deliberately avoided addressing, the link between affluence and energy. This is evident in policies focused purely on energy efficiency without constraints on total energy use and greenhouse gas emissions. As we have shown, once the cost of energy efficient infrastructure is paid back, the dollar savings allow for more production or expenditure, thus setting in train the requirement for more energy, either directly or indirectly.

It is understandable why policy has more readily embraced supporting technological change rather than promoting lifestyle change. After all, what can be achieved by new technology is easy to sell to the consumer: no one has to give up their habits, and governments do not need to risk losing votes, because they do not need to initiate a potentially painful and difficult public discourse, let alone intervention into consumers' choices. In a functioning democracy, convincing the public to forgo certain types of unsustainable consumption now for the sake of future generations is without doubt a formidable challenge, prone to resistance, requiring respected leadership (Beekman 1997).

Decades of unabated and unrestrained economic growth, nurtured by advertising affluent, material lifestyles to an ever-growing portion of the world population makes one wonder whether some sort of lifestyle change is indeed unavoidable. One would hope for such changes to be brought about by conscious and collective decisions rather than by involuntary and unilateral force, or perhaps worse, by natural and socio-economic circumstances.

A compelling reminder of the latter option is Ronald Wright's recount of the fallibility of human civilizations in his book *A Short History of Progress* (Wright 2004, p. 79) as he describes the collapse of the Empire of Ur around 2000 BC located in what is now Southern Iraq:

The short-lived Empire of Ur exhibits the same behaviour we saw on Easter Island: sticking to entrenched beliefs and practices, robbing the future to pay for the present, spending the last reserves of natural capital on a reckless binge of excessive wealth and glory. Canals were lengthened, fallow periods reduced, population increased and the economic surplus concentrated on Ur to support grandiose building projects. The result was a few generations of prosperity (for the rulers) followed by a collapse from which Southern Mesopotamia has never recovered.

In spite of their impressive technological underpinnings, modern cities such as Sydney display many of the fragilities that could in the long term lead to their decline, and then the collapse of what were pre-eminent civilizations in their day. The challenge for the modern city-state is to learn from lessons of the past millennia, and to begin the transition process at least one century before physical problems become intractable. The one thing that has changed from the Empire of Ur is that the problems of tomorrow will be driven by global dynamics rather than local or regional ones.

Acknowledgements

Christopher Dey from the Centre for Integrated Sustainability Analysis at the University of Sydney prepared the maps shown in section 4.3.2.

References

Ang, B.W. (2000). A survey of index decomposition analysis in energy and environmental studies. *Energy*, **25**, 1149–1176.

Australian Bureau of Agricultural and Resource Economics (2006). *Energy update.*

Australian Bureau of Statistics (2000). *1998–99 Household Expenditure Survey – Detailed Expenditure Items, Confidentialised Unit Record File. Electronic file* (unpublished).

Australian Bureau of Statistics (2004). *Australian National Accounts, Input–Output Tables, 1998–99.* ABS Catalogue No. 5209.0. Canberra, Australia: Australian Bureau of Statistics.

Australian Bureau of Statistics (2007). *2001 Census Data by Location, Community Profiles.* Internet site http://www.abs.gov.au/websitedbs/d3310114.nsf/home/Census+data.Canberra,Australia:Australian Bureau of Statistics.

Beekman, V. (1997). Government intervention in non-sustainable lifestyles, in *Environmental Justice, Papers from the Melbourne Conference* (ed. N. Low). Faculty of Architecture, Building and Planning, The University of Melbourne, Melbourne, Australia.

Bin, S. and Dowlatabadi, H. (2005). Consumer lifestyle approach to US energy use and the related CO_2 emissions. *Energy Policy*, **33**(2), 197–208.

Bourque, P.J. (1981). Embodied energy trade balances among regions. *International Regional Science Review*, **6**, 121–136.

Bullard, C.W. and Herendeen, R.A. (1975). Energy impact of consumption decisions. *Proceedings of the IEEE*, **63**(3), 484–493.

Caldwell, L.K. (1976). Energy and the structure of social institutions. *Human Ecology*, **4**(1), 31–45.

Chang, Y.F. and Lin, S.J. (1998). Structural decomposition of industrial CO_2 emissions in Taiwan: an input-output approach. *Energy Policy*, **26**(1), 5–12.

Chen, C.-Y. and Rose, A. (1990). A structural decomposition analysis of changes in energy demand in Taiwan: 1971–1984. *Energy Journal*, **11**(1), 127–146.

Chung, H.-S. and Rhee, H.-C. (2001). A residual-free decomposition of the sources of carbon dioxide emissions: a case of the Korean industries. *Energy*, **26**, 15–30.

Cohen, C.A.M.J., Lenzen, M. and Schaeffer, R. (2005). Energy requirements of households in Brazil. *Energy Policy*, **55**, 555–562.

Cole, M.A., Rayner, A.J. and Bates, J.M. (1997). The environmental Kuznets curve: an empirical analysis. *Environment and Development Economics*, **2**(4), 401–416.

Common, M.S. and Salma, U. (1992). Accounting for changes in Australian carbon dioxide emissions. *Energy Economics*, **14**(3), 217–225.

Dietzenbacher, E. and Los, B. (1997). Analyzing decomposition analyses, in *Prices, Growth and Cycles* (eds A. Simonovits and A.E. Steenge), pp. 108–131. MacMillan, London, UK.

Ehrhardt-Martinez, K., Crenshaw, E.M. and Jenkins, J.C. (2002). Deforestation and the Environmental Kuznets Curve: a cross-national investigation of intervening mechanisms. *Social Science Quarterly*, **83**(1), 226–243.

Ekins, P. (1997). The Kuznets curve for the environment and economic growth: examining the evidence. *Environment and Planning*, A **29**(5), 805–830.

Folke, C., Jansson, Å., Larsson, J. and Costanza, R. (1997). Ecosystem appropriation by cities. *Ambio*, **26**(3), 167–172.

Foran, B. and Poldy, F. (2002). *Future Dilemmas – Options to 2050 for Australia's population, technology, resources and environment*. Working Paper Series 02/01. Canberra, Australia: CSIRO Sustainable Ecosystems.

Garcia, R.A., Manegdeg, F. and Raneses, N.O. (1994). Household energy consumption surveys in three Philippine cities. *Energy*, **19**(5), 539–548.

Grossman, G.M. and Krueger, A.B. (1995). Economic growth and the environment. *Quarterly Journal of Economics*, **110**(2), 353–377.

Hamilton, C. and Turton, H. (1999). Population policy and environmental degradation: sources and trends in greenhouse gas emissions. *People and Place*, **7**(4), 42–62.

Hastings, M. (2007). Addressing binge flying is vital for the climate. *Guardian Weekly*, **176**(20), 19.

Herendeen, R. (1978). Total energy cost of household consumption in Norway, 1973. *Energy*, **3**, 615–630.

Herendeen, R.A. (1974). Affluence and energy demand. *Mechanical Engineering*, **96**(10), 18–22.

Herendeen, R. and Tanaka, J. (1976). Energy cost of living. *Energy*, **1**, 165–178.

Herendeen, R., Ford, C. and Hannon, B. (1981). Energy cost of living, 1972–1973. *Energy*, **6**, 1433–1450.

Heyes, A.G. and Liston-Heyes, C. (1993). US demilitarization and global warming. *Energy Policy*, **21**(12), 1217–1224.

Hills, P. (1994). Household energy transition in Hong Kong. *Energy*, **19**(5), 517–528.

Hoekstra, R. and van den Bergh, J.C.J.M. (2002). Structural decomposition analysis of physical flows in the economy. *Environmental and Resource Economics*, **23**, 357–378.

International Energy Agency. (2007) *Energy Balances*. Internet site http://www.iea.org/Textbase/stats/prodresult.asp?PRODUCT=Balances. Paris, France: OECD/IEA.

Johansson, B. (2003). Transportation fuels – a system perspective, in *Handbook of Transport and the Environment* (eds D.A. Hensher and K.J. Button), pp. 141–157. Elsevier BV, Amsterdam, Netherlands.

Kalma, J.D. and Newcombe, K.J. (1976). Energy use in two large cities: a comparison of Hong Kong and Sydney, Australia. *International Journal of Environmental Studies*, **9**(1), 53–64.

Kalma, J.D., Aston, A.R. and Millington, R.J. (1972). Energy use in the Sydney area. *Proceedings of the Ecological Society of Australia*, **7**, 125–142.

Kempton, W. (1993). Will public environmental concern lead to action on global warming? *Annual Review of Energy and the Environment*, **18**, 217–245.

Kulkarni, A., Sant, G. and Krishnayya, J.G. (1994). Urbanization in search of energy in three Indian cities. *Energy*, **19**(5), 549–560.

Leach, G.A. (1998). Residential energy in the Third World. *Annual Review of Energy*, **13**, 47–65.

Lenzen, M. (1998). The energy and greenhouse gas cost of living for Australia during 1993–94. *Energy*, **23**(6), 497–516.

Lenzen, M. (2001). A generalised input-output multiplier calculus for Australia. *Economic Systems Research*, **13**(1), 65–92.

Lenzen, M. and Murray, J. (2001). The role of equity and lifestyles in education about climate change: experiences from a large-scale teacher development program. *Canadian Journal of Environmental Education*, **6**, 32–51.

Lenzen, M. and Dey, C.J. (2002). Economic, energy and emissions impacts of some environmentally motivated consumer, technology and government spending options. *Energy Economics*, **24**, 377–403.

Lenzen, M., Dey, C. and Hamilton, C. (2003). Climate change, in *Handbook of Transport and the Environment* (eds D.A. Hensher and K.J. Button), pp. 37–60. Elsevier BV, Amsterdam, Netherlands.

Lenzen, M., Pade, L.-L. and Munksgaard, J. (2004). CO_2 multipliers in multi-region input-output models. *Economic Systems Research*, **16**(4), 391–412.

Lenzen, M., Dey, C. and Foran, B. (2004). Energy requirements of Sydney households. *Ecological Economics*, **49**(3), 375–399.

Lenzen, M., Wier, M., Cohen, C., *et al.* R. (2006). A comparative multivariate analysis of household energy requirements in Australia, Brazil, Denmark, India and Japan. *Energy*, **31**, 181–207.

Lin, X. and Polenske, K.R. Input-output anatomy of China's energy use changes in the 1980s. *Economic Systems Research*, **7**(1), 67–84.

Mélanie, J., Phillips, B. and Tormey, B. (1994). An international comparison of factors affecting carbon dioxide emissions. *Australian Commodities*, **1**(4), 468–483.

Mazur, A. and Rose, E. (1974). Energy and life-style. *Science* 186, 607–610.

McKibben, B. (2003). Worried? Us? *Granta*, **83**, 7–12.

Melasniemi-Uutela, H. (1999). The need to heed the changing structure of household consumption, not only technical energy efficiency – the Finnish case, in *Energy Efficiency and CO_2 Reduction: The Dimensions of the Social Challenge* (ed. European Council for an Energy-Efficient Economy). Mandelieu, France: ADEME Editions.

Mukhopadhyay, K. and Chakraborty, D. (1999). India's energy consumption changes during 1973/74 to 1991/92. *Economic Systems Research*, **11**(4), 423–438.

Munksgaard, J. and Pedersen, K.A. (2001). CO_2 accounts for open economies: producer or consumer responsibility? *Energy Policy*, **29**, 327–334.

Muradian, R., O'Connor, M. and Martinez-Alier, J. (2002). Embodied pollution in trade: estimating the 'environmental load displacement' of industrialised countries. *Ecological Economics*, **41**, 51–67.

Nakagami, H. (1996). Lifestyle change and energy use in Japan: household equipment and energy consumption. *Energy* 21(12), 1157–1167.

Newcombe, K. (1975). Energy use in Hong Kong: Part I, an overview. *Urban Ecology*, **1**(1), 87–113.

Newcombe, K. (1975). Energy use in Hong Kong: Part II, sector end-use analysis. *Urban Ecology*, **1** (2–3), 285–309.

Newcombe, K. (1976). Energy use in Hong Kong: Part III, spatial and temporal patterns. *Urban Ecology*, **2**(2), 139–172.

Newcombe, K. (1979). Energy use in Hong Kong: Part IV. Socioeconomic distribution, patterns of personal energy use, and the energy slave syndrome. *Urban Ecology*, **4**(3), 179–205.

Newman, P.W.G. (1982). Domestic energy use in Australian cities. *Urban Ecology*, **7**, 19–38.

Pearson, P.J.G. (1994). Energy, externalities and environmental quality: will development cure the ills it creates? *Energy Studies Review*, **6**(3), 199–216.

Peet, N.J., Carter, A.J. and Baines, J.T. (1985). Energy in the New Zealand household, 1974–1980. *Energy*, **10**(11), 1197–1208.

Pongsapich, A. and Wongsekiarttirat, W. (1994). Urban household energy consumption in Thailand. *Energy*, **19**(5), 509–516.

Proops, J.L.R., Faber, M. and Wagenhals, G. (1993). *Reducing CO_2 Emissions*. Springer-Verlag, Berlin, Germany.

Qiu, D., Ma, Y., Lu, Y., *et al.* (1994). Household energy consumption in Beijing and Nanning, China. *Energy*, **19**(5), 529–538.

Rees, W. and Wackernagel, M. (1996). Urban ecological footprints: why cities cannot be sustainable – and why they are a key to sustainability. *Environmental Impact Assessment Review*, **16**(4–6), 223–248.

Roberts, P.C. (1975). Energy analysis in modeling, in *Aspects of Energy Conversion* (eds I.M. Blair, B.D. Jones and A.J. Van Horn), pp. 759–771. Pergamon Press, Oxford, UK.

Sanchez, M.C., Koomey, J.G., Moezzi, M.M., *et al.* (1998). Miscellaneous electricity in US homes: historical decomposition and future trends. *Energy Policy*, **26**(8), 585–593.

Sathaye, J., Tyler, S. and Goldman, N. (1994). Transportation, fuel use and air quality in Asian cities. *Energy*, **19**(5), 573–586.

Schipper, L. (1993). Global climate change: linking energy, energy efficiency, human activities and climate change. *Environmental Science Research*, **45**, 37–44.

Schipper, L. (1998). Life-styles and the environment: the case of energy. *IEEE Engineering Management Review*, **26**(1), 3–14.

Schipper, L. and Ketoff, A. (1985). Explaining residential energy use by international comparisons. *Annual Review of Energy*, **10**, 341–405.

Schipper, L., Marie-Lilliu, C. and Lewis-Davis, G. (2001). Rapid motorisation in the largest countries in Asia: implication for oil, carbon dioxide and transportation. *Asia Pacific Journal of Energy*. Internet site http://www.iea.org/pubs/free/articles/schipper/rapmot.htm

Schipper, L. and Fulton, L. (2003). Carbon dioxide emissions from transportation: trends, driving factors and forces for change, in *Handbook of Transport and the Environment* (eds D.A. Hensher and K.J. Button), pp. 203–225. Elsevier BV, Amsterdam, Netherlands.

Schipper, L., Bartlett, S., Hawk, D. and Vine, E. (1989). Linking life-styles and energy use: a matter of time? *Annual Review of Energy*, **14**, 273–320.

Schipper, L., Ting, M., Khrushch, M. and Golove, W. (1997). The evolution of carbon dioxide emissions from energy use in industrialized countries: an end-use analysis. *Energy Policy*, **25**(7–9), 651–672.

Scholl, L., Schipper, L. and Kiang, N. (1996). CO_2 emissions from passenger transport. *Energy Policy*, **24**(1), 17–30.

Selden, T.M. and Song, D. (1994). Environmental quality and development: Is there a Kuznets curve for air pollution emissions? *Journal of Environmental Economics and Management*, **27**, 147–162.

Shafik, N. (1994). Economic development and environmental quality: an econometric analysis. *Oxford Economic Papers*, **46**, 757–773.

Smith, A. (1776). *An Inquiry into the Nature and Causes of the Wealth of Nations*. Strand, UK: Published 1904 by Methuen & Co.

Smith, H. and Raemaekers, J. (1998). Land use pattern and transport in Curitiba. *Land Use Policy*, **15**(3), 233–251.

Smith, K.R., Apte, M.G., Yuqing, M., *et al.* (1994). Air pollution and the energy ladder in Asian cities. *Energy*, **19**(5), 587–600.

Stern, D.I. (2003). The rise and fall of the Environmental Kuznets Curve. Working Papers 0302. Troy, NY: Rensselaer Polytechnic Institute.

Stern, D.I., Common, M.S. and Barbier, E.B. (1996). Economic growth and environmental degradation: the Environmental Kuznets Curve and sustainable development. *World Development*, **24**(7), 1151–1160.

Stokes, D., Lindsay, A., Marinopoulos, J., *et al.* (1994). Household carbon dioxide production in relation to the greenhouse effect. *Journal of Environmental Management*, **40**, 197–211.

Trainer, T. (1995). *Towards a Sustainable Economy: The Need for Fundamental Change*. Envirobooks, Sydney, Australia.

Trainer, F.E. (1997). Can renewable energy sources sustain affluent society? *Energy Policy*, **23**(12), 1009–1026.

Treloar, G. (1997). Extracting embodied energy paths from input-output tables: towards an input-output-based hybrid energy analysis method. *Economic Systems Research*, **9**(4), 375–391.

Tyler, S. (1994). Household energy use and environment in Asian cities: an introduction. *Energy* 19(5), 503–508.

UN-Energy (2007). *Sustainable Bioenergy: A Framework for Decision-Makers*. Internet site http://esa. un.org/un-energy/pdf/susdev.Biofuels.FAO.pdf. New York, USA: United Nations.

Unander, F., Ettestøl, I., Ting, M. and Schipper, L. (2004). Residential energy use: an international perspective on long-term trends in Denmark, Norway and Sweden. *Energy Policy*, **32**(12), 1395–1404.

United Nations Statistics Division (2007). *UN comtrade – United Nations Commodity Trade Statistics Database*. Internet site http://comtrade.un.org/. New York, USA: United Nations Statistics Division, UNSD.

VandeWeghe, J.R. and Kennedy, C. (2007). A spatial analysis of residential greenhouse gas emissions in the Toronto Census Metropolitan Area. *Journal of Industrial Ecology*, **11**(2), 133–144.

Various Authors (2007). Special issue on industrial ecology and the global impacts of cities. *Journal of Industrial Ecology*, **11**(2), 1–144.

Vringer, K., Aalbers, T. and Blok, K. Household energy requirement and value patterns. *Energy Policy*, **35**(1), 553–566.

Vringer, K. and Blok, K. (1995). The direct and indirect energy requirements of households in the Netherlands. *Energy Policy*, **23**(10), 893–910.

Wachsmann, U. (2005). A structural decomposition analysis of Brazilian energy consumption and CO_2 emissions, 1970–1996. Ph.D. thesis. Rio de Janeiro, Brazil: Universidade Federal do Rio de Janeiro.

Wade, M., Dale, D. *et al.* (2007). Census 2006 – Snapshot of a Nation. *Sydney Morning Herald* 1–6.

Weber, C. and Perrels, A. (2000). Modelling lifestyle effects on energy demand and related emissions. *Energy Policy*, **28**, 549–566.

Wei, Y.-M., Liu, L.-C., Fan, Y. and Wu, G. (2007). The impact of lifestyle on energy use and CO_2 emission: an empirical analysis of China's residents. *Energy Policy*, **35**, 247–257.

Wiedmann, T., Lenzen, M., Turner, K. and Barrett, J. (2007). Examining the global environmental impact of regional consumption activities – Part 2: Review of input–output models for the assessment of environmental impacts embodied in trade. *Ecological Economics*, **61**(1), 15–26.

Wier, M. (1998). Sources of changes in emissions from energy: a structural decomposition analysis. *Economic Systems Research*, **10**(2), 99–112.

Wier, M., Lenzen, M., Munksgaard, J. and Smed, S. (2001). Environmental effects of household consumption pattern and lifestyle. *Economic Systems Research*, **13**(3), 259–274.

Wilby, P. (2007). Grandma's 80 W bulb will kill her grandkids. *Guardian Weekly*, **176**(25), 24.

Wilhite, H., Nakagami, H., Masuda, T. *et al.* (1996). A cross-cultural analysis of household energy use behaviour in Japan and Norway. *Energy Policy*, **24**(9), 795–803.

Wolvén, L.-E. (1991). Life-styles and energy consumption. *Energy*, **16**(6), 959–963.

Wood, R. (2003). *The structural determinants for change in Australia's greenhouse gas emissions. Honours thesis.* The University of Sydney, Sydney, Australia.

World Resources Institute (2006). *Climate Analysis Indicators Tool (CAIT) on-line database version 3.0.* Internet site http://cait.wri.org

Wright, R. (2004). *A Short History of Progress*. Text Publishing, Melbourne, Australia.

Chapter 5
Energy Development and Sustainable Monetary Systems

SHANN TURNBULL[1]

Principal, International Institute for Self-governance, PO Box 266, Woollahra, Sydney, Australia 1350

5.1 Introduction

This chapter explains how the generation of renewable energy, the creation of sustainable communities and making capitalism sustainable can be designed to become self-reinforcing. This state of affairs can be achieved by economic activities being financed by a local currency whose value is defined in terms of kilowatt-hours generated from sustainable sources to create 'energy' dollars. Energy dollars provide a basis for replacing the current monetary regime that creates compelling incentives to discount the future to jeopardize the existence of humanity on the planet.

To avoid an energy rich community being dependent on external finance or a financially self-sufficient community importing energy, the two resources need to be integrated to create sustainable communities (Swann 1997; Bennello 1997). An important reason for linking the two is that renewable energy sources become more economically attractive than non-renewable energy when a custom designed local currency is introduced that eliminates the cost of interest. Islamic banking achieves this objective. Another way, consistent with Islamic banking, is to establish a local currency that has a user fee as discussed later in this section.

---

[1] Shann Turnbull, Dip. Elec. Eng. (Hobart), BSc (Melbourne), MBA (Harvard), PhD (Macquarie), began his career as an electrical engineer in the Hydro-electric Commission of Tasmania. After attending Harvard Business School he became the founder/CEO of a number of businesses, three of which became listed in Australia. He participated in the identification, acquisition and reorganization of a dozen listed Australian companies as their chairman and/or CEO or advisor from 1967 to 1974. In 1975 he became a founding author of the first educational qualification in the world for company directors and used his PhD research to create an MBA course for designing governance systems in the public, private and non-profit sectors. He first promoted the idea of energy dollars in 1977 and included this in his contributions as co-author of *Building Sustainable Communities*: *Tools and Concepts for Self-Reliant Economic Change* and author of *New Economics for a New World Democracy*. He is also the author of *Democratising the Wealth of Nations*, *A New Way to Govern: Organisations and society after Enron*, 23 book chapters, and hundreds of popular, professional and academic articles mainly on economic reform, see links at http://www.linkedin.com/pub/0/aa4/470 and http://www.aprim.net/associates/turnbull.htm

Interest costs from capital intensive renewable electrical power generation can be an order of magnitude greater than interest charges from fuel intensive power generation. It is the interest cost that makes renewable power much more expensive than burning carbon. An indicative comparison is provided in Table 5.1.

Table 5.1.   Indicative cost comparison between renewable and carbon burning power generation.

Indicative cost comparison of electrical power between: (Assumes sales price=value of investment in plant/kWh)	Renewable energy	Carbon burning energy
Value of investment in electrical generation plant (=sales price)	$20/kWh	$10/kWh
Operating life of plant	20 yrs	20 yrs
Cost of writing investment off over its operating life (5% per year)	$1/kWh	$0.5 kWh
All other operating costs before finance charges – (includes fuel)	$1/kWh	$7.5 kWh
Earnings Before Interest and Tax per kWh (EBIT/kWh)	$2/kWh	$8/kWh
Cost of finance charges and any tax	$18/kWh	$2/kWh
Cost of electrical power with demurrage money/Islamic banking	**$2/kWh**	**$8/kWh**

The investment cost for unit of output of renewable energy ($20/kWh) is taken to be twice that of carbon energy ($10/kWh). The value of the output each year is taken to be equal to the value of the investment to make renewable power twice as expensive. The life of each plant is assumed to be 20 years requiring 5% of the plant cost to be written off each year. The operating cost of the renewable energy plant is mainly maintenance and this is assumed also to be 5% a year ($1/kWh). However, the operating cost of a carbon burning plant is 7.5 times greater as it includes the cost of fuel. The resulting Earnings Before Interest and Tax (EBIT) for a carbon burning plant becomes $8/kWh four times greater than renewable energy with an EBIT of $2/kWh. Renewable energy becomes twice as expensive than burning carbon because its finance costs of $18/kWh is nine times greater than the $2/kWh finance cost of burning carbon. Eliminate the finance cost and carbon generated power becomes four times more expensive than renewable energy.

In addition, the cost of non-renewable fuels like coal, oil and gas do not recognize the cost of nature creating the fuel or the environmental costs when it is used. Economists describe these costs as 'externalities'. The external cost of non-renewable power generation includes the cost of global warming. This illustrates how markets can 'fail'. They operate in a counterproductive manner, misallocating resources to increase costs instead of efficiently sustaining civilization.

To avoid burning carbon fuels and contributing to global warming local renewable energy sources need to be tapped such as sun, wind, waves, tide, geothermal, bacteria generated hydrogen (hydrogenases) and hydrogen fusion. Every community on the planet has access to some type of sustainable sources of energy. This means that every community on the planet can establish its own independent unit of value based on the cost of converting sustainable sources of energy into electricity measured in kilowatt-hours. New technologies like hydrogen fusion and hydrogenases will ensure that significant sustainable energy sources are universally available. Hydrogenases could become especially attractive in urban communities that lack access to water as this is produced by burning hydrogen to generate power.

It is not desirable for sustainable energy dollars to also carry out one of the other conventional functions of money – to be a store of value. For various reasons discussed later a currency needs to be designed on ecological principles to have limited life like all living things. Such ecological forms of money were created in hundreds of US communities in the 1930s to replace the shortage of official money at that time (Fisher 1933).

These non-official 'shadow' currencies were created by local councils or chambers of commerce. The money was given away to citizens to invigorate the local economy. However, the money lost all its value unless a demurrage charge was paid to the issuer. This usage cost varied from community to community. But typically each note with a face value of $1.00 would lose all its value after seven days unless a one cent stamp was affixed to the back of the note. The notes would be cancelled after two years when the issuer had collected one cent each week over 104 weeks. In this way the issuer of the note made a 4% profit from giving away its money!

The cost of the stamp is described as a 'demurrage' charge. A cost of 1% of the face value of a note *per week* would be more attractive to merchants and consumers than modern credit cards that charge around 2% *per transaction*. The more transactions the notes financed each week the smaller would become the average demurrage cost per transaction. For example, if a note was used in 20 transactions in a week the average demurrage cost per transaction would be reduced to one twentieth of 1%, which would represent 0.05% of the average value of each transaction.

Demurrage has been a traditional feature of money; it is not a radical new idea! In past centuries when commodities such as gold, silver, copper, tea, tobacco or grains were used as money, large depositors were charged a fee to cover the cost of storage and insurance. Many banks now charge small depositors a service fee that can reach a similar cost. A demurrage charge between 0.05% and 0.25% of foreign exchange transactions described as a 'Tobin tax' has been proposed to inhibit currency speculation (OECD 2000). The revenues raised would be allocated to finance global projects that could include the costs of reducing climate change.

Locally based renewable energy sources create a self-reinforcing virtuous sustainable economic process. The renewable energy can be used to create a locally determined but globally relevant unit of value while the local currency on which it is based eliminates the cost of interest to make renewable energy sources economically attractive. In addition, renewable energy would plug the drain of economic value flowing out of a community by reducing the need to import fuel for power generators or pay external suppliers for power. Nor would there be a need to export economic value from the payment of carbon taxes.

Currencies with a demurrage charge atrophy with age like living things. The idea that money should increase with age goes against the laws of nature as recognized by Islamic banking. Yet the idea that the value of money increases with the passing of time is widely accepted outside the Islamic community. When money has a time value it results in future values becoming discounted on a compounding basis. It explains why modern investment analysis is inconsistent with sustaining humanity on the planet.

Ecological systems are never static, perpetual or exclusive but subject to continuous change by being renewed through death and birth to obtain a better fit to changing environmental conditions. Property rights to realty and corporations can also be designed to take on ecological characteristics so that they can instigate change, die and/or be renewed. The introduction of ecological characteristics to the ownership and control of money, realty and corporations provides a basis for making urban communities economically sustainable as well as achieving energy sustainability. Ecological ownership and control provides a way of reforming the nature of capitalism to make it more efficient, sustainable, equitable, responsive, and democratic as explained later.

Section 5.2 considers how ecological property rights provide a way to plug the economic drains in urban communities to assist them to become self-financing and so sustainable. Section 5.3 considers how ecological property rights facilitate the ability of urban precincts to become self-financing and so self-governing. The governance architecture of realty with ecological property rights is considered in section 5.4. The financial architecture for a sustainable energy monetary system is considered in section 5.5, with concluding comments in section 5.6.

5.2 Plugging the Economic Drains from an Urban Precinct

This section considers how economic value can drain out of an urban precinct and how this can be minimized by the introduction of ecological property rights.

The adoption of ecological forms of money changes the way a market economy allocates resources as described by Jacobs (1985a). One way as noted above is that it makes renewable energy sources more attractive. The adoption of ecological forms of property rights creates a way of transferring economic value *without money* to introduce an additional way for allocating resources to create self-sustaining, self-financing and self-governing urban precincts as described in section 5.3.

Urban communities, be they towns or precincts of city, can be viewed as micro-economies. To further their financial and hence political independence communities need to eliminate or minimize the loss of economic value from: (i) imports exceeding exports; (ii) migration of its citizens; (iii) wages, salaries, and fees paid to guest workers; (iv) rents, profits, dividends, royalties, and fees paid to external property owners; and (v) interest payments to external lenders. Local generation of renewable energy is but one way of reducing imports.

However, urban planners do not usually consider the architecture of these invisible activities when designing urban precincts. A famous exception was Howard (1946) who designed both the visible and invisible structures for the first Garden City of Letchworth in England at the beginning of twentieth century. However, even Howard did not design ways for the invisible structures of ownership and control to be localized to facilitate self-governance. As a result the Letchworth Corporation became the target of a takeover bid from a corporate raider in the 1970s. To prevent this, the Corporation was nationalized by the UK central government.

The most important and insidious way urban precincts lose value is through external payments of mortgage interest and rent. These payments can drain away up to a third of householder income. Another way economic value can be drained away – preventing urban communities from becoming self-financing – is through external owners of property capturing the windfall gains created by public and private infrastructure investment servicing their property.

Both the inequity and potential of public investment in infrastructure to make urban precincts self-financing are illustrated by the building of the Jubilee underground train line in London. The line involved building 11 new stations in 1999 at a cost of £3.5 billion. Riley (2002) reports that the aggregate uplift in the value of land within 1000 yards of the new stations was £13 billion, 3.7 times greater than the investment of the whole project. This illustrates how public money creates private profit, and reveals how inefficient and inequitable the current system of urban land ownership is. It also illustrates how the ecological land tenure system described in section 5.3 allows windfall gains to be captured by tenants and home owners alike (but not by commercial interests or those living outside the precinct) while also using the gains to borrow the funds to build community developments.

While a local ecological currency can minimize the export of interest, ecological property rights are needed to limit the export of value through rental payments and windfall gains. Windfall gains can be very substantial as illustrated by the building of the Jubilee Line. Surprisingly windfall gains are not part of the traditional calculus of economists. Windfall gains are not typically measured over an urban precinct – and what is not measured and reported is not seen and hence not managed.

Even when windfall gains are measured and reported, they are not necessarily considered part of economics. Economists traditionally limit the extent of their analysis to the 'production and exchange of goods and services' and ignore 'balance sheet' transactions like the exchange and transformation of assets and liabilities. As a result economic textbooks do not identify the concept of 'surplus profits' (Turnbull 2006a). When they see the words 'surplus profit' they assume that it is a component of what their jargon refers to as 'economic rent'. Economic rent is typically described as the revenues required to create or maintain production. Surplus profit is a complementary concept because it describes the revenues that are *not* required for an investment in either productive or non-productive assets.

Surplus profits are income cash flows in excess of the incentive to invest. They arise after a period described as the investment 'time horizon', a point after which investors do not expect any return on their investment. The lack of expectation may arise from perceived risks in production or product failure or obsolescence, emergence of competitive products, technological change, social and political uncertainty and/or the discounting of future income because of the immediate lost opportunity to earn interest without risk today. Accounting doctrines do not require investor time horizons to be identified and so surplus profits are not measured or reported. They can be a major source of wealth inequality and a drain of wealth from a precinct to create poverty in resource-rich communities (Turnbull 2006b).

Ecological property rights can minimize value loss from a community through external owners capturing surplus profits that include windfall gains. Like ecological currencies, ecological property rights have limited life. They follow the rule found in squatter settlements: 'if you do not use it you lose it'. Incentives can be used to change the existing system of perpetual, static and inclusive rights to a system that has time limited, dynamic and inclusive rights.

Ecological property rights could be introduced by making the tax deductibility of investment property conditional upon the ownership rights being written off at the same rate as the tax deduction. In this way the profitability of investment property would not be changed as the cost of the ownership loss is being recognized in any event. Ownership would be transferred directly or indirectly to residents to stop the export of rents and capital gains as explained in greater detail in the following section.

To stop corporations exporting surplus profits outside an urban precinct, tax and other incentives are required for shareholders to adopt ecological properties rights. The incentive would provide shareholders with a bigger, quicker and so less risky profit in return for giving up smaller, slower and more risky returns over the long run.

The tax incentive required is surprisingly small as shown by Turnbull (2000a). In return shareholders would change corporate constitutions to create 'stakeholder' shares as well as the shares owned by investors that would now be described as 'investor' shares. The constitution would transfer the rights to corporate assets, earnings, dividends and votes from the investor shares to the stakeholders shares at say 5% per year so that in 20 years the corporation would become 100% owned by resident stakeholders. Residents, who were natural persons, could receive stakeholder shares in a similar manner to frequent flyer points according to their contribution to the business as a supplier, employee and/or customer (Turnbull 1997a).

In this way, ownership transfer corporations (OTCs) would be created that would transfer the control of corporations owned by investors anywhere in the world to individuals in the community that hosts its operations. OTCs limit the size of firms because ecological property rights force firms to continually distribute most of their cash flow to shareholders. Shareholders can then reinvest their dividends in other firms or in offspring enterprises formed by the parent OTC to expand its operations (Tricker 2000a; Turnbull 2001a). As a result a profound change results in the ecology of firms as their size becomes limited to human scale while the extent of their activities continues to grow in a way demonstrated by Mondragón firms (Turnbull 2000b). Growth is achieved by OTCs raising new funds by giving birth to many 'offspring' enterprises that take over and/or develop part of their operations to form productive networks that can expand globally.

Besides plugging the economic drain created by firms being externally owned, OTCs also localize the control of businesses in urban precincts to make them accountable to residents. In this way ecological property rights enrich democracy and the ability of communities to become self-governing. But the ability of a precinct to become self-governing also depends upon residents owning and controlling the land and buildings of the precinct as noted earlier. How this can be achieved is considered in the next section.

5.3 Establishing Self-Financing, Self-Governing Precincts

This section explains how ecological ownership can be introduced to urban communities to stop economic value draining away as referred to in the previous section. The result is to create a more equitable and efficient and so more sustainable form of capitalism.

For cities to maximize their ability to become energy and financially self-sufficient, their component precincts and/or suburbs need to become energy and financially self-sufficient. Ecological ownership and control of urban precincts introduces a political structure for governing energy conservation initiatives at the local level on a self-enforcing bottom-up basis. This includes renewable energy storage facilities within a precinct, energy trading between precincts, providing an authority to issue energy dollars and managing their integrity. These considerations indicate the multidimensional processes required to be integrated and managed at the local level for establishing financially and energy self-sufficient sustainable urban communities.

The value of urban land is created by how well it is serviced with: water, sewerage, power, roads, transport, communication, hospitals, schools, and places of employment, entertainment and recreation. The value is not in the land but how well the site is serviced by external public and private investment described as 'externalities'. The site also obtains value from the improvements on it which may be a dwelling: home unit, shop, office, factory or entertainment facility. To establish equity and efficiency, property rights need to be designed to separate the externally created values in a site from those created by improvements on the site as indicated in Fig. 5.1.

The separation of private and community property rights is a common feature of condominium, company or 'strata title' systems and in community land trusts (CLTs). However, these 'duplex' ownership systems do not provide separate publicly negotiable title deeds to each type of property right. Nor do they operate over an area sufficiently large to capture most of the values generated externally to any single site.

What are required are two separate title deeds with one deed being represented by a share in a corporation that owns all the sites in a contiguous viable precinct. The other

Value of urban property has two components:
1. Site or 'Land value'
2. Value of improvements

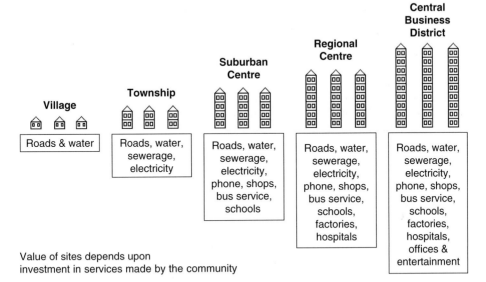

Value of sites depends upon
investment in services made by the community

Two different types of property rights are required to create efficient equitable
markets for the private ownership of urban property:

1. Dynamic Lease (DL) or 'Strata Title' for improvements on the site/land,
2. Shares in the co-operative which owns all sites/land in the community.

Dynamic Lease (DL) 'Strata Title' captures value of improvements	Value of DL determined by market value of improvements (not by windfall gains from community)
Shares in Community Land Bank (CLB) captures value of site ('Land Title')	Issue value of shares by CLB determined by average market price for all community land. Buyback price from vendors of DL's proportionally discounted for rates not paid over a 25 year period.

CLB captures for all resident home owners and tenants:
- **Equity values in all land not occupied by dwellings including all commercial and industrial sites;**
- **Windfall gains in all sites from public investment in infrastructures services;**
- **Windfall gains from private investment in local amenities and facilities servicing the site;**
- **Surplus profits from private investment in buildings and other improvements;**
- **Profits from buying back its own shares at a discount from short term residents;**
- **Market value of quality of life created by community sites, services and governance.**

Fig. 5.1. Duplex tenure.

title deed would provide negotiable rights to a specific volume in space like a lease or an Australian 'strata title'. It is referred to as a 'Dynamic Lease' (DL) in Fig. 5.1 for reasons described below.

One share in the land owning corporation could be issued for every square metre occupied by each *residential* DL whether or not the dwelling was on the ground floor or in a high rise. In this way only residents would own all the land occupied by non-residents, trusts, partnerships, corporations and higher levels of government.

Another way of thinking about the arrangement is that it represents an incorporated unit of urban government that issues voting shares *only* to its residents, be they owners or tenants. Unlike a CLT and other forms of duplex tenure, the scale of operations needs to be sufficient to establish a public market for the two different types of urban property rights. A basis is then established for conventional lenders to use the property rights as collateral to finance home ownership.

There is no necessity to introduce any new law to create the duplex system of property rights that form a cooperative or Community Land Bank (CLB). This is because corporate constitutions possess replaceable rules and the rules can be designed to provide the most desirable property rights for the particular structures built in each precinct. Competition for investment between precincts/suburbs would provide a way of determining the most efficacious design for the rules.

A CLB provides a framework to introduce 'use it or lose it' property rights to both DLs and the CLB shares. Surprisingly, such dynamic ecological property rights would become more attractive to investors in apartment buildings or in other commercial improvements than with the existing system. The attraction for investors is that they would not need to purchase the site they occupied. The cost of a site in advanced economies is typically half the cost of a dwelling as reported in the US (Davis and Palumbo 2006). For pioneer home owners in a CLB this creates half price housing as sites become self-financing from the value added by development and only the cost of the dwelling needs to be financed by its owner (Turnbull 1976).

Commercial investors in a CLB would significantly reduce the size of their investment as they would not need to purchase a land site. However, they would be required to relinquish their ownership rights of their investment at the same rate that they wrote it off for tax purposes. Their rate of profit would not be reduced as the cost of losing ownership is offset by the depreciation cost that would be incurred in any event. In this way the residual value of investments in shopping centres, office blocks, recreational facilities and factories would become owned by the CLB and so by all residents be they home owners or tenants. Tenants are included as they acquire CLB shares associated with their dwelling without charge at the rate ownership of their dwellings is depreciated.

Citizens of CLB in which OTCs were operating would acquire asset ownership in both productive enterprises and in realty of their community. Ecological property rights in CLBs and OTCs introduce a mechanism for mass asset transfer to citizens without the use of money, taxes, welfare or the associated government bureaucracy involved in traditional ways of distributing wealth. *Democratizing the wealth of nations* (Turnbull 1975a) in this manner is neither identified nor explained by orthodox economic analysis. One reason is because the nature of property rights is assumed to be fixed rather than being a variable. (Turnbull 2007a,b)

Only residents in the precinct can own and so vote CLB shares to control their precinct. In this way external ownership and control can be almost eliminated. The ownership of the DLs in investment dwellings transfers, as they are written off for tax purposes, to the tenants rather than the CLB. The CLB transfers ownership of the shares 'stapled' to the DLs to tenants at the same rate. If investors wrote off the cost of their investment over 25 years, their tenants would obtain 100% ownership of both their dwelling and the CLB shares during this time without paying any more than a normal competitive rent as indicated in Fig. 5.2. The transfer provides an incentive for the tenants to take over the maintenance cost of their dwellings to increase the return to investors.

The incentive for buying a home rather than renting for pioneer residents in a CLB arise from obtaining half cost housing. If they leave their home and rent it out then they would lose ownership rights in both their DL and the associated CLB shares at 4% per year to become co-owners with their tenants. This creates an incentive for non-user home owners to sell their property rights.

The price paid for the DLs on the open market would take into account the cost of buying the associated shares from the CLB who would price them in the same manner as for a real estate investment trust (REIT). The CLB would purchase its shares back from the

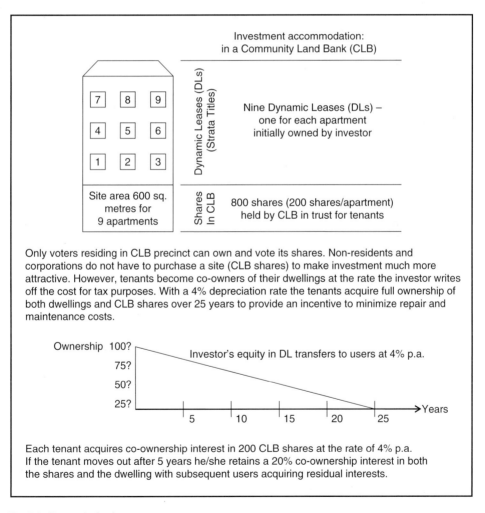

Fig. 5.2. Dynamic duplex tenure.

seller at a discounted price to recapture some of the windfall gains created in the community by either public or private investment and/or by improvements in the quality of life created by how the CLB is governed.

The values recaptured from trading in its shares assist in making the CLB self-financing in a way not available to CLTs. CLTs also do not borrow money secured by the equity created from uplift in its land value like a CLB. This denies CLTs from becoming self-financing to force them to be dependent upon obtaining gifts of land to eliminate its cost. For this reason the introduction of CLTs is very restricted. They cannot provide either a widely reproducible or sustainable solution to the inefficiencies and inequities in current urban tenure systems.

Ideally, the CLB precinct will include a rich mix of commercial activities to provide rent/rates to service any borrowings to finance its infrastructure and/or cross subsidize residents and/or pay a dividend to residents. Ideally also, the number of dwellings in the precinct would be sufficient to support educational facilities up to a basic tertiary level with supporting health care services to sustain its mix of residents over generational changes. This would typically mean a population of from say 50 000 to 100 000 residents that might involve, say 1000 to 5000 acres. With a density of around 18 people per acre residing in London there would be about 12 000 individuals residing on average within a 1000 yards of each Jubilee line station. If a CLB owned the land then each individual would obtain a windfall gain of £101 000.

CLBs have in total six mechanisms for transferring wealth to their residents without including the use of OTCs. These are: (i) pioneer home owners acquiring shares without cost; (ii) tenants acquiring ownership over time of their dwelling without the need to make a purchase; (iii) tenants acquiring ownership of CLB shares without cost; (iv) all residents capturing a proportion of any gain from improvements and windfall values in their dwellings; (v) all residents capturing a share in ownership values of all non-residential land and depreciated improvements in the CLB precinct through ownership of CLB shares; (vi) all residents acquiring a proportion of the windfall gains captured by the CLB when it buys back its shares at a discount from residents selling their dwellings.

As only residents can vote CLB shares, CLBs promote self-governance. The ability of CLBs to become self-financing, unlike CLTs, facilitates their political independence. Both the self-financing and self-governing abilities of CLBs are promoted by the introduction of OTCs as they localize both the ownership and control of productive activities hosted in the precinct. In these ways ecological property rights to money, realty and corporations make capitalism more efficient, sustainable, equitable, responsive and democratic.

The economic and political independence introduced by ecological property rights provides a basis to promote and protect the creation of an independent community banking system based on sustainable energy dollars. The strengths and weaknesses of creating energy dollars are considered in the next section.

5.4 Evaluation of Sustainable Energy Dollars

This section extends the evaluation of energy dollars developed by Turnbull (1997a) to include national currencies. Most types of money used throughout history have been connected with reality by being redeemable into physical assets such as gold, silver, copper, tea, tobacco, rum, wampum shells, wheat, corn, cattle, slaves and even wives (Galbraith 1976). Modern money is no longer related to reality since President Nixon took the US dollar off the gold standard in 1971 (Galbraith 1976).

Modern currencies are described as 'fiat' money as they are defined to exist by the force of law (Greenspan 1997). The monetary tokens issued by governments in the form of notes and coins are given a face value not related in any way to the material used in the token or paper money that may have negligible intrinsic value. Monetary tokens typically represent less than 5% in value of all the money in modern societies. Most money is represented by bank deposits and other credits. The value of fiat money in the form of tokens, deposits and other credits is not defined by any specific goods or services but by the totality of all items traded and invested in one national currency compared with the totality of all transactions in other national currencies. This means the value of modern money is indefinable in terms of any one or defined combination of goods and/or services.

As modern money is not redeemable into any specified goods or services, there is no limit on how much is created (Greenspan 1997). Monetary tokens, such as notes and coins, are produced by governments to generate a profit, referred to as 'Seigniorage', which is the difference between the cost of manufacturing paper money and coins and their face value. The rest of the money supply is created by private banks who earn 'special profits' from creating money in the form of credit.

As Galbraith (1976) observed: 'The process by which banks create money is so simple the mind is repelled. With something so important, a deeper mystery seems only decent.' Banks create money by issuing loans to borrowers who deposit the funds back with the bank. In this way banks create both an asset and a liability on their balance sheet simultaneously. The loan is recorded as an asset of the bank with the deposit by the borrower becoming a matching liability of the bank. When the deposit is drawn down, to be spent on the purchase of goods and services, deposits are placed back in the bank by the vendors of the goods and services. If the vendor deposits are at other banks then the money can be lent back to the bank that created the money. The creation of money in this way allows banks also to create 'special profits' from the difference between the interest charged on the loan and the interest they may pay on the deposits also created. For non-religious folk the creation of modern money is the second biggest confidence trick perpetuated in the history of civilization.

If the banks creating money are outside an urban precinct then they will drain value away to external regions. According to Huber and Robertson (2000) the value of the special profits earned by US private banks in 1998 was US$114 billion and £48 billion in the UK. These authors reported that this would have represented respectively 4.5% and 15% of central government tax revenue in each country!

Modern money carries out the role of being a 'medium' of exchange and a 'store of value' but it no longer carries out its historical role in providing a physically definable 'unit of value' like a pound weight of sterling silver or a defined weight of gold. In earlier centuries money was defined in terms of specific commodities as noted earlier. There was no need for a government to get involved in creating or defining the value of money or what could be used as money. It was an era of highly decentralized and mostly non-government controlled 'Free Banking' (White 1993). However, the need for governments to raise money resulted in decentralized banking being replaced by central banking which is a form of central planning.

A problem in using commodities to define units of value is that the characteristic of the commodity has also to be defined and measured. The purity of metal commodities can be more easily defined, measured and maintained than the characteristics of tea, tobacco or cattle and so on. By using a kilowatt-hour of energy as a unit of value the definitional and management problem can be overcome as energy can be measured as precisely as required. In this way energy dollars provide advantages not available in modern forms of money- or commodity-based currencies.

Table 5.2 provides a number of other criteria for comparison. No quality testing is required for national currencies as quality is not defined as noted in row 2 of the table. Tokens of fiat money have negligible intrinsic value while gold can be used in industry to some degree as suggested in row 3 of the table. Another special feature of energy dollars is that they have an intrinsic consumable value that is little shared by gold and not at all with fiat money as indicated in row 4.

The definition of what is considered national currency is determined by governments, as noted in row 5. The sources for gold are globally distributed in limited and far-flung pockets, while, as noted earlier, renewable energy is globally available. With a gold-backed currency, countries that are well endowed with gold obtain an international competitive advantage. Row 6 recognizes that renewable energy currencies are democratic, being available around the world. As noted in closely related rows 7, 8 and 9, the volume of national currencies made available is typically controlled indirectly by interest rate and fiscal policies but some government may also introduce direct controls as noted in row 9 of the table. The availability of gold to back a currency in an economy can very much depend upon external factors as noted in the table. The amount of power made available is closely related to consumer demand so the volume of renewable energy currencies can become automatically related to the level of economic activity or GDP. However, not shown in the table, the volume of gold and energy currencies could also be controlled by political interventions.

The use of a physical commodity like gold as the unit of value or 'reserve' currency introduces storage and insurance costs as noted in rows 10 and 11. These costs are avoided with the national currencies and renewable energy dollars. This does not mean that some storage devices are not required for some forms of renewable energy services. Both gold and energy suffer costs in being moved as noted in row 12. While the cost of transporting gold is relatively minor the cost of distributing energy across a nation can be very significant and can rise to over 30% of power generated. However, this cost is an advantage as it provides the incentive for urban precincts to become self-sufficient to promote their financial and political independence.

Table 5.2.   Comparison of National currencies with gold and renewable energy dollars.

No.	Criteria for comparison	National dollars	Gold dollars	Energy dollars
1	Unit of value	Not defined	Ounces/grams	Kilowatt-hours
2	Quality testing	Not required	Density	Not required
3	Intrinsic value	Negligible	Say 10%	100%
4	Subjective value	100%	Say 90%	Nil
5	Source of currency	Government decree	Haphazard locations	Globally
6	Democratic availability	Depends on Gov.	Depends on location	Globally
7	Changes in production of money	Controls & interest rates	Little related to consumption/GDP	Related to consumption/GDP
8	Volume of money controlled	Indirectly by interest rates	Geography, trade and government	Related economic value/GDP
9	Rate of change in production of money	Fiscal and monetary policies	Fluctuates with region and time	Relatively stable by region and in time
10	Cost of storage	Not required	1% of value per year	Not required
11	Cost of insurance	Not required	1% of value per year	Not required
12	Cost of distributing reserve currency	Negligible with electronic transfers	Changes little with distance	Increases with distance
13	Ecological features	None	Yes	Yes

Both gold and renewable energy depend upon nature while national currencies are not connected to nature in any way as indicated in row 13. Indeed, the ability of modern money to increase its value from earning interest over time is inconsistent with natural processes and is not sustainable.

The importance of having an ecological local currency connected to environmental conditions can be profound. The nature of a currency determines how resources are priced and markets allocate resources according to prices. To sustain humanity on the planet it is the environment that should influence how resources are allocated and governed. In other words society needs to become an 'environmental republic' with feedback mechanisms to allow it to be automatically controlled by nature. This cannot occur with modern currencies that are controlled by governments and their monetary institutions in order to maintain political power, a problem exacerbated by the current type of money that creates compelling short-term political incentives to exploit nature through its ability to earn interest.

The importance of having a local currency to allocate resources was highlighted by Jacobs (1985b) who stated that 'Because currency feedback information is so potent, and because so often the information is not what governments want to hear, nations go to extravagant lengths to try and block off or resist the information.'Jacobs (1985c) went on to explain:

> *Individual city currencies indeed serve as an elegant feedback control because they trigger specifically appropriate corrections to specific responding mechanisms. This is a built-in design advantage that many cities of the past had but which almost none have now. Singapore and Hong Kong, which are oddities today, have their own currencies and so they possess this built-in advantage.*

As outlined in Table 5.2 renewable energy dollars have quite different operating characteristics to national currencies. This means that changes need also to be made in the architecture of a local monetary system. These issues are considered in the following section.

5.5 Designing a Local Real Monetary System

An autonomous monetary system in an urban precinct requires four elements as considered in this section. Many well-known local exchange systems are not autonomous as they tie their unit of value to the national currency. Local Exchange Trading Systems (LETS), commercial barter systems and stamped scrip are examples that define their unit of value in terms of the national currency. While such 'shadow' currencies are useful in providing a way to identify and so support local activities they do not provide a way for a local community to establish its own priorities on how its resources are valued. As shadow currencies are tied to the national unit of value their value is not definable in any real goods or services as discussed earlier. An internally generated unit of value is required to insulate local resource allocation from those governed by currencies without ecological attributes. A local unit of value also insulates the local monetary system from the perversity of forces produced by globalization.

The four elements traditionally found in an autonomous monetary system described in greater detail by Turnbull (1997c) are: (i) savings institutions, (ii) money multiplying institutions, (iii) risk management institutions for loss of value or liquidity and (iv) money-changers. Commercial banks are money-multiplying institutions: they expand the volume of money by making loans as described earlier. Commercial banks today may also carry out the other three activities. This has become possible because money is no longer definable in physical units to deny it being subjected to any reality test for its creation or availability.

An energy dollar-based monetary system would be subjected to the reality test of both the physical capacity to generate power and its availability as and when required. For this reason basic principles need to be considered in designing a local energy monetary system and managing risks to its integrity. It also requires consideration of the practicalities of the technology. To illustrate these points, an urban precinct using solar cells to obtain energy and financial independence will be considered.

We will assume that a solar cell has a 20-year life and so must recover 5% of its cost each year in the value of the output it generates to pay for itself. We will also assume that its annual maintenance cost is 5% so that to become self-financing over its life the solar cells must produce at least 10% of their value each year. A cost of 10% per year could be recovered with a demurrage charge of 0.19% per week, less than a fifth of the 1% cost per week of the stamped scrip example considered earlier.

The 5% maintenance cost could be reduced and even avoided if individual owners of solar cells undertook the maintenance themselves. In this way home owners could obtain an additional payback, income and/or 'sweat equity' from their solar cells. Tenants, apartment management associations and/or other community associations, and businesses could also become directly involved in minimizing solar cell maintenance costs. The point being that private ownership provides opportunities and incentives to service solar cells or other renewable energy sources even if their purchase and set-up costs were financed by lease or rental arrangement with a third party.

Contracts are then created between the owner/agents of the solar cells to sell the minimum expected kilowatt-hours produced each year to the energy management agency of the CLB. This allows the CLB to take over the 'lender of last resort' activity that used to be the most important function of central banks in past eras when money was redeemable into a specified commodity. This rationale for central banking no longer exists as national currencies are no longer redeemable into any specific goods or services.

Any short fall in delivery of kilowatt-hours from one contractor in the urban precinct could be compensated with extra output produced by other contract suppliers. If more aggregate output was produced than required by demand within the precinct then the CLB could export power to external communities or store it by generating hydrogen from water or by other means. The hydrogen could then fuel energy cells or be burnt to generate electricity by conventional means without any pollution as its only residue is very pure hot water. If the CLB did not possess an energy storage system to average out shortages or surpluses within the CLB precinct, it could import or export power from or to external sources.

This simplistic illustration on how sustainable energy dollars can be created hides the devil in the detail. A typical problem of establishing a local currency is how to get it issued, used and distributed. This problem is minimized by a commodity-backed currency when there is a universal demand for the commodity like electrical power. The power authority could require or provide a compelling incentive for its energy bills to be paid with energy dollars. In addition, the CLB, like any other local government authority could require or provide a compelling incentive for some or all of its rates, taxes and rents to be paid in energy dollars.

Another problem of defining money and making it redeemable into a single real resource is that insufficient money will be created to service all the transactions involved in modern capitalistic societies. In societies with private ownership of land, firms and productive assets, the value of all monetary transactions becomes many times the value of goods and services produced and exchanged. An autonomous community monetary system will need institutional arrangement to multiply the amount of money created to service all transactions.

Commercial banks carry out a money multiplying function by simultaneously creating loans and deposits as described earlier. However, an autonomous banking system requires a sound basis for managing the risks introduced from money multiplication. Commercial banks are involved in managing two types of risks: (i) credit risk of a loan loss and (ii) liquidity risk of all depositors wanting to withdraw their funds.

There is also a system risk that excessive money creation will produce inflation if it is not matched by increases in productivity. As noted by Nobel Prize winning economist Professor Lawrence Klein, 'the expansion of Federal Reserve credit will not be inflationary if the funds made available flow into investment that raises national productivity' (Speiser 1986). However, current banking practices do not provide a mechanism to differentiate between expanding credit to finance increased productivity or increases in consumer demand. This problem can be overcome with introduction of credit insurance as is commonly available for housing loans. Unlike home loans, the cost of obtaining the insurance can be used to create a market mechanism for allocating credit expansion selectively for increasing productivity. It also provides advantages in managing credit risk.

Another way of controlling inflationary pressures is by the introduction of competing currencies as described by another Nobel Prize wining economist Professor Friedrich von Hayek (1976a). The introduction of energy dollars achieves this objective. Intellectual creditability for introducing an alternative urban currency system is also provided by von Hayek (1976b) in his suggestion for the 'Denationalization of Money'.

Credit and liquidity risks are best managed by separate and different types of financial institutions. Credit risks are best managed by organizations without any debt, and liquidity risks are best managed by organizations that can leverage their equity with a large amount of debt. For instance, secure investments like government bonds may allow a bond trading organization to have a 20:1 debt to equity ratio. This could be over twice that of a commercial bank with say a 10:1 debt to equity ratio that trades in more risky commercial loans. An investment bank trading in even more risky publicly traded equities might have only a 5:1 debt to equity ratio. Risks that have limited liquidity as accepted by insurance companies may mean that they have little or no debt on their balance sheet.

These observations raise questions regarding the desirability of commercial banks accepting both credit and liquidity risks together, especially when they act as money multipliers for a redeemable currency. With modern money, these risks can be accepted as national currencies are not defined in terms of real resources and not subject to any requirement that it is redeemed into anything specific. But in an autonomous community monetary system, it would be much safer and efficient for differently designed institutions to manage the two different types of risk.

A CLB could license commercial banks to multiply the volume of energy dollars in its precinct on condition that all credit risk was insured by organizations in the non-bank sector and their service fees would replace interest payments on energy dollars. In this way the cost of credit insurance would replace the cost of interest. The elimination of interest costs would make a CLB precinct a highly attractive locality for establishing and operating businesses. It would more than halve the cost of infrastructure investment like toll ways, tunnels, rail services, hospitals and schools etc. as the cost of interest can more than double the repayments required over their operating life of 20, 30 or more years. This would mean the unit cost of essential services would be halved to directly reduce inflationary pressures. A proposal for the US Federal Reserve system to provide interest-free funds for states and local government was introduced into the US Congress in 2004 (HR 2004).

The CLB would control the volume of money created through its purchases of insured loans from the commercial banks. In other words, it would be the CLB that created new energy dollars by purchasing the loans issued by the banks to their borrowers. The commercial banks could also be making loans in the national currency to introduce competition between currencies as described by von Hayek (1976). However, the special profits obtained by multiplying the volume of money and credit would be captured by the community through the CLB rather than any external private shareholders of the commercial banks. This is another way in which an independent monetary system can reduce the loss of value from its precinct to enhance financial independence.

The cost of energy dollar loans would be determined not by the interest rate policy of the central government but by the cost of obtaining loan insurance. The cost of loan insurance would then take over the role that interest is supposed to carry out in allocating resources most efficiently.

The introduction of ecologically sustainable energy dollars combined with ecological property rights introduces a profoundly different type of economic system. It introduces quite different technical concepts, language and method of analysis as described by Turnbull (1992) with the institutional arrangements described in Turnbull *et al.* (1975b).

Some of the implications of this new type of political economy created from ecological property rights are considered in the concluding next section.

5.6 Governance of Sustainable Urban Communities

In concluding our consideration of how renewable energy dollars can play a role in transforming the production and consumption of energy, there is also a need to consider community governance. The introduction of ecological property rights to money, realty and corporations introduces quite a different type of political economy.

Ecological capitalism introduces continuous change into the power structures of society. It is only through change that progress can be achieved. Evolution is likewise predicated upon change and ecological property rights follow the example of nature. However, instead of change being progressive it can also be regressive and self-destructive. Checks and balances are required to constrain and mitigate regressive and self-destructive changes. However, checks and balances are not typically found, or are not present in sufficient richness, in contemporary societies.

Without checks and balances power can be become concentrated. It has been long recognized that power corrupts and absolute power corrupts absolutely. An appropriate division of power then becomes a condition for the building of sustainable communities. An appropriate division of power is also required to manage complexity. This is specified by the science of communication and control that developed in the mid twentieth century. The application of this new science to social organizations at the beginning in the twenty-first century provides criteria for designing the governance architecture of sustainable communities (Turnbull 2000a: 401–13; 2000b; 2002a; 2003b: 256–72).

The science of governance explains why nature adopts network governance. The reason is to economize the volume of data, information, knowledge and wisdom that social animals need to receive, process, store and transmit to sustain their species. By economizing data, biota economize the associated matter and energy involved. Network governance is most compellingly illustrated by the network of firms located around the Spanish town of Mondragón. The firms have over 60 000 workers governed by their stakeholders (Turnbull 2000c).

OTCs provide a way to transform existing corporations with a centralized command and control governance architecture to network firms governed and owned by their stakeholders like Mondragón firms. CLBs are also owned and controlled by their stakeholders and so become another type of multi-stakeholder mutual (Turnbull 2001b). Stakeholder governance requires firms adopting constitutions that introduce a rich separation of powers as illustrated by the Mondragón firms that introduce sustainable competitive advantages.

An analysis of Mondragón firms was undertaken by Thomas and Logan (1982) who reported:

> *During more than two decades a considerable number of cooperative factories have functioned at a level equal to or superior in efficiency to that of capitalist enterprises. The compatibility question in this case has been solved without doubt. Efficiency in terms of the use made of scarce resources has been higher in cooperatives; their growth record of sales, exports and employment, under both favorable and adverse economic conditions, has been superior to that of capitalist enterprises.*

The reason why the rich and apparently complex architecture of network firms provides competitive advantages is that it paradoxically simplifies the role of individuals (Turnbull 2000d). Other reasons why network firms provide operating advantages is provided by the laws of communication and control (Turnbull 2002a).

The science of corporate governance (Turnbull 2002b) identifies how competitive advantage of a stakeholder mutual is achieved without the necessity of it being publicly traded. This is because with an appropriate form of network governance, competition for corporate control is created internally between competing constituencies of stakeholders. Continuous increases in efficiency are achieved in this way (Tricker 2000c; Turnbull 2001a).

Competition for existence also occurs in nature without the use of markets dependent upon money or prices. The evolution of life forms through natural selection is governed by feedback messages obtained from the local environment. Sustainable communities also require feedback information from their host bio-region to govern how they can adapt appropriately. Ecological capitalism provides a much richer, more sensitive and responsive way to initiate and facilitate changes in how the economic system operates. It also changes how society is governed as the ownership and control of realty and corporations becomes dynamic, time limited and inclusive. This limits the concentration of power of individuals and their ability to exploit each other or the environment.

The possibility arises for local communities to be partly governed by the characteristics of their host environment. This would assist in reversing the current trend of humanity governing their environment rather than allowing their environment to govern society as it did in primitive societies. Societies like traditional Australian aboriginals could be described as environmental republics. For a modern society to be at least partly governed by its local environment, a bottom-up political economy is required. CLBs as a local governing system create a basic building block for a global system of governance (Turnbull 2003). In the language of system science, CLBs would become the bottom level of a hierarchy of 'holons' described as a 'Holarchy' (Turnbull 2000b).

The changes required to improve the ability of society to be sustainable are not as great as those that have occurred in the past. Table 5.3, details many of the changes that have occurred over the last couple of millennia and compares the changes with the current situation with that proposed in this article for a sustainable future as tabulated in the last column of Table 5.3. However, for humanity to avoid extinction from global warming, the more modest changes proposed are required in a much shorter time. Global warming has created the need to transforming society with elements of environmental republicanism for governing society.

Table 5.3. Dominant characteristics of social change: past, present and sustainable future.

	Characteristic	Past	Present	Sustainable future
1	People treated as	Property	Resource	Potential
2	Role of women	Breeding	Cheap labour	Full partners
3	Purpose of work	Sustenance	Income distribution	Fulfilment
4	Distribution of national income	Employment	Employment & government transfers	Employment & transfer of property income
5	Relationship to the environment	Subservient	Dominant	Stewardship
6	Natural resources	Use	Exploit	Sustain
7	Source of land acquisition	Conquest or inheritance	Purchase or inheritance	Use and/or purchase
8	Period of land ownership	Time of use	Perpetual	Time of use & so limited
9	Source of business ownership	Start-up or inheritance	Purchase/start-up & inheritance	Start-up, investment and stakeholder rights
10	Business owners	Proprietors	Shareholders	Stakeholders
11	Period of business ownership	Life of owner	Perpetual	Limited
12	Property rights	Discretion of sovereign	Static, monopoly and perpetual	Ecological: dynamic, inclusive, time limited
13	Structure of business	Paternal and centralized	Hierarchal and centralized	Decentralized stakeholder mutuals
14	Monopolies	Granted to private interests by Sovereign	Prohibited or controlled by government	Eliminated by time limited dynamic rights
15	Institutions	Perpetual	Evolving	Dynamic
16	Basis of money	Commodities	Fiat of government	Goods or services
17	Creation of money	De-centralized competitive basis by private sector	Centralized government controlled monopoly	De-centralized competitive basis by private sector
18	Cost of money	Cost of storage & quality control	None, earns interest	Demurrage cost
19	Allocation of resources	Command and control	Markets & hierarchies	Use, benevolence, semiotics & markets
20	Value system	Absolute	Materialistic	Humanistic
21	Distribution of economic values	Autarchic	Market forces	As to stakeholder contributions & need
22	Accumulation of economic value	Limited by political power	Not limited	Limited by ecological property rights
23	Economic and political power	Centralized in Sovereign	Government & big business	Decentralized to communities
24	Power architecture	Hierarchy	Oligarchy	Holarchy as in nature

Financing urban communities with sustainable energy dollars provides one way to begin transforming society to facilitate the survival of humanity. The transformation of the current system of ownership and control of money, realty and property provides complementary approaches to reinforce the establishment of sustainable communities. However, to avoid the extinction of humanity these relatively more modest changes are now required to be achieved without delay.

References

Bennello, C.G. (1997). Community financing resource optimization. In Morehouse, Ward (ed.), Chapter 23, *Building Sustainable Communities: Tools and concepts for self-reliant economic change*. Intermediate Technology Development Group of North America Inc, pp. 184–191.

Davis, A.M. and Palumbo, M.G. (2006). The price of residential land in large U.S. cities. *Finance and Economics Discussion Series*, Divisions of Research & Statistics and Monetary Affairs, Federal Reserve Board, May 2006. Retrieved 9 April 2007 from http://www.federalreserve.gov/pubs/feds/2006/200625/

Fisher, I. (1933). *Stamped Scrip* Adelphi Press.

Galbraith, J.K. (1976). *Money: Whence it came, where it went*. Penguin Books.

Greenspan, A. 'Central Banking and Global Finance' presented to the Catholic University of Leuven, Belgium 1997. Retrieved 9 April 2007 from www.fame.org/PDF/Gsjan14_R2.PDF.

Howard, E. (1946). *Garden Cities of To-Morrow*. Faber and Faber.

HR (2004). State and Local Government Economic Empowerment Act, introduced into the House of Representatives of the 108th Congress, May 6. Retrieved 9 April 2007 from http://thomas.loc.gov/cgi-bin/query/z?c108:H.R.4310.IH

Huber, J. and Robertson, J. (2000). *Creating New Money: A Monetary Reform for the Information Age*. New Economics Foundation, London, available at: http://www.neweconomics.org/gen/z_sys_PublicationDetail.aspx?PID=81

Jacobs, J. (1985a). *Cities and the Wealth of Nations: Principles of economic life*. Vintage Books.

Jacobs, J. (1985b, p. 161). Op. cit.

Jacobs, J. (1985c, p. 163). Op. cit.

OECD (2000). Tobin tax: could it work? *OECD Observer*, March. Retrieved 9 April 2007 from http://oecdobserver.org/news/fullstory.php/aid/664/Tobin_tax:_could_it_work__.html

Riley, D. (2002). *Taken for a Ride*. Centre for Land Policy Studies.

Speiser, S.M. (1986). *The USOP Handbook: A guide to designing universal share ownership plans for the United States and Great Britain*. The Council on International and Public Affairs.

Swann, R. (1997). Building a community banking system. In Morehouse, Ward (ed.), Chapter 22, *Building Sustainable Communities: Tools and concepts for self-reliant economic change*. Intermediate Technology Development Group of North America Inc, pp. 178–183.

Thomas, H. and Logan, C. (1982). *Mondragón: An economic analysis*. George Allen and Unwin.

Tricker, R. (2000a) and Turnbull, S. (2001a).

Tricker, R. (ed.) (2000, pp. 401–413). Op. cit.; Turnbull, S. (2000b). Op. cit.; Turnbull, S. (2002a). *A New Way to Govern: Organisations and society after Enron*. New Economics Foundation; Turnbull, S. (2002b). The science of corporate governance, *Corporate Governance: An international review*, **4**, 256–272; Turnbull, S. (2007) (forthcoming).

Tricker, R. (2000, pp. 401–413). Op. cit.; Turnbull, S. (2001a). Op. cit.

Turnbull, C.S.S. (1975a). *Democratising the Wealth of Nations*. Company Directors Association of Australia. Republished 2000 available from http://cog.kent.edu/lib/TurnbullBook/TurnbullBook.htm and Turnbull 2006 op cit

Turnbull, C.S.S. (1975b). Op. cit.; Turnbull, S. (1983). Financing world development through decentralised banking, in M. Mtewa (ed.), *Perspectives in International Development*. Allied Publishers; and Turnbull, S. (1996). Financing a social economy, in U. Reifner (ed.), *Bank Safety and Soundness – The Bergamo Report*. Institut für Finanzdienstleistungen e.V. (IFF).

Turnbull, S. (1976). Land leases without landlords', June. Retrieved 9 April 2007 from http://ssrn. com/abstract=630861; and Cooperative land banks for low cost housing, in Angel, S., Archer, R.W., Tanphiphat, S. and Wegelin, E.A. (eds) (1983). *Land for Housing the Poor*. Select Books, available at http://ssrn.com/abstract=630861

Turnbull, S. (1992). New strategies for structuring society from a cash-flow paradigm. Retrieved 9 April 2007 from http://ssrn.com/abstract=936868; Turnbull, S. (2006). Op. cit.; and Turnbull, S. (2007). A framework for designing sustainable urban communities. Working Paper. Retrieved 14 April 2007 from http://ssrn.com/abstract_id=960193

Turnbull, S. (1997a). Stakeholder cooperation. *Journal of Co-operative Studies*, **29**(3), 18–52, available from http://ssrn.com/abstract=26238; Turnbull, S. (2001). The case for introducing stakeholder corporations, Working Paper, Institute for International Corporate Governance and Accountability, George Washington University Law School. Retrieved 9 April 2007 from http://ssrn.com/abstract=436400; and Turnbull, S. (2001). The competitive advantage of stakeholder mutuals, in Birchall, J. (ed.), *The New Mutualism in Public Policy*. Routledge available from http://ssrn.com/abstract=242779

Turnbull, S. (1997b). Creating a community currency, in Morehouse, W. (ed.), Chapter 21, *Building Sustainable Communities: Tools and concepts for self-reliant economic change*. Intermediate Technology Development Group of North America Inc, pp. 167–77.

Turnbull, S. (1997c). Elements of autonomous banking, in Morehouse, Ward (ed.), Chapter 20, *Building Sustainable Communities: Tools and concepts for self-reliant economic change*. Intermediate Technology Development Group of North America Inc, pp. 159–166.

Turnbull, S. (2000a). Stakeholder governance: a cybernetic and property rights analysis, in *Corporate Governance: The history of management thought* (ed. R.I. Tricker) Ashgate Publishing.

Turnbull, S. (2000b). *The governance of firms controlled by more than one board: theory development and examples*. PhD Thesis, pp. 199–225. Retrieved 9 April 2007 from http://ssrn.com/abstract=858244

Turnbull, S. (2000b, p. 130). Op. cit.

Turnbull (2000c, pp. 199–225). Op. cit.

Turnbull, S. (2000d, p. 245). Op. cit.

Turnbull, S. (2001a). The case for introducing stakeholder corporations, working paper, Institute for International Corporate Governance and Accountability, George Washington University Law School Retrieved 9 April 2007 from http://ssrn.com/abstract=43600

Turnbull, S. (2001b). Op. cit.

Turnbull, S. (2002a). Op. cit.

Turnbull, S. (2002b). Op. cit.

Turnbull, S. (2003). Emergence of a global brain: for and from world governance. Working Paper, May 2003. Available at http://papers.ssrn.com/abstract_id=637401

Turnbull, S. (2006a, p. 455). Grounding economics in commercial reality: A cash flow paradigm', in Kreisler, P., Johnson, M. and Lodewijks, J. (eds), *Essays in Heterodox Economics: Proceedings, refereed papers, Fifth conference of Heterodox Economics*. University of New South Wales, pp. 438–461, available from http://ssrn.com/abstract=946033.

Turnbull, S. (2006b, p. 451). Op cit.

Turnbull, S. (2007a), 'Affordable Housing Policies: Not identifiable from orthodox economic analysis' in *Essays in Heterodox Economics: Proceedings, referred papers, Sixth conference of Heterodox Economics*, University of New South Wales, Sydney, December, 2007, available from http://ssrn. com/abstract=1027864.

Turnbull, S. (2007b), 'A framework for designing sustainable urban communities', *Kybernetes: The international journal of systems, cybernetics and management science*, Special Issue on "Cybernetics and Design", Guest Editor, Ranulph Granville, Emerald Group, pp. 1543–57, October, available at http://papers. ssrn.com/abstract_id=960193.

von Hayek, F.A. (1976a). Op. cit.

von Hayek, F.A. (1976a). *Choice in Currency: A Way to Stop Inflation*. The Institute of Economic Affairs, Occasional Paper 48.

von Hayek, F.A. (1976b). *Denationalization of Money: An Analysis of the Theory and Practice of Concurrent Currencies*. The Institute of Economic Affairs, Hobart Paper Special 70.

White, L.H. (ed.) (1993). *Free Banking: Modern Theory and Practice*. Volumes I, II, and III, Edward Elgar.

PART II
Policy and Practice Dynamics

Chapter 6
Renewable Energy Policymaking in New York and London: Lessons for other 'World Cities'?

STEPHEN A. HAMMER

Center for Energy, Marine Transportation and Public Policy, Columbia University, MC3366, 420 West 118th St, Room 1105, New York, NY 10027 USA

6.1 Introduction

Over the last 10–15 years, there has been growing interest in renewable energy use at the city level. In response, municipalities around the world have taken up the challenging task of trying to track local energy use, rein in demand, and switch to more environmentally benign energy sources. Interest in local action on these topics stems from a variety of explanations. One factor is growing attention to the larger issue of urban sustainability, a subject highlighted at the 1992 Earth Summit in Rio de Janeiro. One policy initiative coming out of Rio was Local Agenda 21, which specifically focused on the role cities can play in reversing global climate change (UNCED 1992). Groups like Energie-Cités and the International Council for Local Environmental Initiatives (ICLEI) have taken up the banner of local action and are fostering information sharing between communities interested in enacting more 'climate-friendly' policies.

Urban-level interest in energy policymaking is occurring for other reasons as well. Over the past few years, the damaging financial consequences of energy price volatility have led many cities to fear for their local economy (Benson 2002). Energy conservation programs and the local deployment of renewable power schemes can thus provide a hedge against price spikes (Wiser *et al.* 2005), greatly benefiting the local economy during peak electricity demand periods. Other cities are responding to concerns about energy security, believing it makes sense to generate power locally using technologies that do not rely on imported fuel sources. Post-September 11 fears of terrorist attacks on nuclear power plants and liquid natural gas terminals near cities raise a completely different set of energy security and public health concerns (Hall Hayes 2005; Hebert 2005; Lyman 2004). Finally, cities are also concerned about the localized emission impacts of power production, including air quality in the vicinity of power plants. In many cities, these facilities are found near low income or predominately minority communities, giving rise to claims of environmental racism (San Francisco Department of the Environment 2002).

Urban renewables policymaking also deserves our attention by virtue of the fact that cities are an important part of the global energy equation. The Organization for Economic Co-operation and Development (OECD) estimates cities account for 60–80% of the total energy demand in OECD member countries (Capello *et al.* 1999). As the world is growing increasingly urbanized, the proportion of global energy use consumed in cities will likely rise as well. Improving our understanding of the dynamics of local energy policymaking takes on new relevance given these trends. Few academic studies have been undertaken in this area, and policymakers and practitioners seeking to develop citywide strategies would thus benefit from information on the logic behind other cities' approaches to these issues.

This chapter attempts to explain the underlying dynamics of renewables policymaking in London and New York City, two leading world cities (Sassen 2001). There are many parallels between the two cities, both in terms of size, and how each holds considerable financial and cultural sway over the countries and continents where they are located. Both cities have also undertaken significant local energy planning efforts in recent years, although the factors driving action and the policy approaches pursued in each city are vastly different.

There are symmetries between the two cities' energy plans, linked to the subordinate role each city plays on most energy planning matters. This situation is not unique to New York and London – indeed, most cities similarly have a constrained capacity to act, limited by state or national laws or rules specifically designed to rein in the energy policymaking powers of local authorities. This chapter begins by briefly discussing different ways cities can engage on energy policy matters. It then moves on to examine local renewable energy policymaking in London and New York City, and concludes by distilling common themes and lessons relevant to other large cities around the world.

6.2 Urban Renewables Policymaking: What Role for Cities?

As an institution, local government comes in many flavours, reflecting the unique roles assigned to it by state or national-level government. The remit of local government may be quite vast, with local government serving as the local agent charged with carrying out state or national level policies or programs. Local government also takes on roles or responsibilities imposed on it by the community, providing some service or developing policies that address needs unique to that particular community or region.

It is thus inappropriate to speak in terms of some standard set of renewable energy policymaking responsibilities common to all municipalities. Energy responsibilities undertaken by one set of local policymakers may be handled by central government or a multi-city public authority in another. What can be done is to look at generic functions commonly carried out by local government and assign them an energy context. By doing so, we can better understand different ways that cities engage on renewables policymaking and program matters:

- *Direct service delivery* – In many cities local government is a direct service provider, delivering services that are public goods or are otherwise underprovided by the marketplace. Municipally owned electric and/or gas utilities are one example where cities provide direct energy services to the public. Neither New York nor London operate their own public utility, but many large American cities (including Los Angeles, Austin, and Seattle) do, giving local leaders more direct control over local energy policy matters. Management is generally left in the hands of experienced industry professionals rather than local bureaucrats, but municipal officials often have substantial input into major policy decisions. In particular, operating a municipal utility increases a city's ability to: (1) change the power portfolio delivered to local customers (such as emphasizing renewably generated power); (2) subsidize or provide technical support to customers who wish to install renewable energy systems; and (3) structure rates in a way to discourage electricity use or promote the use of clean on-site power. Whether a municipality is allowed to own and operate a public utility may be a function of local, state or national laws.

- *Regulator* – Cities may exert several types of local regulatory powers that have implications for renewable energy. Land use controls typically represent one of the most important energy-related powers of local government. Zoning rules dictate the height of structures in a neighbourhood, facilitating or preventing the construction of rooftop and free-standing wind turbine systems. Zoning rules also dictate how closely buildings are clustered, affecting the solar access of solar photovoltaic and solar thermal systems (Knowles 2003; Roberts 1987).

  Direct regulatory control over renewables deployment decisions can also be achieved through the strategic use of local building construction requirements. Although building codes are often set at the state or national level, many cities have established local versions that reflect local needs or interests. In Barcelona,[1] planning rules require all new or rehabilitated buildings exceeding a certain size to generate at least 60% of their domestic hot water needs using solar thermal collectors (Barcelona Energy Agency 2004). In the United States, the cities of Boulder, Colorado, and Aspen, Colorado, have also incorporated renewable energy requirements into their local building codes. Boulder requires sensitivity to the solar envelope of a structure, referring to a building's placement on its lot so passive solar systems can achieve their maximum performance. Aspen requires structures with a high energy load due to excessive size or other amenities (e.g. pools, spas, snow melting systems) to offset part of their electricity demand through the installation of on-site renewable energy systems. Home owners can also pay a one-time mitigation fee to buy their way out of this obligation (ICLEI 2003).

  Oakland, California has gone in a different direction, focusing on ensuring the regulatory process is not an impediment to local deployment. The city has streamlined its permitting process for new photovoltaic and small wind turbine systems: all reviews by Buildings Department staff must occur within a five-day period, and property owners are given a self-certification checklist to document compliance with local codes. Proposals deemed in compliance are approved and have all application fees waived (ICLEI 2002).

- *Buyer* – Cities purchase equipment, vehicles, and energy services necessary for their own government operations. In cities like London and New York, these expenditures run into the tens or hundreds of millions of dollars. Policies can require the use of certain types of renewable fuels for city fleet vehicles, such as biodiesel or ethanol. Procurement specifications can also be used for electricity. In the UK, a local authority requires procurement officials to seek renewable electricity supplies and accept the bid providing it costs no more than 5% above the best standard electricity supply offer (London Borough of Southwark 2004). Incorporating such a caveat is one technique officials can use to circumvent traditional rules preventing them from purchasing anything other than the least cost bid.

  Cities can also attempt to aggregate purchases of renewable power by consumers or neighbouring communities, thereby commanding more competitive prices or creating the critical mass necessary to allow new renewables projects to move forward (Navigant Consulting 2005).

- *Land owner/developer/building operator* – A significant portion of the energy consumed by local government occurs in its buildings. Passive solar design elements can reduce energy

---

[1] In 2006, the Spanish government adopted rules making this a national standard.

use, while on-site renewable power systems can displace grid-based sources. The use of renewable energy to heat or power government buildings is often an early element of a city's energy plan, because costly energy expenditures can crowd out other important government programs. Local government buildings can also be used as showcases, introducing businesses and home owners to new technology and proving that it can be utilized locally in a cost-effective fashion.

• *Public health/environmental protection* – Power plant emission standards and fuel input requirements are often established at the state or national level, to ensure that stakeholders within that geographic area face a relatively even playing field. Some cities have unique geographic or atmospheric conditions that require different environmental quality or operating standards to protect local health. For example, local air quality concerns led to coal burning curbs in New York City in the early 1970s (Narvaez 1972). Such standards can improve the prospects for renewable fuels or power systems, as they tend to force local power plants to internalize the cost of mitigation measures, making renewable power projects more cost competitive.

• *Advocate/educator* – One of the most common policy approaches employed by local government is to use its highly visible platform to promote energy initiatives or policies. In some cases these efforts are simply aimed at information dissemination, helping local consumers understand the benefits of green energy use and who to contact in the local marketplace for more information or installation assistance. Municipalities can also play an important role in serving as a conduit for information about state energy subsidy or financing programs. Local authorities can target their advocacy efforts upwards as well, using their position as the representative of thousands or millions of people to advocate for renewable energy policies at the state or national level.

London and New York both employ many of these strategies in their energy programming and policy work. As will become clear, each city's use of these approaches varies quite a bit, reflecting local conditions and preferences on these issues.

6.3 London – A Strategic Vision on Renewable Energy Supply and Use

London's energy planning picture is complex, bound up in the recent reformation of regional government linking together the 33 local authorities that make up London.[2] Unlike prior forms of regional government, for the first time London now has a popularly elected Mayor operating in a strong mayor form of government (Travers 2004). First elected mayor in 2000, and then again in 2004, Ken Livingstone has made energy planning a priority since he took office. In doing so, the mayor inserted regional government squarely into the local energy policymaking picture, something that had not occurred in many decades.

Prior to 1947, individual London boroughs and London regional government were actively involved in energy matters, issuing franchises to private firms or directly managing

---

[2] These 33 local authorities include what is referred to as the City of London, the one-square mile area where the city was originally founded. London regional government has existed in one form or another since 1855, but it was disbanded by the British Parliament in 1986, at the urging of then – Prime Minister Margaret Thatcher. When the Labour Party reclaimed power in 1997, they set out to reconstitute regional government, creating what is now known as the Greater London Authority (GLA).

electricity supply and distribution utilities themselves. After the industry was nationalized that year, local government was largely removed from the picture as electricity planning, power generation, and transmission and distribution responsibilities were shifted to the UK government. This changed in the late 1970s, when local government began to re-engage on this issue out of concern over the high cost they were paying for energy due to the oil crisis. Since then, various entities have established working groups to look at electricity issues in London, quantified energy use in the Greater London region, and developed locally oriented action and policy plans they believed were necessary complements to central government policies (Greater London Council 1978; Greater London Council 1981; Greater London Council 1982; Greater London Council 1983; London Research Centre 1993).

The push for local action is partly explained by the sheer scale of local energy use. In 1999, London's total energy use exceeded that of all of Ireland, and was roughly equivalent to that of Portugal or Greece (Greater London Authority 2004). This level – and the corresponding environmental impact associated with this energy use – was one reason that London mayor Ken Livingstone announced shortly after assuming office in 2001 that he would develop an overarching energy strategy for London.

The mayor's decision to voluntarily develop an energy strategy was also related to the fact that the 1999 Parliamentary legislation creating the GLA required the new mayor to prepare a 'state of the environment' report and eight distinct strategic plans for the city, on topics ranging from biodiversity to culture to land use to transportation (HMSO 1999). Energy use was a recurring theme in many of these strategic plans, leading the mayor to conclude there was a need for a ninth strategic plan. This report ultimately became known as *Green Light to Clean Power – The Mayor's Energy Strategy*.

In developing his energy strategy, the mayor laid out three overarching objectives (Greater London Authority 2004):

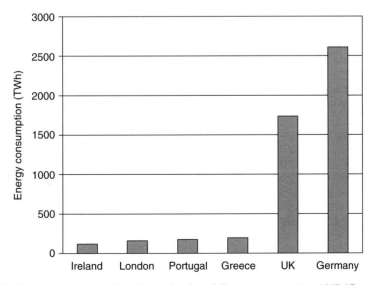

Fig. 6.1. Final energy demand in London and selected European countries, 1997 (Greater London Authority 2004).

1. Reduce London's contribution to global climate change.
2. Help eradicate fuel poverty.
3. Contribute to London's economy by delivering sustainable energy and improving London's housing and building stock.

It took nearly four years to develop and refine the Energy Strategy. The task of authoring the first draft fell to an in-house Energy Team, managed by one of the mayor's principal environmental policy advisors, and staffed by civil servants and political appointees, several of whom had a background on local energy issues. Outside support and feedback came from the general public and a 13-member Energy Strategy Advisory Group, consisting of academics, local authority representatives, developers, and energy and public health experts. Drafts were issued in early 2002 and 2003 to solicit public comment and feedback from the London Assembly. The final report was released in February 2004, making 33 distinct policy statements and 70 programmatic proposals endorsing (Greater London Authority 2004):

- The expanded use of CHP/district heating in built-up areas of London.
- Energy recovery from the waste stream.
- Initiatives to reduce fuel poverty by expanding energy efficiency and energy conservation programs around London.
- Land-use policies that result in increased public transport and bicycle use and pedestrian access to shops and businesses.
- Energy-related business development in London.
- The conversion of Transport for London's vehicle fleet to cleaner power sources.
- The development of planning policies that leads developers to consider the environmental impacts of their projects.
- The development of public/private partnerships to guide action on energy issues in London.
- The use of hydrogen and sewage gases as fuel sources.
- An expansion of the local public transport system, and the implementation of a congestion charging scheme, both designed to reduce automobile use in Central London.

Some of the most noteworthy aspects of the mayor's Energy Strategy relate to its focus on renewable power. Currently, 40% of London's power is generated at large (primarily natural gas-fired) power plants located in the city; the balance is imported via the high voltage transmission grid connecting London to the rest of England, Wales, and Scotland. London thus shares the general power profile with the rest of the UK, meaning its electricity is heavily derived from nuclear, natural gas, and coal-fired power plants (see Fig. 6.2).

In his Energy Strategy, Mayor Ken Livingstone set out to change that, establishing a goal that 2.2% (or 665 GWh) of London's electric power should be derived from renewable sources by 2010. By 2020, the mayor hopes to triple that figure. The targets were based on a central government-funded assessment[3] examining how London could contribute to the UK's overall target of 10% renewable power by 2010 (ETSU/AEA Technology 2001). A breakdown of the renewables targets is found in Box 6.1.

---

[3]The mayor's renewable targets are somewhat misleading, as the study found that achieving 665 GWh of power generation would require significant amounts of power to be derived from waste incineration, pyrolysis, gasification, or anaerobic technologies. Many commentators do not see these technologies as comparable to other 'new' renewable technologies (i.e. solar, wind, tidal) which rely on naturally repeating energy flows. Using this narrower definition, the study found the renewable technologies cited in Box 6.1 would be capable of generating only 0.88% of London's energy needs by 2010.

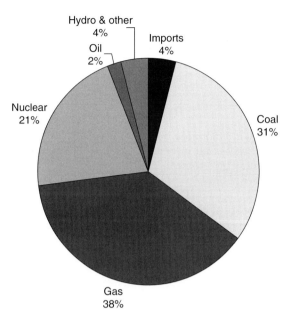

Fig. 6.2. UK electricity generation fuel mix 2000 (Greater London Authority 2004).

Box 6.1.
London Energy Strategy 2010 renewable energy targets (to be tripled by 2020)
(Greater London Authority 2004)

- 7000 domestic solar PV systems (15 MW peak capacity)
- 250 commercial and public building solar PV systems (12 MW peak capacity)
- 500 small wind generators
- 25 000 domestic solar water heating schemes
- 2000 solar water heating systems for swimming pools
- More anaerobic digestion plants with energy recovery and biomass-fuelled CHP plants.

To deliver these targets, the mayor proposed to use the full range of powers at his disposal:

- *Planning control powers* – By statute, certain types of development projects must obtain mayoral approval in addition to approval from the borough with planning jurisdiction over the project. In the Energy Strategy, the mayor explicitly announced projects requiring his approval must abide by certain design and energy standards, such as generating at least 10% of the site's power and heat needs on site from renewable power sources

(Greater London Authority 2004).[4] He also encouraged local authorities to impose similar requirements on projects over which they have exclusive planning control.

- *Leading by example* – The Energy Strategy called on the four mayoral agencies (i.e. Transport for London, the London Development Agency, the London Fire and Emergency Planning Authority, and the Metropolitan Police Authority) to seek to power all their buildings from renewable electricity. They mayor also asked each agency to investigate the feasibility of directly deploying renewable energy technology on their buildings.

- *Bully pulpit* – The mayor announced he would use his high visibility to draw attention to renewable energy issues and installations in London and lobby for policy changes and funding from central government.

- *Partnerships* – The Energy Strategy proposed active use of public/private partnerships, reflecting both the GLA's limited budgetary powers and recognition of the fact that attainment of the strategy will only come with widespread participation of the local business community. London Renewables, a partnership established by the mayor in 2003 that has since been absorbed by the larger London Energy Partnership, consisted of local university researchers, key policymakers, and businesses involved in different aspects of the renewables industry. From early 2003 through the summer of 2004, London Renewables conducted surveys on what various stakeholders had to say about renewable energy in London (Brook Lyndhurst 2003), and developed background information and policy guidance for a variety of audiences, including borough councillors and planners, architects, developers, and housing associations around London (London Renewables/London Energy Partnership 2004). Both documents were seen as laying the foundation for future renewables policy and program efforts around London.

One important player arriving on the scene after the Energy Strategy was released is the new London Climate Change Agency (LCCA). Established by the mayor to replicate the work of the Thamesway Energy Services Company (ESCO) in the village of Woking,

---

Box 6.2.
Profile of current London energy marketplace (Greater London Authority 2004)

- Local electric grid managed by two firms
- 8 firms compete to supply power to Londoners and local businesses
- Peak demand = ~9000 MW; peak demand occurs during winter months
- In-city central station generation capacity = 1392 MW
- Approximately 140 CHP systems in place around London (total = ~175 MW capacity)

---

[4] In 2006, the mayor proposed to change these rules, suggesting that new developments should be required to reduce CO_2 emissions by 20% through the use of on-site renewables. This change in metrics (i.e. energy needs vs. emission reductions) is based on two reasons – it better fits national building regulations assessing the energy performance of buildings and central government's Energy White Paper goal of reducing carbon emissions by 60% by 2050. Whether this rule change will be adopted will not become clear until 2008 when revisions to the mayor's spatial development strategy (i.e. *The London Plan*) are finalized.

southwest of London, the LCCA has the potential to dramatically change the face of London's current energy marketplace. By marrying alternative power generation technologies, energy efficiency, and energy conservation schemes together into a single business model, ESCOs are a cost-effective way to reduce the environmental impacts of local electricity production and use. Woking's Thamesway operation began in 1991 with a £250 000 investment in energy efficiency upgrades in council-owned facilities. Over time, the savings from these installations have been 'recycled' and parlayed into £2.7 million of new investment in Woking. By 2004, Woking was home to more than 10% of all of the solar photovoltaic capacity deployed in the entire UK (Jones 2005), much of which is deployed on council-owned buildings. As of 2005, the annual savings to Woking Council from its original investment approximated £885 000/year.

In London, the LCCA will seek to establish neighbourhood-specific ESCO projects, partnering with local authorities, housing groups, and the utility giant EDF, which will be responsible for designing, installing, and financing the combined heat and power (CHP) systems, renewables schemes, and energy efficient equipment likely to be part of each project. To direct the work of the LCCA, the mayor recruited Allan Jones, the architect of the Thamesway system in Woking (Muir 2004).

6.4 Key Influences on London's Energy Policy

There are several factors influencing both the content and form of London's Energy Strategy and the mayor's energy policy decision-making. First and foremost is the tremendous influence of Parliament and central government. Since 1855 Parliament has abolished and then reformulated regional government in London four times. Travers attributes this to ongoing struggles over who should control the destiny of the UK's capital city (Travers 2004). During the crafting of the GLA Act 1999, he noted Whitehall bureaucrats deliberately authored the legislation in a way so their own powers were not diminished. The implications for the GLA's renewable energy policymaking have been profound.

Some of the most significant impacts relate to the division of authority between regional government and the 33 boroughs that make up Greater London. Although the mayor retains planning control powers over the 250–300 largest development schemes proposed in London each year (Government Office for London 2004),[5] the boroughs are responsible for many more, totalling tens of thousands of planning decisions each year. Thus, their decisions can support, ignore, or contradict the mayor's energy goals. The work of London Renewables was a direct consequence of the mayor's relatively weak position on local authority planning decisions. First, the stakeholder survey (Brook Lyndhurst 2003) was intended to defuse potential opposition by local authority officials claiming Londoners do not care about renewable energy.[6,7] Second, the guidance for local authority planners is particularly detailed, educating them about all aspects of renewable energy, energy

---

[5] The Town and Country Planning (Mayor of London) Order 2000 states that the Mayor of London has significant powers in relation to individual planning applications of strategic importance to London. Boroughs receiving planning applications meeting these detailed criteria must notify the mayor, who has the option of deciding whether to comment on and support these applications, or, if he deems it necessary, direct the borough to refuse planning permission. The mayor cannot direct boroughs to approve applications otherwise opposed by the borough.

[6] Personal communication with GLA Energy Team (21 July 2003).

[7] The Brook Lyndhurst study found that Londoners are big supporters of green power with 81% believing renewable energy is a good idea, a figure in line with previous national surveys.

conservation, and green building design so they can both understand the need for effective energy planning and assess how well new planning applications address such matters. The mayor's on-site renewables requirement for projects subject to his approval is reviewed in detail in the guidance.

Another key area where central government exerts significant control over the GLA is in the budget arena. The budget controlled by the GLA is quite large, principally because the GLA budget includes the huge budget of Transport for London, the local mass transit agency. In general, however, the GLA's fiscal powers are quite limited. The mayor can raise funds only through a precept charged to local authorities, through miscellaneous service charges, and through direct grants from the government. The GLA does not have the powers to directly levy income, property, or sales taxes on London residents or businesses (Loveland 1999). Even the size of the precept charges can be constrained, if central government sees them as excessive (HMSO 1999). The bottom line is the GLA has little of its own funding available to spend on renewable energy projects. The mayor could allocate funds to such projects directly out of the general GLA budget, but these funds would displace funding for other GLA activities and services and be subject to scrutiny by the London Assembly, where its value would be compared to other expenditures. As a result, the mayor rarely imposes energy requirements on the GLA functional bodies, more often than not using exhortatory language (e.g. the mayor encourages the GLA functional bodies . . ., the mayor requests . . ., etc.) rather than outright mandates.

The new Climate Change Agency presents the mayor with one vehicle for circumventing this funding problem, through its capacity to provide direct energy services delivery. As previously noted, the Thamesway model on which the LCCA is based recycles financial savings attributable to past energy system investments into new projects. The level of renewable power system investment pursued by a GLA-affiliated ESCO could be huge, given the large number of buildings owned by the GLA, the various functional bodies, and the 33 local authorities around London. Because Thamesway was so effective at deploying renewable energy systems in Woking, Allan Jones may similarly prioritize the deployment of renewable power systems in London.

Political influences on the new Energy Strategy have come from several different sources. In the run-up to the first GLA election in 2001, several green groups attempted to develop an environmental agenda for the city. In this election manifesto, the Sustainable GLA Coalition explored how an environmental approach to economic development could bring jobs and wealth creation opportunities to London (Sustainable GLA Coalition 2001). The report examined urban regeneration, transport, energy, waste, eco-technologies, and biodiversity, advocating specific policies and programs the Coalition believed could improve the quality of life in London, create jobs, lower the cost of doing business, and address social exclusion. Less than a year later, energy-related business creation was identified as a core theme of the mayor's Energy Strategy, and a key focus of the various public/private partnerships the mayor has established to deliver his strategy. Mayoral staff hedge when asked to ascribe credit for the mayor's focus on energy job creation, noting his strong personal commitment to these issues (Sawer and Bar-Hillel 2004).[8] More evidence of the mayor's personal commitment is his decision to partner with former US President Clinton's charitable foundation to establish the Large Cities Climate Leadership Group. This group will provide information and technical support to advance climate change

---

[8] Personal communication with John Duffy, Director of Environment, Greater London Authority (19 May 2004).

mitigation and adaptation initiatives in large cities around the world (Eilperin 2006); London was instrumental in getting this project off the ground.

Several commentators have made note of the growing influence on the business community on policymaking in London (Travers 2004; Thornley *et al.* 2002). Among local groups, London First has the strongest record of action on local energy policy matters, publishing a report on the mayor's congestion charge plan (London First 2002); a footprint study examining London's global environmental impact (London Remade/London First 2003); and providing feedback on all of the mayor's strategies. London First also partnered with London Renewables to launch the release of new guidance documents for developers, consultants, local authority planners, and councillors (London First 2004).

Evidence of London First's influence on the development of the Energy Strategy can be found by examining how the mayor's mandate regarding on-site renewables generation changed between the early and final drafts. By the time *Green Light to Clean Power* was released, the mandate had been softened by the addition of the modifier 'where feasible', reflecting London First's contention '. . .that the generation of at least 10% energy needs is not viable in all cases in the short term. Renewable Energy does not as yet provide an entirely reliable constant energy supply, nor is it always feasibly installed in all locations of this size'(London First 2003). GLA Energy Team members confirm this change was largely due to business community arguments that these requirements could put London at an economic disadvantage compared to other UK and European cities.

Thus far, London First's concerns appear somewhat overblown. Despite 'grumbling from developers when they first submit their applications', development of large projects

---

Box 6.3.
Comparison of renewables language in different drafts of the London Energy Strategy (Greater London Authority 2002; Greater London Authority 2003; Greater London Authority 2004)

DRAFT 1	**DRAFT 2**	**FINAL LANGUAGE**
PROPOSAL 11: To contribute to meeting London's target for the generation of renewable energy, the mayor will expect applications for new developments of strategic significance (i.e. those referable to the mayor) to generate a proportion of the site's energy needs (electricity and heat) from renewables. The mayor encourages boroughs to expect the same of developments of a significant size.	PROPOSAL 20: To contribute to meeting London's targets for the generation of renewable energy, the mayor will expect applications referable to him to generate at least ten per cent of the site's energy needs (power and heat) from renewable energy. Boroughs should expect the same of new commercial and industrial developments over 1000 m² and new residential developments of ten dwellings or more.	PROPOSAL 13: To contribute to meeting London's targets for the generation of renewable energy, the mayor will expect applications referable to him to generate at least ten per cent of the site's energy needs (power and heat) from renewable energy on the site **where feasible**. Boroughs should develop appropriate planning policies to reflect this strategic policy. [*emphasis added*]

in London has not screeched to a halt since the energy strategy went into effect in 2004.[9] New development plans continue at a rapid pace, in all parts of the city. What has changed is a growth in developer's recognition that energy concerns must now be factored into the project design. By the time the back-and-forth review process is complete, significant changes have ensued, typically resulting in a far more energy-efficient development project than the one first conceived. Even the grumbling may be disappearing; in early 2006 the GLA Energy Team reported receiving a proposal that would generate 14% of its own power on-site from renewable sources.

6.5 New York City – A Comprehensive but Less Renewables-Focused Energy Path

Roughly the same size as London, and similarly situated as a centre of commerce and culture, New York has an even more voracious appetite for energy. Like London, energy use is expected to rise in the coming decade, along with the local population. New York and London also share a similar energy policy storyline, involving varying levels of local engagement on energy planning issues over the last three decades. Current Mayor Michael Bloomberg's administration has increasingly adopted a forward-thinking approach, ending a 25-year drought when other New York City mayors dwelt on it only in times of crisis, such as in the aftermath of a blackout.

Though the range of energy initiatives currently undertaken by the city of New York is quite comprehensive, there has historically not been a coordinated strategy guiding New York's efforts. This may be changing, however, thanks to work under way on *plaNYC2030*, a long-term growth and sustainability plan scheduled to be released by the city in mid-2007. This plan is discussed in greater detail below.

Box 6.4.
Profile of current New York City energy marketplace (New York City Energy Policy Task Force 2004; Resource Insight Inc., 2003; US Census Bureau 2000)

- Consolidated Edison, a private firm, manages the local electrical grid serving 99% of New York City
- 13 firms compete to supply power to local households and businesses; the city of New York procures its power from the state-run New York Power Authority
- Peak electricity demand = ~11 630 MW; peak demand occurs during summer months
- State law requires sufficient in-city generation capacity to meet 80% of peak demand
- Current in-city central station generation capacity = ~10 305 MW
- New York City households consume approximately 4370 kWh of electricity/year, the lowest total of any major US urban area
- New York City has the highest public transportation ridership of any large US city, while just 46.5% of city households own a car. In Manhattan, this figure drops dramatically, as just 23% of households own a vehicle

---

[9] Personal communication with GLA Energy Team, 8 February 2006.

Transportation is one area where New York City has an illustrious energy and environmental record. New York City's huge mass transit programs are a well-known feature of the city, the mode of choice for one of every two commuting trips made each day Schaller Consulting (2002). Less well known is the fact that New York City is also a global leader in the area of alternative fuel vehicle deployment. The state-level public authority[10] responsible for local bus operations has purchased nearly 1000 compressed natural gas-powered (CNG) and diesel-electric hybrid buses to reduce energy consumption and improve air quality. New York City government agencies have collectively purchased another 2100 CNG and gasoline-electric hybrid passenger vehicles, representing approximately 16% of the city government's total passenger vehicle fleet. The New York City Sanitation Department has piloted natural gas-powered waste collection vehicles and street sweepers, and is working with equipment manufacturers to design a diesel-hybrid collection truck suitable for local operations. Sanitation also has 450 trucks powered by E85, an ethanol/diesel blend, refuelling them at six department fuelling stations around the city.[11] A new local law signed by the mayor in 2005 is driving the city's alternative fuel efforts, requiring city agencies to purchase vehicles that are the cleanest in their vehicle class (City of New York 2005).[12]

The city has also targeted fuel use by privately owned vehicles over which it has regulatory control, incentivizing the conversion of taxicabs to run on natural gas and selling discounted new taxi licences specifically allocated to hybrid-electric vehicles (Considine 2006). By the end of 2006, 281 hybrid taxis operated on city streets (Gross 2006). No special provisions or policies exist to promote biofuel use by local taxicabs.

City-owned buildings are the other focus of significant energy policy activity. The logic behind these efforts is simple – on its own, local government consumes approximately 1100 MW of power each day, or just under 10% of all electricity consumed citywide (New York City Energy Policy Task Force 2004). Although the city pays lower-than-market rates for power from the New York Power Authority (NYPA), this high level of usage nonetheless constitutes an enormous expense, totalling approximately $430 million in fiscal year 2006 (New York City Department of Citywide Administrative Services 2006). Since 1989 New York City agencies have been under a mayoral directive to conserve energy, and local officials collaborate with the New York Power Authority's Energy Cost Reduction Program (ENCORE) on approximately 25 projects per year. Between 1998 and 2004 more than $163 million of energy conservation projects were undertaken, generating agency cost savings that amount to tens of millions of dollars each year (New York City Energy Policy Task Force 2004; New York City Comptroller 2006).

In 1999, the city began to promote high performance design principles on new city-owned construction projects as another means of reducing government building-related energy consumption. The guidelines were not intended to be prescriptive; instead, they were presented as a menu of ideas consultants and architects advising the city could voluntarily draw from when designing a new building or retrofit project. Because of the

---

[10] The Metropolitan Transportation Authority is a regional organization established by the state of New York in 1965. It is governed by a 17-person board, four of whom are appointed by the Mayor of New York City. The MTA operates completely independently of New York City government, although it does consult closely with the city on various transport matters.

[11] Personal communication with Mark Simon, Director, Alternative Fuels Program, New York City Department of Transportation, 3 November 2005 and 29 September 2006.

[12] Ironically, this law penalizes ethanol use, as trucks using E85 are not considered to be the cleanest in class. In the future, the Sanitation Department must obtain a waiver before it purchases any additional ethanol trucks.

gains achieved on several high performance building demonstration projects, in 2005 the city changed course, moving to require Leadership in Energy and Environmental Design (LEED®) certification[13] on all city-funded construction projects exceeding $2 million in value (Hu 2005). Like the high performance design guidelines, LEED® does not require the agency to emphasize on-site renewable power generation or other energy cost-saving strategies. Points toward certification can be obtained from a wide range of project variables, some of which are only tangentially related to on-site energy use.

The city's focus on its own energy consumption has been embraced as one core component of a larger citywide electricity strategy developed by an Energy Policy Task Force convened by Mayor Bloomberg in 2003. Composed of 16 representatives from local utilities, environmental and community organizations, the real estate community, and a range of local and state government agencies, the group was tasked with the following responsibilities (New York City Energy Policy Task Force 2004):

- Assessing the current state of the city's energy resources, with a primary focus on electricity;
- Projecting the energy needs of the city through 2008; and
- Developing recommendations necessary to secure the city's energy future.

Dominating the Task Force report was the idea that New York City's electricity supply picture is rather tenuous, and that absent some sizable reduction in demand or introduction of new supply sources, the city faced a supply gap of thousands of megawatts by 2008. This shortfall stems from two factors – growing demand and the anticipated closure of several in-city power plants. Because New York City is considered a load pocket, with limited power transmission capacity connecting the city to other parts of New York and neighbouring New Jersey, the anticipated plant closures have a disproportional impact on the city's power supply situation. The Task Force also called for additional supply capacity to stabilize or lower local electricity prices, which are among the highest in the nation. (Table 6.1)

Mayor Bloomberg's Energy Policy Task Force was not the first group to identify this supply gap; beginning in 2001, five other reports raised similar concerns (Resource Insight

Table 6.1.   Anticipated New York electricity supply gap (New York City Energy Policy Task Force 2004).

Category	Capacity (MW)
Need to meet demand growth	665
Need to assure market stability	1000
Need to replace aging power plants	2115
Total capacity needed	3780
Less	
Power plants under construction	(875)
Distributed resources (includes renewables)	(300)
Net capacity needed through 2008	2605

[13] LEED® is an internationally recognized set of green design protocols that award points for different design elements or construction processes; to achieve certification, some minimum point threshold must be achieved.

Inc. 2003; New York Building Congress Energy Committee 2001; New York Building Congress Energy Committee 2002; NYISO 2001; NYISO 2002). Recent updates to the Task Force analysis find the urgency of the problem has abated somewhat, but given antici-pated population growth, the problem will continue to grow worse with time, with the city facing a potential supply shortfall of 6600 MW by 2030 (New York City Energy Policy Task Force 2006; Quiniones 2006).

The Task Force recommendations make clear the limits on local government's capabili-ties in this area. For over 100 years, the city of New York has purchased power from other entities rather than generating power itself. There were attempts to build municipal power plants early in the twentieth century, but these all failed (Hughes 1983). As a result, the mayor cannot simply turn to his energy commissioner and tell him or her to build new power plants in the city to solve this problem. Instead, the Task Force put forth an advocacy agenda, offering a series of recommendations emphasizing enhancements to the workings of the local energy marketplace (New York City Energy Policy Task Force 2004):

- *Siting* – The Task Force urged the mayor to advocate for the reauthorization of Article X, the state's expired power plant siting law. The Task Force believes successful reau-thorization of Article X would greatly facilitate the deployment of new large scale, in-city generation capacity, as the Article X process was specifically designed to overcome local opposition to power plant siting.
- *User-based initiatives* – The Task Force urged the mayor to promote the expansion of commercial and industrial demand reduction efforts, focusing particularly on those parts of the city where the grid is overloaded. The Task Force also urged the mayor to support increased private sector investment in energy efficiency and clean on-site gen-eration systems.
- *Project financing* – The Task Force noted that several large power plant proposals exist that would add new supply capacity in the city; some of these projects have already received state permits. They have been slow to move forward, however, because of dif-ficulty obtaining financing. This is not a problem unique to New York City. Prior to electricity market restructuring efforts in the 1990s, such facilities would have been assured a predictable rate of return. Current market structures now place more finan-cial risk on the project developer, making investors wary about the long-term viability of these plants. The Task Force urged the mayor to examine ways to help these projects obtain financing, including using the city's purchasing powers as one means of under-writing these projects.

The Task Force also noted the potential for in-house initiatives, reducing consumption and increasing power generation at city-owned and managed facilities. Recommendations were made to create an energy efficiency plan for city buildings, to emphasize high performance design in city construction projects, and to work with local experts and government agencies to enact other improvements in the city's internal energy manage-ment practices.

Following the release of the report, Mayor Bloomberg asked Task Force members to remain empanelled and they continue to meet on a periodic basis to discuss steps the city is considering or has taken to implement various recommendations. On the anniversary of the report's release, the Task Force issued an update detailing progress made by different energy stakeholders around the city (New York City Energy Policy Task Force and 'Status Report' 2004). A second update was released in June 2006 (New York City Energy Policy Task Force 2006).

Many of the initiatives reported on in the 2005 and 2006 updates were undertaken by private sector members of the Task Force, and it is unclear whether one can attribute causality to the firm's involvement in the Task Force, or whether such measures would have been made regardless. There have been several initiatives taken by the city that do appear to be directly linked to recommendations made by the Task Force. These include:

- *Con Edison rate case settlement* – The city benefited greatly from the timing of Con Edison's decision to apply to state regulators for an increase in the rates charged to New York City residential and commercial electricity customers. As part of their assessment of whether to approve the rate increase, state regulators invited feedback from interested parties (including the city of New York) on the pros/cons and alternatives to the proposed increase. The city's top energy advisors leveraged this opportunity to press Con Edison for several initiatives that built on Task Force recommendations, including:
  - A three-year, $224 million investment in distributed resources development around the city, including demand reduction initiatives, energy efficiency, and clean on-site generation. The goal is to achieve 675 MW in new supply or demand reduction resources citywide; 150 MW are to be achieved in load constrained areas, where current demand is near the carrying capacity of the electric distribution system in that neighbourhood. By promoting demand reduction and on-site generation, the city believes Con Edison can delay or forego significant new infrastructure investments in that area, while simultaneously addressing a portion of the anticipated citywide supply gap. This investment requirement creates a big opportunity for renewable energy deployment, although renewables will be competing against other technologies that may have a shorter payback period.
  - Con Edison also agreed to develop Energy Infrastructure Master Plans for major redevelopment zones around the city, analysing what role district heating and cooling systems and other types of distributed resources (including renewables) could play in serving these areas, in lieu of traditional energy infrastructure upgrades. The first plan developed covers the 360 acre Hudson Yards area in Manhattan, while future plans will examine lower Manhattan and certain neighbourhoods in Brooklyn.
- *NYPA contract renewal* – The city's long-term power contract with NYPA also came up for renewal during this period, and the city used this opportunity to clarify its interest in procuring renewable power. Shortly thereafter, NYPA issued a solicitation seeking 500 MW of new power sources to satisfy its New York City government contract, of which 20 MW were ultimately based on renewable power sources.

In May 2005, Mayor Bloomberg signed the US Conference of Mayors' Climate Change Protection agreement, committing the city to reduce greenhouse gas (GHG) emissions by at least – if not more than – the 7% level called for under the Kyoto Protocol. As part of this undertaking, the city has been working to compile GHG emissions data for local government operations, after which an action plan will be prepared to reduce future city government emission levels (NYC Office of Environmental Coordination 2006). The city has also been compiling data on citywide carbon emissions.

Both datasets are expected to play a critical role in supporting the development of *plaNYC2030*, the long-term growth and sustainability planning initiative first announced by Mayor Michael Bloomberg in early 2006. Perhaps inspired by strategic planning efforts in London,[14] *plaNYC2030* is designed to articulate a long-term vision for New York City that both accommodates anticipated population growth and addresses short- and long-term housing, infrastructure, and environmental pressures facing the city. In December 2006, the mayor laid out ten key *plaNYC2030* goals, four of which have clear or implicit energy links (City of New York 2006):

Goal 2: Improve travel times by adding transit capacity for millions more residents, visitors, and workers.

Goal 6: Provide cleaner, more reliable power for every New Yorker by upgrading our energy infrastructure.

Goal 7: Reduce global warming emissions by more than 30%.

Goal 8: Achieve the cleanest air of any big city in America.

At the time of this writing, there was nothing in the city's published materials or *plaNYC2030* website[15] that explicitly discussed renewable power, so it is difficult to tell whether the city will choose to emphasize these technologies in the final plan.

6.6 Key Influences on Energy Policymaking in New York City

Despite its size and complexity, New York City has frequently been cited for its leadership on sustainability issues (Owen 2004; SustainLane 2006), but renewables deployment is not an area that has drawn much kudos. Several factors appear to be at play.

First, there is no question that concerns over the anticipated energy shortfall dominate local policy debates, because the economic impacts could be so severe. This was made clear during the citywide 29-hour blackout in August 2003, when economic losses to New York City businesses and individuals were in the vicinity of $1 billion (Herman *et al.* 2003). Frequent blackouts or brownouts could discourage local business expansion or drive businesses out of the city, dramatically harming the city's long-term economic prospects. The decision of the mayors' Energy Policy Task Force to emphasize this problem is thus a rather logical one.

It is noteworthy that the Task Force report essentially ignores renewables as a partial solution to this problem. The bulk of the report focuses on new power plant siting and construction, or on the development of new transmission lines which can import power to the city. In the section on distributed resources, renewable power technologies are identified as a potential source of new supply, but their contribution is downplayed, noted merely for their lack of current market share and the fact they are subject to 'detailed and complicated engineering studies . . . [that require] a substantial commitment of time and resources both by Con Edison and the developer . . .'(New York City Energy Policy Task Force 2004).

---

[14] There is reason to believe that New York City's planning efforts were partly inspired by planning efforts undertaken by the Greater London Authority. In October 2005, London Deputy Mayor Nicky Gavron visited New York City to attend several meetings regarding an upcoming urban climate change conference. While in New York, she met with New York Deputy Mayor Dan Doctoroff, speaking at length about the 'London Plan', London's strategic land use plan. Shortly thereafter, the New York City Economic Development Corporation's Energy Office was asked to prepare a briefing for Deputy Mayor Doctoroff explaining the London Plan and the Mayor's Energy Strategy in greater detail. In January 2006, Mayor Michael Bloomberg announced his administration was undertaking a long-term growth management plan, covering a range of issues very similar to those addressed by the London Plan; eventually this effort came to be known as *plaNYC2030*.

[15] See www.nyc.gov/planyc2030.

Con Edison's ambivalence about renewables may be one reason the report was written this way. Con Edison has long expressed concern about renewable energy systems in the city, dating back to 1977, when urban homesteaders had to sue to force the utility to permit the interconnection of a small wind turbine atop an abandoned tenement in lower Manhattan. Con Edison had refused fearing power surges from this two-kilowatt system would damage their ten million-kilowatt network. After a long back-and-forth debate, the New York State Public Service Commission intervened, ruling that Con Ed must allow this windmill and up to 24 others to connect to its grid (Energy Task Force 1977; Greenhouse 1977).

Since then, Con Edison has developed formal interconnection guidelines to facilitate deployment, but the underlying message remains one of caution, emphasizing the need to ensure 'there is no adverse effect on the Company's other Customers, equipment or person-nel, while maintaining the quality of service' (Con Edison 2002). Con Edison has updated their interconnection guidelines several times in recent years, most recently in 2006, but their focus on system reliability remains unchanged (Con Edison 2006). Such an orienta-tion is understandable – Con Edison is required by regulators to deliver power wherever and whenever New Yorkers need it (New York State Public Service Commission 1991). Failure to achieve these Electric Service Standards can result in fines, such as the $1.9 million penalty imposed on Con Edison for a 1999 blackout in upper Manhattan (New York State Public Service Commission 2002).[16]

Against this backdrop, energy systems that plug seamlessly into the existing electric grid – such as the new transmissions lines and central station power plants advocated by the Energy Policy Task Force report – hold great appeal. Speaking off-the-record, several Task Force members observed Con Edison offered a robust endorsement of such systems in their meetings, but spoke less favourably about other alternatives.

Lacking significant technical expertise, Task Force members may have felt a need to defer to the utility's concerns on such matters, believing that if renewable power and other alternatives didn't work out as planned, the citywide financial consequences could be grim. The *Electricity Outlook* report's emphasis could therefore be viewed as a form of tech-nology path dependency or lock-in (Berkhout 2002; Unruh 2000), where stakeholders find it difficult to change course because embedded system assets like wires, relay switches, and substations do not easily accommodate alternative technological approaches.

The second key factor influencing local policymaking is New York City's constrained capacity to act on local energy matters. This situation dates back almost a century, when state rules revoked local control over the electric utility industry, placing it in the hands of state regulators. Until that time, local government had primacy, reflecting the inherently local nature of the DC-power distribution systems then in use. Local authorities deter-mined which utilities could operate in certain parts of the city, and the rules they must follow. In 1907, the New York State Public Service Commission was created to drive out corruption linked to the issuance of these franchise rights (Hirsh 1999; Read 1998). This move had huge consequences, placing local government in a subservient position with regards to rate-setting, power plant location decisions, and what type of power sources are encouraged or discouraged. The city can (and does) provide input to state regulators on these issues, but as seen in Fig. 6.3 has limits on its ability to establish market rules and structures emphasizing renewables.

---

[16]More recently, Con Edison has been criticized for system and management failures associated with a several days-long blackout in central Queens in August 2006. At the time of this writing, news reports were hinting fines levied against Con Edison could reach $100 million.

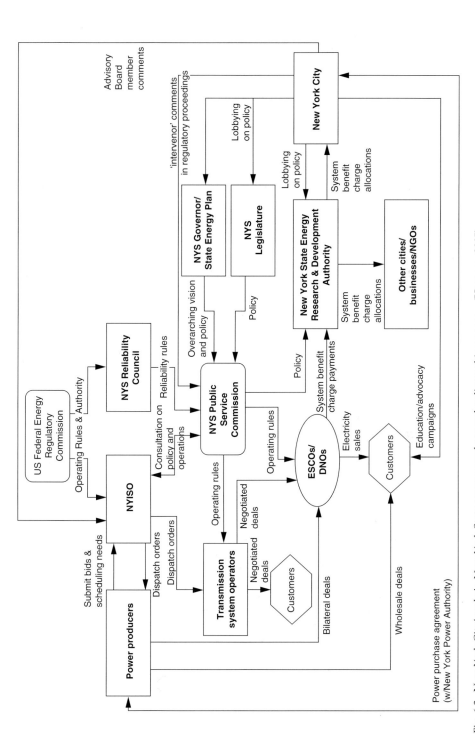

Fig. 6.3. New York City's role in New York State energy markets and policymaking process (Hammer 2005).

How has the city's disadvantaged policymaking powers influenced its renewables agenda? The New York State Energy Research and Development Authority (NYSERDA) has lead responsibility for renewables planning in the state, and over the years has gained a national reputation as a leader in this field. Several members suggested the mayor's Task Force was hamstrung in developing a renewables policy because NYSERDA had yet to promulgate rules on a significant new statewide renewables initiative, the Renewables Portfolio Standard (RPS).[17] Rather than risking policy moves that might subsequently be undercut by the RPS, the Task Force chose to focus on other issues. In effect, the institutional environment led the Task Force to behave reactively, waiting for others to act before deciding whether or how to respond.

A third factor influencing local renewables policymaking is an apparent preference for voluntary action on renewables deployment rather than the use of mandates. This policy has permeated the city's approach with both the public and private sectors. As discussed earlier, the high performance buildings guidelines developed by the city's Department of Design and Construction were intended to support voluntary decisions by city agencies constructing new buildings. Local Law 86, the 2005 law requiring LEED certification for new city agency buildings, continues this approach, giving designers and agency officials considerable flexibility when determining how to achieve the minimum point threshold; renewables are again not required.

The city has a limited record of advocacy regarding renewables use by the residential and business sectors. The 2005 announcement that the city would purchase renewable power for the city-owned Brooklyn Army Terminal complex (Con Ed Solutions 2005) was the Bloomberg administration's first public statement on the subject of green power. Because this purchase was touted as an example for others to follow, we can infer the mayor's preference that households and businesses arrive at similar decisions on their own.

The current high cost of renewable energy in New York City is likely driving this preference for voluntary action. Two recent university studies noted renewable power systems in New York City generate electricity at a cost higher than that available through grid-based electricity sources (Center for Sustainable Energy 2007; Columbia University Workshop on International Energy Management and Policy 2006). For most energy buyers and policymakers, this puts renewables at a significant disadvantage. Mayor Bloomberg's principal energy advisor made exactly this point in 2005 when testifying before the New York City Council about a proposal mandating higher rates of renewable power use by city agencies. Although highly sympathetic to the goals of the legislation, he noted that green power comes at a cost premium, and that the city's electricity costs were already expected to increase by over $50 million a year due to recent rate hikes. In his view, to increase costs further simply didn't make sense.

The mayor's advisor also made another important point at the hearing, noting 'The cost premium to purchase green power must also be evaluated in terms of competing energy investments . . . Presently, each dollar invested in energy efficiency upgrades will conserve more fossil fuel and therefore protect the environment more than the same dollar invested in renewable energy sources.'(Quiniones 2005)

---

[17] New York State's renewable portfolio standard (RPS) seeks to increase the amount of electricity generated from renewable sources to no less than 25% of the state's power supply by 2013. When the RPS was first established, renewables already accounted for approximately 19% of the state's supply. It is estimated that the RPS will result in 3700 MW of new renewables capacity statewide.

Given limited budget resources, the city's rate case settlement strategy forcing Con Edison to pursue distributed resources deployment makes great sense because it shifts programmatic and budgetary responsibility away from local government coffers and onto the utility. Such a move also offers the city political coverage, with Con Edison bearing the brunt of any ratepayer backlash rather than local government.

6.7 Distilling Policymaking Lessons for Other Cities

London and New York are two cities that are both highly engaged on energy matters, pursuing a wide range of strategies, yet with different policy approaches that reflect local circumstances and perspectives on renewable energy deployment and use. To be sure, neither city has historically been a hotbed for renewables activity, as evidenced by the amount of installed capacity, which pales in comparison to the amount of installed fossil fuel-based generation capacity in each city.

The past five years hint at change, a situation that bears watching.

In London, a new mayor operating under a mandate to develop a strategic vision for the city has put forth an expansive vision of London's energy future. Environmental sustainability is an explicit goal of that plan, and renewable energy is one way the mayor intends to deliver this goal. Since the London Energy Strategy was released in 2004, the GLA has

Table 6.2.   Current renewable power system deployment levels.

	London	NYC
Solar PV	~1.0 MWp	1.1 MWp[18]
Biogas-based power	35.3 MWp[19]	1.6 MWp[20]
Wind power	3.6 MWp[21]	–
Landfill gas	20.0 MWp[22]	–[23]
Total	59.9 MWp	2.7 MWp
Total installed fossil-fired powerplant capacity within city limits	2194.7 MWp	~10 300.0 MWp

[18] As of the end of 2005, there were 45 solar PV systems in New York City with a combined capacity of just over 1.1 MWp.

[19] Personal communication with Helen Burton, Generation Manager, Thames Water, 20 October 2006. Approximately 3.3 MWp = biogas combusted directly, 16.0 MWp = biogas combusted in the presence of a small amount of fuel oil, and 16.0 MW = power generated by sludge combustion triggered by a small quantity of natural gas.

[20] The NYC Department of Environmental Protection has deployed eight 200 kW fuel cells powered by digester biogas at five sewage treatment plants around New York City. *Source*: Yan Kishinevsky, Program Manager, Research and Technology Development, New York Power Authority, presentation at Columbia University School of International and Public Affairs, 23 February 2006.

[21] Ecotricity has deployed two 1.8 MWp wind turbines at the Ford Motor Co. plant in Dagenham, East London. Several small wind turbines have also been deployed around London, but their total combined generation capacity is negligible.

[22] Power generation system is installed at the now-closed Rainham landfill.

[23] New York City does not have any power generation capacity currently installed at local landfills, but methane gas is captured at the now-closed Fresh Kills landfill, cleaned, and sold to residential and commercial customers on Staten Island.

focused on implementation, instigating the deployment of new renewable power systems and undertaking market-building activities that should speed the adoption of these technologies by home owners and the private sector. In New York, local officials have been less focused on renewables, devoting most of their policy attention to resolving a supply gap that could seriously harm the city's economy.

In one sense, it is unfair to compare the two approaches, because presumably London would have a similar emphasis if transmission lines entering the city could not keep up with growing demand. Moreover, the city of New York has not been sitting idly by, implementing an extensive array of energy conservation, green building and green vehicle initiatives. Most have targeted internal government operations, although the city has also pushed for policies that move two of the biggest energy stakeholders in the city – the local distribution network operator and the city's own energy provider – to expand their efforts at energy efficiency and clean power generation and sales. *plaNYC2030*, the city of New York's new growth and sustainability plan, is another ambitious new initiative, partly focused on reducing the city's greenhouse gas emission levels and upgrading the local power infrastructure.

It is thus difficult to make a normative judgement that one city's approach is better than the other, although we can draw important lessons relevant to cities considering their own energy future.

First, cities must recognize the extent to which their local policymaking powers are shaped by laws or policies at the regional, state, or national level. These powers, best characterized as a municipality's capacity to act, can be either strong or weak. We saw examples of both types in London and New York. Weak powers are exemplified by clear limits on what a locality can do; in London, these manifest themselves in the mayor's budgetary powers and in the rules that give the mayor planning permit authority over just 250–300 planning applications each year. That these applications represent the largest development schemes in London is not insignificant. The total number pales, however, when compared to the tens of thousands of applications handled by the 33 London boroughs whose planning control powers were protected by the crafters of the law imposing these limitations (HMSO 2000). Similarly, in New York, laws passed 100 years ago stripped the city of its regulatory power over utilities, leaving it in the much weaker role of intervenor on regulatory matters.

A strong capacity to act manifests itself in areas where the municipality does have direct line authority or regulatory control. In New York, the purchasing powers of government have been used to procure clean energy vehicles, greener buildings, and grid-based renewable power. In London, regulatory powers were used to impose energy analysis and on-site power generation requirements on new development projects answerable to the mayor.

Although both cities have exploited their capacity to act, London has historically been more systematic in its approach. According to GLA Energy Team members, prior to crafting their draft energy strategy, the team undertook a full assessment of the GLA's policymaking powers, and then designed the strategy around that reality to maximize their policy leverage. Thus, where the GLA had the ability – and the preference – to impose a mandate, it did so. Where mayoral powers were weaker, the GLA has adopted advocacy or educational strategies that further the GLA's goals.

Does London's strategy of 'joined-up' thinking (Bundred 2006) give it an advantage over New York's loose-knit plans? Probably. Because London's many strategic planning documents were written in roughly the same time frame, they all closely link with one another, keying off the same core principles, cross-referencing ideas, and expanding the number of stakeholders who are encouraged or empowered to act on these issues. In this regard, London's energy strategy broadens the action agenda beyond a narrowly crafted policy silo, forcing others who normally ignore energy considerations to now take heed. London's joined-up approach also allows it to systematically and explicitly incorporate central government and local authority advocacy strategies into the local action plan, thereby ensuring that every possible opportunity is exploited. Whether New York will follow London's joined-up approach in its *plaNYC2030* initiative will only become apparent once the report is released in 2007.

A second lesson from London and New York is that municipalities must determine which sectors will be the target of their policymaking efforts, which policy mechanisms are politically palatable, and which policies will best deliver these goals. Is the focus primarily on internal government operations, or do local officials also wish to target residential or commercial energy use? London and New York have significant stylistic differences in this area. The GLA makes clear its intention to directly engage households and businesses on energy issues. Educational initiatives, leading-by-example, and mandates are all techniques employed by the GLA to promote action. The mandates are aggressive – and apparently rather effective. Researchers at Southbank University investigating the energy component of dozens of planning applications approved by the Mayor's Office found the renewables component of the average project results in a 5% decline in the building's overall CO_2 emissions.[24,25]

The GLA's policy leadership is producing results elsewhere as well. In the London Borough of Croydon, one of three boroughs[26] that independently established similar on-site renewables requirements for building projects over which it has sole planning jurisdiction, 120 solar photovoltaic, solar thermal, and small wind systems have been approved since the mandate first went into effect in December 2003. When completed, these projects are expected to result in over 2 MW of new solar PV and 8500 square metres of solar water heating capacity,[27] significantly increasing the total amount of renewables capacity installed citywide. Whether an aggressive incentive strategy would have proven equally effective is an open question requiring further analysis.

In New York, the city has adopted a lead-by-example approach, but beyond that, has been generally silent on the subject of what households and businesses should do as regards energy use. Until recently, the city's best-known energy education initiative was a voluntary demand reduction program, asking large businesses to reduce power use when local energy assets are stressed, typically on the hottest days of the year. The city also reports it has been collaborating with housing and real estate firms working on city-related development projects, encouraging them to 'green up' their designs (New York City Energy Policy Task Force 2006).

---

[24] *Source*: Presentation by Dr Antony Day, Southbank University, to Society of Operations Engineers, Birmingham, UK, 7 June 2006.

[25] Energy efficiency initiatives incorporated into these projects resulted in even greater impacts, responsible for anticipated building CO_2 emission reductions of 25%.

[26] According to www.TheMertonRule.org, an independent website tracking the adoption of this type of renewables mandate among UK local authorities, three London boroughs (Croydon, Hammersmith and Fulham, and Merton) have formally adopted some type of requirement on new development schemes, while 21 others (Barking and Dagenham, Bexley, Bromley, Camden, Corporation of London, Ealing, Enfield, Greenwich, Haringey, Harrow, Havering, Hillingdon, Islington, Lambeth, Lewisham, Newham, Redbridge, Southwark, Sutton, Tower Hamlets, and Westminster) are considering or are close to finalizing such a rule. The Merton Rule is so named because it was the London Borough of Merton which first implemented this policy; it was embraced by Mayor Livingstone several months later in the second draft of his citywide Energy Strategy. (List accurate as of 6 February 2007.)

[27] Personal communication with Eddy Taylor, Environment and Sustainability Manager, Croydon Council, 3 October 2006.

plaNYC2030 is the city's most noteworthy effort to engage New Yorkers on local energy and climate issues, but at the time of this writing the program's focus has largely been on soliciting public input on the topic, rather than laying out any specific action plan. In early 2007, New York City Deputy Mayor Dan Doctoroff said the city would incentivize clean power generation to help meet its *plaNYC2030* greenhouse gas reduction targets (Smith 2007). Whether this means *plaNYC2030* will ultimately target existing large power plants or promote new clean on-site power installations is not yet clear.

If the city does emphasize voluntary action, its approach will stand in stark contrast to London's tactics. It also contrasts with policies in place on a 90-acre piece of state-owned land known as Battery Park City, located just five blocks from New York's City Hall in lower Manhattan. As in London, new commercial and residential buildings in Battery Park City are required to generate some of their power from on-site renewable power systems (Battery Park City Authority 2002; Battery Park City Authority 2003). New York City officials have made no effort to replicate these mandates on a citywide basis thus far, and given the deputy mayor's comments it appears it will remain this way. There is historical precedence for a voluntary approach, as the city has a long history of using financial incentives to prod developers to voluntarily deliver some social objective (Bressi (Ed.) 1993; Kayden 2000). It could also reflect government or developer concerns that green mandates could exacerbate the already high construction costs plaguing the city. (Satow 2007).

The final lesson holds relevance both for local energy policymakers and officials in state and national government. It stems from London and New York's efforts to advocate energy policy or program changes at the state or national level that would improve circumstances locally. Such efforts are important because they highlight areas where local policymaking powers are weak, requiring state or national level intervention. They also raise the question of whether the current allocation of policymaking and program responsibilities between different stakeholders are appropriate, or whether changes to this system are necessary.

There are sound reasons why certain energy policies or regulatory functions are more appropriately handled at a macro level. There could be market chaos if each municipality had its own unique renewables interconnection guidelines or performance standards for biofuels. On the other hand, state and national policies do sometimes inhibit desirable local action. State programs promoting renewable technology installations are one paradoxical example, because many have historically prioritized least-cost approaches, resulting in a disproportionate focus on systems less suitable to an urban context (e.g. large wind farms). New York State's new renewables portfolio standard thus deserves special note for the way it allocates funds for smaller scale technologies most likely to be deployed in cities. (New York State Department of Public Service 2004).

New York's RPS shows that state regulations can be structured in ways that are urban-friendly. But is this the best solution – to fine-tune our traditional state or national government-led policymaking approach with special urban carve-outs – or is it time to rethink our fundamental regulatory and market systems so they better address the unique challenges of influencing energy use in cities?

The latter would likely involve the rescaling of energy governance (Bulkeley 2005), requiring new institutional approaches that give local authorities:

1. An expanded role in energy and climate policymaking;
2. A concrete set of relevant powers; and
3. Adequate resources to deliver the results.

Moving from our current situation to this new institutional environment would be no easy feat; London's battles over power sharing make this point abundantly clear. The key will be determining whether it is possible to better optimize roles and resources among market stakeholders and different tiers of government. The answer will depend on how a city uses energy, where the energy comes from, which policy or program levers are under the control of key stakeholders, and the strength of these powers.

Table 6.3 represents the beginning of such a capacity-to-act analysis, detailing which stakeholders exert influence over energy use in New York City's residential sector, and how this influence manifests itself.[28]

In New York's liberalized energy marketplace, residential customers have the right to select their energy supplier (A), allowing households to express preferences for the pricing and sources of the energy they use. These decisions are constrained by choices made by power generators (B) and state and federal authorities, with the latter group holding significant influence thanks to their emission limits and rate-setting capabilities (C). State and national-level officials can also establish technology standards (D) for home appliances and residential energy systems. Consumers have limited say over such matters, although they do have the ability to minimize their use of energy-intensive technology or opt to deploy certain household energy systems (E). A variety of stakeholders can attempt to influence such behaviour (F) through educational programming.

At the local government level, the most significant policymaking powers relate to those determinants of urban form known as zoning rules and building codes (G). Zoning rules have long been the province of local officials (Johnson 1996), giving policymakers the capacity to determine building density and size in different neighbourhoods. Building codes are important because they dictate a variety of energy-related standards, such as how much insulation is required in a building, the minimum energy efficiency rating of windows, whether renewables must be deployed on site, etc. Local government may be limited in its ability to develop a city-specific building code, as such codes are generally developed at the state or national level and these tiers of government will dictate whether localities can impose their own version on local building projects.[29]

A comprehensive assessment would require that similar analyses be conducted for transportation-related energy use or building-related energy use in the commercial, industrial or government sectors.[30] Once this basic system description is completed, the real challenge comes in assessing which aspects of the system will deliver the greatest results, and on what time schedule. Actions by some stakeholders will quickly touch most energy users in a city, while actions by others may deliver longer-term or geographically targeted results.

It is this more detailed analysis that ultimately helps us understand whether the current policymaking and market system is logically structured, and whether responsibilities should be devolved or shifted in a way that enhances society's ability to manage local energy use. Politics may intervene to make such change difficult to achieve, but this knowledge is helpful regardless of whether we seek to effect change for economic, energy security, or environmental reasons.

---

[28] This analysis intentionally goes beyond this chapter's narrow focus on renewables to explicate the full range of policy and program powers relevant to residential energy use.

[29] New York City's first building code guidelines date back to 1647, when the city was known as New Amsterdam. Over the years, these requirements have grown in detail, with the approval of state government, a recognition that the city's dense built environment brings with it unique health and safety considerations.

[30] Buildings and transportation are generally considered to be the two primary categories of energy use. Power plants and on-site heating and cooling systems are seen as necessary to meet the needs of occupants of these buildings, so energy consumed to satisfy the power and thermal demand is allocated to residential, commercial, industrial, or government sector buildings.

Table 6.3. Capacity to Act analysis of the residential energy sector in New York City.

Policy control/ program function	City of New York	State of New York	Federal government	Private sector	Households	NGOs etc.
Energy supplier	– prohibited by state law from owning generation capacity	– sells power to NYC Housing Authority via New York Power Authority (state utility)	C	– Con Edison – ESCOs – Power plant owners	– install self generation power systems	
Regulatory control **G**	– zoning rules – building codes – power plant siting	– all aspects of utility regulation (rates/ emissions/ interconnections) – emission standards – power plant siting – technology and fuels standards	– emission standards – power plant siting – technology and fuels standards D		A	
Procurement	– demand aggregation – green power from NYPA for city-run residential facilities	– fuel choice decisions	B	– fuel choice decisions – demand aggregation	– select power supplier – make decisions over household energy usage – technology decisions	
Advocacy/ education	– lead by example – info on technology – info on local rules/installers/ manufacturers	– lead by example – info on technology – info on local rules/ installers/ manufacturers	– lead by example – info on technology – info on local rules/ installers/ manufacturers	– lead by example – info on technology – info on local rules/ installers/ manufacturers	– lead by example – info on technology – info on local rules/ installers/ manufacturers	– lead by example – info on technology – info on local rules/installers/ manufacturers
Research and development/ policy research	– local energy policy research	– state energy R&D programs	– federal R&D programs F	– technology R&D		– policy research
Landowner/ developer/ operator	– technology decisions – building design decisions – operating practice decisions	– technology loans – subsidies/ tax credits – energy efficient mortgages	– subsidies/tax credits	– energy efficient mortgages – technology decisions – building design decisions – operating practice decisions	– technology decisions – building design decisions – operating practice decisions	

References

Barcelona Energy Agency (2004). *Ordinance on the Incorporation of Solar Thermal Energy Collection in the Buildings.*

Battery Park City Authority (2002). *Hugh L. Carey Battery Park City Authority Commercial/Institutional Environmental Guidelines.*

Battery Park City Authority (2003). *Hugh L. Carey Battery Park City Authority Residential Environmental Guidelines (Version 4).*

Benson, M. (2002). Light switch: as woes mount for utilities, cities try to take charge. *Wall Street Journal*, **4** November, A1.

Berkhout, F. (2002). Technological regimes, path dependency, and the environment. *Global Environmental Change*, **12**, 1–4.

Bressi, T.W. (ed.) (1993). *Planning and Zoning in New York City.* Center for Urban Policy Research, Rutgers University.

Brook Lyndhurst (2003). *Attitudes to Renewable Energy in London: Public and Stakeholder Opinion and the Scope for Progress.* London Renewables.

Bulkeley, H. (2005). Reconfiguring environmental governance: towards a politics of scales and networks. *Political Geography*, **24**, 875–902.

Bundred, S. (2006). Solutions to silos: joining up knowledge. *Public Money and Management*, **26**(2), 125–130.

Capello, R., Nijkamp, P. and Pepping, G. (1999). *Sustainable Cities and Energy Policies.* Springer-Verlag.

Center for Sustainable Energy (2007). *New York City's Solar Energy Future (Part II) – Solar energy policies and barriers in New York City.* City University of New York (Bronx Community College).

City of New York (2005). *Local Law 38 – Cleaner Vehicles.*

City of New York, 'About plaNYC2030: 10 key goals' (2006). Retrieved 20 December 2006 from http://www.nyc.gov/html/planyc2030/html/about/10-goals.shtml

Columbia University Workshop on International Energy Management and Policy (2006). *Powering Forward: Incorporating Renewable Energy into New York City's Energy Future.* Columbia University School of International and Public Affairs.

Con Ed Solutions (2005). New York City makes largest purchase of 'green power' in New York City Government history [press release 9 February 2005]. Retrieved 15 May 2005 from www.conedsolutions.com/news_events/pressreleasFeb92005.htm

Con Edison (2002). *Specification EO-2115 (Revision 5): Handbook of General Requirements for Electrical Service to Dispersed Generation Customers.*

Con Edison (2006). *Specification EO-2115 (Revision 8): Handbook of General Requirements for Electrical Service to Dispersed Generation Customers.*

Considine, A. (2006). Green Tech: Is that a tinge of green on New York's Yellow Cabs? *New York Times* 21 May, B1.

Eilperin, J. (2006). 22 cities join Clinton anti-warming effort. *Washington Post*, 2 August, A3.

Energy Task Force (1977). Windmill Power for City People: A Documentation of the First Urban Wind Energy System.

ETSU/AEA Technology (2001). *Development of a Renewable Energy Assessment and Targets for London: Final Report to Government Office for London and the GLA (Volume 1 – Main Report).*

Government Office for London (2004). Roles in Planning and Development Control – the London Boroughs, the Mayor of London and the Secretary of State. Retrieved 5 June 2004 from http://www.go-london.gov.uk/planning/land_use_planning/rolesindevcontrol.asp

Greater London Authority (2002). *The Mayor's Draft Energy Strategy: Assembly and Functional Bodies Consultation Draft.*

Greater London Authority (2003). *Green Light to Clean Power: The Mayor's Draft Energy Strategy.*

Greater London Authority (2004). *Green Light to Clean Power: The Mayor's Energy Strategy.*

Greater London Council (1978). *Energy Policy and London.*

Greater London Council (1981). Energy Use in London (*Statistical Series: No. 10*).

Greater London Council (1982). *Electricity Supply in London: Part 1 – A Review (Reviews and Studies Series: No. 13)*.

Greater London Council (1983). *The Energy Economy – Building Employment by Investing in the Rational Use of Electricity (Strategy Document No. 5)*.

Greenhouse, L. (1977). State tells Con Ed to buy 2 kilowatts – from a windmill. *New York Times* May 6, 1.

Gross, S. (2006). Taxi medallions for 'green' vehicles receive record bids. *Staten Island Advance* 22 June.

Hall Hayes, K. (2005). Possible disasters at LNG site spelled out. *The Providence Journal* 14 April, E4.

Hammer, S. (2005). *Urban Policy for Renewable Energy: Case Studies of New York and London*. PhD thesis, London School of Economics and Political Science.

Hebert, H.J. (2005). Potential of catastrophic fire from terrorist attack worries LNG opponents. *Associated Press*, 19 January.

Herman, E., Saltonstall, D. and Siemaszko, C. (2003). City hurt, but it's back in business; loss estimated range past $1b. *New York Daily News*, 19 August, 5.

Hirsh, R.F. (1999). *Power Loss: The Origins of Deregulation and Restructuring in the American Electric Utility System*. MIT Press.

HMSO (1999). *Greater London Authority Act 1999 (Chapter 29)*. Her Majesty's Stationery Office.

HMSO (2000). *Town and Country (Mayor of London) Order 2000*. Her Majesty's Stationery Office.

Hu, W. (2005). Mayor approves construction rules. *New York Times*, 16 September, B5.

Hughes, T.P. (1983). *Networks of Power: Electrification in Western Society 1880–1930*. Johns Hopkins University Press.

ICLEI (2002). Permits and Ordinances for Wind Development. Retrieved 1 December 2002 from http://www.greenpowergovs.org/wind/Oakland%20wind%20permit.html

ICLEI (2003). ICLEI Case Studies. Retrieved 18 February 2003 from http://www3.iclei.org/iclei/casestud.htm

Johnson, D. (1996). *Planning the Great Metropolis: The 1929 Regional Plan of New York and its Environs*. E&FN Spon.

Jones, A. (2005). *Local Sustainable Community Energy*. London Climate Change Agency.

Kayden, J. New York City Department of City Planning and the Municipal Art Society (2000). *Privately owned public space: the New York City experience*. John Wiley & Sons.

Knowles, R. (2003). The solar envelope: its meaning for energy and buildings. *Energy and Buildings*, **35**, 15–25.

London Borough of Southwark (2004). *Delegation of award of energy supply contracts* (meeting notes).

London First (2002). *Getting the best from London's roads*.

London First (2003). *The Mayor's Draft Energy Strategy: The London First Response*.

London First (2004). Sustainable Development Initiatives. Retrieved 29 December 2004 from www.london-first.co.uk/improving_london/initiatives.asp?L2=85

London Remade/London First (2003). Towards a Sustainable London: Reducing the Capital's Ecological Footprint (Phase 1 Report).

London Renewables/London Energy Partnership (2004). *Integrating Renewable Energy into New Developments: Toolkit for Planners, Developers and Consultants*.

London Research Centre (1993). *London Energy Study: Energy Use and the Environment*.

Loveland, I. (1999). The Government of London. *The Political Quarterly*, **70**(1) , 91–100.

Lyman, E.S. (2004). *Chernobyl on the Hudson? The Health and Economic Impacts of a Terrorist Attack at the Indian Point Nuclear Plant*. Union of Concerned Scientists.

Muir, H. (2004). Global quest for 'Mr Green' ends just down the road: the London Mayor's new climate change agency has found its leader – the man who made Woking a pioneer of energy efficiency. *The Guardian* 20 October, 10.

Narvaez, A. (1972). PSC backs city on coal-use curb. *New York Times*, January **22**, 32.

Navigant Consulting (2005). Community Choice Aggregation – Base Case Feasibility Evaluation. City of Berkeley.

New York Building Congress Energy Committee (2001). *A Matter of Urgency: New York City's Electric Supply Needs*.

New York Building Congress Energy Committee (2002). *Electricity Outlook: A Call to Action*.

New York City Comptroller (2006). *Report to the Mayor and City Council on City Comptroller Audit Operations Fiscal Year 2005.*

New York City Department of Citywide Administrative Services (2006). DCAS Office of Energy Conservation. Retrieved 14 June 2006 from http://www.nyc.gov/html/dcas/html/resources/dcas_oec.shtml

New York City Energy Policy Task Force (2004). *New York City Energy Policy: An Electricity Resource Roadmap.*

New York City Energy Policy Task Force (2006). *NYC Energy Policy Task Force 2006 Status Report.* NYC Economic Development Corporation.

New York City Energy Policy Task Force, 'Status Report' (2004). Retrieved 14 May 2005 from http://www.newyorkbiz.com/About_Us/eptf2004statusreport.pdf

New York State Department of Public Service (2004). *CASE 03-E-0188 – Proceeding on Motion of the Commission Regarding a Retail Renewable Portfolio Standard. (Order regarding retail renewable portfolio standard.)*

New York State Public Service Commission (1991). *Case 90-E-1119 (Proceeding on Motion of the Commission to Consider Establishing Standards on Reliability and Quality of Electric Service) – Order Adopting Standards on Reliability and Quality of Electric Service (Issued and Effective July 2, 1991).*

New York State Public Service Commission (2002). *Case 99-E-0930 (Proceeding on Motion of the Commission to Investigate the July 6, 1999 Power Outage of Con Edison's Washington Heights Network) – Order Approving Settlement (Issued and Effective November 13, 2002).*

NYC Office of Environmental Coordination, 'Greenhouse Gas Inventory' (2006). Retrieved 20 June 2006 from www.nyc.gov/html/oec/html/sustain/green_gas_inv.shtml

NYISO (2001). *Power Alert: New York's Energy Crossroads.* New York Independent System Operator.

NYISO (2002). *Power Alert II: New York's Persisting Energy Crisis.* New York Independent System Operator.

Owen, D. (2004). Green Manhattan: Everywhere should be more like New York. *The New Yorker*, 18 October, 111.

Quiniones, G. (2005). *Testimony of Gil Quiniones, Senior Vice President, New York City Economic Development Corporation before the City Council Environmental Protection Committee (February 15, 2005).*

Quiniones, G. (2006). *Presentation by Gil Quiniones, New York City Economic Development Corporation – NY Academy of Science Panel Discussion (Energy System Research & Innovation: New York City Electricity – Past, Present, Future).* New York Academy of Science.

Read, H.J. (1998). *Defending the Public: Milo R. Maltbie and Utility Regulation in New York.* Dorrance Publishing Company.

Resource Insight Inc. (2003). *City Energy Plan for the City of New York (working draft).*

Roberts, S. (1987). Debating the right to the sunshine and a solar grid. *New York Times* 1 June, B1.

San Francisco Department of the Environment (2002). *The Electricity Resource Plan – Choosing San Francisco's Energy Future.*

Sassen, S. (2001). *The Global City* (2nd ed.). Princeton University Press.

Satow, J. (2007). NYC developers resist the push to go green. *Crain's New York Business*, 29 January, 2–3.

Sawer, P. and Bar-Hillel, M. (2004). Red Ken goes green. *The Evening Standard*, 26 January.

Schaller Consulting (2002). *Commuting, Non-Work Travel and the Changing City.*

Smith, C. (2007). Uncool New York: this city is behind several others – even Chicago! – in its efforts to combat global warming. Mayor Bloomberg can do much better. *New York Magazine*, 22 January, 24–26.

SustainLane (2006). The SustainLane 2006 US City Rankings. Retrieved 29 September 2006 from http://www.sustainlane.com/article/895

Sustainable GLA Coalition (2001). *Making London Work: Gaining Economic and Social Advantage from a Better Environment.* Forum for the Future's Sustainable Wealth London project.

Travers, T. (2004). *The Politics of London: Governing an Ungovernable City.* Palgrave MacMillan.

Thornley, A., Rydin, Y., Scanlon, K. and West, K. (2002). *The Greater London Authority: Interest Representation and the Strategic Agenda (Discussion Paper #8).* LSE London (Metropolitan and Urban Research).

UNCED (1992). *Agenda 21 – Action Plan for the Next Century.* United Nations Conference on Environment and Development.

Unruh, G.C. (2000). Understanding carbon lock-in. *Energy Policy*, **28**, 817–830.
US Census Bureau (2000). H041. *Tenure by Vehicles Available – Universe Occupied Housing Units (Data Set 2000 Supplementary Survey Summary Tables)*.
Wiser, R., Bolinger, M. and St Clair, M. (2005). *Easing the Natural Gas Crisis: Reducing Natural Gas Prices through Increased Deployment of Renewable Energy and Energy Efficiency*. Environmental Energy Technologies Division, Ernest Orlando Lawrence Berkeley National Laboratory.

Chapter 7

Climate Change and Cities: The Making of a Climate Friendly Future

SHOBHAKAR DHAKAL

Executive Director, Global Carbon Project, National Institute for Environmental Studies,
16-2 Onogawa, Tsukuba, Japan 305 8506

Over the last few decades the world experienced urbanization to an unprecedented extent. The urban development path adopted by urbanizing regions, more specifically by cities, is a crucial factor in determining the scale and choice of energy use and the accompanied emissions of local air pollutants and greenhouse gases. Such pathways also impact on urban carbon sinks and play a significant role in determining the vulnerabilities of cities to the impacts of climate change. The connections between cities and climate change are not straightforward; understanding the drivers that explain such pathways is important to finding appropriate solutions. Above all, governing climate change at the city level and implementing measures for climate change are daunting tasks given that cities are primarily focused on local issues. However, climate change is already occurring and there is an urgent need to clarify and communicate how cities in developing and developed countries can contribute to lessening the causes and impacts of climate change from their respective levels.

7.1 Examining the Connections

Cities and climate change interlinkages manifest themselves in several forms. For a long time, the focus of such interactions in climate policy debate has been one-sided, i.e. the impact of cities on climate through emissions of energy-related greenhouse gases. This did not provide a complete picture, especially since climate change has been scientifically confirmed as already occurring: cities are increasingly affected by the physical impact of climatic change such as the heightened frequency of adverse climate effects – flooding, hurricanes, infrastructure damage, etc. – and the socio-economic impacts of such consequences. In addition to the implications of impacts of physical climatic changes, cities will be increasingly affected economically, socially and environmentally, though indirectly, due to the requirements of international climate regimes, increasingly low emission technologies, growing carbon markets, technology transfers, and financial mechanisms. Such interactions are shown in Fig. 7.1.

From a natural science perspective, cities affect climate through changes in a number of other parameters in addition to emissions but these are largely not considered in most of the literature dealing with urban climate change issues. However, this is not surprising

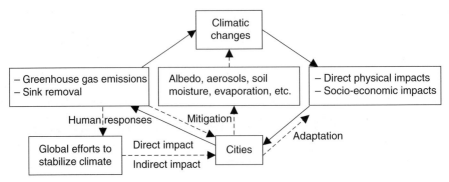

Fig. 7.1. Cause/impact interactions between cities and climate change.

as such studies at urban levels are largely led by scholars from the energy-environment field. In reality, changes in urban vegetation can affect the amount of carbon sequestration by a large city. Of all organic carbon pools in cities, perhaps trees are the most intensively studied because of their links to other concerns such as urban heat island, air pollution and urban aesthetics. McPherson *et al.* (2005a–c) show that in six US cities – Fort Collins, Berkeley, Boulder, Bismarck, Cheyenne and Minneapolis – the amount of carbon dioxide absorbed by each city tree – whether in streets or parks – is in the range of 79 to 107 kg annually. Urban trees not only sequester carbon but also lower the cooling burden of air conditioners and thus reduce emission of CO_2 related to electricity production (Akbari 2002; Taha 1997; Dhakal 2002). The heat discharged by air conditioning equipment is a significant outcome of urban energy use in many cities. In Tokyo, studies show that such urban energy uses are responsible for a maximum of about 3.4°C changes in temperature in downtown Tokyo on a typical summer day (Dhakal and Hanaki 2002; Ichinose *et al.* 1999). Apart from explaining cities' links to climate change through emissions and seques-trations, radiative forcing-based examinations of such linkages are generally ignored. Cities – and urbanization in general – boost the emission of aerosols, and alter the carbon balance in soil, evaporation, moisture and albedo value (the degree of surface reflectiv-ity) that affect radiative forcing. Cities critically influence the dynamics of climate change through many such factors outside the realm of greenhouse gas emissions. Although important, these are largely ignored.

However, cities are often identified as sources of climate change problems in broader terms. Fossil fuels are reported to supply 80.3% of global commercial energy (IEA 2006) and a large share of this is used to support urban areas. In OECD countries, cities con-sume between 70 to 80% of fossil fuel nationally (OECD 1995). Svirejeva-Hopkins *et al.* (2004) mention that 'although the total area of urbanised territories is relatively small (2% of the global land area in the 1990s) (Grubler 1994), they play a more important role in the carbon cycle because these territories emit up to 97% of all anthropogenic carbon emis-sions (Grubler 1994) and urbanised territories transform the structure of the local carbon flows over considerably larger areas than what they occupy'. We doubt such high share of emissions by urbanized territories but regardless of the accuracy of such ambitious number, it highlights the key role that cities play through their emissions. We should raise the subsequent question then: are rural areas doing any better? In the developing world, which has a low level of urbanization and a small fraction of people living in cities, the per capita end-use of energy is perhaps higher than in rural areas. Previous studies have already established that income has a strong correlation with energy intensive lifestyles up to a certain level. Thus, pertaining to large income gaps between urban and rural areas, it seems reasonable that cities are responsible for higher end-use and embodied energy use per capita than rural areas. Additionally, fossil energy availability in cities is usually better and thus supply problems do not limit the use of commercial energy, as is more common in rural areas. End-use energy consumption in many rural areas is indeed supply constrained, and often does not involve a 'modern', i.e. fossil or nuclear, form of energy. Being dominated by traditional sources, the end-use efficiency of energy use in rural households is expected to be low and therefore it is unlikely that per capita primary energy use in rural household surpasses urban for that reason. So we accept here that per capita energy use and carbon emissions in rural households are lower because these units tend to use traditional energy sources that largely come from renewable energy sources – biomass – unless they cause deforestation, releasing greenhouse gas and affecting carbon sinks. Since urbanization is rapidly increasing in the developing world, the contribution of cities to the emission of greenhouse gases will further increase due to the greater demands of energy despite continuously improving technologies.

In the developed world, the per capita incomes in urban and peri-urban areas are already higher and energy cost makes a smaller fraction of total disposable income compared to the developing countries. It does not make income elasticity of energy demand as significant a factor as in the developing world. As income plays a lesser role in energy use, population density effect becomes more important. As a result, per capita energy use in cities might be smaller than suburban and other peri-urban areas due to the fact that economy of scale persists in cities due to efficient production and distribution systems, and that people in cities live in smaller houses, commute less by private car, and use public transport more than in rural areas. It is generally accepted that a denser city consumes less energy per capita and thus emits less carbon than a sprawled city if other conditions remain unchanged. However, the extent of density difference matters too because the definition of 'dense' in Asian cities is often different from sprawled North American cities. When it comes to rural areas, a study in Japan shows that per capita total energy use in rural areas is more than in urban areas (e.g. Ichinose *et al.* 1993), primarily due to excessive automobile dependence in rural areas. However, there are arguments that if we consider total energy use, people in cities opt for more services and consume more, similar to those in cities in developing countries. Lenzen *et al.* showed in 1997 that in Australia urban total energy use was higher than in rural or country areas – simply due to higher total consumption, including that of energy.

In the year 2005, about half of the world's population, or about 3.2 billion people out of 6.5 billion people on earth, lived in urban areas, and urbanization rates, particularly in developing countries of Asia and Africa, are rising rapidly.[1] Projections show that urban populations will grow twice as fast as compared to the total population resulting in 4.9 billion people living in urban areas by 2030. This is about 60% of the projected world population of 8.2 billion. That means that 1.8 billion people will be added to urban populations in the next 25 years, of which 1.1 billion alone will be added to Asia (UN, 2006). Such an urban expansion would mean an enormous increase in energy demand due to rapidly rising per capita energy use, brought about by industrialization and rising urban incomes

---

[1] World urbanization rate is projected to reach 50% by 2008. Asia hosts the largest urban population of 1.6 billion, Europe 0.5 billion, Africa 0.3 billion, North America 0.3 billion, and others 0.4 billion. China, India and the USA have the largest urban populations as countries (UN 2006).

and consumption, and increases in the urban population itself. It is very likely that all gains in technological efficiency will be overshadowed by the very scale of energy use in cities. This puts an enormous pressure on efforts to control greenhouse gas emissions through the introduction of renewable power sources, while preserving the local environment and natural carbon sinks. As mentioned earlier, urban areas already consume large amount of global commercial energy and are therefore likely to be responsible for a comparable share in fossil fuel-related CO_2 emissions. Studies show that cities embody considerable amounts of energy in their buildings and infrastructure, as well as carbon emissions due to the flows of goods and services that are consumed in the city. A large city's actual carbon responsibilities are hence much higher than they appear from accounts of direct energy use or direct carbon emissions, say from electricity use or transport. Urban areas are among the most significant systems to be considered when developing a human world that is low in carbon dioxide emissions. This does not mean that we should or indeed can turn away from cities. Historically, cities have played a crucial role in human modernization, industrialization, innovation, and technology development, and urban communities are key to finding solutions to a variety of human problems. They should be viewed as places that offer great economic and socio-cultural opportunities to mitigate emissions. The denser settlement in cities provides a great opportunity to create an efficient production and distribution infrastructure, make public transportation feasible, and encourage people to use efficient technology in addition to allowing the conservation of energy relating to shelters, lifestyle, mobility and other key human needs (Lebel *et al.* 2007).

Urban greenhouse gas emissions are generated by three kinds of activities. The first type of activity produces *direct emissions* within the physical boundary of the city. Coal burnt in boilers in the city, natural gas burnt in homes, gasoline or diesel combusted by vehicles are only a few of such examples for direct emission sources. Such emissions generally result from the use of fossil forms of energy sources such as coal, gas, oil – but not electricity – and activities from non-energy sectors in the city such as methane emanating from urban rice plantations, or the collection, transfer, treatment or combustion of solid waste, and other activities. The second activity type, producing *indirect emissions*, encompasses those types of energy use in the city where energy is used inside a city with the actual emissions taking place outside its boundaries. Most electricity generation, in particular, still involves primary energy sources such as coal, oil or gas being burnt outside a city's boundaries. In addition, emissions from energy are often accounted for from the perspective of end-use without considering the energy lost in transportation and conversion, or from the *primary energy supply side* after accounting all such losses in the long chain of energy production, conversion and distribution. In practice, such direct and indirect emissions are well accepted in policy discussions, policy design, and policy instruments in many cities. However, the question of how to evaluate the responsibility of a particular city and for what we should make the city accountable remains to be answered. Cities have ecological 'footprints' that are many times their size – meaning that the city survives due to goods and services, in both the physical and ecosystem sense, from outside of the city – and a large waste stream requires massive ecological space to be absorbed. Similarly, a large volume of emissions takes place outside of the city to produce the goods and services consumed by the city dwellers, and large areas of forest and oceans are relied on as emission sinks. The question of how to account for this third type of activity when estimating greenhouse gas emissions is unresolved and has not yet reached the point of being widely acknowledged or resolved in policy discussions, let alone international agreements. If we consider such *embodied emissions*, the per capita emission of greenhouse gases allocated to a city will likely increase significantly; it can increase several times if the city is big and well developed but resource-scarce with a predominance of service and commercial activities (Kaneko *et al.* 2003). Whether embodied emissions will be considered in any future policy response in cities is a different matter and a complicated one, but consideration of such factors would help to advance cities towards societal development that is not as focused on the overconsumption of materials, which in itself has multiple benefits. In addition, this concept would aid in moving towards the development of the so-called *Sound Material-Recycle Society* that the Japanese government has been touting of late, enhance the concept of a *Circular Economy in China*,[2] and assist in the conservation of natural resources and friendly material substitution policies in Europe (MOEJ 2006b; Yuan *et al.* 2006).

Cities are also victims of climate change and environmental problems. Historically cities have played a key role in the world's industrialization and faced air and water pollution and other environmental adversities. Through the heat island effect the heat balance in cities is altered, contributing to heat stress, increasing trends of rising smog and poor visibility, raising ozone levels and lowered life expectancy – and growing energy bills for greater cooling demands (Dhakal 2002). Coastal cities face flooding and hurricane problems and their infrastructures are increasingly being compromised in their capacity to protect urban dwellers. For example, the capital of Bangladesh, Dhaka, which is very prone to drainage congestion, periodic flooding and hurricanes, is 2 to 13 metres above sea level with the majority of urban areas only 6 to 8 metres above sea level (Alam and Rabbani 2007). The 1988 flood in Dhaka inundated 85% of the city. In the greater Cape Town area, water demand exceeds total potential yield (Mukheibir and Ziervogel 2007). The downtown part of the city of Mumbai sits on a flood-prone and poorly drained area. The devastating flood of July 2005 in Mumbai that killed more than 1000 urban dwellers was triggered by 944 millimetres of rainfall in a 24-hour period (Sherbinin *et al.* 2007). In Shanghai, severe flooding of the Yangtze River in August 1998 caused 3000 deaths and displaced 16 million people with a total damage estimate of US$36 billion (cited from several sources by Sherbinin *et al.* 2007). In the New York Metropolitan Region, many coastal communities have been identified as being at high risk of sea-level rises and storm surges that affect both affluent as well as low income communities (Gornitz and Couch 2000). Poor, inner city communities are seriously affected by summer heatwaves due to the prevalence of high risk populations such as the elderly poor (Kinney *et al.* 2001). Hurricane Katrina's impacts on New Orleans in the United States in 2005 is a reminder to us that cities are prone to natural disasters irrespective of whether they are located in developed or developing parts of the world and thus future climatic changes are going to increase the burden and exacerbate these already pressing problems in cities.

7.2 Underlying Emission Drivers

What drives emissions of greenhouse gases in cities? Different academic lenses see these drivers differently. System analysts prefer to identify and quantify the immediate and proximate drivers, and assign them their respective share of the emissions problem. They typically look at contributions of various energy sectors such as residential, commercial, industrial, and transport sectors and the energy mix. Some follow broad aggregation methods using the Kaya Identity – emission is a product of a number of a decomposed factors (see Box 7.1). Economists see the energy use and emissions in income linkages, elasticity

---

[2] China, learning and inspired by Japanese and German recycling-related initiatives, is promoting a Circular Economy (CE) initiative.

Box 7.1.
Different approaches to explaining emission drivers

Emissions = sum (sectoral and subsectoral disaggregations)
Emissions = f (emission/energy, energy/income, income/person, population)
Emissions = f (economic growth, price signal, externality handling, market
     mechanism)
Emissions = f (population, organization, environment, technology, institutions,
     culture)
Emissions = f (mobility, shelter, food, lifestyle)

---

of demand, market role, price signals, drivers that act in such a market, and the handling of externalities. Social scientists see such drivers from a broader perspective, based in the evolution of the social system itself (Canan and Schienke 2006; Scholz 2006). To help create a balanced understanding, we mention a few major drivers propounded by these groups that reasonably describe the key drivers following the author's previous work (Dhakal 2004) on such drivers in Asian megacities.

Urban demography: Urban demography can affect emissions through size, growth, composition and distribution of population. While urban population growth has levelled off in many developed countries, and indeed is declining in some European cities, it continues to rise in the rapidly developing and least developed cities in the developing world (UN 2006). High population growth means bigger population size, and such population size, combined with other factors, becomes a direct measure of the emissions volume of a city. However, since household size has an important bearing on energy uses, such that per capita energy use in smaller size households is expected to be different from bigger size households, the number of households provides a better indicator than population itself in a city. In Japan, the 2006 Annual Report of the Ministry of the Environment shows that per capita energy use in a one-person household is about 17 GJ/person/year in comparison to about 11 GJ/person/year in a four-person household (MOEJ 2006a). Similarly, it shows that the amount of household waste, which is a good proxy of consumption in a one-person household, in 2003 was 941 grams/person/day compared to 440 grams/person/day in a four-person household in Kawasaki City. We do not have empirical evidence to show that it prevails only in the same income groups or across different income groups, or whether this is due to the fact that smaller households tend to be more affluent and hence more wasteful or if it is due to the scale effect that leads to reduced per unit impact.

When it comes to population composition, the exact impact of different shapes of urban population pyramids[3] to emissions is largely unknown, and how different households with different age structures use energy differently in various settings is also an open question. At the national scale, developed countries have a higher share of senior citizens – the aging society syndrome – but whether this can be generalized to cities is unknown because young people often move to cities for better opportunities, leaving an increasingly aging population in small towns and the rural areas. Japan is one example of this trend. In general, in developed cities, senior citizens are expected to use more energy per capita and in a society

---

[3] A distribution of male and female population by age groups with population in X-axis and age group in Y-axis.

of nuclear families it might be easier to see that effect than in others. More than age structure, the population difference from daytime to the night-time has a well-established bearing on emissions. The Tokyo population, for example, has stabilized since the early 1970s, and the population of the 23 wards constituting downtown Tokyo is decreasing, but Tokyo attracts one third of its total workforce from surrounding cities and prefectures utilizing its well-developed surface rail and subways. The ratio of daytime to night-time population was 1.25 in 1999 in Tokyo; in the 23 wards, the ratio is as high as 1.41 (TMG 2000).

Economic development: As mentioned above, economic growth has a very clear relation to emissions because higher growth usually requires more use of energy, and post-fossil fuel sources are still vastly underused. Economic growth has been regarded as the most influential factor for the rise in emissions in Tokyo, Seoul, Beijing and Shanghai over the last three decades (Dhakal 2004). It is also a key driver in Mexico City, Mendoza, Buenos Aires, and Santiago (Romero Lankao *et al.* 2005). However, the extent of the influence of economic growth on emissions depends also on the structure of the economic activities prevailing in a city. If a city is industrial in nature and is dominated by primary and secondary industries, the emission intensity of the economic activities will be higher such as in Beijing, Shanghai, Ho Chi Minh and so on. However, if the city is heavily commercial in nature, dominated by service industries, the direct emission intensity of its economic activities is lower such as in Tokyo, and perhaps Seoul. Does slow or negative economic growth result in slowing or reducing emission growth to the same extent as income does? We do not have published data to answer this question with quantifiable precision but studies in Tokyo and Seoul have shown that there are time lags between the slowing of economic activities and the slowing of emissions, and the extent of reductions in emissions may be smaller than the extent of reduction in economic growth (Dhakal 2004). In other words, within commercial or service-oriented cities, economic failure may not lead to an equivalent drop in emissions although its indirect impact on the embodied carbon stream would be larger because economic activities in such cities are not energy intensive. Along the same lines, it remains to be seen whether a rise in economic growth leads to a proportional rise in emissions, but it is assumed that this will not be the case since technological improvements and diffusions lessen environmental impacts over time. The learning effect prevails not only for technology but for society at large. Decoupling economic growth from emissions is a key issue among scientific and policy communities worldwide and this debate needs to be accelerated in the attempt to find options that will lead to low carbon societies.

Infrastructure and technology: Energy demand itself is use derived; urban residents do not demand energy directly but they procure and use services such as lighting, heating, cooling, motive power and so on. In the energy literature, a clear distinction has been made between service demand, useful energy demand, end-use energy demand, secondary energy demand and primary energy demand. There are a number of different pathways through which a particular service demand can be met. Since the emission impacts of different energy pathways are different, choice of such pathways determine the extent of emissions. Dhakal (2004) describes one such example in which the per capita energy consumption of Tokyo, Beijing, Seoul, and Shanghai are converging over time but their per capita emissions are diverging due to rising energy use, improvements in energy efficiency, and the coal dependency of the energy system in Beijing and Shanghai (Fig. 7.2). Urban physical infrastructure and technology play crucial roles in shaping such alternative emission pathways. They are related to energy supply and choice, buildings, modes of mobility, and energy and process efficiencies of technologies in household, commercial, transportation and industrial sectors in a city. Tokyo has lower CO_2-equivalent per capita emission levels

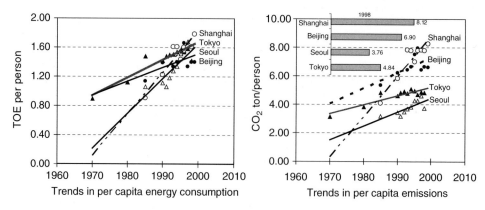

Fig. 7.2. Per capita energy and emissions in Tokyo, Beijing, Seoul and Shanghai.
Source: Dhakal (2004).

than Beijing and Shanghai, due to its efficient urban infrastructure, greater reliance on somewhat lower emitting sources, notably nuclear power, natural gas and hydro-electricity, and more efficient end-use technology, in addition to other factors (Dhakal 2004). If a city draws energy from coal, the city does not perform well in terms of emissions. A city that heavily relies on private cars and has an underdeveloped or underutilized mass public transport system is not emission friendly. In commercial and residential buildings, the insulation of the buildings, the efficiency of heating and cooling equipment, the type of lighting, and the efficiency of appliances determine the extent of energy use. Urban residents' desire to take services from an efficient infrastructure leads to low emission pathways. In many cities, the rate of private motorization is becoming an increasingly serious problem. In denser and bigger cities in the developing world, industrial relocation is pushing primary and secondary industries beyond the city limits over time and thus motorization is outpacing other sectors for emissions (Dhakal and Schipper 2005). Continuing economic growth, rising motorization and the lack of efficient and adequate public transport systems are causing a bleak emissions outlook for rapidly developing cities in Asia, Latin America, and Africa.

Urban form and function: Urban forms are an important determinant of emissions. City size, density, shape and distribution of functions affect emissions through setting the conditions under which infrastructure and urban services are demanded, provided and consumed. The compactness of urban settlements influences the demand for energy for transportation and other areas such as district heating and cooling using co-generation systems. Asian cities, in contrast to European and North American cities, are continuously getting denser while expanding – starting from a less dense base. Because many North American metropolitan areas are low in density and sprawled over large areas they require large amounts of energy to run their transportation and distribution systems; their public transportation is not cost effective. Urban sprawl, in which low density suburbs depend on extensive distribution systems, undermines efficient energy use. Mixed land use results in different energy use than does segregated land use. Urban zoning and industrial relocation from city centres to peri-urban areas in cities significantly influence travel demand and energy use. A combination of mixed land use and high density self-reliant settlement clusters served by efficient mass transportation is generally regarded to help make a city emission reducing, particularly if further opportunities existed to use non-motorized modes, and pedestrian friendliness in those clusters was assured. In essence, determining precisely which urban forms are emission friendly is a difficult question to answer. Given the widespread failure of cities to successfully control development through planning across many developing countries, identifying what opportunities do exist to restructure urban forms and functions is a critical challenge.

Behavioural and societal factors: Cultural and social contexts shape the behaviour of urban residents towards carbon emission reducing pathways, in combination with the supply-side conditions, such as infrastructure, technology, urban forms, etc., that are imposed on the urban system. Choosing cars with smaller engines, opting for fuel efficient cars, rational use of energy (switching lights off when not needed in homes, using air conditioning and heating moderately), maintaining greenery around houses, combating mass consumerism by living a more self-sufficient life are results of behavioural changes that are subject to public policy and civic leadership and affect energy use and emissions levels positively. One example of behavioural dogma in need of adjustment is the fact that a car is promoted and perceived as a symbol of social class and lifestyle status in many developing countries, and typically associated with city life and achievement. How precisely such behaviour is guided is little studied today, and what actions are likely to induce positive behavioural changes are equally difficult to ascertain. By and large, the prevailing social value system affects behaviour; many public efficiency campaigns bank on the assumption that awareness, education, correct information, and proper incentives can induce positive behaviour.

Globalization: Not many past studies have examined whether rapid globalization is a key factor in emission growth, but it is plausible, even self-evident. Globalization is taking place along economic and cultural fronts and is shaping the conditions under which the physical infrastructure of cities and the behaviour of urban residents are developed. Foreign direct investments and trade agreements affect the location and technology of manufacturing and commercial activities and labour reorganization (Romero Lankao *et al.* 2005). The prevalence of too many multinational corporations and their strong political leverage may create a situation where carbon relevant decisions cannot easily be locally decided or controlled. Romero Lankao *et al.* (2005) show that due to strong lobbying by the automobile industry, emission regulations in Mexico have lagged far behind federal regulations in the United States. In China, individual cities are competing with each other to attract foreign direct investments and are compromising their local environmental conditions and tax policies (Dhakal 2005). Such competitions are also observed in other countries such as Vietnam and India. In addition to economic globalization, global media such as television and the internet increase global connectedness and influence individual choices and behaviour. This is not to say that globalization in itself is either detrimental or beneficial to efforts to stem emissions but also illustrates that it is an important player in a variety of direct and indirect ways.

Institutional and political factors: Beyond any doubt, the role of political and institutional factors is important for emissions as well as mitigation. Not only decision-making in cities but also provincial and national decision-making processes and actors affect emissions from cities because a city is governed in a complex way by many different layers of governments. For urban environmental issues, in most Asian cities, municipal governments manage solid waste while only in relatively big cities in a few countries do municipal governments substantially manage air pollution (Dhakal 2005). Who makes the policies? Are the policies well consulted and applied? Are policy instruments reasonably robust? Are the implementing authorities well resourced? These questions are important. Decentralization of responsibilities to local agencies without adequately developing their capacity only defers the responsibility. Therefore, a multi-scale style of governance with clearly defined institutional mechanisms is propounded for issues like greenhouse gas emissions (Gustavsson *et al.* 2006; Bulkeley and Betsill 2005).

Natural factors: We cannot discount the natural phenomena of cities. Climate factors directly affect energy use, and thus emissions, due to greater demand of heating or cooling services. A city in the tropics uses more air conditioning energy. High latitude cities in Northern Europe, North America, China, and Mongolia, for example, need more heating energy during the winter than cities in temperate climate zones. A coastal city could be more vulnerable to the impacts of climate change than others.

Above, we discussed many drivers that influence emissions from a city. The impacts of some of these drivers on emissions are well established while others are not clear. In reality, a number of such drivers act collectively on an urban system and we often tend to miss existing inter-relations among drivers and rebound effects of interventions in those drivers in a larger context. This is one reason why devising policies for urban carbon management becomes complex. In addition, we do not know clearly what combination of these drivers applies to different city types – by size, function, geography, income, etc. – and what different clusters of drivers lead to a particular defined emission pathway. This leads us to the next point, which is to identify which of these drivers is most important. We leave this unanswered question to future researches through further examination of the facts.

Before we conclude this section, it is important to discuss the complexities involved in comparing carbon emissions among cities. It is not possible to say that one city is doing better than another simply by comparing the volume of carbon dioxide emission given their local situations. Emissions per capita is a better indicator than emissions alone because it internalizes some of the equity debates. This chapter proposes a framework of indicators to assess a city's carbon policy friendliness through the collective effect of the rate of change in emissions per unit economic activity and the rate of change in emissions per capita scaled by appropriate climate factor (Fig. 7.3). In Dhakal *et al.* (2002), we outlined a graph where the X-axis is emissions per capita over time and the Y-axis is emission intensity of economic activities over time plotted for a few Asian megacities. The direction, vertical gradient and horizontal gradient of the graph over time show the performance of a city over time. However, we could not scale for climate factors at that time. While this

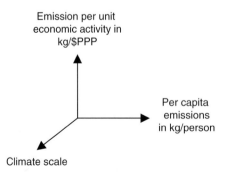

Emission per unit economic activity in kg/$PPP

Per capita emissions in kg/person

Climate scale

What to note in the figure:
– General direction of graph (towards right: unfavourable in general; towards left: favourable; downward: favourable; upwards: unfavourable)
– Vertical gradient of graph (speed of carbon intensiveness of economic activities)
– Horizontal span of graph (speed of rise or fall in per capita emissions)

Fig. 7.3. Comparing cities' carbon performance over time.

framework has yet to be tested in quantitative terms, we believe it is a better indicator than others for now.

7.3 Confronting the Challenges of Mitigation and Adaptation

Carbon management is a challenging task. Carbon management is in practice difficult to implement across nations because of the very nature of the climate change problem. Climate change challenges, unlike other local environmental issues, involve delays between emissions and impacts, spatial divides between who emits and who suffers from impacts, and problems taking place on a global level but with the required action having to take place locally. While we have no control over these issues, there are a number of carbon management challenges over which we have reasonable degrees of control, such as: defining the climate change problem appropriately across multiple levels and scales to give wide ownership of the problem, minimizing the conflicts between energy choice and demand with economic development, and engaging in rational international politics devoid of excuses by both developed and developing countries. Unfortunately, more challenges add up when we opt for managing carbon at the city level. These new challenges start right from the definition of a city (administrative/political, agglomeration) to its spatial and functional 'reach' across various scales (temporal, spatial, institutional) and the levels of those scales. Internal dynamics within a city as well as the external dynamics of city in a regional context make a complex web of difficult-to-understand cross-scale linkages involving trade, energy, material, mobility and services. As a result, it is a daunting challenge to 'govern' carbon in urban areas: this requires us to answer few seemingly simple but actually complex questions. Who should govern – municipal authorities, national or state governments, everybody or nobody? Who is responsible and who should be managed – producer or consumer? What are the temporal dimensions of such management? Can cities do it locally? Accordingly, Box 7.2 highlights some of the key challenges for managing carbon at an urban level.

Box 7.2 only shows a list of generic challenges that urban decision-makers are facing (Bai 2007; Lebel *et al.* 2007; Dhakal 2004; Bulkeley and Betsill 2005). The combination of such challenges differs from one city to the next. We recognize that it is premature to categorize cities into certain groups for similar policy approaches and that there are no such clear distinctions. But broadly, we can characterize cities into three groups that can be used to understand what type of approaches might make sense for urban carbon management (see Box 7.3).

Physically, the areas that need intervention for mitigating direct emissions are quite clear; these are the different urban sectors that are responsible for energy use, waste management and those activities that lead to a reduction in vegetation and urban forestry. The transport sector appears to be a 'time bomb' in many large cities in developing countries due to rising incomes and the lack of efficient public transportation systems (Sari 2004). Even in many cities in developed countries the share of public transportation is dwindling over time (Table 7.1). There is no doubt that the generation side should be the first target of any such intervention, and that it should be followed by the consumption or demand side. Such interventions should lead to better energy efficiency, new innovations in the process efficiencies on the production side as well as demand reduction and changes in behaviour and lifestyle on the consumption side. It is often inevitable that the additional cost of such intervention would ultimately be passed down to consumers.

After Russia ratified the Kyoto Protocol, it went into effect in February 2005. Cities in Annex I countries that have ratified the protocol – most developed countries with a few exceptions, such as the United States – are under pressure from their respective national

Box 7.2.
Challenges for urban decision-makers

Scale: The carbon problem has not been defined as a package of related problems across multiple scales (temporal, spatial, institutional, etc.) and levels, nor have the solutions been synergized with existing realities at multiple scales. Institutionally, if carbon management needs multi-level governance such a structure is underdeveloped.

Readiness: Local institutions are not ready to address carbon issues due to low knowledge, awareness, technical capacity, and priority in local institutions.

Fit: Environmental units in local institutions are not designed to address carbon issues in an adequate manner.

Free riding: There are no firm commitments for cities to act except in the case of cities in Annex I countries that have ratified the Kyoto Protocol.

Public response: Except in some world regions, even serious local public pressure is unlikely to introduce carbon into policy agendas given the state of public awareness.

Resources: Local institutions' financial resource base is weaker not only in cities in developing countries but also in many developed countries – appropriate tax reforms, revenue collection, green budgeting and objective decentralization remain challenging.

Jurisdiction: Urban decision-makers' jurisdiction is often limited and their control over policy instruments is minimal hindering decision-making and their capacity to implement their decisions.

Box 7.3.
An indicative sketch of status quo in different cities

- *Cities in less developed countries:* Cities exhibit far less of a carbon emission issue, the capacity of local policymakers is weak, resources are scarce, institutions for urban environmental governance are less developed and capable, involvement of stakeholders is less pronounced, local issues dominate, and implementation of measures is relatively poor.

- *Rapidly developing/industrializing cities:* The capacity of local policymakers is improving rapidly, resources are scarce but starting to build up, local institutions are being built up, local issues dominate but there has been growing awareness of the need to consider newly emerging issues such as global warming.

- *Developed/matured cities:* Cities and local governments are under growing pressure to tackle emerging global environmental issues due to rising local awareness or to fulfil national (legal or voluntary) commitments to international environmental regimes. Many cities have implemented such policies but it is not yet clear how capable these cities are of truly managing their emissions.

Table 7.1. Trends in mode share of public transportation in selected cities.

City	Earlier year	Public transport as % of motorized trips	Later year	Public transport as % of motorized trips
Bangkok	1970	53	1990	39
Buenos Aires	1993	49	1999	33
Kuala Lumpur	1985	34	1997	19
Mexico City	1984	80	1994	72
Moscow	1990	87	1997	83
Sao Paolo	1977	46	1997	33
Seoul	1970	67	1992	61
Tokyo	1970	65	1990	48
Shanghai	1986	24	1995	15
Warsaw	1987	80	1998	53

Source: WBCSD (2001).

governments to mitigate greenhouse gas emissions. In Japan, the law requires prefectures and cities to formulate plans to reduce greenhouse gas emissions (Dhakal 2004). However, Japanese cites have refrained from setting clear numerical goals and their actions are much behind European cities in developing and implementing local climate policies. In Sweden, where the national goal is to reduce GHG emissions by 4% of the 1990 level (Government Bill 2001/02:55), a survey made by the Swedish Society for Nature Conservation in 2004 (Rylander 2005) shows that 62 of the 135 Swedish municipalities that answered the survey had their own local climate goals, and some of them even went further in reducing GHG levels than the national climate goal (Gustavsson 2006). Despite the fact that they are not bound by the Kyoto Protocol, combined with less enthusiasm in federal government, a number of cities in United States have targets and policies for mitigating greenhouse gases and they have even formed alliances with each other. The US Mayor's Climate Protection Agreement has already attracted 330 mayors as of November 2006 and there were about 160 US cities in ICLEI's (International Council for Local Environmental Initiatives') Cities for Climate Protection (CCP) campaign by 2006 (Betsill 2006). Europe has been at the forefront of these endeavours and their cities are much ahead in implementing local climate policies in comparison with other regions. In Europe, some of the limitations for carbon governance stem from often contradictory European Union policy goals in the areas of liberalization and privatization, climate change policy, and thematic strategy on the urban environment (Kern 2006). The financial crisis of many cities in Europe, notably in Germany, has imposed limitations on how far climate policies can go. As a result, despite leading climate agenda politically, Europe has not had an easy time garnering more aggressive political support of municipal leaders than before. At least three key transnational networks have been observed in Europe for climate policy: Climate Alliance (over 1000 memberships), ICLEI's Cities for Climate Change Protection (mainly in the UK and Finland), and Energie-Cités (Kern 2006). In summary, cities in developed countries tend to make comprehensive climate policies locally and in many cases have specific targets. However, there are large differences in climate policy approaches used by these cities and generalization is not possible. In addition, there are a numbers of cities that remain passive.

The way to confront the challenge of mitigating emissions in cities in developing countries is different from that in developed countries because these cities, first of all, do not acknowledge that they are obliged to take serious action. Their per capita emissions are still far lower due to a lower level of economic activity and there are reasonable arguments that they should be reducing emissions from the baseline future scenarios rather than the absolute value of emissions at present. Following an equity argument (Byrne and Yuan 1999), these cities should perhaps be entitled to actually increase their emissions somewhat in the short term but policy measures should dampen the 'rate of increase' in emission. Their priorities are overridden by an urgency of tackling local environmental problems and the need for focusing on economic growth to meet basic human services such as food, shelter, sanitation, education and health. Invariably, cities in developing countries do not have locally developed and implemented climate policies. Rapidly developing cities who feel that addressing climate issues raises their international profile often show climate policy linkages through 'co-benefits' in relation to activities in air pollution and health, transport management and energy efficiency. Cifuentes *et al.* (2001) show that key policies aimed at mitigating greenhouse gases can provide immediate air pollution benefits in Santiago, Sao Paulo, Mexico City and New York City. Dhakal (2006) shows that key local policies to control particulate matters, traffic congestion, and energy saving simultaneously in the transport sector would reduce a large amount of carbon emissions in Kathmandu Valley. While energy-related policy interventions often reduce carbon emissions, one should note that it is not necessarily true that all local actions in transportation management and air pollution actually do reduce carbon emissions. The synergies and conflicts of local actions with greenhouse gas mitigation depend on many factors and require careful evaluation. Many cities in developing countries are interested in embracing climate policies for boosting local revenue sources by generating Certified Emission Credits (CERs) through Clean Development Mechanism (CDM) projects and selling them to the Kyoto-deficient or other intermediate parties. Many urban projects related to energy efficiency, fuel switching, and the waste management sector provide such promises. However, a major urban sector that is key to reducing emissions – urban transportation – is not a very popular sector for Clean Development Mechanisms. CDM strictly follows a project-based approach which imposes many methodological complexities, making the transport sector an unattractive target unless a new type of CDM project, such as a program-based approach, is considered in future Kyoto negotiations.

A number of transnational networks assist cities in the area of climate change in developing countries but they are relatively weak. ICLEI, which has a strong presence in North America, Australia and Europe, is under-represented in developing countries primarily due to the fact that ICLEI is a fee-based organization. Cities in the developing world cannot afford to pay large membership fees. In many instances, a particular city's interest in actively pursuing ICLEI-defined causes is minimal since they are locally preoccupied and the level of their jurisdiction cannot keep up with the agendas of transnational networks. A number of other intercity networks supported by multilateral and international groups and some countries are operational in Asia[4] but they carry out limited capacity-building activities and information sharing. Such networks were established by various interest groups with good motives but are not organically formed with a clear need, a shared cause, ownership and governance.

---

[4]Kitakyushu Initiative for Clean Environment (established by the 4th Ministerial Conference on Environment and Development in Asia and the Pacific in 2000 and supported by the Japanese Government), CITYNET (started by UN organizations and supported by Japan), Asian Pacific Mayors' Summits, etc.

Reducing carbon emissions is not a local environmental priority for all cities but adaptation to the impacts of climate change is a key priority. As mentioned in the first section of this chapter, cities need to adapt to (a) the physical impacts of climate change that take place locally, and (b) the impacts of international environmental regimes and the market forces brought about by such regimes and physical climate change elsewhere. For the first category of such impacts, cities primarily need to adapt in the area of precipitation change – mainly water resources management and urban drainage management, heat stress, and sea-level rise – mainly coastal protection and flood protection, depending on the cities' location. Coastal cities in particular will be vulnerable. In 1990 there were nine costal cities with populations of more than eight million people and rapid urbanization is expected to produce 20 such coastal cities by 2010 (Nicholls 2004). Many coastal cities sit on geologically young sedimentary strata and are prone to subsidence in case of excessive groundwater withdrawal (Nicholls 2004). Adaptation requires improvements in physical infrastructure as well as appropriate institutional arrangements and policy tools. Box 7.4 shows some of these examples. Their aim should be to reduce vulnerability in general by negating anticipatory impacts and improving reactive capacity through emergency preparedness for natural disasters. For the second category of impacts, cities may lose their competitive advantage in economic activities due the accompanied changes in market forces and urban residents may be subject to additional taxes for energy, price hikes and other effects. If a more stringent international regime would be in place after the Kyoto Protocol expires in 2012, it, together with voluntary carbon markets, could affect market forces and technology substantially, although such effects are yet to be seen.

Naturally, coastal cities such as Dhaka, Malé, London, New York, Tokyo and many European cities are more serious about pre-empting such impacts. Casualties due to

---

Box 7.4.
A list of measures for adaptation to climate change occurring locally in cities

Water	*Sea levels*
River basin management	Remove subsidies to sensitive coastal lands
Contingency planning for drought	Buffer zoning of coastal land
Conserve water	Increase marginal height of coastal development
Use of market-based water allocation	
Urban water drainage for flood control	Preserve coastal wetlands
Pollution control	
Improve disaster preparedness	
Heat stress	*Ground subsidence*
Promote cool construction materials	Control ground water withdrawal
Urban greening initiatives	
Building energy management	

Source: By author based on Smith and Lenhart (1996), Akbari (2002), Nicholls (2004) and other sources.

heatwaves in Europe in 2003 and 2006 forced improvements in emergency preparedness. France was severely affected and that resulted in many deaths in 2003. In many cases, such impacts are also used as a rationale to promote policies for emission mitigation. One such example is mitigation of the urban heat island effect in Tokyo which is related to improving urban carbon sinks and using less energy (Dhakal 2003). However, how much adaptation is enough and what it takes to achieve it are questions that are still difficult to answer due to many scientific uncertainties as well as financial realities. However, it is important to note that not all adaptations will necessarily be capital intensive. Despite unclear distinctions between the impacts of climate change and some regional or local effects, cities need to build their capacity in a timely way and implement simple and practical measures to respond to the impacts because scientists are already delivering the message that the impacts of climate change have already been observed and the frequency of such impacts will be increasing in the future.

7.4 Making Cities Climate Friendly

Science is robust and unequivocal in saying that anthropogenic contributions are causing climate change and that adverse impacts of such climate change have already been observed. Given the unabated rise in emissions at present and what is expected in the future, climate change will have an increasingly adverse effect on humans. Cities have a considerable stake in both controlling emissions and lessening impacts and, therefore, have to play a leading role in mitigation and adaptation activities. Cities may not find it easy to deal with climate change – but the rise in awareness in decisions-makers and society at large has been phenomenal in the last ten to 15 years and increased awareness in the coming years will create better opportunities for action. The present challenges lie in two areas: further creating a sound knowledge base, and implementing mitigation and adaptation practices based on what is already known. Better scientific knowledge is a prerequisite for understanding the various causes and impacts of emissions and for implementing sequestration and full cycle carbon accounting following the framework presented in the beginning of this chapter. A number of today's decisions relating to urban forms and functions, transportation, physical infrastructures, energy supply systems, coastal development plans, and water management systems, among others, lock urban systems into a distinct path that is difficult to change for a long time to come. Deferring action is not an option any longer under the guise of accumulating better scientific knowledge and similar reasoning. Incorporating climate change concerns into various urban sectors at an earlier stage of development is essential.

Despite many challenges discussed earlier in this chapter, carbon management in cities plays an important role in global climate policy. Decarbonizing cities has promising prospects given that cities are designed and operated in sub-optimal fashion in relation to emission optimization and climate impacts. Existing prospects in cities to further optimize energy use and supply and a close relationship between a number of key local priority issues – especially urban infrastructures – and mitigation of greenhouse gas emission widen such prospects further. A key need in these efforts is to develop an integrated assessment framework for cities where mitigation, land use changes, and adaptation measures are addressed in an integrated manner and to explore alternative development pathways for cities. Dawson *et al.* (2007) have shown one such example of a framework for London. This also calls for new and innovative approaches to planning that are locally suitable. Roberts (2006) presents one such planning approach – open space management – for the city of Durban in South Africa and an integrated approach that the city is adopting to mitigate emissions, sequester more carbon, and adapt to climate change impacts. Such integrated approaches are very important from the research as well as the actual planning viewpoint but are greatly underdeveloped at this time. Scientific and policy communities need to develop a comprehensive and integrated knowledge base and planning tools. Transnational scientific and policy networks, bilateral donors and multilateral agencies should assist in developing and disseminating such an integrated approach in research and planning and transfer best practices.

Cities in developed countries need a far reaching numerical target-based policy and more and more cities should adopt such an approach. In addition to required actions in cities of developed countries, cities in developing countries cannot shrink from the mitigation of greenhouse gas emissions. Interventions in the area of energy efficiency, energy choice and energy demand reduction provide an easy entry point for these cities to streamline climate policy concerns into local activities. The second largest energy consuming and carbon emitting nation in the world, China, set an ambitious target to achieve a reduction of 20% in the energy intensity of its economic activities in its eleventh five year plan. A large number of such measures deal with cities. Multiple benefits can be ascertained from such activities. Air pollution control, transportation management, urban planning and many other local activities can provide an easy entry point to climate change mitigation through careful planning and in many cases through nominal additional financial investment. As a first step, local decisions-makers need to have a better understanding of the synergies, conflicts and trade-offs between emission mitigation, adaptation measures and local development issues. That would enable them to make a valuable contribution to a more integrated climate policy and the effective climate-change 'proofing' of our cities (McEvoy *et al.* 2006). Tapping co-benefits of climate change mitigation locally requires integrated strategies such as reducing air pollution and carbon emission, waste management and reducing methane leakage, reducing carbon emission from transportation and level of road congestions, etc. Cities in developing countries need multi-level support to streamline climate concerns into local development planning and actions. How that multi-level support is framed globally or nationally is a key to determine the extent to which cities in developing countries will want to be engaged in climate policy issues in the near-term.

Such need for multi-level support fits well with the growing realization for the need for multi-level governance of carbon issues which has long been discussed and, to some extent, practised. Various actors at multiple levels can influence carbon decisions locally – international environmental regimes, multilateral actors, transnational networks, etc. International financial mechanisms such as the Clean Development Mechanism and the Global Environment Facility have been helpful tools in realizing co-benefits for cities. The Clean Development Mechanism (CDM), in particular, has a relatively smaller influence globally in comparison with the other carbon markets but has been a key mechanism in the context of developing countries.[5] It has provided additional financial support for a number of projects on waste management, renewable energy, energy efficiency, fugitive emissions and HFC-23 (Point Carbon 2007). Some of the successful sectors in CDM such as waste management are strongly urban related while other key urban sectors such as urban transportation are facing serious methodological issues and remains unattractive. Larger scale energy efficiency programs which can play an important role in emission reduction in cities are essentially program based and are not suitable to make CDM projects. Reforms are necessary for Clean Development Mechanisms in post-2012 climate regimes, such as

---

[5]CDM market value in 2006 was at 3.9 billion euros in a 22.5 billion euros carbon market (Point Carbon 2007).

implementing program based approach and making it less methodologically complicated and reducing transaction costs, so that they can play an important role in urban carbon management in developing countries. There is a growing consensus that strictly project-based approaches have to give way to program-based efforts that are capable of engaging governments, institutions and community organizations. Flexible and simplified baseline, verification and operation procedures have to be devised. And last not least, the transaction costs of developing and implementing both projects and programs will have to be dramatically lowered. The question is still open of whether COM is in fact capable of such improvements, and even able to reach a significant scale to be constructive in a positive sense.

Another mechanism, the Global Environmental Facility, neither favors nor excludes the urban sector. In summary, existing frameworks – including activities of bilateral and transnational networks – were useful in raising awareness but have not been adequate to provide structural support to cities. Efforts are neither far-reaching nor focused enough. The inherent limitations of CDM aside, this is because they do not yet see 'city' as a unit of concern and address mitigation and adaptation in cities on a sectoral or piecemeal basis. Due consideration to the city as 'one of the key entities' for global climate policy in bilateral arrangements and future negotiations for international climate regimes beyond 2012 will be instrumental in encouraging cities in developing countries. Moreover, they should provide mechanisms to provide crucial multi-level support.

Locally, a forward-looking stewardship of the political arena – engaged championship – can greatly help to move cities towards climate friendly paths since level of public awareness is so far minimal. The role of political champions in transforming cities is evident. International regimes, transnational networks, bilateral agencies, multilateral agencies, and national government need effort to create such a favourable environment for local authorities to champion the climate cause. A long-held notion that climate policies will have adverse implications on the economic development of cities needs further clarification in the context of an integrated framework of mitigation and adaptation and considering short- and long-term costs and benefits. While there could be adverse implications for economic development on a case-by-case basis, it is not necessarily true that climate policy and economic development have to contradict all the time and this is important for political leadership to understand.

Acknowledgements

The author is grateful to Peter Droege and Shaney Crawford for their very useful comments and suggestions which have helped to improve the earlier version of this chapter.

References

Akbari, H. (2002). Shade trees reduce building energy use and CO_2 emissions from power plants. *Environmental Pollution*, **116**, S119–S126.

Alam, M. and Rabbani, M.G. (2007). Vulnerabilities and responses to climate change for Dhaka. *Environment & Urbanization*, **19**(1). International Institute for Environment and Development (IIED), London. pp. 81–97.

Bai, X. (2007). Integrating global environmental concerns into urban management. *Journal of Industrial Ecology*, **11**(2), 15–29.

Betsill, M. (2006). Carbon management in US cities and states. Presentation to International Workshop on Institutional Dimensions of Carbon Management at the Urban and Regional Levels, December 5, 2006, Grand Hyatt Bali, Nusa Dua, Bali, Indonesia.

Bulkeley, H. and Betsill, M. (2005). *Cities and Climate Change: Urban Sustainability and Global Environmental Governance*. Routledge. ISBN: 9780415359160, 256 pp.

Byrne, J. and Yun, S. (1999). Efficient global warming: contradictions in liberal democratic responses to global environmental problems. *Bulletin of Science, Technology & Society*, **19**(6, December), 493–500.

Canan, P. and Schienke, E.W. (2006). Responsibility, opportunity, and vision for higher education in urban and regional carbon management. *Carbon Balance and Management*, **1**, 13.

Cifuentes, L., Borja-Aburto, V.H., Gouveia, N., *et al.* (2001). Assessing the Health Benefits of Urban Air Pollution Reductions Associated with Climate Change Mitigation (2000–2020): Santiago, São Paulo, Mexico City, and New York City. *Environmental Health Perspectives*, **109**(3, June), 419–425.

Dawson, R. *et al.* (2007). A blueprint for the integrated assessment of climate change in cities, Tyndall Center Working Paper 104, February, UK.

Dhakal, S. (2002). The urban heat environment and urban sustainability, in *Future Cities: Dynamics and Sustainability* (eds F. Moavenzadeh, K. Hanaki and P. Baccini), pp. 149–172. Kluwer Academic Publishers.

Dhakal, S. (2003). Assessment of local strategies for countering greenhouse gas emissions: case of Tokyo. Working Paper, Urban Environmental Management Project, Institute for Global Environmental Strategies, Kitakyushu, Japan.

Dhakal, S. (2004). *Urban Energy Use and Greenhouse Gas Emission in Asian Mega-cities: Polices for a Sustainable Future.* Institute for Global Environmental Strategies, Hayama, Japan.

Dhakal, S. (2005). Strengthening urban environmental management in Asia. In: *Sustainable Asia 2005 and Beyond – In pursuit of innovative policies*, pp. 97–107. Institute for Global Environmental Strategies, Hayama, Japan, 2005, ISBN 4-88788-017-0.

Dhakal, S. (2006). *Urban Transport and Environment in Kathamandu Valley Nepal: Integrating Global Carbon Concerns into Local Air Pollution Management.* Institute for Global Environmental Strategies, Hayama, Japan.

Dhakal, S. and Hanaki, K. (2002). Improvement of urban warming by managing heat discharges and surface modifications in Tokyo. *Energy and Buildings*, **34**(1), 13–23. Elsevier Science BV.

Dhakal, S., Kaneko, S. and Imura, H. (2002). An analysis on driving factors for CO_2 emissions from energy use in Tokyo and Seoul by Factor Decomposition Method. *Environmental Systems Research*, **30**, 295–303. Japan Society of Civil Engineers (JSCE).

Dhakal, S. and Schipper, L. (2005). Transport and environment in Asian cities: reshaping the issues and opportunities into a holistic framework. *International Review for Environmental Strategies*, **5**(2), 399–424.

Gornitz, V. and Couch, S. (2000). Sea level rise and coastal hazard. In: C. Rosenzweig, W. Solecki (Eds.), *Climate Change and a Global City: An Assessment of Metropolitan East Coast Region.* US National Assessment of the Potential Consequences of Climate Variability and Change. US Global Change Research Program. Columbia Earth Institute.

Grubler, A. (1994). Technology, In: *Changes in Land Use and Land Cover: A Global Perspective* (eds B.M. William and B.L. TurnerII) Cambridge University Press, Cambridge. 287pp.

Gustavsson, E., Elander, I. and Lundmark, M. (2006). Multilevel governance, networking cities and climate change. Paper presented at the Sixth European Urban and Regional Studies Conference 21–24 September 2006, Comwell Hotel, Roskilde, Denmark.

Ichinose, T., Hanaki, K. and Matsuo, T. (1993). International comparison of energy consumption in urban areas. *Proceedings of Environmental Engineering Research, Japan Society of Civil Engineers (JSCE)*, **30**, 371–381. (in Japanese)

Ichinose, T., Shimodozono, K. and Hanaki, K. (1999). Impact of anthropogenic heat on urban climate in Tokyo. *Atmospheric Environment*, **33**, 3919–3930.

IEA (2006). *Key World Energy Statistics 2006.* International Energy Agency, Paris.

Kaneko S., Nakayama, H. and Libo, W. (2003). Comparative study on indirect energy demand, supply and corresponding CO_2 emissions of Asian mega-cities. Proceedings of International Workshop on Policy Integration towards Sustainable Urban Energy Uses of Asian Mega-cities (4–5 February 2003, East West Center, Hawaii), Institute for Global Environmental Strategies, Hayama, Japan.

Kern, K. (2006). Carbon management and institutional issues in European cities. Presentation to International Workshop on Institutional Dimensions of Carbon Management at the Urban and Regional Levels, December 5, 2006, Grand Hyatt Bali, Nusa Dua, Bali, Indonesia.

Kinney, P. *et al.* (2001). Climate change, air quality and public health: an impact analysis for New York City, in *Climate Change and a Global City: An Assessment of Metropolitan East Coast Region*, C. Rosenweig and W. Solecki eds. US National Assessment of the Potential Consequences of Climate Variability and Change. US Global Change Research Program. Columbia Earth Institute.

Lebel, L. *et al.* (2007). Integrating carbon management into the development strategies of urbanizing regions in Asia. *Journal of Industrial Ecology*, **11**(2), 61–81.

Lenzen, M. (1997). Energy and greenhouse gas costs of living for Australia during 1993/4. *Energy*, **23**(6), 487–586. Pergamon.

McEvoy, D., Lindley, S. and Handley, J. (2006). Adaptation and mitigation in urban areas: synergies and conflicts. *Municipal Engineer*, **159**(4), 185–191. Print ISSN: 0965-0903.

McPherson, E.G., Simpson, J.R., Peper, P.F., *et al.* (2005a). City of Boulder, Colorado Municipal Resources Analysis. USDA Forest Service, Pacific Southwest Research Station, Center for Urban Forest Research.

McPherson, E.G., Simpson, J.R., Peper, P.F. *et al.* (2005b). City of Minneapolis, Minnesota Municipal Resources Analysis. USDA Forest Service, Pacific Southwest Research Station, Center for Urban Forest Research.

McPherson, E.G., Simpson, J.R., Peper, P.F. *et al.* (2005c). Municipal forest benefits and costs in five US cities. *Journal of Forestry*, **103**, 411–416.

MOEJ (2006a). *Annual Report on the Environment in Japan 2006*. Ministry of Environment, Japan.

MOEJ (2006b). Sweeping Policy Reform Towards a 'Sound Material-Cycle Society' Starting from Japan and Spreading Over the Entire Globe – The '3R' loop connecting Japan with other countries. Ministry of Environment Japan, 2006 (downloaded from http://www.env.go.jp/en/wpaper/smc2006/index.html on 1 June 2007).

Mukheibir, P. and Ziervogel, G. (2007). Developing a Municipal Adaptation Plan (MAP) for climate change: the city of Cape Town. *Environment & Urbanization*, **19**(1). International Institute for Environment and Development (IIED), London, pp. 143–158.

Nicholls (2004). Coastal megacities and climate change. *GeoJournal*, **37**(3, November), 369–379. Springer, Netherlands.

OECD (1995). *Urban Energy Handbook*. Organization of Economic Cooperation and Development, Paris.

Point Carbon (2007). Carbon 2007 – A new climate for carbon trading. Røine, K. and Hasselknippe, H. eds, 62 pages, Point Carbon. Accessed http://www.pointcarbon.com/getfile.php/fileelement_105366/Carbon_2007_final.pdf on 17 June 2007.

Roberts, D. (2006). The Potential Role of Municipal Open Space in Carbon Management at the Local Government Level. Presentations to International Conference on Carbon Management at Urban and Regional Levels: Connecting Development Decisions to Global Issues, Mexico City, September 4–8, 2006.

Romero, L. *et al.* (2005). *Can Cities Reduce Global Warming? Urban Development and Carbon Cycle in Latin America*. Universidad Autonoma Metropolitana Unidad Xochimilco, Mexico City. 92 pp.

Rylander, Y. (2005). *Kommunernas Klimatarbete. Klimatindex for Kommuner 2005*. Stockholm Svensaka Naturskyddsforeningen.

Sari, A. (2004). Case Study – Jakarta. In Integrating Carbon Management into Development Strategies of Cities – Establishing a network of case studies of urbanisation in the Asia-Pacific, Final report for APN project 2004-07-CMY-Lasco, Asia Pacific Network for Global Change Research, Kobe, Japan.

Scholz, S. (2006). The POETICs of industrial carbon dioxide emissions in Japan: an urban and institutional extension of the IPAT identity. *Carbon Balance and Management*, **1**, 11.

Sherbinin, A.D., Schiller, A. and Pulsipher, A. (2007). The vulnerability of global cities to climate hazards. *Environment & Urbanization*, **19**(1). April, pp. International Institute for Environment and Development (IIED). pp. 39–64.

Smith, J.B. and Lenhart, S. (1996). Climate change adaptation policy options. *Climate Research*, **6**, 193–201.

Svirejeva-Hopkins, A., Schellnhuber, H.J. and Pomaz, V.L. (2004). Urbanised territories as a specific component of the Global Carbon Cycle. *Ecological Modelling*, **173**(2–3).

Taha, H. (1997). Urban climates and heat islands: albedo, evapotranspiration, and anthropogenic heat. *Energy and Buildings*, **25**, 99–103.

TMG (2000). Transport Demand Management Action Plan. Tokyo Metropolitan Government.

UN (2006). *World Urbanization Prospects: The 2005 Revision*. United Nations, Department of Economic and Social Affairs, Population Division.

WBCSD, (2001). *Mobility 2001: Mobility at the End of the Twentieth Century*. World Business Council for Sustainable Development, WBCSD, Geneva.

Yuan, Z., Bi, J. and Moriguichi, Y. (2006). The circular economy: a new development strategy in China. *Journal of Industrial Ecology*, **10**(1–2), 4–8. Winter/Spring, MIT Press.

Chapter 8
City Energy Networking in Europe

MARCO KEINER AND ARLEY KIM

Urban Environment Section, UN-HABITAT, Nairobi, Kenya
Institute for Spatial and Landscape Planning, ETH Zurich, 8093 Zurich, Switzerland

8.1 Introduction

According to the medium projection scenario of the Population Division of the Department of Economic and Social Affairs of the United Nations (UN 2004) the world population will rise from 6.1 billion persons in 2000 to a maximum of 9.2 billion persons in 2075 and decline thereafter to reach 8.3 billion in 2175. By far, the most rapidly growing continent will be Africa, nearly tripling its overall population from 0.8 to 2.3 billion. However, in 2100, with 55.4%, Asia is still supposed to hold the biggest share in world population. By contrast, the population is to decrease from 1.2 billion in 2000 to 1.1 billion in more developed countries. For Europe, the number of residents is projected to decrease from 728 million in 2000 to 538 million in 2100. China, India and the United States are and will continue to be the most populous countries of the world. By 2050, India is expected to surpass China in population size and continue to be the most populous country in the world thereafter – and in 2100, out of 9.1 billion earthlings the lion's share will live in less developed countries (7.9 billion, which is 3.0 billion more than in year 2000).

Today, more than half of the world's population and three quarters of Europe's population live in cities and these numbers are on the rise. Due to the outlined population growth, the share of the developing world's population living in cities is expected to rise from 32% in 1985 and 40% (2000) to a projected 57% in 2025 (UN 2004). Paradoxically, this amazing world population growth and urbanization, if realized, would have to be sustained with continually diminishing resources, if recent consumption patterns were to continue. Today's urban structures have been developed in the era of fossil fuels, everything from suburban landscapes of detached houses based on the notion of individual mobility to the engines and motors running our industries depend on finite resources. These, in turn, can only be maintained by exploiting and often destroying the natural environment, and their use generates a number of environmental externalities such as pollution and climate change, which negatively affect human health, ecological stability, and economic development. Increased levels of motor vehicle use and urbanization are the main causes for increased energy consumption (Imran and Barnes 1990). This increased fossil fuel consumption, in turn, exacerbates urban-based pollution, increasing related environmental impacts. These effects will be increasingly felt all the way from the neighbourhood to the global level. Ultimately, these impacts seriously threaten urban environments and their liveability, a problem that will only worsen with time if no mitigating measures are introduced. And to make matters worse, global population growth fuels the demand for energy in all sectors of the urban economy and partially explains why the energy demand of the less developed countries is projected to rise from current consumption of one third of that of OECD countries to parity with OECD demand by the year 2015.

If the above-mentioned prognostics on demographic development are accurate, economic development continues at its current pace, production methods and behavioural patterns do not profoundly change, and oil, gas and coal were to continue as the most important energy sources – despite the impending oil peak and the increasing prices of these resources due to higher demands – global temperatures will continue to increase. UNDESA's urbanization rates imply the continuation of the status quo despite the impending oil peak: minimal concessions on the part of developed countries, continued massive migration from rural areas to cities in the south, and OPEC's disregard of their charge to manage their resources responsibly. The urbanization trends in developing countries and concomitant fossil fuel demand are counteracted only to a minor extent with changing practices and technological improvements of developed countries. At the moment, urbanization in the south 'costs' less than that in the north, hence trends can continue despite warnings of a global energy crisis. But this will change as these developing societies grow, as the current picture in China gives us. The problem is made worse when fossil fuels become too expensive for even the richest countries to be dependent on; technological and urban infrastructures will already have been established that can hardly run on anything else. We will have built up an extraordinary but deeply increasingly dysfunctional urbanization machine at the cost of our non-replenishable natural resources. Already today, the impacts of warmer climates in different regions of the world are obvious: melting glaciers in the mountains and the poles, more intensive flooding, erosion and desertification, higher risks for agriculture through changing variability of precipitations, drought periods, storms and hurricanes, rising sea levels putting small islands, lowlands and megacities at the seashore in danger, and much more. The risks to human life grow.

If sustainability is to be more than a catchword, and the quality of living conditions are not to diminish for future generations, a radical change regarding energy use is a must. This is not a new insight. However, since Meadows *et al.*'s 1972 Limits to Growth report, and despite the Rio Conference and its follow-ups (such as the Kyoto Protocol), there have only been some superficial changes in energy consumption but not the necessary U-turn. A wide-ranging shift towards sustainable energy systems in our cities is urgently required. This can be achieved by taking three measures in parallel: reducing energy consumption, improving energy efficiency, and increasing the share of renewable energy – solar, wind, hydropower, and biomass. Many innovative concepts have been developed in this area and are already in limited use. The future task will be to significantly and efficiently increase the renewable energy contribution in densely built-up areas and efficient energy distribution strategies and communal energy management systems. Global sharing of technical or other energy change knowledge and the exchange of cutting-edge high-tech solutions and insight into implementing best practices in sustainable energy production, distribution, consumption and reuse assist in achieving change at the local level.

8.2 Networking

In response to the urgent need to address the threat of climate change, thousands of local governments around the world have implemented local climate protection and sustainable energy programs in close consultation with local citizens through Local Agenda 21 efforts. In addition, many have joined city networks and associations, adopting joint commitments and approaches, exchanging experience, and providing mutual support. Indeed, transnational networking can be considered an expression of the globalization of knowledge sharing and a new form of governance based on local components (Kern 2001).

Hall and Pfeiffer (2000) emphasize the potential of networks for facilitating mutual learning processes through the transfer and transformation of sustainable innovations (technical and organizational) and knowledge, a process involving both public and private stakeholders. Thus, networks in this context are understood not as physical connections (as, for example, energy networks in terms of power supply systems or interconnected electrical power grids between countries) but as intentional communities of members facing the same problems and looking for similar solutions. Many cities network with one another, in order to strengthen their capacity to solve major environmental and social problems, deliver urban services to their residents, and develop more effective urban governance and management structures. In so doing, city administrators can stimulate local sustainable innovative practices, leading to a myriad of economic, environmental, and social benefits (Jensen 2004).

Networks primarily function as clearinghouses of information. Some concentrate on this role of information dissemination and exchange (i.e. best practice databases), while others are more involved in actively producing new information. The elements of the network architecture involve a formalized organizational and communication structure that facilitates the smooth functioning of decentralized and democratic processes, the primary task of which is managing resources, including finances, and network activity. The extent of networks, their effectiveness, their objectives and agendas are a product of the decentralized power of the individual members.

Resource complementarity is the key to the success of networks. Networks do not merely aggregate resources, but are structured to take advantage of the fact that each participating entity brings different aspects to the fore. This pooling of know-how and exchange of expertise in unforeseeable and unexpected ways makes networks the productive and flexible workshops of the twenty-first century. They also facilitate the forms of trisectoral collaboration – e.g. public sector, private sector, and civil society – increasingly needed to address transnational challenges (Reinicke *et al.* 2000).

The intangible outcomes of networks – building trust between participants and the creation of a forum for raising and discussing other new issues – are often as important as the tangible ones (i.e. negotiating and setting of global standards, gathering and disseminating knowledge, and acting as innovating implementation mechanisms for traditional intergovernmental treaties). Decision-making in networks is structured with a minimum of organization to guarantee the freedom of democratic processes. For example, meetings, events, and training sessions providing communication and information exchange platforms are often planned on a regular basis. There seems to be great consensus as to the necessary components, because the majority of formal networks show great similarity in their architecture, with minor variations.

8.3 A Short History of Networks Supporting Sustainability in Urban Development

Agenda 21, one of the main outcomes of the conference at Rio de Janeiro in 1992, was a key driver for the founding and furthering of sustainability-oriented city networks. Agenda 21 stipulates that all cities, particularly those characterized by severe sustainable development problems, should develop and strengthen programmes aimed at addressing such problems and guiding their development along a sustainable path.

In Chapter 34 the 'Transfer of Environmentally Sound Technology, Cooperation and Capacity-building' is named as a primary objective (Agenda 21), calling for access to scientific and state-of-the-art technologies, particularly to developing countries. In addition, the development of international information networks which link national, subregional, regional and international systems is proposed as an integral part of the implementation process:

Existing national, subregional, regional, and international information systems should be developed and linked through regional clearing-houses covering broad-based sectors of the economy such as agriculture, industry and energy . . . [which] disseminate information on available technologies, their sources, their environmental risks, and the broad terms under which they may be acquired . . . [and help] users to identify their needs and in disseminating information that meets those needs, including the use of existing news, public information, and communication systems. The disseminated information would highlight and detail concrete cases where environmentally sound technologies were successfully developed and implemented. In order to be effective, the clearing-houses need to provide not only information, but also referrals to other services, including sources of advice, training, technologies and technology assessment. The clearing-houses would thus facilitate the establishment of joint ventures and partnerships of various kinds. (Agenda 21, 34.15)

Agenda 21 provided a much-needed impulse for supporting local efforts in sustainability all around the world, including the development of city networks dedicated to the cause even though it lacked a clear definition of the term and disregarded the primary hindrance: global fossil fuel dependence. In Europe a number of city networks with specific sustainability-related scopes were founded from 1989 to 1994, for example on climate (Climate Alliance); culture (Les Rencontres); urban policy (Congress of Local and Regional Authorities of Europe, Eurotowns); communication and technology transfer (TeleCities); transportation (POLIS); urban development (QeC-ERAN); and energy (Energie-Cités, Brundtland City Energy Network). The IT revolution of the 1980s and rise of the internet in the 1990s set the stage for many of these networks to take advantage of the World Wide Web for the furthering of their efforts, an effect particularly relevant for sustainability-related networks due to the global scale involved (see Fig. 8.1). Parallel was also an increase in the initiative of individual cities to act on the sluggishness or inaction of their national governments.

This interest and activity in sustainability-related networks culminated in the First European Conference of Sustainable Cities and Towns in Aalborg, Denmark (1994), where the European Cities and Towns Campaign was launched. Initially, 80 authorities signed the Charter of European Cities and Towns towards Sustainability (Aalborg Charter), committing themselves to long-term action plans toward sustainability and implementing Local Agenda 21 (LA21) processes. To date, more than 2000 local and regional authorities from 34 European countries have signed. This makes the campaign the largest European initiative of local sustainable development.

The Aalborg Charter, more concrete in its definition of sustainability and its various components, is directed specifically to 'major local authority networks in Europe' (Aalborg Charter).

We foresee the principal activities of the Campaign to be to:

- *Facilitate mutual support between European cities and towns in the design, development, and implementation of policies towards sustainability;*
- *Collect and disseminate information on good examples at the local level;*

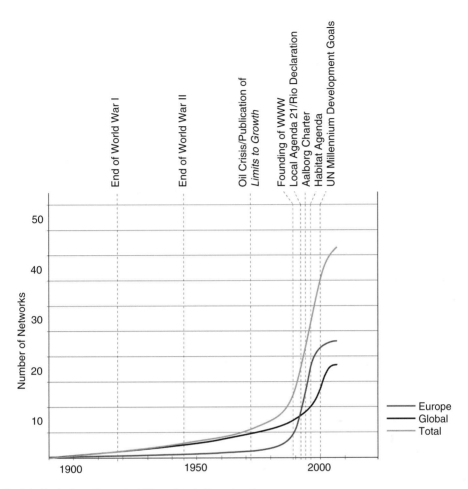

Fig. 8.1. Evolution of sustainability-related city networks.

- *Promote the principle of sustainability in other local authorities;*
- *Recruit further signatories to the Charter;*
- *Organise an annual 'Sustainable City Award';*
- *Formulate policy recommendations to the European Commission;*
- *Provide input to the Sustainable Cities Reports of the Urban Environment Expert Group;*
- *Support local policy-makers in implementing appropriate recommendations and legislation from the European Union;*
- *Edit a Campaign newsletter. (Aalborg Charter).*

Today, a wide variety of networks aimed at promoting sustainable urban development exist (Keiner and Kim 2007). Every network takes a different focus on sustainability themes, although the majority, especially the larger global networks, tend towards a more general approach (i.e. focused on urban development) with smaller sub- or project-oriented networks. For example, one of the largest, ICLEI, uses Cities for Climate Protection (CCP) as its flagship campaign, a network with a particular focus. There are many other smaller independent special focus networks (idem; see Fig. 8.2), such as the European Forum for Urban Safety (EFUS).

According to the scope of each network, the balance between learning, lobbying and branding varies considerably. For example, for some networks like the Congress of Local and Regional Authorities of Europe and the International Union of Local Authorities (now UCLG) the emphasis is not on learning, but on gaining influence on political decisions at national and supra-national levels through the network. Through CLRAE, representatives from individual cities find an appropriate forum to further their interests within the EU. Other networks are primarily focused around maintaining information databases concerned with sustainability (i.e. Best Practices and Local Leadership Programme, or the former Cities Environmental Reports on the Internet).

Spatial scope provides the broadest differentiating category among urban sustainability networks. National networks follow well-established geographic boundaries. However, there is a growing number of transregional and regional networks, which confirm an increased prominence of territories sharing cultural, climatic, and/or larger geographic contexts (i.e. Alliance of the Alps, Union of Baltic Cities). The acknowledgement that such overriding factors provide a better basis for cooperation and exchange as opposed to traditional lines has manifested itself in more focused and successful networking efforts.

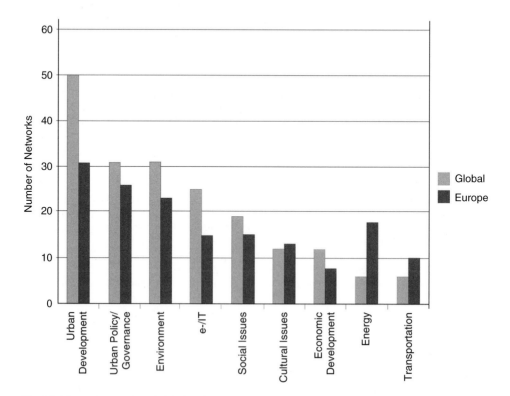

Fig. 8.2. Thematic agendas of global and European sustainability networks.

An omnipresent factor is language as the basis for communication, one that connects as well as continues to isolate regions of the world where English is not spoken, the dominant language of the internet and the primary language of most networks. Global networks do tend to communicate through an English-oriented platform (translated into other major languages such as French and Spanish). Our research, being focused on global and European city networks, did not take into account regional networks spanning Latin American or Asian countries. Very often, their websites are only available in Spanish (as, for example, Mercociudades and FLACMA) or Chinese. Arab city networks, however, also offer information in English. The internet is only one part of sustainability-oriented city networks, except for those focused on information technology and its interfaces with municipal governance. But other means of communication such as formal meetings, working groups, seminars and publications also depend on language and the accessibility of venues.

Research on global networks has shown that under-represented regions of the world tend either to lack access to modern networks entirely or participate in stronger regional networks (Keiner and Kim 2007). But perhaps this is a moot point, considering the very factors differentiating between groups (language, regional culture) are also those which determine the relevance of information in general. It would be erroneous to assume that cities in Europe can profit from best practices in China to the same degree as other Chinese cities, although they might have a marginal interest and may, indeed, learn from the experience. Some networks, for instance, take the issue of regional culture as a main focus. Regional cooperation and exchange networks, for example, connect local communities and actors of two different cultural backgrounds, as is the case for European and Arab countries (CEDARE), geographic regions (i.e. MEDCITIES: countries bordering the Mediterranean Sea) and continents (i.e. City-net: Europe and Asia). The objective of such networks is to disseminate expertise and best practices as well as build up contacts to open formerly closed or new markets to specialized businesses.

Networks based on regional cooperation tend to be primarily project-specific with limited time spans and focused objectives. Constructive work in one project can very well lead to successive projects with the same or other network members. Important, however, is the very well-defined and specific nature of these project partnerships. It differs from the more open-ended aims of sustainability networks, whose agendas are defined over longer time frames and objectives are broader, more generally formulated. The ability to quantify the success or impact of individual networks will be addressed toward the end of this chapter.

Sectoral or thematic issues form another criterion differentiating urban sustainability networks. Here one becomes aware of the depth of sustainability-related interests, which we have attempted to group into larger categories: ecology/environment, energy, governance and urban policy, urban management, society, culture, and communication/technology transfer. Jensen (2004) argues that networks oriented towards one specific technical subject or sector (i.e. energy, urban safety) are more successful in their outreach and learning transfer, because technical subjects and innovations are easily transferable and more or less independent of context, and members are relatively homogeneous. Indeed, most of the networks analysed in our study show a particular thematic bias, and larger networks with a more general focus tend to have smaller thematic or project-oriented working groups.

8.4 Network Types

City networks can also be differentiated with respect to membership characteristics, including community size and type of constituency: cities/local governments, scientific institutions/universities, businesses/the private sector, NGOs, and individuals. In general, it can be stated that networks profit from broad-based inputs that, in turn, are structured by their organizational and communication frameworks. In other words, the constituency, its make-up, and the formal relationships established between various parties and interest groups determine the kind of information exchange furthered by networking. For example, websites whose content is more or less exclusive to members may reduce the overall impact and outreach of their resources, but also make membership in the network more attractive. Also membership requirements or access fees (free of charge, flat fee, or dependent on the population size of a local community) will also influence the make-up of network membership.

Founding or leading organizations also significantly influence the focus and orientation of city networks. In addition to networks sponsored by private companies there is a battery of sustainability-oriented networks organized by international organizations such as the UN. Within Europe, the EU takes an overriding role as network builder and meta-network supporter (Behringer 2003). Universities, research institutions, and think-tanks also sponsor or house networks, one example being the Globalization and World Cities Study Group and Network (GaWC), based in three universities in the UK, USA, and Belgium. Network activity can also be influenced by membership in larger meta-networks or partner networks with complementary interests.

Other characteristics differentiating sustainability-oriented networks can be divided into issues of organization (year founded, organizational structure, location of headquarters, budget) and content (website information, databases/resources, type and frequency of communication/information exchange, projects/awards). The latter are particularly insightful about the various strategies utilized by networks to achieve their aims. Some organs achieve sustainability related networking goals without a formal network. For example, prizes and performance awards (i.e. EEA, UNESCO Cities for Peace Prize, Bremen Initiative, Dubai Award, Stockholm Partnerships for Sustainable Cities) tend to accomplish objectives similar to structured sustainability networks but within a totally different framework. Here, an attempt is made to publicize the work of particular cities fulfilling certain criteria or those making extraordinary efforts to further the goals of the prize or award in question. These prizes and awards are usually established by larger organizations, which have a vested interest in furthering these goals and publicizing these efforts, such as the UN, EU, and private corporations.

8.5 Energy Network Overview

Parallel to diminishing fossil fuel resources and increasing concern for climate change is the increasing influence of city networks dealing with energy issues. Energy sustains the urban system, from meeting basic needs, driving industry, guaranteeing mobility to achieving socio-economic advancement. Thus, energy is a core issue and crucial for sustainable development, the economy, social welfare and the environment, in other words, the future of the globe (REN21 2004). The extent of energy use is determined by a number of factors, such as urban settlement structures and functions, the transportation system make-up, socio-cultural issues such as income levels and lifestyles as well as climate.

With regard to geographical spread and political influence as well as membership structure, several different kinds of energy networks exist:

- Non-profit associations of individuals, companies, organizations, and research institutions (i.e. World Council for Renewable Energy, Eurosolar, World Wind Energy Association, International Solar Energy Society).

- Research institution networks, to facilitate cross-disciplinary research.
- Project networks initiated by international organizations or agreements.
- City or local government networks.

Our research focused on networks primarily comprised of cities to concentrate on one genre. However, a number of networks on energy issues fall under the category of non-profit associations, perhaps due to the strong link to technology advancements furthered by research institutions and the private sector. Often national energy agencies are actively involved in these kinds of networks as opposed to individual cities, as these issues are typically handled at this level of government.

8.5.1 *World Renewable Energy Network and European Energy Network*

One example of a transnational energy network is the World Renewable Energy Network (WREN), a non-profit organization established in 1992 and based in Brighton, UK (www. wrenuk.co.uk). It is affiliated with the UNESCO and maintains links to many United Nations, governmental, and non-governmental organizations. WREN's objectives are to promote the communication and technical education of scientists, engineers, technicians and managers in the field of renewable energy and to support and enhance the utilization and implementation of renewable energy sources that are both environmentally safe and economically sustainable. In order to achieve this objective, WREN runs a worldwide network of almost 100 national agencies, laboratories, university and research institutions, private companies and individuals, all working together towards the international diffusion of renewable energy technologies and applications. Representing countries in the world, it promotes the communication and technical education of scientists, specialists and managers of the field and addresses itself to the energy needs of developed and developing countries. To this aim, WREN organizes international congresses (WREC), regional meetings, workshops on renewable energy topics, and competitions and awards promoting renewable energy. It also publishes a journal *Renewable Energy*, scientific publications, books, and a newsletter. WREN's honorary president is the deputy director general of UNESCO, and it is led by a governing council, an executive committee and a director general.

The European equivalent of WREN is the European Energy Network (E^nR; www. enr-network.org), an association of European organizations taking the responsibility for the planning, management or review of national research, development, demonstration or dissemination programs in the fields of energy efficiency and renewable energy. E^nR comprises one member from each of the 18 EU countries, plus Croatia. E^nR acts as an informal communication network, aiming at creating added value from joint activities in the promotion of pan-European technical support to European organizations on matters of energy policy, strategies and programmes. Projects include a networking/twinning programme for energy efficiency organization, the development of databases, giving assistance to energy agencies, etc. The funding for E^nR's activities comes primarily from national resources; the network has no own budget. However, specific projects within the seven different working groups (among others on energy efficiency, renewables, and energy behaviour) can receive financial contributions from outside sources. The administration of E^nR is in the hands of a yearly rotating presidency and secretariat.

8.5.2 *EU Project Networks*

In the 6th Framework Programme for Research and Technological Development, funded by the European Union, and also in the EU's structural fund INTERREG IIIC, there are several noteworthy projects that are based on a kind of network cooperation between the participating institutions and cities, such as the POLYCITY project and the RUSE project.

The POLYCITY project (running from 2005 to 2010) focuses on large-scale urban developments where living and working areas are integrated to result in sustainable city quarters with short distances and low transport energy consumption. The project treats different aspects of urban conversion in three project cities. In the outskirts of Barcelona, new constructions for residential and industrial/scientific uses on underdeveloped sites encompassing an area of over $2\,000\,000\,m^2$ will be accompanied by innovative polygeneration systems for the year-round use of renewable energy. Especially for commercial and industrial areas with high cooling demand, a comparative demonstration of various thermal cooling technologies will be utilized, making the Barcelona project the largest European solar and biomass powered thermal cooling installation by far.

In Turin, an old city quarter of $875\,000\,m^2$ is rehabilitated through the implementation of efficient district heating. The urban conversion project near Stuttgart covers $178\,000\,m^2$ of newly constructed surface area for 10 000 people and provides very high building standards combined with biomass co-generation on a former large military base. The three projects are integrated in a regional sustainable development strategy and are accompanied by a large Eastern European and Canadian observer network.

The INTEREG IIIC-sponsored RUSE operation (www.sea.gov.sk/english/inter_programs/interreg.htm) aims at directing the development in urban areas in new European Union member states (Poland, Slovakia, Czech Republic and Lithuania; Bulgaria, Romania and Serbia are also expected to participate as associate members) towards the use of sustainable energy. To this end, RUSE aims at progressing towards a better integration of sustainable energy issues, based on improving capacity building in collective structures (city networks, agencies) and individual bodies (municipalities). The network Energie-Cités (see below), a network of European municipalities promoting local sustainable energy policies, coordinates its operation, supported by city networks of the target countries as well as by ten municipalities and municipal energy agencies and two national energy agencies from the former EU-15.

Other networks on sustainable energy sponsored through the 6th Framework Program are Sustainable Energy Systems in Advanced Cities (Sesac), organized by the community of Växjö (Sweden), including the cities of Delft, Grenoble, Kaunas and partnering with Energie-Cités and ICLEI; and the Action of Cities to Mainstream Energy Efficient Building and Renewable Energy System Across Europe (ACT2), managed by the city of Hanover, partnering with Nantes, Koszalin, Newcastle and Malmö. Both projects focus on mitigating greenhouse gas emissions and innovative technologies, processes and systems for sustainable energy use.

An example of European Community-driven cooperation in the framework of the Intelligent Energy–Europe program is the Campaign for Sustainable Energy. Running from 2005 to 2008, it is the successor of the Campaign for Take-Off (2000–2003) with its Renewable Energy Partnership concept, which has involved 130 renewable energy programs and projects with more than 700 partner organizations (i.e. regions, cities and municipalities, industries, agencies and universities) in the EU. The Campaign for Take-Off served as a tool to encourage and enhance the visible commitment of public authorities, which benefited from a series of promotional aids such as conferences, an annual award competition, and communication and media schemes.

8.5.3 City Networks on Energy

Local governments play a key role in changing energy consumption habits, especially for decentralized applications of renewable energies. While national governments are in the position to set up proper frameworks, local governments and communities can facilitate concrete investments in renewable energy technologies, thereby promoting and supporting policies aimed at an increased share of renewables. Given the decentralized nature of a sustainable renewable energy system, national policies and measures will only be successful in the long run if they are joined by local-level commitment and action, and vice versa (REN21 2004). In turn, the switch to renewable energies, together with the significant energy efficiency improvements that this switch requires, is linked to the prospects of reducing the health and environmental impacts of energy production and consumption, fostering regional economic development and job creation, and minimizing dependence on energy imports. A number of city networks in the area of energy focus on the implementation of sustainable energy practices.

8.5.4 Cities for Climate Protection

Energy issues are also a theme in a number of city networks that primarily aim at climate protection, as, for example, the Climate Alliance and Cities for Climate Protection (CCP). CCP (www.iclei.org/index.php?id=800) is a program run by ICLEI since 1993, aimed at educating and empowering local governments worldwide to take action on climate change. For integrating climate change mitigation into their decision-making processes, cities are supported by CCP with case studies, publications, and a tool kit in adopting policies and implementing measures to achieve quantifiable reductions in local greenhouse gas emissions, improving air quality, and enhancing urban liveability and sustainability. CCP is structured around five milestones that local governments commit to undertake. The milestones provide a simple, standardized means of calculating greenhouse gas emissions, of establishing targets to lower emissions, of reducing greenhouse gas emissions and of monitoring, measuring and reporting performance. This framework allows local governments to understand how municipal decisions affect energy use and how these decisions can be used to mitigate global climate change while improving community quality of life. Today, more than 670 local governments in 30 countries worldwide participate in the program. Two thirds of the partnering cities are located in the USA, Canada and Australia. In Europe, mainly cities from Finland and the UK are members.

8.5.5 Brundtland City Energy Network

The Brundtland City Energy Network (BCEN), founded in 1990 and coordinated by the Municipality Nørre-Rangstrup, Toftlund (Denmark), is a group of cities working together to improve levels of sustainability, in part focused on energy. The purpose of this network is to promote a strategic approach to the process of establishing higher energy efficiency and the stronger use of renewables, and to enable the exchange of information and experiences between European towns and cities. Technical and administrative support for development of the network is provided by ISES Europe (the European Unit of the International Solar Energy Society) and by Esbensen Consultants, Denmark. The network includes both local Brundtland team contacts and development contacts. Brundtland Cities are expected to organize a Brundtland team of local stakeholders to work together in the city to introduce energy efficiency measures and renewable energy solutions. The teams prepare a survey of the present use of energy in the city and create a future energy plan for the city, highlighting energy efficiency and renewable energy measures, which may be introduced. There is emphasis on citizen involvement, on creating new business opportunities and on exchanging experiences with other Brundtland network members in Europe. At network meetings, members of local Brundtland teams represent established Brundtland Cities. So-called development contacts, appointed by the network coordinator, report on cities and towns which are in the process of joining the network. At present, BCEN has 17 member cities, most of them relatively small in population size, from 15 European countries.

8.5.6 Renewable Energy Strategies for European Towns (RESETnet)

RESETnet (www.resetters.org), founded 1995, is at the same time a net of individuals which can sustain and reinforce the official communication between cities, a net for the energetic treatment of ideas, developing doable solutions by working on the smaller stepping-stone implementation processes, and it is committed to encourage new ideas, 'no matter how remote they may seem'.

Currently, RESETnet is running three projects called RESTART (Renewable Energy Strategies and Technology Applications for Regenerating Towns), RESET Project, and CLICK. RESTART aims at providing the public authorities, the institutions and the professionals of European industrial cities with some exemplary urban projects concerning innovative energy integration on the built environment. The RESET Project was intended to verify the feasibility of the penetration of renewable energies in four European metropolitan areas with peculiar conditions, such as large industrial infrastructure to be strongly reconverted in the future; need for a major change in labour perspectives; and high pollution levels and environmental vulnerability. The project is based on the Action Plan for Renewable Energy Sources worked out by the Commission of the European Communities for the European Parliament, which stimulated the cities involved in the RESET Project to develop an investigation on the feasibility of substituting 15% of the real primary energy consumption with renewable energy sources, within the year 2010. The third project of RESETnet, CLICK, aims at promoting and extending the renewable energies action planning to the cities which constitute the RESET network, as an accompanying measure of the RESTART targeted demonstration project. RESETnet has eight member cities spread out among eight European countries.

8.5.7 Association of European Local Authorities Promoting a Local Sustainable Energy Policy (Energie-Cités)

Energie-Cités, founded 1990 and based in Besançon (France), focuses on valorizing renewable energy and local resources, and studying the energy budgets of municipal structures. This network aims at developing its member's sustainable energy initiatives through exchange of experiences, the transfer of know-how and the organization of joint projects, providing expertise in local energy strategies, promoting renewable and decentralized energy sources and protection of the environment, as well as influencing the policies and proposals made by European Union institutions in the fields of energy, environmental protection and urban policy. The 116 members of Energie-Cités are mainly municipalities, but also intermunicipal structures, local energy management agencies, municipal companies, representing 466 cities or organisms from the European Union, Romania and Ukraine. The activities of this network include annual conferences, training sessions, internships, technical visits, study tours to member cities as well as publications on integrated action, energy efficiency, renewable energy, and urban mobility. On its website (www.energie-cites.org), Energie-Cités also offers more than 500 case studies.

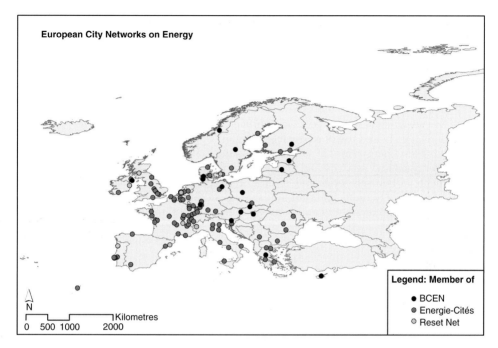

Fig. 8.3. European cities involved in energy networks.

A look at the geographical spread of cities active in the above-mentioned energy networks in Europe (Fig. 8.3) shows a concentration in industrialized areas in Western Europe but also an even distribution across Eastern Europe. While the Brundtland City Energy Network tends to concentrate around Germany and Scandinavia and Eastern Europe, Energie-Cités is more dominant in the Benelux countries, France and Southern Europe. Evidentially, energy network membership is regionally defined, which leads one to conclude that cultural factors come into play. It is also interesting to note that smaller to mid-sized cities dominate in these networks – the regional capitals as opposed to the global metropolises. RESETnet, for example, with Barcelona, Glasgow, Lyon, Turin, Rotterdam, Copenhagen, Porto, and Dublin is comprised of only such cities. In contrast, Moscow, London, and Madrid are rarely on the membership lists, with Rome and Berlin being the two significant exceptions to the rule.

8.5.8 Forum European Energy Award (eea®)

The Forum European Energy Award (www.european-energy-award.org), which in some ways functions as a network, is based on previous experience with programs in Switzerland (Energiestadt Schweiz) and Austria (e5 Programme for Energy Efficient Municipalities). Founded in 2003, it was developed within the 5th EU Framework Programme for Research and Technological Development and has been successfully implemented in a number of communities in Switzerland, Austria and Germany. Currently the SAVE project EURENA extends the implementation of the program to include other European countries and regions: Basque region/Spain, Ireland, Liguria/Italy, Lithuania and Slovakia.

The Forum European Energy Award coordinates the activities of all regional and national organizations using the eea® instrument based on jointly agreed standards, evaluates applications and bestows the European Energy AwardGOLD, is responsible for providing information both internally and externally, seeks out suitable organizations in other countries, and cultivates contacts with municipal networks in Europe and with the relevant EU committees.

The aim of the European Energy Award® is to support communities that want to contribute to a sustainable energy policy and urban development through the rational use of energy and an increased use of renewable energies. To achieve this aim the main elements of the European Energy Award® are: a total quality management system (TQM) for communal energy-related services and activities, certification and award for energy-related achievements in two categories and control of success through regular audits, qualification and assistance to municipal staff in planning and implementation of measures, and a network within the community and between communities through increased cooperation and communication.

The certification process is supported by different eea® tools, such as an audit tool (a checklist for carrying out the energy reviews, an aid for planning activities, and a measurement tool for benchmarks and use during external audits and certification), a guidebook, and a good-practices catalogue showing good examples of local practice.

8.6 Effectiveness and Problems of City Energy Networking in Practice

Power in networks comes from the resources, including monetary ones, and wealth of information accessed through connections. Here, the most interesting questions are: who pulls the strings, or provides monetary support or incentives to keep networks alive? Are connections broad based? In the ideal case, members come from a wide range of backgrounds. Interestingly enough, most energy networks we studied were either oriented towards industry and the private sector with only a few cities involved, or city networks where information from the private sector must be brought in externally by one of the members. For us, this attests to the difficulty of resolving disparate issues in networking: a decentralized information base is needed to maximize networking potential even though the relevance factor of a common background is often a prerequisite for constructive networking impulses. Like most city networks working on sustainability issues, energy networks tend to be smaller in terms of the number of members, and more focused in order to serve as an effective organ of small to mid-sized cities of regional importance. This is not to say that energy issues are not a theme in larger global metropolises (i.e. represented in the Large Cities Climate Leadership Group), only that these cities seem go about them in other – more independent – ways. Smaller regional metropolises can profit more from pooled resources and knowledge and gain more in the perceived significance of their interests. It is also noteworthy that places with the greatest potential for making an impact with respect to energy are in the developing and far less intensively networked world. On the international scale, the rapidly growing metro-polises of Africa, Asia, and Latin America are represented among city networks for sustainability in general (particularly Dakar, Quito, Kathmandu, and Riga; see Keiner and Kim 2007) but with respect to energy networks, more or less non-existent. This could suggest that energy-based information networks are regionally focused on the Western world due to the access to computer and telecommunication access, skills and resources required to manage and maintain networks, budgets to sustain content provision, seen as ineffective or irrelevant, and/or that renewable energy or other forms of sustainable energy action are seen as a low priority theme for cities in the developing world, due their amply documented entrapment in the fossil energy economy.

If we focus on city energy networks in Europe, two main problems hinder the effectiveness of city networking. The first is the communication divide between the great number and diversity of internal offices responsible for sustainability issues, running the gamut from the department of the environment and spatial development, to transportation and mobility and building departments. The second, perhaps greater, problem is despite the seemingly limitless potential of global communication in the age of the internet, networking is still mostly dependent on the personal initiative of the individuals involved as well as the time and resources they are allowed to devote, including access to equipment. In addition, the specific and irreproducible details of local context make the transfer of best practices impossible or difficult at best.

Evaluating the work or 'success' of energy networks in quantifiable terms remains difficult at best. The decentralized dissemination of information is impossible to trace and so other parameters are used: membership size, database size, internal project/research results and support of successful policy initiatives or efforts. It is arguable to what extent the internet has boosted the success of urban networking, and the contribution of networking efforts to initiating political change, in fossil fuel use reduction or otherwise, is difficult to pinpoint.

The pressure to measure the effectiveness of networks based on membership size has unleashed a trend towards consolidation, creating giant but relatively inflexible organizations, whose productive component is only a small part of the whole. Often the number of cities involved is hailed as the most important asset, even though a great number may not be actively involved, or only indirectly through another umbrella organization, such as community associations. Initial interviews with random cities have confirmed the supposition of a high rate of passive members in such networks. These networks are reduced to their lobbying capacity for cities in the international arena.

Using city networks for sustainability to support the cause of cities is not only negative (e.g. puts more power into lower and more local levels of government closer to the heart of problems and their solution) but significantly diminishes their potential as a whole. Within networks with a particular focus, however, growth can actually be attributed to the conviction or persuasiveness of a method and the actual dissemination of specific knowledge. For example, the Forum European Energy Award (www.european-energy-award.org) promotes an instrument for evaluating, steering and controlling communal energy policy. Although a number of such benchmarking tools exist, it represents the consolidation of experience of smaller regional networks in Switzerland, Austria, and Germany has expanded to a number of other European countries attests to the apparent success of this method. All networks are to an extent promotional and activist. Depending on their communication efficiency (i.e. website organization, public outreach), they can assess and compile good practices and cutting-edge knowledge. Whether this knowledge is then accessed and used by members and others is open to question. However, we are able to observe that sometimes networks are able to 'package' information and communicate more effectively than conventional methods, i.e. through government agencies. Because sustainability issues often lie at the crossroads of a number of different disciplines and government bodies, the structure or nature of this 'information landscape' often has a lot to do with its ability to channel and connect people to information most efficiently. To some extent, networks are also in competition with each other in this respect, spurring each other on to more visually inviting websites or innovative strategies to further their aims and gain active and productive members. The attractiveness of networking lies in its inherent dynamic nature and flexibility to changing framework conditions. In contrast, networks that are either entirely or in part dedicated to the creation of new knowledge in terms of projects, specific initiatives, or private investment generate more easily quantifiable progress. To measure concrete progress in energy issues through networking, an independent organ would have to document progress in this area and be able to distinguish the part attributable to networks from the rest. In our continued research, we will turn to the primary agencies initiating and supporting networks for sustainability (i.e. UN-Habitat) to ascertain how they concretize or evaluate networking success.

8.7 Conclusion

In the near future, urban settlements are projected to house more than half of the growing population of our planet. In other words, the causes of energy problems today and tomorrow are mainly situated in cities as well as the solutions. The nature of energy consumption and its associated development infrastructure significantly affects the way cities function and grow. History has shown us the significant changes energy consumption patterns have made on human societies: one only has to compare the pre-industrial and post-industrial city, their infrastructure, their housing structure, and their size. Our current problem is that we are building cities based on outdated infrastructures relying on an abundance of fossil fuels. Moreover the most rapidly growing cities are in developing countries with priorities other than the efficient use of energy resources to maximize their future return. This exacerbates existing problems of rapid population growth, finite and secured resources, and inadequate governance.

Networks on energy issues promise to yield solutions at several levels: in their ability to produce new information, disseminate best practices, and further the cause by supporting changes in public policy and industry on a global scale. New information may include the results of project partnerships among cities as well as public/private initiatives and campaigns, while the dissemination of best practices is dependent on the active participation of a broad-based membership as well as an attractive and effective website as communication platform. Finally, in the age of the internet, networks are able to bring global players together who would not normally collaborate with each other. Supported by 'real' events bringing people physically together, modern communication channels are used to enrich the collaborative process.

Major hindrances to energy networks include already established communication paths, for example within governments, which are inconducive to cross-disciplinary themes and thinking. Dealing with the issue of rethinking a better future for our cities requires the ability to put scientific information in a governance context; only then can technological advancement and innovation be put into practice. This can only happen when the communication paths at the level of city governments are capable of channelling the wealth of information onto most fruitful paths. Energy impacts many different levels of the city in different ways and when the people working in these offices are aware of this fact and its far-reaching implications, they can contribute to networking effectiveness as a whole.

In theory, networking profits from the broad base of decentralized information. But in practice, networking tends to be most effective among members with similar profiles or within the limited context of a focused objective or project framework. This attests to the regional specificity of information and the great extent to which language and cultural barriers come into play. Within the context of city networks, this tends either to produce strong regional networks which share similar political or cultural contexts, or evolve into gigantic apparatuses formed to further the interests of the city in general and increase their influence and clout as a group.

Finally, modern communication methods have changed the face of networking, but at its heart it remains the same. Personal contacts and individual initiative determine the success of networking efforts. Here individual initiative also comes to play in terms of how cities jockey for position and prominence among their peers. The number of network memberships and hosting of conferences, or the day-to-day of network management, tends to overshadow concrete gains in international energy practices and policy.

References

Aalborg Charter (2006). Retrieved 12 February 2006 from http://www.aalborgplus10.dk

Agenda 21 (2006). Retrieved 12 February 2006 from http://www.un.org/esa/sustdev/documents/agenda21/index.htm

Behringer, J. (2003). Nationale und transnationale Städtenetze in der Alpenregion. *WZB Discussion Paper SP IV 2003-104.*

Hall, P. and Pfeiffer, U. (2000). *Urban Future 21 – A Global Agenda for Twenty-First Century Cities.* E & FN Spon.

Imran, M. and Barnes, P. (1990). Energy demand in the developing countries: prospects of the future. World Bank Commodity Working Paper 23.

Jensen, O.J. (2004). Networks as tools for sustainable urban development. *Paper presented to the International Conference Innovation, Sustainability and Policy,* Munich, Kloster Seeon.

Keiner, M. and Kim, A. (2007). Transnational city networks for sustainability. *European Planning Studies* (forthcoming).

Kern, K. (2001). Transnationale Städtenetzwerke in Europa, in *Empirische Policy- und Verwaltungsforschung – Lokale, nationale und internationale Perspektiven* (ed. E. Schröter). Leske + Budrich.

Meadows, D.H., Meadows, D.L., Randers, J. and Behrens, W.W. (1972). Limits to Growth – A Report for the Club of Rome's Project on the Predicament of Mankind. Universe Books.

Reinicke, W. *et al.* (2000). Critical Choices – *The United Nations, Networks, and the Future of Global Governance.* IDRC.

Renewable Energy Policy Network for the 21st Century (REN21). Retrieved 9 July 2006 from http://www.renewables2004.de/en/related/LocalRenewables_Final_Declaration.pdf

UN Department of Economic and Social Affairs (2004). World Population 2300. Retrieved 21 November 2006 from http://www.un.org/esa/population/publications/longrange2/2004worldpop2300reportfinalc.pdf

Chapter 9

Energy Use and CO$_2$ Production in the Urban Passenger Transport Systems of 84 International Cities: Findings and Policy Implications

JEFFREY R. KENWORTHY

Professor in Sustainable Cities, Institute for Sustainability and Technology Policy, South Street, Murdoch University, Murdoch Western Australia 6150

Summary

The urban transport systems of most cities are particularly vulnerable to the 'big rollover' in world oil production and still unprepared for the inevitable transition to a post-petroleum world. Likewise, global warming is placing additional pressure on urban transport to reduce its energy use and CO$_2$ output. This chapter provides a review of private and public transport, urban form, energy use, modal energy efficiency and CO$_2$ emissions patterns in an international sample of 84 cities in the USA, Canada, Australia, Western Europe, high and low income Asia, Eastern Europe, the Middle East, Africa, Latin America and China. It concentrates on factors such as urban density, transport infrastructure and car ownership and use, public transport and non-motorized mode use, in order to better understand patterns of passenger transport energy use and CO$_2$ emissions in different cities. It finds that average per capita energy use in private passenger transport is about 24 times higher in the study's US cities than in the Chinese cities. CO$_2$ emissions from passenger transport follow a similar pattern. For example, Atlanta, the most car-dependent city in the study, produces 105 times more CO$_2$ per capita than Ho Chi Minh City. Some policy implications are outlined to reduce urban passenger transport energy use and CO$_2$ production. Wealth, for example, is not found to be a fundamental explanatory variable in understanding car use and energy use patterns in urban transport systems. Physical planning and infrastructure differences on the other hand are found to be fundamental. This means that urban and transport planners have a key role to play in shaping the direction of passenger transport energy consumption and CO$_2$ production in cities. This includes programs to limit growth in car and motorcycle ownership and usage, especially in developing cities and to protect and enhance the roles of public transport, walking and cycling, which are being decimated by motorization. Urban rail modes are found not only to be the most energy efficient, but also result in higher public transport use, offering speeds that are more competitive with

211

the car. They also fit best with the need for strategically increasing urban densities through transit-oriented development, another key policy conclusion from the research. Traffic congestion is shown to act as a break on growing car use and energy use and urban policy needs to recognize this by a cessation of freeway building in cities and prioritizing infrastructure for public transport and non-motorized modes. Likewise, high parking levels in central cities encourages greater car and energy use and needs to be curtailed and reduced. Finally, cities need to strategically focus denser, mixed use urban development into nodes, including traditional CBDs, linked by high quality public transport operating on its own right-of-way to create more transit-oriented, polycentric metropolises.

9.1 Introduction

World oil production is likely to peak some time in the first decade of this century. Thereafter total production will decline and the highly oil-dependent transport sector will have to make a series of sweeping adjustments, which will be felt acutely in cities (Andrews and Udall 2003; Fleay 1995). Likewise, the imperative to reduce CO_2 emissions and minimize the possibility of destructive global warming will likely be felt more acutely throughout the world. It is important therefore to understand global patterns of urban transport, how dependent different cities are on energy to run their transport systems, how they produce different levels of CO_2 emissions, and some of the underlying reasons for these patterns.

Sound comparative data on metropolitan areas around the world by which our understanding of these issues can be increased are, however, difficult to find. This chapter helps to fill this gap by providing an overview of the patterns of automobile dependence, passenger transport energy use and CO_2 emissions across 84 cities in nearly all regions of the world and some underlying reasons for these patterns. The discussion points to a series of policy implications.

The chapter commences with a brief description of the methodology, data sources and cities covered by the study. Results are then presented for a wide range of urban characteristics, summarized by 11 different high and low income regions in the world. Data covered include urban form and wealth, vehicle ownership, private and public transport infrastructure and usage, public transport service and modal split. Transport energy use and detailed modal efficiency data are presented for private and public transport, along with the resulting CO_2 emissions. Some overall conclusions and perspectives are drawn.

9.2 Methodology and Data Sources

The data in this chapter are drawn from the *Millennium Cities Database for Sustainable Transport* compiled over three years by Kenworthy and Laube (2001) for the International Union (Association) of Public Transport (UITP) in Brussels. The database provides data on 100 cities on all continents. Data summarized here represent averages from 84 of these fully completed cities (Table 9.1) in the USA, Australia and New Zealand, Canada, Western Europe, Asia (high and low income areas), Eastern Europe, the Middle East, Latin America, Africa and China.

The database contains data on 69 primary variables. The methodology of data collection for all variables was strictly controlled by agreed-upon definitions contained in a technical booklet of over 100 pages and data were carefully checked and verified by three parties before being accepted into the database. A detailed discussion of methodology is not possible in this chapter.

Table 9.1. Eighty-four completed cities in the Millennium Cities Database for Sustainable Transport by Region.

USA	Canada	Aust/NZ	Western Europe	High Income Asia	Eastern Europe	Middle East	Africa	Latin America	Low income Asia	China
Atlanta (2.90)	Calgary (0.77)	Brisbane (1.49)	Graz (0.24)	Osaka (16.83)	Prague (1.21)	Tel Aviv (2.46)	Dakar (1.94)	Curitiba (2.43)	Manila (9.45)	Beijing (8.16)
Chicago (7.52)	Montreal (3.22)	Melbourne (3.14)	Vienna (1.59)	Sapporo (1.76)	Budapest (1.91)	Teheran (6.80)	Cape Town (2.90)	S. Paulo (15.56)	Bangkok (6.68)	Shanghai (9.57)
Denver (1.98)	Ottawa (0.97)	Perth (1.24)	Brussels (0.95)	Tokyo (32.34)	Krakow (0.74)	Riyadh (3.12)	Jo'burg (2.25)	Bogota (5.57)	Mumbai (17.07)	Guangzhou (3.85)
Houston (3.92)	Toronto (4.63)	Sydney (3.74)	Copenhagen (1.74)	Hong Kong (6.31)		Cairo (13.14)	Harare (1.43)		Chennai (6.08)	
Los Angeles (9.08)	Vancouver (1.90)	Wellington (0.37)	Helsinki (0.89)	Singapore (2.99)		Tunis (1.87)			K. Lumpur (3.77)	
New York (19.23)			Lyon (1.15)	Taipei (5.96)					Jakarta (9.16)	
Phoenix (2.53)			Nantes (0.53)						Seoul (20.58)	
San Diego (2.63)			Paris (11.00)						Ho Chi Minh City (4.81)	
S. Francisco (3.84)			Marseilles (0.80)							
Washington (3.74)			Berlin (3.47)							
			Frankfurt (0.65)							
			Hamburg (1.70)							
			Dusseldorf (0.57)							
			Munich (1.32)							
			Ruhr (7.36)							
			Stuttgart (0.59)							
			Athens (3.46)							
			Milan (2.46)							
			Bologna (0.45)							
			Rome (2.65)							
			Amsterdam (0.83)							
			Oslo (0.92)							
			Barcelona (2.78)							
			Madrid (5.18)							
			Stockholm (1.73)							
			Bern (0.30)							
			Geneva (0.40)							
			Zurich (0.79)							
			London (7.01)							
			Manchester (2.58)							
			Newcastle (1.13)							
			Glasgow (2.18)							
Av. Pop. 5.74	Av. Pop. 2.30	Av. Pop. 2.00	Av. Pop. 2.17	Av. Pop. 11.03	Av. Pop.1.29	Av. Pop. 5.48	Av. Pop. 2.13	Av. Pop. 7.85	Av. Pop. 9.70	Av. Pop. 7.19

Note: Population sizes are shown next to each city in millions with the average population size per city for the group shown at the bottom of each column.

From these primary variables, some 230 standardized variables were calculated. Cities can thus be compared on urban form, private and public transport performance, overall mobility and modal split, private and public transport infrastructure, the economics of urban transport (operating and investment costs, revenues), passenger transport energy use and environmental factors, including CO_2 emissions. For this overview, which is focussed on transport energy and CO_2 in cities, only a selection of salient features was chosen for comment. Tables 9.2–9.7 provide these data summarized according to the 11 areas shown in Table 9.1, divided into higher and lower income parts of the world.[1] The data are for the year 1995. Data collection on these cities commenced in 1998 and was completed at the end of 2000. Thus data for 1995 provides the latest perspective that can reasonably be expected for a study of this magnitude.

The following discussion summarizes the results of how the 11 areas compare on factors related closely to urban passenger transport energy use and CO_2 emissions.

9.3 Characteristics of Urban Transport Systems

9.3.1 Urban transport and the wealth of cities

Rising wealth is a factor that is nearly always associated with increasing motorization and energy use, so a brief examination of wealth patterns in cities is provided here. The relative wealth of metropolitan regions in this chapter is measured by the gross domestic (or regional) product (GDP) per capita in US dollars of the actual functional urban region (Tables 9.2 and 9.5). This factor is the basis for the split in the sample of cities between higher and lower income regions. The five higher income areas have average GDPs between US$20 000 and US$32 000, while the six lower income areas range from US$2400 to US$6000. As will be seen later from the patterns of private and public transport, wealth alone does not provide a consistent or satisfactory explanation of metropolitan scale transport patterns. This is despite claims by a number of commentators that increasing wealth automatically tends towards higher auto dependence (Gomez-Ibañez 1991; Kirwan 1992; Lave 1992). In fact, within the 58 higher income cities, there is no significant statistical correlation between per capita private transport use (or energy use) and metropolitan GDP per capita. Rather, for metropolitan scale analyses, the data point towards deeper underlying physical differences between cities, which will become clearer in later discussions.

---

[1] The key to regional abbreviations used in Tables 9.2–9.7 is as follows. The specific cities comprising the regional averages are found in Table 9.1.

HIGHER INCOME

USA   US cities
ANZ   Australia/New Zealand cities
CAN   Canadian cities
WEU   Western European cities
HIA   High income Asian cities

LOWER INCOME

EEU   Eastern European cities
MEA   Middle Eastern cities
LAM   Latin American cities
AFR   African cities
LIA   Low income Asian cities
CHN   Chinese cities

Table 9.2.   Land use, transport infrastructure and private transport system characteristics in higher income regions, 1995.

Indicators	Units	USA	ANZ	CAN	WEU	HIA
Urban form and wealth						
Urban density	persons/ha	14.9	15.0	26.2	54.9	150.3
Metropolitan gross domestic product per capita	USD	$31 386	$19 775	$20 825	$32 077	$31 579
Private transport infrastructure indicators						
Length of freeway per person	m/person	0.156	0.129	0.122	0.082	0.020
Parking spaces per 1000 CBD jobs		555	505	390	261	105
Length of freeway per $ of GDP	km/$1000	4.97	6.52	5.85	2.56	0.65
Public transport infrastructure indicators						
Total length of reserved public transport routes per 1000 persons	m/1000 person	48.6	215.5	55.4	192.0	53.3
Total length of reserved public transport routes per urban hectare	m/ha	0.81	3.41	1.44	9.46	5.87
Ratio of segregated transit infrastructure versus expressways		0.41	2.00	0.55	3.12	3.34
Total length of reserved public transport routes per $ of GDP	km/$1000	1.55	10.90	2.66	5.99	1.69
Private transport Supply (cars and motorcycles)						
Passenger cars per 1000 persons		587	575	530	414	210
Motorcycles per 1000 persons		13	13	9	32	88
Passenger cars per $ of GDP	cars/$1000	18.71	29.09	25.43	12.90	6.66
Motorcycles per $ of GDP	mc/$1000	0.42	0.68	0.46	1.00	2.78
Private mobility indicators						
Passenger car passenger kilometres per capita	p.km/person	18 155	11 387	8645	6202	3614
Motorcycle passenger kilometres per capita	p.km/person	45	81	21	119	357
Passenger car passenger kilometres per $ of GDP	p.km/$1000	578.44	575.80	415.15	193.35	114.44
Motorcycle passenger kilometres per $ of GDP	p.km/$1000	1.43	4.11	1.0	3.70	11.32
Total private passenger kilometres per $ of GDP	p.km/$1000	579.86	579.91	416.14	197.05	125.76
Traffic intensity indicators						
Total private passenger vehicles per km of road	units/km	98.7	73.1	105.8	181.9	144.4
Total single and collective private passenger vehicles per km of road	units/km	98.9	73.3	106.1	183.1	149.6
Average road network speed	km/h	49.3	44.2	44.5	32.9	28.9

Table 9.3. Public transport system characteristics and modal split in higher income regions, 1995.

Indicators	Units	USA	ANZ	CAN	WEU	HIA
Public transport supply and service						
Total public transport seat kilometres of service per capita	seat km/person	1557	3628	2290	4213	4995
Total public transport seat kilometres per $ of GDP	seat km/$1000	49.60	183.46	109.95	131.33	158.17
Rail seat kilometres per capita (tram, LRT, metro, suburban rail)	seat km/person	747	2470	676	2609	2282
% of public transport seat kilometres on rail	%	48.0	68.1	29.5	61.9	45.7
Overall average speed of public transport	km/h	27.4	32.7	25.1	25.7	29.9
* Average speed of buses	km/h	21.7	23.3	22.0	20.2	16.2
* Average speed of metro	km/h	37.0		34.4	30.6	36.6
* Average speed of suburban rail	km/h	54.9	45.4	49.5	49.5	47.1
Ratio of public versus private transport speeds		0.58	0.75	0.57	0.79	1.04
Mode split of all trips						
* Non-motorized modes	%	8.1%	15.8%	10.4%	31.3%	28.5%
* Motorized public modes	%	3.4%	5.1%	9.1%	19.0%	29.9%
* Motorized private modes	%	88.5%	79.1%	80.5%	49.7%	41.6%
Public transport mobility indicators						
Total public transport boardings per capita	bd./person	59	84	140	297	431
Rail boardings per capita (tram, LRT, metro, suburban rail)	bd./person	22	42	44	162	238
Proportion of public transport boardings on rail	%	36.7%	50.7%	31.7%	54.6%	55.4%
Proportion of total motorized passenger kilometres on public transport	%	2.9%	7.5%	9.8%	19.0%	45.9%

Table 9.4. Transport energy use and CO_2 characteristics in higher income regions, 1995.

Indicators	Units	USA	ANZ	CAN	WEU	HIA
Overall transport energy indicators						
Private passenger transport energy use per capita	MJ/person	60034	29610	32519	15675	9556
Private passenger transport energy use per $ of GDP	MJ/$1000	1913	1497	1562	489	303
Public transport energy use per capita	MJ/person	809	795	1044	1118	1423
Public transport energy use per $ of GDP	MJ/$1000	26	40	50	35	45
Energy use per private passenger vehicle kilometre	MJ/km	4.6	3.9	5.0	3.3	3.3
Energy use per public passenger vehicle kilometre	MJ/km	26.3	14.9	22.0	14.7	14.4
Energy use per private passenger kilometre	MJ/p.km	3.25	2.56	3.79	2.49	2.33
Energy use per public transport passenger kilometre	MJ/p.km	2.13	0.92	1.14	0.83	0.48
Overall energy use per passenger kilometre	MJ/p.km	3.20	2.43	3.52	2.17	1.40
Public transport energy use per vehicle kilometre by mode						
Energy use per bus vehicle kilometre	MJ/km	28.8	17.0	24.1	16.3	17.3
Energy use per tram wagon kilometre	MJ/km	19.1	10.1	12.1	13.7	7.9
Energy use per light rail wagon kilometre	MJ/km	17.1	–	13.1	19.5	11.7
Energy use per metro wagon kilometre	MJ/km	25.3	–	10.6	11.6	10.0
Energy use per suburban rail wagon kilometre	MJ/km	49.9	12.1	48.8	15.4	10.4
Energy use per ferry vessel kilometre	MJ/km	846.5	170.5	290.8	120.7	601.7
Public transport energy use per passenger kilometre by mode						
Energy use per bus passenger kilometre	MJ/p.km	2.85	1.66	1.50	1.17	0.84
Energy use per tram passenger kilometre	MJ/p.km	0.99	0.36	0.31	0.72	0.36
Energy use per light rail passenger kilometre	MJ/p.km	0.67	–	0.25	0.69	0.34
Energy use per metro passenger kilometre	MJ/p.km	1.65	–	0.49	0.48	0.19
Energy use per suburban rail passenger kilometre	MJ/p.km	1.39	0.53	1.31	0.96	0.24
Energy use per ferry passenger kilometre	MJ/p.km	5.41	2.49	3.62	5.66	3.64
Greenhouse indicators						
Total passenger transport CO_2 emissions per capita	kg/person	4405	2226	2422	1269	825
Total private transport CO_2 emissions per capita	kg/person	4322	2107	2348	1133	688
Total public transport CO_2 emissions per capita	kg/person	83	119	74	134	162
Percentage of total passenger transport CO_2 emissions from public transport	%	1.9	5.3	3.1	10.6	19.7

Note: In this table and Table 9.7, the energy use of electrically powered modes is based on end use or actual delivered operating energy. The CO_2 emissions calculations for electrically powered modes take account of the fuel sources for electrical energy generation (hydro, nuclear, different grades of coal, gas, etc.) in each country, as well as electrical energy generation efficiency in each country.

Table 9.5. Land use and private transport system characteristics in lower income regions, 1995.

Indicators	Units	EEU	MEA	LAM	AFR	LIA	CHN
Urban form and wealth							
Urban density	persons/ha	52.9	118.8	74.7	59.9	204.1	146.2
Metropolitan gross domestic product per capita	USD	$5951	$5479	$4931	$2820	$3753	$2366
Private transport infrastructure indicators							
Length of freeway per person	m/person	0.031	0.053	0.003	0.018	0.015	0.003
Parking spaces per 1000 CBD jobs		75	532	90	252	127	17
Length of freeway per $ of GDP	km/$1000	5.26	9.59	0.62	6.41	3.99	1.17
Public transport infrastructure indicators							
Total length of reserved public transport routes per 1000 persons	m/1000 pers	200.8	16.1	19.3	40.2	16.1	2.3
Total length of reserved public transport routes per urban hectare	m/ha	10.67	2.18	1.15	2.39	2.50	0.32
Ratio of segregated transit infrastructure versus expressways		9.11	3.54	3.36	3.16	1.33	0.77
Total length of reserved public transport routes per $ of GDP	km/$1000	33.74	2.93	3.92	14.25	4.30	0.96
Private transport supply (cars and motorcycles)							
Passenger cars per 1000 persons		332	134	202	135	105	26
Motorcycles per 1000 persons		21	19	14	5	127	55
Passenger cars per $ of GDP	cars/$1000	55.78	24.49	41.04	47.89	28.08	11.03
Motorcycles per $ of GDP	mc/$1000	3.50	3.49	2.91	1.96	33.90	23.30
Private mobility indicators							
Passenger car passenger kilometres per capita	p.km/person	2907	3262	2862	2652	1855	814
Motorcycle passenger kilometres per capita	p.km/person	19	129	104	57	684	289
Passenger car passenger kilometres per $ of GDP	p.km/$1000	488.57	595.37	580.35	940.48	494.13	344.05
Motorcycle passenger kilometres per $ of GDP	p.km/$1000	3.13	23.57	21.17	20.09	182.20	122.34
Total private passenger kilometres per $ of GDP	p.km/$1000	491.70	618.94	601.53	960.57	676.33	466.39
Traffic intensity indicators							
Total private passenger vehicles per km of road	units/km	168.8	180.7	144.1	58.4	236.1	117.2
Total single and collective private passenger vehicles per km of road	units/km	170.9	197.1	146.2	60.0	249.1	131.8
Average road network speed	km/h	30.8	32.1	31.5	39.3	21.9	18.7

Table 9.6. Public transport system characteristics and modal split in lower income regions, 1995.

Indicators	Units	EEU	MEA	LAM	AFR	LIA	CHN
Public transport supply and service							
Total public transport seat kilometres of service per capita	seat km/person	4170	1245	4481	5450	2699	1171
Total public transport seat kilometres per $ of GDP	seat km/$1000	700.80	227.16	908.78	1932.77	719.01	495.01
Rail seat kilometres per capita (tram, LRT, metro, suburban rail)	seat km/person	2479	126	316	1715	402	45
% of public transport seat kilometres on rail	%	59.4	10.1	7.1	31.5	14.9	3.8
Overall average speed of public transport	km/h	21.4	20.9	18.4	31.4	18.0	13.6
* Average speed of buses	km/h	19.3	18.5	17.8	25.8	16.2	12.5
* Average speed of metro	km/h	29.5	–	32.4	–	33.9	35.4
* Average speed of suburban rail	km/h	37.6	36.6	41.0	34.4	33.0	–
Ratio of public versus private transport speeds		0.71	0.68	0.60	0.80	0.81	0.73
Mode split of all trips							
* Non-motorized modes	%	26.2%	26.6%	30.7%	41.4%	32.4%	65.0%
* Motorized public modes	%	47.0%	17.6%	33.9%	26.3%	31.8%	19.0%
* Motorized private modes	%	26.8%	55.9%	35.4%	32.3%	35.9%	15.9%
Public transport mobility indicators							
Total public transport boardings per capita	bd./person	712	152	265	195	231	375
Rail boardings per capita (tram, LRT, metro, suburban rail)	bd./person	409	18	19	37	40	23
Proportion of public transport boardings on rail	%	57.5%	12.0%	7.2%	19.0%	17.4%	6.1%
Proportion of total motorized passenger kilometres on public transport	%	53.0%	29.5%	48.2%	50.8%	41.0%	55.0%

Table 9.7. Transport energy use and CO_2 characteristics in lower income regions, 1995.

Indicators	Units	EEU	MEA	LAM	AFR	LIA	CHN
Transport energy indicators							
Private passenger transport energy use per capita	MJ/person	6661	10573	7283	6184	5523	2498
Private passenger transport energy use per $ of GDP	MJ/$1000	1119	1930	1477	2193	1471	1055
Public transport energy use per capita	MJ/person	1242	599	2158	1522	1112	419
Public transport energy use per $ of GDP	MJ/$1000	209	109	438	540	296	177
Energy use per private passenger vehicle kilometre	MJ/km	3.1	4.2	3.7	3.7	2.6	2.7
Energy use per public passenger vehicle kilometre	MJ/km	11.8	16.1	16.9	9.5	11.9	10.6
Energy use per private passenger kilometre	MJ/p.km	2.35	2.56	2.27	1.86	1.78	1.69
Energy use per public transport passenger kilometre	MJ/p.km	0.40	0.67	0.76	0.51	0.64	0.28
Overall energy use per passenger kilometre	MJ/p.km	1.31	1.99	1.60	1.26	1.20	0.87
Public transport energy use per vehicle kilometre by mode							
Energy use per bus vehicle kilometre	MJ/km	14.2	20.6	18.0	18.1	14.4	9.8
Energy use per tram wagon kilometre	MJ/km	10.5	2.9	–	–	–	–
Energy use per light rail wagon kilometre	MJ/km	19.7	14.3	–	–	11.7	–
Energy use per metro wagon kilometre	MJ/km	10.4	–	12.7	–	21.3	10.5
Energy use per suburban rail wagon kilometre	MJ/km	5.2	35.8	4.7	17.5	14.7	–
Energy use per ferry vessel kilometre	MJ/km	84.2	54.0	–	–	25.0	297.6
Public transport energy use per passenger kilometre by mode							
Energy use per bus passenger kilometre	MJ/p.km	0.56	0.74	0.75	0.57	0.66	0.26
Energy use per tram passenger kilometre	MJ/p.km	0.74	0.13	–	–	–	–
Energy use per light rail passenger kilometre	MJ/p.km	1.71	0.20	–	–	0.05	–
Energy use per metro passenger kilometre	MJ/p.km	0.21	–	0.19	–	0.46	0.05
Energy use per suburban rail passenger kilometre	MJ/p.km	0.18	0.56	0.15	0.49	0.25	–
Energy use per ferry passenger kilometre	MJ/p.km	4.87	2.32	–	–	2.34	4.90
Greenhouse indicators							
Total passenger transport CO_2 emissions per capita	kg/person	694	812	678	592	509	213
Total private transport CO_2 emissions per capita	kg/person	480	761	524	443	441	180
Total public transport CO_2 emissions per capita	kg/person	214	51	154	149	96	33
Percentage of total passenger transport CO_2 emissions from public transport	%	30.8	6.2	22.7	25.2	18.8	15.5

9.3.2 Private transport

9.3.2.1 Car ownership

Globally there are enormous variations in urban vehicle ownership. Clearly, North American and Australian/New Zealand (ANZ) cities lead the world in car ownership with over 500 cars per 1000 people (US cities nearly 600). Western European cities are, however, closing on 'new world' cities with 414 cars per 1000, while Eastern European car ownership was, in 1995, more moderate at 332, though it is rising rapidly. All other groups of cities average between 100 and 200 cars per 1000 people, except for the three Chinese cities, which in 1995 had a mere 26 cars per 1000 people, though this is also growing at an enormous rate (Tables 9.2 and 9.5).

Car ownership is always associated with wealth in the literature, so it is useful to express car ownership relative to wealth (i.e. cars owned per US$1000 of GDP). The ANZ and Canadian cities are clearly the leaders in the higher income cities (25 to 30 cars per US$1000), while US cities are less at 19. Western European and prosperous Asian cities have only a fraction of the cars relative to their wealth (13 and 6 respectively). Of major concern in lower income regions is the much higher level of car ownership relative to their low income status. Eastern European cities lead the race with 56 cars per US$1000, but African and Latin American cities are not so far behind with 48 and 41 respectively. Less prosperous Asian cities already have a rate of car ownership relative to wealth that is virtually equal to cities in Australia/New Zealand. Chinese cities, despite an average GDP of only US$2400, already had in 1995 almost the same rate of car ownership per dollar of GDP as Western European cities (11 compared to 13), despite the latter cities having an average GDP per capita of US$32000.

9.3.2.2 Motorcycle ownership

Motorcycle ownership is relatively insignificant in all areas (between 5 and 30 motorcycles per 1000 people), except in the Asian cities (Tables 9.2 and 9.5). In the high and low income Asian cities, including China, motorcycles average between 55 and 127 per 1000 people, and they form a significant part of the transport system. The huge take-up of motorcycles in low income Asian cities and in urban China is seen in the fact that they have between 23 and 34 motorcycles per $1000 of GDP, compared to an average across all other regions of just two. Motorcycles are the most manoeuvrable motorized mode for avoiding traffic queues and the most energy-efficient and affordable form of motorized private transport for moderate income people. However, they are a major cause of air pollution, noise, danger and death in these cities.

Urban policy needs to try to limit the growth in car and motorcycle ownership, especially in rapidly motorizing developing cities, through programs such as the Certificate of Entitlement in Singapore (and now in Shanghai), which place very high additional costs on the purchase of new motor vehicles (Ang 1990, 1993, 1996).

9.3.2.3 Car usage

Car usage follows a more extreme pattern than ownership, indicating that while cars may be owned to a similar degree in different regions, the need to use them varies dramatically (Tables 9.2 and 9.5). This in turn relates to urban form factors and the viability of other modes for various trip purposes. US cities 'require' over 18000 car passenger km per capita to meet the access needs and discretionary travel of their inhabitants. By contrast, their high income counterparts in Europe and Asia have only between 20% and 63% of that level of use. In the lower income regions, car passenger km per capita range from a mere 814 (4% of the US figure) in Chinese cities, up to 3300 in Middle Eastern cities (18% of the US figure).

Car use can also be expressed relative to wealth. The US and ANZ cities are the leaders with 578 and 576 car passenger km per $1000 of GDP, with Canadian cities some way behind at 415. Again, the Western European and the prosperous Asian cities distinguish themselves in their low levels of private car mobility relative to wealth (193 and 114 respectively).

In lower income cities the rates of private car mobility per unit of wealth are comparatively high. African cities have some 940 car passenger km per $1000 of GDP, which is close to double the US and ANZ level. This effect seems to come from the South African cities where two clearly distinct transport systems exist side by side (the sizeable automobile-based system for prosperous people and the informal, public transport and walking-based systems for the vast majority of poorer residents). Latin American and Middle Eastern cities are virtually identical to the US and ANZ cities in this factor (580 and 595 respectively). In 1995, low income Asian cities and Chinese cities (494 and 344 car passenger km per $1000) had already far exceeded their wealthy Asian neighbours (193) and even Western European cities in this factor (114).

9.3.2.4 Motorcycle usage

Usage of motorcycles is comparatively small in high income cities. Motorcycle use, as a percentage of total private passenger km, ranges from 0.25% in the US cities up to 9% in the high income Asian cities (Tables 9.2 and 9.5). By contrast, in low income Asian cities and Chinese cities, motorcycle mobility represents 26% of private passenger km, while in the other lower income regions it again is small, at between 0.7% and 3.8%. Again, if we normalize this by wealth, the huge commitment to motorcycles in low income Asian cities and Chinese cities is apparent (152 motorcycle passenger km per $1000 of GDP, while all the other regions average a meagre 10).

Why motorcycles have burgeoned to this extent in most Asian cities and in no other parts of the world (nor in Manila where motorcycle ownership is actually about half the US level) is an interesting policy question. The low penetration of motorcycles in Manila is possibly a result of the extensive and effective jeepney system and paratransit-like motorized tricycles (Barter 1998). The role of motorcycles in urban transport, their potential to facilitate urban sprawl by providing low cost private transport to large numbers of people, and their environmental and human impacts are important issues to understand. This is especially so in cities like Taipei where ownership is some 200 per 1000 people and usage represents 35% of private mobility. Notwithstanding this, they are the most energy-efficient form of private motorized mobility available.

9.3.2.5 Private motorized transport modal split

The final variable that provides insight into private transport patterns is the percentage of all daily trips (all purposes) that are catered for by private transport (Tables 9.3 and 9.6). Not surprisingly, US (89%), ANZ (79%) and Canadian cities (81%) head the list. By contrast, their wealthier counterparts in Europe and Asia have only 50% and 42% respectively of all trips by private transport. In the lower income cities private transport caters for only between 16% (Chinese cities) and 36% (Asian cities) of all trips. The exception is the Middle Eastern cities where the proportion rises to 56%.

Despite the overwhelming visual and sensory impacts of traffic and its capacity to rapidly saturate the public space of a city, private transport is a minority player, relative to public transport and non-motorized modes, in seven out of the 11 areas in this study. Because of their size, cars and other private transport vehicles have a huge impact, even at relatively low ownership levels, in urban environments not designed for them. This is true in most rapidly developing cities in the world and, of course, it has enormous social justice and equity implications. If urban transport priorities are primarily directed towards facilitating car and motorcycle travel through new freeways, parking facilities and so on, then this can threaten public and non-motorized mobility, which operate at high energy efficiency and low CO$_2$ output and provide effective transport services to a high proportion of people.

9.3.2.6 *Non-motorized modal split*
The most egalitarian and sustainable modes of urban transport are foot and bicycle. These modes have few fossil fuel or CO$_2$ generation implications, outside of the embodied energy and CO$_2$ output involved in human food production, bicycles and pedestrian and bicycle infrastructure and likewise for CO$_2$ generation. There is an extraordinary range in the use of these energy-efficient modes (Tables 9.3 and 9.6). In US cities, only 8% of all trips are made by foot and bicycle. Other auto cities are a little higher (respectively, 10% and 16% in Canadian and ANZ cities). Eastern and Western European, high and low income Asian, Middle Eastern and Latin American cities all have very similar levels of non-motorized mode use, ranging from 26% to 32% of all trips. The African cities have 41% walking and cycling, due to the majority of low income populations who rely heavily on walking, while the world leader in 1995 was still the Chinese city with 65%.

It would appear very sensible for social, environmental and economic reasons, and certainly from an energy and CO$_2$ perspective, to prioritize the use of non-motorized modes by ensuring that facilities for them are actively promoted and not eroded by motorization. This is especially urgent in rapidly developing cities, especially in China where their pedestrian and cycling advantage appears to be under increasing threat from policies against bicycles and the sheer scale of motorization (de Boom *et al.* 2001; Kenworthy and Hu 2002; Kenworthy and Townsend 2002).

9.3.3 *Energy use*

9.3.3.1 *The oil problem in transport*
Urban automobile dependence has large implications for energy use and CO$_2$ production. In this new century, when the world is being affected by rapidly escalating oil prices and the fallout is being felt everywhere from trucking industry blockades and protests to the significant effect on many household budgets of rising fuel prices, energy is back on the policy agenda. World oil production is predicted to peak by 2010 ('the big rollover') and then to enter a phase of irreversible decline, leading to shortage and supply interruptions, rapidly rising prices and a greater concentration of oil power in the Middle East. This will have profound implications for transport, which is utterly dependent upon conventional oil and cannot restructure overnight (Campbell 1991; Campbell and Laherrere 1995; Fleay 1995). The relative certainty of peaking oil can be seen historically in the accuracy of M. King Hubbert's original prediction of the US oil production peak in 1970 (Hubbert's bubble) (Hubbert 1965). He predicted in the 1950s and 1960s that world oil production would peak shortly after the year 2000, which is now being confirmed by many people, based on much more comprehensive data.

9.3.3.2 *Private passenger transport energy use*

Tables 9.4 and 9.7 reveal a very large range in transport energy consumption, with US cities leading the world at over 60 000 MJ per person per year of energy used for cars and motorcycles. This is twice as high as their nearest rivals, the Canadian and Australian cities, and four to six times more than their biggest competitors in the global economy, the Western European and wealthy Asian cities, such as in Japan. Even cities in the oil-rich Middle East only use 10 600 MJ per person, despite some relatively conspicuous consumption in cities such as Riyadh (25 000 MJ per person). The three Chinese cities consume an average 2500 MJ per person in private passenger transport, which means that an average US city of 400 000 people consumes in one year the same amount as a Chinese megacity of 10 million people.

Three groups of cities stand out as being the most intensive in passenger transport energy use relative to their wealth. These are the US and the Middle Eastern cities (1900 MJ/US$1000 of GDP), and the African cities (2200), again highlighting the high consumption and private transport orientation of a wealthy minority against a backdrop of widespread poverty in African cities. The outstanding cities are again the Western European cities and wealthy Asian cities who consume only 489 and 303 MJ/US$1000 respectively. All the other regions fall between these extremes with an average of 1364 MJ/US$1000.

As countries stake out their claims on ever diminishing and more costly conventional oil, especially those who so far have not yet significantly shared the benefits that flow from this valuable non-renewable resource, oil is likely to become a major destabilizing geo-political and economic issue.

9.3.3.3 *Public transport energy use*

The use of energy in urban public transport systems (Tables 9.4 and 9.7) is small compared to private transport, regardless of the significance of the transport task undertaken by public transport (see later). Also, where rail is extensively used electricity is a major energy source for public transport, which is often generated without oil (gas, hydro-electric, nuclear), and of course can also come from renewable sources. In the US, ANZ, Canadian, Western European and Middle Eastern cities, public transport energy use per capita does not exceed 7% of the combined private and public passenger transport energy use (average of 4%). The biggest contribution is in Latin American cities (23%), with the other five regions averaging 17%.

9.3.3.4 *Modal energy consumption differences between private and public transport*

Tables 9.4 and 9.7 show how relatively energy-inefficient private transport is compared to public transport. Energy consumed per passenger km in public transport in all cities is between one fifth and one third that of private transport, the only exception being in the US cities where large buses dominate public transport and attempt to pick up thinly spread passengers in suburbs designed principally around the car. In US cities, public transport energy use per passenger km stands at 65% that of cars. Part of the reason for this is that in US cities the public transport vehicles have the highest use of energy per vehicle km of all cities (26 MJ/km, with most other regions under about 16 to 17 MJ/km, or as low as 10 MJ/km in African cities).

Examining the overall modal energy consumption of motorized transport in cities (private and public transport combined), Canadian cities are the least efficient at 3.5 MJ per passenger km, followed closely by US cities at 3.2 MJ per passenger km. This reflects the large vehicles in use in North American cities, especially 4WD sports utility vehicles, their low use of motorcycles and their high levels of private versus public mobility. The private vehicles in US and Canadian cities consume about 5 MJ/km, whereas most other regions are under 4 or even 3 MJ/km, despite generally worse levels of congestion in these latter areas.

By contrast to North America, ANZ cities average 2.4 MJ per passenger km for their total motorized passenger transport system, while all the lower income regions range between 0.9 (China) and 2.0 MJ per passenger km. All these lower income cities have a more significant role for energy-efficient public transport, some have high use of motorcycles and many operate fleets of minibuses, which are relatively energy efficient (especially with high loadings).

9.3.3.5 Energy consumption by different public transport modes

Modal energy use can be examined on a per vehicle km or per passenger km basis. The former is an indication of the inherent energy use of the particular vehicle, the technology it exploits and the environment in which it operates (congestion, etc). In the case of rail modes, the data are reported on a per wagon km basis, not train km. Energy use per passenger km is an indication of the mode's efficiency in carrying people, based on the kind of loadings that the mode achieves in different cities. Tables 9.4 and 9.7 contain these data for buses, trams, light rail (LRT), metro systems, suburban rail and ferries. Not all modes are present in some regions and the averages for a particular mode are taken from the cities in the region where the mode is found. All energy data are based on end use or actual delivered operating energy.[2]

It is difficult to discuss the energy use per vehicle km for public transport modes in any detail because of the huge variety of vehicle types, sizes and ages that lie behind the averages. A few general points can be made.

- As with cars, buses in US and Canadian cities are the most energy consumptive (between 24 and 29 MJ/km, compared to an average of 16 MJ/km in all other regions and only 10 MJ/km in Chinese cities).
- Big differences occur in vehicular energy use in suburban rail operations depending on whether higher consumption diesel systems are present.
- There are 29 cases where rail modes are represented in the two tables and in 24 cases the energy use per vehicle km for the rail systems is lower than that of the respective bus system in the region.
- Ferries clearly have the highest use of energy per km due to the frictional forces involved in operating through water. However, there is a huge variation based on vessel size (e.g. double-deck ferries in Hong Kong and small long tail boats in Bangkok) and speed of operation. The average operational energy use across the nine regions where ferries exist is 277 MJ/km, but figures range from 846 in US cities to only 25 in low income Asian cities.

---

[2] The primary energy use for electric rail modes in each city will vary according to the overall efficiency of electrical generation in each country, including power station efficiencies and transmission losses. The use of primary energy in modal energy consumption for electrical modes would have necessitated a fuller accounting of the energy used in producing and delivering petrol, diesel and gaseous fuels, if a genuine comparison were to be made.

More meaningful results can be obtained from energy use per passenger km because this takes into account vehicle loadings and is a measure of the success in public transport operations. It is also the only way to fairly compare public and private transport modal energy use.

- Except for trams and light rail in Eastern European cities, rail modes use less energy than buses per passenger km in each region.
- Across all regions buses average 1.05 MJ per passenger km. This is compared to 0.52 for trams, 0.56 for LRT, 0.46 for metro and 0.61 for suburban rail. There is, on average, not a huge difference in energy consumption between the different rail modes, and on average rail systems in cities use about half the energy of buses per passenger km.
- Urban rail modes, taken together across regions, are on average 4.6 times less energy consuming than the average car (0.54 compared to 2.45 MJ/passenger km).
- The above averages do, however, mask some exceptional energy performance by specific rail modes in particular regions. For example, light rail in low income Asian cities and metro systems in Chinese cities consume only 0.05 MJ/passenger km. This is 57 times more efficient than an American urban bus and 76 times more efficient than a Canadian car per passenger km. These high efficiencies are mainly due to some exceptional loading levels on Chinese systems.
- In every region, ferries are by far the most energy consumptive public transport mode. In fact, in six out of the nine regions where ferries are featured, their energy use per passenger km exceeds that of private transport.

In policy terms, rail modes are clearly the most energy efficient; they have the greatest potential to run on renewable energies and should be prioritized in urban transport infrastructure development where cities are facing a coming oil crisis. They are also best suited to serving dense nodes and linear strips of urban development and thus fit well with increasing urban densities, discussed later in the chapter.

9.3.4 CO_2 emissions from passenger transport

Tables 9.4 and 9.7 show the average per capita emissions of CO_2 from passenger transport in each of the regions. These have been calculated from the detailed energy data on private and public transport through standard grams of CO_2 per MJ conversion factors. For electrical end use energy in electric public transport modes in different countries, reference was made to UN energy statistics showing the contribution of various energy sources to electricity production (i.e. thermal, nuclear, hydro, geothermal). The data also showed the relative contribution of different feedstock to the thermal power plants and the overall efficiency of electrical energy production in the country (http://unstats.un.org/unsd/energy/balance/default.htm). This combination of data was used to ensure the correct multiplier for end use electrical energy and to calculate the kilograms of CO_2 from end use electrical energy consumption by the transit systems in each city.

The results show a similar pattern to that of private passenger transport energy use per capita because of the dominance in most cases of energy use for private transport in cities. US cities generate an average of 4405 kg of CO_2 per person from passenger transport, while the next highest group, the Canadian cities, produce roughly half that level (2422 kg). Australian cities are a fraction lower (2226 kg). Thereafter the figures are much lower, starting with the Western European cities (1269 kg) and followed by the high income Asian cities (825 kg). In the lower income cities, the figures range from 812 kg in Middle Eastern cities down to a mere 213 kg in Chinese cities. US cities are producing 21 times more CO_2 per capita from passenger transport than are the Chinese cities.

The other interesting factor in these tables is the proportion of CO_2 that is attributable to public transport. Again not surprisingly, the US are the lowest at only 1.9%, while the Eastern European cities are the highest at 30.8% due to the fact that they have the most extensive and well-utilized public transport systems in the world. The other high income regions, excluding the USA, have cities where public transport contributes on average 10% to passenger transport CO_2 emissions. The average for the lower income cities, excluding Eastern Europe, is 18%. In most cities public transport is by far the minor player in CO_2 emissions, due partly to its comparatively low share of trips, but also because of its greater efficiency in moving people.

There are, however, exceptions to this, as revealed in Fig. 9.1. This shows the total per capita emissions of CO_2 from passenger transport in all 84 cities, divided into private and public transport (public transport is the top portion of the graph). The data show an extraordinary range in CO_2 emissions from a low in Ho Chi Minh City of 71 kg per capita per annum up to Atlanta's figure of 7455 kg per capita (a 105-fold difference). In addition, in a handful of cities, public transport is one half or more of the total figure, whereas the average for the entire sample of cities is 13%. Figure 9.2 shows this more clearly with Manila, Dakar, Bogota and Cracow all having between 51% and 78% of total CO_2 emissions from passenger transport coming from public transport. After these, Hong Kong is the only exceptional city with over 40%, while the rest of the sample plunges to below 30%, ending with 0.5% in Riyadh and 0.6% in a handful of US cities.

9.3.5 Public transport patterns

Having examined the broad patterns of private transport and the resulting energy use and CO_2, we need to better understand some of the factors that lie behind these patterns.

9.3.5.1 Public transport service levels

Public transport service supply in annual seat km per capita measures the amount of service provided by public transport, taking into account the different capacity of public transport vehicles (from mini-buses to double-deck trains). Public transport service levels are by far weakest in Chinese, Middle Eastern and US cities. Chinese cities still rely very heavily on non-motorized modes (though this is falling rapidly) and their public transport systems have consequently never been very well developed (only 4% of service was by rail in these three cities in 1995). Public transport in Middle Eastern cities relies quite heavily on mini-bus systems that restrict supply capacity (only 10% of public transport service is rail based). US cities, although having had some extensive public transport systems earlier in the twentieth century (e.g. Los Angeles' large rail system), have had a long history of decline. This has only begun to change a little over recent years (Pucher 2002) and 48% of service is now rail based in the US cities in this study. The Western and Eastern European, high income Asian, Latin American and African cities provide the highest levels of public transport service. There are qualitative differences, however, with the European and Asian cities being more rail oriented and offering public transport services that compete with cars in quality, reliability and speed (46% to 62% is rail based). By contrast, African cities have 31% of service on rail, while Latin American cities have only 7%, notwithstanding some fine busway systems in Curitiba and São Paulo.

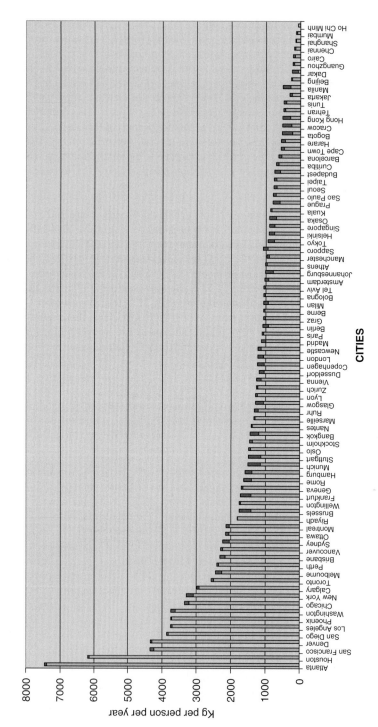

Fig. 9.1. Per capita passenger transport emissions of CO_2 in 84 cities worldwide.

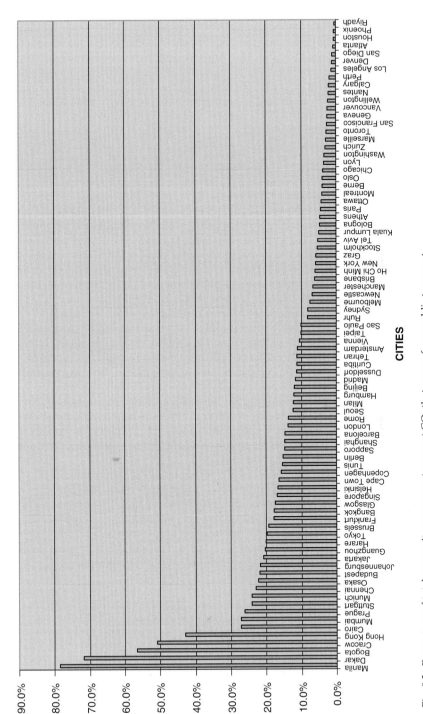

Fig. 9.2. Percentage of total per capita passenger transport CO$_2$ that comes from public transport.

The overall average seat km of service per capita across the high and low income regions is almost identical (3336 and 3203 respectively). However, the quality of service provided, as measured by the proportion of service that is by rail modes, is very much higher in the richer regions (51% compared to 21%). This helps to explain the data in the next section on public transport use. The data also show that relative to wealth, low income cities provide a very much larger amount of public transport service. High income cities provide 126 seat km per $1000 of GDP, while low income cities provide 831, or more than 6.5 times greater service levels. The fact that low income cities do not generate exceptionally high public transport use from the services they provide is an important point discussed in the following sections.

9.3.5.2 Public transport usage levels

There are two clear extremes in public transport use. The US cities stand out globally with the lowest rate of public transport trips per capita (59 per annum), while the Eastern European cities are clearly the world leaders with 712 trips per person per annum (12 times more). This is also reflected in the modal split for all trips, where US urban residents use transit for only 3% of daily trips and Eastern European city residents use transit for 47% of all trips. The other high users of public transport, either in terms of trips per capita or modal share of trips (but not always both), are high and low income Asian cities, Western European, Latin American, African and Chinese cities. For example, Chinese cities, despite poor service, have high per capita usage due to captive riders (375 trips per capita), but the overall share of total trips is low (19%), due to very high walking and cycling (65% of total trips). African cities have only mediocre trips per capita on public transport (195), but the share of trips is quite high at 26%, due to the lowest overall daily trip rate of all cities.

ANZ, Canadian and Middle Eastern cities are comparatively low users of public transport, regardless of the measure used. However, Canadian cities distinguish themselves within the most auto-oriented cities with public transport use that is essentially double that of the US and ANZ cities. This relative success of Canadian cities in public transport is discussed elsewhere (Raad and Kenworthy 1998).

Although some low income cities enjoy comparatively healthy public transport use for their very low level of service (e.g. in China), this will not last. The crowded, unreliable and poor safety of public transport is something that people in such cities will try to escape at their first opportunity, given access to a car or motorcycle. Many low income cities may be unable to resist further rapid motorization unless they improve the quality and speed competitiveness of their public transport systems. This will generally mean developing more extensive rail systems, or fully segregated busways. If not, transport energy use and CO_2 emissions will rise rapidly.

9.3.5.3 Importance of rail and comparative modal speeds

The data in this chapter highlight the importance of urban rail systems in developing competitive public transport systems, reducing energy use and minimizing CO_2 from transport. Not only are they the most energy-efficient modes, but they are the most effective at capturing modal share from private transport. In the high income cities, only the European and Asian cities have public transport systems that capture a healthy share of the overall transport market and these are the cities where urban rail systems are most developed, especially in relation to their private transport equivalent, the urban freeway. The ratio of fully segregated public transport rights-of-way to urban freeways in these rail-oriented cities is over 3 compared, to 0.4 and 0.5 in US and Canadian cities and 2 in ANZ cities. In the lower income sample, by far the healthiest public transport, by whatever measured used, is in the Eastern European cities where segregated public transport right-of-way is some nine times greater than urban freeways (see next section).

While there are cities in Latin America, Africa and China that achieve significant public transport use with little or no rail systems, they rely mostly on poor 'captive' riders. As already suggested, as incomes rise and car ownership levels grow, the public transport systems of these low income, bus-based cities will tend to lose market share, because they cannot compete in speed or comfort with private transport. Motorcycles in particular tend to compete heavily with bus systems that are engulfed in traffic (Barter 1998).

This is seen in the comparative modal operating speeds. There are no regions where the average speed of bus systems exceeds 26 km/h and the overall average across the 11 regions is only 19 km/h. In Chinese cities buses operate at an average 12.5 km/h, or about the same speed as cycling. On the other hand, metro systems operate between 30 and 37 km/h (average 34 km/h). Suburban rail systems across the regions average 42.9 km/h. When these speeds are compared to general road traffic speed, which averages 34 km/h across all regions, it can be seen that only rail systems can compete.

In policy terms cities, wishing to protect, rebuild and grow their public transport base, save energy and reduce CO$_2$ emissions as incomes rise and competition from motorcycles and automobiles escalates, need to seriously consider some form of segregated rail system. At the very least, cities need effective busways or Bus Rapid Transit (BRT) systems as demonstrated in Curitiba and São Paulo, and more recently the Transmilenio BRT system in Bogota. Competitive transit speeds appear to be critical for public transport in any city.

9.3.6 Transport infrastructure provision

Underlying the patterns of urban transport, energy use and CO$_2$ emissions are significant differences in the extent and type of infrastructure for private and public transport. Different priorities in transport infrastructure facilitate different movement patterns.

9.3.6.1 Public transport infrastructure

Western European, high income Asian and Eastern European cities, and to a lesser extent the ANZ cities, are the only regions that have significant reserved alignments for public transport. This consists mainly of railways, but also a few physically segregated busways. All of the lower income city groups, apart from Eastern Europe, have comparatively scarce reserved public transport right-of-way. Chinese cities stand out as being particularly low in this factor and in the wealthier auto-dependent groups, US cities are clearly the lowest in segregated public transport right-of-way, though Canadian cities are not far behind. The data again highlight that the wealthier regions have an edge in the provision of better quality public transport systems, with an average of 113 metres per 1000 persons of reserved public transport routes compared to 49 metres in lower income cities. On the positive side, relative to their wealth, poorer cities have more than double the reserved rights-of-way of their wealthier neighbours (10.0 km versus 4.6 km per $1000 of GDP).

9.3.6.2 Private transport infrastructure

Urban freeways: The private transport corollary of fixed route public transport is the urban freeway. US cities, without any surprise, have the highest availability of freeway per person in the world, followed by ANZ and Canadian cities with 83% and 78% as much respectively. Outside of these three regions freeway provision falls away rapidly, especially in Latin American and Chinese cities (only 2% of the US level). The other eight regions altogether average only 0.028 metres of freeway per capita, compared to 0.156 in US cities (18% of the US level).

It is not surprising that cities with the highest freeway provision also have the highest average speed of general traffic (44 to 49 km/h in US, ANZ and Canadian cities). The other cities with considerably lower freeway provision achieve only 29 km/h average road system speed. It has been understood since the 1970s how urban freeway provision is directly associated with higher car and energy use in cities (e.g. Watt and Ayres 1974). The mechanism for this, in terms of longer travel distances rather than savings in time, has been explained elsewhere (e.g. Newman and Kenworthy 1984, 1988, 1999b).

Lower income cities have marginally higher provision of freeways per $1000 of GDP than high income cities (4.5 km compared to 4.1 km). In fact, Eastern European, Middle Eastern and African cities exceed the US figure in kilometres of freeway per $1000 of GDP. Poorer cities appear to be giving priority to freeway construction.

Congestion: Freeways and congestion issues are often linked together in discussions, especially in relation to efforts to save energy and reduce emissions. The tables provide some different measures of congestion in cities, since increasing urban traffic congestion is frequently cited as being responsible for huge wastage of energy resources and extra emissions in cities. The results in this study run directly counter to this assertion. They suggest that as congestion increases, there is less car use, more motorcycle use, more public transport use and more use of non-motorized modes. Conversely, lower congestion reflected in higher average speed of traffic is associated with the direct opposite of this.

In policy terms, the results suggest that congestion acts as a brake on car use. Congestion encourages greater motorcycle use, and works in favour of public transport, but only where these options offer speed advantages over cars in congested conditions, and where parking is limited. This is often the case in more congested high income cities, since they commonly have urban rail systems, or less frequently, busways. The results suggest that rather than saving energy and lowering emissions, reduced congestion, reflected in higher traffic speeds, increases energy use and emissions (including CO_2). Higher traffic speeds favour cars, increase urban sprawl and travel distances, and reduce the viability of other modes. Cities should not be in any rush to reduce congestion in order to decrease energy and CO_2. Rather they should be strategically improving the public transport and non-motorized mode alternatives to help travellers avoid congestion.

Parking: Parking in the central business districts (CBD) of cities is another indicator of private transport infrastructure, which varies dramatically across regions. Parking supply in the CBD is an important factor in determining modal split for trips to these space-constrained areas of cities. The availability of parking, much more than price, tends to determine the attractiveness of car commuting to the central city. High levels of parking will encourage the most energy consuming trips to work, while well-loaded, radial rail systems will use the least energy.

The highest providers of parking are the US, ANZ and, perhaps surprisingly, the Middle Eastern cities, all having over one parking space for every two jobs in the CBD. Middle Eastern cities are greatly affected by Riyadh, which is a world extreme, having some 1883 spaces for every 1000 jobs, due to its huge on-street parking supply. Teheran has only

22 spaces per 1000 jobs and Tel Aviv, the next highest after Riyadh, has 467. By contrast, Chinese, Eastern European and Latin American cities average 17, 75 and 90 spaces per 1000 jobs respectively, with Canadian cities at 390.

In summary, the experience of rapidly motorizing low income cities with dense urban forms, such as in China, suggests that orderly urban transport development depends upon controlling the rate of motorization, while building public transport infrastructure to effectively balance investment in new roads and parking areas. No city in the world, not even Los Angeles, has ever been able to supply enough roads and parking to meet demand. High density cities have particularly severe limits on how much space they can devote to cars without degrading their public environments. Finally, the evidence suggests that reducing per capita transport energy use and CO$_2$ by building more private transport infrastructure is a vain hope.

9.4 Urban Form

The importance of urban form in helping to explain the macro patterns of urban transport, especially the level of auto dependence and transport energy use, has been widely discussed (Cervero 1998; Kenworthy and Laube 1999; Newman and Kenworthy 1989, 1999a). Tables 9.2 and 9.5 provide data on urban density, the most significant measure of urban form that has been found in the above studies. The data show how the higher car and energy use cities, and the highest CO$_2$ producers, are low in population density, while the higher density cities have reduced car and energy use per person and lower passenger transport CO$_2$. Average densities range from lows of 15 per ha in the US and ANZ cities up to 150 to 200 per ha in the Asian cities, including Chinese cities. On average, the lower income cities are more than double the density of the wealthier cities (109 versus 52 persons per ha).

In the high income cities, 82% of the variance in car passenger km per capita and 78% of the variance in per capita private passenger transport energy use are explained by urban density. In the low income cities, where other factors such as extreme variations in income affect the outcome, still 47% of their variation in per capita car use and 44% of the variance in per capita private passenger transport energy use are explained by urban density (Kenworthy 2001).

Any city wishing to better manage the automobile, minimize car and energy use and reduce CO$_2$ output must address urban form and its effects on urban transport. The best policy response seems to be one of selective density increases and mixing of compatible land uses, especially around areas of high public transport accessibility – transit-oriented development or TOD. As well, centralization of jobs in the CBD, but also in satellite sub-centres built at public transport nodes (decentralised concentration), appears to be an effective strategy. Urban growth boundaries or green belts to minimize spread of the city also appear to be important (Newman and Kenworthy 1999a).

The above approach develops a polycentric city connected by excellent public transport, i.e. a 'transit metropolis'. The tightly integrated dense, mixed-use development around stops on a fixed-route transit network maximizes walk-up patronage and encourages multiple public transport trip making. This is the approach in Curitiba and Ottawa using busways and in European and Japanese cities, as well as Hong Kong and Singapore, focused on rail systems (Cervero 1998). Bus or light rail feeders to the main fixed-route system are also a critical part of this strategy.

Cities in North America and Australia that are attempting to reduce their car use and increase their public transport and non-motorized mode use, are faced with trying to increase densities in strategic locations, as well as extending and improving the infrastructure for these modes. Many lower income cities, especially those in Asia, already have dense and centralized patterns of urban land use but require better public transport and non-motorized mode infrastructure to exploit this advantage. However, much recent development is more dispersed and dependent on private motorized transport. Enforceable land use policies are needed to curb this trend (Kenworthy and Hu 2002; Kenworthy and Townsend 2002).

9.5 Conclusions

This chapter has highlighted some major differences in transport patterns in metropolitan areas in 11 different regions around the world and focused attention on passenger transport energy use and CO_2 emissions. US cities, followed a long way behind by Canadian and Australian cities, are the heaviest passenger transport energy users and CO_2 producers from transport in the world, while Chinese cities are the lowest (US cities are 24 times higher than Chinese cities in per capita passenger transport energy use and 21 times higher in CO_2 emissions).

The patterns demonstrate the energy and CO_2 reduction potential of reducing the role of the automobile in urban transport systems and enhancing the role of public transport and non-motorized modes. Motorcycles are also relatively energy- and CO_2-efficient modes, but they are associated with high local smog emissions, noise and traffic danger, which can detract from their energy and greenhouse benefits.

Public transport energy use per capita is always a fraction of private transport energy use, never exceeding an average of 23% of the total passenger transport energy use in any region, regardless of the extent of service and usage. For CO_2 emissions per capita from passenger transport, Eastern European cities experience the highest contribution from public transport at 31%.

Urban passenger transport energy use and CO_2 emissions can be linked directly and indirectly to a host of factors. Some significant factors found internationally include the extent and quality of the public transport system, especially the kilometres of dedicated public transport right-of-way and the amount of service provided by urban rail systems. Lower income cities provide comparatively high levels of public transport service, but little of it is rail. Most service consists of buses that operate within general road traffic congestion, which are apt to lose market share to cars and motorcycles. Their public transport use is low compared to many wealthier cities that are focused strongly on speed-competitive rail systems.

Freeways and high levels of parking in the CBD are associated with higher energy use and CO_2 emissions in cities. Minimizing such infrastructure in cities will have energy and CO_2 benefits. Attempting to get rid of congestion through freeway building and other means, rather than building up the non-auto modes to help people avoid congestion, will not save energy or reduce CO_2 emissions.

Low density, heavily zoned land use is particularly strongly associated with high transport energy use and CO_2 emissions. This is primarily because such land uses generate high levels of car use and low use of public and non-motorized transport. Compact, mixed land uses, extensive public transport systems operating on a backbone of rail and attractive environments for walking and cycling will minimize transport energy use and CO_2 emissions.

Passenger transport energy use and CO_2 are of course dependent upon the modal share between private, public and non-motorized modes, but also upon the relative energy use of different modes. Energy consumption per passenger km is significantly higher for cars than public transport in all groups of cities. Rail modes (trams, light rail, metro and suburban rail) are in virtually every instance less energy consuming than buses in their respective regions.

In most cases, except selected suburban rail systems, rail operates on electric power, which can be and is generated from non-fossil fuel and renewable energy sources. Ferries are the highest energy consuming modes, exceeding even that of cars in some regions.

Overall, this chapter provides useful guidance to policymakers concerned about the energy and greenhouse implications of urban planning and transport policy at a time when oil supplies and prices and global warming are high on international and national political and economic agendas. The key overall policy points are:

- Wealth is not found to be a fundamental explanatory variable in understanding car use and energy use patterns in urban transport systems. Physical planning and infrastructure differences on the other hand are found to be fundamental.
- Urban and transport planners have a key role to play in shaping the direction of passenger transport energy consumption and CO$_2$ production in cities.
- Programmes are needed to limit growth in car and motorcycle ownership and usage, especially in developing cities (but also of course in wealthier cities) and to protect and enhance the role of public transport, walking and cycling, which are, or have been, already decimated by motorization.
- Urban rail modes are found not only to be the most energy efficient, but also to result in higher overall public transport use in cities, offering speeds that are more competitive with the car.
- Urban rail also fits best with the need for strategically increasing urban densities through transit-oriented development. Cities need to strategically focus denser, mixed-use urban development into nodes, including traditional CBDs, linked by high quality public transport operating on its own right-of-way to create more transit-oriented, polycentric metropolises. Bus Rapid Transit also has a particular role to play here in developing cities, though a transition to urban rail should generally feature in the planning to cope with expanding use and growing customer expectations about the quality of public transport as incomes rise.
- Traffic congestion is shown to act as a break on growing car use and energy use and urban policy needs to recognize this by a cessation of freeway building in cities and the prioritizing of infrastructure for public transport and non-motorized modes. This means transport investment spending needs to change from being road oriented to being focused on 'green modes'.
- High parking levels in central cities encourages greater car and energy use and needs to be curtailed and reduced, through parking restriction policies and a programme of re-urbanization, building new residential, mixed-use development on commuter car parks. This also works synergistically with the development of better public transport systems, especially high capacity urban rail, to deliver large numbers of people to such constrained sites.

Acknowledgement

The author wishes to acknowledge the very significant contribution of Dr Felix Laube, co-author of the Millennium Cities Database for Sustainable Transport, in collecting and testing the data that lie behind the analyses in this chapter.

References

Andrews, S. and Udall, R. (2003). Oil prophets: looking at world oil studies over time, in *Second International Workshop on Oil Depletion*, French Institute of Petroleum, Paris, May 2003. www.peakoil.net/iwood2003/paper/AndrewsPaper.doc

Ang, B.W. (1990). Reducing traffic congestion and its impact on transport energy use in Singapore. *Energy Policy*, 871–874.

Ang, B.W. (1993). An energy and environmentally sound urban transport system: the case of Singapore. *International Journal of Vehicle Design*, **14**.

Ang, B.W. (1996). Urban transportation management and energy savings: the case of Singapore. *International Journal of Vehicle Design* 17, 1–12.

Barter, P.A. (1998). An International Comparative Perspective on Urban Transport and Urban Form in Pacific Asia: Responses to the Challenge of Motorisation in Dense Cities. Murdoch University PhD Thesis, Perth.

Campbell, C.J. (1991). *The Golden Century of Oil 1950–2050: The Depletion of a Resource*. Kluwer Academic Publishers, Dordrecht.

Campbell, C.J. and Laherrere, J.H. (1995). *The World's Oil Supply 1930–2050*. Petroconsultants, Geneva.

Cervero, R. (1998). *The Transit Metropolis: A Global Inquiry*. Island Press, Washington, DC.

de Boom, A., Walker, R. and Goldup, R. (2001). Shanghai: the greatest cycling city in the world?. *World Transport Policy and Practice*, **7**(3), 53–59.

Fleay, B.J. (1995). *The Decline of the Age of Oil – Petrol Politics: Australia's Road Ahead*. Pluto Press, Australia, Sydney.

Gomez-Ibañez, J.A. (1991). A global view of automobile dependence. *Journal of the American Planning Association*, **57**(3), 376–379.

Hubbert, M.K. (1965). Energy resources, in *Resources and Man*. National Academy of Sciences, Freeman, San Francisco.

Kenworthy, J. (2001). Insights into some key transport and land use policy questions in cities: a statistical analysis of the Millennium Cities Database for Sustainable Transport. A report to the UITP, Brussels, Institute for Sustainability and Technology Policy, Murdoch University, Perth.

Kenworthy, J. and Hu, G. (2002). Transport and urban form in Chinese cities: an international comparative and policy perspective with implications for sustainable urban transport in China. *DISP*, **151**(4), 4–14.

Kenworthy, J.R. and Laube, F.B. (1999). *An International Sourcebook of Automobile Dependence in Cities, 1960–1990*. University Press of Colorado, Boulder.

Kenworthy, J. and Townsend, C. (2002). An international comparative perspective on motorisation in urban China: problems and prospects. *IATSS Research*, **26**(2), 99–109.

Kenworthy, J. and Laube, F. (2001). The Millennium Cities Database for Sustainable Transport, International Union of Public Transport (UITP), Brussels and Institute for Sustainability and Technology Policy (ISTP), Perth.

Kirwan, R. (1992). Urban form, energy and transport – a note on the Newman-Kenworthy thesis. *Urban Policy and Research*, **10**(1), 6–23.

Lave, C. (1992). Cars and demographics. *Access*, **1**, 4–11.

Newman, P.W.G. and Kenworthy, J.R. (1984). The use and abuse of driving cycle research: clarifying the relationship between traffic congestion, energy and emissions. *Transportation Quarterly*, **38**(4), 615–635.

Newman, P.W.G. and Kenworthy, J.R. (1988). The transport energy trade-off: fuel-efficient traffic versus fuel-efficient cities. *Transportation Research*, **22A**(3), 163–174.

Newman, P.W.G. and Kenworthy, J.R. (1999a). *Sustainability and Cities: Overcoming Automobile Dependence*. Island Press, Washington, DC.

Newman, P.W.G. and Kenworthy, J.R. (1989). *Cities and Automobile Dependence: An International Sourcebook*. Gower, Aldershot.

Newman, P.W.G. and Kenworthy, J.R. (1999b). 'Relative speed' not 'time savings': a new indicator for sustainable transport, in: *Papers of the 23rd Australasian Transport Research Forum*, **23**(1), 425–440, Perth.

Pucher, J. (2002). Renaissance of public transport in the United States?. *Transportation Quarterly*, **56**(1), 33–49.

Raad, T. and Kenworthy, J.R. (1998). The US and us. *Alternatives*, **24**(1), 14–22.

Watt, K.E.F. and Ayres, C. (1974). Urban land use patterns and transportation energy cost, in *Proceedings of the Annual Meeting of the American Association for the Advancement of Science*, San Francisco.

PART III
New Aspects of Technology

Chapter 10
Storage Systems for Reliable Future Power Supply Networks

DIRK UWE SAUER

Electrochemical Energy Conversion and Storage Systems Group, Institute for Power Electronics and Electrical Drives (ISEA), RWTH Aachen University

Summary

As pointed out in many international studies, it is necessary to decrease global CO_2 emissions drastically. On the other hand worldwide economic growth should be maintained. This requires more electricity-based energy systems where the electricity is produced mainly from renewable energy sources. Electricity can replace gas and oil consumption, e.g. by using electrically driven heat pumps in well-insulated houses for space heating and by using plug-in hybrids or full electrical vehicles at least for individual city driving.

Electricity is supplied via complex power supply networks whose operation is costly and difficult, especially if large quantities of stochastic power generators such as wind turbines and solar power by means of photovoltaics or solar thermal processes are fed into the grid. However, the latter will be necessary to replace fossil and nuclear power generation in the coming decades for a sustainable future energy supply. To assure reliable power supply via the networks and to provide customers with the power quality needed for their respective business, significant storage capacities in terms of power and energy are required.

Various storage technologies for high power and high energy are available. Even though several technologies still need improvements with regard to costs and lifetimes, from a technical point of view it is clear that the storage problem can be solved. However, an efficient and economic power supply will be achieved only if all options for an efficient management of loads, storage and generators are used. Heat and power co-generation with intelligent control, space heating in well-insulated houses by electrical power and incorporating local individual transportation by means of plug-in hybrid and full electric vehicles offer the possibility for an efficient management of the energy supply system and a minimization of investment costs.

10.1. Introduction

Storage systems in grids are as old as the idea of grids itself. Power generation and consumption must be in equilibrium at any given time. Due to rapid and unexpected changes in load demand and unexpected changes in generation capacity (e.g. caused by a sudden shutdown of a power plant) highly flexible units are necessary to compensate for differences between generation and demand. For an optimal match between generation and demand, load and generation forecasts are used to prepare schedules for the operation of power plants 24 hours in advance. This allows using a mixture of base, medium and peak load power plants. However, schedules are never perfect and therefore several measures are necessary to ensure the equilibrium between generation and demand at any time, ranging from seconds to hours. Pumped hydro power plants are used today for these purposes for any event beyond one minute; events shorter than a minute are typically taken from the rotational kinetic reserve of the turbines.

In addition to their use at the power transport level (high voltage grid), several storage systems are used to serve critical loads with the required power quality and reliability directly to the customer's site. Uninterruptible power supply (UPS) systems can react within less than 10 ms to level out voltage dips or blackouts. This is essential for all computer-based systems. Large UPS systems in the MW range are the heart of central communication centres or internet servers but they can be found also in hospitals or at lower power levels in private households for PCs. Today, UPS systems are equipped almost 100% with lead-acid batteries.

Today the requirement for load levelling on the power generation level is typically limited to times below one hour because the various types of power plants in use allow for a flexible adaptation of power generation. Peak power plants, such as combined cycle gas power plants, can easily increase their power generation to full power within a quarter of an hour. The situation is changing completely with the increasing production of power from wind generators. In Germany the installed capacity of wind turbines now totals approximately 20 GW and about 7% of electricity was produced by wind turbines in 2006. At times of lowest load in the German grid, power from wind generators is already 50% of load demand. Offshore wind parks could add another 20 to 60 GW installed power (Fig. 10.1). In Denmark it is already the case that wind power regularly exceeds the national load. In this situation storage systems become a relevant alternative to shutdowns of wind generators or selling the electricity at very low cost to other countries. The latter is only possible if sufficient grid transport capacities are available.

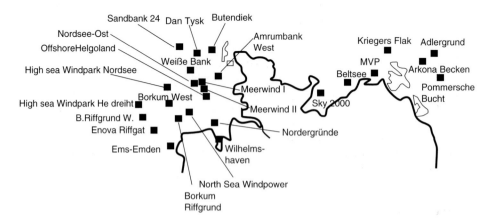

Fig. 10.1. Plans for off-shore wind parks in Germany for the coming 20 years with an installed capacity totalling 60 GW (*Source*: Jahrbuch Erneuerbare Energien 02/03).

To keep the worldwide temperature increase below 2°C, as highly recommended by the Intergovernmental Panel on Climate Change (IPCC) on 2 February 2007, different scenarios have been developed. Fig. 10.2 [Plate 12] shows a scenario aiming at this ambitious goal, taking also into account that nuclear power is not an option for reliable and sustainable power supply. Other scenarios show that the CO_2 emissions per capita (worldwide) must be as low as 1 metric tonne per year in 2050 to keep the CO_2 concentration in the atmosphere below 450 ppm, which is necessary to keep the temperature increase below 2°C according to the climate models. All scenarios state that a significant increase in the share of renewable energies is necessary. While hydro power, geothermal energy and biomass can be stored easily and used at any time, increasing amounts of wind and solar power are contributing to the solutions needed for handling volatile energy production.

When discussing the need for energy storage it is necessary to look at regions like Germany or Europe as a whole as so-called 'copper plate' where power can be distributed to any point without any limitations. This is, however, the assumption in many scenario calculations. In reality we have a grid which provides a physical limit to the energy distribution and there is a balance between load and power generation. Load and power generation must be balanced in any grid segments. Therefore, a need for storage systems for unloading the grids can arise also under local conditions, which need to be examined in detail in every case. A wind turbine installed at the end of a grid extension can cause excessive voltage in the grid segment and needs a balancing system, even though there might be a generation deficit at the other end of the grid segment.

Two kinds of problems must be solved in grids with a significant penetration of wind and solar power:

1. Matching the differences between load forecast and real load as well as between power generation and the power generation forecast.
2. Supplying electrical energy to the grid in periods with low power generation from wind or solar systems.

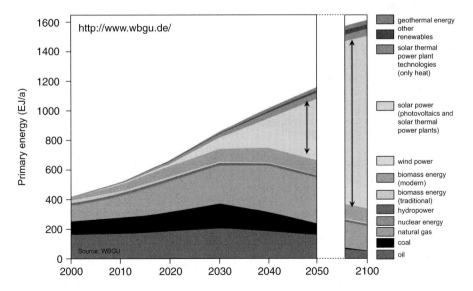

Fig. 10.2. Scenario for a worldwide energy mix in the next decade. [Plate 12]

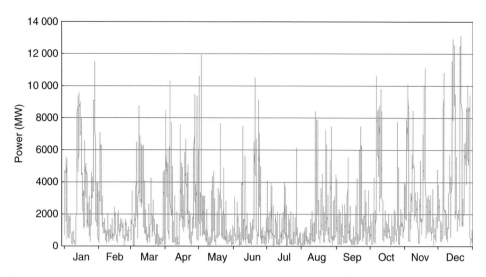

Fig. 10.3. Power generation from wind turbines in Germany in the year 2003.

The first problem can be solved perfectly by different storage technologies such as hydro power, compressed air, or electrochemical storage systems of various kinds. High power is needed for periods of up to 4 hours. The second problem is more complicated. Today, it is solved by still keeping full reserve capacity available in conventional power plants, which can be used in periods of low power generation from wind and sun. However, this is obviously an unfavourable situation because investments for renewable power generators and conventional power plants are needed in parallel. In the medium to long term, other strategies are needed. One solution is increased energy transport via grids within Europe and maybe even beyond because wind and solar radiation never fail to be available in all areas at the same time. The costs will be very high and for the moment it is unclear who should invest in such a grid, but from a technical point of view it is a very interesting option. Another solution is the installation of very large storage systems to supply loads in periods of low power generation.

The requirements concerning power, energy and discharge times are very different and so are the characteristics of the storage technologies (Fig. 10.4 [Plate 13]). The good news is that an appropriate storage technology is available for all practical needs, ranging from power quality systems and off-grid power supply systems to large scale energy storage in the medium and high voltage grids for load-levelling purposes.

For an economic and efficient solution to the problem, intelligent energy management concepts are needed that take into account all kind of loads and power generation units on all voltage levels. There is a significant potential in future energy systems and urban infrastructure for a growing synergy in the transportation and mobility infrastructure, space heating and power generation. A discussion of future scenarios is given in section 10.3 Future Trends in Urban Energy Supply.

We define a 'storage system' as a device that can be connected directly to an electrical grid or an appliance, thus for the user it could be a black box with a single- or three-phase AC or DC connector (Fig. 10.5). A 'storage technology' is the entire device which stores the energy. Thus, a storage system consists of a storage technology utilizing charge and

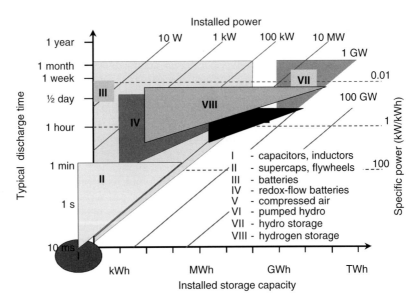

Fig. 10.4. Typical systems sizes for different storage technologies as a function of the installed energy capacity, the available charge/discharge power and the typical discharge times. [Plate 13]

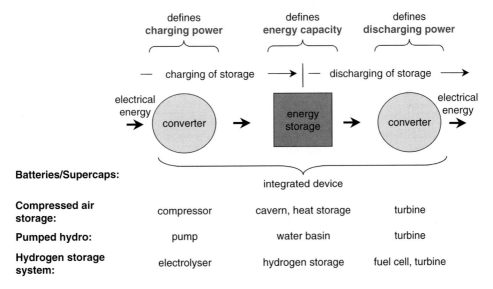

Fig. 10.5. General description of a storage system and examples of some storage technologies.

discharge devices including the power electronic interface to the grid or appliance, and the monitoring and diagnostics to achieve any necessary information for a safe, reliable and long-lasting operation of the storage technology.

There are various requirements for storage systems concerning power, energy and discharge times (Fig. 10.4 [Plate 13]). Obviously the optimization of a storage technology with regard to certain performance parameters often results in performance decrease on another parameter. As an example, an increase in energy density typically results in a decrease of power density. On the other hand, an increase in power density is typically also associated with an increase in safety risks due to a higher degree of concentration of energy in a certain volume. The most important performance parameters are listed below.

- *Energy and power density (volumetric and gravimetric):* In applications with limited available space a high volumetric energy density is required. For mobile applications the weight of the battery is a major parameter of concern and therefore high gravimetric energy densities are required.
- *Lifetime (years and energy throughput):* For most storage technologies lifetime in months and years must be distinguished from cycle lifetime (equivalent to energy throughput). Electrochemical storage technologies in particular can serve only a limited number of cycles but on the other hand the storage systems age even if they are not used. The operation time depends therefore on the operation profile of the battery, but is limited definitively by the calendar lifetime of the storage technology.
- *High power capability:* In an application with short charge or discharge pulses the high power capability is of major concern. It defines how much energy can be charged or discharged from the storage system in a short time, e.g. during regenerative braking in cars.
- *Fast response time:* Especially for power quality devices which are used for voltage and frequency control or uninterruptible power supply (UPS) systems a very fast reaction is mandatory. UPS systems must be able to support the grid in less than 10 ms (half-wave at 50 Hz grid frequency) with nominal power. This requires high performance storage technologies and power electronics and controls.
- *Low requirements on charging strategies and charging electronics:* A charger or a bi-directional power electronic device is the interface between the storage itself and the grid. This interface device must fulfil on the one hand the requirements defined by the grid (voltage, frequency, fast response time, grid failure detection, e.g. islanding) and, on the other hand, the storage technology requirements of safe and efficient operation and long lifetimes. Special charging strategies with regard to current rates, voltage limits and temperature control are needed. The lower the requirements of the storage technology, the lower are the costs for the power electronics interface and the higher the reliability.
- *Wide temperature range:* Storage systems in outdoor applications or if placed close to heat emitting devices such as internal combustion engines or power electronics must cope with a wide temperature range. The performance needed by the application must be available in the full temperature range occurring during the use time.
- *Low safety risk:* The concentration of energy in a small volume entails a safety risk. As an example the gearing of a flywheel may fail and cause damage to its surroundings. Electrochemical systems may explode and cause significant damage. It is noteworthy that, e.g., a lithium-ion battery cell pack holds more energy than a piece of dynamite of the same volume but fortunately does not discharge as fast as the dynamite. Explosions may cause mechanical or thermal damage or release environmentally incompatible materials.
- *High efficiency:* Energy losses cost money, because more energy is needed to charge the storage system than is available from the storage system. In many cases the losses also result in a temperature increase which may limit the maximum throughput to avoid overheating, shorten the lifetime or make an active cooling system necessary.

- *Low maintenance:* Storage systems in rural areas or relatively small units such as in cars should not require significant maintenance, otherwise costs will be extremely high or the storage systems will fail due to insufficient maintenance.
- *Easy estimation of 'state of* function': Parameters such as state of charge, state of health, and cranking or high power capability must be known by the system management systems to ensure proper system operation or to arrange for a replacement of the system in time. Whereas in mechanical storage systems the measurement of, e.g., the state of charge can be made easily from pressure, speed or level measurements, it is often quite difficult for electrochemical storage systems, especially batteries.
- *Easy and cost-efficient recycling:* If storage technologies with limited and rather short lifetimes are used (especially electrochemical systems) an efficient recycling is needed to remove and unload these often polluting materials from the environment. A recycling process that results in materials which can be used directly again for the manufacturing of the same product from which they originate is preferred.

When selecting a storage technology for a certain application, a detailed list of the requirements that includes a prioritization with regard to the above-listed features is needed. On this basis a comparison and selection among different storage technologies can be made.

10.2. Storage Technologies for Electrical Energy

There are various options for storing electrical energy. In most cases a reversible conversion of the energy takes place. Fig. 10.6 shows a classification of the storage technologies with regard to the internal occurrence of the stored energy. Technologies can be distinguished into systems using mechanical energy (potential or kinetic), electrical fields (electromagnetic or electrostatic), or chemical energy.

Furthermore, storage technologies can be separated into so-called 'high power' or 'high energy' storage systems. While high power storage systems deliver energy for typically less than 10 seconds at very high rates, high energy storage systems can provide energy for hours. A clear cut between the two classes is difficult, but a useful parameter is to use the ratio of peak power to stored energy. High power systems should provide 20 kW per kWh of stored capacity or more (see also Fig. 10.4 [Plate 13]). Often the power is limited by the sizing of the power electronics rather than by the storage technology itself.

For the selection of a suitable storage technology it is important to know the typical frequency of charge/discharge cycles (see also Fig. 10.4 [Plate 13]). The relevance of parameters such as efficiency, self-discharge or cycle life depends significantly on the use profile. As an example, self-discharge is of almost no importance for storage systems with a high

Electrical	Mechanical	Electrochemical
• superconductive coils	• pumped hydro	• accumulators with internal storage (e.g. Pb, NiCd, Li-Ion)
• capacitors (various technologies)	• fly wheel	
	• compressed air	• accumulators with external storage (e.g. hydrogen storage system, redox-flow battery, primary batteries with external regeneration (e.g. Zn-Air))

Fig. 10.6. Classes and technologies for storing electrical energy.

charge/discharge frequency, whereas in uninterruptible power supply systems with less than one charge/discharge event per month the energy losses are determined by the self-discharge rate.

For comparison of the energy densities of electrochemical storage systems, different storage technologies are listed below without taking into account conversion efficiencies:

- The energy density of diesel fuel is approx. 12 kWh per litre.
- Energy density of pumped hydro power storage systems: 10 kWh can be stored by pumping $10 m^3$ water from a lower to an upper level with 360 m height difference.
- Storing 10 kWh electrical energy by means of hydrogen requires a volume of 3000 l at 1 bar. At 200 bars pressure 15 l of storage volume are required.
- Storing 10 kWh electrical energy by means of lead-acid batteries requires 130 l battery volume.
- Storing 10 kWh electrical energy by means of lithium-ion high energy batteries requires 50 l volume.

The examples show that the energy density of electrochemical systems is low compared with diesel fuel. Compressed hydrogen (200 bars) is one order of magnitude better compared with lead-acid batteries. The energy density of pumped hydro storage systems is very low.

10.2.1. *Mechanical and electrical storage technologies*

10.2.1.1. *High energy storage systems*
High energy storage systems typically provide power in the range of hours. The typical number of discharges per day is one to two, e.g. for load-levelling, or less in seasonal storage systems or uninterruptible power supply systems.

Compressed air storage systems (CAES):   Compressed air storage systems use electrical power to compress air by means of electrical driven compressors. The compressed air is stored and can be used later for power generation, e.g. with turbines.

Two different concepts can be distinguished:

1. Diabatic compressed air energy storage.
2. Adiabatic compressed air energy storage.

Globally there are two large systems installed and both are of the diabatic type. These systems are so-called hybrid systems because they combine the energy storage system with a conventional gas power plant. The compressed air is used in the turbine process. The air is stored in caverns at pressures between 50 and 100 bar. The maximum efficiencies for these storage systems are around 55%. In addition to a relatively low efficiency, the technology has two characteristics which could be considered as disadvantages: (1) appropriate geological formations are necessary with existing caverns or possibilities to build caverns at reasonable costs; (2) besides the storage system an additional power plant must be installed. This is especially true in those disadvantageous cases where transport capacities in the grid are limited.

The adiabatic concept is in an early stage of development. Fig. 10.7 shows a schematic of the system. No gas power plant is needed but additional heat storage is included in the system to take up the heat given off during the compression of the air. It seems that a

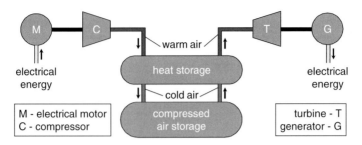

Fig. 10.7. Schematic of an adiabatic compressed air storage system.

cavern will be needed as well for the compressed air. This heat is applied to the air during expansion. Finally, a turbine with a generator is used to generate electrical power.

Theoretical calculations show that efficiencies of 70% can be achieved with the system. In addition no gas power plant in needed. Several challenges must be solved before a large scale prototype can go into operation. Among them are high temperature storage systems and high temperature compressors. Both require new materials and concepts. The installation of a large scale prototype is not expected before 2015.

However, compressed air storage systems are a technology which is suited to build storage systems in the range of several 100 MWh or more. Their characteristics are comparable with pumped-hydro storage plants. In Germany a diabatic compressed air storage system (CAES) has been in operation in Huntdorf since 1978 with an installed charging capacity of about 150 MW, a storage capacity of 580 MWh and a discharge power of 290 MW. Two salt caverns with a volume of 150 000 m³ each and a pressure between 50 and 70 bars are used.

Pumped-hydro storage system: Pumped-hydro storage systems are the backbone of the large power supply grids. With installed capacities in the GW range they are the most important storage systems. During periods of energy, surplus water is pumped from a lower basin to an upper basin, thus the energy is stored by means of potential mechanical energy. When additional power is needed in the grid, the potential energy is used to drive water turbines. Depending on the rated power, the height difference between the lower and the higher basin, and the water flow rates, different types of turbines are available.

In pure pumped-hydro storage systems the upper basin is often an artificial lake with no natural intake of water from rivers, glaciers or streams. These artificial lakes typically can supply water to run the turbines for approximately 8 hours, thus the typical use profile is one to two cycles per day.

The power of the turbines can be controlled over a wide range but the pumps can operate typically only at rated power. Therefore a pump could be either off or on at nominal power. When partial load operation is needed, the net power consumption from the grid is controlled by operating the pumps at rated power and the turbine/generator set in partial load. This definitely results in a low efficiency but gives the flexibility needed to stabilize the grids.

The newest generation of pumps can now work also at partial load thanks to modern electrical motors, drives and power electronics. The newest pumped-hydro storage system in Germany is located in Goldisthal/Thuringia and consists of four machine sets with approximately 250 MW each with the modern pump design. The upper basin provides energy at the full rate of 1 GW for 8 hours, thus 8 GWh. The roundtrip efficiency for pumped-hydro storage systems is between 75 and 80%.

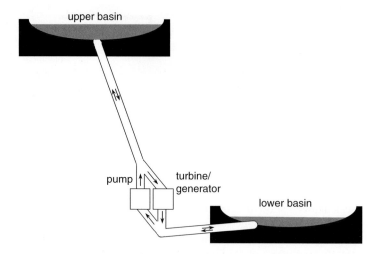

Fig. 10.8. Schematic of a pumped-hydro system with upper and lower basin.

Pumped-hydro storage systems have very long lifetimes; up to 100 years can be achieved, provided regular maintenance is performed. However, investment costs and legislative restrictions are high today and therefore planning and construction typically takes at least a decade. Full power (pumping or generation) can be achieved typically within 1 to 3 minutes. This is faster than any conventional thermal power plant, but slow compared with battery storage systems, flywheels or SMES.

Typically pumped-hydro storage systems are used for load-levelling on a daily basis, thus serving peak power demands typically during noon and evening hours and pumping the water during night hours. In addition, they also serve as an emergency reserve in case of sudden shutdowns of power plants, unscheduled peak power demand or significant deviation between the actual and the forecasted wind power generation.

However, these specialized pumped-hydro storage systems cannot store energy for days or weeks to level out, e.g. weeks of low wind speed or seasonal variations in solar radiation. This can only be solved by using conventional hydro power plants with large dams, which are used today only for power generation from natural feeders (rivers, glaciers, streams). Lakes with more than 60 GWh of stored energy are available even in Europe. These systems would need a retrofit with an additional pump. A major problem in many cases is the lack of a lower water basin or any other sufficient water resource for filling the lake by the pumps.

Based on modern high voltage direct current (HVDC) transmission lines, even storage systems with some distance to the power generation are possible at sufficient efficiencies. This allows, e.g., the use of the mountain areas in the Alps or in Scandinavia for storing energy from power generators almost all over Europe.

10.2.1.2. *High power storage systems*
Strictly speaking, the term 'power storage' is not correct as only energy can be stored. Nevertheless, the term describes very well a class of storage systems which are able to deliver energy at very high rates but typically for short times. Typical discharge times for high power storage systems are below 10 seconds.

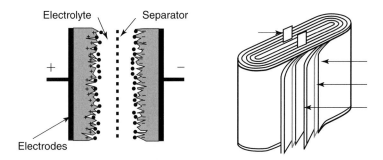

Fig. 10.9. Schematic design of electrochemical double layer capacitors (*Source*: Montena (Maxwell), Epcos).

Electrochemical double layer capacitors: Electrochemical double layer capacitors (EDLC) that close the gap between classical capacitors with almost unlimited cycle number and very high power ratings as used in electronics and power electronics on the one hand and secondary (rechargeable) batteries with several orders of magnitude higher energy densities on the other hand. For EDLCs several names are used, mainly based on the brand names of different manufacturers. Among them, 'supercap', 'boostcap' or 'goldcap' are the most common names.

Energy is stored without an electrochemical reaction. Therefore more than 500 000 full cycles can be achieved. High capacities (up to 5000 farad per litre) compared with conventional capacitors result from the very high effective surface given by the porous active material and the very small distance between the two plates of the capacitor. The capacitor plates are formed from the porous carbon material and the ions from the electrolyte. The distance is only in the order of 10 nm. In a discharged EDLC the ions are located in the bulk electrolyte; in the charged state the ions are located next to the carbon electrode surface. Fig. 10.9 shows the schematic design of electrochemical double layer capacitors.

The nominal voltage of commercial EDLCs is typically between 2.5 and 2.7 V. Organic electrolytes, which are stable up to these voltage limits, are used. High power EDLCs are characterized by very small internal resistances. This allows for very high power densities of up to 18 kW/kg while the energy densities are between 2 and 5 Wh/kg. The cost per kWh is still in the order of some 10 000 €/kWh; however, the costs per kW are in the order of 10 €/kg. Mass production will allow for some further cost reductions.

EDLCs are most suited for applications with a large number of charge/discharge cycles and discharge times in the order of 10 s to make use of the high power capability and the very good cycle lifetime. EDLCs are not suited to store large amounts of energy for long periods.

Flywheels: Energy can be stored also by means of kinetic energy. For compact system designs rotational energy is the only option. The stored energy depends on the momentum of the rotational body and the rotational speed.

Three classes of flywheels can be distinguished: slow turning flywheels at approx. 5000 rpm, medium turning flywheels (approx. 20 000 rpm) and high speed flywheels (approx. 100 000 rpm). The maximum speed of flywheels depends on the mechanical strength of the material which is used for the rotational bodies. Today, composite materials known from the aircraft industry are used. The forces depend on the radius of the rotational body and the rotational speed.

Flywheels are typical high power storage systems. Charging and discharging at very high rates is possible and typically limited only by the motor/generator and power electronic unit. The number of charge/discharge cycles is not really limited by the technology and different manufacturers promise several million charge/discharge cycles. This can be achieved, for example, in the public transport sector (buses, trams, subways) but also in lifting devices such as container terminals in harbours for the recovery of the braking energy. The flywheels can be installed either on the vehicles themselves or in the power supply stations. Flywheels with installed power of several MW are commercially available.

A major disadvantage of flywheels is their high self-discharge rate. Without acceleration, the rotational speed of the flywheel decreases continuously due to the friction losses caused by the remaining air in the evaporated systems and the bearings. Depending on the technology, a flywheel loses its complete energy within less than a day. When there are high numbers of charge/discharge cycles per day this is of no relevance, but for uninterruptible power supply systems with very few discharge events this is a significant disadvantage and cost factor. In mobile applications gimbal mounting and special burst protection are necessary to minimize safety risks in case of failure or accident.

Supra-magnetic energy storage (SMES):   While capacitors store electrical energy in a static electrical field, coils store energy in an electro-dynamic field. The stored energy depends on the inductance of the coil and the current flowing through the coil. To store energy for long periods of time the ohmic resistance of the coil must approach zero. This can be fulfilled especially by supra-magnetic materials. Below a certain characteristic temperature supra-magnetic materials lose any ohmic resistance and current can flow without energy loss. Supra-magnetic coils are used in many technical applications such as medical devices that require very high magnetic fields.

However, the large scale use of supra-magnetic coils is limited for two main reasons. Supra-magnetic conditions occur only at very low temperatures. Depending on the materials, liquid helium or liquid nitrogen is needed. This requires very good thermal insulation and energy to compensate for thermal losses. Besides this efficiency problem, large storage systems would create strong magnetic fields. The impact of these magnetic fields on human beings or other biological material is not yet fully understood.

Small storage systems are available on the market as prototypes. Typically they are used in applications where very high power is required for short times (low energy content, high discharge currents). Short circuit current sources based on supra-magnetic coils are sold as commercial devices for providing sufficient currents in case of grid failures.

The electrical characteristic of supra-magnetic storage systems is comparable with double layer capacitors and flywheels. However, the system technology, including the cooling systems for the liquid nitrogen, is more complex and the efficiency including the cooling systems is relatively low. Therefore, supra-magnetic coils are currently not realistic storage systems for high power storage applications in grids.

10.2.2. *Electrochemical storage systems*

Electrochemical storage systems can be classified according to Fig. 10.10. Two major classes can be distinguished: Storage systems with integrated energy storage and storage systems with external storage.

In systems with integrated storage the electrochemical reaction takes place directly on the surface or in the active material and no spatial separation between the charged active material and the charge/discharge reaction is possible. As a result an increase of the

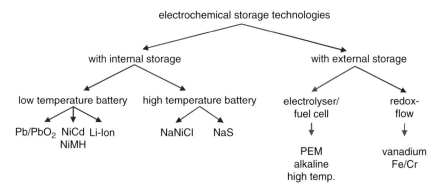

Fig. 10.10. Classification of electrochemical storage technologies.

charge/discharge power always requires an increase in the energy storage capacity. All classical batteries belong to this group. Batteries can be distinguished further into batteries operating at room temperature and those which need higher temperature for proper operation. In systems with external storage a separation of the energy conversion unit and the stored active material is possible. This allows for an independent sizing of power and energy of the storage system. The power conversion unit and storage tanks can be separated physically. The most important representatives of this class are redox-flow batteries and hydrogen storage systems (see below).

10.2.2.1. Rechargeable batteries with internal storage

Electrochemical batteries can be made from an almost infinite number of material combinations. Several of them are used in niche markets for special applications with very specific requirements concerning energy densities, self-discharge rate or operating temperature. These technologies are not suitable for large scale application and therefore they are not discussed here.

Commercially available battery technologies for large scale energy storage systems in stationary applications are today lead-acid and NiCd batteries. Lithium-ion and NiMH batteries, and for high temperature sodium sulphur (NaS) or sodium nickel chloride (NaNiCl) batteries, have very good potential for commercialization. These battery technologies will be discussed in the following sections.

Lead-acid batteries: The most important storage technology according to the installed battery capacity worldwide is the lead-acid accumulator. Fig. 10.11 shows the design of a battery system starting from the porous active materials with high internal surface area. Different plate designs are available for different applications. Positive and negative plates together with an electrolyte form a battery cell and several battery cells connected in series or parallel result in battery systems with capacities and voltages that are compatible with the technological requirements.

The most important technical parameters of lead-acid batteries, which are mainly formed from lead, sulphuric acid, plastic housings and separators, are gravimetric energy densities around 25 Wh/kg or 40 kg/kWh and volumetric energy densities around 75 Wh/l or 131/kWh. The efficiency depends on the load profile and is between 80 and 90%. High

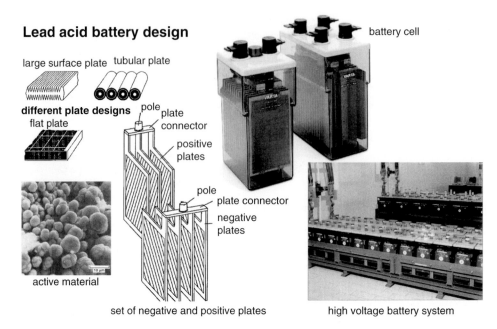

Lead acid battery design

large surface plate   tubular plate

different plate designs   pole plate connector
flat plate

positive plates

pole
plate connector
negative plates

active material

set of negative and positive plates

battery cell

high voltage battery system

Fig. 10.11. Design of battery storage systems from lead-acid batteries. Porous active materials with high internal surface are used in different plate designs. The plates are integrated into a cell housing together with an electrolyte. A cell is the smallest technical unit with a nominal voltage of 2 volts. Cells can be connected in series to achieve necessary voltage and power levels.

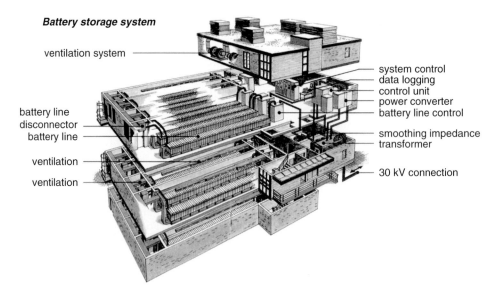

Battery storage system

ventilation system

battery line disconnector
battery line

ventilation

ventilation

system control
data logging
control unit
power converter
battery line control

smoothing impedance
transformer

30 kV connection

Fig. 10.12. Battery storage system for frequency and voltage stabilization for the West Berlin power supply grid before the reunification of Germany. (Start of operation in 1986. *Source*: Exide Technologies).

quality stationary batteries can achieve lifetimes of six to 12 years and 2000 full cycles. With partial-state-of-charge cycling, more than 10000 times the nominal capacity can be charged and discharged from the battery. The costs depend on the quality of the battery and for stationary lead-acid batteries the range is between 100 and 300€/kWh. Starter batteries for the automotive industry are sold below 40€/kWh, but their lifetimes are limited with regard to energy throughput. Almost 100% of industrial batteries are collected and recycled in Europe. The lead recovered in the recycling plants can be used for the production of new batteries.

Valve regulated batteries (VRLA) with internal gas recombination are used as well as flooded batteries which need refilling with water regularly and require sufficient air flow in the battery compartment to avoid the build-up of critical concentrations of hydrogen and oxygen. VRLA batteries are easier to handle but are more expensive, and flooded batteries often show better cycle lifetimes.

Battery storage systems based on lead-acid batteries are in operation in many countries worldwide to solve local power quality problems. Applications include the stabilization of grid extensions or providing frequency and voltage stabilization. The largest battery system which has been operated in Germany was a 17MW/14MWh system in Berlin. It went into operation in 1986 when West Berlin was not connected to the Western European grid. The battery was used for frequency and voltage control. The battery system provided more than 7000 equivalent full cycles in a partial-state-of-charge operation. There were plans to build more systems of the same type, but the German reunification followed by the connection of Berlin to the Western European grid made this unnecessary.

Table 10.1 shows a selection of battery storage systems with a maximum of 70MW installed power and up to 40MWh. Power is limited typically by the installed power electronics rather than by the battery itself. Batteries can supply very high power for short times. Start pulses in conventional cars are in the order of 5kW from a battery with a capacity of 0.5kWh.

Lithium-ion batteries: Lithium-ion batteries dominate the market in the field of mobile applications such as laptop computers, mobile phones, or cameras. Compared with lead-acid or NiCd batteries, the energy density is very high and reaches 200Wh/kg in high energy cells. High specific costs are acceptable in these applications. Today production costs for lithium-ion batteries in mass production are below 300US$/kWh. Custom-made battery packs that include electronic battery controls and safety equipment are sold at significantly higher costs.

When discussing lithium-ion battery technology, it is important to be aware of the fact that this describes a class of batteries that use lithium. However, there is a very wide

Table 10.1. List of large battery storage systems which have been operated or which are still in operation (selection).

Unternehmen	Leistung	Energie
BEWAG, Berlin, Germany	17MW	14MWh
Kansal Electric Power Company, Tatsumi, Japan	1MW	4MWh
Southern California Edison Company, Chino, CA, USA	10MW	40MWh
Vall Reefs, Godmine, South Africa	4MW	7.4MWh
Hawaii Electric Light Company, Hawaii, USA	10MW	15MWh
Puerto Rico Electric Power Authority, San Juan	20MW	14MWh
Chugach Electric Assn., Anchorage, Alaska, USA	20MW	10MWh
Golden Valley Electric Assn., Fairbanks, Alaska, USA	70MW	17MWh

variety of electrolytes and electrode materials in use today in commercial cells and which for the future promise lower costs, longer lifetime and increased intrinsic safety. Different materials result also in different properties such as lifetime, open circuit voltage, intrinsic safety and electrical performance. Therefore it is very difficult to make general statements concerning these parameters. Two major developments can be distinguished: high power and high energy cells. The materials used are not necessarily different, but high power cells are made from very thin electrodes and very thin separators to keep the internal resistance as low as possible. As a result the energy density of high power cells is significantly less compared with high energy cells. High power cells are available today with more than 2500 W/kg for a 10 second charge or discharge pulse at 50% state of charge (SOC). The energy density for these cells is between 80 and 100 Wh/kg. The high energy cells as used in laptops or mobile phones are typically limited to currents which discharge the battery within one hour. However, this current can be used continuously.

Lithium-ion batteries, which are most common today, have a nominal voltage of 3.6 V/cell. This reduces the number of cells connected in series for a certain voltage by a factor of three compared with NiMH battery technologies. But an individual cell equalization and control electronic is necessary to ensure that no cell is overcharged.

Today a high quality, high energy cell can achieve 5000 cycles with 80% DOD with energy densities between 150 and 200 Wh/kg. High power cells with more than 2500 W/kg can achieve cycle lifetimes of several 100 000 cycles with 2% DOD. The efficiency is between 90 and 95%. The biggest threat is the safety of lithium-based batteries. Many different approaches are under development to solve this problem. Among them $LiFePO_4$ is a very promising material with a high intrinsic safety.

Lithium-based battery technology has the potential to solve most problems in the field of portable and mobile applications and will also be an alternative to lead-acid batteries in the stationary sector. However, this market is still under development. Cells with more than 100 Ah are hard to find on the market.

Nickel-metal-hydride (NiMH) and nickel-cadmium (NiCd) batteries: Nickel-cadmium (NiCd) batteries have been available as commercial products for more than 100 years and represent the most important storage technology for stationary and traction applications beside lead-acid batteries. Nickel-metal-hydride batteries have been developed to replace NiCd because cadmium is a poisonous heavy metal. The cadmium electrode has been replaced by a metal hydride electrode which adsorbs hydrogen during discharge. The NiMH technology was very successful and achieved superior performance compared with NiCd batteries, especially with regard to the gravimetric energy density. Therefore NiMH batteries were for several years the dominant technology for portable applications such as laptops or mobile phones. Today NiMH batteries are used in all commercially available hybrid electric vehicles. However, the lithium-ion battery technology is replacing NiMH batteries. In portable applications this has already taken place and for hybrid electric vehicles the replacement is expected to start before the end of this decade. The main reason is that NiCd and NiMH batteries have only a limited cost reduction potential because nickel is an expensive raw material and contributes to the costs significantly. Furthermore lithium-based batteries are superior in almost all technical performance parameters such as energy density, voltage per cell and efficiency.

The efficiency of the nickel-based batteries is in the range of 70 to 75% and therefore significantly less compared with lead-acid or lithium-based batteries. NiCd batteries can operate with sufficient power even at −40°C and are therefore, among all other battery technologies, most suited for low temperature applications. Charging above 45°C is almost impossible for NiCd and NiMH batteries.

The largest stationary NiCd battery provides more than 40 MW for 20 minutes and was installed a few years ago in Alaska as a local back-up battery in a weak grid with long grid extensions. The battery is bridging the time gap until diesel generators can supply the grid in case of grid failure.

Sodium-based high temperature batteries:　Sodium nickel chloride (NaNiCl, also called zebra battery) and sodium sulphur (NaS) batteries are different to the other secondary batteries discussed above. The battery is based on liquid active masses and a solid state electrolyte. The electrolyte, which serves also as the separator, is typically an ion-conducting ceramic. To achieve a sufficient conductivity of the electrolyte and to melt the active masses, temperatures between 270 and 350°C are necessary.

The temperature must remain high and this can be achieved by using both a very good thermal insulation and the heat produced during charging and discharging. However, during stand-still periods some external heating is required, which reduces the overall energy efficiency.

In the 1980s NaS batteries were considered an interesting technology for full electric passenger vehicles. Due to the safety risk of liquid sodium and sulphur in case of a crash the technology did not succeed. NaNiCl batteries are offered by one manufacturer in Europe for the transport sector. However, as the batteries should cool down as seldom as possible, they are most suited for vehicles in fleet operation or many operational hours per day such as in buses, taxis, police cars, urban transport, etc. High temperature batteries are well suited for high energy applications rather than for high power applications. NaS batteries are currently produced in Japan by one manufacturer and they are used for load-levelling applications (Fig. 10.13). Systems of up to several 10 kWh are in operation. The lifetime results are very good.

Generally high temperature battery technology has a potential for low costs and long lifetimes in mass production. To achieve long lifetimes, high quality, fully automated production lines are needed.

Quelle: Tokyo Electric Power Company

Fig. 10.13. Sodium sulphur (NaS) high temperature battery for load-levelling with 6 MW, 48 MWh (Tokyo Electric Power Company, Tsunashima).

10.2.2.2. Electrochemical storage systems with external storage
As shown in Fig. 10.5 storage systems with external storage have separated energy converters and storage systems. They can be designed independently from each other according to the needs for power and energy. Especially for storing large quantities of energy this is an interesting option. Redox-flow batteries and hydrogen storage systems will be discussed as typical representatives of this technology. In addition, the zinc-bromine battery will be discussed in this section, even though it is, to a certain extent, a technology between conventional batteries and storage systems with external storage.

Redox-flow batteries: The active material of redox-flow batteries is dissolved salt ions in fluid electrolytes. The electrolyte with salts is stored in tanks and is pumped into a reaction unit for charging and discharging with pumps. The solubility of the salts is typically limited and therefore the energy densities are in the range of the lead-acid battery. During charging and discharging the charge of the ions is changed (e.g. $V^{5+} + e^- \rightarrow V^{4+}$ and $V^{2+} \rightarrow V^{3+} + e^-$ for the two electrodes during discharging in a vanadium redox-flow battery). This process happens also, e.g., in lead-acid batteries (e.g. $Pb^{\pm 0} \rightarrow Pb^{2+} + 2e^-$) but the charged and the discharged materials appear as crystals. Currently different material combinations are under investigation such as the vanadium battery (V/V), Fe/Cr, Br_2/Cr, or $NaBr + Na_2S_4/Na_2S_2 + NaBr_3$ (Regenesys). The vanadium battery is very interesting because both electrodes contain vanadium at different charged states (V^{5+}/V^{4+} and V^{2+}/V^{3+}). In this case crossing over through the membrane in the reaction unit does not poison the other electrode (Fig. 10.14).

The central reaction unit operates in a similar way to fuel cells or electrolysers. A membrane separates the positive and negative electrolytes (Fig. 10.14). The charge/discharge reaction takes place on the membrane by means of catalysts. The size of the reaction unit defines the charge/discharge power; the size of the tank determines the energy content of the storage system.

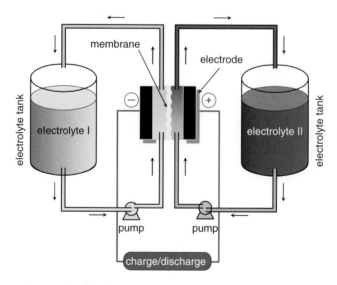

Fig. 10.14. Schematic of a redox-flow battery.

Fig. 10.15 shows the prototype of a vanadium-redox battery and a concept study for a large scale redox-flow battery. Generally the technology is suited for stationary application with low to medium power requirements. Increasing the capacity can be achieved very simply by increasing the volume of the tanks. The electrolytes can be delivered easily and efficiently with trucks.

Redox-flow batteries have already been investigated in the 1970s and 1980s for stationary applications. Various material problems caused a decline in the activities; recently the activities have started to increase again and, e.g., new membrane materials are promising solutions to the old problems.

Reliable field data on the lifetime of the central reaction unit and the electrolyte are very rare. As none of the components undergoes any structural change during charging and discharging, the cycle life is very long. For vanadium batteries cycle lifetimes of more than 13 000 full cycles have been presented in the literature. Especially in the case of vanadium batteries the electrolyte can be recycled without any losses. Even though commercial systems based on the vanadium technology are now offered by different manufacturers, it is still difficult to estimate the costs for a mass product. Costs below 200 €/kWh seem to be realistic. The vanadium redox-flow battery has achieved the most mature status and therefore it is the only commercially available redox-flow battery to date. Other technologies such as Fe/Cr are cheaper with regard to the materials, but technological problems with a so-called crossing over through the membrane prevent these materials today from being commercialized. But vanadium is a relatively rare element and the costs are difficult to estimate for a large scale market introduction.

Efficiencies of 80 to 85% for vanadium redox-flow batteries have been demonstrated. Taking into account electronics and pumps, the overall system efficiency drops to approx. 75%. This is well below the efficiency of lithium batteries (>90%, see lithium-ion batteries, above) but well above hydrogen storage systems (<40%, Fig. 10.18). Self-discharge is almost zero.

Zinc-bromine batteries: Zinc-bromine batteries represent an intermediate between conventional batteries and redox-flow batteries. Bromine is used as active material for the positive electrode. The bromine is dissolved in an electrolyte. Zinc is the active material of the

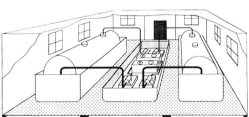

Fig. 10.15. Left-hand picture: prototype of a vanadium-redox battery with the tanks for the negative and the positive active mass in the background and the stacks for the energy conversion in the foreground (picture: Fraunhöfer ISE); concept study for a large scale redox-flow battery.

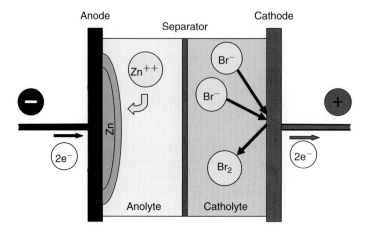

Fig. 10.16. Schematic of the reactions in a ZnBr battery.

negative electrode and is dissolved in the charged state in an electrolyte as well. However, during discharge the zinc is deposited as metallic zinc on the anode. Therefore, the overall capacity is limited by the amount of zinc that can be deposited on the anode. An increase in the capacity is not possible by simply increasing the electrolyte tanks. Fig. 10.16 shows a schematic of the reaction of the zinc-bromine battery.

Initial commercial products based on the zinc-bromine technology are available with energy densities in the order of 70 Wh/kg. A five year lifetime with one daily cycle has been demonstrated; a doubling of the lifetime should be achieved in the coming years. Expected costs per kWh stored energy are in the order of €0.05.

Hydrogen energy storage systems: Using hydrogen for storing electrical energy requires three components (Fig. 10.5):

1. An electrolyser to convert electrical energy into chemical energy by means of hydrogen production from electrical power.
2. Hydrogen storage to store the chemical energy.
3. A fuel cell or hydrogen gas turbine to reconvert the chemical energy stored in the hydrogen into electrical power.

Only this configuration is equivalent to a battery system. Fuel cell systems on their own are energy converters, such as internal combustion engines or turbines, because they need to be fed with fuels continuously.

Hydrogen production by electrolysis is a well-established process and large commercial systems are available. Simply speaking, electrolysis can be achieved by inserting two electrodes in an ion-conducting electrolyte and by applying a potential of more than 1.23 V to the electrodes. Oxygen is formed on the positive electrode, hydrogen on the negative electrode (Fig. 10.17).

The technology development aims at high rate throughput electrolysers at high efficiencies. This requires noble catalysts such as platinum and palladium for low temperature electrolysers. Low temperature electrolysers (below 100°C) based on polymer electrolyte membranetechnology can achieve efficiencies between 75 and 85%, depending on the current rate. High temperature electrolysis at temperatures above 1000°C gives the highest

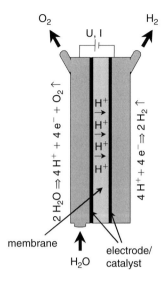

Fig. 10.17. Schematic of a low temperature polymer electrolyte membrane electrolyser for the production of hydrogen from electrical power and water.

electrical efficiencies of up to 93%. This technology uses a solid state electrolyte which can be compared with high temperature batteries. Additional energy can be fed from heat sources.

A reduction in the power consumption of polymer electrolyte membrane (PEM) electrolysers can occur within milliseconds. This is an interesting feature for grid controls. In case of a rapid increase of loads or a rapid decrease of power generation, the electrolysers can be turned off to level out demand and supply immediately. Nevertheless it is necessary to understand that hydrogen from electrolysers is only CO_2 free if the electricity comes from CO_2-free sources such as hydro, wind, solar or nuclear power.

While the hydrogen is collected, dried and cleaned, the oxygen typically is released to the environment and is taken back from the atmosphere if needed for power production. Only in special cases is the oxygen used directly onsite for chemical processes or to stimulate biogas production. Fuel cells have higher efficiencies when operated with pure oxygen compared with air with 21% oxygen, but the additional effort in transporting and storing the oxygen typically does not pay for the increase in efficiency.

Today the most suited technologies for storing and transporting hydrogen are:

1. Compression
2. Liquefaction
3. Adsorption (e.g. in metal hydrides)

Compressed hydrogen is a well-established technology. Hydrogen can be transported in gas bottles, tanks or pipelines. Compared with natural gas (CH_4) the energy density of hydrogen per volume is only approximately one third. Therefore the storage volumes and transport capacities are significantly higher for hydrogen. Storage tanks made from composite materials for up to 700 bar are under development; however, large storage tanks for several $100\,000\,m^3$ are operated at pressures well below 100 bar.

Liquefaction of hydrogen requires cooling down to $-253°C$. There is no way to liquefy hydrogen at room temperature by increasing the pressure. The liquefaction process needs approximately one third of the energy which is stored in the hydrogen. Transportation by ship over long distances in well-insulated tanks is possible. The losses per day to keep the hydrogen cool are between 0.1 and 1%.

The best ratio between energy density and energy consumption for storing the hydrogen is provided by metal hydrides. Spongy and highly porous metals are used which can adsorb hydrogen. A metal hydride storage device at 10 to 20 bar can store in the same volume as much hydrogen as compressed hydrogen storage tanks at pressures between 200 and 700 bar. Because of the lower pressure the energy efficiency is significantly better. The major disadvantages are high costs for the metals (e.g. nickel-based alloys with rare earth metals) and the high weight of the storage systems. The hydrogen content per weight is only 1 to 2%; 98 to 99% are needed for the metal alloy. Even though it is the most efficient and safest storage technology, it is not suited for mobile applications nor for large storage capacities due to the high costs.

Several other materials using the adsorptions process are under development. Among them carbon nano-tubes are one of the promising technologies.

To supply electrical energy from a hydrogen storage system, a conversion of the hydrogen into electricity is necessary. Hydrogen turbines can be used in large systems, but fuel cells have the best medium- to long-term perspective. Various types of fuel cells are available with different characteristics. Two interesting technologies are polymer electrolyte fuel cells, which operate at temperatures around 80 to 100°C, and solid electrolyte fuel cells operating at around 1000°C. Conversion efficiencies are between 40 and 60%.

Fig. 10.18 shows the efficiency chain for a hydrogen storage system. The chain goes from electrical energy to electrical energy. Only small losses are taken into account for the hydrogen storage. High pressure or liquefied hydrogen storage systems would reduce the storage efficiency to 70 to 80% from 95%. The figure compares a calculation of the efficiency

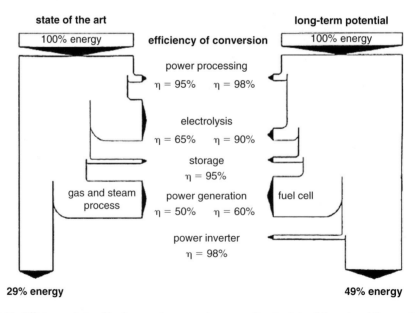

Fig. 10.18. Efficiency chain of hydrogen storage system according to state of the art and the maximum long-term potential.

assuming today's best efficiencies and a calculation assuming the best efficiencies that might be achieved in the coming decades. The overall efficiencies amount to only 29 to 49%. Maturity and reasonable costs for all components still need to be demonstrated. Nevertheless, the complexity of a hydrogen storage system is significantly higher and the efficiency is just about 50% compared with battery storage systems.

10.3. Future Trends in Urban Energy Supply

In the following section two additional approaches to the task of energy storage are discussed. Storing heat instead of electrical power by just using different control strategies for combined heat and power generation reduces the costs significantly (Section 10.3.1). The same can be achieved if existing storage systems can be used also for grid stabilization purposes. These storage systems will be available in a future plug-in hybrid of full electric vehicles for personal transport (Section 10.3.2). Finally a discussion on the usefulness of the so-called hydrogen economy is given because it is related very closely to the question of the types of storage systems needed for electrical power (Section 10.3.3).

10.3.1. Heat and power co-generation (CHP)

Heat and power co-generation (CHP) allows for a highly efficient use of primary energy sources including biomass. This is achieved by using heat and power instead of wasting the heat as in large central power plants. However, the transport of heat is efficient only over relatively short distances; therefore CHP is most effective if the CHP devices are located where the heat is needed. Today, most CHP units have an installed electrical capacity ranging from several kW to some 100 kW. They are controlled by the need for heat. However, this results in electrical power generation in times when heat is needed rather than in times where power is needed. A typical example for a mismatch between heat and power demand is the early morning. Many space heating systems 'wake up' around 5 to 6 am to ensure comfortable room temperature after the thermostat was turned down to save energy during the night. However, this is also a time of minimum power demand, because people are still sleeping. To store the electrical energy before feeding it to the grid in peak load time is inefficient and expensive. But if the system is equipped with a heat storage system a decoupling between heat demand and heat generation is possible and gives the freedom to operate the CHP unit on the basis of power demand rather than heat demand. This gives a significantly increased value and helps to resolve the peak power demands. The CHP unit can be operated even more as part of virtual power plant (VPP) by a central control. This might be operated either by the local utility, the grid operator or an independent power producer (IPP).

Thermal storage systems are more efficient and cheaper than electrical storage systems. Low temperature heat (60 to 100°C) can be realized, e.g. with water. For high temperature storage systems with temperatures well above 200°C, which are needed for several industrial processes, more research is necessary. Different melting salts or phase change materials (typical from the solid to the liquid phase) are interesting options for such heat storage systems.

10.3.2. Growing synergy of the power supply system and the transport sector

Beside the rapid growth of renewable energy generation capacities in the classical power supply sector, the upcoming transition process in the transportation sector is important for a sustainable future. Transportation is responsible for a quarter to a third of the overall CO_2 emissions and it depends today to a very high degree on oil. Further discussion will focus on private passenger traffic. Airplanes, ships and trucks are a different story.

The reports on climate change have forced politicians all over the world to call for more efficient cars. In the European Union a fleet average of $130\,g/km$ CO_2 by 2012 is under discussion; however, this can only be a first step. Furthermore, particle emissions, especially from diesel engines, are also under discussion. There are different technologies available to realize these goals for the coming five years. A consistent use of biofuels or synthetic biofuels and very efficient diesel engines or hybrid electric vehicles is the most promising short-term option. Fuel cell cars will not enter the market in large quantities at acceptable life cycle costs before 2015 or 2020. Full electric vehicles are not more efficient with regard to CO_2 emissions compared with highly efficient diesel engines as long as electricity is produced mainly from fossil fuels, as, e.g., in Germany.

A very likely scenario is the market penetration of plug-in hybrids and full electric vehicles in large quantities until 2020. Plug-in hybrids will have electrical storage systems of 5 to $10\,kWh$ for 30 to $60\,km$ of pure electrical driving. Due to the hybrid concept the vehicles can also travel long distances with the support of an internal combustion engine powered by biofuels. Fuel cells might replace the internal combustion engines later on, whereas it is unclear if they will use biofuels or hydrogen. Full electric vehicles probably will have electrical storage systems with a capacity to serve the vehicle for 100 to $200\,km$.

Thus, especially in urban areas, electricity will become the major 'fuel' for individual transport. According to the scenario depicted in Fig. 10.19 the vehicles will not only need electricity, but also provide excellent large scale, decentralized energy storage. They become an active part of the grid. Charging of the battery can be done during times of excess electrical energy production from renewable energy sources. Together with intelligent management, the storage capacity can be used also for stabilizing and supply the grid for load-levelling purposes. Especially plug-in hybrid technology offers a high degree of flexibility because the vehicles can be operated even when the energy storage is not fully charged due to an unexpected need to use the vehicle.

10.3.3. *Hydrogen economy – a solution for future energy supply in urban areas?*

Beside the fact that using oil and natural gas results in CO_2 emissions, which need to be reduced drastically to protect the earth from a further significant temperature increase, they are also limited resources. Therefore, oil and natural gas will not be available for many decades to the same extent as today for powering cars for individual transport or space heating in urban houses.

While trying to maintain the general infrastructure and technologies as they are today, different alternatives for oil and natural gas as the primary energy carrier have been discussed for decades. Hydrogen is the most popular alternative among them and is often seen as the solution for the future energy infrastructure.

In the following some aspects about a possible hydrogen economy are discussed to understand how hydrogen could become a major energy carrier in future urban energy supply scenarios.

No doubt, hydrogen has some desirable properties, which make it a clean and universal energy carrier. Hydrogen can be used as fuel for fuel cells or turbines for power generation at efficiencies as high as 50%. It can be used also in internal combustion engines (ICEs) in cars instead of gasoline or diesel, or for space heating and cooking by using catalytic burners. All processes are emission free in terms of CO_2, SO_2 or particles, and result mainly in steam or water.

Example: Plug-in hybrid serving as grid storage systems

- Assuming a middle class car with 6 l/100 km fuel consumption.
- Taking into account the energy content of fuel and an average efficiency fuel to wheel of 25% results in approximately 15 kWh of energy needed for 100 km.
- 10 kWh$_{el}$ from a battery is sufficient for 60 km full electric driving, taking into account efficiencies of electric converters, drives and motors as well as regenerative braking.
- Assuming an average usage of a car according to long term statistics between 10 000 and 20 000 km/year or approximately 30 to 60 km/day results in less than 2 hours usage of the car per day on average (statistical even significantly less).
- Assuming an installed power for an external charge/discharge unit of 5 kW, which is possible in Germany for any household, allows recharging of 80% of the capacity in 1.6 hours
- Even under the conservative assumptions made above no more than four hours a day for driving and recharging the battery are needed.

The battery of plug-in hybrids or full electric vehicles can be used 20 hours a day for any other purpose. According to the example 10 kWh of electrical storage capacity with 5 kW charge/discharge power is available for 20 hours a day for purposes besides driving.

- The average electrical energy consumption of a German standard household is approximately 10 kWh.
- In Germany approximately 40 Million cars are in operation.
- Today, more than 20 GW wind power are installed in Germany.
- Having 10% of the cars equipped with a battery as given above results in 20 GW installed charge/discharge capacity for a period of 2 hours.
- Taking into account an electrical peak load of close to 80 GW in Germany this is 25% of the peak load, or 40% of the cars are needed to equal the peak load.
- The average electrical energy consumption per day in Germany is approximately 1 TWh.
- 40% of the cars with 10 kWh storage capacities can serve the full electrical load in Germany for close to 4 hours.

Fig. 10.19. Example on how plug-in hybrid or full electric vehicles can support the grid with peak power (charging and discharging).

Beside safety aspects, over which there is some controversy and which are not necessarily negative for hydrogen compared with gasoline or natural gas, there are other aspects which need to be taken into account. The energy density of hydrogen is only approximately one third compared with natural gas. Thus pipelines and tanks for hydrogen must be either enlarged by a factor of three or the compression must be increased accordingly. Liquefaction of hydrogen is only possible after cooling down to −253°C. Compressing the gas at roomtemperature will not result in liquefaction. The process is energy consuming and needs in the order of 30% of the energy stored in the hydrogen. Even without liquefaction, energy losses during storage, transport and distribution of compressed hydrogen still amount to 10 to 20%.

Most importantly, hydrogen is not a primary energy source such as natural gas, oil or coal; it is an energy carrier and can be compared in its function with electricity. Hydrogen must be generated either by reforming of different kinds of fossil fuels or biomass, by biochemical or solar-chemical processes, or by electrolysis from water using electrical power. Biochemical and solar-chemical processes have not yet shown their ability to produce hydrogen in large quantities; nevertheless this is an important field for R&D with potential for significant innovations. Producing hydrogen from fossil fuels will not solve the CO_2 problem nor help to overcome the shortcomings of fossil fuels in the coming decades. Therefore, this process cannot be the basis for a sustainable future energy supply. This

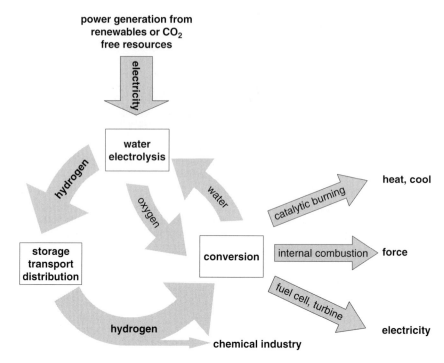

Fig. 10.20. Solar hydrogen circle.

leaves the production of hydrogen from electrical power with electrolysers. As discussed under hydrogen energy storage systems above, several different electrolyser technologies are available; among them high temperature electrolysis (Hot Elly) is the most efficient. Today commercial electrolysers of the membrane and alkaline type are available with up to $100\,MW_{el}$. But electrical energy is needed and it has to come from somewhere. Only CO_2-free power generation is worthy of consideration as a solution to the CO_2 problem. This leaves power generation from any kind of renewable energy source or nuclear power plants (Fig. 10.20). The latter still leaves the questions concerning safety of nuclear power plants, handling of radioactive waste, proliferation of nuclear material and a sufficient supply of uranium in the coming decades.

Besides, if electricity is already available, the question is why should we transfer the energy at its highest energy level into hydrogen? Surely, there are boundary conditions where transport of electrical energy via grids is not possible due to limited transport capacities. This could be the case with off-shore wind parks where transmission lines for the rated capacity would be necessary if no alternative is available. Production of hydrogen directly off-shore and transporting the hydrogen via pipelines or ships might be an interesting alternative to expensive power grids. But energy efficiencies are always very low when using hydrogen as an intermediary energy carrier. Other storage technologies with higher efficiency, such as pumped hydro or electrochemical storage systems, can do the job at lower costs. Electrical energy transport by means of high voltage direct current (HVDC) transmission grids is reliable and efficient. Losses are assumed to be in the order of 5% per 1000 km.

All future energy needs in the urban environment (not taking into account large scale industry, especially energy intensive or chemical production plants) can be served perfectly with electrical energy. The main areas of energy consumption in the urban environment are:

- Energy for lighting, communication and electrical appliances in private households and public buildings.
- Energy for space heating and cooling in private households and public buildings.
- Energy for individual transport.
- Energy for public transport.
- Energy for small and medium business enterprises.

By increasing the efficiency of energy use the overall energy consumption per resident can be reduced easily by a factor of 2 without any reduction in the standard of living.

The need for space heating can be reduced to a minimum by using appropriate measures for insulation. The remaining heat (for the Central European climate a reduction below $10\,kWh/m^2/year$ is possible already with today's technology) can be supplied from electrical driven heat pumps. The transport sector can make use of electricity for urban traffic as well; long-distance journeys may be performed with biofuels.

Thus, from today's point of view, there is no need and no efficient or economic business case for a large scale introduction of hydrogen into urban areas. Electricity produced from renewable energy sources with various technologies (wind, photovoltaics, solar thermal power plants, geothermal energy, wave energy, tidal energy, biogas and biomass) can serve all energy needs in the urban environment.

Electricity becomes increasingly prominent and this increases the need for efficient and cost-effective electrical energy storage units at sufficient efficiencies.

10.4. Conclusion

A phasing out of the use of fossil fuels for power generation moving towards 2050 is necessary to meet the required CO_2 levels to stabilize the world climate. A large variety of renewable energy technologies are available to fill the gap at acceptable costs. However, the variability of power generation will increase due to the fluctuating nature of energy sources such as wind and solar.

This will require additional storage capacities for electrical energy to level out power generation and loads at any one time. In addition to the grid level there is a need to stabilize weak grid segments and to maintain the power quality of critical loads directly at the customer's site. To serve these storage needs a large variety of storage technologies are available. As the requirements range from systems for very high power for a few seconds to very large energy storage systems, there is also a need for many different storage technologies. No single storage technology can fulfil all requirements.

Flywheels are a promising technology for high power for a few seconds. For large scale energy storage in the range of many 100 MWh or GWh, pumped-hydro storage systems will remain the most efficient technology. Adiabatic compressed air storage systems with thermal storage are an interesting option but the concept needs to be proved. For energy storage systems ranging from several 10 kWh to several 10 MWh, battery systems based on lead-acid, lithium-ion and sodium high temperature battery technology are suitable solutions. Further development must aim mainly at an increase of lifetimes and a reduction of cost. Energy and power densities are not of major concern in stationary applications. Redox-flow batteries are a technology for filling the gap between batteries and pumped-hydro storage systems. But as with the adiabatic compressed air storage systems the redox-flow batteries must be developed further and their suitability for large scale cost and energy efficient energy storage has to be demonstrated.

High storage efficiencies are a prerequisite. Otherwise the costs for the energy which is wasted in the storage systems mounts up significantly. Therefore it is unlikely that hydrogen will play a major rule as a storage technology for electrical energy. Nevertheless, hydrogen will play a role as a basic element for the chemical industry and maybe also in those transport sectors which haven't been discussed here, namely trucks, ships and planes.

Especially in the urban environment there are options to reduce the need for electrical storage systems and for using existing storage systems for grid purposes. These are namely CHP units where the energy storage task is shifted from the electrical storage system to the thermal storage systems, and the use of battery storage system in plug-in hybrid and full electric vehicles. The vehicles can amount to a charge/discharge power capacity in the range of the peak load in the grids. Together with the CHP units and appropriate thermal storage systems, any fluctuations within a day can be easily solved.

Today it is still difficult to estimate how much energy storage capacity would be needed if power generation were to be based mainly on renewable energies. This question can be answered only on the basis of a detailed scenario for a mix of technologies. But even more important is the question of transmission grid structures in the future. Will we have something like a super-grid, which allows the transfer of large quantities of electrical energy throughout Europe? The larger the transport capacities, the smaller will be the need for long-term storage systems for periods beyond a few days. Storage systems always become uneconomic if they are only very seldom needed. The investments are high and the amortization requires a high throughput.

Nevertheless, there is no doubt that there are storage technologies available for any level of renewable energy penetration. Specific costs will depend significantly on the boundary conditions as discussed above.

Chapter 11
The Media Laboratory City Car: A New Approach to Sustainable Urban Mobility

WILLIAM J. MITCHELL, RYAN CHIN AND ANDRES SEVTSUK[1]

With the exception of electric trains, trams, and trolleys which are continuously powered from rails or overhead wires, and elevators, cable cars, and ferries which are pulled along by motors located off board, mechanically powered vehicles must carry their energy supplies with them. When onboard supplies run out, they are immobilized. This, as we shall demonstrate, has some fundamental effects upon urban spatial structures and mobility patterns.

Furthermore, these effects have major impacts upon urban energy consumption. They can create barriers to achieving energy efficiency, or when appropriately understood, they can open up new opportunities for significant energy savings. Here, therefore, we will show how integrated design of mobility systems, energy distribution systems, and urban spatial patterns can significantly enhance overall energy efficiency and reduce carbon emissions. In particular, we will introduce the Media Laboratory City Car – a new concept that illustrates this design strategy, and that opens up exciting opportunities for creation of clean, energy-efficient, cities for the twenty-first century.[2]

11.1 The Geography of Refuelling

The basic tradeoffs that are involved here are very familiar to mechanical engineers and transportation system designers. The more energy a vehicle can carry, the greater its range. But carrying more energy increases bulk and weight, and reduces energy efficiency. It is always necessary, in vehicle design, to find an appropriate balance.

---

[1] Contributors to the City Car project have included Mitchell Joachim, Patrik Künzler, William Lark, Jr, Phil Liang, Michael Chia-Liang Lin, Raul-David Poblano, Peter Schmitt, and Franco Vairani.

[2] There has been little systematic study of the overall energy efficiency of cities. The pioneering work of Abel Wolman ('The Metabolism of Cities', *Scientific American*, **213**(3), 179–190 (1965)) introduced the term 'urban metabolism' and conceived of complete urban systems as organisms, the inputs and outputs of which could be modelled. For a general introduction to the topic of overall urban energy efficiency, see Droege, Peter, *Renewable City: A Comprehensive Guide to an Urban Revolution*, Chichester, Wiley-Academy, 2006. The BP-Imperial College Urban Energy Systems Project is an ambitious current attempt to develop 'a systematic, integrated approach to the design and operation of urban energy systems, with a view to at least halving the energy intensity of cities' (*First Annual Report*, London, Imperial College, December 2006).

Fuels and energy storage media vary in their energy densities, and this profoundly affects the balance point. For example, fuel supplies for early steam-powered vehicles – coal or wood – were bulky and difficult to manage; this was not a fatal flaw for very large steamships and steam locomotives, but it limited the development of steam-powered loco-motives. Gasoline, however, has very high energy density, and its liquid form makes it relatively easy to manage, so it became key to the development of the modern automobile – enabling fairly compact vehicles with ranges of 300 miles. Batteries, unfortunately, have low energy density (so far, at least) and other limitations, and this has traditionally pre-sented designers of electric cars with a difficult dilemma. Either the car carries few batter-ies and consequently has very limited range (as with golf carts), or else it carries sufficient batteries for long range – but is bulkier, heavier, and less efficient.

Gasoline-electric hybrid vehicles cleverly combine the advantages of gasoline's high energy density with the benefits of electric drive – providing an attractive combination of fuel efficiency, good performance, and normal automobile range. However, there is another tradeoff. They do so at the cost of vastly increasing the vehicle's complexity.

Vehicles that carry their energy supplies also require infrastructures of energy supply points, such as coaling stations for steamships and trains, gasoline stations for automo-biles, and charging stations for battery-electric automobiles. Clearly these supply points must be strategically located within transportation networks, and the density with which they are distributed will depend upon vehicle ranges. Short-range vehicles will either be limited to use within the immediate vicinities of their supply points, or will require many, closely spaced supply points. Longer-range vehicles will be less local to supply points in their use, and will require less dense distributions of supply points.

The effects of this simple logic can be traced in settlement patterns. In the pre-industrial era, settlements often grew up around points on stagecoach runs where horses were replenished. Similarly, railroad towns developed at locations where coal and water could conveniently be stored. Later, the automobile produced the emergence of gas stations at busy intersections, highway off-ramps, and so on. Frequently, energy supply points have a 'village well' effect – providing an attraction that creates opportunities for other businesses, and for social space.

Finally, energy supply points must themselves be supplied in bulk – and this can be dif-ficult and expensive when they are located remotely from sources. Thus railway towns were supplied by coal trains – often running on the same tracks as the passenger and goods trains that they supported. Gas stations are now supplied by elaborate global networks involving pipelines, tanker ships, and tanker trucks. At remote stations, as, for example, in the Australian outback, the cost of getting gasoline there makes it particularly expensive.

This logic applies not only to today's fuels, but also to the proposed use of hydrogen and hydrogen fuel cells. In a hydrogen economy, hydrogen would need to be transported in bulk from production points, stored at hydrogen filling stations – much like gasoline stations, and carried in hydrogen fuel cell-powered vehicles.

11.2 Mechanical, Pipe, and Wire Distribution Networks

For powering fixed rather that mobile machinery, there is an alternative tradition of mechanical, pipe, and wire distribution networks. For example, the mills of the early industrial revolution clustered around central sources of water or steam power, and net-works of shafts and pulleys distributed power to the machines on the mill floor. These forms of mechanical transmission could not operate effectively over long distances, so mills were compact. (Furthermore, the worker housing of the mill towns tended to cluster tightly around the mills.) Occasionally, this principle was extended to powering vehicle movement at a small scale – notably, for example, in the cable car system that still operates in San Francisco, and in ski lifts.

During the nineteenth and early twentieth centuries, many cities acquired elaborate pipe networks for distributing coal gas to gaslights and gas stoves.[3] The infrastructure consisted of centrally located gasworks, large gas storage tanks, and the necessary piping. These networks could service considerably larger areas than shaft and pulley systems, but they were still limited in scale by the cost of constructing and maintaining the piping systems, and by the effects of transmission losses. In principle, these systems might have incorporated stations for refuelling coal gas-powered vehicles, but this never became popular – although, in later years, compressed natural gas has found some use as a vehicle fuel.

Eventually, electric lighting replaced gaslights, and electrical supply networks developed at urban scale. Early systems simply had generators at fairly central locations and wires distributing power to wherever it was required – much as with the gasworks and pipe networks that they partially supplanted. (Mostly, gas systems did not go away, but modernized, and shifted their roles to supplying heating and cooking devices.) There was a small zone of electrification around a source, much like an oasis around a well. Remote settlements and isolated dwellings, with diesel generators and the like, still follow this simple spatial logic.

As the scale of urban electrical use grew, it became evident that it was often more attractive, from many points of view, to transport electricity rather than fuel over long distances. Hydro electricity, for example, *must* be generated at points of availability of falling water, and these are often in locations remote from cities. Similar considerations apply to large-scale solar, wind, and wave power. Electricity is cheaper to transport than coal, so it made sense to locate coal-fired plants near to sources of coal rather than near to the demand for electricity. And considerations of safety, air quality, and political acceptability have often mandated the location of fossil fuel and nuclear power plants remotely from population centres. Thus the now-familiar infrastructure of long-distance power lines developed. However, physically inevitable transmission losses continued to impose some scale limits.

Over time (this simplifies a very complex story), individual power systems linked up to form vast electrical grids that extend across nations and continents.[4,5,6] These incorporate numerous power plants, serve many cities, and have redundant transmission paths. Their behaviours, and the associated management strategies, are complex.

One of the problems with traditional power grids, as they have developed, is that they do not incorporate batteries or other storage devices on a large scale. There is, for example, considerably higher demand during the day than during late night and early morning hours, so it is not possible to run power plants at optimal output all the time. A second problem is that, while grids can efficiently service predictable base loads, random peaks – such as those that occur on hot summer days when all the air conditioners are turned on – often require the expensive measure of bringing standby generators online for relatively short periods. And a third problem is that clean but intermittent power sources, such as wind and solar,

---

[3] Schivelbusch, W. (1998). *Disenchanted Night: The Industrialization of Light in the Nineteenth Century*, Berkeley, University of California Press.

[4] Hughes, T. P. (1983). *Networks of Power: The Electrification of America*, Baltimore, Johns Hopkins University Press.

[5] Nye, D. E. (1990). *Electrifying America: Social Meanings of a New Technology*, Cambridge, MIT Press.

[6] Schewe, P. F. (2007). *The Grid: A Journey through the Heart of Our Electrified World*, Washington, DC, Joseph Henry Press.

obviously cannot be relied upon to produce when needed, since they respond to Mother Nature, not the energy market. All this leads to strategies of fast-paced buying, selling, and moving electricity around in grids, and of managing demand through pricing – in particular, by making off-peak power less expensive.

An emerging alternative strategy, and one that is made increasingly feasible by inexpensive embedded intelligence, is to think of cities as distributed virtual power plants. Under this model, each building would function both as a consumer of electricity, and – through incorporating solar panels, wind turbines, fuel cells, etc. – as a small-scale producer and seller. Smart devices would buy electricity when prices were low (for example, dishwashers would switch on in off-peak hours), and fuel cells would come online to sell electricity back into the grid when prices were high. Transmission losses would be reduced, since electricity production and consumption would be closely co-located. And the whole system would be decentralized and redundant – much like the internet – offering potential reliability and service advantages.

Such a system would be even more like the internet, and even more effective, if it incorporated storage everywhere. This could be accomplished by storing hydrogen for fuel cells in distributed fashion, or through batteries or capacitors. Feasibility clearly depends upon technological advances that reduce the cost of these measures, finding good ways to pay for them, or some combination of the two; we shall return to this.

11.3 The Geography of Battery Recharging

Generally, in the interiors of buildings, we opportunistically recharge mobile phones and laptop computers whenever we need to. This works out reasonably well, since power outlets are never far away. And it means that very long battery life, while desirable, is not essential – allowing designers of these devices to minimize battery bulk and weight.

However, this strategy does not currently work for battery-electric vehicles on the exteriors of buildings. This is partly because the power outlets aren't provided. Partly because there aren't ways of charging users for power that is tapped in this way. And partly because the larger amounts of energy that must be transferred tend to make the process lengthier, more difficult, and more dangerous.

But these problems don't seem to be insoluble. Imagine, then, a city that provided automatic recharging of battery-electric vehicles whenever they came to rest, and maybe even when in motion. Vehicles would function much like electric toothbrushes, which recharge whenever they are put back in their holders. Without loss of functionality or reduction of effective range, they could carry far fewer batteries, which would make them simpler, lighter, and more energy efficient.

One requirement for this is the development of effective, reliable, automatic charging mechanisms – perhaps through automatic mechanical connection, through inductive charging (as with electric toothbrushes), or through emerging technologies of wireless charging. A second requirement is for some combination of battery and capacitor technology that allows sufficiently rapid charging, and that works well under this pattern of charging and demand. And a third requirement is an architectural and urban design strategy for providing ubiquitous charging points in vehicle-accessible zones – most obviously (but this is not the only way) by incorporating them into parking stalls.

11.4 Dual-Use Battery-Electric Vehicles

When battery-electric vehicles are connected to the grid in this way, in large numbers, they become dual-use devices. When they are in motion they provide mobility in the usual way, and when they are at rest they provide the grid with storage capacity – creating the benefits discussed earlier.

Imagine a city, then, in which all cars are electric, and all parked cars are connected to the electric grid. If the cars have sufficient intelligence, and they have knowledge of electricity price patterns and their own use patterns, they can play the electricity futures market. When electricity prices are low, and they can anticipate that they will soon need to drive, they can draw electricity from the grid to charge their batteries. Thus, for example, a commuter car that is parked at home and will be used in the morning can be charged during late-night off-peak hours. Conversely, when prices are high, batteries are charged, and they can anticipate that they will be parked for a while, they can sell electricity back to the grid. There is already a body of research on car-to-grid electrical systems, and the concept seems very promising.[7,8,9]

A system of this type should be able to make particularly effective use of intermittent solar and wind power, simply by taking the opportunity to charge batteries when extra power from these sources is available. Solar power output, for instance, usually peaks around solar noon, while mobility demand peaks a few hours later during the evening commute. Storage in the batteries of parked cars can effectively bridge the temporal gap.

The vision that emerges from all this is of a city that has small-scale electrical generation, electrical storage, and points of consumption scattered throughout. It makes particularly effective use of clean, renewable power sources, and utilizes simple, clean, silent electrical vehicles that produce no tailpipe emissions. It gains efficiencies through co-location of power production, storage, and consumption. And, like the internet, it achieves robustness through high levels of redundancy.

All this depends upon embedded intelligence everywhere, and the capability to effectively manage complex, dynamic electricity markets. There are many technical and policy issues still to be explored in depth. But this type of urban energy system does seem increasingly feasible, and promises a major advance in urban energy efficiency and sustainability.

11.5 The Role of Private Electric Vehicles

In most cities today, most of the vehicles on the road – cars, motor scooters, etc. – are privately owned. Private vehicle ownership allows realization of many, though not all, of the benefits that we have described.

A private ownership model would be based upon providing charging capabilities at homes, and in parking facilities at workplaces, shopping facilities, subway stations, airports, and other popular urban destinations. Over time, vehicles would be able to learn the habits of their owners, and so adjust their electricity buying and selling patterns to match them optimally – although, obviously, override capabilities would be needed to allow for breaks in routine.

---

[7] Kempton, W. and Letendre, S. E. 'Electric Vehicles as a New Power Source for Electric Utilities', *Transportation Research D*, Vol. 2, No. 3, 1997, pp. 157–175, www.udel.edu/V2G/docs/Kempton-Letendre-97.pdf

[8] Kempton, W. and Dhanju, A. 'Electric Vehicles with V2G: Storage for Large-Scale Wind Power', *Windtech International*, Vol. 2, No. 2, March 2006, pp. 18–21, www.udel.edu/V2G/docs/KemptonDhanju06-V2G-Wind.pdf

[9] Letendre, S. Denholm, P. and Lilienthal, P. 'Electric and Hybrid Cars: New Load, or New Resource?' *Public Utilities Fortnightly*, December 2006, pp. 28–37, www.udel.edu/V2G/docs/LetendDenLil-LoadOrResource06.pdf

Like private vehicles today, most of these electric vehicles would spend most of their time parked. However, unlike today's cars and scooters, they would not be uselessly taking up space and consuming materials and embodied energy while doing so; they would be serving as the city's battery packs.

The private vehicle ownership model has the advantage of fitting with the current business goals of automobile manufacturers, and of allowing incremental introduction. A system could begin with fairly sparse distribution of recharging points, and correspondingly large battery packs in cars to provide sufficient range. Then it could evolve, over time, into a system of much denser recharging points – allowing correspondingly lighter, simpler, shorter-range electric vehicles.

11.6 The Role of Shared-Use Vehicles

An alternative to private ownership, and one that has some significant advantages, is to provide urban mobility services through shared-use vehicles.[10] This has been demonstrated on a significant scale, with bicycles, in the French city of Lyon.

In Lyon, bicycles are available, at numerous locations throughout the city, in specially designed bicycle racks (Fig. 11.1). A customer walks to a nearby rack, unlocks a bicycle by swiping an electronic identification card, rides it to the destination, and deposits and locks it at another rack. Rental, then, is one-way; customers do not have to return bicycles to where they obtained them. Prices are very low, since this particular system is designed as a public utility that substitutes for mass transit; use is free for the first half hour, and equivalent to subway fares thereafter. Patterns of demand are carefully monitored, and there is a system of retrieving bicycles from locations where there are excess bicycles for the demand to locations where demand is higher.[11] Users can access a simple online map that provides real-time information about the current availability of bicycles at any location in the city (Fig. 11.2).

A profit-making, car-based system with many relevant features is that of ZipCar – which began in Cambridge, Massachusetts, and has now successfully spread to many other cities. ZipCars are available in parking spaces scattered throughout the city. (At the time of writing, there were about 500 of them in the Boston/Cambridge area.) They are rented by the hour. Customers locate and reserve them online, go to the chosen location, unlock the car with an RFID card, drive it away, and eventually return it to the same location. This is still two-way rental (as with traditional car rentals), but it provides access distributed throughout the service area, minimizes the complexity and time of the rental transaction,

---

[10] There is now a growing technical literature on shared-use vehicle systems. For a useful introduction see Shaheen, Susan A., 'Shared-Use Vehicle Systems' (September 2, 2004), *Institute of Transportation Studies*, UC Davis, Paper UCD-ITS-RP-04-13, http://repositories.cdlib.org/itsdavis/UCD-ITS-RP-04-13. See also Shaheen, Susan A., and Adam P. Cohen, 'Worldwide Carsharing Growth: An International Comparison' (December 2006), *Institute of Transportation Studies*, UC Davis, Research Report UCD-ITS-RR-06-22, and Katzev, Richard, 'Car Sharing: A New Approach to Urban Transportation Problems', *Analyses of Social Issues and Public Policy*, Vol. 3, No. 1, 2003, pp. 65–86, www.asap-spssi.org/pdf/katzev.pdf

[11] The 'retrieval problem' in shared-use vehicle systems has been quite extensively studied, and a variety of solutions have been proposed. See, for example, Barth, Mathew, Michael Todd and Lei Xue, 'User-Based Vehicle Relocation Techniques for Multiple-Station Shared-Use Vehicle Systems' (November 2003), *Transportation Research Board*, Paper 04-4161, www.communauto.com/images/TRB2004-002161.pdf

Fig. 11.1. Lyon bicycle racks (Image: Andres Sevtsuk).

and allows vehicles to be rented for short periods at fairly low prices. It is particularly popular, for example, with students who cannot afford to own cars but occasionally need them for supermarket shopping, weekend trips, and so on.

In general, shared-use systems have the potential to reduce the number of cars on the road. It is claimed, for example, that one ZipCar typically substitutes for about seven personal cars.

11.7 The City Car

The MIT Media Laboratory's City Car concept proposes extremely convenient, flexible, one-way rental in the context of an electric vehicle, distributed urban energy and mobility system as discussed above. There are many potential synergies between these two ideas.

In a City Car system, shared-use electrical cars are available at stacks densely located throughout the service area. Cars can either be stacked in parallel, as with the Lyon bicycles, or in a first-in-first-out (FIFO) arrangement, like luggage carts in a cart vending machine at an airport. Cars recharge while they are parked in the stacks. Parallel stacks take up a fixed amount of space, independent of the number of cars actually parked, and allow a customer to choose any available vehicle – important if vehicles vary in style and price, and if some may not be sufficiently charged. FIFO car stacks, on the other hand, are very compact, can vary dynamically in length, allow cars to charge as they progress through

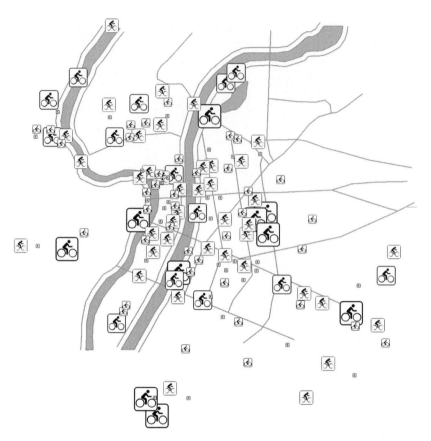

Fig. 11.2. Location of bicycle racks and relative numbers of bicycles in Lyon, France (image: Andres Sevtsuk).

the stack (arriving at the back of the stack discharged, and reaching the front charged), and work particularly well when cars are uniform (Fig. 11.3); like taxis, you take the first one that comes along. Both arrangements have advantages and disadvantages, and probably find appropriate roles in different contexts.

Providing acceptable levels of mobility service with a City Car system is a matter of efficient queue management within a network (there are some parallels with packet switching computer networks), and can be investigated using queueing theory models. Pedestrians walk to stack locations. If there are cars available at a stack, waiting time is zero. If there are no cars available, pedestrians must queue and wait for cars to arrive – much as with taxi queues. Customers will drive for varying periods, and then want to deposit their cars at stacks near their destinations. If there are spaces available in the chosen stack, then the car can be deposited immediately, but if not, the driver must wait or go to another stack. The dynamics of the system are basically determined by patterns of mobility demand (as expressed, in the usual way, by origin-destination data), and by network capacity. The key measure of mobility service quality is total time from origin stack to destination stack, including any waiting time. Systems with more cars available will provide better mobility service than systems with fewer cars, but the cost will be higher. System performance will

Fig. 11.3. City Car stack – first-in-first-out (FIFO) system (image: Franco Vairani).

be affected by spatial factors – particularly number and location of stacks relative to the time-varying spatial distribution of the demand, and congestion points in the road network. It will also be sensitive to vehicle characteristics such as charging time, pricing strategies, and to fleet management strategies – particularly to strategies for moving vehicles around to meet demand. Monitoring the performance of the system and collecting information on the daily trends can allow 'experience-based' improvements to the allocations of vehicles on a daily or weekly basis.

Mobile connectivity can significantly enhance the performance of the system. Unlike traditional taxi dispatching systems that can be hard to contact at peak hours, every vehicle can be equipped with an accurate real-time information system that reports the availability of vehicles and parking spaces throughout the city. This information can also be made available through cellular phones. Knowing the exact location and availability of parking at each stack, the system can solve parking queues in a much more efficient manner than a traditional 'race' for parking spots, where several drivers compete for the same place, by managing several cars at a time and suggesting alternative parking or renting solutions to drivers according to their current locations and queue situation.

A well-designed and effectively managed City Car system should provide the equivalent of instant valet parking everywhere, reducing the presently excessive amounts of gasoline, time and congestion spent on the search for cheap street parking.[12] Preliminary studies suggest (although much remains to be done) that, in many contexts, this can be done at a sufficiently low price to make it attractive, while allowing a sophisticated operator to run a profitable service business.

From a sustainability perspective, this system has many attractions. It achieves much higher utilization rates for vehicles – so reducing the amount of material and embodied energy that must be allocated to urban mobility. It is environmentally benign (small footprint and light weight, silent, no tailpipe emissions), and meshes well with the distributed, renewable-friendly

---

[12] Shoup, D. C. (2005). *The High Cost of Free Parking*, Chicago, Planners Press, American Planning Association.

Fig. 11.4. City Car urban energy system (image: Franco Vairani).

urban energy systems discussed earlier (Fig. 11.4). And, through embedded intelligence, sensing, networking, and use of optimization algorithms, it lends itself to effective management for energy efficiency and optimal use of available parking space and road real estate.

It is hardly to be expected that large automobile and scooter manufacturers would initially welcome the shared-use model. After all, the goal under this model is to minimize the number of cars on the road, while auto companies want to maximize. But these companies currently market products that generally aren't much different under the skin, have low profit margins, and face intense price competition. Furthermore, in many parts of the world, cities are reaching automobile saturation point – beyond which it is not feasible to increase the number of cars on the road – leading to dramatically restrictive responses such as the London Congestion Zone. In contexts such as the cities of China, where incomes and aspirations are growing, and private automobile ownership offers the seduction of expressing new-found freedom, wealth, and prestige, it seems likely that producing and selling automobiles will remain an attractive business. But, in areas that are approaching automobile saturation, and where considerations of energy efficiency and carbon emissions translate into consumer demand and political pressure, it may be increasingly tempting to move from an old-fashioned commodity product business into an innovative mobility, energy, and information service business.

11.8 Shared-Use Electric Scooters and e-Bikes

In some contexts – particularly in the rapidly growing cities of Asia – electric scooters and e-bikes may make more sense, in this role, than electric cars. They are simpler than cars, considerably less costly, and even more compact. In China, for example, the popularity of e-bikes has been growing rapidly in recent years.[13]

---

[13] Weinert, J. X., Chaktan Ma and Xinmiao Yang, 'The Transition to Electric Bikes in China and its Effect on Travel Behavior, Transit Use, and Safety' (October 2006), *Institute of Transportation Studies*, UC Davis. Working Paper UCD-ITS-RR-06-15.

Fig. 11.5. Electric folding scooter (design: Michael Chia-Liang Lin).

Fig. 11.6. Electric scooter charging stations (design: Michael Chia-Liang Lin).

As with electric City Cars, as another design from the Media Laboratory's Smart Cities group demonstrates, electric scooters and bicycles can be designed to fold into very compact configurations (Fig. 11.5). When folded, they can be pulled along like wheeled luggage. When parked, they can be locked into special racks and recharged (Fig. 11.6).

11.9 Combination with rapid transit

It is conceivable that this sort of personal vehicle could substitute for mass transit (as the Lyon shared bicycle system seems to do), or substitute for private automobile ownership

Subway stations per square mile in New York City

Zone	Subway stations	Radius (miles)	Area (sq. miles)	Stations/sq. mile
1	37	1	3.14	11.8
2	66	3	25.13	2.6
3	102	5	53.40	1.9
4	103	7	100.53	1.0
5	92	9	153.94	0.6

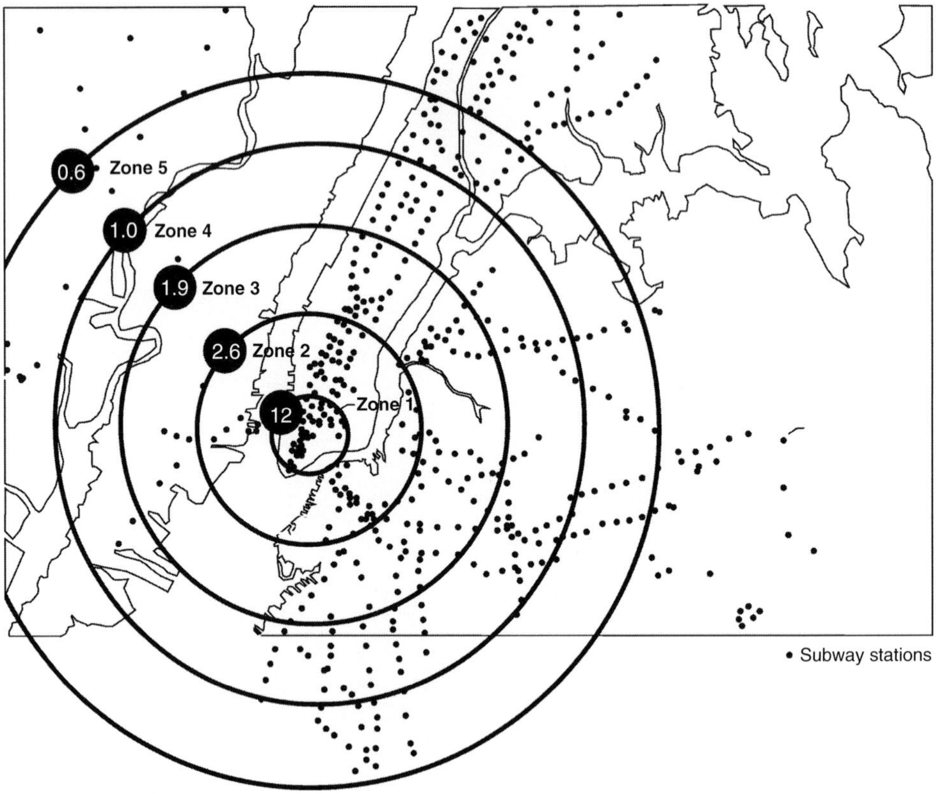

Fig. 11.7. Subway station density in New York City (image: Ryan Chin).

(which seems to be the primary effect of ZipCar). Probably there are some substitution effects. But the most likely scenario, in our view, is that City Cars would complement rapid transit – essentially by locating car stacks at transit stops to solve the 'last kilometre'or 'last ten kilometre' problem.

Rapid transit systems are unbeatably efficient for fast, high volume transport between fixed points, and in recent years advances such as high speed trains and bus-rapid-transit systems have increased their usefulness. But they suffer from the inherent problem that departure points are rarely exactly where you need them, and destination points are only an approximation to where you want to be. For example, the large subway network (over 500 subway stations) of New York City provides for a significant portion of mobility needs of the city; however, like most radial organized cities the density of the subway network decreases towards the periphery of the city (Fig. 11.7). Correspondingly, the number of

Fig. 11.8. Solving the 'last kilometre problem' – Taiwan's high speed rail system (image: Ryan Chin).

households with private automobiles increases as access to the subway network decreases. With the notable exception of the upper east side of Manhattan, an overwhelming percentage of registered automobiles reside in these outlying areas.[14]

Private automobiles, on the other hand, offer the flexibility of arbitrarily chosen departure and destination points, at arbitrarily chosen times – but at comparatively very high cost, and with undesirable side-effects. A combination of City Cars with rapid transit offers the best of both – with City Cars providing flexibility at both ends, and rapid transit providing speed and efficiency in between.

In the context of linear transit systems, such as Taiwan's new high speed rail system, City Cars (or even smaller and lighter electric scooters) could provide personal mobility

---

[14] Neuman, W. (2007).'Cars Clogging New York? Most are from the City', *New York Times*, January 12.

Suburban City Car stacks
Placed at the end of subway and commuter rail lines

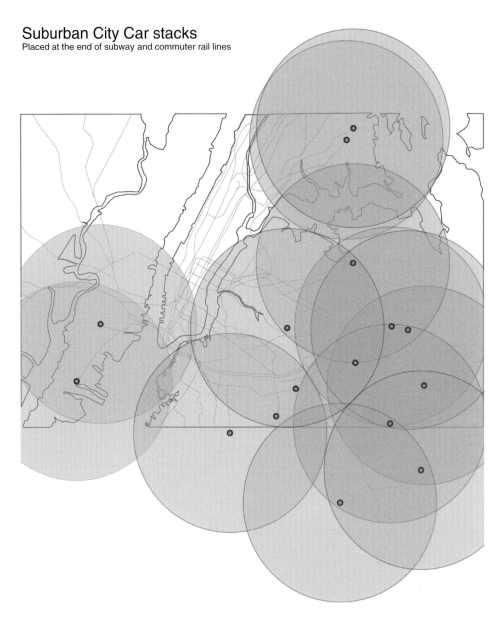

Fig. 11.9. Placing stacks on the last stop of the subway system in New York City. Circles represent a five-mile radius (image: Ryan Chin).

City Car stacks in Congestion Pricing Zones (London)
Stacks placed at Underground stations

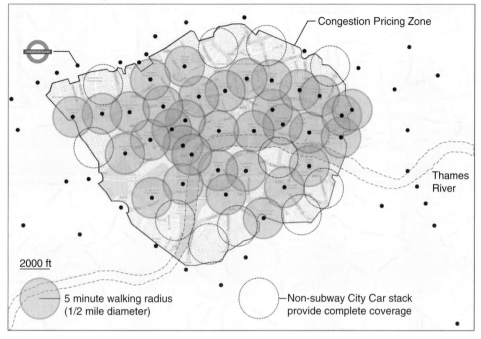

Fig. 11.10. Placing City Car stations at Underground stations within the London congestion pricing zone (image: Ryan Chin).

within a zone around each stop (Fig. 11.8). In radial systems, such as that of the New York metropolitan area, City Cars could provide personal mobility at the sparsely served extremities, where it is mostly needed (Fig. 11.9). In dense transit grid systems, such as that provided by the London Underground, City Cars could provide personal mobility 'infill' between Underground stations (Fig. 11.10).[15]

Radial transit systems often require travel to the city centre and out again in order to move circumferentially. City Car systems, with stacks of cars at transit stops, can reduce this problem by creating 'virtual rings', as illustrated in Fig. 11.11.

11.10 Conclusion

Strategies for achieving urban energy efficiency frequently focus upon reducing overall transportation demand, and upon shifting demand from automobiles to public transportation. But there is abundant evidence that high levels of interconnectivity are crucial to

[15] Litman, T. 'London Congestion Pricing: Implications for Other Cities', *Victoria Transport Policy Institute*, January 10, 2006, www.vtpi.org/london.pdf.

Eliminating city centre travel requirements

Neighbourhood City Cars reduce virtual rings in radially centred mass transit systems
*Radius 5 miles

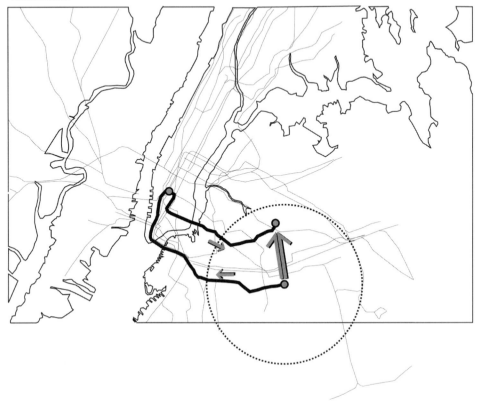

Fig. 11.11. Tackling the virtual ring problem in radial centred cities (image: Ryan Chin).

vibrant, flexible, creative cities, and that the inhabitants of cities greatly value personal mobility, so these traditional strategies have significant downsides.

Here, then, we have proposed an alternative – an urban mobility system that takes advantage of ubiquitous networking and embedded intelligence to enable combination of compact, simple, lightweight, battery-electric vehicles with ubiquitous recharging infrastructure and sophisticated electrical grids and markets. It is inherently energy efficient, it makes optimal use of clean, renewable energy sources, it keeps mobility costs low, and it does not compromise convenience or quality. By solving the 'last kilometre' problem, it can work in effective combination with transit systems. And there are no insurmountable technological barriers to implementation in the near future.

Chapter 12
Towards the Intelligent Grid: A Review of the Literature

JOHN GARDNER AND PETA ASHWORTH

Contributors: Naomi Boughen, Simone Carr Cornish, Anna Littleboy, Sam MacAulay, Simon Niemeyer, Australian National University, Anne Pisarski, University of Queensland CSIRO

12.1 Background

Cities in energy transition can be thought of as moving from centrally supplied, one-way consumption modes to open systems of distributed, ubiquitous providers and users of renewable and other energy streams. This places entirely new demands on the intelligent networking of energy supply, storage and management devices and arrangements. Indeed, any future energy system capable of helping combat climate change demands an electricity network that balances supply and demand for the benefit of consumers and utilities but minimizes greenhouse gas emissions, losses and price. One such vision is an 'intelligent grid' which foresees an evolved energy system with greater levels of embedded or distributed energy than are currently in place. A critical challenge for the delivery of this vision is to determine the value of an intelligent grid system and the triggers that will bring it about. The case of Australia and CSIRO's initiatives are the background to this investigation.

Distributed energy systems are defined as local, small-scale, modular technologies for on-site, grid-connected or stand-alone energy conversion and delivery. The energy production units may, for example, be located in the distribution network of a building or factory. Distributed energy systems typically include heat production and cooling systems, and frequently use renewable energy sources. Production can be independent, mobile, and/or connected to a power or a heat network.

Distributed energy systems include technologies for both supply-side management (distributed generation, small scale, close to load) and demand-side management (consumer loads, energy efficiency, peak shaving, load deferral). Specifically, our underlying definition of distributed energy incorporates any demand supply management (DSM) technology, and any distributed generation (DG) technology that produces less than 30 MW, and is located close to load. The 30 MW threshold represents the minimum size for power production units bidding into the National Energy Market (NEM).

Embedding greater levels of distributed energy technologies goes hand in hand with changing patterns of consumption and consumer expectations. Interest in changing the consumption behaviour of energy consumers has its origins in the search for a policy response to the 1970's energy crisis. From early on in the policy response, distributed energy systems have been seen as having the potential to improve the energy efficiency of power generation and distribution. However, even in an era that saw dramatic oil price rises and a presidential call for Americans to reduce demand, little sustained behavioural change towards the adoption of these technologies occurred. It is clear that the availability of distributed energy technology alone is not sufficient to achieve impact.

As the world refocuses on the twin issues of energy supply and global warming in the twenty-first century, the potential for distributed energy systems to contribute to solving this problem remains significant. Engendering the large-scale behavioural change required to adopt and support the large-scale technological deployment of distributed energy is an essential requirement for its success. The aim of this literature review is to outline the likely drivers of public uptake behaviour, and useful theoretical frameworks that will help identify options for delivering an intelligent grid. The review also describes potential strategies to inform the acceptance and uptake of distributed energy systems.

Initially, this chapter presents a conceptual model of how social research fits into the framework of an intelligent grid (section 12.2). Subsequently, literature describing factors likely to affect the uptake of distributed energy technology is reviewed. For the purposes of the intelligent grid the highly complex society-individual-technology system has been structured into a discussion of the *processes* affecting technology uptake (section 12.3), the *characteristics* of individuals that affect their choice to adopt technology (section 12.4) and the influence of social *situations* on the acceptability of different courses of action with regard to technology (section 12.5). In section 12.6, theoretical frameworks for intervening in the society-individual-technology system and thereby changing attitudes and behaviours are discussed. Section 12.7 describes an array of external factors which may influence the adoption of distributed energy generation, and presents some of the policy incentives for distributed energy technologies already in place around the world. Section 12.8 incorporates all of the preceding information into a discussion on the implications of this review for an intelligent grid system.

12.2 A Framework for Integrating Social Research into the Intelligent Grid

By understanding attitudes and behaviours that can reduce energy demand and increase the acceptance of distributed energy systems, various interventions can be implemented to help maximize technology diffusion and the likelihood of distributed energy technology uptake. Our proposed approach is to identify appropriate theoretical frameworks from the social research literature and to apply them in an energy context.

The central premise of the combined framework (see Fig. 12.1) is that people's values, attitudes and beliefs will drive their intentions, subsequent action and eventual long-term acceptance of distributed energy and reductions in consumption. Although a wide array of psychological research supports this central premise, it is important to acknowledge that people's decisions are made within a broader context, where a range of external influences also have an impact. These external influences include *economic factors*, such as changes in the demand for renewable energy, and the costs of implementing distributed energy generation; *physical/technological factors*, such as the development of and access to technology; and *societal factors*, such as community support for reduced emission technology, government incentives and industry reactions. The impact of societal factors on the adoption of distributed energy is particularly relevant, since some distributed energy technology is likely to be implemented at the community level, rather than in individual households.

Once the psychological and contextual drivers relevant to the intelligent grid concept are understood, a range of interventions can be designed to either directly or indirectly promote

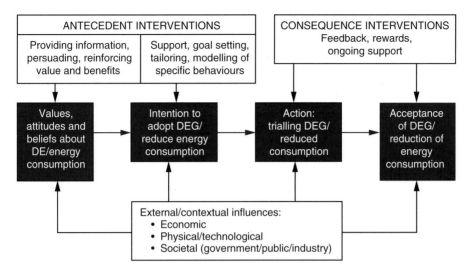

Fig. 12.1. A conceptual framework for the intelligent grid.

uptake and acceptance of distributed energy and reduction in energy consumption. Such interventions might provide information about potential economic and environmental benefits, provide examples of specific types of distributed energy uptake, or provide feedback about particular levels of energy usage in a household or a broader community/organization. This literature review suggests such interventions would be informed by a combination of social theories implemented via a participatory action research approach. Evaluation and ongoing tracking of these interventions will help to identify what mix of components of distributed energy is acceptable to Australian society, and what methods of promotion are liable to be effective in the long term.

12.3 Adoption, Diffusion and Acceptance: Processes Affecting the Uptake of Distributed Energy Technology

Although there is limited research that examines specific adoption, diffusion and acceptance of distributed energy systems, a large body of research pertains to the uptake and adoption of information and communication technologies. Examining this literature together with research into social risk can shed light on the potential dynamics for the introduction of distributed energy. Several relevant theories are discussed in more detail below (see, for example, Bagozzi *et al.* 1992; Burkman 1987; Rogers 1962, 1995).

12.3.1 *Consumer uptake and societal acceptance*

The dynamics of adoption differs depending on the characteristics of the technology involved (Niemeyer 2004). An important distinction has to be drawn between:

- *consumer technologies,* which involve the adoption of small-scale or individual (firm or household) technologies, such as the hybrid car, where decisions are often made by the individual consumer; and

- *social technologies,* which involve the adoption of much larger-scale technologies, such as a centralized energy generation system, where consumers are not directly involved in adoption, but are involved as part of a broader system (Niemeyer and Littleboy 2005).

Distributed energy technologies can be considered to be both a consumer technology *and* a social technology. In the case of distributed energy, consumers decide whether or not to use distributed energy systems or demand-side management tools. However, much of the motivation to change current social and environmental conditions requires a collective decision. Hence, issues of individual uptake as well as collective acceptance need to be considered.

12.3.2 Theory of diffusion and adoption of innovations

In his original theory of diffusion and adoption of innovations, Rogers (1962) proposed that the timing of adoption would fit a bell curve, such that individuals could be categorized on the basis of their speed of adoption as innovators (2.5% of the population), early adopters (13.5%), early majority (34%), late majority (34%) or laggards (16%). A central implication of this model is that technological innovations will diffuse into a market in an S-curve, with slow initial penetration, followed by a period of rapid growth ('take-off'), and then a period of slowed growth as the technology becomes commonplace (a 'saturated' market; see Fig. 12.2). An individual's willingness and ability to adopt an innovation is dependent on their awareness, interest, evaluation and trialling of the technology (Rogers 1995). The speed of adoption is influenced by the acceleration at the beginning of the adoption phase and the speed of later growth. This speed may be affected by factors such as price and the relative benefits of the technology (Rogers 1995).

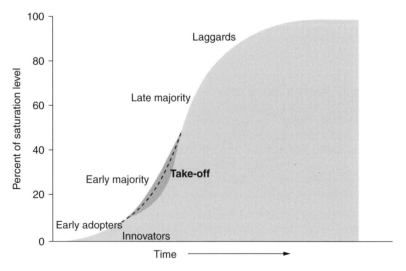

Fig. 12.2. Stylized diffusion process.
(Summary graphical representation of Rogers' (1995) theory).

In Rogers' (1995) five-stage model, potential adopters first learn of the existence and function of an innovation. By evaluating the perceived attributes, either positively or negatively, they form an attitude towards the innovation. This evaluation of perceived attributes leads them to a decision to accept or reject the innovation. If an individual is persuaded to adopt, they implement the innovation to a point where it becomes a regular part of their life. Subsequently, the individual seeks supportive messages to reinforce their decision. If the adopter gains support or confirmation of their decision, then continued adoption will occur, but if no support is gained, adoption may be discontinued. Similarly, if the decision-maker initially rejects the innovation and finds support for this decision, the decision-maker will continue to reject adoption; however, if there is no support to reject, adoption may occur at a later stage.

Rogers (1995) also suggested through his Theory of Perceived Attributes that the adoption and diffusion of a technology will also be judged according to trialability, observability, its relative advantage over other technologies, its complexity and its compatibility with other technologies in the situation in which it is to be adopted.

12.3.3 The Technology Acceptance Model

Another model developed to explore the factors affecting the uptake of information technology in the workplace is the Technology Acceptance Model (TAM; Bagozzi *et al.* 1992; Davis 1989; Davis *et al.* 1989; Venkatesh and Davis 2000). The TAM is based on the assumption that two main drivers of technology uptake are perceived usefulness and perceived ease of use (Fig. 12.3). Perceived usefulness relates to a subjective assessment of the benefits of the technology. Perceived ease of use is the assessment of the level of effort needed to master the technology (Davis *et al.* 1989). The TAM draws heavily on the theory of reasoned action, which is described below.

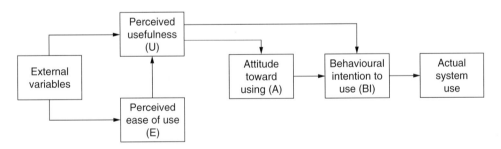

Fig. 12.3. Technology Acceptance Model.
(*Source*: Davis *et al.* (1989)).

12.3.4 The Theory of Reasoned Action

The Theory of Reasoned Action is strongly dependent on the concept of behavioural intention – the commitment to a certain action or behaviour (Fishbein and Ajzen, 1975, Ajzen and Fishbein, 1980). The theory asserts that behaviour is a deliberate act based on the beliefs of the individual and the norms imposed by society. Therefore, when an individual is positively predisposed toward a particular behaviour, and when they perceive support for that behaviour from people around them, then they will form a positive behavioural intention towards that behaviour. Behavioural intention, in turn, leads to actual performance of the relevant behaviour (Ajzen and Fishbein 1980).

The application of this theory can be illustrated in the context of recycling behaviour. In Australia over the past decade there has been an increasing expectation for individuals to recycle their household rubbish. In other words, a subjective norm exists that recycling is an appropriate and reasonable thing to do, but this norm alone is unlikely to be sufficient to produce recycling behaviour in all individuals. Only individuals who also hold a positive attitude towards recycling are likely to form an intention to recycle, and then to actually perform the behaviour.

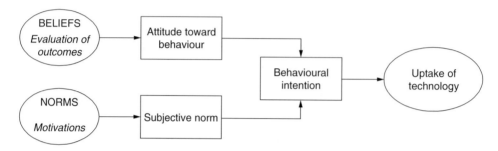

Fig. 12.4. Theory of Reasoned Action.
(*Source*: adapted from Fishbein and Ajzen (1975)).

12.3.5 *Theory of Planned Behaviour*

The theory of reasoned action was later modified to develop Ajzen's (1989) Theory of Planned Behaviour, which incorporates the person's belief about how easy or difficult it is to perform the behaviour, based on their abilities, opportunities and resources (Fig. 12.5). Ajzen and Fishbein's (2005) research has demonstrated that incorporating this measure of perceived control improves the prediction accuracy for both intentions and behaviour. For

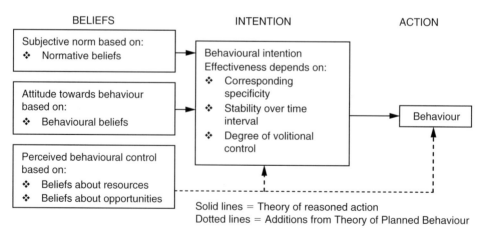

Fig. 12.5. Theory of Planned Behaviour
(*Source*: adapted from Madden *et al.* (1992)).

example, if a person holds a positive attitude towards distributed energy technology, and if they perceive support for the technology from within their community, their intention to adopt the technology will be higher. In addition, if they feel they possess the capability and opportunity to adopt the technology, their intention to act will increase, as will the likelihood of actual adoption.

12.3.6 Risk as a moderator of technology uptake and adoption

In addition to the theories for adoption, diffusion and acceptance of technologies outlined in this section, society also has a fundamental affect on the market for a technology depending on its perceived risk. Oskamp (2000) identified that perceived risk can delay or halt the diffusion and implementation of new technology, which raises a key question about ways to encourage individuals, communities and organisations in the uptake of distributed energy systems.

In this sphere, the literature on risk perception comes to the fore (Cvetkovich and Löfstedt 1999). Risk perception has historically been attributed to problems of public information, but more recently has been understood as part of the relationships within the technological system (Grove-White *et al.* 2000). Trust in the messenger is an important part of the communication process, thus influencing technology acceptance and/or uptake (Marks 2001; Pueppke 2001).

Trust is not something that can be decreed, but emerges as part of genuine engagement with the public, to address its concerns and aspirations. A key conclusion in the literature is that engaging the public in meaningful dialogue can foster significant trust building and reshape the trajectory of technology uptake (Löfstedt and Frewer 1998). Not only can technology be reshaped to address citizens' concerns, but also factors that amplify social risk can be addressed by concerted efforts such as trust building and targeted information. Additionally, in true dialogue the information gained from these processes can be used to shape the technologies or to assess their viability.

12.4 Attitudes and Behaviour: Characteristics Affecting the Uptake of Distributed Energy Technology

The theories in the preceding section demonstrate the importance of understanding attitudes and behaviours for the uptake of distributed energy technology. A critical factor for the success of the intelligent grid is therefore the extent to which behaviours, attitudes and attitudinal change towards distributed energy technologies can be understood and predicted. Although the theoretical frameworks describing technology uptake and acceptance (e.g. Bagozzi *et al.* 1992; Burkman 1987; Rogers 1962, 1995) can be criticized for assuming attitudes are rational and that socially significant behaviours are intentional (Abelson 1972; Wicker 1969), they are still valuable for understanding the relationships between attitudes, intentions and subsequent behaviours (Doll and Ajzen 1992; Smith and Stasson 2000). To deepen our understanding of attitudes and behaviours towards technology uptake, it is also useful to consider broader perspectives from social psychology.

The challenge of mobilizing society to change behaviours relevant to energy and distributed energy systems has been recognized as a global problem (IEA 2002; UK DTI 2006a). The approaches of Dake (1992) and Hallman and Wandersman (1992) demonstrate ways of encouraging individual or group-level attitude and behaviour change using applied social psychology techniques. These techniques include applied behavioural analysis and processes of social interaction and persuasion at the individual level, and deliberate shifting of social identity at both the individual and group level (Hogg and Cooper 2002; Oskamp 1983). Specific interventions at the individual level include information campaigns, prompts, modelling, incentives and feedback. Social interaction interventions observable at the individual and group level include such things as social incentives and the use of cognitive dissonance (Oskamp 2000).

However, recent research has demonstrated that along with advantages, there are difficulties with using attitudinal and behavioural change strategies within the general area of energy conservation and efficiency (Poortinga *et al.* 2004; Poortinga *et al.* 2003; Renn *et al.* 1995). These difficulties are not only isolated to energy, but have also been encountered across areas of science and technology transfer, for example resistance to the introduction of genetically modified foods in some countries, and the failed siting of hazardous waste disposal facilities in Canada and the United Kingdom (Renn *et al.* 1995). In this regard, there is evidence that people's values can contribute to the success of strategies aimed at changing their behaviour (Poortinaga *et al.* 2004).

12.4.1 *Understanding values*

Values are judgements or ideas people hold about what is right, good or desirable and contain elements that demonstrate the strength or intensity of these judgements and together create an individual's value system (Gutman 1982; Maniywa and Crawford 2001; Rokeach 1973; Zhao *et al.* 1998). Values tend to be inflexible, relatively stable and enduring, and influence our attitudes and behaviour (Rokeach 1973). The seminal work on values by Rokeach (1973, p. 5) argued that values are 'an enduring belief that a specific mode of conduct or end-state of existence is personally or socially preferable to an opposite or converse mode of conduct or end-state of existence'. Rokeach (1973) also proposed values are likely to be established in childhood through culture, personality, and societal experiences, and should be considered more enduring than their consequences – attitudes, motives, preferences and interests.

Rokeach's (1973) assertion that values can be classified into two components, terminal and instrumental values, has been supported by subsequent research (e.g. Gutman 1982; Maniywa and Crawford 2001; Zhao *et al.* 1998). Terminal values represent our preferred end-state existence, such as the goals we wish to achieve over the course of our lives, while instrumental values are the means or ways in which we intend to achieve our terminal values (Rokeach 1973). Although individuals will vary in their adopted value systems, groups or cohorts in society may share values (Munson and Posner 1980). The cohorts may be based on occupation, generation (age), interests, cultures or subcultures (Munson and Posner 1980; Poortinaga *et al.* 2004).

Research suggests that people's pre-existing values can explain their attitudes to energy-saving measures (Poortinaga *et al.* 2004), so it is important to understand what specific values influence people's reactions to distributed energy technology. Values may sometimes be changed through a process of questioning, which may then in turn change our attitudes and behaviours; however, it is more likely that questioning values will reinforce those values (Rokeach 1973). Values and attitudes are interrelated, but attitudes are more amenable to change (Eagly and Chaiken 1998; Rokeach 1973).

12.4.2 *Attitude formation and behaviour change*

An attitude has been described as a tendency towards an object based on an organization of beliefs and feelings (Eagly and Chaiken 1998; Himmelfarb and Eagly 1974). Attitudes

have been found to endure across time and situations and are bound to socially significant events or objects. They involve a degree of abstraction and can be generalized
(Himmelfarb and Eagly 1974).

Thus, a one-off event, such as the breakdown of a new solar hot water system, generally
is not sufficient to produce attitude formation. However, if the event is significant enough
to promote a distrust towards solar power in general, then this mistrust can emerge as
an attitude (Himmelfarb and Eagly 1974). Early research attempted to predict people's
behaviour from their attitudes but found little relationship between surveyed measures
of attitude and measures of actual behaviour (Abelson 1972; Wicker 1969). Later research
showed that the relationship between an attitude and behaviour can vary depending on
how accessible the attitude is (Doll and Ajzen 1992), whether the attitude is expressed
publicly or in private, individually or in a group, and the degree of identification with a
particular group (Smith and Stasson 2000).

12.4.3 *Attitude strength and accessibility*

Attitude strength is thought to be a multidimensional construct which includes our direct
behavioural experience, combined with our expressions and thoughts about the object
(Krosnick *et al.* 1993). The stronger the attitude and the more accessible it is, the more
impact this attitude will have on our behaviour (Fazio *et al.* 1992; Fazio and Williams 1986).
Accessible attitudes focus our attention, facilitate decision-making and assist categoriza-
tion, which are extremely useful functions in organizing data and reducing stress (Fazio
and Powell 1997; Olsen and Zanna 1993). However, changes in the attitude object may
also be associated with accessible attitudes becoming dysfunctional and an insensitivity to
change (Fazio *et al.* 2000). The association between attitude object and attitude accessibility
can be demonstrated by the example of solar power.

Consider an individual whose attitude towards solar power is strong and positive; that
is, they strongly believe solar power is a useful and green source of energy. Then when
that individual is discussing the merits of distributed energy, solar power should emerge
as a preferred source of energy. However, for a person who does not have a strong opinion on solar power, they are less likely to associate solar power as a useful ingredient in a
distributed energy system. In addition, if an individual has a strong negative attitude to
solar power, even when given information that suggests solar power has advanced to be
a useful alternative source of domestic energy, they may have difficulty deciding whether
it has changed for the better or worse (see Fazio *et al.* 2000). In general, previous research
suggests that strong attitudes exert a more consistent influence over our behaviour and are
more accessible than are weak attitudes (Eagly and Chaiken 1998; Fazio *et al.* 2000; Fazio
and Williams 1986).

12.4.4 *Personality as a moderator of attitude and behaviour*

Research has shown that, although situational factors can influence attitude and behaviour, personality can also have a significant influence. Personality characteristics, also
referred to as traits, are thought to remain fairly stable across a person's life. Examples of
personality characteristics are extraversion or introversion and positive or negative affect
(i.e. optimism or pessimism; 'seeing the glass half full or half empty'; Vaughan 1977).
Moods, habits, perceived degree of control over the situation and cognitive biases are

personality variables also known to influence attitudes and behaviours (Paglia and Room 1999; Semmler and Brewer 2002; Triandis 1977, 1980a, 1980b).

Acknowledging the potential influence of personality characteristics permits us to further understand attitudes and behaviours towards distributed energy technologies. For example, we could infer that individuals with a pessimistic view of the world and an external locus of control may strongly feel they cannot do anything about climate change, seeing it as inevitable, or as something only the government can address. We might also infer that people who run their air conditioners at 21°C may be acting out of habit, rather than on the basis of a consciously made decision (see Triandis 1977, 1980a, 1980b). Understanding these moderators can assist in the development of a range of strategies to change people's behaviour (Eagly and Chaiken 1998; Vaughan 1977). For example, legislation or regulatory control may be the most effective means of bringing about change among pessimistic people with an external locus of control and high energy demand habits. People with an optimistic world view and an internal locus of control may respond best to reason and persuasion (see Triandis 1977, 1980a, 1980b).

12.5 Society and Community: Situations Affecting the Uptake of Distributed Energy Technology

The discussion in section 12.4 focused on individual attitudes and behaviours. The theories of reasoned action and planned behaviour, in particular, acknowledge the influence of societal 'norms' on individual intentions and behaviours. In consequence, technology uptake is affected both by individual action and societal acceptance (Echabe and Castro 1999). An action or behaviour will be more likely to occur when there are supportive social norms, that is, when a person believes significant others (their peers, friends, family, and society at large) support the particular behaviour. The effects of supportive social norms can also be examined via notions of social identity and group membership.

12.5.1 Social Identity Theory and influencing change

The concept of Social Identity Theory (SIT) is relevant to understanding social relations within communities and groups and is helpful for the focus of this research. Social Identity Theory can be described as a theory of the dynamic relationship between individuals and the groups to which they belong (Hogg and Abrams 1999). In short, SIT argues that an individual's membership of a specific social group, together with the emotional and value significance they place on that group membership, influences their behaviour in social settings (Hogg and Abrams 1999; Gallois *et al.* 1998; Tajfel and Turner 1986). Social Identity Theory focuses on explaining individual behaviour in interactions both with other members of the same group (in-group members) and with members of competing groups (out-group members).

Knowledge of the groups that people identify with can help us to understand people's reactions to and perceptions of social issues (Gardner *et al.* 2001) including climate change and distributed energy. When people identify strongly with a particular group their perceptions of individual differences are minimized, and the in-group behavioural norms become more relevant or salient (Terry *et al.* 1999, 2000). For example, the more an individual identifies as an environmentalist (their in-group), the more he or she will feel distinct from out-group members, who might be defined as 'environmentally unfriendly'. Even if a person doesn't hold any specific pre-existing view about distributed energy (i.e. a weak attitude), strongly identifying with environmentalists as a group should lead them to be more supportive of distributed energy, providing they label distributed energy as environmentally friendly. In a study by Siero *et al.* (1996), which examined energy consumption behaviour in a metallurgy company, an intervention based on SIT and group membership resulted in sustained energy conservation, regardless of people's original attitudes to energy consumption.

In some situations, people with weak attitudes will act in a manner that is inconsistent with their attitudes. Such inconsistency is dependent on aspects of the situation, especially the presence of an implicit social norm (Calder and Ross 1973; Lavine *et al.* 1998; Terry *et al.* 1999). For example, a person with a faintly negative attitude towards energy conservation might actually engage in energy-conserving behaviour at work, if there is a strong pro-efficiency norm in their workplace. More generally, Terry and her colleagues (Terry *et al.* 1996, 1999, 2000), drawing on SIT, demonstrated that a person's attitudes are more likely to be translated into actual behaviour if those attitudes are supported by norms within a social group that the person identifies with.

12.5.2 *Collective choice and social negotiation*

The impact of social identity on distributed energy will be seen in difficulties in making choices about potential generating technologies. In the late twentieth century, 'expected utility theory' dominated expert literature describing decision-making processes (Savage 1990). This assumes that all rational people will make the same 'rational' choice when given the same information (Cayford 2001). However, the theory is based on individual choices (either that of an institution or an individual) and that this single decision-maker has a single view of the desired or best outcome. These ideas break down when applied by a collective.

Distributed energy generation requires a collective approach, which challenges the applicability of expected utility theory. More complex still, Stirling (2002) argues that many aspects of decisions involving environmental considerations are *incommensurable*, in which case a typical assessment approach of resolving conflicts analytically in an objective manner needs to be rethought (Stirling 2002). The implication is that choices on behalf of a community cannot be made using the same rules that govern individual choice. This is illustrated below.

Three people each have preferences for heating their water. They need to choose a system to share but, as can be seen in the following table, there is no clear preference which relates to the group. This doesn't mean they won't make a collective choice – other things will come into play. Mary may detest solar, Elizabeth doesn't really mind and George had an electric hot water system in his last house. So they choose gas. There may be a number of other sensible choices. However, these will only be uncovered through a process of 'social negotiation' to uncover deeper weightings and values.

Person	1st choice	2nd choice	3rd choice
George	Solar	Gas	Electric
Mary	Gas	Electric	Solar
Elizabeth	Electric	Solar	Gas

In the context of the intelligent grid, this presents some challenges for assessing the viability of distributed energy generation. A deep investigation of values and preferences will be required to enable comment about the acceptability of different scenarios and technologies

12.6 Changing Attitudes and Behaviours

Sections 12.4 and 12.5 identify the numerous interrelated factors influencing the chances of technology uptake, that when considered together can determine the acceptability of distributed energy technologies. The conceptual model presented in section 12.2 indicates the potential for knowledge-based interventions to increase the intention to adopt distributed energy technologies. Such interventions also require an understanding of factors that can influence attitudinal and behavioural change such as situation, habits, and experience (Vaughan 1977). The processes used to accomplish change in people's attitudes and behaviours can include either persuasion, reason, or a combination of both (Hogg and Cooper 2002). Theoretical frameworks describing the dynamics of knowledge intervention processes are discussed below, together with a summary of research exploring behavioural change specifically within the context of energy technologies.

12.6.1 Cognitive dissonance theory

The theory of social cognition suggests that if we can change a person's beliefs about something we are likely to change their attitudes as a result (Manis 1978). More specifically, Festinger's (1957) cognitive dissonance theory states that when we are exposed to information that is incongruent with previously or currently held beliefs, it is likely to bring about discomfort or dissonance and therefore attitudinal and behavioural change. One simple example of cognitive dissonance influencing behaviour is that of a mother who asks her children to recycle because it is good for the planet, yet due to tiredness and against her stated beliefs she throws out a can that could be recycled. If her children observe this action, the mother may feel a degree of guilt or discomfort that prompts her to shift the can to the recycling bin, in order to reduce her level of cognitive dissonance.

Cognitive dissonance theory has been applied in the development of persuasive messages designed to prompt behaviour change. Two mediating factors are likely to effect the attitude and behaviour change via such messages: the degree of cognitive effort that is applied to the message (Echabe and Castro 1999; Petty and Cacioppo 1984), and the extent to which the user perceives they already exhibit desirable behaviour (Bittle *et al.* 1980; Kantola *et al.* 1984). A persuasive message is more likely to drive attitude change if there is strong effort invested in processing the message's content (Petty and Cacioppo 1984). The strength of cognitive effort is thought to increase in discussion groups, because in groups a stronger effort is exerted compared to individual thinking (Echabe and Castro 1999). The strength of cognitive effort is also likely to increase when processing of the message leads to predominantly favourable thoughts, for example the resolution of a problem or greater understanding of an issue such as climate change (Petty and Cacioppo 1984).

Applications of cognitive dissonance to household energy consumption have demonstrated the effectiveness of dissonance interventions in prompting attitudinal and behavioural changes and may be compromised depending on how the person targeted perceives their behaviour (Bittle *et al.* 1980; Kantola *et al.* 1984). Research has found people who are higher users of energy are more likely to reduce consumption when made aware of their behaviour. But when low energy users were given feedback, their energy consumption increased (Bittle *et al.* 1980; Kantola *et al.* 1984). As a result, it may be inferred that information interventions intending to create change on the basis of dissonance should be specifically targeted to people who perceive their behaviours as positive when they are not, in order to create the opportunity for positive attitude and behaviour changes (Bittle *et al.* 1980; Festinger 1957; Kantola *et al.* 1984).

12.6.2 Participatory action research

Participatory action research (PAR) is a process which can be used to direct intervention and change in communities and groups. The process is context dependent, focusing on human values, experiences and beliefs, and brings about knowledge that is conducive for generating social improvement (Greenwood and Levin 2003; Wadsworth 1998). The PAR process is underpinned by the theory of action (Argyris and Shon 1978 cited in Dick and Dalmau 2003), which suggests that thoughts and actions are enacted and can be changed through cyclical learning events that are designed to engage stakeholder groups (Dick and Dalmau 2003; Wadsworth 1998).

Participatory action research is a useful framework for guiding the engagement of communities through an investigation into potential distributed energy implementation. The topic of distributed energy is a complex issue, encompassing a number of interrelated concepts including energy supply and demand, energy efficiency, effects on CO_2 emissions and climate change. Further, the sheer complexity required for implementation of distributed energy means it has an influence on a large and diverse range of stakeholders. Hemmati (2002, p. 2) describes stakeholders as 'those who have an interest in a particular decision, either as individuals or representatives of a group. This includes people who influence a decision, or can influence it, as well as those affected by it.' For the purpose of the intelligent grid, the stakeholders of distributed energy technologies include not only domestic consumers (individuals and local communities), but also small to medium enterprises, industry groups, government (both as a consumer of electricity and as a policy-maker), the regulators of energy, retailers of distributed energy technologies, and energy generators.

Effective PAR involves promoting a number of conditions. The situation or context is fundamental, as it needs to allow for the stakeholders of distributed energy to share their values and preferences, in a deliberative and iterative process that identifies concerns and also initiates action (Wadsworth 1998). The theory of action suggests the consequences and the effectiveness of these actions should be measured throughout the PAR process (Dick and Dalmau 2003).

Initially the research should identify concerns about issues specific to the research participants (Wadsworth 1998). For distributed energy these might be the current concerns people have about their energy use, and the reliability of distributed energy technologies, or the cost of the technologies. There should also be inquiry into how the process intends to challenge the knowledge of participants and the subsequent action (Wadsworth 1998), for example measuring attitudes before and after a discussion on how to reduce household energy consumption. Once a PAR process around distributed energy technology is initiated, it should be maintained over a number of meaningful interactions, in order to sufficiently challenge and hence change people's attitudes and behaviours (Wadsworth 1998; Dick and Dalmau 2003).

12.6.3 *Reducing household energy consumption*

Research into household energy conservation suggests there are two broad types of interventions that have been used to alter household energy consumption: antecedent and consequence interventions (Abrahamse *et al.* 2005; Geller 1981, 2002; Stern 1992). Antecedent interventions are indirect, designed to influence the *causes* of the desired behaviour. Such interventions can be based on gaining a commitment to change behaviour, goal setting, providing information through workshops, mass media, or home audits, and exposure to exemplar behaviour. Consequence interventions are more direct, designed to influence the behaviour itself by providing positive (or negative) outcomes for the desired (or undesired) behaviour. There are two primary types of consequence intervention strategies, feedback and rewards. The most effective approaches that could be employed to limit consumption are likely to involve a combination of antecedent and consequence interventions (Abrahamse *et al.* 2005).

Behaviour change has been demonstrated with a number of interventions that could be applied to the conservation of energy and the adoption of distributed energy technology. An antecedent intervention based on commitment involves establishing a commitment to change behaviour, and making this commitment in public is likely to result in greater influence on the behaviour (Abrahamse *et al.* 2005). A commitment intervention can easily be combined with goal setting, for example making a pledge to reduce energy consumption by a certain percentage, and then increasing the difficulty of the goal (e.g. saving 20% rather than 10%) and providing feedback has been shown to increase conservation behaviour (Abrahamse *et al.* 2005). The effectiveness of feedback (a consequence intervention) has been shown to increase when the feedback is given closely after the behaviour and feedback is frequent (Abrahamse *et al.* 2005). However, feedback may induce negative behaviour in low and moderate users (Bittle *et al.* 1980). Comparative feedback has not been found to be beneficial unless combined with rewards (Abrahamse *et al.* 2005). Rewards generally reduce energy consumption in the short term, but are unlikely to be associated with long-term behaviour use, thus are more applicable to short-term strategies (Abrahamse *et al.* 2005).

Studies of antecedent interventions suggest that when interventions are employed individually, they are most likely to influence knowledge and attitude, while few are associated with actual behaviour change (Abrahamse *et al.* 2005). The use of general information such as workshops and mass media campaigns has been demonstrated to be effective for increasing attitudes and knowledge, but are unlikely to directly change behaviour (Geller 1981; Luyben 1982; Hutton and McNeill 1981; Staats *et al.* 2004). Behaviour change using information requires information to be tailored to the recipient, for example conducting household energy audits (Abrahamse *et al.* 2005). Demonstrations of model behaviour, such as being shown ways to save energy, has been effective in changing behaviour, though it is unlikely that such change will persist in the long term (Winnet *et al.* 1985).

Overall, the literature suggests that applications of antecedent and consequence intervention strategies to change knowledge, attitudes and behaviour towards energy consumption need to consider how long the change is desired, and whether change in knowledge, attitudes, or behaviour is the primary goal (Abrahamse *et al.* 2005). A limitation of research into these intervention strategies is that results are often based on self-perception measures of change, which may be inaccurate. However, advances in energy metering technology now offer a means of measuring actual change (Abrahamse *et al.* 2005; Poortinaga *et al.* 2003). Another limitation arises because the influence of external factors such as income and regulatory controls is often not considered (Clark *et al.* 2003; Guagnano *et al.* 1995).

12.6.4 *Reducing organisational energy consumption*

Although distributed energy is 'increasingly becoming a mainstream corporate option' (WADE 2006), studies that focus on the implementation and evaluation of energy conservation measures are more abundant at the household level than studies at the organizational level (Horowitz 2004; Siero *et al.* 1996). There is some evidence to suggest that programs targeted at commercial electricity users are effective. For instance, over a 12 year time period from 1989 to 2001, demand-side management programs across 42 states in the United States reduced consumption by 1.9%, and programs changing both demand and supply accounted for a 5.8% reduction (Horowitz 2004). The lack of programs specifically designed to influence organizational energy consumption is probably due, in part, to the complexity of assigning cost savings to particular agents and actions with any precision (Siero *et al.* 1996).

There is some research that has examined the application of antecedent and consequence intervention strategies to commercial energy conservation. These studies have taken place in a transport company (Siero *et al.* 1989) and in a metallurgical company (Siero *et al.* 1996). Because of the organizational context, energy can refer both to energy from vehicle fuel (Siero *et al.* 1989) and energy from power grids (Siero *et al.* 1996). Both studies have drawn on social identity theory, with Siero *et al.* (1989) also applying the theory of reasoned action and Siero *et al.* (1996) using cognitive dissonance theory. The earlier study (Siero *et al.* 1989) tested the influence of three intervention strategies on reducing energy consumption from driving activities; these interventions were providing information, providing task assignment and control, and providing feedback on fuel consumption. The results showed there was a significant increase in self-reported measures of attitudes, social norms, and actual behaviour, and that energy savings also increased (Siero *et al.* 1989).

The later study (Siero *et al.* 1996) applied four intervention strategies: information about conservation, goal setting, feedback about actions, and comparative feedback. The comparative feedback intervention is a specific form of consequence intervention and the study suggests feedback comparing groups had the greatest influence on energy conservation (Siero *et al.* 1996). The comparative feedback is also suggested to be the reason for achieving behavioural changes and only slight attitude and intention changes (Siero *et al.* 1996). The findings from both studies suggest that a combination of intervention strategies is desirable for bringing about behaviour that conserves energy in an organizational context. However, because of the combination of interventions it is difficult to establish the influence of individual interventions, though the findings that conservation increased with comparative feedback suggests that the social environment plays a significant role in influencing attitudes and behaviours related to organizational energy conservation (Kurz 2002).

12.7 External Influences on the Uptake of Distributed Energy Generation and the Reduction of Energy Consumption

As outlined in section 12.2, there are a range of external factors which are relevant to an intelligent grid. Although the psychological models presented in earlier chapters have demonstrated value, such models alone are insufficient to fully explain people's behaviour

(Clark *et al.* 2003; Stern 1992). Rather, individual behaviour occurs within a broader context of non-psychological, external variables.

A wide range of frameworks have been used to organize these external factors (e.g. Abrahamse *et al.* 2005; Guagnano *et al.* 1995; Kurz 2002). Typically, such frameworks include economic factors, technological/physical factors, and broader societal factors, and these broad categories are adopted here. In Table 12.1, a range of factors are identified, relevant to the reduction of energy consumption and the adoption of distributed energy generation. These factors are also categorized as *drivers* (underlying elements that might promote change), *barriers* (elements that may prevent inhibit change) and *incentives* (likely advantages realized once change has occurred). An exhaustive description of all these factors is beyond the scope of the present review, but the remainder of this chapter reviews some specific elements that may be particularly relevant.

Historically, the opening of energy markets, the importance of energy security and environmental values have led to a growing demand for renewable energy sources especially those that combine heat and power (CHP). The spread of the industrial revolution and its associated transition to mass production and electrification was one of the primary catalysts of the drastic reduction in traditional distributed energy generation (e.g. wind and water mills, steam power) seen in most developing countries between the late nineteenth century and mid-twentieth century (Chandler 1994; Hughes 1983; Nuvolari 2006). While this reduction in distributed energy generation (DEG) continued relatively unabated into the twenty-first century, a small number of countries (e.g. Denmark, Germany and the Netherlands) began to make major shifts in public policy towards distributed energy with the emergence of the energy crisis in the 1970s (Munson 2006).

More recently, an increasing number of countries are beginning to make substantial investments in distributed energy systems and have begun instituting major public policy programs to facilitate its development (IEA 2002; WADE 2006). The International Energy Agency (2002) argues that this shift has primarily been driven by five major factors: developments in distributed generation technologies; constraints on the construction of new transmission lines; increased customer demand for highly reliable electricity; the electricity market liberalization; and concerns about climate change. In addition, an improvement in the efficiency of distributed energy production and general system performance has also played a role in improving the competitiveness of distributed energy technologies. These increases, partially due to macro changes in manufacturing technology, have enabled the traditional relationships between scale and scope to be reformatted in favour of more flexible energy solutions (Chandler 1994; Essletzbichler 2003). Case studies of a number of these countries and their public policy programs aimed at distributed energy are examined below.

12.7.1 International case studies

Distributed energy generation plays an important part in Denmark's energy production and supply system (van der Vleuten and Raven 2006). After the energy crisis of 1973 firm policy measures were instituted to reduce Denmark's dependence on foreign energy (van der Vleuten and Raven 2006). The introduction of a variety of measures, such as thermostatic valves, insulation of radiator pipes, electronic heating control systems, wind power and CHP, has meant that the share of distributed generation in gross electricity production increased from 1% in 1980 to over 35% in 2001, made up of decentralized CHP (24%) and wind turbines (11%) (DEA 2004). New regulations recently introduced focus on energy consumption in buildings, where the largest and most cost-effective potential lies.

Table 12.1.   Potential drivers, barriers and incentives for adoption of distributed energy and reduced consumption.

Drivers	Barriers	Incentives
Economic issues		
• The opening of energy markets • Emissions/carbon trading schemes • Competitiveness of DEG technologies • Declining access to energy sources such as oil or gas • Dependency on overseas energy suppliers • Demand for sustainable and renewable energy • Demand for reliable energy	• Lack of access to energy storage and distribution grids • Price of DEG technology • Costs of reducing demand • Compliance costs • Lack of incentives for implementation • Taxation systems • Regulatory practices such as building codes and standards, environmental permits • Government ineligible to enter international emission schemes • Lack of reliability of supply	• Access to energy markets • Admittance to carbon trading schemes • Long-term cost savings • Increased consumer choice • Selling energy to a state or national grid • Provision of incentives such as tax benefits, exemption from levies, etc.
Physical/technological issues		
• Development of DEG technologies • Improvements in efficiency and system performance • Increased spending on research and development • Need for ease of implementation and use	• Lack of technology transfer partnerships • Lack of technical performance metrics • Lack of incentives for development • Lack of awareness, interest, evaluation and trialling of the technology • Lack of perceived ease of use of the technology • Lag in development of economically viable alternatives • Discrepancies between ability to supply energy and the demand for energy	• Incentives from government to develop technology • Development of technology transfer partnerships • Ease of implementation • Ease of use of the technology • Compatibility/synergy with existing technology
Societal issues (government, public and industry)		
• Increasing concerns in relation to climate change • Increases in energy consumption • Demand for higher standards of living • Increasing demand for sustainable and renewable energy • Desire for convenience of energy supply	• Lack of government, industry and public knowledge and acceptance of environmental issues • Aesthetics of DEG • Lack of understanding of the values, attitudes, beliefs and behaviour of the public and industry • Resistance to change • Lack of commitment to short-term sacrifices that may be needed such as reduced demand • Lack of government, industry and public acceptance and perceived usefulness of DEG • Public and industry energy consumption • Public lack of knowledge and understanding of DEG	• Reduced environmental impact and emissions • Increased efficiency of energy use • Flexibility and reliability of supply • Ownership of supply • Reduced reliance on external energy suppliers

The most important initiatives are a tightening of energy provisions in building regulations, a new and improved energy-labelling scheme, enhanced supervision of boilers and ventilation installation regulations and finally a separate initiative within the public sector (DEA 2005). While the Finnish government see environmental drivers for distributed energy adoption as important, the real force behind its push into this arena seems to be a perceived economic opportunity that can be exploited with targeted industry policy (Laine 2004).

The United Kingdom government views the prospective environmental and social benefits envisaged to stem from micro-CHP distributed energy systems as the primary driver for their policy support to the distributed energy industry (Watson 2002). In an effort to bring about this change, the UK government is using a broad range of tools to break down barriers to distributed energy and to facilitate its integration into local and national electricity grids. These measures include providing tax benefits (e.g. potential tax reductions), changing regulatory controls (e.g. new processes for awarding permits to potential distributed energy sites), encouraging emissions trading (e.g. EU Emissions Trading Scheme), introducing various carbon-based taxes (e.g. Climate Change Levy) and expanding support for research and development (UK DTI 2006a).

An interesting example of these measures in action is in the field of micro-CHP units. The Carbon Trust is running a field trial of micro-CHP units, monitoring the energy and financial savings (UK DTI 2006a). The UK government, subject to these field trials results, intends to reduce the rate of value added tax on micro-CHP units from 17.5% to 5% (Cogen 2005a). Also, for non-residential energy users paying the Climate Change Levy, a levy on electricity and gas sales, any CHP that is certified as 'good quality' is exempt from the levy. This strategy appears to focus on eliminating the need for planning permission for low impact distributed energy (e.g. solar, micro-wind, etc.) and attempts to reduce market uncertainty. For example, the UK 2006 Budget allocated an additional £50 m of capital grants for micro-generation. Using this fund, the government is working to set up a framework agreement whereby a number of suppliers agree to provide micro-generation installations at reduced prices, secure in the knowledge that they will have access to the market created by the £50 m grant funding. This is designed to provide a level of certainty for suppliers in return for reduced retail prices for their products. It is hoped that participation in the framework will encourage suppliers to invest in larger-scale production, bringing down prices on a permanent basis and stimulating demand yet further (Cogen 2005a). Expressions of interest were published mid-June 2007, and the government aims to have the program up and running by the end of the year.

The UK government is currently undertaking research into distributed energy generation for a government White Paper, which is to be published in the second half of 2007 (UK DTI 2006b). Its primary focus is: the economic and other incentives on suppliers to buy electricity from distributed generators; the economic costs and benefits, and other incentives on distribution network operators (DNOs) to connect new generators and to invest in upgrading distribution networks in order to accommodate increasing amounts of distributed generation; the incentives on DNOs to engage in innovation aimed at minimizing the costs and capturing the benefits of distributed generation; options for resolving potential barriers to the sale of electricity from small generators, for example, licensing procedures; and technical standards for connection and for network operation (UK DTI 2006a).

In the USA, investments in distributed natural gas technologies like advanced turbines and microturbines, cooling, heating and power systems, fuel cell systems, hybrid systems and natural gas engines are developing (US NEPDG 2001). The general aim of the policy within the US is to use these investment funds first to support the development of enabling technologies, like fuel processing, hydrogen energy systems, materials and manufacturing and power electronics and sensors; second to develop energy generation and delivery systems architecture for distributed energy resources, including district energy, energy storage, grid interconnection, modelling and simulation tools, power parks, mini-grids and superconducting materials for electric systems and transmission and distribution; third to coordinate activities in renewable energy development, like solar, geothermal and photovoltaic projects – in order to facilitate this process, the government is aiming to establish collaborative technology transfer partnerships between industry, state agencies, universities and national laboratories; and finally to conduct systems integration, implementation and outreach activities aimed at addressing infrastructure, institutional and regulatory needs, including building codes and standards, environmental permitting and siting and other initiatives (US NEPDG 2001).

The Netherlands has one of the highest distributed energy generation penetration rates in the world (Strachan and Dowlatabadi 2002). In order to encourage the development of the distributed energy generation market the Netherlands government used the liberalization of the national energy market to encourage electricity distribution utilities to support distributed energy production and provided extensive subsidies (Blok and Farla 1996). For example, businesses purchasing micro-CHP units are entitled to tax breaks on their investment; however, at present there are no incentives for households to purchase or use micro-CHP (Cogen 2005b).

12.7.2 The Australian context

As indicated in Table 12.1, within Australia, the primary drivers of distributed energy technologies are viewed as being energy conservation, efficiency and capital deferment (MacAulay and Ashworth 2006). While there are important drivers of distributed energy technologies, there are also significant barriers and incentives that may obstruct or facilitate the wide-scale adoption of these technologies. Primary among these barriers is the need to change the consumption behaviour of actors within each respective market and create social acceptance of distributed energy generation. While there has been extensive investigation into the technical performance metrics associated with distributed energy occurring around the world (e.g. Hoff *et al.* 1996; Outhred and Spooner 2002; Strachan and Farrell 2006), little research to date has examined its implementation or social acceptance.

12.8 Conclusions and Application for an Intelligent Grid in Australia

This review confirms the need for understanding social perspectives toward distributed energy within the Australian context, as part of a vision for an intelligent grid. The conceptual model (Fig. 12.6) proposed suggests a theoretical way forward to enable the future vision for an electricity network that balances supply and demand for the benefit of consumers and utilities, and minimizes greenhouse gas emissions, losses and price.

The model is based on a causal chain from knowledge, beliefs and attitudes, through intention to act, actual behaviour and longer-term acceptance and uptake of distributed energy technology. In addition, the model acknowledges external (non-psychological) influences on individual behaviour including physical, social, economic and political factors. It also recognizes that a supportive socio-political environment (via appropriate

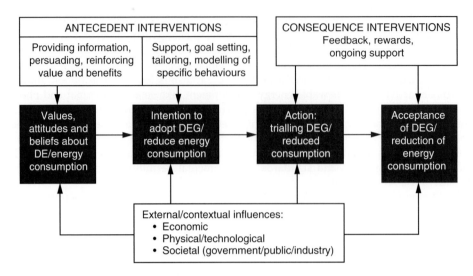

Fig. 12.6. A conceptual framework for the intelligent grid.

public policy, regulatory changes, access to distributed energy technology, etc.) is a vital component of any successful broad-scale adoption of distributed energy technologies.

The review highlights that no single body of literature is fully and solely relevant for the public acceptance of distributed energy generation, and empirical data is required to advance our understanding in this area; however, the theoretical basis for this model has been confirmed through this review. While the model draws heavily on two specific theories, those being Rogers' Theory of Perceived Attributes and Ajzen's Theory of Planned Behaviour, it allows for the highly complex interrelationships between value, attitudes and behaviours evidenced in other bodies of literature. Because the convenient distinction between 'consumer' technologies (requiring individual decisions to purchase a particular system) and 'social' technologies (requiring collective decisions about the acceptability of a particular course of action) is blurred when it comes to distributed energy technologies, both individual and group behaviour must be taken into account in determining the viability of the intelligent grid.

Within the model two distinct types of social/psychological interventions are identified. The first are early interventions which are designed to increase knowledge and promote positive attitudes and intentions towards distributed energy generation. The second are interventions based around incentives to increase the likely uptake of specific distributed energy technologies, either through distributed energy generation or demand-side management. Both of these interventions identify potential research areas to be explored in more detail. This review also highlights the importance of interventions through policy frameworks supporting and encouraging the uptake of distributed energy technologies and ultimate implementation of an intelligent grid.

References

Abelson, R.P. (1972). Are attitudes necessary? in *Attitudes, Conflict and Social Change* (ed. B.T. King), pp. 19–32. Academic Press, New York.

Abrahamse, W., Steg, L., Vlek, C. and Rothengatter, T. (2005). A review of intervention studies aimed at household energy conservation. *Journal of Environmental Psychology*, **25**(3), 273–291.

Ajzen, I. (1989). Attitude structure and behavior, in *Attitude Structure and Function* (eds A.R. Pratkanis, S.J. Beckler and A.G. Greenwald), pp. 241–274. Erlbaum, Hillsdale, NJ.

Ajzen, I. and Fishbein, M. (1980). *Understanding Attitudes and Predicting Social Behaviour*. Prentice Hall, Englewood Cliffs, NJ.

Ajzen, I. and Fishbein, M. (2005). The influence of attitudes on behavior, in *The Handbook of Attitudes* (eds D. Albarracín, B.T. Johnson and M.P. Zanna), pp. 173–221. Erlbaum, Mahwah, NJ.

Argyris, C. and Schon, D. (1978). *Organizational Learning: A Theory of Action Perspective*. McGraw-Hill, New York.

Bagozzi, R.P., Davis, F.D. and Warshaw, P.R. (1992). Development and test of a theory of technological learning and usage. *Human Relations*, **45**(7), 660–686.

Bittle, R.G., Valesano, R. and Thaler, G. (1980). The effects of daily cost feedback on residential electricity usage as a function of usage level and type of feedback information. *Journal of Environmental Systems*, **9**, 275–287.

Blok, K. and Farla, J. (1996). The continuing story of CHP in the Netherlands. *International Journal of Global Energy Issues*, **8**(4), 349–361.

Burkman, E. (1987). Factors affecting utilisation, in *Instructional Technology: Foundations* (ed. R.M. Gagne) Lawrence Earlbaum, Hillsdale, NJ.

Calder, B.J. and Ross, M. (1973). *Attitudes and Behaviour*. General Learning Press, Morristown, New Jersey.

Cayford, J. (2001). Informal risk perception and formal theory. In: *Proceedings of the 2nd VALDOR Conference on Transparency in Risk and Decision Making*.

Chandler, A. (1994). The competitive performance of U.S. industrial enterprises since the Second World War. *Business History Review*, **68**(1), 1–72.

Clark, C.F., Kotchen, M.J. and Moore, M.R. (2003). Internal and external influences on pro-environmental behaviour: participation in a green electricity program. *Journal of Environmental Psychology*, **23**, 237–246.

Cogen Europe (Cogen) (2005a). *Micro-CHP fact sheet United Kingdom*. Available: http://www.cogen. org/Downloadables/Publications/FactSheet_MicroCHP_UK.pdf. Retrieved 28 November 2006.

Cogen Europe (Cogen) (2005b). *Micro-CHP fact sheet Netherlands*. Available: http://www.cogen.org/ Downloadables/Publications/FactSheet_MicroCHP_Netherlands.pdf. Retrieved 28 November 2006.

Cvetkovich, G. and Löfstedt, R. (1999). *Social Trust and the Management of Risk*. Earthscan, London.

Dake, K. (1992). Myths of nature: culture and the social construction of risk. *Journal of Social Issue*, **48**(4), 21–37.

Danish Energy Authority (DEA) (2004). *Energy in Denmark 2003*. Danish Energy Authority, Copenhagen.

Danish Energy Authority (DEA) (2005). *Energy policy statement 2005*. Available: http://www.energistyrelsen.dk/graphics/Publikationer/Energipolitik_UK/Energy_Policy_Statement_2005/index.htm. Retrieved 28 November 2006.

Davis, F.D. (1989). Perceived usefulness, perceived ease of use, and user acceptance of information technology. *MIS Quarterly*, **13**(3), 319–340.

Davis, F.D., Bagozzi, R.P. and Warshaw, P.R. (1989). User acceptance of computer technology: a comparison of two theoretical models. *Management Science*, **35**(8), 982–1002.

Dick, B. and Dalmau, T. (2003). Argyris and Schön: some elements of their models. *Resource Papers in Action Research*. Available: http://www.scu.edu.au/schools/gcm/ar/arp/argyris2.html. Retrieved 11 January 2007.

Doll, J. and Ajzen, I. (1992). Accessibility and stability of predictors in the theory of planned behavior. *Journal of Personality and Social Psychology*, **63**(5), 754–765.

Eagly, A.H. and Chaiken, S. (1998). *Attitude Structure and Function*. McGraw-Hill, New York.

Echabe, E. and Castro, L. (1999). Group discussion and changes in attitudes and representations. *The Journal of Social Psychology*, **139**(1), 29–43.

Essletzbichler, J. (2003). From mass production to flexible specialization: the sectoral and geographical extent of contract work in U.S. manufacturing. *Regional Studies*, **37**(8), 753–761.

Fazio, R.H., Blascovich, J. and Driscoll, D.M. (1992). On the functional value of attitudes: the influence of accessible attitudes on the ease and quality of decision making. *Personality and Social Psychology Bulletin*, **18**(4), 388–401.

Fazio, R.H., Ledbetter, J.E. and Towles-Schwen, T. (2000). On the costs of accessible attitudes: detecting that the attitude object has changed. *Journal of Personality and Social Psychology*, **78**(2), 197–210.

Fazio, R.H. and Powell, M.C. (1997). On the value of knowing one's likes and dislikes: attitude accessibility, stress, and health in college. *Psychological Science*, **8**(6), 430–436.

Fazio, R.H. and Williams, C.J. (1986). Attitude accessibility as a moderator of the attitude-perception and attitude-behavior relations: an investigation of the 1984 presidential election. *Journal of Personality and Social Psychology*, **51**(3), 505–514.

Festinger, L. (1957). *A Theory of Cognitive Dissonance*. Row, Peterson, Oxford, England.

Fishbein, M. and Ajzen, I. (1975). *Belief, Attitude, Intention and Behaviour: An Introduction to Theory and Research*. Addison-Wesley, Reading, MA.

Gallois, C., Giles, H., Jones, E. *et al.* (1995). Accommodating intercultural encounters: elaborations and extensions, in *Intercultural Communication Theory* (ed. R.L. Wiseman), pp. 115–147. Sage Publications, Thousand Oaks, CA.

Gardner, M.J., Paulsen, N., Gallois, C. *et al.* (2001). Communication in organisations: an intergroup perspective, in *The New Handbook of Language and Social Psychology* (eds W.P. Robinson and H. Giles), pp. 561–584. John Wiley and Sons Ltd.

Geller, E.S. (1981). Evaluating energy conservation programs: is verbal report enough? *Journal of Consumer Research*, **8**, 331–335.

Geller, E.S. (2002). The challenge of increasing proenvironment behavior, in *Handbook of Environmental Psychology* (eds R.G. Bechtel and A. Churchman), pp. 525–540. Wiley, New York.

Greenwood, D.J. and Levin, M. (2003). Reconstructing the relationship between universities and society through action research, in *The Landscape of Qualitative Research: Theories and Issues* (eds N.K. Denzin and Y.S. Lincoln) Sage Publications, Thousand Oaks, CA.

Grove-White, R., Macnaughten, P. and Wynne, B. (2000). *Wising Up: The Public and New Technologies*. Lancaster University, Centre for the Study of Environmental Change.

Guagnano, G.A., Stern, P.C. and Dietz, T. (1995). Influences on attitude-behaviour relationships: a natural experiment with curbside recycling. *Environment and Behaviour*, **27**(5), 699–718.

Gutman, J. (1982). Means-end chain model based on consumer categorisation process *Journal of Marketing*, **46**(Spring 1983), 60–72.

Hallman, W.K. and Wandersman, A. (1992). Attribution of responsibility and individual and collective coping with environmental threats. *Journal of Social Issues*, **48**(4), 101–118.

Hemmati, M. (2002). *Multi-Stakeholder Processes for Governance and Sustainability: Beyond Deadlock and Conflict*. Earthscan Publications, London.

Himmelfarb, S. and Eagly, A.H. (1974). *Readings in Attitude Change*. John Wiley and Sons, Oxford.

Hoff, T.E., Wenger, H.J. and Farmer, B.K. (1996). Distributed generation: an alternative to electric utility investment in system capacity. *Energy Policy*, **24**(2), 137–138.

Hogg, M.A. and Abrams, D. (1999). Social identity and social cognition: historical background and current trends, in *Social Identity and Social Cognition* (eds D. Abrams and M.A. Hogg), pp. 1–25. Blackwell Publishers In, Oxford.

Hogg, M.A. and Cooper, J. (eds) (2002). *Sage Handbook of Social Psychology*. Sage Publications, London.

Horowitz, M. (2004). Electricity intensity in the commercial sector: market and public program effects. *The Energy Journal*, **25**(2), 115–138.

Hughes, T.P. (1983). *Networks of Power: Electrification in Western Society 1880–1930*. Johns Hopkins University Press, Baltimore.

Hutton, R.B. and McNeill, D.L. (1981). The value of incentives in stimulating energy conservation. *Journal of Consumer Research*, **8**, 291–298.

International Energy Agency (IEA) (2002). *Distributed Generation in Liberalised Electricity Markets*. International Energy Agency, Paris.

Kantola, S., Syme, G. and Campbell, N. (1984). Cognitive dissonance and energy conservation. *Journal of Applied Psychology*, **69**(3), 416–422.

Krosnick, J.A., Boninger, D.S., Chuang, Y.C. *et al.* (1993). Attitude strength: one construct or many related constructs? *Journal of Personality and Social Psychology*, **65**(6), 1132–1151.

Kurz, T. (2002). The psychology of environmentally sustainable behaviour: Fitting together pieces of the puzzle. *Analyses of Social Issues and Public Policy*, **2**(1), 257–278.

Laine, J. (2004). *Results of the distributed generation technology programme in Finland.* Presented at the International Energy Agency conference on distributed generation: key issues, challenges and roles. Available: http://www.iea.org/dbtw-wpd/Textbase/work/2004/distgen/Laine.pdf. Retrieved 28 November 2006.

Lavine, H., Huff, J.W., Wagner, S.H. and Sweeney, D. (1998). The moderating influence of attitude strength on the suseptibility to context effects in attitude surveys. *Journal of Personality and Social Psychology*, **75**, 359–373.

Löfstedt, R. and Frewer, L. (1998). *The Earthscan Reader in Risk and Modern Society.* Earthscan Publications Ltd, London.

Luyben, P.D. (1982). Prompting thermostat setting behavior: public response to a presidential appeal for conservation. *Environment and Behavior*, **14**(1), 113–128.

MacAulay, S. and Ashworth, P. (2006). Summary report of interviews with key stakeholders. CSIRO Energy Transformed: National Research Flagship.

Madden, T.J., Ellen, P.S. and Ajzen, I. (1992). A comparison of the theory of planned behavior and the theory of reasoned action. *Personality and Social Psychology Bulletin*, **18**(1), 3–9.

Manis, M. (1978). Cognitive social psychology and attitude change. *The American Behavioural Scientist*, **21**(5), 675–690.

Manyiwa, S. and Crawford, I. (2001). Determining linkages between consumer choices in a social context and the consumer's values: A means-end approach. *Journal of Consumer Behaviour*, **2**(1), 54–70.

Marks, L.A. (2001). Communicating about agrobiotechnology. *AgBioForum*, **4**(3 and 4), 152–154.

Munson, R. (2006). Yes, in my backyard: distributed electric power. *Issues in Science and Technology*, **22**(20), 49–55.

Munson, J.M. and Posner, B.Z. (1980). The factorial validity of a modified Rokeach value survey for four diverse samples. *Educational and Psychological Measurement*, **40**(4), 1073–1079.

Niemeyer, S.J. (2004). Deliberation for the wilderness: displacing symbolic politics. *Environmental Politics*, **13**(2), 342–347.

Niemeyer, N. and Littleboy, A. (2005). Societal uptake of energy technologies: a framework for examining social responses to energy technologies in Australia energy futures. Report No. P2006/30. Brisbane: CSIRO.

Nuvolari, A. (2006). The making of steam power technology: a study of technical change during the British industrial revolution. *The Journal of Economic History*, **66**(2), 473–485.

Olsen, J.M. and Zanna, M.P. (1993). Attitudes and attitude change. *Annual Review of Psychology*, **44**, 117–145.

Oskamp, S. (1983). Psychology's role in the conserving society. *Population and Environment: Behavioral and Social Issues*, **6**(4), 255–293.

Oskamp, S. (2000). A sustainable future for humanity? How can psychology help? *American Psychologist*, **55**(5), 496–508.

Outhred, H. and Spooner, E. (2002). Network issues associated with distributed generation. *Proceedings of Solar 2002, Australian and New Zealand Solar Energy Society.*

Paglia, A. and Room, R. (1999). Expectancies about the effects of alcohol on the self and on others as determinants of alcohol policy attitudes. *Journal of Applied Social Psychology*, **29**(12), 2632–2651.

Petty, R.E. and Cacioppo, J.T. (1984). The effects of involvement on responses to argument quantity and quality: central and peripheral routes to persuasion. *Journal of Personality and Social Psychology*, **46**(1), 69–81.

Poortinga, W., Steg, L. and Vlek, C. (2004). Values, environmental concern, and environmental behavior: a study into household energy use. *Environment and Behavior*, **36**(1), 70–93.

Poortinga, W., Steg, L., Vlek, C. and Wiersma, G. (2003). Household preferences for energy-saving measures: a conjoint analysis. *Journal of Economic Psychology*, **24**(1), 49–64.

Pueppke, S.G. (2001). Agricultural biotechnology and plant improvement: setting the stage for social and economic dialogue. *American Behavioral Scientist*, **44**(8), 1233–1245.

Renn, O., Webler, T. and Widemann, P. (1995). *Fairness and Competence in Citizen Participation: Evaluating Models for Environmental Discourse*. Kluwer Academic Publishers, Dordrecht, Germany.

Rogers, E.M. (1962). *Diffusion of Innovations* (1st edn.). The Free Press, New York.

Rogers, E.M. (1995). *Diffusion of Innovations* (4th edn.). The Free Press, New York.

Rogers, E.M. (2003). *Diffusion of innovations* (5th ed.). New York: Free Press.

Rokeach, M. (1973). *The Nature of Human Values*. The Free Press, New York.

Savage, L.J. (1990). *The Foundations of Statistics*. Wiley, New York.

Semmler, C. and Brewer, N. (2002). Effects of mood and emotion on juror processing and judgments. *Behavioral Science and the Law*, **20**, 423–436.

Siero, F., Bakker, A., Dekker, G. and van den Burg, M. (1996). Changing organizational energy consumption behaviour through comparative feedback. *Journal of Environmental Psychology*, **16**, 235–236.

Siero, S., Boon, M., Kok, G. and Siero, F.W. (1989). Modification of driving behaviour in a large transport organization: a field study. *Journal of Applied Psychology*, **74**, 417–425.

Smith, B.N. and Stasson, M.F. (2000). A comparison of health behaviour constructs: social psychological predictors of AIDS-preventive behavioural intentions. *Journal of Applied Social Psychology*, **30**(3), 443–462.

Staats, H., Harland, P. and Wilke, H. (2004). Effecting durable change: a team approach to improve environmental behaviour in the household. *Environment and Behavior*, **36**(3), 341–367.

Stern, P.C. (1992). What psychology knows about energy conservation. *American Psychologist*, **47**(10), 1224–1232.

Stirling, A. (2002). Risk at a turning point. *Journal of Risk Research*, **97**, 109.

Strachan, N. and Dowlatabadi, H. (2002). Distributed energy and distribution utilities. *Energy Policy*, **30**, 649–655.

Strachan, N. and Farrell, A. (2006). Emissions from distributed versus centralized generation: the importance of system performance. *Energy Policy*, **34**, 2677–2689.

Tajfel, H. and Turner, J.C. (1986). The social identity theory of intergroup behaviour, in *Psychology of Intergroup Relations* (eds S. Worchel and W.G. Austin), pp. 7–24. Nelson-Hall, Chicago.

Terry, D.T., Callan, V.J. and Sartori, G. (1996). Employee adjustment to an organisational merger: stress, coping and intergroup differences. *Stress Medicine*, **12**, 105–122.

Terry, D.T., Hogg, M.A. and Duck, J. (1999). Group membership, social identity and attitudes, in *Social Identity and Social Cognition* (eds D. Abrahams and M.A. Hogg), pp. 280–314. Blackwell, Oxford.

Terry, D.T., Hogg, M.A. and White, K.M. (2000). Attitude behaviour relations: social identity and group membership, in *Attitudes, Behaviour and Social Context: The Role of Norms and Group Membership* (eds D.J. Terry and M.A. Hogg), pp. 67–93. Erlbaum, Mahwah, NJ.

Triandis, H.C. (1977). *Interpersonal Behaviour*. Brooks, Monterey, CA.

Triandis, H.C. (1980a). Social factors in science education. *Instructional Science*, **9**(2), 163–181.

Triandis, H.C. (1980b). Values, attitudes, and interpersonal behavior. *Nebraska Symposium on Motivation*, **27**, 195–259.

United Kingdom Department of Trade and Industry (UK DTI) (2006a). *The Energy Challenge*. Available: http://www.dti.gov.uk/energy/review/page31995.html. Retrieved 28 November 2006.

United Kingdom Department of Trade and Industry (UK DTI) (2006b). *Distributed Energy*. Available: http://www.dti.gov.uk/energy/review/implementation/distributed-energy/page35076.html. Retrieved 28 November 2006.

United States National Energy Policy Development Group (NEPDG) (2001). *National Energy Policy*. Available: http://www.whitehouse.gov/energy/National-Energy-Policy.pdf. Retrieved 28 November 2006.

van der Vleuten, E. and Raven, R. (2006). Lock-in and change: distributed generation in Denmark in a long-term perspective. *Energy Policy*, **34**(18), 3739–3748.

Vaughan, G.M. (1977). Personality and small group behaviour, in *Handbook of Modern Personality Theory* (eds R.B. Cattell and R.M. Dreger), pp. 511–529. Academic Press, London.

Venkatesh, V. and Davis, F.D. (2000). A theoretical extension of the technology acceptance model: four longitudinal field studies. *Management Science*, **46**(2), 186–205.

Wadsworth, Y. (1998). What is participatory action research? *Action Research International*. Available: http://www.scu.edu.au/schools/gcm/ar/ari/p-ywadsworth98.html. Retrieved 11 January 2007.

Watson, J. (2002). Renewables and CHP deployment in the UK to 2020. Tyndell Centre Working Paper. Report No. 21. Available: http://www.tyndall.ac.uk/publications/working_papers/wp21.pdf. Retrieved 28 November 2006.

Wicker, A.W. (1969). Attitudes versus actions: the relationship of verbal and overt behavioral responses to attitude objects. *Journal of Social Issues*, **25**(4), 41–78.

Winett, R.A., Leckliter, I.N., Chinn, D.E. *et al.* (1985). Effects of television modeling on residential energy conservation. *Journal of Applied Behavior Analysis*, **18**, 33–44.

World Alliance for Distributed Energy (WADE) (2006). *World Survey of Decentralized Energy*. World Alliance for Distributed Energy, Edinburgh.

Zhao, J., He, N. and Lovrich, N.P. (1998). Individual value preferences among American police officers: the Rokeach theory of human values revisited. *An International Journal of Police Strategies and Management*, **21**(1), 22–37.

Chapter 13
Innovations Promote Rural and Peri-Urban Electrification in Developing Countries

NANCY E. WIMMER

microSOLAR, www.microsolar.com

13.1 Introduction

The electrification of rural and peri-urban areas in developing countries is in transition. Electric grids in poor countries grow so slowly that they will not reach many young villagers in their lifetime. But windmills and photovoltaic systems already generate electricity in thousands of remote places – island and mainland, steppe and farmland. They will increase as oil and kerosene prices soar, and the next generation of renewable energy systems enters the market. While this course of rural electrification seems predictable to some, a closer look reveals how deeply innovation is changing rural life. New information and communication technologies, new services for finance and insurance, health and education are impacting rural markets. Many of them seem unrelated to rural electrification – but will push it forward nonetheless.

Rural electrification is a rapidly evolving field, spurred by a mixture of innovations that interact and reinforce each other, offering villagers new opportunities. Significant change lies ahead because innovation encompasses much more than technology and products. It is about applications, business models and new markets – far different from those in the industrialized urban world. The combined force of innovations can speed up rural evolution and helps meet the untapped energy needs of some 1.6 billion rural customers. Moreover, they can serve to bridge the gap between rural and urban communities and provide models to stabilize poor areas in megacities.

If current trends continue, the demand for electricity will soon go beyond those for light, TV, water pumps or appliances. An innovative initiative by the Massachusetts Institute of Technology (MIT) Media Lab highlights this trend (OLPC 2006). If it succeeds, millions of children in poor countries will receive cheap computers in the next several years. This $100 laptop is specifically designed for young students living in harsh environments, featuring advanced technologies like wireless communication, sunlight readability, open source software, and a user interface adapted to many local languages.

As an educational tool it could impact significant change in rural areas, where schools are known to be poor and where most young people in developing countries live. Eighty per cent of all school-age children in China live in rural areas. Half of Bangladesh's 110 million villagers are under 18 years of age. Developing countries have young populations, and children are known to take to computers readily, even in slums and remote villages.

The *One Laptop per Child* idea is that governments provide the laptops free of charge to children in much the same way they provide text books. Argentina, Brazil, Ethiopia, India, Libya, Nepal, Nigeria, Pakistan, Peru, Romania, Russia, Rwanda, Thailand, Uruguay and USA are candidates to receive the first wave of laptops (OLPC 2006).

But what will power computers in areas where few people have access to an electric grid, car batteries and diesel generators? The laptops can be 'human powered', whereby one minute of cranking a generator allows for ten minutes of use. Clearly, it is an interim solution where batteries and generators are expensive to fuel, charge, maintain and replace, or where the electric grid is notoriously unreliable.

13.2 Wireless Networks – New Forms of Connectivity

More than laptops, communication networks will drive rural development and they are booming. Wireless communication for mobile telephony has spread throughout the developing world more quickly and deeply than any previous technology. In less than a decade, mobile phones are being used in each of the 70000 villages of Bangladesh. Africa, so often left behind, has seen the world's most rapid growth in mobile phone penetration (The World Bank 2006). An estimated 80% of Tanzanians can now access a mobile phone. The rural poor in much of Southern Asia readily devote 5–7% of their income to communication access.

Underestimated in the beginning, wireless communication meets a basic demand. A mobile phone can allow farmers and fishermen to find out the day's prices for their products and make more profit. Already there is evidence that the growth in mobile telephony is reducing the variations in prices between markets in rural and urban areas. Small businesses can shop more efficiently for supplies. A handyman living in a village can advertise in a large town nearby for work and travel only when he is informed by phone that there is a job available. Any innovation that saves unnecessary travel and reduces journeys is of great value in rural areas with underdeveloped transport infrastructures.

Hammond and March (2006) argue that powerful wireless networks, low-cost technology and entrepreneurship blend to create community-based communication networks. A range of value-added information and communication services promise to help expand economic capacity and commerce of rural areas in developing countries.

13.3 Pervasive Mobile Phone Applications

Mobile phones are widely used to maintain family ties to workers abroad and to cut the cost of money transfers from overseas. According to the World Bank international remittances are worth $225 billion a year, and constitute as strongly growing business. The African economy is increasingly supported by its worldwide diaspora, with communication technology spurring the process.

The importance this remittances flow constitutes for developing countries cannot be underestimated. Remittances account for more than 10% of the gross domestic products (GDP) of 15 developing countries studied by the International Monetary Fund (IMF) and are growing at a double-digit annual rate. The number of participants in this market is unprecedented. Estimates are that 200 million transnational labour migrants sent money home to their families during 2005, and the number of fund-dispatching domestic labour migrants is even larger (Hastings 2006). A multi-million dollar industry devoted to money transfers has emerged, with profit margins as high as 30%. In 2007, the remittance market is expected to generate more than $15 billion in annual revenue (Hastings 2006).

Wire transfer companies such as Western Union or MoneyGram have traditionally been the common formal means of sending remittances, but this is changing as new services and service providers appear. Cell phones have become electronic wallets, using SMS technology for electronic money transfer and payment (Rhyne and Otero 2006). Table 13.1 lists some examples.

Funds transfer via mobile phones has dramatically reduced the cost of sending remittances, even for people who are not or inadequately served by banks.

Many more innovative applications will build on cell phone technology as they are applied to solve local problems. The *Village Phone* program in rural Bangladesh pioneered loans to poor women to buy a mobile phone and earn an income by selling telephone services to the surrounding villages. Within a decade, there are now over 300 000 *village telephone ladies*, many of whom have already diversified their business, providing telephony, telefax and other services based on information and communication technology (ICT).

More powerful solutions in wireless communication make it likely that even remote and poor villagers will experience the internet by using a cell phone.

13.4 Information for Villagers

ICT is taking root in rural areas and enabling entirely new applications which will grow as they meet the needs of rural customers. Already, the rural way of life seems to be in transition, influenced by new applications for education, health, farming, trading, financing and connecting dispersed family members. *Will it help me to generate income?* This is the key question villagers ask. If the answer is positive, even poor villagers are ready to invest. Centres of application innovation are emerging, notably in Southern Asia.

Consider the e-Choupal phenomenon in India, where millions of farmers gather not at their village meeting place, a *choupal* in Hindi, but at places where a PC is linked to the internet. A small-scale soya farmer can visit his village e-Choupal, check the website of the

Table 13.1. Electronic money transfer and payment services.

Area	Service
The Philippines	*Globe Communications* and *SMART* have introduced cell phones to store and transfer electronic cash for payments.
Kenya	*Pesa* has partnered with *Vodafone, Faulu-Kenya*, a local microfinance institution (MFI), and *CBA*, a local bank. The M-Pesa system supports money transfers, cash withdrawal and deposits at retail outlets, and disbursement and payment of loans.
South Africa	The *Wizzit* banking facility allows low income account holders to use their mobile phones to remit money to a friend, buy airtime, or pay accounts.
Zambia	*Celpay* has developed a mobile phone payment system, which enables money transfers via phone and without a bank account.
Latin America	*Motorola* recently unveiled *M-Wallet*, a downloadable software application that allows users to pay bills, purchase products, and/or transfer money using their cell phones. The company is targeting the remittance market between the US and Latin America.

Table 13.2.   Selection of ICT-based services.

Area	Service type	Access	Organization
India	agricultural information and marketplace portal	Internet PC	ITC Ltd, www.echoupal.com
Kenya	agricultural information portal	mobile phone-based SMS system	The Kenya Agricultural Commodity Exchange (KACE) www.kacekenya.com/marketinfo/sms.asp
East and Central Africa	agricultural information portal	mobile phone-based SMS system	FOODNET www.foodnet.cgiar.org
The Philippines	agricultural marketplace portal	mobile phone-based SMS system	B2Bpricenow www.B2Bpricenow.com
India	agricultural information and marketplace portal	Internet PC	Agriwatch http://agriwatch.com/
South Africa	health services	phone-based SMS system	On-Cue Compliance (www.on-cue.co.za) Cell-Life (www.cell-life. org)

Chicago Board of Trade and local markets on prices for his crop, the weather forecast, and tips on increasing his yield – all in the local language. If he chooses, he can sell his crop online and even buy seeds and fertilizer. In only six years, 6500 e-Choupals have been installed, covering 38 000 villages and serving 3.5 million farmers (ITC 2006).

e-Choupals were launched in 2000 by the Indian agricultural trading company, ITC, as an e-commerce platform to capture more of the soybean crop and to lower transaction costs. e-Choupals were an immediate success, because they solved a local problem. Both ITC and the soya farmers have long been dependent on the same archaic system. Farmers sold to village traders, settling for whatever price was offered. ITC then had to buy from the traders, with little quality control and high transaction costs. e-Choupals allow the company to buy directly from farmers and let farmers check prices to decide whether they want to sell – all of which increases farmer incomes and *excludes* the village soybean traders.

If information technology is designed to meet business needs and give easy access, even poor populations can benefit: illiterate farmers can use e-Choupals easily for a small fee, because each e-Choupal is run by a local, trusted farmer, who is trained by ITC and obligated by oath to serve the entire community.

e-Choupals – like telecentres, village internet kiosks and cabinas publicas elsewhere – often combine conventional technology with a killer application. Table 13.2 indicates a variety of service types and access methods (Hammond and March 2006).

Special software platforms are being developed to set up portals for audio-based information in local languages for fast and easy access (WORDTALK 2006). More information services will appear to meet demands for housing, health, markets, legal rights, employment and crime prevention – both in rural and peri-urban regions.

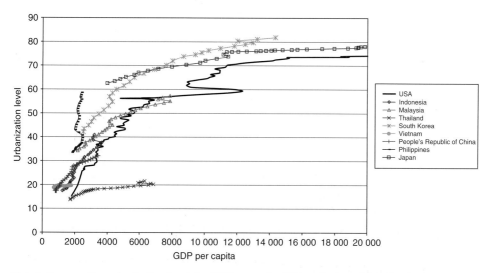

Plate 1 Comparative urbanization levels by GDP per capita, USA and selected Asia-Pacific economies.

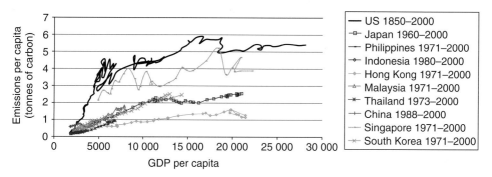

Plate 2 Comparative changes in carbon emissions by GDP per capita, USA and selected Asia-Pacific economies.

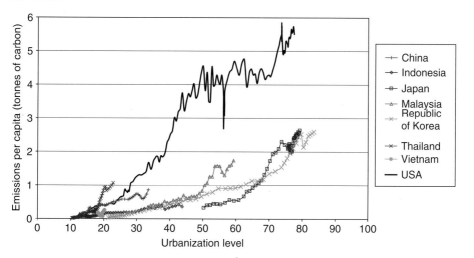

Plate 3 Comparative carbon emissions per level of urbanization, USA and selected Asia-Pacific economies.

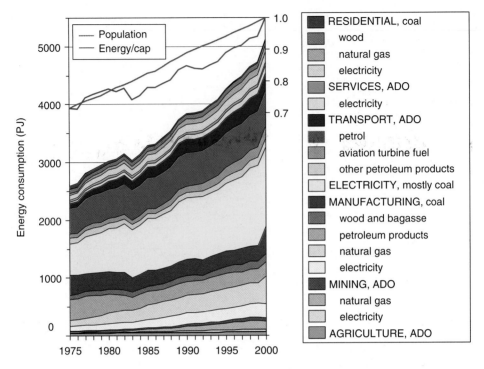

Plate 4 Energy consumption in Australia 1975–2000 (Australian Bureau of Agricultural and Resource Economics 2006). Areas represent energy consumed minus derived fuels produced. Brown curve: Australian population; red curve: per capita energy consumption (both indexed to 2000 = 1). ADO = Automotive Diesel Oil.

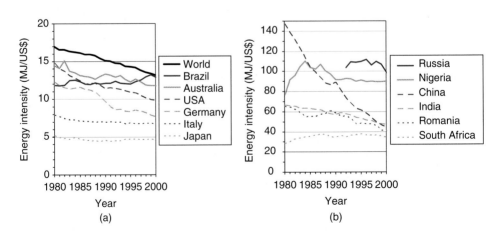

Plate 5 World energy intensity trends (MJ/US$) for selected countries (World Resources Institute 2006): (a) energy-efficient economies, (b) energy-intensive economies.

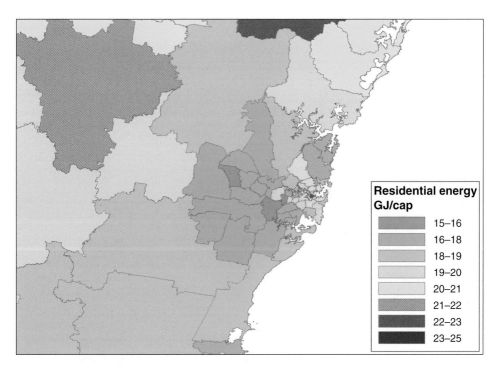

Plate 6 Per capita residential energy in Greater Sydney SLAs.

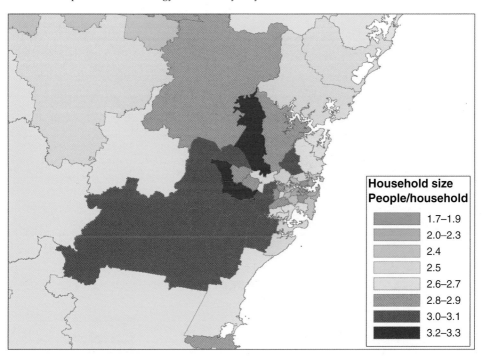

Plate 7 Household size in Greater Sydney SLAs (Australian Bureau of Statistics 2007).

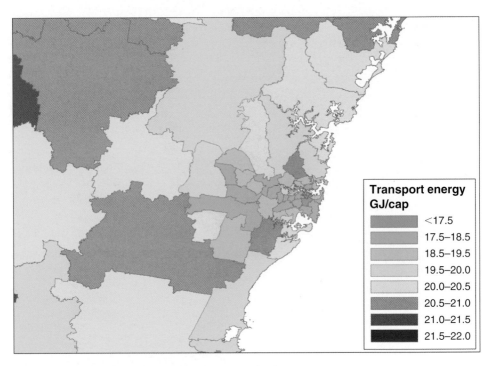

Plate 8 Per-capita transport energy in Greater Sydney SLAs.

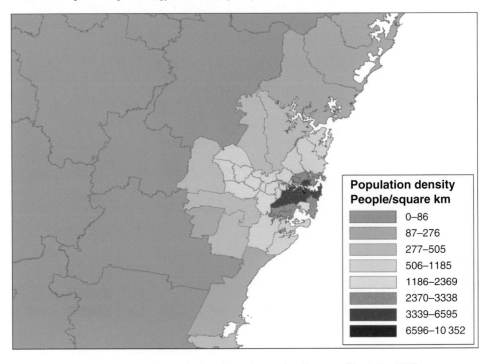

Plate 9 Population density in Greater Sydney SLAs (Australian Bureau of Statistics 2007).

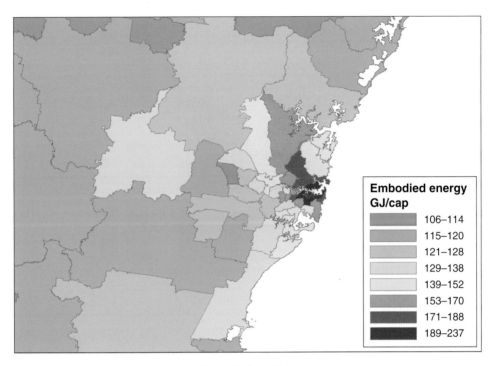

Plate 10 Per capita embodied energy in Greater Sydney SLAs.

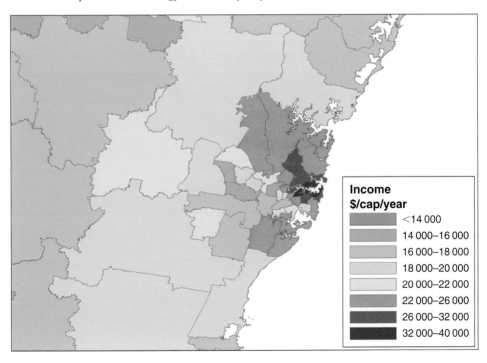

Plate 11 Per capita annual income in Greater Sydney SLAs (Australian Bureau of Statistics 2007).

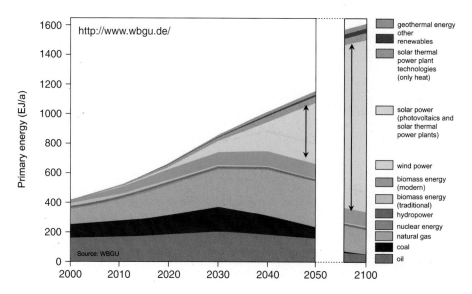

Plate 12 Scenario for a worldwide energy mix in the next decade.

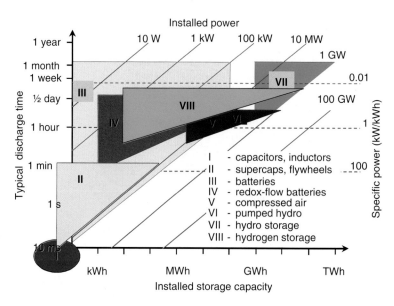

Plate 13 Typical systems sizes for different storage technologies as a function of the installed energy capacity, the available charge/discharge power and the typical discharge times.

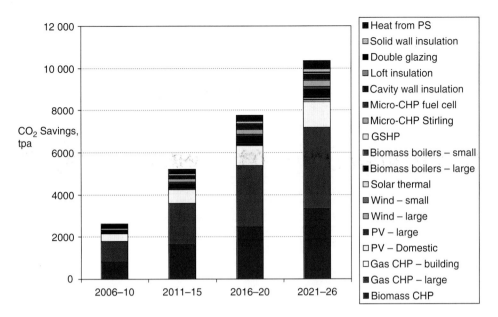

Plate 14 Breakdown of CO_2 savings for scenario 5. (London Carbon Scenarios 2026).

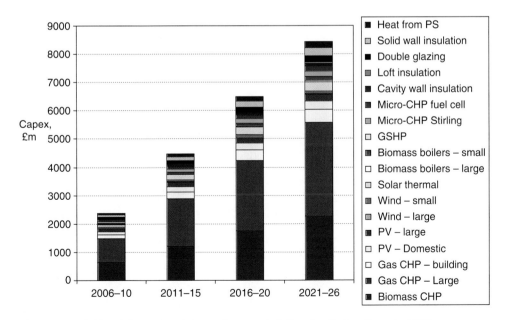

Plate 15 Detailed breakdown of capital costs for scenario 5. (London Carbon Scenarios 2026).

Plate 16 Representation of annual CO_2 emissions in the form of GIS thematic maps using DECoRuM® for dwellings in north Oxford. The dwelling footprints in red and pink indicate the hotpots of energy use.

Plate 17 Estimating the solar potential of dwellings in north Oxford using DECoRuM®.

13.5 Rapid Digital Evolution

The digital revolution in India is largely an information revolution, because technology can deliver *reliable* information the rural poor desperately need. Take, for example, the *Bhoomi* land records program. Karnataka state has digitized 20 million deeds that farmers can get for just 30 cents in the local language from any of 200 computer kiosks. Until the records were computerized, the deeds were used and controlled by powerful village accountants, who charged farmers as much as $22 for documents needed to sell property or to get bank loans for fertilizer and seeds.

Computerizing land records is even impacting social change. No longer can accountants and upper-class landlords trick farmers into signing away their property. Deed fraud, which according to World Bank estimates once cost poor farmers in Karnataka $20 million a year, has been virtually wiped out.

Rural India, where 72% of the population lives, is gearing up to be part of the information society. What for a decade were merely small-scale experiments are now aiming for economies of scale to meet the needs of the rural population. Telecentres offer services ranging from e-government, education and finance to telemedicine. A cheap wireless handheld computer may soon be available for the Bhoomi land records program. And ITC plans to install 20 000 more e-Choupals to cover 100 000 villages in the next ten years. Expectations are that internet will reach 25 million farmers by 2010 and 100 million farmers during the next decade.

Some of the Indian internet projects already operating in rural areas are (Hay 2005):

- ITC e-Choupal
- N-Logue
- MS Swaminathan centre in Pondicherry
- Akshaya in Kerala
- Gyandoot in MP with focus on e-governance
- Bhoomi in Karnataka
- E-seva in Godavari District of AP
- Warana in Maharashtra by NIC
- Aksh Broadband
- Jagriti in Punjab

13.6 Village Banking

Still, millions of impoverished villagers, slum dwellers and tenant farmers have yet to benefit from ICT. Jobs are scarce, cash is scarcer. Yet, decades before the advent of village internet kiosks, a local experiment in Bangladesh pioneered ways to supply working capital for small-scale rural business. Village banking. This bold new form of banking succeeded in serving even poor villagers without collateral, who need *micro-credit* and want to save, but are traditionally exploited by moneylenders. The idea that financial services can be extended to poor people to build assets, increase income and enjoy a better life has in the course of 30 years experienced unprecedented growth as a tool for economic development. Today, institutions providing microfinance, MFIs, have spread to all corners of the developing world, and evolved into a billion dollar industry.

But what is now generating so much interest is less the discovery of some entirely new way to deliver financial services to the poor than the effect of the rapid innovation that has taken place in the past three decades. The Grameen, *village*, Bank in Bangladesh, which began by loaning $27 to 42 villagers, has since loaned $6.5 billion to more than 7 million microentrepreneurs. It has computerized and networked 2300 of its 2400 branch offices and created over 20 Grameen companies. To increase income for its borrowers via new technology, the bank created Grameen Telecom, a provider of rural communication services, and the *Village Phone* program mentioned above – now so successful that it is being replicated by MFIs in Cameroon Uganda and Rwanda.

Microenterprise loans spawned new businesses and housing loans afforded families their own homes, but couldn't bring reliable power to rural Bangladesh where only an estimated 15% of the population has access to electricity. To meet this need, Grameen Shakti, *village energy*, started business in 1996 as the first rural energy service company (RESCO) in Bangladesh. By 2007 Shakti will have installed 100 000 Solar Home Systems (SHS) to power lights, motors, pumps, TV, mobile phones and computers.

Grameen is a good example of the initiative and entrepreneurism of village banks. Their financial capabilities and expertise play a crucial role not only in fostering microenterprise, self-employment and income generation for the poorer parts of the population, but also in developing new markets.

In 2003, 140 microfinance insitutions were surveyed (Wimmer and Barua 2004) to assess their interest and activities in emerging rural and peri-urban renewable energy markets. Several MFIs were already active in this field and a significant share of the sample were considering new financing methods for renewable energy generation devices and even including them in their home improvement loans.

Still a young industry, nobody knows exactly how many institutions are providing microfinance in some form, but they are growing fast and serving a vast number of people both in rural and urban areas.

Village banking is done by a variety of institutions: by banks, regulated and unregulated non-bank financial intermediaries, regulated and unregulated NGOs, financial cooperatives and credit unions. An estimated 10 000 of these institutions exist, but comprehensive information on all of them is still lacking. However:

- According to a Microcredit Summit report (Daley-Harris 2006) in 2006, 3133 village banks reached 113 million clients, 82 million of which are poorest (see Table 13.3) A person is considered poorest if he or she is among the poorest 25% of the part of a nation's population which lives below the national poverty line.
- Current information can be found in (MIXMARKET 2006), a portal on village banking. Its database lists 90 Country Profiles. For details about Latin America see (Miller and Martinez 2006).

13.7 Village Technologies

Demand and potential markets for services like village banking are immense. But reaching out to remote rural customers is a formidable challenge. Bank workers must have local knowledge and be willing to live in rural areas with poor infrastructure. To collect the loan installments they navigate miles of difficult terrain everyday and risk theft and attacks. They must keep precise records on each borrower and do most of the bookkeeping by hand at night, with a kerosene lamp and pocket calculator often their only technology.

Village banking is especially problematic in sparsely populated rural regions, as is the case in most African countries. Eighty-five per cent of the Ugandan population lives in rural areas, of which only 10% have access to microfinance. The World Bank estimates that

Table 13.3. Outreach of microfinance in 2005.

Region	Number of microfinance programs reporting	Number of clients	Number of poorest clients
Sub-Saharan Africa	959	7 429 730	5 380 680
Asia and the Pacific	1652	96 689 252	74 330 516
Latin America and Caribbean	439	4 409 093	1 760 405
Middle East and North Africa	30	1 287 318	387 951
Developing world totals	**3080**	**109 815 393**	**81 859 552**
North America and Western Europe	35	55 707	13 318
Eastern Europe and Central Asia	18	3 390 290	76 166
Industrial world totals	**53**	**3 445 997**	**89 484**
Global totals	**3133**	**113 261 390**	**81 949 036**

of the 500 million potential clients worldwide, only 100 million have access to the services offered by microfinance institutions. Further, most of these 100 million customers live in urban or peri-urban areas and MFIs lack means of reaching out to those who live in isolated rural areas. MFIs are now looking to innovative applications in technology to help expand their outreach and reduce operational costs.

In India, paper-based, manual keeping of ledgers is being replaced by entering records into a handheld computer fitted with smartcards to record transactions. In Bolivia MFIs are using automatic teller machines, wireless communication, *chipcards* and user interfaces in several local languages. Similar systems are being tested in India and Uganda to lower the cost of expansion. Other MFIs are working through merchants in small towns, who use mobile phones for MFI clients to pay loans, make deposits and get cash. Vodafone is testing such services in Kenya, and Hewlett Packard has just finished the Remote Transaction System (RTS) pilot project with MFIs in Uganda using a point-of-sale (POS) device with mobile phone technology.

And still the problem of electric power remains. SKS, for example, a fast-growing MFI in India, has had to build back-office systems that can work on two hours of power a day to compensate for India's unreliable power supply. It must closely monitor voltage when its computers are running and keeps a diesel generator on hand in case of power failure. As new technologies are developed to help lead the microfinance industry into the next stage of its growth, access to reliable power will be as important as the technology itself.

13.8 Reaching Out to a Rural Clientele

In a recent survey by CGAP (Ivaturi 2006), 62 microfinance institutions in 32 countries report using technology channels to handle transactions for poor people. These technologies include automated teller machines (ATM), point-of-sale devices and wireless communication.

The fact that most survey respondents use ATMs suggests that they target customers in urban and semi-urban areas. These locations are more likely to have reliable access to electricity and the 'always-on' telecommunication connections required by most ATMs to connect to a bank's central server. In addition, because ATMs must be regularly refilled or emptied of cash manually, it is most cost effective to place them in densely populated areas.

The success of ATMs in India has encouraged the ICICI Bank to pilot test a low cost ATM that can withstand high temperatures and handle soiled and crumpled notes. Citigroup has begun installing ATMs in microfinance institutions to enable clients to open savings accounts with no minimum balance. The machines are designed to identify clients by their fingerprints and to speak to them in any of six Indian dialects. Point-of-sale devices, in contrast, are typically used to handle payment transactions and often placed in rural shops and businesses. The device can consist of a card reader, mobile phone, personal computer (PC), barcode scanner, or any hardware that can identify customers and receive instructions for the transfer of value. Although most POS devices are card-reading terminals, PCs may be used where transaction volume is expected to be high, or where wireless internet access is available. Each POS device uses a telephone line, mobile phone connection, or the internet to send instructions for transferring value from one account to another. For example, after swiping a card through the POS device, the merchant presses a button on the POS terminal authorizing payment from the customer's line of credit (credit card) or funds available in the customer's current account (debit card). If the POS device is a mobile phone, the customer uses her mobile phone to send a text message authorizing payment from her bank account, or from her account with the mobile phone company to the merchant's phone.

POS technology could provide MFIs with a cost-effective way to scale up services in rural areas even where they do not have branch networks. It enables loan payments, cash deposits and withdrawals by microfinance clients through a network of agents that can either be microfinance staff or third parties. Moreover it serves demands still unmet in remote areas: rural micro-entrepreneurs lack credit and reliable banks for their savings, and MFIs need greater outreach to increase revenues, lower costs and bring in economies of scale.

The Remote Transaction System mentioned above was pilot tested in Uganda because of its large rural population and low population density, active MFI networks and strong national GSM network coverage. RTS objectives were not only to demonstrate its technical viability, but to improve the transparency and efficiency of banking processes for those MFIs seeking to become commercially viable. Lending money to the poor is a financially and socially sound investment, but the annual capital requirements to operate the business are a continuous challenge.

Support is coming from new technologies to reduce operation costs and new legislation allowing MFIs to lend from the deposits of their clients. The recently enacted Microfinance Deposit Taking Institutions Act (MDI) in Uganda, for example, commits an MFI to computerize and network its branches to qualify for an MDI licence. MFIs were thus motivated to join the RTS pilot both as a low cost means of expanding their branch network into rural areas and as a means to capture data electronically to meet the MDI commitments.

In the wake of improving access to financial services is the push to create new products and better ways for low income households to access insurance. Providing *micro-insurance* has all the problems associated with providing credit, i.e. high transaction costs and data collection, and there has yet to be a breakthrough innovation of a kind parallel to those in microfinance. And yet a growing movement within microfinance is pushing to provide insurance on top of loans and deposit services. Life insurance has been most successful to date, but health insurance plans are being tried, as well as property and crop insurance. Microinsurance is offered in the Philippines, Zambia, Uganda, Malawi, India, Bangladesh and elsewhere (Leftley and Mapfumo Shradek 2006). To operate an efficient distribution network, insurance companies may rely on networks built for financial services and make clever use of modern ICT.

In future, every service organization will eventually rely on ICT – on POS devices, laptops, handheld computers (PDA), PCs, ATMs, back-office servers, cell phones, base stations for wireless communication or internet kiosks. The demand for services in rural areas will bring such ICT machinery to most villages, where growing demands for electricity will have to be met.

13.9 Technical Trends

Little doubt we will soon see even more electronic devices for rural populations in the future. Intel is building cheap and rugged PCs for rural markets, which are resistant to power failure and tailored to community use (INTEL Community PC 2006). Other approaches favour individual use. AMD's Personal Internet Communicator PIC (AMD Personal Internet Communicator 2006), for example, is a low cost internet access device engineered for low income markets. They compete with an increasingly popular and unconventional solution for $10 per month.

Here all applications and user data remain in a network, accessed by a *frontend* consisting merely of display, keyboard and network connector. Many such systems will implement advanced user interface technology, which masters a variety of languages, and can both recognize and generate voice and script (HP Labs India 2006). Important for rural areas is how these innovations will reinforce each other to melt down the barriers to village computing and to spawn new applications.

As promising as this sounds, the benefits of information technology can be widespread *only if* information technology reaches a large percentage of the population, the reason why wireless communication is and continues to be the most important part of rural ICT.

Wireless telephony led the way and more powerful technologies will enter the market full scale in 2007. A Wimax (*Worldwide Interoperability for Microwave Access*) station may bring broadband connections to several hundred villagers living as far as 20 kilometres away. Wimax will render internet telephony, email and information services fast and cheap (WIMAX Forum 2006). Moreover, experts predict that breakthroughs in communication technology will bridge today's digital divide between urban and rural areas, thus enabling villagers to access needed services, education and jobs.

Several key communication technologies will impact rural areas:

- Internet – packet switching replaces the industry-standard circuit switching, substantially reducing costs and allowing for additional value-added services such as internet or VoIP telephony.
- WiMAX and Wi-Fi – standards-based fixed wireless networks for distributing the internet to rural areas and for last-mile community networks. They operate at lower cost and with higher capacity than cellular telephone networks. Cheap Wi-Fi phones and Wi-Fi-enabled cell phones are already in the market. In Vietnam, USAID is piloting the use of a $10 VoIP cordless phone.
- Satellites – where terrestrial solutions are not available, satellites will allow remote communities to reach outside their network. New satellites, such as the IPStar and BGAN, are providing IP services at a substantially lower cost than has been possible in the past.

If the internet goes to the countryside wirelessly, it may well be linking primarily phones and not computers. And many new applications will be voice and local language based.

Wireless technologies may be the most novel, powerful and even disruptive technologies to be considered, but there are many more relevant technical innovations:

- User interface technology – Many villagers have a low level of literacy and live in countries where numerous languages, dialects and scripts are used, all of which are different from the Roman alphabet and English, the lingua franca of the internet. The ICT industry is remedying the situation through software that recognizes speech and handwriting. This allows even a low literate clientele to use public email services.
- Collaboration technology – Western industry employs this mature technology extensively. Call centres and various business functions like human resource processing are being outsourced to distant countries. Distributed engineering or sales teams can cooperate synchronously, seeing the same document on many computer screens and discussing it in a voice conference.

  Professional software for business administration and engineering can be used via Internet on a pay-per-use basis. Users are thus relieved from the costs and complexities of software acquisition, installation and maintenance. *Social software* like Blogs and Wikis may or may not bring the secret revolution envisioned by some, but it will facilitate interaction and collaboration of teams of any size.

  Little doubt, remote resources – software and information, many skills and expertise – are becoming accessible flexibly and ubiquitously. And while today urban areas are more effectively accessible than rural areas, this will change with time.
- Free software, free information – Rich urbanites and poor villagers can access most of the World Wide Web for free. Only an internet browser is needed to write text, produce graphics or compute a spreadsheet.

The list of ICT-related innovations impacting rural areas could be extended. But ICT is not the only field of innovation. Breakthroughs have been achieved in other vital areas such as water purification (HANS HUBER AG 2006). Small decentralized water purifiers turn contaminated surface water into drinking water, and recycle used water where water is in short supply.

In addition to information and communication services, finance and insurance services we will witness new clean water services. Together they can provide a foundation to develop tourism and new local economies. This brave new rural world, however, relies on electricity and must be powered reliably.

13.10 Education in Rural Areas

The Massachusetts Institute of Technology (MIT), a leading US university, is pioneering the *Opencourseware* approach and making its course material public. Seventeen hundred courses from business and engineering to physics are downloadable free of cost by any user of the internet. More than 100 higher education institutions in the USA, France, China and Japan have joined the *Opencourseware* movement, offering open-access to high quality educational materials in a variety of different languages to educators and students every where. From the start, *Opencourseware* has been well received and is even being used in rural areas, where access to academic institutions is problematic.

Opencourseware seems to have been inspired by the Open Source Software movement, which has enabled millions of programmers around the world to create free software. The aim of this popular form of self-organized collaboration is to have as many people

Table 13.4. MIT *Opencourseware* disciplines.

Aeronautics and Astronautics	Foreign Languages and Literatures
Anthropology	Health Sciences and Technology
Architecture	History
Athletics, Physical Education and	Linguistics and Philosophy
Recreation	Literature
Biological Engineering	Materials Science and Engineering
Biology	Mathematics
Brain and Cognitive Sciences	Mechanical Engineering
Chemical Engineering	Media Arts and Sciences
Chemistry	Music and Theatre Arts
Civil and Environmental	Nuclear Science and Engineering
Engineering	Physics
Comparative Media Studies	Political Science
Earth, Atmospheric, and Planetary	Science, Technology, and Society
Sciences	Sloan School of Management
Economics	Special Programs
Electrical Engineering and	Urban Studies and Planning
Computer Science	Women's Studies
Engineering Systems Division	Writing and Humanistic Studies
Experimental Study Group	

as possible writing, improving and distributing software for public use. Its success has empowered enthusiastic programmers everywhere. Schools in Brazil, Malaysia and Namibia are using free software to teach young people how to use computers to support their education. User groups from all over the world collaborate to evolve the programs. So there is a good chance that the user interface of the $100 educational laptop will be adapted to local languages and the educational needs of children in developing countries.

That *Opencourseware* is well received in developing countries by students and teachers alike is well documented by their comments to the *Opencourseware* portal. Education is in demand. A major part of the World Summit on the Information Society in Tunis in 2005 focused on education in developing countries for good reason.

Consider Bangladesh, where, in 1997 the Grameen Bank created a new loan product, which it would not have thought possible 20 years earlier: the higher education loan. Its 12000 bank workers conducted a household survey to determine how many children in Grameen member families attended school. Ninety-nine per cent was the result. Now the children want to go on to university. To date Grameen has given 13000 higher education loans to cover tuition and other school expenses to students of engineering, medicine, computer science, education and other professions. Over 7000 students are now added to this number annually. A social triumph for poor, illiterate parents. What is more, the higher education loan has become so popular among the rural population that it is attracting many new Grameen clients.

Other new activities have been introduced to educate and train the 56 million people in rural Bangladesh under 18 years of age. Grameen Shikkha, *village education*, awards 30000 scholarships every year to encourage children to get better grades and not to drop out of school – a common problem in rural areas. And the primary reason, half of the scholarships are awarded to young girls. BRAC, a leading NGO has created a country-wide network of schools and even BRAC university. Education via computer assisted learning is being explored with success in many developing countries.

Distance education has made it possible for students anywhere in the world with internet connections to enroll in online courses. Pioneering efforts by the African Virtual University (AVU) illustrate how ICT applies to higher education: in 2002 AVU requested Australia's Royal Melbourne Institute of Technology to help build a computer science program for sub-Saharan African students. ICT enabled all students to attend their lectures via video conferencing with Australia.

According to Intel, its *Teach Program* has trained 3 million teachers in 35 countries since 1999. In addition, its *World Ahead Program* (INTEL World Ahead Program 2006) will train an additional 10 million teachers in developing communities on the effective use of technology in the classroom over the next five years. The program will also provide 100 000 free PCs to classrooms in these communities to both improve learning skills and to accelerate the benefits of technology in education.

In China, Intel has pledged to train one million teachers to install 10 000 PCs in rural schools by 2008. China in turn has pledged $5 billion to Africa, to build schools and hospitals and to train thousands of professionals (CNN 2006). Aiming at similar targets, Microsoft has announced a partnership program to benefit 45 million people in Africa by 2010 (MICROSOFT 2006). Whether these pledges materialize or not, all in all, there seems little doubt that the links between education, ICT, financial services and energy supply will become stronger, and that progress in one area will spur progress in others.

13.11 New Rural Electrification Policies and Programs

The new technologies, services and private efforts described above are increasingly supported by new national policies and programs for an estimated 360 million households worldwide, which still lack access to electric power networks. Because private investment alone will not suffice to reach this goal, public donors direct $500 million per year to renewable energy programs in developing countries. Here is a brief overview of rural electrification policies and programs (Martinot 2006; REN21 2005; REN21 2006):

- Brazil plans to electrify 2.5 million households by 2008 under the 'Luz para Todos' program (about 700 000 have already been electrified), and has targeted 200 000, or about 10% of these households for renewable energy.
- Several other Latin American countries have recently launched or revamped new rural electrification programs, including Argentina, Bolivia, Chile, Costa Rica, Guatemala, Mexico, Nicaragua, and Peru.
- China's Township Electrification Program (The Brightness and Township Electrification Program in China 2006), which was substantially completed during 2004, provided power to 1 million people in rural areas with renewable energy.
- The Indian government's 'Remote Village Electrification Programme' has identified 18 000 villages for electrification, partly with renewable energy technologies like biomass gasifiers.
- The Philippines launched a strategy in 1999 to achieve full rural village electrification by 2007, including renewable energy explicitly in that strategy.

Other Asian examples of countries with explicit mandates for renewable energy for rural electrification include Bangladesh, Cambodia, Nepal, Sri Lanka, Thailand, and Vietnam.

13.12 The Art of doing Rural Business

Banks operate in cities. Villagers in rural areas are not creditworthy because they are poor, backward, illiterate, too dispersed, have no assets and are too great a risk – so the conventional wisdom goes. Village banking proved otherwise. Village banking for the rural poor is an *innovation made in Bangladesh*. When the Grameen Bank began 30 years ago, it was not about adapting commercial banking to some new, still unserved clientele in rural areas. Rather it was about doing business in a radical new way, in a change of paradigm. Before, banking simply could not be done profitably in a poor rural environment. The Grameen Bank could – thanks to an ingenious business model, based on the credo: *The rural poor should have a business and make a profit.*

Grameen Telecom follows this business model to promote mobile telephony. A villager buys a mobile phone handset, antenna, battery and bulk airtime, and lets other villagers use the phone for a fee. The Grameen Bank finances the purchase of phone equipment from Grameen Telecom, and trains the village phone operators how to run a viable business. Many of these micro-entrepreneurs make a profit averaging $100 per month, and over 100 million villagers benefit from shared access to affordable phone services in their own village. Grameen Shakti has built up a flourishing renewable energy business providing solar powered mobile phone chargers in off-grid areas, easy credit terms for its customers, and training for solar entrepreneurs to start their own business. An entrepreneur buys a solar home system (SHS) on easy credit terms, installs it in a village market, and then rents solar powered lights to neighbouring shops for a fee. Increased income via solar business has made renewables so attractive to rural entrepreneurs that they now number over 10 000.

But if it is hard to make a success out of something, it is an order of magnitude harder to sustain the success. Grameen Telecom's initial business model had to be revised as the demand for rural phone services grew to the point that some telephone operators could develop monopolies. After over 20 years of successful micro-credit delivery, the Grameen Bank was confronted with repayment problems and complaints from its borrowers. Ironically, some of the innovations that had helped grow the bank to over one million members were now causing some of them to leave the bank. Group lending, the innovation which had allowed Grameen to grow so explosively at the start, was not flexible enough over time to afford successful entrepreneurs access to larger loans on their own repayment terms. To make matters worse, the worst flood in Bangladesh history in 1998 stopped business, damaged homes and required new models for insurance and savings.

The Grameen Bank radically reversed course and completely revised its lending methodology, introducing new loan and savings products, loan insurance and a Grameen pension scheme. Since its inception in 2001, *Grameen II* has enabled the bank to grow from 2.4 million members in 2002 to 7 million in 2007. Grameen's founder, Muhammad Yunus, explains *'if the model is right, the poor always pay back'* and a book (Dowla and Barua 2006) by the same title documents Grameen's transition from a bank pioneering micro-credit to the poor, to a mature financially reliant institution, which has enabled over 58% of its borrowers to cross the poverty line.

13.13 Income Generation and Innovation

Business innovations have opened up unforeseen opportunities in rural markets. The Grameen Bank's recovery rate is 99%, far above the banking industry's average. Rural demand for telecommunication, education and renewable energy systems is high, and villagers readily spend substantial parts of their income to pay for the services. They easily embrace new technologies, be it mobile phones or solar home systems. Even illiterate Village Telephone Ladies had no problem operating a GSM cell phone, despite having no prior experience with phones of any kind. They memorize the area codes and enjoy village esteem as they connect their customers to all parts of the world.

The art of doing rural business is as much about innovation and local knowledge, as enabling the poor to gain leverage with new technology to raise their income and social status. It's a bottom-up approach, creating entirely new businesses by taking advantage of the growth opportunities in rural markets.

Grameen's rural business operations have been researched in depth, and absorbed by areas other than banking. They also helped to inspire an academic discipline dedicated to finding strategies on how to do business at the so-called bottom of the social pyramid.

'If you can conceptualize the world's 4 billion poor as a market, rather than a burden, they must be considered the biggest source of growth left in the world,' says (C.K. Prahalad 2005). Today the economics of this market have been researched, taught and covered by the media. The findings are that if this market is well understood, it can be served profitably.

Several corporations seem to have understood the message: Intel engages ethnographers to study rural markets in Brazil, Egypt, China and India, where as many middle class consumers live in rural as in urban areas. Manufacturers of cars, appliances, food and ICT gear as well as service providers have *gone local* to design products tailored to a poor and rural clientele, e.g. cheap transport vehicle, detergents in small quantities, low cost insurance, rugged community computers, and washing machines which cope with power outages.

Villagers may benefit from corporations in yet another way. *If work can shift from the US to India, it can shift from cities to villages* – this assumption is being tested by GramIT, an outsouring initiative (GramIT 2006). GramIT centres serve as rural back-offices for urban corporations. GramIT employs trained young villagers and enjoys the rural cost advantage.

Computers, today, offer all the means needed for outsourcing and worldwide tele-cooperation of knowledge workers. Computing is also becoming more affordable. Microsoft's FlexGo approach (FlexGo 2006), for instance, lets people pay for computers as they use them. It accounts for the economic realities of emerging markets where many people work in informal jobs with unpredictable paychecks.

13.14 Towards a Rural Business Strategy

New kinds of rural businesses are emerging in developing countries, triggered by many kinds of innovation. They need to mature and prove themselves in differing environments, be it in rural or peri-urban regions in the developing world. Few business rules exist at present. However, something can be learned from a broad range of business efforts put forth as hypotheses on rural business practice, e.g.:

- Even the poor invest if they can earn more or spend less and save money. Such behaviour led to the strong growth of mobile telephony.
- ICT is a catalyst of business innovation. Even simple chipcards provide a secure identity to illiterate or unregistered clients.
- It is economical to share expensive resources, e.g. shared access to village phones and internet kiosks.
- Providing services often generates more revenue than selling products for individual use. Inexpensive services attract numerous customers and encourage frequent use.
- Large networks of small entrepreneurs can evolve through replication. Replication, a simple form of franchising, works particularly well in the service sector.

- Business benefits from cooperation and partnership, which bring down the high cost of infrastructure needed to reach a dispersed rural clientele. ITC, for instance, has part-nered with 28 companies that extend its e-Choupal services.
- Income generation for and training of villagers are key success factors for rural business.
- Subsidies should not be given to households to obtain goods like solar home systems, but to the companies that supply them. In this way subsidies can help to create an local service industry. If, for instance, an SHS supplier is financially rewarded for every installed SHS, installation rate and volume tend to increase. Subsidies, thus, stabilize and increase cash flow and help start-up organizations to reach break-even.
- Public business support programs should provide *smart* subsidies which are limited in time, decreasing over time, and offer incentives for start-ups to reach sustainable oper-ations early.

13.15 Innovations in Renewable Energy Supply for Rural Areas

Hybrid systems can use solar, hydro, wind, biomass and diesel energy in clever combi-nations both off- and on-grid. Already, they supply power day and night, summer and winter to hamlets, villages and townships. Experts expect that the growth of decentralized power generation, liberalized markets, modern power electronics, and the introduction of advanced integrated circuit technologies are leading to a dramatic change in the manage-ment and operation of electricity grids worldwide (Second International Conference on Integration of Renewable and Distributed Energy Resources 2006). Leading institutes like the *Institut für Solare Energieversorgungstechnik* (ISET 2006) focus on the integration of many different electric power generators and energy storage systems into autonomous hybrid systems, micro-networks as well as large island networks. Standardization, test proced-ures, network interfaces, system management and security issues are well advanced.

Issues of networking numerous and regionally distributed power generators are being investigated. Here, electricity is no longer delivered by electromechnical devices, but by electronic power inverters which are minutely controlled. In development are hybrid sys-tems with advanced features, e.g.:

- Automated operation of decentral power generators.
- Economic expandable and grid-compatible standard hybrid systems.
- Use of biomass in hybrid systems as substitute for fossil fuels.
- New energy conversion technologies for economic use like micro-gas turbines and fuel cells.
- Combination of different types of energy storage.
- Automatic operation and remote monitoring, remote control, and remote maintenance.

The flexibility of hybrid systems is an advantage in particular for rural areas. They can exploit the naturally given renewable energy (RE) resources on-site. They can be estab-lished where needed, large or small, stepwise extended and eventually interconnected and integrated. Moreover, they can absorb the latest innovations in wind or solar technologies.

So at least from a technical viewpoint, the energy demand of rural areas can be served. Hybrid RE systems are also interesting under structural and dynamic aspects of rural evolution. It is unlikely that evolution will follow a masterplan, rather the plans of many diverse players. Base stations, internet kiosks, ATMs, branch offices, sawmills and storage houses, coolers, pumps and dryers will be installed in various locations, step by step.

Hybrid systems can be installed and used where needed and grow as needed. What is more, the distributed infrastructure of power supply matches the distributed architecture of ICT systems, which in turn creates new synergies: wireless communication systems can use RE for power supply, while RE systems use wireless communication to control their many remote components.

13.16 Conclusion

A bewildering variety of technical innovations is entering the rural scene of developing countries – wireless networks, hybrid renewable energy systems and many more. Farmers use new information and communication services to find out how to best sell their goods. Villagers in remote places benefit from microfinance and micro-insurance, e-health and e-education. Products, notably in the ICT sector, are being designed specifically for rural markets, new enterprises are offering rural services, and new government programs focus on rural electrification. These promising new developments are no instant panacea. Rural evolution takes time and effort. Still, many innovative approaches proceed in parallel, reinforce each other and are pushed forward by the growing demands of a large rural market. There seems little doubt that life in rural environments is undergoing substantial change, accelerated by innovation and globalization. The innovation, microfinance, has in the course of only 30 years spread to all parts of the developing world. Little more than a decade ago, the World Wide Web began to proliferate. Today, it seems to be ubiquitous and part of everyday life. At about the same time, cell phone technology started its unparalleled spread throughout the developing world. e-Choupals reached three million farmers in merely five years. Powerful rural markets in China and India and a growing, new entrepreneurial spirit are stimulating rural development. Quite likely we will witness significant change in rural and peri-urban areas in less than a decade, notably in South Asia where many experiments are already under way. The demand for connectivity and trustworthy information is rapidly increasing with every innovation. Western minds may not easily understand such innovations. Village banking, for example, means a change of paradigm for anyone attuned to western ways of banking. A bank that serves poor and illiterate women is difficult to comprehend. One needs to understand the local environment and to experience village life to grasp how microfinance evolves a rural society.

But when, where and how rapidly will rural economies and markets evolve? This evolution is, to a large extent, stimulated by new technologies which depend on electricity and hence on rural electrification, the lack of which has slowed down rural development. There are, however, other accelerating developments. Grameen Shakti has demonstrated that a rural renewable energy service company can bring electricity to half a million villagers within a few years. Moreover, Shakti's business model is mature for replication in other developing countries. Replication has proved to be an effective mechanism to spread rural services worldwide, microfinance being an outstanding example. Education and training stimulate rural evolution in an unforeseen way. New hardware, courseware and the internet combine forces to meet a vast and truly basic demand. Poor people eagerly embrace and are willing to make sacrifices for the education of their children.

Questions remain. Will villages eventually be able to produce the energy they consume – a desirable goal as argued in (Scheer 2005)? And if rural areas advance, can the poor areas of megacities benefit – where energy supply and income generation are also problematic?

Rural energy markets offer a unique advantage: renewables must often fight an uphill battle against established power suppliers, lobbies and subsidies – the legacy of generations of power supply. Many rural areas, however, are still void of such legacy systems. Here, renewables can find a level playing field – the rural advantage.

Acknowledgement

I thank Klaus E. Wimmer, microSOLAR, for his valuable contribution and Dipal C. Barua, Managing Director Grameen Shakti and Deputy Managing Director Grameen Bank for sharing his expertise.

References

AMD Personal Internet Communicator (2006). Retrieved 1 November 2006 from http://www.amdboard.com/pic.html

Armendariz de Adhion, A. and Morduch, J. (2005). *The Economics of Microfinance*. The MIT Press.

CNN (2006). Retrieved 1 November 2006 from http://edition.cnn.com/2006/WORLD/asiapcf/11/03/china.africa.ap/index.html

Daley-Harris, S. (2006). State of the Microcredit Summit Campaign Report 2006. Retrieved 1 November 2006 from http://www.microcreditsummit.org/papers/2006papers.htm

Dowla, A. and Barua, D. (2006). *The Poor Always Pay Back*. Kumarian Press.

FlexGo (2006). Retrieved 1 November 2006 from http://www.microsoft.com/whdc/flexgo/default.mspx

GramIT (2006). Retrieved 1 November 2006 from http://www.byrrajufoundation.org/gramit.htm

Hammond, A. and March J. (2006). A new model for rural connectivity. Retrieved 1 November 2006 from http://www.usaid.gov/our_work/economic_growth_and_trade/info_technology/tech_series/Rural_Connectivity_508.pdf

Hankins, M. (2006). Fresh ideas needed: building the PV market in Africa. *Renewable Energy World*, 9(5), 103–116.

HANS HUBER AG (2006). Retrieved 1 November 2006 from http://www.huber.de

Hastings, A. (2006). Remittance market: opportunities and challenges. Retrieved 1 November 2006 from http://www.microcreditsummit.org/papers/2006papers.htm

Hastings, A. (2006). Entry of MFIs into the remittance market: Retrieved 1 November 2006 from http://www.microcreditsummit.org/papers/2006papers.htm

Hay, K. (2005). *Expanding Broadband Access in Rural India*. World Bank Group Global ICT Department.

HP Labs India (2006). Retrieved 1 November 2006 from http://www.hpl.hp.com/india/research/aad-gkb.html

INTEL Community PC (2006). Retrieved 1 November 2006 from http://www.intel.com/pressroom/archive/releases/20060329corp.htm

INTEL World Ahead Program (2006). Retrieved 1 November 2006 from http://www.intel.com/intel/worldahead/index.htm

ISET (2006). Annual Report 2005. Retrieved 1 November 2006 from http://www.iset.uni-kassel.de

ITC (2006). Retrieved 1 November 2006 from http://www.itcportal.com/sets/echoupal_frameset.htm

Ivaturi, G. (2006). Using technology to build inclusive financial systems. Retrieved 1 November 2006 from http://topics.developmentgateway.org/microfinance/rc/ItemDetail.do~1081070?itemId=1081070

JÜHNDE (2006). Retrieved 1 November 2006 from http://www.bioenergiedorf.de/contenido/cms/front_content.php?client=1&idcat=0&idart=0&lang=1&error=1

Landesinitiative Zukunftsenergien NRW (2006). Retrieved 1 November 2006 from http://www.50-solarsiedlungen.de/frame_siedlungen.html

Leftley R. and Mapfumo Shradek (2006). Effective micro-insurance programs to reduce vulnerability. Retrieved 1 November 2006 from http://www.microcreditsummit.org/papers/Workshops/11_MapfumoLeftley.pdf

Martinot, E. (2006). Renewables global status report: investment and capacity soar while support policies continue to multiply. *Renewable Energy World*, 9(4), 34–45.

MICROSOFT (2006). Bill Gates announces Microsofts partnerships and programmes to benefit more than 45 million people in Africa by 2010. Retrieved 1 November from http://www.microsoft.com/emea/presscentre/pressreleases/GLFAfrica2006Day2PR_1172006.mspx

Miller, J. and Martinez, R. (2006). Championship league: an overview of 80 leading Latin American providers of microfinance. *MicroBanking Bulletin* April, 15–21.

MIXMARKET (2006). Retrieved 1 November 2006 from http://www.mixmarket.org/

OLPC (2006). Retrieved 1 November 2006 from http://www.laptop.org/index.de.html

Prahalad, C. (2005). *The Fortune at the Bottom of the Pyramid.* Wharton School Publishing.

Rabinovitch, J. (2000). *Rural-urban linkages: an emerging policy priority.* UNDP Bureau for Development Policy.

REN21 (2005). *Renewables Global Status Report 2005.* Worldwatch Institute, Paris: REN21 Secretariat and Washington, DC.

REN21 (2006). *Renewables Global Status Report 2006 Update.* Worldwatch Institute, Paris: REN21 Secretariat and Washington, DC.

Rhyne, E. and Otero M. (2006). Microfinance through the next decade: visioning the who, what, where, when and how. Retrieved 1 November 2006 from http://www.microcreditsummit.org/papers/2006papers.htm

Roberts, D. and Rocks, D. (2005). Let a thousand brands bloom: multinationals are competing with local companies for a more discerning Chinese consumer. *BusinessWeek,* **17** October, 24–26.

Scheer, H. (2005). *Energy Autonomy. A New Policy for Renewable Energy.* Verlag Antje Kunstmann.

Second International Conference on Integration of Renewable and Distributed Energy Resources (2006). Retrieved 1 November 2006 from http://www.2ndintegrationconference.com/index.asp

SPINTRACK AB (2005). Options for improving backbone access in developing countries (with a focus on sub-Saharan Africa). Retrieved 1 November 2006 from http://www.spintrack.com/itadvice/reports/Spintrack_OpenAccess_infoDev.pdf

The Brightness and Township Electrification Program in China (2006). Retrieved 1 November 2006 from http://www.nrel.gov/docs/fy04osti/35790.pdf

The World Bank (2006). Information and Communications for Development 2006: Global Trends and Policies (World Information & Communication for Development Report). Retrieved 1 November 2006 from http://go.worldbank.org/0SVRFYVD90

UNEP Indian Solar Loan Programme (2006). Retrieved 1 November 2006 from http://www.uneptie.org/energy/act/fin/india/docs/IndSolLoanReview.pdf

WIMAX Forum (2006). Retrieved 1 November 2006 from http://www.wimaxforum.org/home/

Wimmer, N. and Barua, D. (2004). Microfinance for solar energy in rural areas. *Renewable Energy World,* **7**(4), 170–179.

WORDTALK (2006). Retrieved 1 November 2006 from http://www.wordtalk.org.uk/

PART IV
Transforming the Built Environment

Chapter 14
Towards the Renewable Built Environment

FEDERICO BUTERA

Professor, Politecnico di Milano, Italy

14.1 Background

According to the UN, (United Nations 2004) a very fast process of urbanization is taking place, especially in developing countries (Fig. 14.1), with a world population expected to reach 9 billion people in 2050 (medium forecast). It has been estimated that to accommodate the urban population, in the next 45 years the equivalent of a new town of 1 million

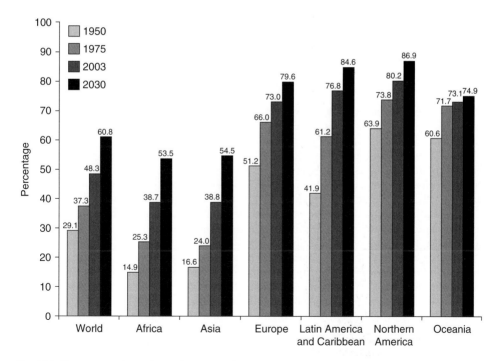

Fig. 14.1. Percentage of population living in urban areas.
(Reproduced from United Nations, *Urban and rural areas 2003*, New York, 2004).

inhabitants will be built every week. China, especially, is experiencing an extraordinary expansion of its building stock; the construction sector grows at a rate of about 1.8 billion square metres per year: (Long Siwei 2004) i.e. the equivalent of the entire Italian building stock is built every 2.5 years.

In settlements, on the other hand, is concentrated the world's final energy consumption (for heating and cooling buildings, for lighting, for electric and electronic appliances, for transport), accounting for more than 70% of the total. Thus, more than two thirds of total energy consumption is needed for urban metabolism, and more than two thirds of the CO_2 emissions are due to it.

Apparently, our future is dark: oil and gas are not going to last so much – at present low prices – because of the depletion of the low cost reserves; coal is not an option due to its high carbon emission per unit of energy (neither is carbon sequestration a safe, sustainable and cheap answer); nuclear energy faces insurmountable problems such as the depletion of affordable uranium reserves, the unsolved issue of radioactive wastes and security against terrorism; renewable energy sources cannot have any significant role in the present energy system because the large amount of energy needed to keep the system running cannot be provided by them, due to their low density, territorial distribution and availability.

On the other hand, it is well known that by adopting appropriate techniques and technologies, the same services could be provided with far lower energy consumption. If this is done, then renewable energy sources can easily meet the energy necessary to preserve our civilization and planet ecosystem.

Putting together all these data, the need arises for very strong actions for curbing the fossil energy consumption trend in cities, where most of the energy is consumed. In other words, it is necessary to develop a new urban design approach based on a new urban energy system, to avoid the catastrophic effects of global warming on one side, and to cope with the unavoidable constant increase of oil cost on the other.

14.2 Designing a Low Energy City

The energy system of a city can be considered as a thermodynamic system in which high grade energy (exergy) is transformed into low grade energy (Butera 1998). This process allows the urban metabolism to run by means of thermodynamic transformations that take place at all levels: individual devices such as domestic appliances and cars, systems for heating and cooling buildings, etc. Like any thermodynamic system, an urban energy system can be more or less efficient, i.e. can require more or less high grade energy to perform its tasks: the efficiency of the first power plants was less than 10%; nowadays it is well above 50%. Currently, urban thermodynamic systems are very inefficient, as inefficient as they were two centuries ago, and for this reason we waste a very large amount of the high grade energy contained in fossil fuels – as well as in solar radiation, wind, water heads, and biomass.

To design a renewable built environment means, first of all, that we must maximize its thermodynamic efficiency, i.e. we must minimize the amount of exergy used or, more commonly, the amount of primary energy consumed. Only after this is done, is it possible for a community to run mainly on renewable energy sources.

In low energy urban design, therefore, the main aim is to minimize primary energy consumption, two thirds of which is currently used in residential, commercial and transport sectors. The fulfilment of this aim involves several combined actions, i.e.:

- Optimize the energy efficiency of the urban structure.
- Minimize the energy demand of buildings.

- Maximize the efficiency of energy supply.
- Maximize the share of renewable energy sources.

This list, however, is not exhaustive, since the entire urban metabolism is based on energy; therefore other actions must be taken into consideration, involving water, wastes and mobility, i.e.:

- Minimize primary water consumption and exploit the energy potential of sewage water.
- Minimize the volume of disposable waste generated, and use the energy content of wastes.
- Minimize the need for transport and optimize transport systems.
- Minimize the primary energy consumption of transport systems.
- Maximize the share of renewable energy sources in transport.

The aim is to increase energy efficiency of the urban structure, of individual buildings, of mobility and of energy supply systems and to furthermore maximize the proportion of clean and renewable energy sources.

14.3 Optimize the Energy Efficiency of the Urban Structure

Usually, when urban planners start the design of a new settlement, they look for pre-existing landmarks, such as roads, railways, rivers, etc., and align the new buildings and streets accordingly. Very rarely do they look for the most ancient pre-existing landmarks: the path of the sun and prevailing winds.

14.3.1 Urban layout

Low energy urban design implies that shadow and surface illumination analysis, combined with wind analysis, must be used to optimize the shape, orientation and distances between buildings, in order to obtain maximum solar radiation and wind protection in winter and minimum solar radiation combined with openness to ventilation in summer (Fig. 14.2). The energy demand of individual buildings can be reduced significantly if this design strategy is adopted.

It is not easy to combine these requirements, as they are often contradictory. There are some unmistakable rules, however. Buildings obliged to develop along the north/south axis because of the shape and size of the land, for example, will require more energy both for heating and cooling; or buildings too close to each other will hamper natural lighting, necessitating artificial lighting which increases the internal heat gains: more electricity consumed not only for lighting but also for cooling.

Generally, western urban planning culture does not favour such rules, since they push towards an urban layout based on an east/west, north/south grid which is very rare in European settlements after the fall of the Roman Empire. However, Chinese urban planning culture has used this grid system both in urban and in rural settlements. Such a grid allows full use of solar gains in winter and makes easy solar protection in summer, provided that the ratio of building height to the distance between rows is appropriate, i.e. no shadowing occurs on the south facing facades in the winter season, when the sun is low. These rules have been always followed in Chinese urban planning, but unfortunately they are fading away, under the pressure of property speculation, leading to very tight rows and tall buildings.

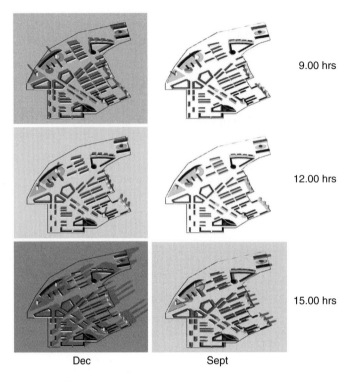

Fig. 14.2. Urban design with sun and wind.

Optimizing the energy efficiency of the urban structure by minimizing its energy demand is a very cost-effective approach, but it must be considered at a very early stage of urban planning. Overlooking this issue has very long lasting consequences, far more than in any other energy-related decision (Fig. 14.3).

14.3.2 Heat island

Temperature distribution in urban areas is highly affected by the urban radiation balance. Solar radiation incident on the urban surfaces is absorbed and then transformed in sensible heat. Walls, roofs and the ground emit long wave radiation to the atmosphere. Because the radiant heat loss is lower in urban areas, the net balance is more positive than in the surrounding rural areas and thus higher temperatures are obtained (Fig. 14.4). This is the so-called heat island effect.

Increased urban temperatures have a direct effect on energy consumption. The heat island effect in warm to hot climates exacerbates cooling energy use in the summer. For large US cities the peak electricity load increases from 3 to 5% for a 1°C increase in temperature; taking into account that urban temperatures during summer afternoons in the US have increased by 1.1 to 2.2°C during the last 40 years, it can be assumed that 3 to 8% of the current urban electricity demand is used to compensate for the heat island effect alone. (Santamouris 2003).

Another perverse effect should also be taken into account: the positive feedback due to the hot air blown by air conditioning units in the urban environment, contributing to the urban

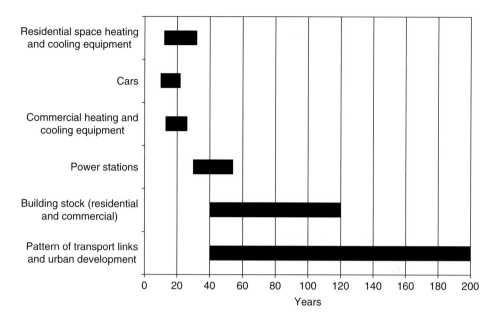

Fig. 14.3. Average lifetime for energy related capital stock.
(Adapted from IEA World Energy Outlook 2000).

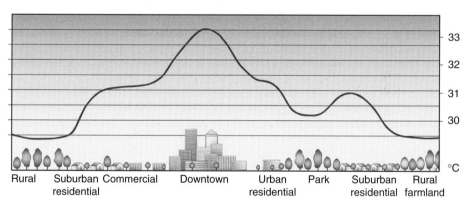

Fig. 14.4. Sketch of an urban heat-island profile.

temperature increase. A modelling study found that, in Tokyo, the waste heat emission from air conditioners is responsible for 1°C warming in summer; a similar study in Houston, Texas, found a 0.5°C increase in daytime and 2.5°C at night (Sawin and Hughes 2007).

The heat island effect, and the consequent energy demand for cooling, can be reduced with appropriate urban design by avoiding large dark surfaces subject to solar gains, by emphasizing the use of green areas and by exploiting the cooling property of water (ponds, fountains, canals, etc.).

In addition, the use of so-called 'cool materials' such as tiles, plasters, and asphalt are capable of reflecting the infrared part of the sun spectrum, reducing surface overheating (Akbari 1998).

By adopting these actions, temperature increase in urban areas is controlled, and the need for air conditioning and related energy consumption can be substantially reduced.

14.3.3 Mixed use

The manner in which the different functions of a settlement are distributed has a strong impact on energy consumption, for several reasons. The first, most obvious, is that if the three main functions, i.e. work, leisure and living, are not closely integrated, the need for transportation is strongly increased.

Another important advantage of compact mixed use developments is that they allow energy and power to be shared between activities in a more efficient way, taking into account their different time patterns, which smooths power peaks.

14.4 Minimize Energy Demand of Buildings

Building design, after urban design, has the second major impact on long-term energy consumption (Fig. 15.3) and new buildings should therefore meet the best energy performance.

14.4.1 Envelopes for low energy buildings

In the past years heating has been the main issue addressed by low energy design, and there are many examples of very well performing buildings, such as the German so-called Passive House, whose energy consumption for heating is less than $15 \, \text{kWh/m}^2$ per year. In Gothenburg, Sweden, 20 terraced houses without heating systems were built, by combining highly insulated walls and south facing low-e triple glazing (Houses without Heating System). Heat from the sun and heat generated by occupants and every day activities such as cooking is sufficient to heat these dwellings to a comfortable temperature.

In recent years rapid increase in space cooling has begun to create worldwide concern, together with its consequent energy consumption. From 1980 to 2000 in EU15 the number of cooled square metres has increased by a factor of five; electricity for cooling was about 51 500 GWh in 2000, and it is expected that it will rise to 115 000 by 2020 (Adnot 2004). Also in China, national room air conditioner (RAC) production is growing fast (Fig. 14.5) and already in 2001 the average number of RACs owned per family was 1.26 in Guangdong, 1.14 in Shanghai and 1.07 in Beijing (Long *et al.* 2003).

Appropriate building shape and orientation, internal layout, the position of openings and sun shielding can enhance ventilation in mid seasons and reduce the need for air conditioning in the hottest periods. Naturally, the implementation of most of these rules is possible or made easier if the layout of urban settings has been properly configured.

Building envelopes designed in such a way to require a very low amount of energy for both heating and cooling is not an impossible task, either technically or economically, as demonstrated by a growing number of examples; but to take a broad view of this result it is necessary to change the architectural language.

14.4.2 Glass architecture

The first change must be architectural fashions such as fully glazed buildings, whose diffusion is the cause of energy loss, especially in mild and hot climates.

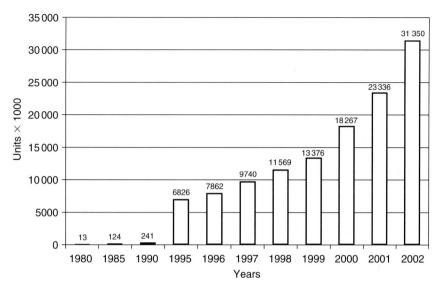

Fig. 14.5. Room air conditioners growth in China.
(Adapted from Ref. 9).

What's wrong with glass facades? There is no doubt that glass envelopes are light and transparent in architectural terms (and this is highly appreciated by architects and their clients). The fact that glass is light and transparent in physical terms affects thermal losses and gains and thermal inertia. But this is not the only problem. Let's analyse how these fully glazed envelopes are used, and their effect on energy consumption and comfort. Since part of the solar spectrum is absorbed, in sunny summer days the glass warms up to 30–40°C, and the infrared radiation emitted makes nearby spaces uncomfortable. On the other hand, during cloudy days or by night in winter, the glass is cold, and for this reason the area close to the glazed surface is uncomfortable. In most cases this undesired effect is reduced or eliminated by blowing a jet of hot or cold air parallel to the glazed surface, whose temperature becomes closer to room air temperature. In this way comfort is improved, but at expenses of higher heat losses.

There is another environmental drawback when tinted glass facades are used, especially if the colour is blue-green, the most appreciated and used by architects. The drawback is evident in such buildings during clear days: in spite of the sun shining and the large amount of natural light available, artificial lighting is being used. The reason is that, even if the illumination level in the rooms reaches or is above the required value, the light coming from the fenestration is too 'cold', due to the colour of the glass, and – as has been well known since more than 60 years – the occupants feel the luminous environment uncomfortable; as a result, artificial light is used to 'warm' the interior, which compensates the cold natural lighting.

Another visual comfort-related issue to consider is when clear glass is used. The benefit of a large aperture that lets in a flood of natural light is entirely cancelled by the effect of glare suffered by a building's occupants: visual comfort is restored by obscuring the glass surface with curtains, venetian blinds or whatever it is available. This is clearly shown in all glazed buildings (Sutter *et al.* 2006) (Fig. 14.6).

Fig. 14.6. Visual comfort in glass architecture.
(Photo credit: Lavinia Tagliabue).

The result on the energy balance of the building is easy to evaluate: high thermal losses through the facade, uncontrolled solar gains in summer (curtains, even if white, absorb solar energy that is transferred to the room) and lights always on.

To moderate this undesirable effect in more recent times some leading architects protect the large glazed curtain walls with external shading devices. It seems a good idea, but unfortunately with light and outside vision cut off, artificial lighting must be on all the time.

In the last decade or so a new envelope technology has been proposed and used: the glass double skin facade. Scientists and experts are very doubtful about its effectiveness from an energy point of view, mainly because to design a glass double skin facade with a good energy performance is a very difficult task that requires time and high skill. (Streicher 2005) This is even more true in case of its adoption in temperate and hot climates, since one of the main recognized drawbacks of such a technology has been found in summer periods, when overheating in higher floors often occurred, even in cold climates. In any case, most of the problems associated with single skin glazed facades apply to double skin, plus some new ones.

Glass is potentially a very effective material for low energy buildings, thanks to the most recent technological developments, but fully glazed buildings exacerbate the problem of air conditioning in hot seasons, increasing the energy demand because of the large solar gains and the induced need for artificial lighting in spite of natural light always being available.

The most effective way to stop the proliferation of energy waste in buildings is to develop and put in force appropriate building regulations dealing with energy conservation. In some countries, such as France, Spain, Portugal in Europe, and California in the US, there are limits to the maximum glazed area allowed in each facade, unless it is demonstrated that the energy required for heating and cooling is less than a pre-set limit value.

14.5 Maximize Efficiency of Energy Supply

Once energy demand has been minimized, with appropriate urban and building design, it is time to evaluate the use of the most energy efficient technologies for providing heating and cooling, hot water production, lighting, etc.

14.5.1 From energy to exergy

To accomplish this, a paradigm shift must take place. For most of the last century energy was cheap, apparently inexhaustible, and the greenhouse effect was not taken into consideration. Even if new – far more efficient – technologies were potentially available, heating was provided by burning coal, oil or gas in a sort of improved wood stove: the boiler, i.e. the most inefficient way to use an energy source to keep room air at 20°C. For a short time there was an exception, however, when initially electricity was produced locally by means of small steam power plants, whose waste heat was distributed to provide space heating in the surrounding buildings. A single plant for supplying light and warmth, with a very high overall efficiency, even if as power plant alone was rather inefficient (nowadays this practice is called combined heat and power, CHP). Because energy sources were very cheap, it was more profitable for electric utilities to produce electricity with large scale power plants, far from the settlements where waste heat could have been used; thus this approach was soon abandoned and we started our current irrational practice of burning fuel separately: for producing electricity – releasing waste heat into the environment – and in boilers for heating.

The second law of thermodynamics deals with exergy, which produces clear statements: on the one hand, burning fuel in a boiler for heating air at 20°C is the most inefficient way to obtain the desired result (use high grade heat when we need low grade); on the other hand, the production of electricity with a thermal power plant implies necessarily the production of some low temperature heat. Since we need low grade heat for heating, why not use this, otherwise it is wasted? In this way the overall efficiency of the system is significantly improved (Fig. 14.7). Moreover, water consumption – in a world in which water shortage is already a threat – is strongly reduced. Approximately three quarters of water consumption in Germany and about 50% in the US are used in the cooling system of fossil and nuclear power stations for extracting low grade heat (Scheer 2006) which is wasted in the atmosphere, in rivers or in the sea causing so-called thermal pollution.

At the same time the second law states that the most efficient way to produce (or subtract) heat for heating or cooling with high exergy content sources such as fuel or electricity is by means of an appropriate use of the heat pump. The heat pump (Fig. 14.8) is a device that 'pumps' a heat flow from a lower to a higher temperature, in the same way as a water pump raises a water flow from a lower to a higher level. When it is used for heating, it pumps heat from the outdoor *to* the indoor environment, heating it; when used for cooling (refrigerators, air conditioners), it pumps heat *from* the indoor to the outdoor environment.

Within this framework, only two technologies are consistent with the second law of thermodynamics for supplying the final services required in buildings to assure our well-being: combined heat and power (CHP) systems for providing electricity, heating, cooling and hot water if fuel is used as the energy source, and heat pumps for heating, cooling and hot water production if electricity is used as the energy vector.

The twentieth century was the century of the first law of thermodynamics, dealing only with quantities of energy, disregarding the issue of its quality. The twenty-first century must be the century of the second law that deals with energy quality. It is a necessary paradigm shift for radically reducing the consumption of exergy and, at the same time, improving comfort. The required technologies have been available for a long time.

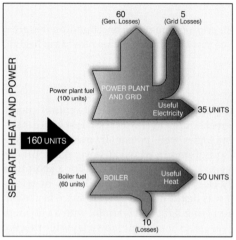
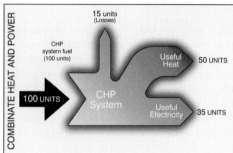

Fig. 14.7. Combined heat and power principle.

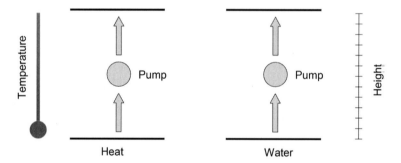

Fig. 14.8. Heat pump principle.

14.5.2 Low exergy technologies for heating and cooling

CHP or co-generation plants are the most efficient energy supply systems, as their waste heat is used not only for heating but also for cooling by means of devices such as the absorption chiller, which uses low temperature heat to produce chilled water.

Small-scale gas fuelled CHP units (micro-CHP) are now available on the market, ranging from 1kW for a small apartment to hundreds of kW for large buildings, but this kind of technology is best exploited with district heating and cooling, since the overall efficiency of the system increases with its size.

Town-scale gas fuelled CHP systems have been present for a long time in many European cities, but almost all use them for heating only. A few, like Helsinki, Finland, have developed a more efficient CHP district heating and cooling system.

The most efficient way to use a heat pump in winter is by minimizing the delivery temperature of water or air in the rooms and, in summer, by maximizing it. In order to obtain this result, the best current technology is radiant heating and cooling, using radiant ceilings or radiant floors.

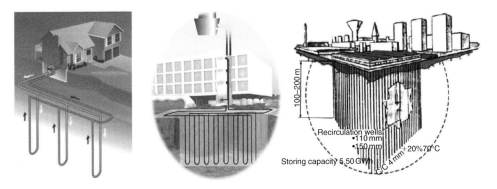

Fig. 14.9. Ground source heat pump.

In cold climates, the outdoor temperature in winter is low; this implies that the difference between outdoor and indoor temperatures is high, and the performance of the heat pump is poor. To overcome this problem, there are several options. If aquifers are available, they can be used as a heat source in winter and heat sink in summer, improving the heat pump performance. In summer, in the periods in which the load is not very high, underground water, usually at about 10–15°C, can be used directly for cooling; in this case the only energy consumption derives from the pumps used for water circulation. Instead of aquifers, rivers, lakes or even sea water can be exploited as heat sink/source.

There are several examples of district heating and cooling using water as heat source/ sink and heat pumps. In Stockholm, both Baltic sea water and aquifers are used. Presently 500 GWh of district cooling are sold per annum. If that cooling had been produced conventionally, it would require five times more electric energy (IEA Annex VIII Showcase of district cooling systems in Europe – Stockholm).

A project of district cooling using deep ocean water is ongoing in Hawaii, (State of Hawaii 2002) and in Geneva, Switzerland, lake water is planned to be used for both heating and cooling with heat pumps (European Commission and Concerto Projects).

If water is not available, the ground itself can be used as outdoor environment, taking benefit from the fact that ground temperature, below 10 m depth, is constant and roughly equal to the average annual air temperature. To exploit this phenomenon, ground heat exchangers located into boreholes 50–150 metres deep are used (Fig. 14.9). This technology, developed both for heating and cooling is an effective alternative for significantly reduce the energy consumption, due to the very high efficiency that can be reached by the heat pump.

Also in this case, the adoption of the technology at district scale could help, since the heat pump efficiency increases with its size.

Either CHP or heat pumps used for district heating and cooling can dramatically reduce the primary energy consumption of the existing building stock.

14.6 Maximize the Share of Renewable Energy Sources

As energy consumption is minimized with appropriate technological systems, renewable energy can have a significant role in the energy balance of an urban settlement. Many technologies are available and already in use.

14.6.1 Biomass

Wood biomass can be used for supplying CHP power plants directly (as pellets or wood-chips) or after gasification. These practices are applied more and more frequently. There is an assortment of biomass CHP plants all over Europe ranging from the 37 MWe CHP plant of the Swedish town of Växjö (52 000 inhabitants) supplying 35% of the electricity and 95% of the heat needed (Energie-Cités and Biomass CHP – Växjö) to the 1.1 MWe CHP plant in the alpine town of Tirano, Italy, supplying heat and electricity to 6900 inhabitants, plus other smaller ones.

Generally, wood biomass is chipped and burnt in boilers to produce steam which supplies one or more turbines coupled with generators. Some CHP systems instead use internal combustion engines fuelled with gas produced by a biomass gasification plant. The Sterling engine is used in smaller (down to 10 kW) biomass fuelled CHP units that are being developed for multifamily housing.

The popularity of biofuels is growing, mainly in cars as a substitute for or as an integration of gasoline or diesel oil; more recently, small internal combustion CHP units running on biofuels have become available.

14.6.2 Solar energy

A consequence of an appropriate urban design is that most buildings are aligned along the east/west axis, making it easy to install solar collectors in their flat or south facing roofs, with limited costs.

The most immediate and cost-effective use of solar energy is for hot water production. In a low energy urban development the use of solar systems for hot water production should be compulsory, as it is, for example, in Israel, Spain and, more recently, in Italy.

Solar thermal systems can also be integrated in the district heating network, as in the community of Ballerup, in Denmark, where 50% of the energy consumption for hot water and heating of 100 apartments is provided by roof mounted solar collectors connected to a gas fired CHP plant (Energie-Cités best practices and Solar District Heating – Ballerup) the heat produced by these systems is injected into the district heating network, which provides the remaining 50%.

Solar thermal district heating with seasonal storage is another approach, implemented in Friedrichshafen, where excess heat in summer is stored in a large and well-insulated underground water tank and used during winter (Energie-Cités best practices and Solar District Heating – Friedrichshafen).

Solar thermal systems equipped with evacuated tubes are also suitable for solar cooling, either by means of absorption and adsorption chillers or coupled to desiccant cooling systems (IEA Task 25, Solar Assisted Air Conditioning of Buildings) (Fig. 14.10).

Solar energy can be used either for DHW production or for cooling, complementing the heat produced with CHP systems, helping to balance the mismatch occurring between the electricity and cooling demand of buildings and electricity and cooling supply systems. Photovoltaic systems, in a low energy consumption settlement, are best used when integrated in the buildings' envelopes. The present drawback is their cost, but it is steadily dropping and the crossover, evident with the rising curve of fossil fuel cost, is predicted to be not that far away.

In a future in which PV systems will be competitive with fossil fuels for electricity production, these systems will be the main actors of the energy system of a low energy urban settlement, providing electricity also for cooling, coupled with heat pumps.

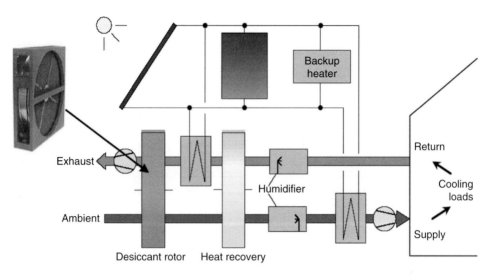

Fig. 14.10. Solar desiccant cooling principle.
(Adapted from IEA Task 25, Solar Assisted Air Conditioning of Buildings).

In the meanwhile, another technology could be exploited for the production of electricity: the solar power station (Fig. 14.11). It is a technology that, after a short period of success in the early 1980s, was largely ignored, but nowadays many new plants are being built. Based on solar concentration, solar power plants are not suitable in hot humid climates because of the low amount of direct solar radiation, but they are very effective in dry climates, where they could become part of the landscape around the settlement.

14.6.3 Wind energy

Wind power is not available everywhere, but in coastal areas it is often significant enough to make the installation of wind turbines cost effective. Also the ridges on hills are suitable locations. Offshore wind farms are becoming attractive options, due to the technological improvements and the lowering of costs; besides being capable of harvesting higher speed winds, these plants have the advantage of reducing the problem of visual impact, which often prevents or slows down the development of wind power.

Not only should large wind turbines be taken into consideration, even small wind generators, either horizontal or vertical axis, are also an option (Fig. 14.12). Although their cost effectiveness is lower than that of large wind turbines, due to the rise in fossil fuel costs, they are a valuable option and could give a considerable contribution to the energy balance of the settlement, owing to the large number that could be installed on building roofs, integrated into the urban landscape.

14.6.4 Geothermal energy

Low and moderate temperature geothermal resources have been discovered in many areas of the world, and are being used increasingly for district heating. The geothermal water acts as a heat source directly heating the network fluid through a heat exchanger (see Fig. 14.14).

Fig. 14.11. Solar power stations.

Fig. 14.12. Mini wind generators.

The scale at which geothermal district heating is used ranges from small to large settlements: in Boise Warms Springs Water District, Idaho (the first district heating system in the United States), heating is provided to about 200 dwellings, while in the Italian town of Ferrara (Energie-Cités best practices and Geothermal Energy – Ferrara) to more than 8000.

Geothermal energy plays a crucial role in Iceland's energy economy. The dominant use is for space heating where almost 90% of the houses are heated with geothermal water. (Gunnlaugsson *et al.* 2000). The first district heating utility was in Reykjavik, which commenced operation in 1930; it is probably still the world's largest district heating system using geothermal water, serving the homes of about 157 000 people, i.e. 99.9% of the population. Gunnlaugsson *et al.* 2000). The heating utility provides geothermal water not only for homes but also for commercial and industrial buildings in its area.

14.6.5 Mini-hydro power

Within the municipal borders of many existing settlements several derelict small hydro power stations can be found, especially in Europe. Most of them were built in the first half of the twentieth century and abandoned in the second half because they were not cost effective. Many of those mini-hydro power plants are now being restored and put into operation, and new sites are also being explored and exploited. Mini-hydro potential is still largely untapped, however, especially in developing countries.

Another potential is that deriving from the water supply; this water often is collected in springs or basins up in the mountains and delivered at the settlement's lower level via forced conduits. The pressure available at the bottom can be exploited by means of water turbines.

14.6.6 Energy storage

The more CHP and renewable energy sources are used in the energy system the more storage technologies become crucial. Thermal storage is relatively easy: a more or less large and well-insulated tank containing water, or ice. Also phase change materials, capable of storing more heat per unit volume than water, can be used.

Less easy is to store electricity. The most common technology is the battery: well established is the lead-acid type. In the last years some other types of battery have been developed, with higher storage density. Well-established means for storing electricity are pumped hydro and compressed air, used by energy utilities.

Other technologies, however, are close to market, such as supercapacitors, superconductive coils and flywheels (Sauer 2006). Unfortunately, up to now very modest resources have been put into research for implementing them. Far more funds are flowing into the hydrogen research. Hydrogen is another storage medium, using electricity for producing it from water (hydrolysis) and then recovering part of this electricity by using the gas for supplying a fuel cell. This system is by far the less efficient for electricity storage (Bossel 2006) (Table 14.1), so the present enthusiasm about a future hydrogen economy for a clean world energy system should be curbed, unless other ways of producing it are found.

Table 14.1.   Energy storage transfer losses (%).

Super capacitors	10
Lithium-ion batteries	14
Lead-acid batteries	23
Pumped water	28
Compressed air	36
Compressed hydrogen	68
Liquefied hydrogen	75

Small-scale advanced batteries, supercapacitors, flywheels, and compressed air will be intrinsic parts of a low energy system, strongly relying on renewables.

14.7 Wastewater and Solid Wastes

Recycling wastewater and solid wastes and using them as energy sources is essential in an energy efficient city.

Solid waste incineration, after selection and pre-treatment, supplying a CHP plant can give a significant contribution to the city's energy balance.

An alternative to the incineration of wastes is their gasification, producing syngas. Such gas can be used either to supply a CHP unit or to be distributed for cooking use. A very good example (Fig. 14.13) of water and solid waste energy content reuse is implemented in Gothenburg, Sweden (Energie-Cités best practices and Urban Energy Planning – Gothenburg).

The sewage water is led to a wastewater purifying plant just outside the city. The sludge with solid content that settles in the bottom of the secondary settlers is thickened and pumped to an anaerobic digester, which is fed also with separately collected organic food wastes. The biogas produced is delivered – according to the city needs – to the city gas network or to a filling station for gas powered vehicles or to a CHP plant, whose heat and electricity production feed, respectively, the district heating network and the electric grid. Part of the digested sludge, after dewatering, is used as fertilizer.

In winter, after the cleaning process and before flowing in the river, the water is pumped through large heat pumps which provide energy for 13% of the annual demand of the district heating. Heat pump performance is very good because it transfers heat to the cleaned water, whose temperature is 10–15°C, i.e. higher than air, river, or sea temperature in the same period. This is due to the fact that part of the sewage water comes from hot water taps and from piping that, being in the heated buildings, is warm; moreover it flows underground, where the temperature corresponds to the annual mean of the air above ground. With this clever technological solution, part of the heat used for heating water is recovered and reinjected in the district heating network.

The waste incineration supplies the energy for running a CHP plant providing electricity equivalent to the annual consumption of 60 000 apartments and additionally heating and hot water for 120 000 apartments.

14.8 Urban Mobility

Transport is a major factor contributing to energy consumption, directly and indirectly. Urban noise, for example, comes mainly from road traffic (80%) and causes higher energy consumption for air conditioning, because windows must be closed, impairing natural ventilation.

In recent years the use of energy in transport has risen dramatically. Forty years ago, in most developed economies, transport's proportion of total energy use was between 15 and 20%. Today it is around 35% of world energy consumption and is still rising (Potter 2003) (according to Exxon forecasts (Exxonmobil global liquid demand for transportation in 2030) will rise by 35%, mainly due to the development of motorization in developing countries). The highest value of CO_2 emissions per capita for passenger transport is found in the USA, with 4.4 tonnes per person: four times higher than in Europe (Kenworthy 2003).

In EU25, of all CO_2 emissions road transport accounts for 22%, (European Commission 2006) nearly half of which is due to urban mobility (European Commission 2007). This

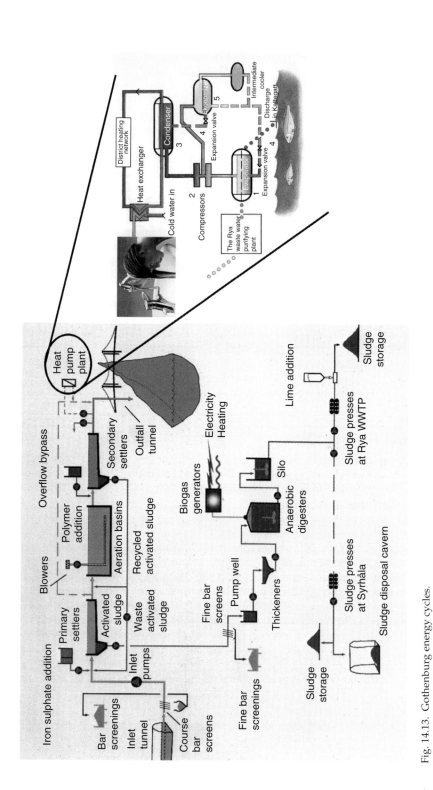

Fig. 14.13. Gothenburg energy cycles.

share will grow because of the progressive reduction of energy consumption in the building sector, as a consequence of new, tighter, energy standards.

If the present trend is to continue, private vehicles will be the most important cause of
CO_2 emissions in urban settlements. This is already the case in cities where, because of the
mild climate, energy consumption of buildings is relatively low and where energy intensive
industrial activities are not present. For example, in Palermo, Italy, energy consumption in the
transport sector accounts for more than 56% of the total (Butera 2000). Technological improvements found in most modern vehicles to improve efficiency can change the picture only marginally; to face and solve the mobility impact on global warming a radical change is needed.

The path to follow, at least in Europe, is easy to detect if available data are considered.
Today, approximately 75% of the EU population lives in urban areas (80% in 2020); (Uhel
2006) on average a European citizen makes 1000 trips per year and half of these are less
than 5 km long (Uhel 2006). For many of these shorter trips walking and cycling could
be a better alternative. Even if public transport is available, the car is by far the dominant
urban means of travel, contributing about 75% of kilometres travelled in EU conurbations.
Cars cause so much congestion that, in some European cities, average traffic speeds at
peak times are lower than in the days of the horse-drawn carriage.

Present mobility, based on private cars, whose average efficiency (useful work done/
primary energy used) is lower than 15%, carrying most of the time a single passenger
(i.e. spending energy to push more than one tonne to move 70 kilos), is incompatible with
energy sustainable urban development. Already a dwelling designed according to the new
energy standards produces less CO_2 every year than an average car travelling in the city
for 12 000 km. The number of circulating vehicles must drastically decrease. A completely
different urban mobility must be developed, i.e. a new generation of cars used in a new way.

14.8.1 Mitigated environment paths

An important issue is to give more priority to pedestrians and cyclists. The aim is to maximize the attractiveness and usability of walking and cycling as alternatives to motorized
transport. The goal should be a dense, high quality, supply-oriented infrastructure network for pedestrians and cyclists.

The main problem connected with walking and cycling is comfort, especially when it is
too cold or too hot or when it is raining. There are means, however, for improving the comfort of outdoor spaces, by mitigating environmental conditions. Arcades giving sheltering
from the rain, wind and sun were very common in many European cities before the advent
of the car. In colder periods some radiant heating could be provided to improve comfort.
For cyclists other kinds of sheltered paths are conceivable. In summer, green shading concepts can be developed and, in a dry climate, water can be used in many ways, as demonstrated in the EXPO 92 in Seville (Alvarez *et al.* 1992) (Fig. 14.14). This is possible by making
use of very limited energy consumption and by using the purified water of the wastewater
treatment plant. Therefore, walking and cycling mobility could be enhanced by creating a
grid of mitigated environment paths, warm and sheltered in winter and cool and shaded in
summer. Even if some energy is needed, it will far be less than that used by cars.

14.8.2 Individual motorized travel

It is not only public transport that should be promoted. Advanced mobility means can also
be envisaged. Car sharing now takes advantage of GPS technologies and the internet. In

Fig. 14.14. Outdoor climate mitigation.

the US, a company provides a car sharing service based on some very simple steps: after paying a membership fee and undergoing a background check, drivers can reserve a vehicle online. They choose the closest pick-up site from a map and the type of vehicle from a list. At the arranged time, the driver can enter the vehicle by swiping a coded card over a windshield reader. The member then starts the car using keys stored inside, drives off and returns the car to the pick-up site.

A substantial improvement to this system is forthcoming. We are already technologically capable to produce cars able to move safely at low speed without a driver, thanks to radar and video cameras; they could be guided to, on request, our front door (or anywhere else) using a GPS system. Cars capable to self drive have been announced already by many car manufacturers. The aim, at the moment, is to leave the car in automatic drive mode in the slow intense city traffic, and in active warning mode on motorways, to increase safety.

An improved car technology, already on the market, is the so-called 'plug-in hybrid', (Plugging into the future 2006) a car with an enhanced battery. A plug-in hybrid car, with its 35 km range as an electric car, is capable of meeting all the mobility needs of an average citizen. If, moreover, the engine is powered with biofuels, which is becoming more popular in Europe and is already well established in Brazil, it is easy to reach the objective of a car running only on renewables. Hybrid cars are a transition technology, since low energy urban mobility should be based on electric cars specially designed for the urban context in which they have to work; not the present electric cars proposed by the most important car manufacturers: they are simply ordinary cars in which the internal combustion engine is substituted with an electric motor. These types of car must be conceived in a different, specialized way; not multi-purpose like today's cars. With the car sharing approach we can use the type of car we need when we need it.

The electric car, if powered with a new generation battery, is the most low emission means of transport, even if the electricity is produced with fossil fuels – this is due to the efficiency of the motor and the electricity storage system. With an urban mobility based on a more or less advanced car sharing scheme it is possible to implement a transport system powered only with renewable energy, where batteries are charged by mini wind turbines, where suitable, or dedicated PV systems are integrated into buildings. Such a mobility system is already cost effective, because powering an electric car with wind or even PV produced electricity is cheaper than powering a car with gasoline or diesel oil, if they are heavily taxed, as in Europe.

Because of the very low efficiency of the chain (electricity from the source, hydrolysis, compression or liquefaction of hydrogen, transport, storage, fuel cell) at present hydrogen powered cars do not seem to be a viable alternative.

14.9 Carbon Neutral Buildings and Settlements

According to the local climate and resources, by combining in an appropriate way the above-mentioned technical and technological means, it is possible to design the energy system of a building or of a settlement with the aim of achieving zero CO_2 emissions. The energy system must be conceived at the beginning according to a new energy paradigm. This implies not only that the architectural design process for the individual buildings has to change, but also – and mainly – that the planning rules of the community have to change: no longer can there be a linear, fossil fuel-based, energy economy, but instead a circular, renewable sources-based, energy economy.

14.9.1 Kronsberg city district (Hannover, Germany)

The Kronsberg Rumming district, comprising 3000 dwellings, is home to around 6600 people and the final municipal plan foresees a total of 6000 homes for 15000 people. Three children's day centres, a primary school, a district arts and community centre, a health centre and shopping complex are in operation (Fig. 14.15).

In designing and creating this new city district at the end of the last century, all the available knowledge of ecological optimization in construction and habitation, along with urban planning and social aspects, was applied.

14.9.1.1 Energy concept

In devising the energy concept for Kronsberg, the aim was to develop generally applicable energy efficiency measures that would be acceptable to developers and residents. Most of the principles described in previous paragraphs were used for designing and implementing the district.

Minimization of heating energy demand in buildings was obtained by obliging all clients and construction companies through the land sale contracts or urban construction contracts to carry out construction work according to precise standards, such as:

- Heating energy index of $50\,kWh/(m^2\,a)$ as a target value.
- The calculation method for the heating energy index defined.
- Monitoring by qualified engineers.
- Penalty payments of €5/m^2 for exceeding the limit value.
- Provision of subsidies by the local authority.

Moreover, a special project for 32 terraced family Passive Houses (heating energy consumption $15\,kWh/m^2\,a$) was developed.

flächendeckend/across the entire district:

– NEH-Bauweise mit Qualitätssicherung
  und Qualifizierung
  LEH construction principles with quality
  assurance monitoring and skilling & qualification
  programme
– Nahwärmeversorgung/district heating provision
– Stromsparen/electricity saving

① Photovoltaik/photovoltaic installation

② BHKW/decentral CHP

③ Passivhäuser/passive houses

④ Mikroklimazone/microclimate zone

⑤ Solare Nahwärme/solar district heating

Windkraftanlagen im Landschaftsraum
wind turbine generators in the countryside

Fig. 14.15. Kronsberg.

Minimization of heat demand for water heating was achieved by fitting water-saving devices and water meters in all apartments and by environmental education measures.

Minimization of electricity demand was obtained by offering to tenants and owner/occupiers incentives for the use of low energy appliances and promoting energy saving awareness, i.e.:

- Free distribution of energy-saving light bulbs.
- Grants for the purchase of electricity-saving appliances (washing machines and dish-washers, refrigerators and freezers).
- Advice in person or by telephone on electricity saving habits.
- Appliance checks with an electricity meter, producing a cost/benefit analysis that showed how much electricity could be saved by buying new appliances.
- Analysis of electricity consumption with a consumption logbook.
- Investigation of further possible energy savings in the apartment, e.g. consumption of standby function, settings for the heating and ventilation systems.

Minimization of primary energy consumption was achieved mainly by means of gas powered CHP district heating. Other measures were, however, implemented, such as the obligation to have washing machines and dishwashers connected to the hot water system by short pipe runs, to avoid the use of electricity for heating water, and using instead the heat delivered by the district heating.

The combination of all these measures, not including passive houses, led to an overall 60% CO_2 emission reduction, compared to current standards for conventional residential buildings.

The action of maximizing renewable energy sources was implemented in two ways:

- Solarcity, a pilot project to demonstrate the suitability of a large social housing complex for extensive solar energy provision. A total of 104 apartments in the Solarcity complex are heated from thermal solar collector panels, which also replace conventional insulation on the south-facing roofs of the housing blocks. Superfluous solar energy in summer is piped to an extremely well-insulated cistern, and thus solar heating is possible from spring through to December. This covers around 40% of the total heating demand, the rest being supplied by the district heating network.
- Wind energy. Two wind turbines of 1.5 and 1.8 megawatts were erected. A limited partnership was founded to finance and run the plants; the operating company's own capital is in large part provided by private investor partners (most are owners of dwellings in Kronsberg).

A 20% reduction in CO_2 emissions was achieved with wind power, and a further 5–15% was derived from Solarcity and passive houses.

The overall result is that emissions are 85–95% lower than they would have been adopting current building and management standards (Fig. 14.16). It is an excellent first stage towards a carbon neutral settlement.

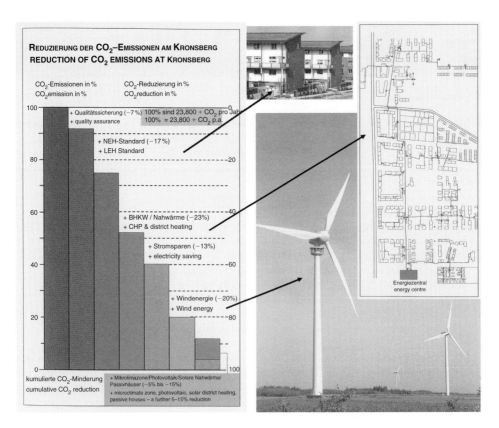

Fig. 14.16. Kronsberg: energy concept.

14.9.1.2 Remarks

The energy consumption of new buildings is, along with municipal buildings, one area over which a local authority has direct influence on the implementation of energy-saving measures. As the Kronsberg example has clearly shown, the possibilities go far beyond setting conditions for the site construction; accompanying measures such as quality assurance monitoring through the planning and construction phases or public relations work are just as much a part of a good energy concept for new settlements as the exploitation of various regulatory instruments (prescriptions in the development plan, bylaws on district heating and legal agreements such as land sale contracts) or supplementary funding programs.

14.9.2 A carbon neutral settlement: BedZed

BedZED (Beddington 'Zero (fossil) Energy Development) (Twinn 2003) is a mixed development urban village, built by a private investor; on a brownfield wasteland site in the London Borough of Sutton; it provides 82 dwellings in a mixture of flats, maisonettes and town houses, and approximately 2500 m^2 of workspace/office and community accommodation including a health centre, nursery, organic café/shop and sports clubhouse. The community is designed to host about 240 residents and 200 workers. BedZED is Britain's first urban carbon-neutral development (Fig. 14.17).

Fig. 14.17. BedZed. (Adapted from pictures of Bill Dunster and Arup).

14.9.2.1 Energy concept
In order to minimize energy demand the houses face south to make the most of the heat from the sun, and are fitted with high levels of insulation, air tight triple-glazed windows and the latest energy saving appliances, including water saving devices. With the combination of super-insulation, heat recovery, passive solar gain stored within each flat by thermally massive floors and walls and internal gains, comfort temperature is obtained for most of the time without any need for backup heating (Fig. 14.18). In summer, workspaces placed in the shaded zones of south facing housing terraces enable all flats to have outdoor garden areas, with good access to sunlight, and at the same time allowing cross-ventilation.

Moreover, wherever possible building materials have been selected from natural, renewable or recycled sources and brought from within a 55 km radius of the site. Primary energy consumption is minimized at two levels: individual building and community; in both cases renewable energy is used to power the energy efficient systems. At individual building level, an innovative device was designed for ventilation heat recovery: the wind cowl. In well-insulated buildings, the main energy demand is due to the need to heat the ventilation air. For this reason the current practice is to use mechanical ventilation combined with efficient heat exchangers for preheating fresh air with the heat of the exhaust one. Conventionally, much high grade fan and pump electricity is consumed to deliver low grade energy for room comfort temperature control and ventilation. Energy consumption tends to be significant because these systems run for extended operating periods. The wind cowl system delivers preheated fresh air to each home and extracts its vitiated air, complete with heat recovery from the extracted ventilation air, using wind power, i.e. only renewable energy.

At community level, a 135 kWe wood fuelled CHP plant meets all the energy requirements of the community, exchanging electricity with the grid. The CHP unit generates electricity, and distributes hot water around the site via a district heating system, (Fig. 14.19) delivering constant heat to oversized domestic hot water cylinders. The cylinders have electric immersion heaters for emergency back-up. They are in cupboards within each home and office, positioned centrally so that they can be opened to work as a radiator in cold spells.

The CHP plant satisfies both the electrical and heat demand of the project due to three interrelated factors:

- Average loads are reduced.
- The design evens out the normal fluctuations between summer/winter and daytime/evening space-heating demands.
- The domestic/commercial mix also evens out the daily electrical demand to more closely match the CHP output.

The generator engine is fuelled by gas produced from woodchips by an on-site gasifier. The woodchips come from tree pruning waste from nearby woodlands – waste that would otherwise go to landfill. The demand is 1100 tonnes per year. London produces 51 000 tonnes of tree surgery waste per year (which could rise if local authorities develop plans for recycling park and garden waste).

It was calculated that if the electricity is used to displace transport fossil fuel, with its high tax levels, the PV financial payback period was about 13 years. With EU/UK grants equating to 50% of capital cost, the theoretical payback period went down to just 6.5 years. On this basis, 107 kWp of PV has been integrated into the south-facing BedZED facades, generating enough solar electricity to power 40 electric vehicles for approximately

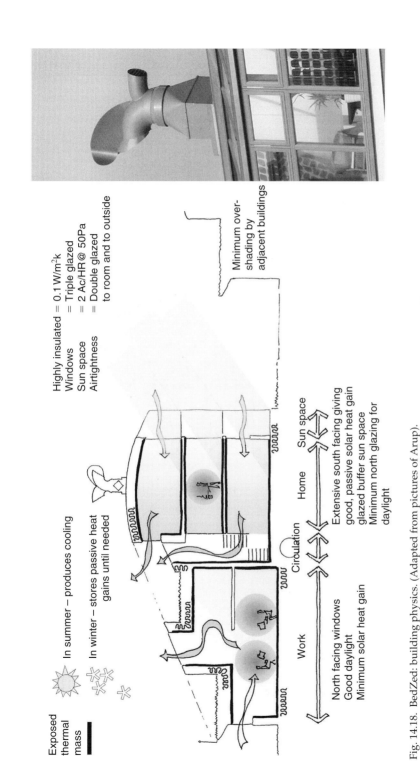

Exposed thermal mass

In summer – produces cooling

In winter – stores passive heat gains until needed

Highly insulated = 0.1 W/m²k
Windows = Triple glazed
Sun space = 2 Ac/HR @ 50Pa
Airtightness = Double glazed to room and to outside

Minimum over-shading by adjacent buildings

Work
North facing windows
Good daylight
Minimum solar heat gain

Circulation

Home

Sun space
Extensive south facing giving good, passive solar heat gain glazed buffer sun space
Minimum north glazing for daylight

Fig. 14.18. BedZed: building physics. (Adapted from pictures of Arup).

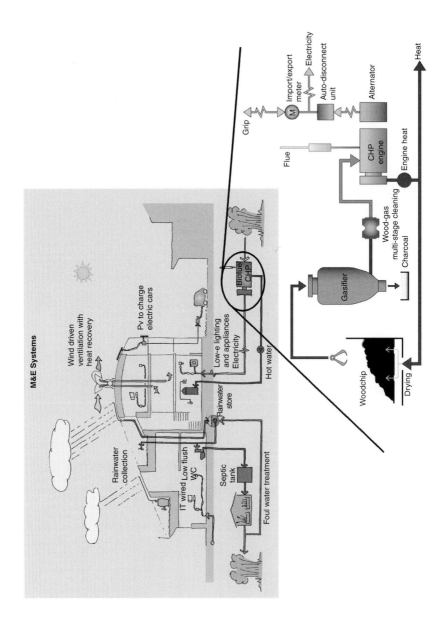

Fig. 14.19. BedZed: energy system. (Adapted from pictures of Arup).

8500km/year. Charging points have been installed and occupants can have free parking and charging if they use electric cars.

Car pool and car hire services with electric cars are provided to reduce car ownership and journeys. The idea of the service is to create carbon neutral transport. As a mixed-use development, BedZED offers the opportunity for residents to live and work on site, therefore eliminating the need to commute to work.

14.9.2.2 Remarks

BedZED seeks to offer to its occupants the opportunity to live and work with a completely carbon-neutral lifestyle, making this choice attractive, cost effective, and appropriate to modern living. It offers solutions to many sustainability lifestyle issues in a practical and replicable way.

One key reason for embarking on the BedZED project was to demonstrate to a sceptical industry how sustainability is possible and cost effective, and how we can really make a difference for society and its future.

14.9.3 Huairou new town concept design

Huai Rou (Butera 2006) is a new town for 100 000 people of which 80 000 permanent residents are to be located north of Beijing (Fig. 14.20). The mandate for the planner was to envisage a green settlement for medium to high income residents, a university, hotels for tourists attracted by the high environmental quality of the surroundings, some light industry and all the consequent services needed, such as schools, shops, etc.

With its 117 611 ha of forest area and its farms, Huairou district has a large biomass potential that could be used for radically reducing dependence on fossil fuels and CO_2 emissions. Solar energy potential is also large and mini-hydro potential is already partially exploited. No information was available about wind at the time of the study but, considering the orography of the territory, its potential should also be significant. The renewable energy potential of Huairou district is reported in Table 14.2, including also solid wastes and sludge produced by its future population.

Taking into account the renewable energy potential, a concept energy system for Huairou was developed.

14.9.3.1 Energy concept

The energy system design aimed to produce an example of the real possibility, with today's technologies and in a context of cost effectiveness, to design a fully renewable energy supplied, carbon neutral new town.

In order to minimize the building stock energy demand, the following assumptions were made:

- All buildings heated and cooled (also residential buildings are supposed to be fully air conditioned, as summers are very hot and humid); heating and cooling energy limits according to Table 14.3.
- Energy demand for electricity according to best practices, i.e. compact fluorescent lamps and presence detectors, A + rated domestic appliances, standby consumption control, variable speed pumps and fans, heat recovery, etc. (Table 14.3).
- Low water consumption taps and showers for reducing energy for hot water production.

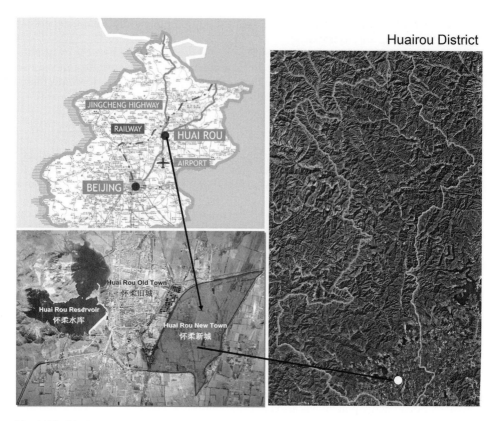

Fig. 14.20. Huairou.

Table 14.2.   Renewable energy potential of Huairou district.

	Source (MWh /y)	Electricity (MWh/year)	Notes
Solar	1 314 000		Only 75% of roof surface has been considered
Biomass	1 140 000		Forests and agricultural residuals; very conservative estimation
Sludge	4 000		
Waste	95 000		
Hydro		10 000 ÷ 15 000	10 000 MWh/year is roughly the present production
Wind		n.a.	No information available
Total	**2 553 000**	**31 600**	

Table 14.3. Energy consumption standards is Huairou.

	Heating (kWh/m²y)	Cooling (kWh/m²y)	Lighting (kWh/m²y)	Equip./appliances (kWh/m²y)
Residential	30	30	3	15
Commercial	25	60	15	20

 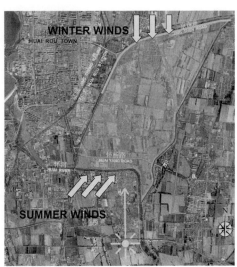

Fig. 14.21. Huairou: layout design; shadows and wind.
(Courtesy Mario Occhiuto Architecture (MOA)).

In order to comply with the energy demand limits of the buildings, an iterative trial and error process was carried out with the urban designer, in order to optimize building types, heights, orientation, functions (mixed use), spatial distribution, reciprocal shadowing, etc. (Figs 14.21 and 14.22). Also green areas to reduce the heat island effect were appropriately distributed and microclimate mitigated paths to favour walking and cycling were designed.

With regard to transport energy demand, it was assumed that in Huairou new town, due to good urban design (easy walking/cycling, the superior network of public transport with hybrid and all electric buses and trams), the use of private cars is significantly reduced (6000 km/year instead of the average 12000 in present European cities, 80% of which is for urban mobility).

The energy demand by sector derived from the assumptions is plotted in Fig. 14.23. It can be noted that the mobility share is far lower than the sum of the other two, and this derives from both urban and energy system design.

Exergy consumption minimization for meeting low grade energy demand was obtained by providing heating and cooling of all buildings except villas with district heating and cooling networks (not a single one but several, according to different demand patterns of different

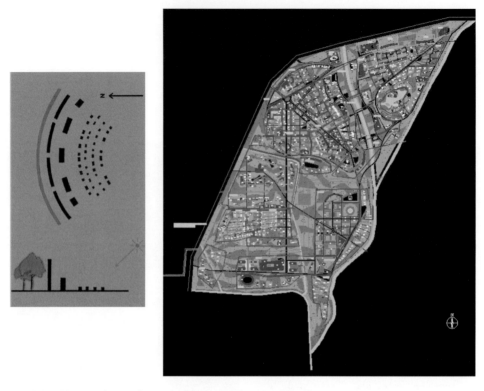

Fig. 14.22. Huairou: layout design; buildings and functions distribution.
(Courtesy Mario Occhiuto Architecture (MOA)).

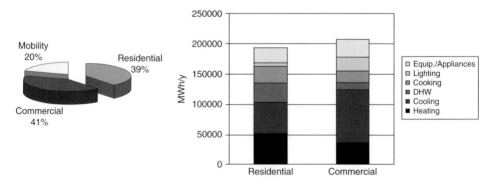

Fig. 14.23. Huairou: energy demand by sector and by end use.

areas); this approach, despite its energy effectiveness, allows progressive construction of the
town, area by area, each with its own energy system, but all of them interconnected.

Hot and cold water circulating in the networks is provided by the waste heat of biomass fuelled CHP units; hot water is directly distributed in the district heating network in

Table 14.4.   Energy sources powering Huairou vehicles.

Type of vehicle	Renewable energy source used
Lorries	Biodiesel
Buses	Biodiesel
Trams	Electricity from CHP units
Vans	Electricity from PV
Cars (urban use)	Electricity from PV
Cars (extra urban use)	Bioethanol

winter, while in summer cold water is produced by means of absorbtion and compression chillers, the latter supplied with part of the electricity produced by the same CHP unit.

Heating and cooling of villas is provided with ground source heat pumps, powered with the electricity produced by their PV roofs.

As much as 50% of hot water demand in residential buildings is provided by solar systems; the remaining share is provided by the biomass fuelled CHP district heating system.

Electricity demand of all buildings, except villas, is provided by biomass fuelled CHP systems, mini-hydro production, a biogas generator (anaerobic digestion of sludge) and a turbine fuelled by a biomass gasifier, acting as an energy storage unit, modulating its power for matching electricity demand and supply of the entire settlement.

Electricity demand of villas is met with the production from their PV roofs. Other PV collectors are also present on part of the roofs of residential and commercial buildings, for powering electric cars.

Cooking energy is provided by gas produced by the biomass gasifier.

According to the calculations, while the heating and cooling demand of the residential sector is exactly matched, the corresponding electricity produced annually almost exceeds demand. This excess electricity is delivered to the electric grid and it is used for street lighting and for balancing the electricity deficit in the commercial and transport sector. The residential sector is a net producer of electric energy.

The Huairou New Town transport system is based on biofuels replacing fossil fuels and electricity produced either from PV systems or biomass powered generators, according to Table 14.4. Private cars for urban use are assumed to be electric, and the plug-in hybrid cars are used for extra-urban use; it is also assumed that 50% of families own an electric car and that a car sharing system is available and used.

In order to produce the energy required with PV, about 110 000 m^2 of panels are necessary, i.e. 80% of the solar potential.

Biodiesel and bioethanol are supposed to be either produced in the district or bought in the market. In the latter case the amount of energy imported would be compensated by an equivalent amount of electricity produced with biomass powered generators and delivered to the national grid.

The overall energy balance of Huairou is shown in Fig. 14.24, in terms of end uses of each energy source. The item 'Electricity from mixed sources' stands for that produced by the waste CHP plant, the biogas CHP plant, the syngas fuelled gas turbines, the mini-hydro plants and the excess from by the biomass CHP systems supplying heat to the residential sector. In the same figure the share of the renewable energy sources used to meet the whole energy demand of Huairou is shown. No fossil energy is used, and biomass is by far the most important contributor to the town energy system.

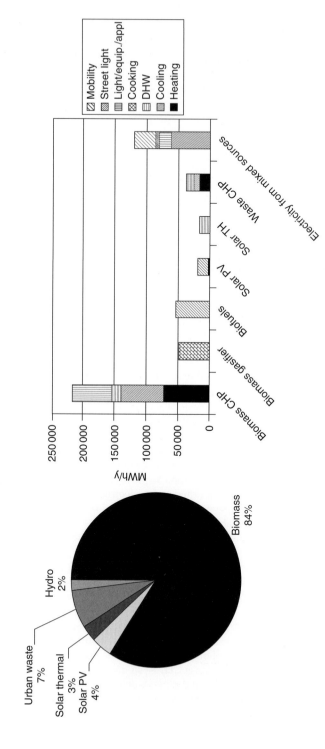

Fig. 14.24. Huairou: energy balance.

14.9.3.2 Remarks
The energy system of Huairou New Town has been designed with the aim of reaching the full sustainability in its operation, i.e. zero fossil energy consumption and zero emissions. Even if more detailed technical and economic evaluations are necessary, the preliminary ones show that this result is fully achievable and:

- can be obtained with presently available technologies and already adopted – in some parts of Europe – building standards for energy conservation,
- requires minor changes of lifestyle,
- guarantees a high standard comfort level,
- leaves some of the district's renewable energy potential still to be used (about 25% of the biomass potential; about 20% of roof area for solar collectors; some spare hydro-power not accounted for; and presumably some wind power potential).

The approach, based on the distributed energy resources concept, besides being energy efficient, also has a practical value, since it is possible to develop the town construction zone by zone, because of the modular nature of the urban subsystems.

Thanks to its centralized heating and cooling systems and to electric cars, Huairou New Town could be a comfortable, clean, noiseless and hospitable place to live.

14.10 Conclusions

When, about two centuries ago, the first gas networks started to be built in our cities, the bases for the present urban energy system were set up. At that time coal was the only fossil energy source used. A little more than a century ago the urban electric grids appeared, and coal was slowly superseded by oil and natural gas. Wood and charcoal soon disappeared; horses were substituted with cars and public transportation systems started to develop. At the beginning of last century the main cities of the western world were provided with a sewage system, a water network and a solid waste collection system. No more hard work carrying water from the fountains, no more epidemics, comfortable interiors with heating, cooling and electric lighting, an easier life at home with domestic appliances, and rapid mobility. A revolution in the quality of life. All thanks to cheap fossil fuels and to the technologies fed by them.

Cities slowly changed, and learnt to metabolize fossil fuels, building up either an urban energy system based on them or an overall metabolic system that left no room for recycling: the saturation of the environment with wastes was not an issue. At the end of the process a new organism, the modern city, was born, fit to an unreal environment imagined as an infinite source and an infinite sink. The present urban energy system is designed on this assumption; for this reason, it is incompatible with the extensive use of renewable energy sources, in the same way that you cannot feed a lion with vegetables – it needs meat. Our task is to transform, with a sort of genetic engineering, our carnivore – the fossil energy-based city – into a herbivore, the renewable energy-based city. Unfortunately we do not have two centuries in front of us, we do not have that much time; we must do it in less than 50 years. Fifty years to redesign the energy system of our existing settlements and less than 20 years to learn how to design in a different way all the new ones.

Starting from the 1970s – after the first oil shock – the energy issue was introduced, by a few pioneers, in architectural design. Guidelines for low energy building design were implemented, and now they are becoming compulsory practice in all European Union and in some other countries of the world.

Now is the time to introduce the energy issue to urban design, since the most significant energy savings can be obtained at this scale, by redesigning the energy system.

This implies a change of priorities in the formal design of the urban layout and in the organization of the urban functions. But the Distributed Energy Resources (DER) approach must also be introduced, made of many small-scale interconnected energy production and consumption units instead of a few large production plants: a new web, the *energy web*. (Droege 2006).

It is the only way to design new settlements (or redesign existing ones) capable of relying mostly on renewable energy sources, and implies an evolutionary jump towards a far more 'intelligent' urban energy system, because it is needed also as a distributed control system – made possible by the present development of information and communication technologies.

The change in the energy paradigm is the only chance we have of coping with the present world trend leading to either an economical or ecological catastrophe, or both. It is not an easy task, because it is a technological change that implies a cultural change. It has to change the culture of architects and city planners, of citizens, of entrepreneurs, of city managers and of politicians.

The fossil-based energy system developed spontaneously, under the contrasting pressures exerted by different social, economical and institutional actors; the new one must be guided by a strong political will. We have no time to wait for the spontaneous cultural change that would lead to the death of present technologies for individual mobility; we have no time to wait for a citizen to spontaneously feel more desirable and self-satisfactory to owe a solar collection, or a mini-CHP, than a famous designer's piece of furniture, if he has to make a choice.

These changes must be guided, by means of new rules in urban planning and building regulations, with incentives and penalties.

Urban services are an integral part of the urban energy system: water and waste cycles, energy supply and transportation are all interconnected, part of a single system. In the past such urban services were, at least in Europe, owned by the municipalities; the present trend, however, is towards private ownership, which is averse to long-term investments. This trend should be reversed, since the change in the energy system cannot be done with a short-term approach. Again, a strong political will is necessary.

The paradigm shift from a fossil-based urban energy system to a sustainable one will imply that there will be, in the market arena, losers and winners (many products must be excluded from the market and new ones substituted for them). This process must be guided too, to avoid dramatic social and economic events.

Actually, such a revolution is not new to mankind, as we know that (Mumford 1934):

'Behind the great materials inventions of the last century and a half was not merely a long internal development of technics: there was also a change of mind. Before the new industrial process could take hold on a great scale, a reorientation of wishes, habits, ideas, goals was necessary.'

References

Adnot, J. (2004). *Performance Certification of Components, Systems and Buildings.* Proc. Forum 'Cooling Buildings in a Warming Climate'. Sophia Antipolis, France, 21–22 June 2004.

Akbari, H. (1998). *Cool Roofs Save Energy.* ASHRAE Proceedings, pp. 791–796, January.

Alvarez, S.A. *et al.* (1992). *Control Climatico en Espacios Abiertos – Projecto EXPO '92.* Secretaria General Tecnica del Ciemat, Madrid.

Bossel, U. (2006). *Physics and Economy of Energy Storage*. First International Renewable Energy Storage Conference (IRES I), Gelsenkirchen, October 30/31.

Butera, F. (1998). Urban development as a guided self-organisation process, in *The City and Its Sciences* (eds C.S. Bertuglia, G. Bianchi and A. Mela). Physica-Verlag, Heidelberg.

Butera, F. (2000). Municipal energy planning – Palermo as a case study, in *World Renewable Energy Congress VI (WREC 2000)* (ed. A.A.M. Sayigh), pp. 1536–1542. Elsevier Science, London.

Butera, F. (2006). Energy for a sustainable settlement, in Mario Occhiuto Architecture (MOA), *Huairou New Town: Aims, Strategies and Tools for a Sustainable Urban Planning*. Interim report to Beijing Municipal Environmental Bureau.

Droege, P. (2006). *Renewable City –A Comprehensive Guide to Urban Revolution*. Wiley-Academy.

Energie-Cités, Biomass CHP – Växjö. http://www.energie-cites.org/db/vaxjo_139_en.pdf

Energie-Cités best practices, Geothermal Energy – Ferrara. http://www.energie-cites.org/db/ferrara_140_en.pdf

Energie-Cités best practices, Solar District Heating – Ballerup. http://www.energie-cites.org/db/ballerup_139_en.pdf

Energie-Cités best practices, Solar District Heating – Friedrichshafen. http://www.energie-cites.org/db/friedrichshafen_139_en.pdf

Energie-Cités best practices, Urban Energy Planning – Gothenburg. http://www.energie-cites.org/db/goteborg_107_fr.pdf

European Commission (2006). DG Energy and Transport, *Energy and Transport Figures*. http://ec.europa.eu/dgs/energy_transport/figures/pocketbook/doc/2006/2006_energy_en.pdf

European Commission (2007). DG Energy and Transport, Clean Urban Transport. http://ec.europa.eu/transport/clean/index_en.htm

European Commission, Concerto Projects. http://www.concertoplus.eu/projects/TetraEner/Geneva.php

Exxonmobil. The Outlook for Energy – A View to 2030. http://www.exxonmobil.com/corporate/files/corporate/energy_outlook_2006_notes.pdf

Gunnlaugsson, E., Frimannson, H. and Sverrisson, G.A. (2000). *District heating in Reykjavik – 70 years experience*. Proceedings of the World Geothermal Congress 2000, Kyushu-Tohoku, Japan, May 28–June 10. http://www.or.is/media/files/0415.PDF

Gunnlaugsson, E., Gislason, G., Ivarsson, G. and Kjaran, S.P. (2000). *Low temperature geothermal fields utilized for district heating in Reykjavik, Iceland*. Proceedings of the World Geothermal Congress 2000, Kyushu-Tohoku, Japan, May 28–June 10, 2000. http://www.or.is/media/files/0436.PDF

Houses without Heating System, http://www.vgregion.se/upload/Regionkanslierna/Milj%C3%B6sekretariatet/Energi/Folder_Lindashusen_EN.pdf

IEA Annex VIII, Showcase of district cooling systems in Europe – Stockholm, http://www.iea-dhc.org/download/Showcases_District_Cooling_Stockholm.pdf

IEA Task 25, Solar Assisted Air Conditioning of Buildings. http://www.iea-shc-task25.org/

Kenworthy, J.R. (2003). *Transport Energy Use and Greenhouse Gases in Urban Passenger Transport Systems: A Study of 84 Global Cities*. Third Conference of the Regional Government Network for Sustainable Development, Fremantle, Western Australia. http://www.sustainability.murdoch.edu.au/

Lang Siwei, Prospect on Energy Efficiency Design Standards for Buildings, International Mayors Forum on Sustainable Urban Energy Development, Kunming, Yunnan Province, P.R. China, November 10–11, 2004.

Long, W., Zhong, T. and Zhang, B. *China: The Issue of Residential Air Conditioning.*, Proc. XXI IIR International Congress of Refridgeration, August 17–22, 2003, Washington, DC, USA. http://www.iifiir.org/en/doc/1056.pdf

Mumford, L. (1934). *Technics and Civilisation*. Harcourt Brace & Company, Orlando, FL.

Plugging into the future. *Economist*, June 8, 2006.

Potter, S. (2003). *Transport energy and emissions; urban public transport*. In: Handbook of transport and the environment, 4. Handbooks in Transport. Elsevier, Amsterdam, Netherlands, pp. 247–262. http://oro.open.ac.uk/4378/01/PT_Energy_and_Emissions.pdf

Rumming, K. (ed.). *The Kronsberg Handbook*. BRE Garston, Watford WD25 9XX, UK. http://ec.europa.eu/energy/res/publications/doc1/the_kronsberg_handbook.pdf

Santamouris, M. Coordinator. *Cooling the City* – Report on Energy Efficient Cooling Systems Techniques for Urban Buildings. European Commission, DG TREN. (2003).

Sauer, D.U. (2006). *The Demand for Energy Storage in Regenerative Energy Systems*. First International Renewable Energy Storage Conference (IRES I), Gelsenkirchen, October 30/31.

Sawin, J.L. and Hughes, K. (2007). Energizing cities, in *State of the World 2007*. The Worldwatch Institute, Norton & Company, New York.

Scheer, H. (2006). *Energy Autonomy*. Earthscan.

State of Hawaii (2002). *Sea Water District Cooling Feasibility Analysis for the State of Hawaii*. Department of Business, Economic Development & Tourism Energy, Resources, and Technology Division. http://www.hawaii.gov/dbedt/info/energy/publications/swac.pdf

Streicher, W. (ed.). (2005) BESTFAÇADE – Best Practice for Double Skin Façades, EIE/04/135/ S07.38652, Intelligent Building Europe. http://www.bestfacade.com/pdf/downloads/Bestfacade_ WP1_Report.pdf

Sutter, Y., Dumortier, D. and Fontoynont, M. (2006). The use of shading systems in VDU task offices: a pilot study. *Energy and Buildings*, **38**, 780–789.

Twinn, C. (2003). BedZed. *The Arup Journal*, **1**. http://www.arup.com/DOWNLOADBANK/download68.pdf

Uhel, R. (ed.) (2006). *Urban Sprawl in Europe – The Ignored Challenge*. European Environment Agency, Copenhagen. http://reports.eea.europa.eu/eea_report_2006_10/en/eea_report_10_2006.pdf

United Nations (2004). *Urban and Rural Areas 2003*, New York.

Chapter 15
Counteracting Urban Heat Islands in Japan

TOSHIAKI ICHINOSE, FUTOSHI MATSUMOTO AND KUMI KATAOKA

National Institute for Environmental Studies, Tsukuba, Japan

15.1 Summer Heat Problems in Urban Settings

Increasing urbanization is generally associated with heat islands – a phenomenon of rising temperatures in urban settings. As a result, more people are affected by higher temperatures for longer periods. Heat islands not only make life uncomfortable for urban residents, the increased temperatures adversely affect people's health (e.g. accelerating air pollution by less ventilation) and the natural ecosystems (e.g. changing flora and fauna) in cities.

With increasing urbanization, ground surfaces have been converted from natural soil or green tracts of land, which function to lower the surrounding air temperature through the cooling effect of *evapotranspiration*, to materials such as asphalt or concrete, which lack any water content and tend to heat the atmosphere. Moreover, in major urban regions, although work is ongoing to develop parks and similar amenities, there has been a substantial decrease in greenery and 'productive green land' in residential areas. Land for roads, public facilities, offices, and high-rise residential blocks is increasingly surfaced with water-impermeable substances such as asphalt paving. Asphalt or concrete ground surfaces reach relatively high temperatures – in the region of 50–60°C – during the daytime on clear summer days, and this daytime heat builds up and is still present during the night.

Exhaust heat from urban energy consumption, such as air conditioners or vehicle traffic, is also a major factor in atmospheric heating (e.g. Ichinose *et al.* 1999). The warmed atmosphere moves according to meteorological or geographic conditions such that the heat not only affects the region where it originated but also produces downwind effects.

Urban environments create a complex set of factors, such as (1) the blocking of heat dispersal through natural wind flows, because of the formation of conurbations; (2) the location of factories and other major heat sources at coastal sites (windward) in seaside cities; and (3) the formation of low-wind areas where the atmosphere tends to stagnate because of the shape of the ground surface, including urban topography and the presence of large buildings.

Japan's Ministry of the Environment (MoE) has recently investigated these issues through various committees (e.g. MoE 2000, 2001), concluding that heat islands are domains of thermal atmospheric pollution, caused by human activities. The Ministry of Land, Infrastructure and Transport (MLIT) and local municipal authorities have also started to debate various measures, while the general public is also increasingly interested in this topic as an environmental problem (Ichinose 2005). We start this chapter with a discussion of the impact of heat islands using actual examples of their assessment as well as measures to mitigate and adapt to summer heat. We then discuss the potential for developing user and environmentally friendly cities that take into account the atmospheric and thermal environments.

15.2 Impact of Urban Heat Islands

It has recently been suggested that urban warming, as symbolized by heat islands (e.g. Landsberg 1981), is affecting ecosystems, including our environment. The impact on ecosystems was pointed out some time ago (Numata 1987), but unusually almost no research has been undertaken to evaluate this effect. Impact assessments are important in terms of providing warnings concerning the issues involved and educating urban residents and as basic information or evaluation tools for use in social policies and to estimate the costs involved (Matsumoto *et al.* 2006). As such, impact assessments can be thought of as barometers or indices for changing climatic environments.

15.2.1 Impact on human activities

Heatstroke and other health issues are the main risks from higher urban temperatures. Figure 15.1 shows Tokyo Fire Department figures for the number of individuals transported to hospital with heatstroke during the summer. Heatstroke often occurs when individuals are working or exercising outdoors, but this figure also includes some examples of individuals requiring emergency transport after being indoors. The results shown illustrate the harsh environment during the summer heat in cities.

Another problem that is becoming more prevalent is the increase in the number of summer 'tropical nights' in urban settings when daily minimum temperatures exceed 25°C

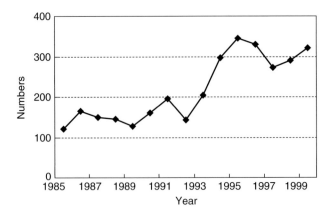

Fig. 15.1. Transported to hospital with heatstroke in Tokyo (running average in 3 years) (*Source*: Tokyo Metropolitan Government 2006).

(Fig. 15.2). This is attributed to heat storage by artificial ground surfaces and its release during the night, which prevents temperatures from falling. Such temperatures produce discomfort, including sleep difficulties, physical fatigue due to inadequate sleep, additional burdens on cardiac function, and psychological stress. There also appears to be an increasing number of heatstroke cases occurring at night. Fatigue due to inadequate sleep affects the individuals concerned the following day and this should not be overlooked as a risk factor associated with the onset of heatstroke during the day.

Other factors behind such hot summer environments involve urban living infrastructure, including the widespread use of air conditioners. The lifestyle of urban residents has become dependent on air conditioning, and this promotes health problems resulting from summer heat. Air conditioning use superficially appears to be a way of preventing or coping with summer heat, but it actually creates a vicious circle whereby the air conditioner is switched on because it is hot, and this then consumes energy and the exhaust heat makes the outside temperatures even hotter. Another recently recognized problem is summer colds caused by air conditioning use.

Of course, such high temperatures are caused not only by urbanization, but can also occur as a result of extreme weather phenomena, such as heatwaves, not directly related to urbanization. References such as the report by the Intergovernmental Panel on Climate Change (IPCC) highlight the possibility of increasingly fierce summers due to global warming. However, urbanization itself definitely plays a role in exposure to high temperatures. When conducting impact assessments of rising temperatures in urban settings, we may need to consider comparisons with the impact from global warming as well as synergistic effects with global warming.

15.2.2 Impact on urban ecosystems

We have recently started to see changes in the vegetation season, with flowers blooming earlier in the spring and leaves changing colour later in the fall (Yoshino and Park 1996; Momose 1998).

Figure 15.3 shows inter-annual changes in average March temperatures and the date of Somei Yoshino cherry tree blossoming in the Tokyo District Meteorological Observatory. The figure demonstrates that the cherry blossoms have recently bloomed earlier each year in line with rising temperatures.

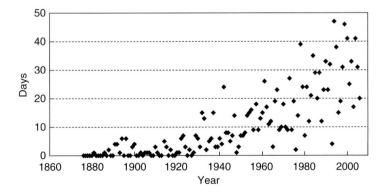

Fig. 15.2. Tropical nights (minimum temperature exceeds 25°C) in Tokyo.

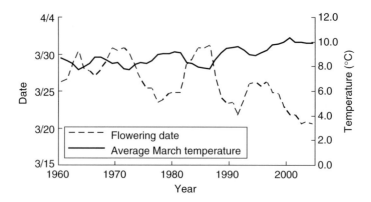

Fig. 15.3. Inter-annual changes in average March temperatures and the date of Somei Yoshino (cherry tree) blossoming in the Tokyo District Meteorological Observatory (running average in 5 years, during 1960–2004).

Figure 15.4 shows an assessment of the impact of heat islands in the Tokyo wards on the date of Somei Yoshino cherry tree blossoming. In heat islands in the heart of Tokyo, the cherry trees are blossoming some five to six days earlier than in the suburbs (Matsumoto *et al.* 2006). Moreover, there have been reports that in Kumagaya City, a small-to-medium sized city in Saitama Prefecture (population: approximately 160 000), cherry trees located in heat islands in the city centre are blossoming around two days earlier than those in the suburbs (Matsumoto and Fukuoka 2003). This suggests that the earlier blossoming is not simply a result of global warming, but may be associated with rising urban temperatures as symbolized by the heat islands. There have also been reports of the leaves on gingko and Japanese maple trees changing colour later in the fall in city centres (Matsumoto and Fukuoka 2002; Matsumoto 2004).

In addition to these effects on vegetation seasons, heat islands are also thought to impact other organisms in various ways. For example, there have been reports of the successful open-field cultivation of tropical aloe plants in Tokyo city centre (Nemoto *et al.* 2001), as well as reports of the disappearance of dragonflies (Shinada *et al.* 1987) and a fall in black-spotted pond frog numbers (Momose 1998).

Of course, we do not know enough about the mechanisms behind the relationship between heat islands and early cherry blossom blooming or tropical plant cultivation, so these examples do not necessarily prove a direct causal relationship. Moreover, we have the impression that these phenomena do not appear to be particularly relevant to urban living. However, ecosystems consist of complex relationships between organisms, one example being the food chain, so a breakdown in the balance of the ecosystem could eventually result in the extinction or, indeed, a plague of a particular organism. We also run the risk that the natural environment in which we live, including the atmosphere, water, or soil that have long been cleaned by other organisms, will no longer function normally. This could also bring about changes in sanitary conditions, such as the emergence of pests or changes to agricultural production in city suburbs. From this perspective, it is becoming increasingly important to evaluate changes not only in the summer, but also in the winter, spring, and fall.

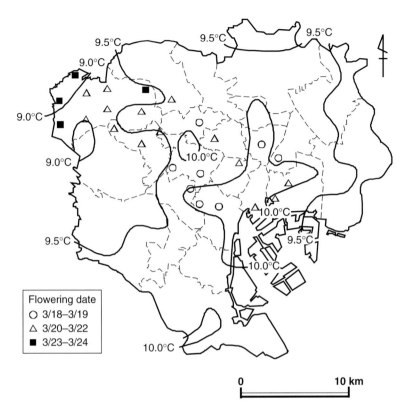

Fig. 15.4. Distribution of flowering dates of Somei Yoshino (cherry tree) and mean temperature in March in the Wards Area in Tokyo in 2004 (based on Matsumoto *et al.* 2006).

In conclusion, it is necessary to conduct careful monitoring and study-based assessments of the impact on natural ecosystems in urban climates. Such assessments need to take a multidisciplinary approach to comprehensive analyses, drawing on expertise from various fields, including meteorology, medicine, ecology, geography, agriculture, and anthropology.

Observations of bio-ecosystems within cities are inexpensive and easily understood methods of appreciating local environmental change and can also be used in observation networks or for monitoring environmental impacts on the citizen level. Moreover, they can act as learning tools for real-life environmental education when schools and other educational establishments are involved.

An understanding of the actual impact of urban warming, as symbolized by heat islands, can play an important role in helping to work out necessary measures (e.g. improving the social environment and preventive measures).

15.3 Measures to Combat Urban Heat Islands

Rising temperatures caused by urbanization can be thought of as anthropogenic climate change and measures to combat such change involve actions to mitigate the effects.

However, the concept of mitigation is not applicable to rising temperatures caused by fierce summer heat unrelated to urbanization. In recent years, we have seen a focus on the concept of adaptation to global warming. Basic measures to combat, or mitigate, heat islands can be understood as adaptation in order to live a cooler, more comfortable, and safer life in hot climates.

15.3.1 Urban heat island mitigation

In Japan, the three main measures currently under consideration to address increasingly high temperatures in urban areas are to (1) reduce anthropogenic heat exhaust associated with energy consumption, (2) change to urban structures and materials on the ground surface that heat the atmosphere less, and (3) ensure airflow through cities. Of these concepts, the measures thought to be the most effective include reducing air-conditioning cooling loads, for example by planting vegetation on buildings (on roofs and walls), use of water-retentive paving, painting walls light colours, and improving the reflective capabilities of roof materials; preserving and developing green tracts of land; developing water channels in parks or flowing small streams through open channels; and relocating large-scale green tracts of land or business facilities (taking into account sea breezes and other prevailing winds) (Fig. 15.5).

Nowadays, regional autonomies have difficulties deciding reasonably (considering their needs and financial situation) their priority on measures for urban heat islands (UHI) they

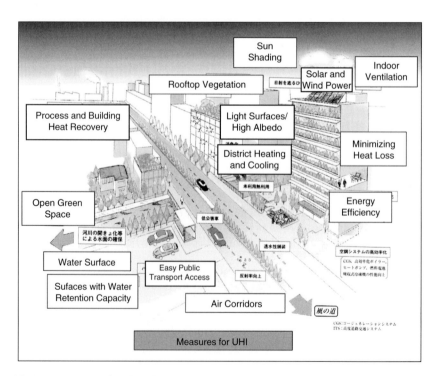

Fig. 15.5. Measures to combat the urban heat island effect (revised from MoE 2000).

actually perform without the information about comparison of measures in terms of their spatial scale, temporal scale, mitigation effect (for nighttime and daytime), cost, and so on. One of the major achievements of the MoE (2001) was a matrix for the evaluation of measures to combat UHI (Table 15.1). It provided basic information on the comparison of several features of each technology as measure. As mentioned above, each technology was classified into three groups of (1) reduction of anthropogenic heat, (2) improvement of anthropogenic coverage, and (3) improvement of urban structure. In general, measures which have larger spatial scale seems to need longer temporal scale, more significant mitigation effect, and higher cost.

Table 15.1.   Example of matrix for evaluation on measures for UHI (MoE 2001).

Measures	Spatial scale	Temporal scale	Effect (nighttime and daytime)	Cost	Responsibility
(1) Reduction of anthropogenic heat					
Energy saving behaviours	building	short	B and B	low	personal-company
Improvement of energy efficiency	building	short-middle	B and B	low	personal-company
Rooftop vegetation	building	short-middle	A and A	low	personal-company
Bright colour painting of building surface	building	short-middle	A and A	low	personal-company
Transportation demand management	district-urban	short-middle	B and C	middle	personal-autonomy
Cascade use of heat	district	middle	B and B	middle	company-autonomy
PV (solar energy use)	building	short-middle	B and B	middle	personal-company
(2) Improvement of anthropogenic coverage					
High albedo and water keeping coverage	district	middle	B and B	middle	company-autonomy
Open space with green	district-urban	middle-long	A and A	middle	company-autonomy
Rooftop vegetation	building	short-middle	A and A	low	personal-company
Arrangement of water surface	district-urban	middle-long	B and A	middle	autonomy
(3) Improvement of urban structure					
Arrangement of urban ventilation path	district-urban	middle-long	B and B	high	autonomy

Effect: A/excellent, B/good, C/fair.

However, its evaluation was not quantitative enough, due to a lack of standard methods to evaluate the mitigation effect of each technology. Which is a more desirable measure, 'Rooftop vegetation' or 'Bright colour painting of building surface'? The high cost of rooftop vegetation in arid climate zones is due to irrigation cost. On the other hand, vegetation has other functions like realizing a better landscape. We conclude, therefore, development of standard methods to evaluate the mitigation effect of each measure is necessary and it has to consider applying conditions.

In 2001, the Japanese government established the Ministries' League (ML) on UHI to promote discussion on the wide-scale mitigation of the urban thermal environment. In 2004, the Fundamental Policy of the Japanese Government on UHI was published. Such action was a world first. These movements of the Japanese government have given much awareness of the local government for UHI issues. In the summer of 2002, the Tokyo Metropolitan Government settled a new monitoring network of UHI with high spatial resolution (120 stations in the 23 wards of Tokyo with a size of 30 km by 30 km). This monitoring network enabled the precise distribution of air temperature and thermal environment to be expressed demographically. It provides high priority information on where measures for UHI in Tokyo are needed.

As yet, there are few actual examples of measures being taken by local municipal authorities in Japan. One of the most famous examples is the recent requirement (since April 2000) in Tokyo for newly constructed buildings to include rooftop vegetation. In Saitama Prefecture, there are also moves to curb urban warming, with a number of trials of rooftop vegetation underway. However, there is still little empirical research on the implementation of such measures and their effects.

Figure 15.6 shows an external view of Keyaki Hiroba (Zelkova Tree Plaza) with rooftop vegetation, built in 2001 in Saitama City, Saitama Prefecture. The facility is located alongside the JR Saitama Shintoshin station. There are 220 Japanese zelkova trees of around 10 m in height growing over an area of approximately 10 200 m^2 on a reinforced concrete

Fig. 15.6. External view of Keyaki Hiroba (Zelkova Tree Plaza) with rooftop vegetation, built in 2001 in Saitama City, Saitama Prefecture, Japan.

roof. Directly under the first floor at ground level, there is a mix of retail and restaurant outlets, while the basement 1 level houses a car park. It is important to evaluate the impact of the rooftop vegetation during summer, when there is strong sunlight and fierce heat. Figure 15.7 shows the temperature distribution across Keyaki Hiroba on 13 August 2001 at noon, the time when the daytime temperature reaches a maximum. Overall, the temperature tends to be higher at the periphery of the square and decreases as you move into the centre. The highest temperature within the square exceeds 33°C at two open-space locations (glass-sided and lawn areas), differing from the area under the zelkova trees by 1–2°C. Therefore, under the trees, we can see the shading effect from the rooftop vegetation. The research also confirms that cool air seeps out into the surrounding area. This research suggests that rooftop vegetation has some effect in mitigating high temperatures. The results can also be interpreted as a method of adaptation, as the shading effect provides the urban residents with a cool and comfortable location.

There is little prospect for a rapid decrease in the use of concrete or asphalt surfaces, or a dramatic increase in green areas, in urban spaces today. In terms of the effective use of limited space, however, there is potential for the improvement of rooftop areas at least.

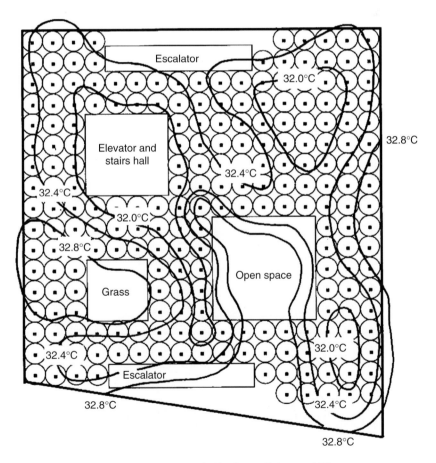

Fig. 15.7. Temperature distribution across Keyaki Hiroba on 13 August 2001 at noon.

15.3.2 Concept of adaptation to heat islands

The concept of adaptation to the summer heat involves the twin objectives of spending time in cooler and more comfortable environments (improving work efficiencies) and minimizing injury from the summer heat (here called 'summer heat injuries').

In terms of the first objective, the Japanese people have a long tradition of protecting themselves from the summer heat through natural adaptation methods. These include the planting of deciduous trees on the western side of homes and watering during the evening. Deciduous plants grown on the west protect from the western sunlight in summer afternoons, but let through the warm sunlight during the winter. This is a useful way of greening walls and can be viewed as a type of adaptation. Figure 15.8 shows wall greening using bitter melon at Ishihara Elementary School in Kumagaya City, Saitama Prefecture. The room where the wall vegetation is used does not require air conditioning at any time of year.

Another type of adaptation is watering in the evening. Any watering done during the day would quickly dry up because of the strong sunlight. Evening watering is an adaptive measure developed by the people to allow them to spend a more comfortable time in the evening, instead of having to endure the daytime heat.

These activities can be thought of as dairy wisdom, but the time may have come when we need to revise our thinking and actively incorporate such concepts.

The other objective of adaptation involves actively avoiding (preventing) risks from summer heat in urban environments, such as heatstroke. If mitigation involves approaches to stop temperatures rising, the concept of adaptation only involves approaches to minimize injury. One example is prevention of heatstroke, which can include the following specific measures:

Fig. 15.8. Wall greening using bitter melon at Ishihara Elementary School in Kumagaya City, Saitama Prefecture, Japan.

1 Adjusting clothing.
2 Cutting back on work outdoors.
3 Changing or shortening exercise times (working times).
4 Providing facilities to allow water to be taken frequently.
5 Caring for one's physical condition, including sleep.
6 Educating people about heatstroke.

Adjusting clothing is significant in that it can also result in savings on air-conditioning costs and a reduction in anthropogenic heat exhaust. Better heatstroke forecasts in actual urban environments are needed in order to pursue the above preventive measures.

Heatstroke forecasts are now provided by the MoE and private meteorological companies. However, these are calculated from forecast temperatures on the lawn in the campus of Meteorological Observatories, which differ from the actual values on artificial surfaces in urban areas. In the future, we may need to develop heatstroke forecasts appropriate for today's urban environments comprising concrete and asphalt surfaces.

If mitigation involves indirect approaches in order to change the surrounding environment to reduce temperatures because of the changing summer heat patterns, then adaptation involves direct approaches on the part of individuals to be personally (actively) aware of summer heat injury.

15.4 Developing Environmentally Friendly Cities

In order to address the worsening problem of heat islands, the MoE, MLIT, Tokyo Municipal Government, and other organizations have started full-scale efforts to promote specific measures by local municipal authorities (Ichinose 2005). These Japanese policy initiatives are the first in Asia and are attracting attention worldwide.

The MoE in particular is aiming to initiate preventive measures, such as the development of environmentally friendly cities or new lifestyles, and is attempting to develop guidelines on such policies.

Such guidelines are revolutionary in terms of providing an official position on mitigating thermal stress for use when local municipal authorities are drawing up urban master plans or environmental plans. However, such thinking on preserving the atmospheric and thermal environments during urban planning has not yet been widely adopted in Japan or across Asia as a whole.

There is a growing need to stress the importance and merits of preserving the thermal environment to urban residents and to get the residents involved in urban planning and development. Ideally, countries and local authorities will provide both urban amenities and summer heat mitigation programs as part of environmental policies and urban planning that actively take into consideration the atmospheric and thermal environments. Examples include:

1 Establishing spaces and cool spots under roadside trees where people can rest (including toilets, arbors, etc.).
2 Riverside spaces where children can play in the water.
3 Ensuring well-ventilated spaces and improved indoor climates.
4 People-friendly paving materials (such as water-retentive materials, building colour schemes that are easy on the eye).

Today, Japan is leading the world in pursuing measures to combat heat islands and, as such, is engaged in various basic research projects to understand the phenomenon and assess its impact and is developing technologies to prevent rising urban temperatures that generate various problems. Some of this research and development has already been applied to actual building construction and basic urban infrastructure. Such moves have picked up pace since the start of the twenty-first century. Here we describe some actual examples from overseas that have triggered these moves in Japan.

15.4.1 Climate analysis for urban planning: measures in Germany

The main environmental problems of Asian megacities arise from the fact that priority is given to economic development more than to environmental protection. It is necessary to control technologies used in creating the urban environment together with methods of urban planning. In Germany, especially in the field of urban planning, many planners take advantage of climatologists' results (Bründl 1988; Horbert *et al.* 1984). In Germany, urban planning with consideration of urban climate, represented by the concept of an urban ventilation lane, is widely practised (e.g. Richter *et al.* 1994). Methods for climatological observations and numerical simulations of the thermal environment are well established in urban planning. In the early 1990s, the Ministry of Environment, the State of Baden-Württemberg, started an air quality control plan (Ministerium für Umwelt, Baden-Württemberg 1991) for its capital city Stuttgart. This plan, also known as the clean air plan for the area of Stuttgart (Luftreinhalteplan Grossraum Stuttgart), includes the concept of the urban ventilation lane as one of its policy options. The plan aims to reduce urban air pollution by natural cold drainage air flows that are to be intensified by suitable alignment of buildings as well as land use zoning based on scientific data (Wirtschaftsministerium, Baden-Württemberg 1998). The plan also aims to improve thermal comfort within this urban area. To realize such a plan, relevant procedures are standardized as VDI (1997). In Japan, many cities are located in coastal zones. The ventilation strategy of Stuttgart is aimed mainly at solving air pollution in winter. Stuttgart is a typical inland basin-bottom city. In the case of Japanese coastal cities like Tokyo, stronger wind speeds are expected. However, they suffer severely from thermal sensation in the summer. We need, therefore, to consider the difference between climatic conditions and natural locations when applying such a German strategy to Japanese coastal cities.

In Germany, there are legal regulations (such as the Baugesetzbuch; Federal Building Code) to prevent localized climate change or atmospheric pollution (Wirtschaftsministerium, Baden-Württemberg 1998). As a result, the country has put into practice programs such as city-scale improvements to the atmosphere, as typified by Stuttgart's urban ventilation lane, as well as feedback on land block design through the mapping of wind-chill temperatures and web-based provision of urban climate information to residents. In particular, Japan is introducing the famous Stuttgart urban ventilation lane (Nemoto 1991; Ichinose 1999). Based on a detailed wind survey, use is being made of Germany's uniquely rigorous urban planning system to promote urban development plans, in order to introduce clean air flows into urban areas. As a result, a flow of cool air generated by nocturnal radiation cooling in the hill country blows through the urban areas and functions as a natural way to mitigate environmental problems such as heat islands and atmospheric pollution.

Many of the examples of Germany's unique environmentally friendly urban planning techniques (especially those involving urban air quality and thermal environmental controls) have been used for inland cities, mainly to decrease atmospheric pollution loads in winter caused by a basin-bottom location and weak solar radiation. Today, however, there is a growing focus on mitigating summer heat. It should be particularly noted here that general procedures for implementing urban plans take into account climate conservation for a particular city by giving due consideration to the work of climatologists and by introducing practical, working-level systems. This could be described as 'climate analysis for urban planning'. Another commonly used term to describe this in German is Klimaanalyse. Such analyses result in the drawing up of Planungshinweiskarte (advice for urban planning communicated in the form of a map or as a diagrammatic plan guide). The planner has legally binding authority and is referred to when drawing up a locally detailed plan (B-Plan) to dictate building forms, with suggestions made to reflect its content.

A Klimaatlas (climate atlas) is a general name describing a collection of thematic maps from various surveys using the above procedures. There are many real-life examples from Northern Germany and the Ruhr region, including Essen (1985), Münster (1992), and Düsseldorf (1995). Currently, these practical systems have been standardized (DIN: German engineering standards) by the Association of German Engineers (Verein Deutscher Ingenieure or VDI), which has also published guidelines. The maps include environmental factors (weather elements or atmospheric pollutants) used in the process of analysis and explanatory remarks to figures (zoning systems, etc.) (VDI 1997). Of the VDI standards, those on urban air quality and thermal environment are particularly well known.

- VDI3784 Blatt 1: Cooling Tower Impact (1986)
- VDI3786 Blatt 9: Visual Weather Observations (1991)
- VDI3789 Blatt 2: Calculation of Short-wave and Long-wave Radiation (1994)
- VDI3787 Blatt 1: Climate and Air Pollution Maps for Cities and Regions (1997)

'Feng Shui' is a discipline or a system of knowledge established in ancient Asia. In 'Feng Shui', there are many statements on the control of the ambient environment through methods of urban or building planning. It also included items on the urban thermal environment. Urban planning with consideration of urban climate practising in German cities seems to be a German form of 'Feng Shui' or modern western 'Feng Shui'. In this case, its application to Asian megacities for mitigating serious urban thermal pollution would be successful because Asia has a long history of traditional 'Feng Shui'. To import the modern German 'Feng Shui' into Asia is a new idea of mitigation. Of course, not only do we need to study the climatological features of Asian cities but we must also compare the legal systems supporting urban planning in Germany and Asian countries.

15.4.2 Cool city surfaces: the US approach

In contrast, the US is developing technologies for individual applications. The most famous are 'cool roofs' and 'cool pavements' as applied to building exteriors and roads. These technologies are being actively researched at the Lawrence Berkeley National Laboratory. The concept is to reduce surface temperatures by applying light colours or highly reflective paints to the surfaces of buildings or road paving. The laboratory is compiling a database of the characteristics of commercially available roof materials (solar radiation reflectance and long-wave radiation emissivity) and is developing new products (Moriyama and Takebayashi 2003). The basic research involves collecting analyses of ground surface material composition on a city-wide scale, targeting cities on the west coast of the US (Akbari *et al.* 1999). The Environmental Protection Agency (EPA) has also issued reports (EPA 1992). Many people associate coastal Mediterranean cities with white-walled buildings. Painting city surfaces in light colours is an effective method in regions such as the US west coast or the Mediterranean coast where there is little rainfall and irrigation to maintain urban green zones is difficult.

Other standards that are known and used globally include new standard effective temperature (SET*) to determine comfortable temperature ranges and the indoor thermal and atmospheric environmental standards from the American Society of Heating, Refrigerating and Air-Conditioning Engineers (ASHRAE).

15.4.3 Cheong-Gye Stream restoration: large-scale stream restoration project in Seoul

To what extent is urban thermal sensation mitigated by the commonly proposed environmentally friendly urban planning initiatives such as large-scale restoration of urban stream? The Cheong-Gye Stream used to run from west to east across central Seoul for around 11 km before flowing into the Han River. However, the density of Seoul's population at the start of the twentieth century created sanitation problems around the Cheong-Gye Stream. In order to address these problems, work was started in the late 1950s to completely cover over the stream, turning it into a culvert under a road, thereby promoting increasing urbanization and traffic flow along the stream. By the early 1970s, some 6 km of the stream had been converted into a four-lane elevated road (double-decked road).

However, in recent years, a number of safety and other problems emerged with the elevated road and the structure of the covering itself. Moreover, the emphasis is now on developing more environmentally friendly cities with a focus on the residents and nature, so the urban residents seized the opportunity and became involved in a campaign to redevelop the Cheong-Gye Stream on a grand scale, restoring it to an open flowing stream that provides an urban space for water lovers. Because of this campaign, the Seoul Metropolitan Government decided to scrap several kilometres of the elevated road and restore the river within the city (the Cheong-Gye Stream) to its original state. The Cheong-Gye Stream restoration project started on 1 July 2003 by removing the elevated road but leaving riverside roads (one-way roads of two to four lanes) in place (Fig. 15.9). The culverts directly under the elevated roads were then dug out; the final step was to create green water facilities.

Various other cities around the world are extremely interested in urban environmental redevelopment with an emphasis on mitigation methods that regenerate a highly natural biotope in an urban setting. However, there are no other examples anywhere in the world of such large-scale river regeneration projects within a city. Of particular interest with this project is its impact on improving the environment, in terms of a cleaner atmosphere from the reduced traffic flow and mitigation of the thermal sensation (curbing rising temperatures) in areas around the river (Ichinose *et al.* 2006).

Fig. 15.9. Landscape around Dongdaemun before and after the restoration of the Cheong-Gye Stream.

The restoration project was completed at the end of September 2005 and many neighbourhood residents noted a significant reduction in dust levels compared with before. Even after the completion of the river restoration project, work is proceeding rapidly to redevelop the areas around the river, which have appreciated in value, into attractive commercial districts, so continued monitoring of atmospheric and thermal environments is necessary in our view. We may see a global paradigm shift in urban development if the restoration of the Cheong-Gye Stream can be shown to have mitigated summer heat.

In Japan, owners of small patches of land have a right to decide how they use their own land. However, the top-down approach in urban development still dominates many Asian countries. Such aggressive social experiments to promote ideal urban construction in a viewpoint of urban climate protection seem to be possible in these countries. Before this restoration project, the author's group has started monitoring the effect of the restoration (Ichinose *et al.* 2006). This is the first chance in the world to notify the atmospheric impact by such drastic change of surface coverage in a real urban space, not in a numerical simulation.

In the future, the spotlight will focus even more on the values of developing environmentally friendly cities and creating amenities. In addressing these issues, we think it will be important to perform more in-depth studies to assess the actual impact measures have on urban thermal environments and to consider and promote effective measures, such as mitigation and adaptation, in response to summer heat, in order to develop cities that take into account the environment from the perspective of preserving the atmospheric and thermal environments.

References

Akbari, H., Rose, L.S. and Taha, H. (1999). *Characterizing the Fabric of the Urban Environment: A Case Study of Sacramento, California.* LBNL-44688.

Bründl, W. (1988). Climate function maps and urban planning. *Energy and Buildings*, **11**, 123–127.

EPA (1992). *Cooling Our Communities: A Guidebook on Tree Planting and Light-Colored Surfacing.* Lawrence Berkeley Laboratory Report; Lbl-31587. US Gov. Printing Washington, DC.

Horbert, M., Kirchgeorg, A. and von Stülpnagel, A. (1984). On the method for charting the climate of an entire large urban area. *Energy and Buildings*, **7**, 109–116.

Ichinose, T. (1999). Klimaanalyse: climate analysis for the urban planning in Germany. *Tenki*, **46**, 709–715 (in Japanese).

Ichinose, T. (2005). Recent counteractions for urban heat island in regional autonomies in Japan, in *Urban Dimensions of Environmental Change: Science, Exposures, Policies and Technologies.* pp. 161–167. Science Press.

Ichinose, T., Bai, Y., Nam, J.-C. and Kim, Y.-H. (2006). Mitigation of thermal stress by a large restoration of inner-city river (Cheong-Gye Stream in Seoul). *Proc. of ICUC-6*, pp. 358–361.

Ichinose, T., Shimodozono, K. and Hanaki, K. (1999). Impact of anthropogenic heat on urban climate in Tokyo. *Atmospheric Environment*, **33**, 3897–3909.

Landsberg, H.E. (1981). *The Urban Climate.* Academic Press, New York.

Matsumoto, F. (2004). Relationship between urban climate and plant phenology in Kumagaya City (2) – leaf-color-change of *Acer palmatum* measured with chlorophyll meter. *Japanese Journal of Biometeorology*, **40**, 337–349 (in Japanese with English abstract).

Matsumoto, F. and Fukuoka, Y. (2002). The relationship between urban climate and plant phenology in Kumagaya City (1) – in the case of leaf-color-change date of *Gingko biloba* and *Acer palmatum*. *Japanese Journal of Biometeorology*, **39**, 3–16 (in Japanese with English abstract).

Matsumoto, F. and Fukuoka, Y. (2003). Effects of urban warming on plant phenology: the *Prunus yedoensis* flowering date in Kumagaya City. *Geographical Review of Japan*, **76**, 1–18 (in Japanese with English abstract).

Matsumoto, F., Mikami, T. and Fukuoka, Y. (2006). Effects of heat island on the flowering dates of *Prunus yedoensis*: case study in the wards of Tokyo. *Geographical Review of Japan*, **79**, 322–334. (in Japanese with English abstract).

Ministerium für Umwelt, Baden-Württemberg (1991). *Luftreinhalteplan Großraum Stuttgart 1991. Teil 1 – Emissionen, Immissionen, Wirkungen*. Ministerium für Umwelt, Baden-Württemberg, Stuttgart (in German).

MoE (Japan's Ministry of the Environment) (2001). *Report of UHI Council on Analyses* (author's translation). MoE, Tokyo (in Japanese).

MoE (Japan's Ministry of the Environment) (2000). *Report of UHI Council on Methods for Counteractions* (author's translation). MoE, Tokyo (in Japanese).

Momose, N. (1998). *The Four Seasons – Animals and Plants Seasonal Behavior* (author's translation). Gihodo Shuppan (in Japanese).

Moriyama, M. and Takebayashi, H. (2003). Heat island mitigation measures in Germany and the U.S.A. *Environmental Information Science*, **32**, 37–41 (in Japanese).

Nemoto, T. (1991). Germany: ecologically developed country (author's translation). *Nikkei Architecture* 9–30, 206–208 (in Japanese).

Nemoto, M., Murayama, H. and Suzuki, K. (2001). The influence of urban climate on the growth and development of *Aloe arborescens* Mill. at the outdoor culture. *Papers on Environmental Information Science*, **15**, 255–260 (in Japanese with English abstract).

Numata, M. (1987). *Urban Ecology* (author's translation). Iwanami Shoten (in Japanese).

Richter, C.J., Röckle, R., Goßmann, H. and Todt, T. (1994). Berücksichtigung klimatischer Belange bei einem städtebaulichen Ideenwettbewerb. *UVP Report* 1994 Nr. 5 (in German).

Shinada, Y., Tachibana, N. and Sugiyama, K. (1987). *Human Environment of Urban Area* (author's translation). Kyoritsu Shuppan (in Japanese).

Tokyo Metropolitan Government (2006). Transported to hospital with heatstroke in Tokyo. <http://www2.kankyo.metro.tokyo.jp/heat/heat1.htm> (in Japanese).

VDI (1997). *Umweltmeteorologie. Klima- und Lufthygienekarten für Städte und Regionen*. VDI-Richtlinien VDI3787 Blatt 1. VDI, Düsseldorf (in German and in English).

Wirtschaftsministerium, Baden-Württemberg (1998). *Städtebauliche Klimafibel. Hinweise für die Bauleitplanung*. Wirtschaftsministerium, Baden-Württemberg, Stuttgart (in German).

Yoshino, M. and Park, H.S. (1996). Variations in the plant phenology affected by global warming, in K. Omasa, K. Kai, H. Taoda, Z. Uchijima and M. Yoshino) (eds), *Climate Change and Plants in East Asia*. Springer Verlag.

Chapter 16
Ecodesign and the Transition of the Built Environment

KEN YEANG

Ecodesign constitutes a fundamental redefinition of building and urban design. The abandoning of the design paradigms inherited from the age of oil is distinguished from the previous performance and hardware-driven approaches to building augmentation, to fundamentally re-examine the climatic, material and biological framework of cities, infrastructure and buildings.

Ecodesign is design for bio-integration. It can be regarded as having three facets: *physical, systemic* and *temporal*. Addressing each of these facets successfully is the challenge of building and urban designers.

Ecodesigners look at nature for guidance. Nature in its entirety without humans exists in stasis. Can our cities, communities, businesses and our built environment imitate nature's processes, structures, and functions? Natural ecosystems have no waste; everything is recycled within. Thus by imitating ecosystems the built environment should produce no waste; all emissions and products would be continuously reused or recycled and eventually reintegrated with the natural environment. Designing to imitate ecosystems is *ecomimesis*. This is the fundamental premise for ecodesign: the built environment can imitate ecosystems in all respects.

Buildings are enclosures erected to protect humans, their assets and livestock from inclement weather and enable activities – whether residential, office, manufacturing, warehousing, etc. – to take place. Seen from an ecological point of view, a building represents a high concentration of materials extracted and manufactured, using largely non-renewable energy resources, from some distant place in the biosphere and transported to a particular location and assembled into a built form or an infrastructure – road, bridge, sewer etc. – whose subsequent operations create further environmental consequences and whose eventual after-life must also be accommodated.

There is a great deal of confusion and misperception as to what exactly constitutes ecological design. It is easy to be misled by technology and to think that if we assemble enough eco-gadgetry such as solar collectors, photovoltaic cells, biological recycling systems, building automation systems and double-skin facades in one single building that this can automatically be considered ecological architecture. Although these technologies are commendable applications of low energy systems they are merely useful components leading towards ecological architecture; they represent some of the means of achieving an ecological end product. Ecological design is not just about low energy systems; to be fully effective these technologies need to be thoroughly integrated into the building fabric; they will also be influenced by the physical and climatic conditions of the site. The nature of the problem is therefore site specific: there will never be a standard 'one size fits all' solution.

The other misperception is that if a building achieves a high score on a green rating scale then all is well. Green rating systems are useful in publicizing certain goals – however, they should be considered as threshold standards that designers should aim at achieving and exceeding.

Ecodesign should be viewed as the design of the built environment as just one system within the natural environment. The system's existence has ecological consequences; the way it functions and interactions with other systems over its entire life cycle must be benignly integrated with the natural environment. In this way it is the life cycle analysis of the system, rather than its value at any one particular point in time, that gives a better idea of its cumulative effect on its neighbouring systems.

Ecosystems are definable units in a biosphere; as such they should contain both biotic (living) and abiotic (non life-supporting) constituents acting together as a whole. Following this model our businesses and our built environment should be designed analogously to the ecosystem's physical content, composition and processes. For instance, besides regarding buildings as currently done as artistic aspirations or serviced enclosures, we should regard them as artifacts that need to be operationally integrated with nature. It should be self-evident that the material composition of our built environment is almost entirely inorganic, whereas ecosystems contain a complement of both biotic and abiotic constituents, i.e. organic and inorganic components.

The enormous number of existing buildings as well as our current manufacturing and processing activities are making the biosphere more and more inorganic and increasingly simplified biologically. To continue doing what we have always done without balancing the abiotic with the biotic content means simply adding to the biosphere's artificiality, thereby making it increasingly inorganic and reducing its complexity and diversity. We must first reverse this trend by starting to balance our built environment with greater levels of biomass; by ameliorating biodiversity and ecological connectivity in the built forms and by complementing their inorganic content with appropriate organic biomass.

We should improve the ecological linkages between our activities, be they design or business processes, with the surrounding landscape in ways that connect them both horizontally and vertically. Achieving these linkages ensures a wider level of species connectivity, interaction, mobility and sharing of resources across boundaries. Such real improvements in connectivity enhance biodiversity and further increase habitat resilience and species survival. An obvious demonstration of horizontal connectivity is the provision of ecological corridors and linkages in regional planning which are crucial in making urban patterns more biologically viable. Besides improved horizontal connectivity, vertical connectivity within the built form is also necessary since most buildings are not single storey but multi-storey. Design must extend ecological linkages vertically from the foundations to the rooftops.

More important than the enhancement of ecological linkages is the biological integration of the inorganic products inherent in the built environment with the landscape so that the two become mutually *ecosystemic*. In this way we can create 'human-made ecosystems' compatible with nature's ecosystems and by doing so we will enhance the ability of human-made ecosystems to sustain life in the biosphere.

Ecodesign is also about discerning the ecology of a site; any building or business activity should take place with the objective of interacting productively with an ecosystem. In the case of site planning we must first understand the properties of the locality's ecosystem before imposing any intended human activity upon it. Every site has an ecological system with a limited capacity to withstand the stresses imposed upon it; if stressed beyond this capacity the ecology will be damaged irrevocably. Stress can be caused just as much by minimal localized impact – such as the clearing of a small land area for access – as by the total devastation of the entire landscape, such as the clearing of all trees and vegetation, levelling the topography and the diversion of existing waterways.

To identify the capacity of a site to withstand human intervention an analysis of the existing ecology should be carried out; we must ascertain, for example, the structure of the site's ecosystems, energy flow and species diversity. Then we must identify which parts of the site, if any, have different ecosystems and which parts are particularly sensitive. Finally, we must consider the likely impact of the intended construction and use. This is, of course, a major undertaking; however, it needs to be done to better understand and appreciate the nature of a site. To be thorough and effective this type of detailed analysis should be carried out diurnally and seasonally over a period of a year or more. To reduce this lengthy process landscape architects have developed the 'layer-cake' method; this sieve-mapping technique enables designers to map the landscape as a series of separate layers that provide a simplified matrix for the investigation of a site's ecology.

As the layers are mapped they can be overlaid and the interaction of the layers can be evaluated in relation to the proposed land use. The final product of this study is a composite map that can be used to guide the proposed site planning (e.g. the disposition of the access roads, water management, drainage patterns and shaping of the built forms). It is important to understand that the sieve-mapping method generally treats the site's ecosystems statically and may ignore the dynamic forces taking place between the layers within an ecosystem. As mentioned above the separation of the layers is a convenient intellectual construct that simplifies the complex natural interactions between layers. Therefore the comprehensive analysis of an ecosystem requires more than sieve mapping – the inter-layer relationships should also be examined.

As designers we should also look into ways of configuring built forms, the operational systems for our built environment and our businesses as low energy systems. In addressing these systems we need to look into ways of improving the internal comfort conditions of our buildings. There are essentially five ways of doing this: *Passive Mode*, *Mixed Mode*, *Full Mode*, *Productive Mode* and *Composite Mode*, the latter being a composite of all the preceding modes.

The practice of sustainable design requires that we look first at Passive Mode (or bioclimatic) design strategies, then we can move on to Mixed Mode, Full Mode, Productive Mode and Composite Mode, all the while adopting progressive strategies to improve comfort conditions relative to external conditions.

Contemporary expectations for office environment comfort conditions cannot generally be met by Passive Mode or by Mixed Mode alone. The internal environment often needs to be supplemented by the use of external sources of energy, as in Full Mode. Full Mode uses electromechanical systems often powered by external energy sources – whether from fossil fuel derived sources or from local ambient sources such as wind or solar power.

Passive Mode means designing for improved internal comfort conditions over external conditions without the use of any electromechanical systems. Examples of Passive Mode strategies include the adoption of suitable building orientation and configuration in relation to the local climate as well as the selection of appropriate building materials. When considering the design of the facade issues of solid-to-glazed area ratios, thermal insulation values, the incorporation of natural ventilation and the use of vegetation are also important.

Building design strategy must start with Passive Mode or bioclimatic design as this can significantly influence the configuration of the built form and its enclosure systems.

Passive Mode requires an understanding of the climatic conditions of the locality; the designer should not merely synchronize the building design with the local meteorological conditions but optimize the ambient energy of the locality to create improved internal comfort conditions without the use of any electromechanical systems. The fundamental nature of these decisions clearly dictates that once the building configuration, orientation and enclosure are considered the further refinement of a design should lead to the adoption of choices that will enhance its energy efficiency. If, as an alternative, a design solution is developed that has not previously optimized the Passive Mode options then these non-energy efficient design decisions will need to be corrected by supplementary Full Mode systems. Such a remedy would make a nonsense of low energy design. Furthermore, if the design optimizes a building's Passive Modes, it remains at an improved level of comfort during any electrical power failure. If the Passive Modes have not been optimized then whenever there is no electricity or external energy source, the building may be become intolerable to occupy.

In Mixed Mode buildings use some electromechanical systems such as ceiling fans, double facades, flue atriums and evaporative cooling.

Full Mode relies entirely on the use of electromechanical systems to create suitable internal comfort conditions. This is the option chosen for most conventional buildings. If clients and users insist on having consistent comfort conditions throughout the year the result will inevitably lead to Full Mode design. It must be clear now that low energy design is essentially a user-driven condition and a lifestyle issue. We must appreciate that Passive Mode and Mixed Mode design can never compete with the comfort levels of the high energy, Full Mode conditions.

Productive Mode is where a building generates its own energy. Common examples of this today can be seen in the generation of electricity through the use of photovoltaic panels that are powered by solar power and wind turbines that harness wind energy. Ecosystems use solar energy that is transformed into chemical energy by the photosynthesis of green plants which in turn drive the ecological cycle. If ecodesign is to be ecomimetic, we should seek to do the same; however, we will need to do so on a much larger scale.

The inclusion of systems that create Productive Modes inevitably lead to sophisticated technological systems that in turn increase the use of material resources, the inorganic content of the built form, the embodied energy content and the attendant impact on the environment.

Composite Mode is a combination of all the above modes in proportions that vary over the seasons of the year.

Ecodesign also requires the designer to use materials and assemblies that facilitate reuse, recycling and their eventual reintegration with ecological systems. Here again we need to be ecomimetic in our use of materials in the built environment: in ecosystems, all living organisms feed on continual flows of matter and energy from their environment to stay alive, and all living organisms continually produce 'waste'. However, ecosystems do not actually generate waste since one species' waste is really another species' food. Thus matter cycles continually through the web of life. To be truly ecomimetic the materials we produce should also take their place within the closed loop where waste becomes food.

Currently we regard everything produced by humans as eventually becoming garbage: waste material that is either burned or ends up in landfill sites. The new question for designers, manufacturers and businesses is: how can we use this waste material? If our materials are readily biodegradable, they can return into the environment through decomposition. If we want to be ecomimetic we should think, at the very early design stages, how a building, its components and its outputs can be reused and recycled. These design considerations will determine the materials to be used, the ways in which the building fabric is to be assembled, how the building can be adapted over time and how the materials can be reused after the building has reached the limits of its useful life.

If we consider the last point, reuse, in a little more detail we come to an increasingly important conclusion. To facilitate the reuse of, let us say, a structural component, the connection between the components should be mechanical, i.e. bolted rather than welded so that the joint can be released easily. If, in addition to being easily demountable, the components were modular then the structure could be easily demounted and reassembled elsewhere. This leads to the concept of Design for Disassembly (DfD) which has its roots in sustainable design.

Another major design issue is the systemic integration of our built forms, operational systems and internal processes with the natural ecosystems that surround us. Such integration is crucial because without it these systems will remain disparate artificial items that could be potential pollutants. Unfortunately many of today's buildings only achieve eventual integration through biodegradation that requires a long-term process of natural decomposition.

While manufacture and design for recycling and reuse relieves the problem of deposition of waste, we should integrate both the organic waste (e.g. sewage, rainwater runoff, wastewater, food wastes, etc.) and the inorganic waste.

There is a very appropriate analogy between ecodesign and surgical prosthetics. Ecodesign is essentially design that integrates human-made systems both mechanically and organically with the natural host system – the ecosystems. A surgical prosthetic device also has to integrate with its organic host being – the human body. Failure to integrate will result in dislocation in both cases. These are the exemplars for what our buildings and our businesses should achieve: the total physical, systemic and temporal integration of our human-made, built environment with our organic host in a benign and positive way. There are, of course, a large number of theoretical and technical problems to be solved before we have a truly ecological built environment; however, we should draw encouragement from the fact that our intellect has allowed us to create prosthetic organs that can integrate with the human body. The next challenge will be to integrate our buildings, our cities and all human activities with the natural ecosystems that surround us.

The answer to a sustainable green future will not be in designing green buildings alone. It has to be at the urban dimension and at the city and regional level. The challenge is to take the basic principles elucidated here and to apply these at the city level to create ecocities. Action ultimately has to be at the political level. Governments must recognize the reality of the current environmental condition and take action to realize this vision.

Chapter 17

"Energy-Contracting" to Achieve Energy Efficiency and Renewables using Comprehensive Refurbishment of Buildings as an Example

A Guide for Building Owners and ESCos

JAN W. BLEYL-ANDROSCHIN[1] AND DANIEL SCHINNERL[2]

[1]*IEA dsm Task XVI 'Competitive Energy Services' Operating Agent;* [2]*c/o Graz Energy Agency Ltd, Kaiserfeldgasse 13/1, 8010 Graz, Austria, Email: bleyl@grazer-ea.at; schinnerl@grazer-ea.at, Tel.: +43-316-811848-0 Fax: +43-316-811848-9, http://www.grazer-ea.at email:*

17.1 Motivation and Introduction

Residential and commercial buildings are major consumers of final energy – and waste it at an alarming rate. Twenty-one per cent of global greenhouse gas emissions or 8.2 Giga tonnes of CO_{2eq} per year can be ascribed to the operation of the worldwide building stock – construction and disposal of the buildings not accounted for.[1] The share in electricity consumption is even more than double: 53% of the world's total electricity consumption is consumed in buildings, quoting the IEA World Energy Outlook 2006.[2]

While new building construction rates range between less than one per cent in an average city to over 10% in booming regions, only some new buildings benefit from model energy performance. The majority of saving potential must be realized in the vast and already existing building stock. It is here where a major effort in the urban energy transition process must be made.

Economic saving potential for building energy efficiency refurbishment measures are high: According to Vattenfall and McKinsey the greenhouse gas abatement potential in the building sector is 3.7 Giga tonnes of CO_{2eq} per year by 2030 or 45% across all building types with measures such as 'improved building insulation, better heating and cooling efficiency, energy efficiency in lighting and appliances'. And what might it cost? Quoting the same source the

---

[1] Vattenfall (2007). *Global Mapping of Green House Gas Abatement Opportunities up to 2030.*
[2] World Energy Outlook 2006.

'marginal abatement cost curve is negative $(-160 \; \text{€}/\text{t} \; CO_2)$', which means that implementation of the saving measures will result in a net positive cash flow over a term of 25 years.[3]

Countries and organs of the European Union (EU) and other regions of the world have embraced increasingly forceful measures and support programs to aim at improving the performance of the existing building stock. In this context Energy-Contracting[4] is being promoted as an important implementation tool for Energy Efficiency (EE).

In some European countries, Energy Performance Contracting (EPC) agreements between clients and contractors (ESCos) are entered into to implement building refurbishment projects with quantifiable savings and contractual long term guarantees. When implemented properly, they have successfully delivered guaranteed savings since they were first established in Europe in the mid-1990s.[5] The new European Union (EU) Directive on Energy End-use Efficiency and Energy Services[6] (EE + ES Directive) supports the Energy-Contracting concept and views it as an important instrument to implement energy efficiency (also termed Energy End-use Efficiency) based on market instruments.

EPC projects realize demand reduction measures that typically comprise building technologies like heating, ventilation, air conditioning (HVAC), lighting, electrical applications and control systems. In most cases, building construction measures[7] such as building envelope refurbishment or passive solar shading measures are excluded. A comprehensive refurbishment (CR) approach to buildings – examining and treating all energy sensitive aspects – is frequently not aimed at. As a result, large saving potentials are neglected in the refurbishment process and they cannot be tapped until the next building refurbishment cycle comes some 30 years later.

Obstacles such as the absence of full cost calculations, no integrated planning, too long payback periods of the energy efficiency investment measures, procurement problems or a lack of knowledge on implementation models are some of the reasons behind this.

For many building refurbishment projects, improvements in energy efficiency are not the driving force. Non-energy goals and benefits like space use efficiency and expansion, increased access or ergonomic workplace comfort, space use efficiency and expansion, external appearance or other ways of lifting income from rent may be more important to the building owner. Nevertheless minimum performance standards for thermal refurbishment and guarantees for maximum energy consumption should always be written into the terms of reference for any building refurbishment. CR-EPC models as described here are a good means to secure both general and comprehensive energy efficiency improvement goals. They are applicable to Public-Private Partnership like sale and leaseback projects as well.

In this contribution, we describe models and how to integrate building construction refurbishment measures into EPC models in order to achieve a comprehensive refurbishment of buildings as indicated above. We propose to call these 'Comprehensive Refurbishment EPC' models (CR-EPC). Three basic CR-EPC models are introduced: a 'General Contractor' (GC

---

[3] See footnote 1.

[4] Also referred to as "ESCo or Energy Service". We prefer the term "Energy-Contracting" to emphasize the difference to a standard fuel supply or maintenance contract, which does not imply any outsourcing of risks or provision of guarantees for the overall system performance (see also Figure 17.1).

[5] Some references to successful examples in public buildings in Austria and Germany: www.grazer-ea.at www.contracting-offensive.de, www.berliner-e-agentur.de (partly in German)

[6] Directive 2006/32/EC of the European Parliament and of the Council of 5 April 2006.

[7] By *building construction measures* we understand measures like refurbishment of facades, windows or passive shading, whereas standard Energy-Contracting measures are *building technologies* like HVAC, lighting or controls.

CR-EPC), a 'General Planner' (GP CR-EPC) and a comprehensive Refurbishment 'Light' (CR 'light' EPC) model, the latter for a reduced scope of refurbishment measures.

The following key features of the three basic models are described in more detail in this contribution: Typical measures, key actors, responsibilities and basic contractual relationships; public and corporate procurement implications; important requirements on the various project partners; contractual guarantees and quality assurance instruments as well as advantages and disadvantages of the different models. We also give some comments on financing options. To sum up, the contribution gives conclusions and recommendations for the implementation of CR projects and a short outlook on future research and development activities.

Not covered in this contribution are payback periods of different CR building measures, subsidy programs or contractual details. We assume that these aspects have no direct impact on the basic selection of the implementation model and leave these topics (and many others) to further elaborations.

Methodologically, the contribution mainly builds on practical Comprehensive Building Refurbishment and Energy Performance Contracting project experiences, developed and implemented by Graz Energy Agency Ltd, Austria. It is supplemented with EPC experiences from the Berlin Energy Agency and the Austrian Energy Agency (former E.V.A).

The groundwork for this contribution has been laid with a systematic description of six existing and planned CR-EPC projects and an evaluation of the experiences made.[8] Earlier basics for this work have been established in the 'CONZUK' project,[9] which have been summarized by Tritthart *et al.*[10] The latter paper also documents three of the six CR project examples mentioned before. Additionally talks with stakeholders such as real estate owners, ESCos and others have been conducted.

This contribution has received support from a number of institutions and individuals and we thank for financial assistance the Intelligent Energy – Europe Programme[11] and the Austrian 'Lebensministerium'.[12] Friedrich Seefeldt and the EUROCONTRACT partners,[13] especially Vollrad Kuhn, Berlin Energy Agency and Marton Varga, Austrian Energy Agency have given helpful comments. The work has been continued within Task XVI 'Competitive Energy Services' run by the IEA (International Energy Agency) Demand Side Management Implementing Agreement (http://dsm.iea.org/).

The findings of this contribution have to be considered as a work in progress, due to the limited practical experiences collected so far. The authors at Grazer Energy Agency Ltd expressly invite feedback and inquiries, please contact Jan W. Bleyl-Androschin (bleyl@ grazer-ea.at).

17.2 Energy-Contracting: Implementation Tool for Energy Efficiency and Renewables. Extended to Comprehensive Refurbishment of Buildings

We focus on some key elements and definitions here, assuming that the reader has a basic knowledge of the Energy-Contracting concept and building energy efficiency. Some

---

[8] Bleyl, Jan W., Kuhn, V. and Schinnerl, D. (2007). *Comprehensive Refurbishment of Buildings with Energy Performance Contracting*. EUROCONTRACT manual. Graz Energy Agency.

[9] Bucar, G., Baumgartner, B., Tritthart, W. *et al.* (2004). *Contracting als Instrument für das Althaus der Zukunft*. Graz Energy Agency.

[10] Tritthart, W., Bleyl, Jan W., Bucar, G. and Bruner-Lienhart, S. (2007). *Contracting and Building Renovation – Does it Work Together?* In ECEEE Summer Study 2007 proceedings, paper id. 5200.

[11] http://www.ec.europa.eu/energy/intelligent/index_en.html

[12] http://www.umwelt.lebensministerium.at/

[13] www.eurocontract.net

further references can be found here: www.grazer-ea.at, www.contracting-portal.at, 'Leitfaden Energiespar-Contracting' published by dena[14] or from the brochure 'Die Energi esparpartnerschaft. Ein Berliner Erfolgsmodell'.[15]

Generally any design approach should first of all focus on energy conservation by evaluating all possible demand reduction opportunities, including the building envelope. Only afterwards the remaining demand should be supplied as efficiently as possible – including renewable supply options. This requires an integrated planning concept. A good example for this approach is the reduction of all electrical and thermal cooling loads including solar shading options before assessing an air conditioning unit.

The Energy Performance Contracting (EPC) concept shifts the focus away from the sale of the units of fuel or electricity towards the desired benefits and services derived from the use of the energy, e.g. the lowest total cost of keeping a room warm, air conditioned or lit (useful energy). The EPC-model, aims at providing useful energy at minimal project- or life cycle or project cost to the end user. And it achieves environmental benefits due to the associated energy and emissions savings.

The EC Directive on 'Energy End-use Efficiency and Energy Services' defines Energy Service as

> *the physical benefit, utility or good derived from a combination of energy with energy efficient technology and/or with action, which may include the operations, maintenance and control necessary to deliver the service, which is delivered on the basis of a contract and in normal circumstances has proven to lead to verifiable and measurable or estimable energy efficiency improvement and/or primary energy savings.*

The Directive also defines an 'Energy Service Company' (ESCo) as an organization that

> *delivers energy services, energy efficiency programmes and other energy efficiency measures in a user's facility, and accepts some degree of technical and sometimes financial risk in so doing. The payment for the services delivered is based (either wholly or in part) on meeting quality performance standards and/or energy efficiency improvements.*

At Energy-Contracting, facility owner and ESCo enter into a long-term contractual relationship. Short-term focusing on profit will not lead to success for either of the parties involved. The term 'Energy Saving Partnership', which has been given to the energy performance contracting campaign of the Berlin Senate mentioned above, expresses this well.

As for Energy-Contracting, transfer of technical and commercial implementation and operating risk as well as takeover of function, performance and price guarantees by the ESCo play a crucial role. These elements create added value compared to in-house solutions and are guaranteed in the EPC contract. In other words: contracting is more than putting together individual components. The contracting concept incorporates incentives and guarantees, that – throughout the contract term – the entire system performs according to specifications.

The central elements of an Energy-Contracting package are summarized in Fig. 17.1.

Most projects are unique in one way or another and require an individual adaptation of the model. Energy-Contracting is a service package that can and must be arranged specifically

---

[14] Deutsche Energie Agentur, 4th edition, December 2004.
[15] Senatsverwaltung für Stadtentwicklung des Landes Berlin, April 2002.

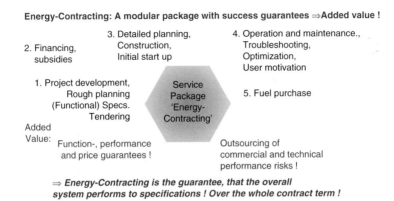

Fig. 17.1. Energy-Contracting: A modular energy service package with success guarantees and outsourcing of risks.

to the needs of the building owner and thus is a modular system. This means the client defines what components they want to outsource and what components they carry out themself. For example, financing can be provided either by the ESCo or the building owner. Or with a (leasing) finance institution as third party. Critical to decision-making is which service provider can offer better financing conditions. This means the contracting package does not automatically include external financing.[16] Other partial tasks, such as ordinary operation management or fault clearance, can readily be assumed by the building owner themself.

Figure 17.2 illustrates an energy added value chain from primary to useful energy, with the respective business models and indications of typical measures carried out. The figure shows the two basic Energy-Contracting models: Energy Supply Contracting (ESC) and Energy Performance Contracting (EPC), the latter expanded to the Comprehensive Refurbishment (CR-EPC) concept:

At Energy Supply Contracting, efficient supply, of final energy such as heat, steam or compressed air is contracted and measured in Megawatt hours (MWh) delivered. The model includes purchasing of fuels and is comparable to district heating or cogeneration supply contracts.

As for Energy Performance Contracting, which is the basis for our models, the focus is on reducing final energy consumption through energy efficiency measures as indicated in Figure 17.2. The business model (see Figure 17.3) is based on a savings guarantee compared to a predefined baseline,[17] also labelled as Negawatt hours (NWh).

At Comprehensive Refurbishment-EPC (CR-EPC) projects, building refurbishment measures are integrated into standard EPC models in order to achieve a comprehensive refurbishment of the buildings. Depending on the CR-EPC model, a general contractor, a

---

[16] This topic has been elaborated in more detail in: Bleyl, Jan W. and Suer, M. (2006). Comparison of different finance options for energy services, in *light+building*. International Trade Fair for Architecture and Technology. Frankfurt a. Main.

[17] In case of new buildings, increased comfort levels or to account for Non-Energy-Benefits, calculatory baselines can be used as long as agreed upon beforehand.

Energy added value chain, two basic Energy-Contracting models and typical efficiency measures

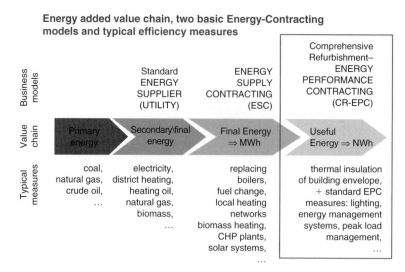

Fig. 17.2. Energy Service models, energy added value chain with typical efficiency measures.

Performance Contracting-Business Model

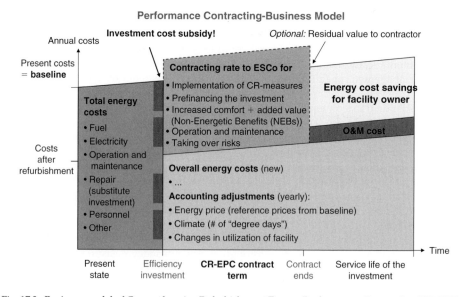

Fig. 17.3. Business model of Comprehensive Refurbishment Energy Performance Contracting (CR-EPC).

general planner or an ESCo will implement a service package encompassing project coordination, overall optimization, detailed planning, implementation of measures, operation and maintenance, subcontracting, fulfilment of energy savings, comfort and other guarantees and may also provide or facilitate financing and acquisition of subsidies.

The CR-EPC business model is shown in Fig. 17.3.

Energy-Contracting models can not decrease pay-back times of energy efficiency investments. Building technology measures can mostly be refinanced from the future energy cost savings within a project period of ten years. This is generally not true for building construction measures, such as building envelope insulation, with today's energy prices. Therefore, the building owner has to co-finance the building measures, e.g. by means of a building cost allowance (which may, e.g., be taken from maintenance reserve funds or subsidies) and/or paying a residual value at the end of the contract (see Fig. 17.3). Another option is longer contract terms of 20 to 25 years, as is common for Public-Private-Partnership contracts.

An important difference between a CR-EPC-model and an in-house refurbishment is the long-term guarantee for the results and quality of the measures taken, which goes clearly beyond the standard legal liability or implied warranty. If there are problems after the refurbishment, such as unexpectedly high energy consumption levels or problems with the formation of mould, the responsibility for remedying these is devolved to the contractor during the contract period. In the case of an in-house refurbishment, the building owner is the person responsible.

It is important to mention here that problems such as lack of quality assurance at the construction site or formation of mould are not related to the contracting model itself. Quality assurance requires controlling and depends on the motivation of the construction company to deliver long term quality. The same is true for problems with formation of mildew. It occurs because more advanced building technologies, such as better sealed building shells, for example, require different or more sophisticated operation and maintenance procedures, which means in this context increased manual or mechanical ventilation.

A CR contracting model offers an instrument to provide incentives to optimize life or project cycle performance, including the operation phase of the building, because the ESCo is not only responsible for the construction but also for the operation and maintenance of the building at a guaranteed price. Thus the ESCo has an inherent interest to take care of quality assurance at the construction site and perform proper maintenance.

In summary, the key features of the CR-EPC model are:

- A CR-EPC partner plans and realizes energy efficiency measures including building construction measures and is responsible for their performance, operation and maintenance throughout the contract term.
- Depending on the implementation model, the contracting partner that will implement the measures is either a general contractor (GC), a general planner (GP) or an energy service company (ESCo).
- The ESCo has to guarantee energy cost savings compared to a predefined energy cost baseline. Further guarantees and quality assurance instruments can be included such as thermal comfort conditions, operation and maintenance or emission reduction guarantees.[18]
- Typical EPC contract terms amount to ten years. Investments for CR-EPC projects – depending on their magnitude – can be refinanced only partially from future energy cost savings. The building owner has to directly pay part of the investments, e.g. with a building cost allowance. Another option is extended contract periods of 15–25 years. After termination of the contract, the entire savings will benefit the client.

---

[18] For more details see Bleyl, Jan W., Baumgartner, B. and Varga, M. (2007). *Quality Assurance Instruments for Energy Services* which has been compiled in a EUROCONTRACT manual. Graz Energy Agency.

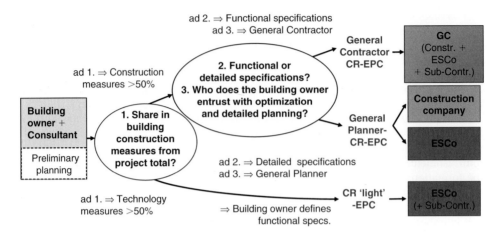

Fig. 17.4. Comprehensive Refurbishment-EPC model selection flow chart.

- The ESCo's remuneration is the contracting rate and depends on the savings achieved. In case of underperformance the ESCo has to cover the shortfall. Additional savings are shared between the partners.

Based on the previous remarks, we define comprehensive refurbishment energy performance contracting (CR-EPC) as:

A comprehensive energy service package including building construction measures aiming at the guaranteed improvement of energy performance and cost efficiency of real estate objects. A general contractor, a general planner or an energy service company (ESCo) implements a customized package of energy efficiency and refurbishment measures and services such as planning, building, operation and maintenance, (pre-)financing or user motivation and takes over technical and commercial performance risks and guarantees for the project. The measures are partially repaid out of guaranteed future energy cost savings, but with (substantial) contributions by the facility owner.[19]

17.3 Three Basic Models to Implement Comprehensive Refurbishment Measures through Energy Performance Contracting

17.3.1 Overview

Energy-Contracting and comprehensive refurbishment activities can be combined in a number of ways, depending among other factors on the scope of the building construction measures, if functional or detailed planning is applied, and on public entities' procurement obligations as contracting authorities (see Fig. 17.4). We present three different approaches to integrate comprehensive refurbishment measures into the standard EPC model. They can be applied in both the private and public sectors.

---

[19] Following Seefeldt, Leutgöb (2003) 'Energy performance contracting – success in Austria and Germany, dead end for Europe?' eceee paper id #5158.

The three approaches can be summarized as follows:

1. **General Contractor model (GC CR-EPC)**
   In this model, the majority of the CR works and services are not described with detailed specifications. Instead the building owner provides functional specification defining the project's technical, financial, organizational, legal and economic performance requirements and the framework conditions for implementation of the measures.

   All services, ranging from overall optimization, detailed planning, construction through operation, maintenance and user motivation, and compliance with the Energy-Contracting quality guarantees over the contract term are contracted to a general contractor (GC, which can be one company or a consortium).

2. **General Planner model (GP CR-EPC)**
   In this model the building owner can specify detailed solutions for the CR measures (e.g. design of the facade). The building owner commissions a general planner who is responsible for overall project optimization, detailed planning, specifications, supervision and quality assurance. Typically, the GP tenders building construction measures (e.g. building envelope) on the basis of detailed specifications; whereas Energy-Contracting are tendered with functional specifications. Hence building construction works and ESCo services are awarded in separate contracts. This model is basically a combination of a standard construction procedure (independent planner + construction company) combined with the Energy-Contracting concept.

3. **Comprehensive Refurbishment 'light'-EPC model (CR 'light'-EPC)**
   Within this model, individual building construction measures (such as top floor ceiling insulation) can be realized with a standard EPC contract. If less than half of the total project cost can be attributed to construction works, the building owner can define detailed specifications for these for the tendering process. An ESCo is awarded an EPC contract and realizes overall optimization, detailed planning, and operation and maintenance and provides all guarantees. The main difference to the GC model lies in the smaller extent of building construction measures. Because only simple building construction measures are involved, we propose to call this the comprehensive refurbishment 'light'-EPC model (CR 'light'-EPC).

In any case, as a first step, a preliminary planning of the CR-EPC project is necessary. For the selection of a suitable implementation model the flow chart in Fig. 17.4 can be used as a guide.

If the project is predominantly comprised of building technology measures, the Comprehensive Refurbishment 'light' model can be used. However, if the project mainly involves construction works, as is the case in most full scale refurbishment projects the 'General Planner' or the 'General Contractor' model must be applied. The 'General Planner' model should be used if the building owner wishes to specify detailed solutions. Otherwise the building owner must decide whom they wish to entrust with the optimization and detailed planning: GC or GP?

The following aspects and implications of the three basic models are described in more detail in the following sections:

- Key features of the models (summary).
- Key actors, responsibilities and contractual relationships.
- Procurement implications (especially for public sector clients).

- Contractual guarantees and quality assurance instruments.
- Advantages and disadvantages of the different models.

Not covered here are aspects such as financing options,[20] subsidy schemes, payback periods of different CR building efficiency measures, details on contractual guarantees and quality assurance instruments[21] or CR model contracts. Of course each project requires an adaptation of the implementation model and contract to the individual project conditions (compare Fig. 17.1).[22]

17.3.2 General Contractor CR-EPC model (GC CR-EPC)

In this model, the majority of the CR works and services are not described with detailed specifications. Instead the building owner provides functional specification defining the project's technical, financial, organizational, legal and economic performance requirements and the framework conditions for implementation of the measures.

All services, ranging from overall optimization, detailed planning, construction through operation and maintenance and user motivation, and compliance with the Energy-Contracting quality guarantees over the contract term are contracted to a general contractor, which can be one company or a consortium. (Table 17.1)

Table 17.1.   Key features of General Contractor CR-EPC model.

Key features	General Contractor CR-EPC model
Example measures:	CR building refurbishment, e.g. in conjunction with sale & Lease back (Public-Private-Partnership) projects
Share of building construction measures:	More than 50% of total project volume[23]
Project specifications and tendering:	Functional specification >50% of project volume ⇒ negotiated procedure
Overall optimization and detailed project planning:	General contractor (individual company or consortium)
Execution of measures:	General contractor (individual company or consortium)
Financing:	Individual combination of EPC savings Guarantee + investment cost allowance + third party financing + subsidy programs

17.3.2.1 Key actors, responsibilities and contractual relationships

In this case, the building owner is not committed or does not have the expertise to plan, optimize and coordinate the overall project in detail. An internal project coordinator or external consultant provides advice and coordinates preliminary project planning, the tendering process, and acceptance and validation of the deliverables. This coordinator in effect represents the interests of the building owner and initiates a negotiated tendering procedure on the basis of functional specifications and selects a suitable general contractor.

---

[20] See footnote 16.
[21] For more details on QAI's see footnote 18.
[22] Some energy agencies and independent consultants have specialized in offering this kind of consultancy, e.g. Graz Energy Agency Ltd. and other Eurocontract partners
[23] More than 50% building construction measures apply for most full-scale comprehensive refurbishment projects.

Figure 17.5 illustrates the contractual relationships for the GC CR-EPC model:

General Contractor CR-EPC Model (GC CR-EPC) *e.g. PPP, sale and lease back of buildings*

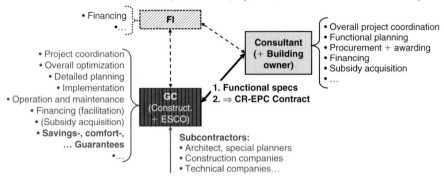

Fig. 17.5. General Contractor CR-EPC model: key actors, responsibilities and contractual relationships.

The general contractor bears the responsibility for the entire project outcome from overall optimization, detailed planning and implementation to operation and maintenance and the coordination of subcontractors. They have to provide energy savings, comfort and other performance guarantees for the results of the project as a whole and may also have to facilitate financing and subsidy acquisition. This requires specialized know-how, experience and a good interdisciplinary understanding of the various project elements and a solid financial background.

The general contractor can be a standard construction company or a standard ESCo. Often, a consortium[24] acts as GC. Most such consortiums comprise a construction company and ESCo and supply the contractual services conjointly. Often, the contractual relationship between the two parties is dissolved after all measures have been implemented, and one of the companies then assumes the remaining contractual rights and obligations. A GC consortium is a viable solution especially if its constituent companies have worked together successfully in the past.

The general contractor must have the statutory permits and authorizations that are required for the project activities. In Austria, for example, the general contractor must have a builder's licence in order to carry out extensive construction activities, and must have a heating/ventilation or gas/plumbing technician's licence to install and service building energy systems.

The general contractor can hire other project partners such as architects, specialized planners, construction companies or technical companies as subcontractors. Since the building owner concludes only one GC contract, they have only one project partner to deal with and thus reduces the number of interfaces for themself.

Financing services are displayed separately in the figure, because experience has shown that for most Energy-Contracting projects it makes sense to differentiate between financing and energy services. ESCos are experts in technical, economic, and organizational matters of Energy-Contracting, which is what they should be commissioned for. Financing is not

---

[24] A consortium is a project partnership with the objective of pooling resources to fulfil a contract, whereby each company is jointly and severally liable for the whole project. One company acts as liaison and represents the consortium externally. An internal partnership agreement governs the relationship between the companies.

necessarily their core business. ESCos can be considered as a vehicle and facilitator for financing. In many cases including a financing institution (FI) as a third party to take over financing matters and risks makes good sense.[25] This holds true for all three models introduced here.

CR-EPC projects with comprehensive refurbishment measures typically have payback times of more than ten years and require either co-financing from the building owner through a partial payment of the investment cost or extended contract terms of up to 25 years.

17.3.2.2 Procurement implications (especially for public sector clients)

In praxis most EPC projects are tendered with a negotiated procedure. Nevertheless, a remark with regard to prerequisites for the applicability of negotiated procurement procedures is appropriate here. The procurement law states that the execution of a negotiated procedure is the exception and not the rule. This exception is subject to prerequisites defined in the public procurement laws. For Energy-Contracting the following prerequisite has to be fulfilled: 'A prior and global pricing is not possible, because of the nature or because of the risks associated with the delivery of the services.'[26]

This translates into procurement praxis as follows:

1. The bidder must be allowed sufficient freedom of scope in formulating their proposal (e.g. selection of EE measures to be implemented). 'Sufficient freedom of scope' requires that a minimum of 50% of the project cost must be subject to negotiations. Project costs are calculated on the basis of preliminary planning (For formula see Section 17.3.4 and Figure 17.8).
2. In order to provide sufficient freedom of scope, the tender documents must be formulated as functional specifications (as opposed to detailed specs), defining the project's technical, financial, organizational, legal and economic performance requirements and framework conditions for the implementation of the measures.
3. The negotiated procurement procedure must actually allow negotiations both for the bidder and for the contracting authority.

If these requirements cannot be met, public sector clients are required to realize the project with the general planner or CR 'light' model.

17.3.2.3 Contractual guarantees and quality assurance instruments

The general contractor takes over technical and commercial implementation and operating risks and gives performance guarantees for the results of the entire project over the contract term.

Typically, the core guarantee is given for the energy cost savings in relation to a reference baseline. At the same time comfort standards such as minima and maxima for room temperatures and humidity are defined and have to be maintained. As a result, it is in the general contractor's very own interest to design and operate the facility's heating, ventilation and other technical systems efficiently, because his remuneration depends on the fulfilment of the savings guarantee given. [27]

---

[25] Bleyl, Jan W. and Suer, Mark (2006). Comparison of different finance options for energy services, in *light + building*. International Trade Fair for Architecture and Technology. Frankfurt.

[26] BVergG 2006 § 30(2) Austrian public procurement law (translation by authors).

[27] Additional quality assurance instruments may be derived from the IPMVP protocols, which can be downloaded from www.evo-world.org

Other guarantees are typically defined for investment costs, for reaction times in case of malfunctions or for quality and (eco-) performance requirements of materials and equipment installed.

In principle, the extent and details of the guarantees in an Energy-Contracting service package can be agreed individually for each project. The goal is to outsource commercial and technical performance risks to the ESCo and demand measurable guarantees as described above (see also Fig. 17.1).

Consortia by their legal nature have to warrant that they will meet all deliverables and warrantees collectively, irrespective of their individual spheres of responsibility. Their relationships within the consortium are regulated by the consortium's bylaws. Companies involved outside the GC contract have a liability which is limited to the legally mandatory (implied) warranties.

17.3.2.4 Advantages of the model

The GC CR-EPC model offers the following advantages in addition to the advantages of Energy-Contracting in general, which are not particularly stated here (see also section 17.4):

- 'One stop shop': The general contractor assumes the coordination and provides guarantees for the entire CR-EPC project including all interfaces and the overall performance. An integrated solution is provided by an expert partner that has at his disposal all the required competencies and can call upon specialized subcontractors as needed.
- The general contractor's performance can easily be judged by evaluating the guarantees agreed in the CR-EPC contract. In addition the general contractor's remuneration is partly performance based with a bonus-penalty system in case of under- or overperformance.
- Financing can be individually arranged from a combination of EPC savings guarantee, investment cost allowance by the building owner, third party financing from a financing institute (or ESCo) and subsidy programs.
- For the building owner, interface problems are reduced since the general contractor is the sole project partner for the realization and operation of all refurbishment measures.

In the case of a consortium only one consort acts as the external contact partner for the building owner, but all corporation partners are responsible (joint and several liability) for providing all deliverables.

17.3.2.5 Disadvantages of the model

The GC CR-EPC model has the following potential disadvantages (in addition to known disadvantages of the Energy-Contracting model):

- Refurbishment measures with detailed specifications, which are typically building construction measures like facades, are limited to less than 50% of the total project value.
- The building owner is highly dependent on the general contractor. Detailed controlling and management options during the project planning and implementation phase are limited. This means that the building owner must have sufficient confidence in the general contractor's capabilities and in addition apply adequate quality assurance instruments. One solution is to require second opinion reports from independent consultants for critical project steps.

- The general contractor usually calculates an additional general contractor surcharge for coordinating and taking responsibility for the overall project. At the same time, pricing pressure for subcontractors is higher compared to a direct contract with the building owner. Generally, the general contractor model favours bigger companies and may be disadvantageous to small and medium sized enterprises (SMEs) or regional companies.
- In the case of a general contractor consortium, project acquisition and long-term contract fulfilment are often with the ESCo and not the construction partner of the consortium, although construction volume exceeds ESCo contract volume.
- Possible conflicts of interest regarding implementation quality may arise between general contractor and subcontractors or consortium partners, because the general contractor is focused on meeting the long-term contractual performance guarantees and to minimize project cycle cost, whereas the subcontractor's horizon is limited to the acceptance directly after the construction period.
- The number of qualified comprehensive refurbishment general contractors, which are familiar with the Energy-Contracting concept, is limited in most markets. Market development activities to familiarize market actors with the CR-EPC concept and with the functional procurement procedure may be necessary.

Applicability and evaluation of the advantages and disadvantages described must be determined on an individual project basis and may need to be adapted to different countries and regions.

17.3.3 General Planner CR-EPC model (GP CR-EPC)

In this model the building owner can specify detailed solutions of the construction measures, such as the detailed design of a building envelope refurbishment. The building owner commissions a general planner who is responsible for overall project optimization, detailed planning, specifications, tendering, supervision and quality assurance.

Typically, the general planner tenders building construction measures on the basis of detailed specifications; whereas ESCo services are tendered with functional specifications. Hence building construction measures and ESCo services are awarded in separate contracts.

This model is basically a praxis-oriented combination of a standard construction procedure with the Energy-Contracting concept. (Table 17.2)

Table 17.2.   Key features of General Planner CR-EPC model.

Key features	General Planner CR-EPC model
Example measure:	CR building refurbishment including facade and building technology
Share of building construction measures:	More than 50% of the total project volume
Project specifications and tendering:	Construction: Detailed specification ⇒ standard procurement ESCo: Functional specs. ⇒ negotiated procedure
Overall optimization and detailed project planning:	General planner
Execution of measures:	Construction company and ESCo
Financing:	Individual combination of EPC savings guarantee + investment cost allowance + third party financing + subsidy programs

From the procurement law perspective, a negotiated procedure can only be applied for the ESCo services, not for the construction measures (please refer to section 17.3.2.2).

The other implications outlined in the GC CR-EPC model with regard to procurement, financing, pay back times and statutory permits for the actors apply here as well.

17.3.3.1 Key actors, responsibilities and contractual relationships

In this case, the building owner wishes to provide detailed specifications for the majority of the refurbishment measures. In praxis this is typically a detailed plan for the building envelope refurbishment. A second reason to choose the GP CR-EPC model may be that the building owner prefers to put an independent planner in charge of the overall optimization, detailed planning and supervision of the refurbishment measures.

The general planner (e.g. a civil or industrial engineer or an architect) represents the interests of the building owner. They are responsible for consulting with the building owner, overall project optimization, detailed and functional planning, procurement and awarding, supervision, acceptance and quality assurance of construction measures and last but not least overall project coordination. This agenda may be expanded to subsidy acquisition or other tasks.

The general planner must possess interdisciplinary competencies and experiences in the overall optimization, realization and coordination of CR and Energy-Contracting projects. Comprehensive refurbishment requires an integrated planning approach, which takes reciprocating effects of the different building technologies into account.

It is the general planner's task to plan and ensure building performance criteria such as air tightness or maximum heat demand $<30\,kWh/m^2/a$, on which the ESCo can base its performance guarantees.

Figure 17.6 illustrates the contractual relationships for the GP CR-EPC model.

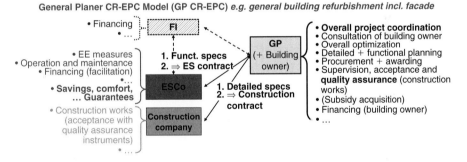

Fig. 17.6. General Planner CR-EPC model: key actors, responsibilities and contractual relationships.

After completion of the planning, building construction measures and Energy-Contracting are awarded in separate contracts. The building construction measures are typically planned in detail and awarded to a construction company on the bases of detailed specifications. Special attention must be given to the definition and control of performance criteria and quality assurance instrument (QAI). The contract must include mandatory QAIs into the specifications such as thermographic pictures, blower door tests, operation manuals and experts' reports or extended liabilities (see Section 17.3.2.3).

ESCo services are typically awarded with a negotiated procedure on the basis of functional specifications. These must also include detailed (performance) data of the building construction measures, which are implemented by other partners, so the ESCo can calculate performance guarantees for the complete refurbishment project. The ESCo takes over operation and maintenance for the entire refurbishment measures. The CR-EPC contract is concluded between the building owner and the ESCo.

For this model, the selection of the general planner is of special importance, taking their scope of responsibilities into account. For the commissioning of the general planner contract, it is possible to define quantitative and qualitative awarding criteria. With the help of a cost/benefit analysis, the qualitative (e.g. consultation fees) and quantitative (e.g. draft concepts or references) criteria can be combined, weighed and evaluated.[28]

17.3.3.2 Contractual warrantees and quality assurance instruments

The responsibility for the entire project is shared between general planner and ESCo. The general planner is responsible for overall coordination, optimization and planning of the project. They are is contractually obligated to the building owner to meet agreed standards and performance criteria. Hence, it is recommended to include measurable success criteria into the general planner contract. As a minimum, the general planner's services must be covered by liability insurance.

The quality of the building construction measures has to be controlled and assured by the general planner. They have to provide evidence of the implementation quality of building construction measures to the ESCo, e.g. by providing thermographic inspections, blower door tests, experts' reports, simulations and similar quality assurance instruments.

The ESCo has to provide performance warrantees for the entire building measures as described in the general contractor model, based on the ESCo as well as the building construction measures.

To secure project cycle optimization and implementation quality it is recommended to integrate the ESCo into the project at an early stage and to allow the ESCo a comment and control status.

The construction company warrants that the materials deployed and methods used to install them meet the quality requirements. Guarantees are typically limited to the legally implied warranty.

17.3.3.3 Advantages of the model

This model is closest to the established standard planning and implementation procedure for building refurbishment measures. In comparison to the general contractor model, the GP CR-EPC model offers the following advantages:

- The building owner themself or via the general planner has more control over the detailed planning and implementation steps of the project.
- Project coordination costs are likely to be lower in comparison to the GC CR-EPC model because there is no general contractor surcharge. On the other hand, costs for the general planner have to be accounted for.
- The standard tendering process with detailed specifications is advantageous for construction companies in that it is easier for them to bid for clearly defined building measures, especially if they are not used to functional call for tenders. In addition, it is easier to contract subtasks to specialized companies for individual or specialized measures.

---

[28] See Bleyl, Jan W. (2006). *Evaluation of Tenders with a Cost-Benefit Analysis.* Eurocontract Training Session. Berlin.

As with the other CR-EPC models, financing can be individually arranged from a combination of EPC savings guarantee, investment cost allowance by the building owner, third party financing from a financing institute (or ESCo) and subsidy programs.

17.3.3.4 Disadvantages of the model

The general planner CR-EPC model has the following potential disadvantages, in comparison to the general contractor model:

- The building owner is highly dependent on the quality and creativity of the general planner, who in return is not responsible for the long-term results and the operation of the building as compared to the general contractor or ESCo. Only the ESCo's performance can easily be judged by fulfilment of guarantees where remuneration is performance based with a bonus-penalty system.
- There are more interfaces with potential problems than in the GC CR-EPC model, e.g. in transitioning from the construction to the operational phase. Especially, the ESCo assuming guarantees for the overall building performance including the building construction measures may be critical.
- Warrantees for the building construction measures are typically limited to the legally implied warranty as opposed to long-term guarantees in a CR-EPC contract over the complete project term.
- The number of qualified comprehensive refurbishment general planners, which are familiar with the Energy-Contracting concept, is limited in most markets.
- In general, there are fewer incentives for innovative solutions, because detailed specifications leave less room for competition of ideas between bidders. Innovation is mostly dependent on the initiative of the general planner.

Applicability and evaluation of the advantages and disadvantages described must be determined on an individual project basis and may need to be adapted in different countries and regions.

17.3.4 Comprehensive Refurbishment 'light'-EPC model (CR 'light'-EPC)

With this model, individual or smaller building construction measures such as top floor ceiling insulation can be realized within a standard EPC contract. If less than half of the total project cost can be attributed to building construction measures, the building owner can define detailed specifications for these for the tendering process.

An ESCo is awarded an Energy-Contracting contract on the basis of functional specifications and realizes overall optimization, detailed planning, and operation and maintenance and provides performance guarantees.

The main difference to the GC CR-EPC model lies in the smaller extent of the building construction measures. Because only simple or 'lightweight' building construction measures are involved, we propose to call this the comprehensive refurbishment 'light'-EPC model (CR 'light'-EPC).

17.3.4.1 Key actors, responsibilities and contractual relationships

The responsibilities and contractual relationships are to a large extent similar to the GC CR-EPC model. Main differences are the extent of the building construction measures. The

Table 17.3. Key features of Comprehensive Refurbishment 'light'-EPC model.

Key features	Comprehensive Refurbishment 'light'-EPC model
Example measures:	Building efficiency technologies + top ceiling insulation, window repair
Share of building construction measures:	Less than 50% of the total project volume
Project specifications and tendering:	Functional specification >50% of project volume ⇒negotiated procedure
Overall project management, optimization and planning:	Energy Service Company (ESCo)
Execution of measures:	Energy Service Company (ESCo)
Financing:	EPC savings guarantee + subsidy program investment cost allowance + third party financing on demand

role of the general contractor can be taken over by a standard ESCo, which may hire a construction company as subcontractor for the building construction measures.

The contractual relationships in the CR 'light'-EPC model are as shown in Fig. 17.7.

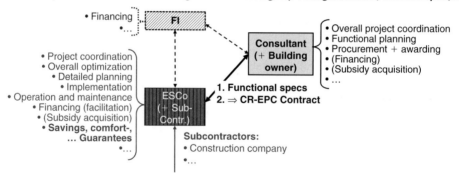

Fig. 17.7. Comprehensive Refurbishment 'light'-EPC model: key actors, responsibilities and contractual relationships.

The ESCo is responsible for overall optimization, detailed planning, implementation of measures, operation and maintenance, subcontracting, fulfilment of energy savings, comfort and other guarantees and may also provide (facilitation of) financing or acquisition of subsidies.

The ESCo must have the statutory permits and authorizations that are required for the project activities. Depending on the amount of construction activities, a heating/ventilation or gas/plumbing technician's licence to install and service building energy systems may be sufficient.

CR-EPC projects with individual building construction measures typically have payback times of 10 years and require only small co-financing from the building owner through a partial payment of the investment cost.

17.3.4.2 Procurement implications (especially for public sector clients)
A CR 'light'-EPC contract can only be awarded with a negotiated procedure if the legal procurement prerequisites – as described in the GC CR-EPC model – are met. The consultant must ensure that more than 50% of the deliverables are tendered with functional specifications, which provide sufficient freedom of scope in formulating proposals. In reality, functional specifications are typically provided for the building technology measures/energy services (>50%) and detailed specifications for the building construction measures.

In order to calculate the value of the measures that allow for negotiations, the value of the building construction and other measures described in detail must be subtracted from the total project value (over the duration of the project). This is done using the formula in Fig. 17.8.

Calculation formula for value of negotiable measures

+ Σ Contracting rates (over project term, excl. interest)	+ Demolition work
	+ Building and roof construction
+ Co-financing by building owner	+ Doors and windows
	+ Thermal and acoustic insulation
+ Third party financing by finance institute (excl. interest)	+ Plastering and painting
	+ Miscellaneous
+ Subsidies	+ All other detailed specs.

Σ total project value   –   **Σ detailed specifications**

= **Σ Value of negotiable measures** (described with functional specifications)

Fig. 17.8. Calculation formula for value of negotiable measures.

The total cost of works and services tendered with functional specifications must account for more than half of the total project value, to meet the legal requirements for a negotiated tendering procedure. The calculation is done on the basis of the preliminary planning results.

17.3.4.3 Contractual warrantees and quality assurance instruments
The ESCo takes over technical and commercial implementation and operating risks (among others) and gives performance guarantees (as described in the GC CR-EPC model) for the results of the entire project over the contract term (as described in section 17.3.2.3).

17.3.4.4 Advantages and disadvantages of the model
The Comprehensive Refurbishment 'light'-EPC model offers the following advantages, in comparison to the other two CR-EPC models as well as the standard basic EPC model: individual building construction measures can be realized within a standard EPC model in the same manner as standard building technology EPC measures.

In addition, the model offers standard-ESCos access to building refurbishment projects including building construction measures in which their energy service expertise can be integrated. This may facilitate access to new and potentially lucrative building refurbishment markets for ESCos.

As a disadvantage of this model, all refurbishment measures with detailed specifications must account for less than 50% of the total project value. Other important advantages and disadvantages are similar to the GC CR-EPC model.

As for the other models, applicability and evaluation of the advantages and disadvantages must be determined on an individual project basis and may need to be adapted in different countries and regions.

17.4 Conclusions, Recommendations and Outlook

Based on the previous sections, the following conclusions and recommendations can be given:

1. The proposed CR-EPC models can facilitate customized packages of building construction and building technology measures combined with the known guarantees of standard EPC models and outsourcing of technical and commercial risks to ESCos.
2. Generally, any building design approach should first of all focus on all possible demand reduction potentials (including the building envelope). Only as a second step, the remaining demand should be supplied as efficiently as possible.
3. An integrated energy efficient planning process is especially necessary, if renewable energy sources are to be applied. For example, solar cooling will hardly be feasible with high cooling loads of more than $40 \, W/m^2$.
4. We propose three different models for the implementation of comprehensive refurbishment through Energy-Contracting: a General Contractor, a General Planner and a CR 'light'-EPC model. All three CR-EPC models presented allow combining (comprehensive) refurbishment measures of buildings with the advantages and long-term guarantees of Energy-Contracting models.
5. The choice of the implementation model (especially for public sector building owners) mainly depends on three factors:
   - The share of building construction versus building technology measures in relation to the total project volume over the contract period. This has implications mainly on the procurement law, (if applicable).
   - Whether functional or detailed specifications for the contracting of the energy efficiently measures are desired and applicable from a procurement law perspective.
   - Who the building owner wants to entrust with the detailed planning, overall optimization and supervision of the project: a general planner or a general contractor.
     The details and implications as well as advantages and disadvantages of each model are described in the main section of this contribution. Naturally, each project requires an adaptation of the implementation model and contract to the individual project conditions.
6. All three models proposed can be applied both in the public and the private sector.
7. Energy-Contracting models as energy efficiency tools will be successful, if the added values compared to an in house implementaion can be communicated. From the perspective of the building owner, the following main advantages of the Energy Service model apply:
   - Guarantees for the results, e.g. for energy cost savings, indoor comfort standards, operation and maintenance, service reaction times and that the overall system performs to specifications. Over the whole contract term.
   - The ESCo's remuneration (contracting rate) is performance based with a bonus-penalty system and depends on the fulfilment of energy savings guarantees.
   - Possibility to save investment costs through part-repayment from future energy cost savings and third party financing.
   - Shifting technical and commercial implementation and operation risks to the general contractor, the general planner or the ESCo.
   - Focusing on the own key business.

- One contact person for all energy matters included in the CR-EPC contract ('one stop shop').
- Increasing comfort and value of the building resulting in long-term increase in the revenue from the property.

     The objective is to create a win-win-win situation for all parties involved. The environment and the building owner's image included.

8. Energy-Contracting models cannot decrease payback times of energy efficiency investments. At current energy prices, the typical guaranteed energy cost savings of a CR-EPC contract cannot repay comprehensive building measures like a complete building envelope refurbishment within ten years. The building owner has to co-finance the investment by way of a building cost allowance (or a residual value payment at the end of the contract). Another option is longer contract terms of 20 to 25 years, as is common for Public-Private Partnership contracts.

9. Financing must be individually arranged from a combination of CR-EPC savings guarantee, investment cost allowance by the building owner, third party financing from a financing institute (or ESCo) and subsidy programs. We recommend differentiating between financing on the one hand and energy services on the other. ESCos are experts in technical, economic, and organizational matters of Energy-Contracting, which is what they should be commissioned for. Financing is not necessarily their core business. ESCos can be considered as a vehicle and facilitator for financing. In many cases including a financing institution as a third party to take over financing matters and risks makes good sense.[29]

10. Comprehensive refurbishment (CR) of buildings is a demanding task in terms of integrating and optimizing all building construction measures and building technologies involved. It requires experienced partners and an integrated planning process taking reciprocating effects of the different EE-measures into account. A good example for this approach is the reduction of all electrical and thermal cooling loads including solar shading options before assessing an air conditioning unit.

11. The necessity for quality assurance at the construction site is not related to the contracting model itself. Quality assurance requires controlling and depends on the motivation of the construction company to deliver long-term quality. Energy-Contracting models offer an instrument to optimise life or project cycle performance, including the operation phase of the building, because the ESCo is responsible not only for the construction but also for the operation and maintenance of the building. Thus the ESCo has an inherent interest to take care of quality assurance at the construction site and perform proper maintenance.

12. In many cases EE is not the driving force behind comprehensive refurbishment of buildings. Nevertheless minimum performance standards for any thermal refurbishment and guarantees for maximum energy consumption should be written into the terms of reference. CR-EPC models as promoted here are a good means to secure these goals and are also applicable to Public-Private Partnering models like sale and leaseback projects.

17.4.1 Outlook

Implementing an Energy-Contracting projects always require dedicated project developers. A future challenge will be standardisation and spreading the concept, initiating further

---

[29] Refer to footnote 25.

projects and collecting more experiences. Our thesis, that the comprehensive refurbishment models introduced are a good instrument to implement building energy efficiency measures still needs more good practice.

To our knowledge, practical experiences with the implementation of CR-EPC models are so far limited to Austria.[30] We would like to learn more about other experiences collected with comprehensive refurbishment of buildings in conjunction with Energy-Contracting or other models and welcome any feedback. Also cooperation models with the facility management community would be of great interest.

Last but not least, the CR-EPC model itself imposes obstacles from a methodological point of view, especially if the cost baseline is difficult to determine or if frequent adjustments of the baseline are necessary due to changes in utilization of the building.[31] The latter problems are not encountered with the Energy Supply Contracting (ESC) model, because for the business model no baseline is needed to measure guaranteed savings. We will carry out research and work on model projects for possible advancements of the ESC model with the objective of integrating demand-side measures and energy saving incentives into the model.[32] For the future we propose, that any energy supply should be coupled with efficiency measures on the demand side. Otherwise our CO_2- and emission reduction goals will not be achievable.

---

[30] Berlin Energy Agency has prepared some projects but not yet reached the implementation phase.
[31] Energy cost and climate adjustments are easy to handle with the yearly final invoice.
[32] Task XVI 'Competitive Energy Services' of the IEA (International Energy Agency) Demand Side Management Implementing Agreement (http://dsm.iea.org/) has recently started research on this topic.

Chapter 18
Sustainability on the Urban Scale: Green Urbanism – New Models for Urban Growth and Neighbourhoods

STEFFEN LEHMANN

School of Architecture & Built Environment, University of Newcastle, Australia; General editor Journal of Green Building

18.1 Introduction

> *Global warming is no longer an inconvenient truth, it is an inescapable reality.*
>
> UN Secretary-General Ban Ki-Moon (2007)

Among the most significant environmental challenges of our time are global climate change, excessive fossil fuel dependency and the growing demand for energy (despite limited fossil fuel reserves) – all likely to be major challenges of the twenty-first century and one of the greatest problems facing humanity (Braungart *et al*. 2002; Flannery 2005).

The increasingly widespread effects of climate change and continuous uncertainty of conventional energy supplies are causing an increasing demand for the transition of our cities through sustainable urban development. However, we are not yet yielding sufficient carbon savings to suggest that adequate action has been taken. The quickest way to reduce greenhouse gas emissions is by being more efficient with the way we consume our energy, by reducing consumption, and by transferring to renewable energy sources. Energy savings require the development and use of more efficient appliances and the retro-fitting of the existing building stock to make the most of natural daylight, rainwater, cooling breezes, and solar exposure – reducing or eliminating the need for artificial lighting, cooling and heating (over the last 20 years, the demand for cooling has increased the most).

Of all the design fields, urban design has the greatest direct impact on the nature of cities and city life. However, urban design and the fundamental principles of how to shape our cities has so far barely featured in the greenhouse debate. The urban dimension and the macro-scale of cities were mostly missing in the debate of the 1980s and early 1990s, as sustainability was predominantly discussed as being about 'alternative lifestyles' or 'ecological houses'.

Much of the more recent debate has circled around ideas about active technology for 'eco-buildings' and sophisticated facade technology – rather than about urban issues.

409

This is surprising since almost half the energy consumed is used by cities and urban built-up areas, and given that avoiding mistakes in the urban design layout at early, conceptual stages could genuinely lead to more sustainable solutions. A large amount of energy consumed in the developed world is used to operate buildings (for instance, in Australia: 44%[1]), whereby the relationship between energy efficiency of buildings and urban design decisions has become a complex issue.

Sustainable architecture is only really effective when set in an urban planning context which itself is based on sustainable principles. D. Gauzin-Mueller (2002)

Several big cities in the developed world have now started large projects and initiatives focused on energy transformation in urban areas, to reduce their dependency on oil and gas sources.

So, what are the practical strategies that focus on increasing sustainability beyond the scope of individual buildings and how can energy transformation be achieved? How can the existing fossil-based energy infrastructure of cities be gradually transformed to confront the environmental challenges of our time; and how can we design urban open spaces and grouping of buildings in such a way that only a minimum of energy is needed to light, ventilate, shade/cool, heat and service them? And by doing so, can we find a poetic architectural response where the building envelope is still informed by simple and strong architectural ideas, rather than by being technologically driven? Will 'Green Urbanism', with its balanced reintegration of landscape, also be fit for hot and humid (tropical) environments, such as in Bangkok, Hong Kong, or Singapore?

This in-progress research deals with cross-cutting issues in architecture and urban design and addresses the question how we can best cohesively integrate all aspects of energy systems, traffic and transport systems, waste and water management, passive and active strategies, climatization (ventilation), etc. into contemporary urban design – and by doing so improve the environmental performance of our cities.

From sustainable urban planning projects conducted over the last years, data show that savings in energy costs of 20 to 50% are possible through integrated planning with carefully considered site orientation and passive strategies (without any further expense); this saving can further increase with on-site renewable energy-producing technologies.

Let's have a closer look at opportunities for regenerating the city centre and the appropriateness of strategies for the various climate conditions, to see how urbanism is affected (and can be expected to be even more affected in future) by the new paradigms of ecology.

18.2 Sustainability Integrated within the Urban Design Process

There are green principles that can creatively support the main design concept. When we study the architecture of Louis Kahn or Alvar Aalto, we find that those architects designed buildings based on what they regarded as 'timeless fundamentals', such as the human experience of space. Both masters designed naturally ventilated office buildings and incorporated climate-responsive design principles long before the notion of 'sustainable architecture' was

---

[1] Australia is the 12th largest emitter of carbon dioxide (CO_2). However, on a per capita basis it is the leading polluter because of its reliance on coal for power; e.g. 93% of electricity in New South Wales is supplied by burning coal. The Australian Business Roundtable on Climate Change (2007) recently showed that it was possible for countries such as Australia to deliver significant reductions in greenhouse gas emissions at an affordable cost, while maintaining strong economic growth.

introduced (e.g. Rachel Carson's book *Silent Spring* was published in 1962). The concepts and knowledge of night-time cooling, evaporative cooling, solar chimneys, cross-ventilation and thermal mass have existed for centuries, even millennia. This supports the notion that sustainability in architecture is about a fundamental attitude of making place and space, and less about the technological solution for 'ventilation'. It is important to recognize that architecture is predominantly about establishing meaning, about the human experience and substance – and not *per se* about technological sophistication.

Of course, a 'green building' is not always automatically a good work of architecture (Lehmann 2006). Architecture and urban design have the potential of re-establishing our relationship with nature, the climate and the experience of the sun, rain and wind. As Scott notes, such environmentally responsible design is at its best 'when it achieves an outcome in which the environmentally sensible elements are closely linked to the design process, go beyond being additive and become meaningful parts of an architectural whole' (Scott 2006). Integration of sustainability aims within the design process demands that the environmental concept and the architecture fully support each other. This requires the identification of environmental strategies that correspond to strong design concepts, support a unique design idea, and reinforce the building's relationship with the landscape and the city. So-called 'exogenous developments' – where improvement comes only from technology, rather than from inside, from society – are only 'quasi evolutionary'.

After more than three decades of environmental debate (starting in 1973 with the oil crisis), there is now an increasing demand to explore and assess the urban design principles and practical strategies for more energy-efficient cities, beyond the scale of the individual building, and to critically examine those strategies for sustainable cities.

Three compelling – and widely published and discussed – contemporary examples of environmentally inspired individual buildings, where ESD principles have successfully informed the design, are:

- The Tjibaou Cultural Centre in Noumea (New Caledonia) by Renzo Piano Building Workshop (1998, www.rpbw.com).
- The Training Centre and Town Hall Mont-Cenis in Herne-Sodingen (Germany) by Francoise-Helene Jourda (1999, www.jourda-architectes.com).
- The Research Centre in Wangeningen (Netherlands) by Guenther Behnisch Architects (1998, www.behnisch.com).

However, it will now be essential to work towards equally persuasive projects on the urban scale, on the level of larger groupings of buildings. The author believes that it is essential and timely to arrive at equally effective proposals of sustainable city projects, even if this may prove to be more difficult.

18.3 The Need for More Comparative Research on Cities

The significance of the research lies in the need for an integration of ecology and sustainable principles in the urban design process, particularly for the growing cities in fast-developing South East Asia, and the general need for a more sustainable city development. The findings are of particular relevance to the rapid urban growth of developing cities undergoing aggressive and largely unplanned and poorly planned transformation, where new towns get virtually built overnight, such as currently in China.

According to the UN, the annual total population increase is around 90 million people (Intergovernmental Panel on Climate Change 2007). In future, more and more people

worldwide will live in cities. In early 2008, for the first time ever, more people are projected to live in urban than in rural areas. Driven by immigration from the poverty-stricken
countryside, soon the number of city dwellers on the globe will reach five billion people.
More than 50% of us live in cities, a figure that will rise to 75% by 2050; therefore, the
twenty-first century will be the urban epoch. This fact poses entirely new challenges on
cities, and the massive influx of rural poor into urban areas in many of the Asian, Indian,
and African cities contributes to this challenge (150 000 people a day leave their rural existence behind to start a new life in the city). As a result, most of this increase in population
occurs in the big cities of developing nations, where most of the world's poverty is already
concentrated, leading to difficulties in water supply.[2] Together with climate change, the
future of the city is the challenge of the twenty-first century.

Traditional European city centres, built prior to and during the fossil fuel era, are the
product of slow growth and complex ownership patterns, but they have not strictly been
laid out on solar principles, with the aim to transfer to solar power. During the nineteenth
and twentieth centuries, the form of those cities was mainly driven by the industrial revolution, to achieve rapid transportation and production, but less sustainability. This leads
to the assumption that it is generally easier to build a new sustainable city or a new sustainable urban quarter from scratch, rather than retrofit an existing one (since the existing,
fossil-based technical infrastructure becomes a restriction to new sustainable development).
Those development patterns that emerged during the industrial revolution, frequently
cluster and network geometries, have not been the product of irrational thought. To the
contrary, governments, developers and citizens have followed a logical course that is the
result of centuries of piece-by-piece outward expansion of land use and the more recent
neglect of the urban city centres. Today, it is very unlikely that a large-scale urban re-plan-
ning on solar principles – where every block could take maximum advantage of passive
solar gain – of those traditional European cities will ever occur.

On the other hand, the Asian 'new towns' – currently being developed and under
construction – could offer unique opportunities for 'total urban design' concepts, to really get
things right: to design and build to the optimized density and ideal day-lighting conditions.
This could be based on master plans including the integration of efficient public transport

---

[2] Increasing environmental damage as a result of humanity's addiction to coal and oil, and declining
resources will further lead to conflicts over water. Desertification, soil erosion and shrinking rainfall
mean that many developing countries will not be able to continue supporting its human and livestock
population.

China's rapid urbanization process makes it an interesting case study. Since the economic reform
in the 1980s, cities in China have undergone phenomenal changes; the sustainability impacts of these
transformations are yet to be assessed. According to its government, around 400 million Chinese people need new houses in the next 12 years. For instance, cities like Chongqing and Shenzen grow annually by around 300 000 people (2006) – at an unprecedented speed. There are also many new towns in
planning: between now and 2020, China will need to build 400 new towns, nearly 30 a year. Two of
such new towns, both supposed to become carbon-neutral cities, are: 'Dongtan Eco-City', a new city
to be entirely based on ecological sustainable principles, with a first stage to be built by 2010 on an
island in the Yangtze River close to Shanghai; the city will grow to 500 000 people by 2040 (planner:
Arup, London; www.arup.com). The other is 'Yimin', a new town for 100 000 people, mainly ethnic
minority groups in Mongolia (planner: Allen, Jack & Cottier, Sydney; www.architectsajc.com). These
rapid growth processes are not the only occurring change: as Dr Peter Herrle (Berlin) points out,
'expansion and shrinking processes of urban areas can occur at the same time in closely neighbouring
city districts. The city of Detroit (USA) is an example for a shrinking downtown area which is accompanied by growth and expansion processes in the outskirts' (Herrle 2007).

(efficient light rail transit systems), where each city block gains maximum solar exposure and uses geothermal technology wherever possible, and decentralized renewable energy generation, which catches the natural breezes for ventilation, collects rainwater and has well-designed shading devices for the western facades. Indeed, by applying sustainable urban planning principles it could all be achieved for those Chinese new towns! China is (again) a place of extremes: we can find some of the most destructive and some of the most hopeful initiatives happening there today (such as Arup's plans for a new zero-carbon development outside Shanghai, 'Dongtan Eco-City' – here, energy-directed urban planning is based on issues of optimized energy flows integrated into the urban form).[3]

So, what exactly is the aim of sustainable design? One major aim is to facilitate the revitalization of pedestrian-friendly city centres, and implement the principles of 'green urbanism'. Most urban designers would generally agree on the following principles:

- Cities and urbanized areas in the developed world need to be our focus, as it will be where (beside the industry sector) most energy is consumed and most waste produced.
- Cities are the main consumers of natural resources and the main producers of pollution.
- Sustainability aims are most effectively pursued when clear principles of sustainable urban development are established.
- Urban patterns, density, public transport, water management, solar orientation, day-light access, construction systems, and supply chains are all essential in the process of guiding urban design decisions towards higher levels of sustainability.
- A 'mixed-use, compact city model' promises the optimum use of space and a new city's land use pattern.

Therefore, comparative research in the urban dimension (macro-scale) of sustainability must be intensified if urban development is to become more sustainable.

Research findings into sustainable urban design support an increased harnessing of the energies embodied in the existing fabric – for instance, through the adaptive reuse of former industrially used sites, inner-city brownfield sites, industrial and military areas, or docklands – and the upgrade and extension of existing building structures, rather than their demolition. The placement of buildings, how compactly they are grouped, can have a profound and direct impact on energy consumption, and determine traffic patterns and, thus, the production of greenhouse gases. In most cities in the US and Australia, there is a large disparity between where people live and where they work, resulting in longer commuting distances than in comparison with citizens in European or Asian cities.

The author suggests that answers to this problem have long been sought in the revitalization of existing city centres, where a higher population density and a more compact community would offer everything within walking and cycling distance. Such high density centres close to public transport nodes are the aim of what has been called 'Transport-Oriented Development' (TOD) (Breheny 1992). Such communities are thought to offer a higher degree of self-sufficiency and, at the same time, good rail links to other areas of the urban region. Densification and Transit-Oriented Design are not always appropriate for all contexts; however, with such strategies, some great improvements were recently achieved in various cities, for instance in Rotterdam, Copenhagen, Barcelona, Lyon and Freiburg

---

[3] The principles developed at Dongtan include: all energy provided by distributed power from a variety of sources – photovoltaic, wind, biomass; use of water recycling and natural treatment systems; a focus on multi-modal public transport and direct and convenient cycle and pedestrian routes as opposed to ring road distribution.

(Hall 2005). The question now is: how is 'best practice' precisely defined in these cities, and which part of the experience can be easily transferred to the other project contexts, e.g. to the US, Asia, or Australia?

Pressure on transport, energy and water systems is likely to increase further. Planning decisions made today in Australian, Indian, or Chinese cities will have immense ramifications in the future. We observe a rapid expansion of neighbouring developing countries and an entirely unprecedented urban transformation process, as currently experienced in the cities in South East Asia, India, along the Pacific Rim, or in the United Arab Emirates. The regeneration of our city centres has become an increasingly pressing task.

18.4 Ramifications of Planning Decisions Made Today

There is a tendency for engineers to favour technology-based solutions for individual buildings, as the author has experienced on many occasions. And indeed, the larger population, too, expects the quick techno-fix to come to their rescue.

With buildings consuming vast amounts of energy in all developed countries, predominantly for cooling, heating, and lighting – almost half of all energy – and directly contributing around 15% of all CO_2 emissions through the construction process, architecture and urban design are often seen as prime technological disciplines in addressing the situation.[4] But the task is too large for architects and urban designers alone; it is a more fundamental issue to be addressed by a series of disciplines, including environmental engineering, urban planning, economics, landscape architecture, sociology, and others. All disciplines working closely together could achieve the goal of the 'City of the Future' which will run entirely on energy drawn from the infinite, unlimited and safe natural sources of renewable energies.[5]

> *While fossil and nuclear industries are fighting to maintain their grip: Renewables are the foundation of the 'City of Tomorrow'. Droege (2007)*

Renewable energy is the fastest growing of all energy industries. Renewable energy industries already employ around two million people worldwide, supplying the globe

---

[4]Roughly half of all the energy consumed in the developed world is used to run buildings (especially for cooling). For instance, in Australia, buildings consume around 44% of all energy (*Source*: OECD Report, Challenges and Policies 2003). A building's air-conditioning system is typically responsible for half the base building's energy consumption; here are the most significant reductions in energy and carbon emissions possible.

The Australian property industry contributes approximately 30% of Australian greenhouse gas emissions. A further 25% of all energy consumed is accounted for by traffic and transport. A large quantity of non-renewable fossil fuel is needed to generate this energy, and the process involved in the conversion of fuel into energy has a lasting negative effect.

[5]At present in Europe, renewable energy sources account for only 7% of the total EU energy mix. In Germany, renewable energy currently contributes 11% to the energy mix, mainly generated by wind power. However, renewable energy already accounts for 25% of the installed capacity of California, and half of Norway's; a new clean energy boom is about to start.

Wind: Off-shore wind parks have probably the largest growth potential at the moment. Wind power is the least expensive of the renewable sources of electricity. Denmark has become a world leader, with wind set to generate at least 25% of Denmark's total energy by 2009. The city of Copenhagen, for instance, recently installed 20 large wind generators off-shore in front of its port.

Germany and Spain are the world's biggest producers of wind energy, slightly ahead of the US (*Source*: Windustry 2007). Both countries adopted the system of 'tariff premiums', which provides energy producers with a guarantee that all the electricity they produce will be purchased by distribution companies at

with the renewable energy technologies which are now available (Australian Conservation Foundation). For instance, over the last ten years, 175 000 new jobs were created in the renewable energy industry in Germany alone. Producing renewable, clean energy does not mean a downturn in employment or industry – quite the opposite: it's about a different way of doing business, about a new industry with innovative products.[6]

A general lack of training, education and research activity in sustainability has been identified across the professions involved – just refer to the recent strategic programs by the Royal Australian Institute of Architects (RAIA) and Royal Institute of British Architects (RIBA) in an attempt to deal with this dilemma. There is clearly a need for more research and education in sustainability. Fortunately, the attitude is now about to change: embracing principles of sustainability has proved for many planners not to be a burden, but a competitive advantage. In future, architects and urban designers are likely to be out of work if they do not know how to deliver environmentally well-performing solutions. Architects must build their own expertise in sustainability, and not simply rely on the expertise of other disciplines. Gaiser has warned of the comfortable mentality to simply hand over responsibility to the building services engineer: 'Architecture could lose or give up its responsibility to perform if we no longer have environmental achievement "per form", but only "per system"' (Gaiser 2007).

The profession has to become an expert in green, energy-efficient planning and design; not only to use more energy-efficient appliances, but to eliminate the future need for

---

premiums over the market price. Both countries are also frontrunners in solar power and bio fuels. Wind turbines now produce more than 25% of Spain's total electricity supply; more than nuclear, coal or any other single energy source.

Europe's biggest solar power plant is outside Seville (Spain): in its centre is a tower 115 metres high that captures the solar energy reflected by a field of giant mirrors and stores it as steam. It already generates enough energy to supply 6000 houses in the first stage. *Solar:* The potential of the solar power source is enormous. For instance, the amount of solar energy that hits Australia in one day is about half the world's total annual energy use. The area required by solar PV panels to generate enough energy for all of Australia's energy needs is 6600 square kilometres in size, roughly the area of greater Sydney.

Covering 4% of the world's deserts with photovoltaic cells would theoretically provide enough energy for the entire global population – without adding any carbon dioxide to the atmosphere. Solar technologies harness the energy either directly as heat (solar hot water systems), or for conversion into electricity, selling the excess power back into the electricity grid. Particularly promising is the Sliver Cells technology and thin-film technology (which is silicium free). Germany is currently by far the biggest user of solar panels worldwide, even more than Japan. Refer also to the carbon-neutral project 'Solar City' in Linz, Austria, at: www.linz.at/solarcity.

The further development and utilization of non-conventional resources like chemical energy from organic waste or thermal energy from wastewater, and the development of decentralized technologies, will come into focus within the next few years.

A study by the German Advisory Council on Climate Change has shown that it is technically feasible for solar energy to provide 25% of all primary energy by 2050.

It is hereby unlikely that large energy companies with business based on fossil fuel (such as Shell or BP) will easily transfer to the renewable energy market. Shell, for instance, spends currently only 3% of its investment on the development of renewable technologies.

[6]According to Sir Nicholas Stern's review (2006), delaying the shift to clean energy will cost us more in the long run. To weaken or delay the clean energy shift would be poor economic management. And there are things we should all do immediately:

- Buy 100% green power (see www.greenelectricitywatch.org).
- Install solar hot water systems (in Sydney, over 80% of the energy required for hot water can be delivered by a solar hot water heater).

artificial heating, cooling and lighting. Increasingly, owners, developers, governments, architects and planners will have to accept, specify or prefer local materials that can be recycled (rather than transporting materials from far away, or using it for landfill),[7] and find ways of reusing and integrating existing buildings over the wish to build new.

18.5 Urban Eco-Systems: Density and Climate are Key Issues

Density is a key issue in planning and influences directly the urban climate. So, to what densities should we build? Is the less structured city with a slowly grown and irregular street pattern better for densification than a grid-planned city structure?

City blocks too close together shade each other – something only desirable in (sub)tropical cities like Singapore, Hong Kong, or Bangkok. The urban design principles that apply to cities in hot, humid, tropical conditions (not only in South East Asia, but also in parts of Australia, India, Africa, and Latin America) are entirely different from the urban design principles in temperate climates. As Bay and Ong have pointed out, the effects of urban canyons, natural ventilation and heat entrapment in the city are entirely different for the tropics: 'While sunlight is welcome in the temperate city and buildings are set back to allow sunlight to penetrate to the road level, shade is preferred in the tropics; (…) more wind and ventilation are welcome in the tropical city' (Bay and Ong 2006). This means, in tropical conditions we want to create shade and overshading between buildings, as long as we ensure the efficient natural ventilation of the spaces between the buildings.

Conventional patterns of urban development that have frequently led to lower densities and greater infrastructure costs are unlikely to remain economically feasible, as the greater dispersion of activity centres leads to an increased automobile dependency.[8] Low density suburbs are incapable of sustaining a public transport infrastructure. Densification is the key – and no inner-city site is too small as to densify and implant good architecture. However, as Gissen 2002 notes, 'there is no one formula for a mix of culture and technology that makes cities vibrant and liveable' based on (Banham 1969).

The aim of a significant amount of new developments is to achieve a rich mix of usage, scale and increased density to create city centres and new urban nodes with a compact and spatially complex model. This model relies heavily on public transport. Hereby, the city centre needs a rich mix of all types of inner-city uses: office buildings, hotels, department stores, university buildings, residential buildings, shops, cinemas, squares, good landscaping, and so on.

---

[7]Many countries, such as Australia, still lack legislation to support recycling. For instance, 'landfilling' with electronic waste has to urgently stop; the old computers and monitors create, among other things, lead pollution.

Ideas about better recycling create considered products, where materials can be used again for a post-life, as eco-effective products. Some manufacturers have started to produce their products using biodegradable materials, e.g. Nike proclaims to be waste free by 2020 – currently, the manufacturer of sports equipment takes used shoes back for disassembly and recycling. Company Trigema now produces a collection of compostable T-shirts. Refer to the 'Cradle-to-Cradle' protocol by William McDonough and Michael Braungart, at www.mcdonough.com and www.braungart.com

[8]The automobile is, unfortunately, not set to disappear yet. However, of all transport methods, the car is the largest CO_2 emitter in the developed world: approx. 45% (whereas trucks contribute around 25%, airplanes 11%, ships 10%, and railway 1.5%). This explains the urgent need to develop smaller, more energy-efficient city car models. The Smart Mercedes was only the beginning. MIT's Media Lab currently develops a new generation of micro cars. The CO_2 emissions from air travel are likely to triple over the next 50 years (refer to the report by Nicholas Stern, October 2006, published January 2007).

Grouping residential units or townhouses together in compact volumes of around four or five storeys – similar to the nineteenth century 'compact city block' model found in Paris, Barcelona, Athens, Amsterdam, or Berlin – would bring considerable environmental benefits, such as:

- Smaller building envelopes, therefore less land use;
- Need for fewer materials, therefore lower construction costs; and
- Sharing fire walls, therefore reduced energy consumption.

18.6 The 'Compact City' Discourse: Social Acceptance of High Density?

The characteristics of a compact city model – such as high density, well-defined boundaries containing city growth, mixed land uses, urban design encouraging walking and cycling, and heavy reliance on public transport – were already advocated by Newman and Kenworthy in 1989, when a compact urban form was argued to be able to reduce much of the environmental vice of low rise and sprawling cities. The attributes, benefits and dis-benefits of compact cities are well discussed in works such as (Newman and Kenworthy 1989; Breheny and Michael 1992; and Burton 2000). In general, there seems to be a superiority of concentrated forms of settlement over the sprawling forms. It has been argued that many of the problems of traffic congestion, noise and air pollution, and greenhouse gas emission in cities in the developing world were caused by car dependency and single occupancy vehicles, partly due to their low density, sprawled expansion and spatial segregation of land use functions (Breheny and Michael 1992).

The environmental claims of compact cities policies are not undisputed. However, the issue of density and compactness is a complex one. 'Densification' and 'intensification' need to be reconciled with the risk of lack of urban green space, rising cost of living (reduced housing affordability) and other expenditures associated with denser centres – which may even result in higher energy consumption and carbon emissions, due to lifestyle or embodied emissions – and negative impacts on the quality of life of the inhabitants. Thus, improved socio-spatial equity is an important consideration of sustainable city form.[9]

Higher housing density *per se* does not deliver sustainable advantages; it is the network efficiency that is crucial. The need for compacting and re-engineering the existing cities (after 100 years of decompacting the nineteenth century city) poses the questions: What form of 'recompacting' should apply? Which density is appropriate?

Great cities have obviously identifiable characteristics offering a high quality network of a public space where it is enjoyable to walk around. A strong focus on how pedestrians move around and the quality of the public domain is therefore essential. Public space is the fundamental basic order of the city, and the author suggests that urban designers need to increase their focus on the public domain, e.g. with a '6-green-star public domain' rating system (which does not yet exist). Quality public space is not just 'what is left over' – it is essential for well-being and social interaction. Given the previously described correlation between compactness and sustainability benefits, higher density requires more careful design of public space within the limited space available.

---

[9]The anti-sprawl critique was raised by Jane Jacobs in the early 1960s leading to a mainly negative image of suburbanization. The implementation of more compact cities needs to deal with emerging issues and challenges that come along with higher densities, such as social changes and conflicts in interneighbourhood relationships, e.g. of multi-apartment housing. In Sydney, for instance, some 25% of the population now live in multi-unit housing. This proportion is still low in comparison with European cities, but is predicted to grow to 45% by 2030.

Fig. 18.1. The block model of the European city: five-storey blocks as found in Paris, Berlin, Barcelona, and in other cities, have proved a solid model for energy efficiency and dense living. These cities possess already the urban characteristics of a compact city model. The strategy of saving from passive-building technology is well accepted: a compact built form with a high level of thermal insulation reduces energy losses through transmission.

18.7 Approaches Towards the Energy-Efficient 'City of the Future'

With a continuous increase in climate change, the world's climate scientists recently unveiled the IPCC Report (report by the UN's Intergovernmental Panel on Climate Change 2007)[10], which forecasts a grim future. According to the findings in the report, design criteria for major infrastructure (such as bridges, ports, campuses, shopping centres, etc.) will need to be urgently redefined to facilitate the transformation of cities towards renewable sources, and readjusted for extreme climatic events, like floods, storm surges, global temperature increases, sea-level rise, water shortages and droughts. This again illustrates the need to implement climate-responsive 'Green Urbanism' and to transform the existing fossil-based energy infrastructure of our cities.

Many of the planning concepts suggested in this chapter are not new; some have traditionally been known as methods of how to plan and build vital cities. However, these concepts simply haven't been much implemented anymore in the mainstream urban design of the twentieth century.

According to the latest IPCC Report, the following three activities will deliver the largest potential to stop climate change and be most effective:

- Saving energy at the point of consumption (energy conservation);
- Increasing energy efficiency; and
- Developing carbon-free renewable technologies, using renewable energy to power, heat and cool cities.[11]

---

[10] Global agreement on scientific evidence confirming the man-made nature of climate change has been established through four reports by the IPCC, the UN's Intergovernmental Panel on Climate Change: reports published between December 2006 and May 2007 (Fourth Assessment Report, May 2007).

[11] Researchers at the ETH Zurich have developed the '2000-Watt-Society' model. This model can only be achieved by increasing efficiency of buildings and machines, and by a shift towards renewable sources of energy. For a country like Switzerland, constant energy needs of 2000 watts per person would be adequate to ensure economic growth and to maintain the current quality of life. In comparison, energy needs in Africa are 500 watts per person, and in the US 12 000 watts.

Fig. 18.2. The impact of industry and fossil-based energy production on the population and land-scape: image of a coal-fired power station in China; recently, the country overtook the US as the world's biggest emitter of greenhouse gases. China is the world's largest consumer of coal, and is likely to continue using massive amounts of coal for many years.

Fossil-based energy generation is the world's largest CO_2 emitter, with almost 30% of all CO_2 and other heat-trapping gas emissions worldwide (data 2005) caused by coal-fired power stations. This is more than through any other source, such as through industry (approx. 20%), or forestry (approx. 18%), through traffic and transport (approx. 13%), or agriculture (approx. 14%). It makes energy production the major cause of global warming. Unfortunately, in countries such as China, India and Australia, coal still accounts for the majority (over 75%) of all energy sources. (*Source*: UN 2005).

Therefore, today, all aspects of the environment have to be considered. In an urban context, the quality of space and light, the compositional questions of urban form and grouping, and the choice of materials all have to be considered as an integral part of an ecological, well-balanced proposal. Sensible, climate-responsive urbanism will result in huge energy savings. Shading the building's western facades with external window awnings or adjustable shading devices, well-placed trees or pergolas will effectively help keep heat gain down.

However, the greatest challenge in terms of sustainable building still lies in the area of energy consumption through operation.

Some of the most effective and available urban design approaches identified are listed below:

- Minimizing the consumption of land through compact development, and densification by containing the footprint;
- Using existing and renewable urban resources such as brownfield sites, integrating underused buildings, structures and sites;
- Designing urban quarters where walking and cycling are highly attractive, with good landscaping and a diversity of uses;
- Utilizing underused sites within walking distance of public transport nodes to reduce reliance on the automobile and to increase pedestrian activity;
- Developing urban areas where efficient infrastructure systems and public transport are already in place, to reduce the need for the automobile;
- Creating a sense of urbanity through density and the design of real places;
- Developing sites in a way that the consumption of energy, non-renewable materials and pollution is reduced;
- Better considering the building's placement on the site, orientation and density issues, with a focus on public space and good landscaping;

Fig. 18.3. A new generation of affordable PV cells (e.g. silicium-free thin film technology) will soon enter the market. There is a wide variety for architecturally integrated installations in facades, on roofs or canopies, to generate the energy decentralized and close to where it is needed. (image: courtesy by Conergy).

- Designing developments in ways that increase access to affordable housing and transportation choices;
- Designing spaces with direct access to natural light and air, and better orientation to take advantage of passive solar design principles;
- Optimizing orientation and solar exposure to maximize the use of renewable resources in the operation of buildings and complexes; and
- Applying the following eight key principles for urban sustainability:
    - low-rise, high density compact communities;
    - functional mix with local and culture-specific uses;
    - eco-buildings which better harness sun, daylight, wind, rain;
    - integration and reuse of existing buildings with elements of local identity;
    - fine grain, with attention to architectural detail and smallness;
    - high quality public space network;
    - reliance on public transport and use of bicycles;
    - variety of urban greenery, integrated in the building.

18.8 Designing Buildings with the Climate – Rather than Against it

So far, our focus has been on urban design, and an understanding that which urban form is most favourable to sustainable development depends much on the different climate zones.

On the scale of energy-efficient buildings, the envelope (facade) obviously plays a crucial role. Most of the time, the facade is the driver for how services in a building are integrated; the skin should allow fresh unconditioned air into the building when possible, optimize daylight to reduce artificial lighting, and minimize heat gains and losses. Such an optimized building can stay comfortable all year long without a mechanical air-conditioning system.

In the past, higher investment costs have been repeatedly identified as the major barrier to green design. But since building components (for instance, the new generation of PV cells, which are increasingly used for city-integrated, decentralized energy generation)[12]

---

[12] Decentralized energy production, for instance with solar PV cells or wind farms, has many advantages compared to large-scale centralized power plants; centralized coal and nuclear power plants require expensive backup in case the entire plant breaks down. Currently developed organic photovoltaic cells, or flexible miniature solar panels, could dramatically reduce the cost of conventional solar power generation.

Fig. 18.4. Sustainable cities integrate brownfield sites and structures: The integration and adaptive reuse of existing buildings, and upgrading of our building stock, are important parts of this strategy (image: London Docklands, 2003; courtesy by MLC).

have become much more efficient and affordable, as they are produced in high quantities, capital investment in renewable technology can pay for itself over a shorter period of around eight to ten years.

High operating costs of buildings require our attention, less so construction costs. Fortunately, renewable energy technologies are now mature and reliable, with components built to last for decades and requiring minimal maintenance. During the life cycle of a building the costs of maintenance and operation can amount to four times the original construction costs. In future, the demand for energy-efficient buildings will inevitably increase. Demonstration projects with a positive energy balance over the year (e.g. new office buildings or large housing complexes) will serve as important pilot projects and are proof that architects have started to assume responsibility beyond aesthetic aspects.

A sensible trend is to shift away from buildings with intensive technical installations to low-tech passive solutions, with structures using efficient but simple facade technology offering operable windows for natural ventilation (as mixed-mode systems). The location for large window openings depends on the orientation and need for protection against overheating and excess solar energy: greater heat gains internally may lead to the use of less energy in winter, but can create problems in summer.

It is crucial to get the relationship between building volume and facade surface (the degree of 'compactness') right. Of course, to a large degree, all urban strategies depend on the geographical location and the local climate. In a temperate climate there is a significant advantage in having long west and east facades, using the summer sun to create a stack effect that draws air from one (cooler) side of the building through to the other (warmer) side. This effect could further be supported by a wind roof such as a 'Venturi spoiler', which uses the prevailing winds to create negative pressure on the facade. A concrete structure can provide better heat storage (thermal mass), while flexible solar shading can protect the interior from too much light and heat gain in summer (in combination with the use of high performance glass), and open up in winter to harness the sun's energy for heating (Daniels 1998; Szokolay 2004).

On the other hand, in a hot and humid (tropical) climate, there are clearly other priorities. Here, it is much more important to ensure good cross-ventilation, and to keep material mass such as concrete floors or masonry walls fully shaded to avoid storage of unwanted heat in the building mass (Koenigsberger *et al.* 1973; Drew *et al.* 1979; Kats *et al.* 2003). But in both cases, from the standpoint of energy consumption and avoidance of heat gain, it is generally seen as an advantage to have compact volumes with a reduced facade surface area.

This is significant for urban planning and suggests that well laid-out subdivisions are ones where homes and offices are oriented to enhance passive solar heating and cooling, for the use of solar hot water heaters, for maximum natural daylight, and for taking advantage of the local prevailing wind direction to catch cooling breezes. When the building facades are made specific to their orientation, with the intentional placement of closed wall surfaces and small window openings, it will result in substantially lower energy consumption – as well as reducing the dependence on air conditioning. For office buildings, this means moving away from the air-conditioned sealed building type and deep plan layout, to buildings with all windows operable and shallow plans (less than 15 metres width) for optimum daylighting. Those office buildings take full advantage of natural daylight for all regularly occupied spaces, which reduces energy consumption and excess heat from electric lighting. As a consequence, there are demonstrable cost benefits in operating such buildings.

The implementation of sustainable design measures frequently implies some additional initial construction costs, although it is well documented that life cycle costs are reduced. Scientist Greg Kats has extensively researched the costs and benefits of green buildings, and has demonstrated conclusively that sustainable architecture and urban design is a cost-effective investment. (Kats and Greg *et al.* 2003).[13]

18.9 Landscape and Building: Reintegrating Green and Maintaining Biodiversity

In recent years, we have seen exciting examples of new public landscapes being developed, combined with increased urban density, and new recreational landscapes utilizing derelict urban sites. Even formerly contaminated sites and rubbish dumps have been successfully rehabilitated and turned into public parkland (e.g. the park at Port Forum in Barcelona 2006).

There is a great potential for urban greenery and gardens to achieve identity and a sense of place. Much research is done into ways to better incorporate landscape as part of urban design, and to integrate new forms of green in the buildings (refer to the recent work by architects and landscape designers such as MVRDV, Ken Yeang or West 8) – in other words, exploring better solutions as to how we can achieve a closer symbiosis between building and nature. These can provide inspiring micro-climates for daily interaction with nature. In this context, landscape can be understood as the 'urban edge' where methods of water management and maintenance of biodiversity are crucial (Johnson 2000).

Despite increased density, the city could still contain small urban gardens for recreation or even food supply, in combination with green roofs to collect the rainwater for irrigation of those gardens. The installation of rainwater tanks, the development of more effective recycling water programs (grey water usage), recycled sewage, and the ability to harvest the storm water runoff, all need to be part of an urban water strategy.

---

[13] Refer to Kats, Greg *et al.* (October 2003): *The Costs and Financial Benefits of Green Buildings*; a report to California's Sustainable Building Task Force. American scientist Greg Kats of Capital E is the nationally known author of the most widely referenced study of the costs and benefits of green buildings. This study has demonstrated conclusively that sustainable building and the use of renewable energies is a cost-effective investment.

It is generally important to ensure the 'hard' appearance of a new development is lessened by soft landscaping, tree planting and vegetation. Innovative ideas of vertical landscaping and the re-creation of ground conditions to roof gardens ('cool roofs') are now being applied by many designers. Today, most cities have established policies on the principle that removal of a potential green area at ground level should be offset by planting the equivalent area at roof level.

18.10 Traffic Planning to Improve Public Transport

Good traffic planning will be essential for the sustainable 'City of the Future', and there are now plenty of innovative ideas and traffic concepts to reduce our car dependency. While it is not possible with buildings, given the appropriate legislation, almost all motorised vehicles could be replaced by environmentally friendly ones within a decade.

Some concepts have become standard in many European cities; however, they have not yet been fully embraced by North American or Australian traffic planners. In fact, inflexible traffic planners seem to frequently be the weakest part in the multi-disciplinary planning chain, hanging on to old ideas, conventions and institutional arrangements. Just to mention a few of the concepts that could be applied:

- Genuine improvement of integrated public transport, where rail/bus/tram/subway/ferry timetable schedules are all coordinated with each other;
- Improved coordination of timing between various red lights to avoid unnecessary halts of traffic flow;
- Step-by-step reduction of inner-city car parking spaces through improvement of parking facilities at the fringe of the historical centre, combined with park-and-ride concepts for rail/light railway stations;
- Street profiles to integrate cycle paths that are sufficiently safe and wide (e.g. Melbourne has now increased the width of cycle paths from 1.5 to 2.5 metres), offering bike stations to park and repair bikes (as is common in the Netherlands and Japan);
- Constructing subway systems is extremely expensive and time consuming. Larger, polycentric cities (cities with several centres but no efficient subway system, such as Sydney, Sao Paulo or Los Angeles) need to find ways to connect their various centres with an efficient, high speed linkage, such as a monorail or light railway system; and
- Adopting a general attitude based on the fact that 'pedestrians are more important than vehicles'.

Australians travel over 7500 kilometres annually by road (RTA data 2006). Their reliance on the private car for transport is having a huge impact on the environment. The situation in the US is similar. Therefore, urban planners need to create more compact, denser cities with a vastly improved and integrated public transport system combined with localized and well-designed pedestrian and cycle networks.

The electrical-driven light railway (tram) system is clearly combining a series of advantages: it only requires a quarter of energy compared with buses, and can transport in one go around four times the amount of people (without blowing exhaust fumes in the air). These are good reasons why currently the tram is running in around 400 cities worldwide and another 250 cities are introducing it as the inner-city public transport system (this includes cities, such as Athens, Sydney, Istanbul, Paris, Lyon, Berlin, Saarbruecken, Edmonton, Bergen, etc.).

Fig. 18.5. The automobile is not the only source of emissions, but a major one. Most CO_2 emitted by global traffic comes from cars (44.5%) and from trucks (25%). Cars will not disappear that soon, therefore
we need to develop more appropriate, energy-efficient models, e.g. electrical or hydrogen fuel-cell powered automobiles (image: study by Media Lab, MIT, 2006; the Media Lab concept relies on electricity).

18.11 Two Examples for Urban Regeneration: The 'City Campus' and 'PortCity' Projects

The following are two case studies that are both based on the findings described
earlier. These two projects are examples for the application of sustainable urban design
principles,[14] both urban design proposals for the city of Newcastle (Australia). 'City
Campus' and 'PortCity' were developed in the fourth year architectural design studio
at the University of Newcastle. The designers were interested to apply the sustainability
criteria previously mentioned and to test these criteria in real projects; the author believes
there are some generic lessons to be learnt from these studies.

The city of Newcastle is located about 70 miles north of Sydney, where the Hunter River
flows into the Pacific. Newcastle is the sixth largest and the second oldest city in Australia,
and the second largest in the state of New South Wales. It is also one of the largest coal export
harbours in the world and boasts coal deposits which cover much of the Hunter region.
Newcastle city has a population of around 150 000 people. It has a temperate maritime climate with four distinct seasons (average maximum temperature is 26°C, minimum 7°C).

Coal mining in Newcastle started in 1804, and the city grew with its mining, steel production and other industrial activities. The city was opened to free settlement in 1823
and quickly expanded into the Hunter Valley area.[15] The post-industrial development
of Newcastle from a former district of heavy industry into a contemporary, twenty-first
century city poses a significant challenge to change management.

Today, most of the steel manufacturing plants and smelters have been closed (e.g. BHP
closed its steel production finally in 1997), and large areas of formerly industrially used

---

[14] In *The Fatal Shore*, R. Hughes tells the story of Newcastle's settlement and development from a tiny
convict outpost to an important port town (Newcastle today: the world's largest coal exporting harbour, 2007). Refer to Hughes, Robert (1987). *The Fatal Shore*. Vintage, London, pp. 425–439.

[15] A definition of 'sustainable' is derived from the Latin verb 'sustinere', which describes relations that
can be maintained for a very long time, or indefinitely (Judes 1996). The idea of 'sustainable urban
development' originated at the 1992 UCED Conference and Earth Summit in Rio de Janeiro. It is based
on the concept of balanced environmental planning instruments and methods, of which a great variety of visions for urban development has been created. However, 15 years after Rio, there is still no
global mandatory commitment to a cap on carbon emissions.

Fig. 18.6. Improving public transport with a modern light railway system to reduce our car dependency. Pedestrian friendly landscaping is possible with integrated light railway tracks.

Fig. 18.7. The alternative to light railway: Upgrading the bus network (Curitiba, Brazil) with efficient, high-tech express buses which can transport up to 120 people.

land – frequently brownfield sites in prime waterfront locations – became available for sustainable urban development or conversion into parkland. This process is accompanied by tremendous environmental challenges, such as heavily contaminated soil, mine subsidence, and the legacy of 200 years of industrial pollution.

The two urban design projects, 'City Campus' and 'PortCity', are described below:

- Project 1: 'City Campus' has a focus on regeneration of the city centre as a key principle of the sustainable city.
- Project 2: 'PortCity' has a focus on transformation and redevelopment of the harbour area as a model for sustainable expansion of the city.

Both projects are based on a balanced approach, high ambition regarding urban ecology and the aim to run the city centre on renewable energy. They offer a great opportunity for Newcastle to grow and regenerate over the next 20 years through energy transition in the right direction.

The City Campus uses careful strategies for urban infill and regeneration of the neglected downtown. The aim is to look beyond the site boundaries at the ecological footprint of the city centre. The brief asks for university facilities for 3000 students, including a new public library, drama theatre, school of architecture and fine arts, faculty of business and law, as well as the integration of existing buildings. The proposal suggests relocating significant parts of the university from its 1960s suburban campus, combined with student accommodation, back to the city centre. A new landscape design for the central park (Civic Park) is part of the project, to get high quality green spaces, green roofs and biodiversity within a sustainable neighbourhood and eco-buildings. 'Sustainable neighbourhood' was hereby defined as 'a compact community cluster using as little natural resources as possible, with careful consideration of public space'. The City Campus will facilitate the revitalization of the centre and stop its further decline.

The second proposal is the 'PortCity'[16] project, which is based on an overall strategy for reclaiming post-industrial waterfront land. It is a mixed-use urban waterfront development of ten hectares of prime waterfront land, of which about half will be dedicated to public parkland. Once the industrial working harbour has moved up the Hunter River (in around six to eight years by 2015), Dyke Point could be connected with the centre by a new pedestrian and cycle bridge, so that the now underutilized land can be turned into a sustainable city precinct.

The 'PortCity' is aimed to be a 90% emission-free development, where all energy is provided by distributed power from a variety of sources – photovoltaics, wind, biomass and possibly geothermal – to demonstrate that it is affordable and achievable today to make all major new urban developments carbon free and based on renewable energy sources. Questions of density, scale, urban public transport (a light railway 'super connector' system was suggested), solar orientation, and the maritime heritage of the working harbour are all critical to the design, as well as the integration of the established Carrington community. The staging of the 'PortCity' development is used to drive the design approach, and activates the existing Pump House as catalyst and starting point.

Newcastle's port is the place where large engineering infrastructure is juxtaposed with the cityscape – a great and poetic inspiration to all architects and urban designers, to reinvent infrastructure and to face up to future challenges.[17] In the post-industrial city there is typically a large potential of undeveloped inner-city areas for the generation of renewable forms of energy.

18.12 Beyond Concerns of Aesthetics: Some Concluding Remarks

During the last century, we spent most of our time developing methods to adjust to the consequences of pollution and global warming rather than dealing with the root causes. The scientific evidence and consensus of international scientists on human-induced global warming is no longer denied.[18] There are also rising public expectations about environmental

---

[16] In preparation of this urban design project, similar port redevelopment projects, such as in Hamburg, Rotterdam, Genoa, Vancouver and Barcelona have been analysed.

[17] The US government is increasingly criticized for having ignored for too long the scientific evidence on climate change impacts. Industry-sponsored greenhouse sceptics – with 'research' funded by the mining, automobile and coal industry – were trying to undermine the credibility of scientists through the 1990s. Today, such censorship is not working anymore, as events like hurricane Katrina (destroying New Orleans in August 2005) lead the American people to their own conclusions. Administration now accepts the mainstream science of climate change, which had been so openly challenged by senior government members only a year ago.

[18] Huge investment efforts will be needed to solve the climate change issue – which is entirely appropriate for saving the environment. In the same way, we invest in all kind of other areas, such as in security, poverty, military, or tourism.

Fig. 18.8. The 'City Campus – Strategies of Urban Infill' Studio, proposal I (2007) by the author and students. The central public parkland and its high quality are an essential part of the urban ecology.

Fig. 18.9. The 'PortCity – Reclaiming the Post-Industrial Waterfront' Studio, proposal (2007) by the author and students. This is a zero-emission city extension with the historical city centre in walking distance.

issues, and with the right leadership we are now in the position to take huge steps forward. Interestingly, the public seems to be ahead of the politicians – the 'public' is really in support.

Today, we understand that the most significant environmental challenges of our time are global climate change and excessive fossil fuel dependency. Related to this are rising greenhouse gas emissions and water, soil and air pollution, all of which have significant environmental, social and economic consequences. Out of this has emerged a new focus of sustainability on large-scale complexes and the need to transform our cities.

Climate change demands action now if we are to make the dramatic and necessary cuts to greenhouse pollution levels. Climate change demands that we change the conventional concepts and layouts of cities, and implement sustainable 'Green Urbanism'. We already have the technologies to commence the rapid transition to an energy future based on renewable energies and new standards in energy efficiency.

By taking early action on climate change and focusing on more sustainable centres, a number of cities have already succeeded in significantly reducing their energy demand

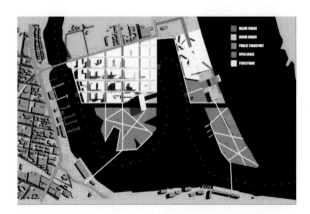

Fig. 18.10. The 'PortCity – Reclaiming the Post-Industrial Waterfront' Studio, master plan (2007) by the author and students.

and, therefore, greenhouse gas emissions (e.g. Freiburg in Germany). While the development industry must take a big share of the responsibility for contributing to the current state of our environment, it is the same industry which is now also in a position to do something about it. Businesses can provide the solution as long as government sets the right standards; however, both together have to face up to the challenge. Several developers and contractors have already started to embrace this paradigm shift and are increasingly changing their practices to incorporate such ideas, as financial investors value ethical and environmentally friendly projects.

We need more intensive investment in sustainability to ensure the rapid application of those new, available technologies – a combined effort like the 'Apollo Program' once triggered 40 years ago.[19] We already possess the scientific and technological know-how to solve carbon and climate problems, and we have practical ways to cut emissions. Those new systems need to be applied in a way that will not only stop the CO_2 emissions that take place, but will also contribute to repairing some of the damage. It is clear that we need to act now, based on practical ways forward.

The city's transition towards renewable energy and compact communities will reduce the dependency on energy and water providers, and reduce the demand on public infrastructure (Stern Sir Nicholas 2006). It is obvious that such sustainable designs need to go far beyond the concerns of aesthetics. Cities are built resources with high primary energy content. Renewable forms of energy, and the strategies outlined earlier in this chapter, are available, achievable and affordable. They present an opportunity to make inner-city life more attractive. Affordable housing and liveable, mixed-use neighbourhoods (with a mix of housing types, prices and ownership forms) are all essential elements of sustainable urban development. Design plays a major role in achieving liveable, affordable inner-city models.

The findings of the research lead to the conclusion that sustainable building development means, first of all, applying technical solutions sparingly and making the most of all passive means provided by the building fabric and natural conditions. In a second step, active and mechanical systems can be integrated. Therefore, for a sustainable city centre, it is imperative first to develop an overarching vision that counters sprawl, curbs commuting, reduces automobile dependency, and promotes improved public space with increased urban density.

There is clearly a need for a better planning and urban design education, combined with increased research activities. Students show a great interest in the subject of sustainable design, and all architecture and planning schools should introduce the subject in their teaching. Education and research are major factors in fighting climate change. Radford points out that design decisions for the sustainable city 'must be based on both an ethical position and a coherent understanding of the objectives and systems involved' (Radford *et al.* 2003). In education, there is a need for a humanistic, socially responsible approach, with environmental sustainability as a non-negotiable underlying ethos. It is important to teach students in architecture and urban design that 'green' is not something to be bolted on once the design is completed, but is to be integrated already in the conceptual phase.

Accompanying the shift towards renewable energy sources, more exchange of experiences between cities is necessary. As we begin to understand that the traditional model of urban development will no longer work for the future, we are still grappling with the difficult task of defining new models, objectives and systems involved. Acknowledging that the overall development patterns of our urban areas are the result of a complex and difficult process, we accept that current patterns of outward-spreading development cannot be sustained economically, environmentally and socially. The increase of densities, combined with urban regeneration strategies for neglected inner-city areas, are therefore valid solutions: it is less environmentally damaging to stimulate growth within established city centres, rather than sprawling into new, formerly unbuilt areas.

Key aspects on our path towards the sustainable city will be an increase of community awareness and incentives for renewables. No doubt we will see a further rise of the renewable energy economy gradually replacing conventional energy technologies with renewable systems.

If we get this right, the city – often thought of as being the most wasteful user of resources – could emerge as the new model of sustainability.

Acknowledgements

Special thanks go to the collaborating students in the design studio: Michael Smith, Tim Hulme, Bede Campbell and Jeremy Pease.

References

Australian Conservation Foundation, www.acfonline.org.au; Also Business Council for Sustainable Energy, Australia, www.bcse.org.au

Banham, R. (1969). *The Architecture of the Well-Tempered Environment*. Chicago Press/Architectural Press.

Bay, J.-H. and Ong, B.-L. (eds) (2006). *Tropical Sustainable Architecture: Social and Environmental Dimensions*. Architectural Press.

Braungart, M. and McDonough, W. (2002). *Cradle to Cradle. Remaking the Way We Make Things*. North Point.

Breheny, M. J. (1992). *Sustainable Development and Urban Form*. Pion.

Burton, E. (2000). The compact city: just or just compact? A preliminary analysis. *Urban Studies*, **37**(11).

Daniels, K. (1998). *Low-Tech. Light-Tech. High-Tech. Bauen in der Informationsgesellschaft*. Birkhaeuser.

Drew, J. and Fry, M. (1979). *Tropical Architecture in the Dry and Humid Zones*.

Droege, P. (2007). *The Renewable City*. Wiley. http://www.renewablecity.org

Flannery, T. (2005). *The Weather Makers. How Man is Changing the Climate*. Atlantic Monthly Press.

Gaiser, W. (2007). Architectural research into environmental performance. *Architecture*, **3**(Spring), 25–26.

Gauzin-Mueller, D. (2002). *Sustainable Architecture and Urbanism: Concepts, Technologies, Examples.* Birkhaeuser.

Gissen, D. (ed.) (2002). *Big & Green: Toward Sustainable Architecture in the 21st Century.* Princeton Architectural Press.

Hall, P. (2005). Keynote speech by Sir Peter Hall for the L'Enfant Lecture on City Planning and Design, on 15 December 2005 in Washington, DC, organized by the American Planning Association and National Building Museum.

Intergovernmental Panel on Climate Change (2007). *Fourth Assessment Report. Contribution of Working Group I: Summary for Policy Makers.* IPCC, UN.

Johnson, C. (2000). *Greening Cities. Landscaping the Urban Fabric.* NSW Department of Commerce.

Kats and Greg *et al.* (2003). *The Costs and Financial Benefits of Green Buildings.* Report to California's Sustainable Building Task Force (October).

Koenigsberger, O.H. *et al.* (1973). *Manual of Tropical Housing and Building, Part One: Climatic Design.* Longman Group.

Lehmann, S. (2006). Towards a sustainable city centre: integrating ecologically sustainable development (ESD) principles into urban renewal. *Journal of Green Building,* College Publishing 1(3)(Summer), 85–104.

Lehmann, S. (2006). Refereed paper presented at the 2nd International Conference on Environmental, Cultural, Economic and Social Sustainability, in Hanoi, Vietnam. Proceedings published in *The International Journal of Environmental, Cultural and Social Sustainability,* CG-Publisher, 2(2), 113–128.

Newman, P. and Kenworthy, J. (1989). *Cities and Automobile Dependence: An International Source Book.* Gower, Aldershot.

Radford, A., Williamson, T. and Bennetts, H. (2003). *Understanding Sustainabile Architecture.* Spon Press.

Scott, A. (2006). Design strategies for green practice. *Journal of Green Building,* College Publishing 1(4)(Fall), 11–27.

Stern, Sir Nicholas (2006, published 01/2007). *The Stern Review: The Economics of Climate Change.* Cambridge University Press.

Szokolay, S.V. (2004). *Introduction to Architectural Science, the Basis of Sustainable Design.* Architectural Press.

PART V
International Urban Agendas

Chapter 19

Barcelona and the Power of Solar Ordinances: Political Will, Capacity Building and People's Participation

JOSEP PUIG

Engineer, vice-president of EUROSOLAR – the European Association for Renewable Energy, former sustainable city councillor of Barcelona, Professor at the Autonomous University of Barcelona, senior consultant at ECOFYS SL (Spain)

19.1 Introduction: Energy and People Living in Cities

In 1975, 4 billion people lived on Earth – and 38% in cities. At the end of 1999, we were 6 billion and 47% of us were living in cities. But in 2006, according the United Nations, our number had risen to 6.5 billion, with half living in urban environments.

But this is an unsustainable path: in the early 1990s the carbon emissions crossed the threshold of 6 billion tons and in 2003 they were 6.7 billion. At the same time the gross global product was US\$ 48.8 trillion (US\$ 44 trillion directly linked to the use of fossil fuels). It means that global carbon efficiency was over \$7302/ton C. Japan, one the most efficient carbon economies in the world, had a carbon efficiency of approximately \$13000/ton C and it has a per capita carbon emission of 2.35 tons C/capita/year. If all the present inhabitants on Earth were emitting the same as in Japan, global carbon emissions would be more than 14.6 billion tons, more than twice the present level of emissions. The IPCC recommends a 60–80% reduction and some authors (Byrne *et al.* 1998) have suggested that 2.35 tons CO_2 per capita as a global sustainable emissions level, which is equivalent to about 1 ton of C.

Let us now take a look at the present energy systems (Fig. 19.1). To increase the efficiency of the energy system we need to increase the efficiency of supply technologies (ST) and also to increase the efficiency of end-use technologies (EUT). But a sustainable energy path requires not only technology but also a sustainable lifestyle.

How much electricity was needed to supply minimum domestic energy services (MDES) to urban populations living in 2006?

Assuming the MDES listed in Fig. 19.2, and not considering the non-electric energy services (cooking and heating water), the electricity necessary to supply each family living in cities depends on the end-use technologies the families were using: conventional or energy efficient technologies (Table 19.1). Each family equipped with conventional technologies requires a supply of 7008 kWh/year, but the use of energy efficient end-use systems

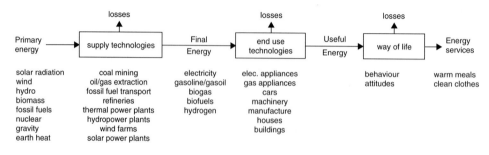

Fig. 19.1. Energy systems: the techno-human chain.
(*Source*: Norgard 1992).

Energy services	level of energy services (electricity)
Lighting	1000 lumen average, corresponding to 6 incandescent lamps, each 60 W, operating 6 hours per day.
Refrigeration	200 litre refrigerator ($+5°C$) and 100 litre freezer ($-18°C$)
Washing	200 laundry washings per year, each 4 kg, in an automatic electric washing machine. Possible need for warm water is assumed to be provided from non-electrical energy
Electronics	several hours of TV watching, radio listening and computer use every day, as well as other minor uses of electronics
Ventilation	supply of fresh air in high rise buildings plus some unspecified ventilation for cooling
Other uses	several other pieces of electric equipment can be added within this category. Equipment with electric heating should generally be avoided

Fig. 19.2. Minimum domestic electricity service per household.
(*Source*: Norgard 1989, 1991).

Table 19.1. Minimum Domestic Electricity Services

Energy services (electricity)		conventional			efficient		
		kWh/ year	W.year	Wyear/ cap	kWh/ year	W.year	Wyear/ cap
Ligthing	6 bulbs*6h/d*365d/y	788	90	23	280	32	8
Refrigeration	refrigerator (200 li.)+ freezer (100 li.)	850	97	24	140	16	4
Washing	washing machine: 200 times/year, 4 kg/laundry washing	400	46	11	70	8	2
Electronics	TV,radio,PC	2483	283	71	351	40	10
Ventilation	fresh air	500	57	14	105	12	3
Other uses	other electric equipment	1987	227	57	280	32	8
Total		7008	800	200	1226	140	35

reduces its demand to only 1226 kWh/year. On a yearly basis the average power choice is between 800 watts or only 140 watts. This means 200 W/capita – or only 35 W/capita.

According to the United Nations the urban population was 3.27 billion in 2006. To cover the MDES for the entire urban population in 2006, it required 5729 TWh/year or only 1003 TWh/year, depending on the type of end-use technology in use.

How is this energy to be supplied? Before answering this question, let us take a look at supply technologies.

Using conventional coal thermal power plants (steam turbine, efficiency 36%) one will need 1022 or 179 units, 800 MW each with a capacity factor – CF – of 0.8, depending on whether consumers were using conventional or energy efficient end-use technologies. The resulting emissions will be the ones listed in Table 19.2.

If the electricity is generated with combined cycle gas turbines (steam and gas turbines, efficiency 53%), in this case one will need 4088 or 715 units, 200 MW each (with a CF of 0.8). The associated emissions are listed in Table 19.2.

If the decision is to supply the electricity with nuclear power plants, in this case one will need 818 or 143 units, 1000 MW each (CF: 0.8), but they will produce thousands of tons of spent fuel (more than 20 000 tons or 3500 tons) containing significant quantities of plutonium (164 tons or 29 tons) and they will introduce huge quantities of radioactivity into the biosphere (54 or 10 quatrillion becquerels). Also they will need to mine huge quantities of uranium ores (212 or 37 million tons) to produce the yellow cake (155 or 37 tons) needed to fuel the reactors. In Table 19.2 are listed the quantities of uranium ores, yellow cake and solid and liquid wastes (containing 85% of radioactivity from the ores).

But if the electricity is generated using renewable energy sources, the electric system will not have any emissions at all. What will be the requirements for producing that electricity with renewable energy technologies? A comparison of technologies produces interesting figures. The selected technologies are established and being commercialized: Solar Energy Generating Systems (SEGS), like those in the Mohave Desert, California, and Wind Energy Conversion Systems (WECS), like ones now operating in many places of the world. In the case of SEGS, it will be necessary to install 1869 GW of capacity or only 327 GW, but 36 063 km^2 or 6311 km^2 will be needed – at a capacity factor of 0.35, and 1.93 Ha/MW. In the case of WECS, it will be necessary to install 2180 GW or 382 GW, and the surface needed will be 87 200 km^2 or 15 260 km^2 (CF=0.3, and 4 Ha/MW). But using modern WECS in the megawatt range with a needed area of 1 Ha/MW, the total surface needed will be 21 800 km^2 or 3815 km^2. See Table 19.2 for more details.

19.2 The Real Case of Barcelona: The Energy Needs and the Energy Supply

But what is happening in a real and concrete city? The example of Barcelona is instructive.[1] Figure 19.3 shows the non-renewable energy flow for Barcelona at the second half of the 1990s. The city imported a large fraction of energy it used in the form of electricity, gas

---

[1] All the data about energy in Barcelona stem from *BarnaGEL – Barcelona Grup d'Energia Local. BarnaGEL* was the local energy agency created with European funds in 1995; it operated until 2000. In that year the councillor responsible for energy policy terminated the support of *BarnaGEL*, apparently because it operated outside the municipality's control. Instead, a new organization under the direct management of the city was created. Since then, the city's performance appears to have weakened, both in terms of efficiency and in the introduction of renewables, illustrating the importance of perseverance and continuity in political leadership. The data about capacities of power plants needed to supply Barcelona derive from the author's lecture notes on energy at the Autonomous University of Barcelona.

Table 19.2. Urban world, 2006: supply and end-use technologies.

Number of fossil fuel power plants, SO$_2$ NO$_x$ and CO$_2$ emissions

| WORLD 3270000000 | | Supply technology | | | | | | |
| | | Coal thermal power plant (steam turbine, 36%) conventional | | | | | | |
		W/cap	TWh/year	number plants	Capacity MW	SO$_2$ mtn	NO$_x$ mtn	CO$_2$ mtn
End-use technology	conventional	200.00	5729.04	1022	800	98.54	7.39	5064.47
	efficient	35.00	1002.58	179	800	17.24	1.29	886.28

| | | Comb. cycle power plant (gas + steam turbine, 53%) efficient | | | | | | |
		W/cap	TWh/year	number plants	capacity MW	SO$_2$ mtn	NO$_x$ mtn	CO$_2$ mtn
End-use technology	conventional	200.00	5729.04	4088	200	0.00	0.57	1976.52
	efficient	35.00	1002.58	715	200	0.00	0.10	345.89

Nuclear power plants, radioactive emissions, spent fuel (Pu content), yellow cake, uranium ore, liquid and solid wastes

| | | Supply technology | | | | | | | | | |
| | | Nuclear power plant | | | | | | | | | |
		W/cap	TWh/any	number reactors	capacity MW	Emissions air + water Bq*10^{12}	Spent fuel ton	Pu spent fuel kg	U$_3$O$_8$ yellow cake ton	Uranium ore ton	Liquid wastes ton	Solid wastes ton
End-use technology	conventional	200.00	5729.04	818	1000	54426	20438	163500	155325	212550000	367875000	245250000
	efficient	35.00	1002.58	143	1000	9525	3577	28613	27182	37196250	64378125	42918750

Number of SEGSth AND WECS, SO$_2$ NO$_x$ and CO$_2$ emissions

| | | Supply technology | | | | | | | |
| | | SEGSth | | WECS | | clean and renewable solar/wind/biomas/hydro | | |
		W/cap	TWh/any	Capacity GW	Surface km^2	Capacity GW	Surface km^2	SO$_2$ mtn	NO$_x$ mtn	CO$_2$ mtn
End-use technology	conventional	200.00	5729.04	1869	36063	2180	87200	0.00	0.00	0.00
	efficient	35.00	1002.58	327	6311	382	15260	0.00	0.00	0.00

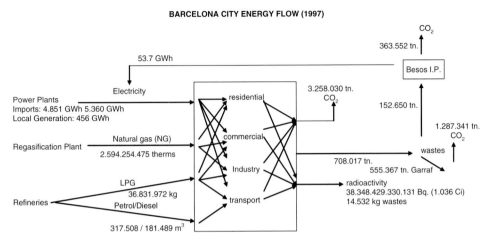

BARCELONA CITY ENERGY FLOW (1997)

Fig. 19.3. Barcelona city energy flow (1997).
(*Source: BarnaGEL* 1999).

(NG and LPG), petrol and diesel. And the city was producing waste, half of it organic. Since 1973 all the waste was deposited in a big landfill (Garraf) and burned in an old incineration plant, only adapted to control emissions in the early twenty-first century. Also, because the electricity the city was using came from nuclear power plants, an enormous amount of radioactive waste was generated. The city also generated copious amounts of greenhouse gases (CO_2 and CH_4).

Looking at Barcelona at the end of twentieth century, one can see that the city was essentially serviced by non-renewable energy inputs. The figures in 1997 were:

- Electricity: 5360 GWh
  - Imports: 4851 GWh
  - Local generation: 456 GWh
  - Incineration plant: 53.7 GWh
- Natural gas (from a regasification plant): 2594 254 475 therms
- Liquefied petroleum gases (from refineries): 36 831 972 kg
- Petrol (from refineries): 317 508 m³
- Diesel (from refineries): 181 489 m³.

At that time Barcelona produced non-renewable outputs, related to energy (1997). The figures were:

- Municipal wastes: 708 017 tons
  - To landfill: 555 367 tons
  - To incineration: 152 650 tons
- CO_2 emissions: 4 980 923 tons
  - From landfill: 1 287 341 tons
  - From incineration: 363 552 tons
  - From energy uses: 3 258 030 tons

- Radioactivity (from electricity generation):
  - Emissions: 1036 curies (38 trillion becquerels)
  - Waste: 14 532 kg.

But Barcelona's land area also receives huge quantities of solar energy: ten times more than the energy the city consumes, or 28 times more than the electricity it uses. The sun over its roofs and terraces alone provides 15 times more energy than the electricity it needs. Also the waste the city produces is a potential source of energy: in energy terms the organic wastes produced by the city represent a quarter of the NG consumption of the city, if the organic fraction of municipal solid wastes mixed with the liquid organic wastes were digested with anaerobic systems in methanization plants. See in Fig. 19.4 the renewable energy flow in the city.

The Barcelona renewable energy inputs are:

- Solar radiation over the city limits: 542 430 150 GJ or 150 676 GWh.
- Over the buildings: 303 156 811 GJ or 84 211 GWh.
- Over the streets: 89 893 439 GJ or 24 970 GWh.
- Over the green spaces and forests: 149 379 900 GJ or 41 494 GWh.
- CH_4 from organic wastes and sewage: 633 506 328 therms.

In the real case of Barcelona, the domestic electricity consumption was 1348 GWh/year (1998). On an annual mean base it represented 102 W/capita, three times more than the electricity required if Barcelona families were using efficient end-use technologies

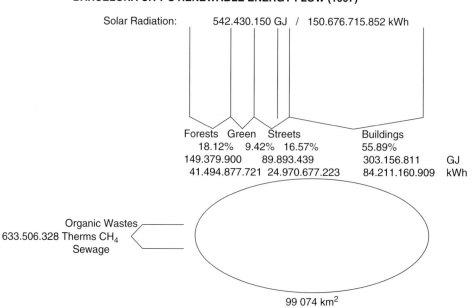

BARCELONA CITY'S RENEWABLE ENERGY FLOW (1997)

Solar Radiation: 542.430.150 GJ / 150.676.715.852 kWh

Forests Green   Streets   Buildings
18.12%   9.42%   16.57%   55.89%
149.379.900   89.893.439   303.156.811   GJ
41.494.877.721 24.970.677.223   84.211.160.909 kWh

Organic Wastes
633.506.328 Therms CH$_4$
Sewage

99 074 km^2

Fig. 19.4. Barcelona city's renewable energy flow (1997).

(35 W/capita) or half than that required if they were using conventional end-use technologies (200 W/capita). But in 2005 the families living in Barcelona crossed the level of 200 W/capita – doubling electricity consumption when compared to 1998, or 2887 GWh/year.

To supply this electricity using conventional coal fired thermal power plants (steam turbine, at an efficiency of 36%) or combined cycle power plants (using both steam and gas turbines at an efficiency of 53%) it would be necessary to have 400 MW or only 70 MW of installed capacity (with a capacity factor of 0.8), depending on whether people were using conventional or energy efficient end-use technologies. The resulting emissions will be the ones listed in Table 19.3.

If Barcelona were to decide to supply the electricity with nuclear power plants, in this case it would need the same capacity, but they would produce several tons of spent fuel containing plutonium and radioactivity into the biosphere. Also they would need to mine hundreds of thousand of tons of uranium ore to produce the yellow cake needed to fuel the reactors. Table 19.3 lists the quantities of uranium ores, yellow cake and solid and liquid wastes – those containing 85% of radioactivity from ore.

But by generating the electricity with renewable energy sources the city of Barcelona would not have any of these problems. What will be the requirements to supply the minimum domestic energy services in Barcelona with SEGS and/or WECS in three different scenarios: the real electricity consumption of the city, if people were using conventional and if people were using efficient end-use technologies?

To cover the real domestic consumption with SEGS it will be necessary to install 942 MW (needed surface: 18 km^2). And with WECS, 1318 MW (needed surface: 13 km^2).

To cover the minimum domestic energy services with SEGS it will be necessary to install 917 MW of capacity or only 161 MW (depending on the end-use technologies the people were using: conventional or efficient), but they will need 18 km^2 or 3 km^2 (capacity factor = 0.35, and 1.93 Ha/MW).

To cover the minimum domestic energy services with WECS, it will be necessary to install 1284 MW or 225 MW, and the surface needed will be 13 km^2 or 2 km^2 (capacity factor = 0.25, and 1 Ha/MW).

The surface required in all the cases is only a small fraction of the city surface (100 km^2). See Table 19.3 for more details.

If all the sectors (not only the domestic sector) of the city of Barcelona were to be supplied with 100% green electricity (7207 GWh/year) it would require 2350 MW of SEGS (45 km^2 or half the city surface) or 3291 MW of WECS (33 km^2 or one third of the city surface). If it were to supply all the electricity consumed by the city with PV solar electricity it would be necessary to install 3268 MW (33 km^2). See in Table 19.4 the requirements to supply different sectors. It is interesting to note that with less than 10 m^2 of PV panels it is possible to supply all the electricity one family consumes in Barcelona.

If the city wants to cover its sanitary hot water needs with the sun, what will be the requirements? In a normal year Barcelona consumes 1.02 billion kWh to heat water (636 million with NG and 384 million with electricity). To heat all the sanitary water the city is using, it will be necessary to cover only 1.7 km^2 of the surface, which means 1 m^2 per person, 2.5 m^2 per apartment or 21 m^2 per building, and all these surfaces are available in most of the buildings of the city.

And the city has the necessary space to make it possible. A study conducted in 1999 across all of Barcelona's neighbourhood typologies, shows that between 19 and 43 m^2 of terrace are available for each existing apartment (Fig. 19.5 and Table 19.5).

The conclusion is that in a Mediterranean city like Barcelona it is possible to reduce the present energy consumption using the most efficient end-use technologies currently available

Table 19.3.   Barcelona, 2005: supply and end-use technologies (domestic sector).

Thermal power capacity, SO₂ NOx and CO₂ emissions 2005

BARCELONA 1 605 502   year 1997

Supply technology — conventional — Coal thermal power plant (steam turbine, 36%)

				Number plants	Capacity MW	SO₂ mtn	NOx mtn	CO₂ mtn
		W/cap	GWh/year					
End-use technology	conventional	200.00	2812.84	1	401	48.38	3.63	2486.55
	efficient	35.00	492.25	1	70	8.47	0.63	435.15
	real (2005) (*)	205.25	2887.30	1	412	49.66	3.72	2552.37

(*) domestic

efficient — Comb. cycle power plant (gas + steam turbine, 53%)

				Number plants	Capacity MW	SO₂ mtn	NOx mtn	CO₂ mtn
		W/cap	GWh/year					
End-use technology	conventional	200.00	2812.84	1	401	0	0.28	970.43
	efficient	35.00	492.25	1	70	0	0.05	169.83
	real (2005) (*)	205.25	2887.30	1	412	0	0.29	996.12

Nuclear power capacity, radioactive emissions, spent fuel (Pu content), yellow cake, uranium ore, liquid and solid wastes

Nuclear power plant

		W/cap	GWh/year	Number reactors	Capacity MW	Surface km²	Emissions air + water Bq*10¹²	Spent fuel ton	Pu spent fuel kg	U₃O₈ yellow cake ton	Uranium ore ton	Liquid wastes ton	Solid wastes ton
End-use technology	conventional	200.00	2812.84	0.40	1000	17.71	26.72	10.03	80.28	76.26	104358	180619	120413
	efficient	35.00	492.25	0.07	1000	3.10	4.68	1.76	14.05	13.35	18263	31608	21072
	real (2005) (*)	205.25	2887.30	0.41	1000	18.18	27.43	10.30	82.40	78.28	107120	185400	123600

Renewable energy capacity, SO₂ NOx and CO₂ emissions

clean and renovable solar/wind/biomass/hydro

		W/cap	GWh/year	SEGSth Capacity MW	Surface km²	WECS Capacity MW	Surface km²	SO₂ mtn	NOx mtn	CO₂ mtn
End-use technology	conventional	200.00	2812.84	917	12.84	1284	17.71	0	0	0
	efficient	35.00	492.25	161	2.25	225	3.10	0	0	0
	real (2005) (*)	205.25	2887.30	942	13.18	1318	18.18	0	0	0

Table 19.4. Supplying green electricity to Barcelona.

	Electricity consumption (2005)	SEGS		SEGSPV		WECS	
	GWh	MW	km^2	MW	km^2	MW	km^2
Total	7206.60	2350	45.36	3268	32.68	3291	32.91
Domestic	2887.30	942	18.18	1309	13.09	1318	13.18
Com./ind.	4319.30	1409	27.19	1959	19.59	1972	19.72
Transport	1051.00	343	6.62	477	4.77	480	4.80
Capacity factor		0.35				0.25	
Occupation Ha/MW		1.93		1		1	
Efficiency				0.15			
Radiation kWh/m^2.year				1470.1			

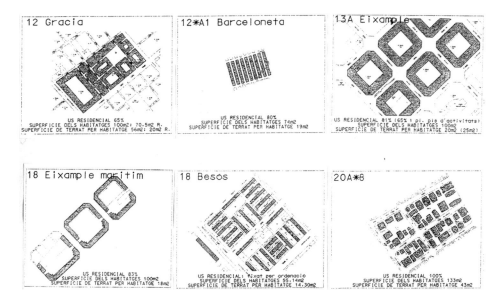

Fig. 19.5. Selection of typologies of different neighbourhoods in Barcelona.

Table 19.5. Typologies of neighbourhoods (sample).

Name of neighbourhood	Surface of house (m^2)	Surface of terrace (m^2)
Barceloneta	74	19
Besos	55.14	14.3
Eixample	100	20
Example marítim	100	18
Gràcia	70.5	20
A*8	133	43

Source: Barcelona City Council.

and it is also possible to generate all the energy the city is using at a local level from renewable energy sources.

19.3 Inefficient and Obsolete Energy Systems

The autonomous community of Catalonia, and its capital Barcelona, inherited a highly inefficient energy system from the Franco regime. But the democratic system built since the dictator's death in 1975 has been unable to modernize the energy system. It is still heavily centralized, based on a few big power stations, fossil or nuclear based, with no cogeneration capacity.

Renewal was started some years ago, building combined cycle gas power plants (CCGPs), but none of these new plants have cogeneration systems to recover the heat lost to the biosphere (40% of all the heat content in the fuel), when this heat could provide heating and cooling to city users. It is stunning that at the beginning of twenty-first century the main Spanish utilities (Endesa, Iberdrola and Gas Natural) are building CCGPs without cogeneration, and even more astonishing is that the public authorities authorize them without resistance. Recently two of these utilities (Endesa and Gas Natural) built two CCGPs, just into the Metropolitan Area of Barcelona (more than 3 million inhabitants), with an electric capacity of 788 MW (394 MW each). Neither plant has a cogeneration system, despite their dumping more than 500 MW of heat to the Mediterranean Sea (this is the equivalent of all the heating needs of Barcelona).

This system has more than 46 000 km of high voltage lines and more than 52 000 km of low voltage lines, and the associated losses are more than 3.5 billion kWh/year (almost 10% of all the electricity generated in Catalonia).

The nascent decentralized energetic system, efficient, clean and renewable it will not be imposed from one day for the other one. There will be a period of transition, the length of which will depend on the political will that those managing the public affairs manifest. And this political will will only materialize if the citizenship exercises actively their energy responsibilities, in a frame where their basic energy rights have been recognized.

In this transition period, fossil fuels will still be used, but they will be only the less pollutant fossil fuel (fossil natural gas), used with the maximum possible efficiency (decentralized technologies of combined heat and power, providing heating and cooling needs), as long as the big thermal power plants based on coal and nuclear continue to be abandoned.

In order to establish a decentralized or distributed energy system in ways that are efficient, safe, clean and renewable, it is important to recognize a set of basic energy rights:

- The right to know the origin of the energy one uses.
- The right to know the ecological and social effects of the manner in which energy is supplied to each final user of energy services.
- The right to capture the energy sources that manifest themselves in the place where one lives.
- The right to generate one's own energy.
- The right of fair access to power networks and grids.
- The right to introduce into power networks energy generated *in situ*.
- The right to a fair remuneration for the energy introduced into networks.

These rights have to be matched by a set of basic responsibilities:

- The responsibility to find out information.

- The responsibility to ask for information.
- The responsibility of generating energy with the most efficient and clean generation technologies available.
- The responsibility to use the most efficient end-use technologies available.
- The responsibility of conservation: of using the generated energy with common sense and avoiding any kind of waste.
- The responsibility of limiting oneself in the use of any form of energy.
- The responsibility of solidarity with those underprivileged societies that have no or limited access to a clean means of energy generation as well as its final use.

Guaranteeing these rights should be one of the tasks to which governments should give absolute priority. Exercising these responsibilities should be considered the fundamental duty of the responsible persons who depend on the sun as a source of energy. Adapting the lifestyles to the solar energy flows (both direct solar energy and its indirect forms) people will discover that fewer costs of every kind will have to be borne to be able to sustain life and prosperity on Planet Earth.

19.4 Realizing Energy Efficiency and Renewable Energy Potentials

What would be necessary to realize both the energy efficiency and renewable energy potentials of the cities of the world? The author's experience as an energy engineer active since the late 1970s, but also as a grassroots activist working with non-government organizations (NGOs), was heightened by his insight into political realities when he served as Sustainable City Councillor in Barcelona (between 1995 and 1999). It showed him that three conditions must come together. The first one is political will. Without political commitment no progress will occur in either energy efficiency or renewable energy transitions. Barcelona's decision to adopt the world's first Solar City Ordinance demonstrates this. How was the political will built that made it possible to adopt a solar ordinance? Many factors contributed. First, personal relations and attitudes are extremely important. These, combined with political leadership, based on experience about the decision it is proposed to adopt, can be a good recipe to succeed. Second, it is necessary to have technical capacity. Political commitment alone does not suffice; superb technical ability and skills to bring about and implement change are also needed. In Barcelona, some people from the city management bodies were not convinced at all when the process started. But the work done within the Civic Table on Energy was essential to convince the sceptics. Also, *BarnaGEL*, organizing the Sustainable Energy Forums, helped to make visible solar and renewable energy companies to citizens and politicians. The third condition is people's involvement and public participation. In Barcelona, some local NGOs have been active in the energy field since the early 1980s, helping to create an environment favourable to solar energy. Without the participation of all the city actors (decision-makers, companies, local citizens), it will be difficult to transform the present energy path (hard energy path) to a soft energy path (efficiency + renewables).

These three preconditions for success mandated the building of partnerships between local politicians, technical expertise, business and ordinary citizens.

In Barcelona, between June 1995 and May 1999, a committed group of people worked hard to open the door to urban energy sustainability. The results of this work were an increase of 93% of the surface devoted to collect the energy from the sun for heating water and an increase of 2400% of the surface devoted to collect the sun's energy and transform it into electricity. In this period of time the solar thermal collectors increased from 700 to

Table 19.6.   Progress on solar thermal in Barcelona.

	1995	1998	1999	Jul-00	Aug.2001	Dec.2001	Jun-02 (m²)	Dec.2002 (m²)	Buildings (number)
Residential						3145	6425.73	8130.41	108
Hotels						1745	2114.72	2071.67	20
Sports facilities						972	1458.00	1822.00	5
Hospitals						307	349.21	545.71	7
Other						152	421.10	1458.10	19
Total	700	1181	1350	1632	5233	6321	10768.76	14027.89	159
Number buildings						65		159	
Residential						3145	6425.73	8130.41	108
Hotels						1745	2114.72	2071.67	20
Sports facilities						972	1458.00	1822.00	5
Hospitals						307	349.21	545.71	7
Other						152	421.10	1458.10	19
Total	700	1181	1350	1632	5233	6321	10768.76	14027.89	159
Number buildings						65		159	

	Dec.2003 (m²)	Buildings (number)	Dec.2004 (m²)	Buildings (number)	Dec.2005 (m²)	Buildings (number)
Residential	12821.22	168	14764	243	19451	315
Hotels	2416.71	24	5641	47	6150	56
Sports facilities	2125.50	7	2200	8	2545	11
Hospitals	545.70	7	549	7	957	9
Other	1684.24	26	1378	22	1975	36
Total	19593.37	232	24532	327	31078	427
Number buildings	232		327		427	

Source: BarnaGEL and Barcelona City Council.

1350 m² and the PV panels from 80 to 2400 m². And this has been only the beginning. All the process resulted in a new trend: the continuous increase of solar energy in Barcelona (see Table. 19.6).

Why did this happen? What were the local circumstances that made possible this step forward to open the door to energy sustainability in a Mediterranean city like Barcelona?

The municipality of Barcelona receives from the sun ten times the energy the city is consuming in a year – or 28 times its yearly electricity consumption. Despite these facts, before 1995 the city never took seriously decisions to push for solar energy.

An interesting set of circumstances came together to make this happen. On the one hand, as a result of the 1995 municipal elections the government of the city adopted sustainability as an issue. On the other hand, some local NGOs active on energy efficiency and renewables had been organizing events around energy (conferences, campaigns, exhibitions, etc.) since early 1990s in complicity with local companies working in the field. In this way, political commitment, together with social pressure and technical capability, united to start walking on the path of soft energy.

The 1995 municipal elections in Barcelona resulted in a new coalition, where a Green Party candidate was appointed for the first time in the city's history. The new government was based on a political agreement that included the creation of a new political post (the Sustainable City Councillor) and the commitment to push programs for the development and diffusion of renewable energies. Also the 1996–1999 Municipal Action Plan included a series of concrete measures to make possible the use of the solar energy in the city. Between the measures there were the following:

- The use of renewable energies in the municipal buildings.
- To give incentives for using renewable energies in the domestic and services sectors.
- To incorporate solar hot water systems in the municipal buildings and sports facilities.

On 22 April 1998 the Plenary Session of the City Council adopted a political decision on energy sustainability. This decision included the promotion of energy efficiency, the use of renewable energies, the provision of information to the citizens and the cooperation with other local energy actors.

To implement these policies the municipality made use of two key instruments: the Civic Table on Energy (*Taula Cívica de l'Energia*) and the Local Energy Agency (*BarnaGEL – Barcelona Grup de Energia Local*).

The Civic Table on Energy was a municipal structure created in 1994 as a result of a public hearing on environment and energy forced by a coalition of local NGOs. This body involved local officials and staff from different departments of the city council with a local representative of an NGO energy platform, Barcelona Saves Energy (*Barcelona Estalvia Energia – BEE*). Its main objective was to build complicities on energy matters from all people in different departments of the city council. Between 1995 and 1999 the author served as chairman of this body.

BarnaGEL was the result of an EU-PERU project (1994) when Leicester City Council (England) and the Metropolitan Area of Barcelona decided to apply for EU funds to create local energy agencies, with the assistance of Ecoserveis (a local NGO working profession-ally on energy and environment) and with the collaboration of *ICAEN* (Regional Energy Agency of Catalonia) and *UAB* (Autonomous University of Barcelona). *BarnaGEL* was par-tially funded by the European Commission between 1996 and 1999. When the project was accepted for funding by the EU the city of Barcelona was not involved in it. Only after the work by the Sustainable City Councillor was done did the city begin working on the project. *BarnaGEL* has been a key facilitator and leader, inspiring all the energy efficiency and renewable energy projects in operation or in construction in Barcelona city. The main objec-tive of *BarnaGEL* has been to build complicities between local actors in order to develop energy projects in the city. *BarnaGEL*, according the European Commission guidelines, was created as an independent body in order to be free to activate all local energy actors.

Also the city of Barcelona was engaged during this time with two more active European networks working on energy in cities: *Energie-Cités* and Climate Alliances.

But the more relevant and well-known energy project has been the Barcelona Solar Ordin-ance, now internationally recognized and awarded with the European Solar Prize 2000.

This very innovative solar law, called 'Barcelona Ordinance on Application of Solar Thermal Energy Systems into the Buildings' or 'Barcelona Solar Ordinance' in brief, enforces all the new buildings to be built in Barcelona (and all the integrally retrofitted buildings) to have solar thermal water systems to cover 60% of sanitary water heating needs.

This Ordinance was adopted by the Plenary Session of the Council in July 1999 and has been mandatory since 1 August 2000. This is a clear example of how to develop renewable energy policies and strategies within a city.

It all started when the Sustainable City Councillor knew that Berlin was working to adopt a Solar Ordinance in 1996. Then the Sustainable City Councillor asked himself a

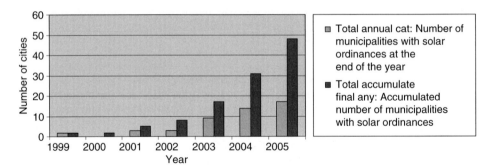

Fig. 19.6. Solar ordinances in Catalonia.

simple question – why was solar energy not widely used in Barcelona, despite the available resource (the sun: 2351 hours/year, 1470 kWh/m²/year), the city's energy needs (the city was heating the sanitary water using natural gas and electricity – 1.02 TWh/year) and its available surface (between 19 and 43 m² of terrace available for each existing apartment)?

When the process to adopt the Solar Ordinance started, some critics said: 'a city like Barcelona has no power to implement a local law on solar energy', but the supporters replied: 'we will proceed and see what will happen'. At the end no one offered strong opposition and the Solar Ordinance was adopted, not only by the city of Barcelona, but by many more cities in Catalonia and in Spain. At the end of 2005, almost 50 municipalities had adopted it in Catalonia (Fig. 19.6) and 25 all over Spain.

All those facts forced the regional government of Catalonia and the Spanish government to react. As a result, in February 2006, the Catalan government enacted a decree on ecoefficiency in buildings (all the new or integrally retrofitted buildings, consuming more than 50 l/day of hot water (60°C), must have a solar heating system to cover between 40 and 70% depending on the climatic zones). And in March 2006, the Spanish Council of Ministers adopted the new Technical Building Code ('CTE – *Código T[wea]cnico de la Edificación*') – it is one of the three documents for implementing European Directive 2002/91/CE on Energy Efficiency on Buildings – which establishes basic quality requirements for buildings, and between them, energy requirements (minimal solar contribution for sanitary hot water in all types of new buildings and minimal PV contribution for electricity in some commercial buildings).

The story of the Barcelona Solar Ordinance is a clear case that shows how a proposal considered 'impossible' became a reality in a short period of time!

Opening the door to solar energy in a city like Barcelona was possible with a broad range of small actions and renewable energy projects in order to visualize real action from a municipality. The following contains a brief description of some projects developed under the Barcelona Sustainable City Councillors' Office which started during the 1995–1999 period:

- A PV solar roof at the two main buildings of Barcelona City Hall. It was an EU Thermie project ('A Grid-Connected and Integrated PV System in the Central Buildings of the Barcelona City Hall') to install 1000 m² of PV, 100 kWp power on top of the two most representative buildings of the municipality. At present, the PV roofs on both buildings are in operation (400 m², 40 kWp in the first one and 600 m² and 60 kWp in the second one). To build the solar roof in the second building it was necessary to deconstruct a 12 floor building and transform it to an eight floor one before covering it with a solar roof. The production of electricity is 170 MWh/year.

- 'Urban ECOTREL – End-user Configuration Offer for Technical solutions on Renewable Energies on-Line'. This was an EU ALTENER project and consisted of a renewable energy information system on urban environments, CD-Rom based, addressed to local officers (political and technical). This package was sent to more than 500 municipalities all around the Catalonia region of Spain and more than 1000 have been given free of charge to all interested people.
- A full-scale demonstration of a solar house during the 14th European Photovoltaic Solar Energy Conference and Exhibition (Barcelona, 30 June–4 July 1997).
- The solar/wind efficient caravan: this is a mobile facility equipped with solar systems (thermal and PV), a small wind system and energy efficient appliances. This equipment is now being used in and around Barcelona to show in real operation how a house could run with renewables and energy efficient appliances.
- The Sustainable City Resources Center: a municipal fixed facility equipped with a permanent exhibition with energy efficient devices and appliances and renewable energy devices. The name of this facility is '*Oficina de Recursos per la Ciutat Sostenible – ORCS*' (Sustainable City Resources Office) and has been open to the public, free of charge, since May 1999.
- An agreement between city and regional governments in order to make possible the construction of the first PV school in Barcelona (30 kWp power, 38 MWh/year of estimated production). At present there are many schools equipped with solar PV systems.
- An agreement between city council, manufacturers and installers of solar thermal systems in order to incorporate solar thermal water systems in all the existing municipal sports facilities. The two first installations (*Poliesportiu Torrent de Melis* with 74 m² of solar collectors and *Piscina Bon Pastor* with 80 m²) are demonstrating that more than 75% of energy consumption is being covered by solar energy. Now all the new sports facilities are equipped with solar thermal systems.
- A project to make it possible for more than 450 new apartments to be equipped with solar thermal water systems (totalling 609 m² of collectors). The apartments were built by the '*Patronat Municipal de l'Habitatge*' (the municipal body charged with social housing construction). This first development has been, and is, quite successful and the users are young people renting (with low rents) the apartments. Also the apartments have incorporated low energy appliances, water saving devices, recycled materials, recycling facilities, etc. By now all the new public owned apartments are incorporating solar thermal systems.
- A municipal policy to give subsidies to solar energy users – thermal and PV – in the city (direct financial subsidies up to 25% of the costs of installation in existing buildings).
- Efficient lighting systems at the City Hall buildings. In October 1995 the City Council decided to replace all incandescent lamps in the City Hall buildings and, in April 1998, all the old fluorescent lighting systems in a main City Hall building – eight flats. The savings have been more than 250 000 kWh/year.
- Biogas valorization from organic waste component: *in-situ* demonstration of the valorization of landfill gas to fuel a car (Garraf landfill, 21 May 1997). Subsequently, the Barcelona Metropolitan Area adopted a Municipal Waste Management Plan that incorporates the construction of three methanization plants (75 000 tons each) to digest the organic fraction of municipal wastes.

All the projects listed above would not be a reality without the complicity of a number of people involved in sustainable energy: local NGOs, experts, companies, etc.

In order to show some of the complicities built around the projects, below is a list of the most relevant actions concerning renewable energy undertaken by local actors with the support of the Sustainable City Office (1995–2000):

- *The Sustainable Energy Forum* organized by *BarnaGEL* – Barcelona Local Energy Group (the local energy agency, EU SAVE program). Three editions have been organized (12–14 March 1998, 25–27 February 1999 and 26 April 2000). During these forums *BarnaGEL* presented the 'Sustainable Energy Green Pages', an electronic catalogue of energy efficiency and renewable energy products and services. *The Sustainable Energy Forum* is currently in its 9th edition.
- Local NGOs Solar Campaign, called *BarnaMIL* (Barcelona thousand) with the goal to achieve $1000\,m^2$ of new solar thermal water systems in private existing buildings before the year 2000. This was a campaign started with the involvement of an NGOs Platform, the professional association of renewable energy professionals – APERCA – and the local energy agency – *BarnaGEL*. Now *BarnaMIL* is a local NGO pushing for renewable energy projects in the city.
- The first small PV installation connected to the local grid. This solar PV system ($18\,m^2$, 2.3 kW) is privately owned by *Fundació Terra*, a Barcelona-based NGO devoted to environmental education. This system was built after having being issued by the Spanish Renewable Energy Feed-in-Tariff (*Real Decreto 28/1998*). The PV system was inaugurated on February 1999 in the presence of Barcelona Mayor Joan Clos and the Sustainable City Councillor. The event had a big impact on the media. In 2007, *Fundació Terra* launched the first project in Spain to build a solar PV roof on top of a municipal market, owned by local citizens.
- The *Sole PV* mobile Greenpeace power plant connected to the City Hall grid: this was the Greenpeace Spain campaign to show how a PV solar system could be connected to the grid. The *Sole PV* station was installed in the square where the City Hall is located and was connected to the City Hall grid. As a result, for the first time in Barcelona the municipal buildings were supplied by 'green' electricity.
- The Xth, XIth, XIIth and XIIIth editions of the *Catalonian Conferences for a Non Nuclear and Sustainable Energy Future*, organized by the Group of Scientists and Technicians for a Non Nuclear Energy Future, a well-established NGO working in the field since the late 1970s. These conferences have the support of a wide range of Catalonian NGOs and local renewable energy companies. In all editions of the conferences (now in the 21st edition) distinguished energy specialists have been invited to contribute.
- The Earth Day activities – principally the Earth Fair where renewable energy equipment is shown in an open space to the general public. The Earth Fair has been organized since 1996 by Earth Day Catalonia (a local NGO) in a public garden-park and each fair has been visited by tens of thousands of visitors. Since the 2004 fair, all the electricity needed by the event has been supplied by renewable energy sources (100% biodiesel). In 2007 the Barcelona Earth Fair was in its 12th year.

19.5 The Future

The experience of Barcelona, like other cities in the world, shows us that putting together political commitment, capacity building and people's participation is the recipe that any city needs to start changing the way energy is generated and used.

Important work commenced between 1995 and 1999 in the city of Barcelona will expand and in the near future we will see Barcelona city and its metropolitan area as a region where the sun will play an important role as an energy source providing the required energy services (with end-use efficient technologies) for living lightly on the earth.

References

BarnaGEL (1997, 1998 and 1999). *L'energia a Barcelona, Regidoria de Ciutat Sostenible, Ajuntament de Barcelona.*

Byrne, J., Young-Doo, W; Hoesung, L; and Jong-dall, K. (1998). An equity- and sustainability-based policy response to global climate change. *Energy Policy*, **26**(4), 335–343.

Norgard, J.S. (1989). *Low Electricity Appliances – Options for the Future.* Energy Group, Physics Laboratory III, Technical University of Denmark.

Norgard, J.S. (1991). *Energy Conservation through Efficiency and Sufficiency.* Physics Laboratory III, Technical University of Denmark.

Norgard, J.G. (1992). *Low Energy Europe. Sustainable Options.* Physics Laboratory III, Technical University of Denmark.

Puig, J. (2004). *Prospectiva energètica. Els contorns d'un nou model energétic i el procés de transició,* in *La tecnologia. Llums i ombres. Informe 2004 de l'Observatori del Risc.* Institut d'Estudis de la Seguretat, 119–149.

Websites

APERCA (Catalan Organization of Renewable Energy Professionals): http://www.aperca.org/

Barcelona Energy: http://www.barcelonaenergia.com/

Barcelona Sustainable City Councillor (1995–1999): http://www.verds-alternativaverda.org/ciutatsostenible/

Earth Day Catalonia: http://www.diadelaterra.org/

Group of Scientists and Technicians for a Non Nuclear Future: http://www.energiasostenible.org/

Terra Foundation: http://www.terra.org/

Chapter 20
Reducing Carbon Emissions in London: From Theory to Practice

A.R. DAY[1], C. DUNHAM[2], P.G. JONES[1], L. HINOJOSA[2], A. DUNSDON[3] AND P. OGUMKA[3]

[1]*London South Bank University (LSBU);* [2]*SEA/RENUE;* [3]*Energy Centre for Sustainable Communities (ECSC)*

20.1 Introduction

This chapter combines the findings of two studies, one conducted for the London Energy Partnership and one for the Greater London Authority, both of which were completed around March 2006. Although independently commissioned, and designed to serve different purposes, they both deal with the practicalities of reducing carbon emissions in London; one through a technical analysis, and the other through the use of policy instruments. The first identifies how best to achieve future carbon savings with different mixes of energy technologies (both supply and demand side), while the other analyses the impact of regional planning policy on renewable energy implementation. Together they provide a picture of what needs to be done to achieve carbon emissions reductions targets in a large city using distributed and embedded energy systems, both technically and through policy frameworks.

The report 'London carbon scenarios to 2026' investigated the possible mix of energy technologies that could be deployed at a local (and regional) level to meet target carbon emissions reductions. The minimum target used in the report was based on the adopted UK target of 60% reduction on 2000 greenhouse gas emissions by 2050 (taken from the Royal Commission on Environmental Pollution (2000) report, 'Energy: the changing climate'. However, the report also discusses the implications of more ambitious targets, based on more recent estimates for avoiding dangerous climate change (for example, see Meinshausen (2006)). The analysis is based on meeting these reductions targets through locally implemented options, as opposed to assuming these may be met through wider national (or even international) technology strategies. These wider options might include, for example, the use of centralized new nuclear power or widespread fuel switching for transport.

The paradigm of local and regional implementation must necessarily assume some local or regional coordination, with some degree of power to influence the energy system. It is becoming increasingly apparent that there needs to be sufficient political will and leadership

451

to help force the pace of change necessary to meet even the minimum targets. In the UK and across the EU there has been a growth of interventionist policies at local and regional government level. In London this has taken the form of the Mayor's Energy Strategy (2004) which sets aspirational targets for embedded renewables, combined heat and power, and energy efficiency. This regional policy also expects local government to adopt similar targets and strategies. In order to meet these targets the mayor adopted a series of measures to tackle the various barriers that were seen as clear obstacles to progress. The key actions were:

1 To engage a wide range of stakeholders in the process. This was done by setting up the London Energy Partnership, an independent body with a remit to take forward the objectives of the Energy Strategy.
2 Setting up the London Climate Change Agency (LCCA). The LCCA was established in June 2005 and is responsible for delivering carbon saving projects and establishing a financial framework that can deliver more widely.
3 The use of the mayor's powers to influence the planning process through the London Plan (2004). This includes a requirement that all new developments that are referred to the mayor for a decision (there are defined thresholds for these) must include 10% on-site renewable energy generation (heat or electricity) and demonstrate significant energy efficiency improvements.

It is the last of these that is the second focus of this chapter. At the time of writing the London Plan was under review, and as a precursor to this process, the Greater London Authority (GLA) commissioned a study into the effectiveness of the energy provisions to that time (Review of impact of the energy policies in the London Plan on planning applications referred to the mayor 2006). While the London Plan was under development some London borough authorities (the tier of local government in London below the GLA) were already putting forward a 10% renewables requirement for developments under their control. Through the adoption of similar targets in the London Plan the mayor also sought a wider take-up of the policy at the London borough level, thus hopefully influencing every new building project in the city.

The implementation of new policies, especially radical ones that have a significant financial price tag, inevitably leads to discontent if the purpose of the policy is unclear or not shared by the stakeholders being targeted. In extreme cases this can cause policy failure, but at the very least there is likely to be a period of time where the objectives are not fully met. It is ultimately through experience of policy implementation, of fine-tuning and of setting up the right systems and flows of information that the success of that policy will be determined. Hence a review was commissioned of the first 18 months of the London Plan to assist with the assessment of where it should be taken in the future.

This chapter synthesizes the findings of the two reports (London Carbon Scenarios to 2026; Review of impact of the energy policies in the London Plan on planning applications referred to the mayor 2006) in order to present a credible route map that can lead a large city towards a low carbon future.

20.2 The London Framework

The Mayor's Energy Strategy (2004) was published in February 2004, with the aim of setting out the mayor's objectives and policies for energy use and delivery in London. Although not part of the mayor's statutory responsibilities at that time, it was an area he felt needed to be addressed at the regional level. Central government now proposes to give him statutory responsibilities in the area of climate change and energy, with a duty to take action to mitigate the effects of climate change.

Proposal 6 of the Energy Strategy sets specific targets for renewable energy technologies in London:

> *London should generate at least 665GWh of electricity and 280GWh of heat, from up to 40 000 renewable energy schemes by 2010. This would generate enough power for the equivalent of more than 100 000 homes and heat for more than 10 000 homes.*
>
> *To help achieve this, London should install at least 7000 (or 15 MW peak capacity) domestic photovoltaic installations; 250 (or 12 MW peak capacity) photovoltaic applications on commercial and public buildings; six large wind turbines; 500 small wind generators associated with public or private sector buildings; 25 000 domestic solar water heating schemes; 2000 solar water heating schemes associated with swimming pools, and more anaerobic digestion plants with energy recovery and biomass-fuelled combined heat and power plants. London should then at least triple these technology capacities by 2020.*

In addition the Energy Strategy calls for an increase in combined heat and power (CHP) capacity, and puts forward the concept of Energy Action Areas that can act as exemplars of how low and zero carbon technologies can be implemented at the community level. It also sets out the channels and mechanisms by which this can all be achieved.

One mechanism under the mayor's powers is the use of the planning system. The London Plan (2004) also known as the Spatial Development Strategy for Greater London, provides the overarching framework for delivering the various strategies for London, such as on transport, noise, waste or air quality. This includes taking forward the objectives of the Energy Strategy. In particular it requires that all large developments referred to the mayor include a proportion of the energy needs to be provided by on-site renewables. This fraction is currently set at 10%, and is expressed in terms of CO_2 emissions avoided – rather than kWh delivered.

20.3 The Low Carbon City

National government sets policies that shape the overall approach to the energy system, and the response to the pressures of climate change, both mitigation and adaptation. In the UK this is leading towards a mixed energy economy (The Energy Challenge, Energy Review 2006) with the likelihood of new nuclear power stations, and fossil fuel plant with carbon capture and storage, together with large-scale renewables. There is also a strong desire to see the use of embedded low and zero carbon technologies, together with improvements in energy efficiency. Indeed these technologies, particularly renewables and CHP, can be deployed much faster than centralized low carbon alternatives and make more rapid inroads into emissions reductions. However, while there are technical challenges, the key problem in implementation is one of economics where the economies of scale and the distributed nature of investments can make these seem unviable or high risk options. There are also regulatory requirements governing supply over private wire networks that may influence investment decisions.

There are a variety of measures in place in the UK that promote the uptake of renewables, whether through first cost grants (e.g. The Low Carbon Buildings Programme (2007), legislation (e.g. building regulations and planning) or through market-based instruments (e.g. the Renewables Obligation (2006). However, on their own these have not resulted in widespread deployment in urban locations where the majority of the demand for energy exists. To make this happen there is a need for a coordinated and structured approach, in which there is an important role for local and regional government. To be effective the authority must have sufficient powers to have real influence, and sufficient information to know how to direct those powers. The London Carbon Scenarios study was commissioned to provide information that can inform policymakers of what technology mix might be the best option to meet future targets.

There are significant benefits from the supply of heat and electricity from sources close to the demand. The obvious ones are the reduction in transmission and distribution losses and the more efficient use of fuel (especially where CHP is used). There are less obvious benefits such as community engagement, increased job opportunities for installation and maintenance, and new business opportunities in the energy sector. As for community engagement there is evidence that locally installed renewables raise local consciousness on issues of energy efficiency and climate change. Once people understand the relationship between their energy use and where it comes from, they are more likely to change their behaviour. Centralized and remote electricity generation has led people to become disconnected (in the social rather than technical sense) from the energy system.

There are drawbacks. Some technologies can be mutually exclusive in some circumstances, for example CHP and solar thermal, and there can be difficulties in managing supply and demand mismatches say between solar availability and electricity consumption. These problems can be dealt with by careful planning and design – significant challenges in their own right for which the skills and experience necessary are only now beginning to be established.

20.3.1 Carbon targets

The London Carbon Scenarios study considered five technology mixes that could meet specified carbon emissions reductions targets. The minimum target was based on the nationally adopted 60% reduction on 2000 emissions by 2050, Royal Commission on Environmental Pollution's (2000) with the assumption that reductions occurred linearly between 2003 (for when baseline figures were known) and 2050. Allowing for anticipated population growth and new developments this gave a 2026 emissions target of 10 343 ktonnes CO_2 pa.

20.3.2 Technology options and scenarios

Table 20.1 shows the technology options considered for the study, together with some of the assumptions used in the model. These represent the key technologies likely to be taken up in London over the next 20 years. There may be other technologies, such as low head mini or micro hydro, but the potential for these is too small to make a significant contribution to overall savings.

The analysis was conducted to see which of the possible combinations of these technologies could provide the target savings most cost effectively. The initial assumption was that a mix of technologies would be employed (given the specific targets in the mayor's Energy Strategy), but that some may be more dominant than others. Understanding which mix is more cost effective, and generally better at saving carbon, is important as it informs what policy measures or support mechanisms may become necessary to promote the uptake

Table 20.1.   Technologies analysed in the London Carbon Scenarios study (2026).

Technology	Description
Biomass CHP	It includes a mixture of different biomass technologies, such as anaerobic digestion, gasification, steam turbines, biodiesel. This technology is measured in number of MW_e installed.
Large gas CCGT CHP	Large CHP (several MW_e) plant that could be built within London, and connected to heating networks. This technology is measured in number of MW_e installed.
Gas CHP – building	These are smaller CHP engines that would be building integrated. Heating networks are not included within this option. This technology is measured in number of MW_e installed.
PV – domestic	Domestic systems of 2.5 kWp are assumed per dwelling. This technology is measured in number of systems installed.
PV – large	Commercial and industrial systems are measured in number of MW_e installed.
Wind – large	Large systems are measured in number of MW_e installed.
Wind – domestic	Domestic systems of 1 kW_e are assumed per dwelling. This technology is measured in number of systems installed.
Solar thermal	Domestic systems of 2.8 kW_{th} are assumed per dwelling. This technology is measured in number of systems installed.
Biomass boilers – large	Commercial and industrial systems are measured in number of MW_{th} installed.
Biomass boilers – domestic	Domestic systems of 20 kW_{th} are assumed per dwelling. This technology is measured in number of systems installed.
Ground source heat pumps (GSHP)	Domestic systems of 5 kW_{th} are assumed per dwelling. This technology is measured in number of systems installed.
Micro-CHP Stirling engine	Domestic systems of 1.2 kW_e are assumed per dwelling. This technology is measured in number of systems installed.
Micro-CHP fuel cell	Domestic systems of 1 kW_e are assumed per dwelling. This technology is measured in number of systems installed.
Cavity wall insulation	This is measured as the number of dwellings that could insulate the unfilled cavity wall.
Loft insulation	This is measured as the number of dwellings that could insulate or increase the levels of insulation of the loft.
Double glazing	This is measured as the number of dwellings that could replace single for double glazing.
Solid wall insulation	This is measured as the number of dwellings that could insulate their solid walls.
Heat from power stations	There are a number of power stations within London that reject heat to the atmosphere. This heat could be potentially used for heating and DHW purposes. It has been estimated that the potential heat that could be exported from Barking power station is in the order of 250 MW_{th}, and another 30 MW_{th} could be taken from SELCHP. In this technology, the amount of potential heat from PS for the scenario under analysis should be entered.

of the right technologies. For example, the installation of large- or medium-scale CHP requires guaranteed connected loads (particularly for heat) and significant infrastructure; this can only be delivered by engaging a number of community stakeholders and ensuring the business model is attractive to investors.

Five technology scenarios were considered: large CHP led; buildings and micro-CHP led; renewables led; insulation led; a hybrid scenario. Table 20.2 summarizes the key attributes of each scenario.

20.3.3 Economic aspects

The economics associated with each technology were considered with respect to capital and investment costs, running costs, and support mechanisms and subsidies. Likely capital cost reductions for immature technologies were taken from published data as production volumes increase. However, the economic model assumed no cost feedback on increased volumes of installations within London (i.e. it was assumed that the *relatively* small London market would have little impact on cost). Current support mechanisms, shown in Table 20.3 were assumed to remain in place until 2026. However, the analysis

Table 20.2.   Summary of technology scenarios.

	Lead technology	Assumptions
Scenario 1	Large CHP led	CHP and community heating will be provided for all new build and that district heating will be widely implemented across London. Minimal insulation measures to existing buildings are considered, mainly in the form of cavity wall and loft insulation. In conjunction with the above, the requirements for renewables are set to at least comply with the 10% requirement.
Scenario 2	Buildings and micro-CHP led	Micro-CHP and small-scale CHP will be provided for all new build and existing buildings. Minimal insulation measures to existing buildings are considered, mainly in the form of cavity wall and loft insulation. In conjunction with the above, the requirement for renewables is also set at 10% as in the previous scenario.
Scenario 3	Renewables led	Renewable energy will be the main driver for reducing CO_2 emissions. Minimal insulation measures to existing buildings are considered, mainly in the form of cavity wall and loft insulation. Micro-CHP and small-scale CHP are also considered.
Scenario 4	Insulation led	Insulation strategies will be implemented widely across London in the form of cavity wall, solid wall, loft insulation and others. In conjunction with the above, the requirement for renewables is set at 10% and micro- and small-scale CHP are also considered.
Scenario 5	Hybrid scenario	Combines the measures of all the previous scenarios, but giving large gas CHP a more important role. This means less large CHP than scenario 1, but more renewables and more energy efficient measures across London.

was also undertaken using a flat rate carbon value across all technologies. The phasing of the installations was assumed (rather than based on some more rigorous modelling) and the life cycle costs of each technology calculated to 2026.

20.3.4 Modelling

A variety of approaches to modelling energy systems is available. The choice largely depends on the type of questions being asked, and what the model is intended to show. Energy systems modelling inevitably requires a large number of variables to be considered, many of which change with increasing uncertainty over time. For example, technology investment costs often depend on market share and rate of take-up; for a confined regional study these costs will be a function of a larger system than that under consideration.

The model was developed by SEA/RENUE in the Microsoft Excel spreadsheet software, using Visual Basic for Applications programming language. Baseline CO_2 emissions for the latest year available were calculated from the GLA Local Energy and Carbon Inventory 2003. This uses real gas and electricity consumption for the year 2003 provided by UK utilities. It was assumed that there had been little variation in London CO_2 emissions between 2000 and 2006. The national target of a 60% reduction by the year 2050 was then used to derive an interim target for the year 2026. A linear reduction from 2006 to 2050 was

Table 20.3. Summary of UK renewable energy support mechanisms London Carbon Scenarios to 2026.

Mechanism	Acronym	Summary	Value	Future variation	Total value	Lifetime
Energy Efficiency Commitment	EEC	Subsidy to domestic energy saving	0.7–0.9p/kWh	Unknown beyond 2008	£148m for EEC1	EEC2 2005–8
Levy Exemption Certificates[a]	LECs	CHP and renewable electricity can claim	0.43p/kWh	Increase with RPI	Unlimited	
EU Emissions Trading Scheme	EU ETS	Cap and trade system on 20 MW_{th} plant	Current price £6/tonne CO_2	Market instrument – value of carbon	Limited by	Phase II to 2012
Low Carbon Buildings Programme	LCBP	Capital support for building integrated renewables and CHP	Varies depending on technology	Unknown beyond 2009	£80m	2006–9
Renewable Obligation Certificates	ROCs	Awarded to generators	4.5 p/kWh	Market instrument	Limited by % obligation	2016

[a] The Clime Change Levy (CCL) is a tax on commercial electricity and fuels (excluding transport), with exemptions for energy from renewables and good quality CHP. Levy Exemption Certificates (LECs) are awarded to renewable or sustainable electricity, which can then be passed on to the end user. LECs are used to show the energy used is CCL exempt.

assumed. This 2026 target was then adjusted for the likely increase in CO_2 emissions in the building sector due to the planned demolition and new build as set out in the London Plan. While new buildings (assuming building regulations are met) would have much lower carbon emissions than the buildings they replace, the net increase in floor space means that emissions will rise by 2026 in the absence of concerted action. This projected increase in emissions is simply added to the 2026 to give a new target.

The process of assessing what mix of measures would meet the target was done manually (the model has since been developed in the course of further use and further work is planned to optimize and automate this process). A menu of 17 measures is available and the user then enters the number or MW of each of these 17 technologies that are expected to be installed to create a scenario. In this case four five-year periods, from 2006 to 2026, were entered.

Behind each measure sit data on capital cost, operation and maintenance cost, heat or electricity generated or saved, and what subsidy a particular measure will qualify for per unit of installed capacity for each year to 2026. These data are then used to calculate the net present value (NPV) and internal rate of return (IRR) for each measure and aggregated for any scenario. The user may select a discount factor against which to assess the economics of any scenario. For the purpose of this study both a public sector rate (3.5%) and a private sector rate (10%) were selected. The model relies on the current UK Building Regulations emission factors for gas and electricity displaced to produce a carbon saving for any given measure.

A further option allows the user to select either the current subsidy regime or a flat rate value of carbon. Currently different technologies attract different rates of subsidy, but a flat rate value allows a level playing field. This flat rate value was based on the UK Treasury figure for the social cost of carbon Clarkson and Deyes (2002). The study predates the publication of the Stern report The economics of climate change (2007).

20.3.5 Results

To achieve a 60% reduction in London by 2050 means a 27% reduction target by 2026 – a reduction of 9077 ktpa on 2003 levels. Taking into account the growth in emissions from new housing and non-residential floor space this target increases to 10 343 ktpa. Initially four scenarios were constructed to (approximately) achieve this target (savings varied less than 1% of this target). A fifth 'hybrid' scenario was then added. What was striking about all four of the original scenarios was that they all required a substantial component of CHP to achieve the target. Scenarios 1 and 2 are both CHP led scenarios, just using different scales of CHP plant. The renewables led scenario contained a high proportion of renewable (biomass) CHP. For the insulation led scenario there simply weren't enough unfilled cavities and lofts to achieve a significant reduction. Many homes in London have solid walls or already have some, albeit inadequate, level of insulation.

Table 20.4 gives an overall summary of the results for the five scenarios.

Figure 20.1 shows the energy delivered (heat and power) by each scenario, while Fig. 20.2 shows the capital expenditure and net present values.

It was clear that of the four original scenarios, the large-scale CHP option presented the best economic case. At a public sector discount factor of 3.5% London could achieve a 2026 carbon reduction target and achieve a positive NPV from an investment of just over £8bn. However, it was the view of the London Energy Partnership and partner organizations that this should not be selected as the preferred scenario, rather a new scenario (5).

Table 20.4. Summary of overall results for the London Carbon Scenarios study(2026).

Scenarios	Description	Heat (GWh/y)	Power (GWh/y)	CO_2 savings (ktpa)	Capital cost (£m)	NPV (£m)
Scenario 1	Large CHP	30296	23587	10442	8392	1192
Scenario 2	Building and micro-CHP	58478	22799	10285	7455	−531
Scenario 3	Renewables	21852	13380	10414	14591	−4237
Scenario 4	Insulation	38177	14526	10362	10797	−1429
Scenario 5	Hybrid	29843	18184	10344	8427	678

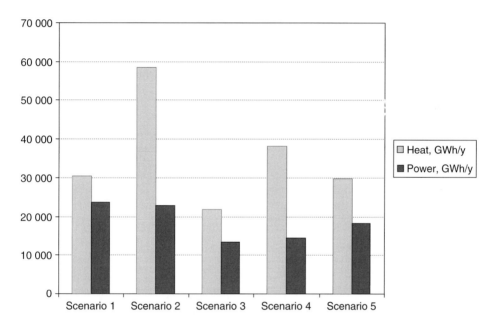

Fig. 20.1. Heat and power delivered for each scenario (2026). (London Carbon Scenarios 2026).

Essentially this scenario arose from discussion of the original scenarios modelled and a desire to see a hybrid based around the large-scale CHP-based scenario 1 but incorporating some elements of the other three. The superior economic performance of large-scale CHP notwithstanding, it was felt that other factors would prevail in practice that would encourage a higher uptake of other measures – despite their poor economic performance over the 20 year period analysed. It was felt that micro-generation technologies in particular might see greater installed capacities than envisaged by scenario 1 since householders do not necessarily make judgements on a purely economic basis. It was also considered that a higher uptake of biomass CHP might be achievable.

Figures. 20.3 [Plate 14] and 20.4 [Plate 15] show the detailed results for the selected scenario (5). It is obvious from Fig. 20.3 [Plate 14] that the greatest share of carbon savings is achieved through the use of biomass, large-scale and building-based CHP. This is also reflected in the capital cost (Fig. 20.4 [Plate 15]). Building CHP requires slightly less capital investment proportionately since no district heating network is required.

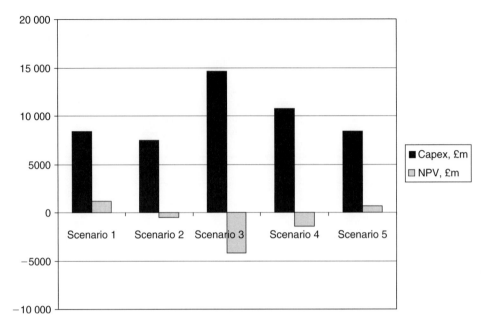

Fig. 20.2. Overall capital costs and net present values for all five scenarios. (London Carbon Scenarios 2026).

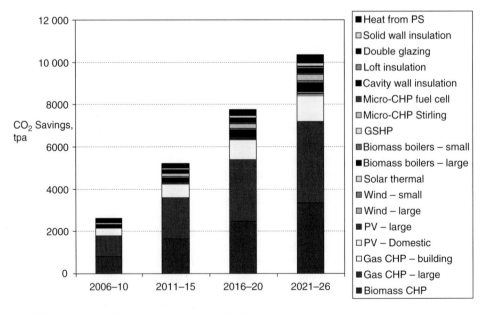

Fig. 20.3. Breakdown of CO_2 savings for scenario 5. (London Carbon Scenarios 2026). [Plate 14]

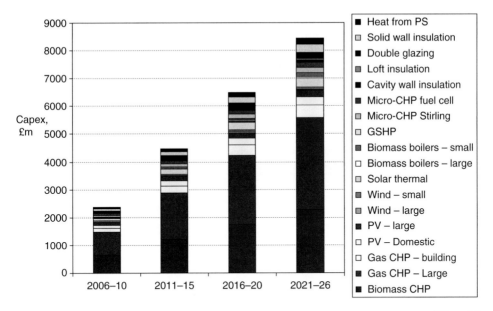

Fig. 20.4. Detailed breakdown of capital costs for scenario 5. (London Carbon Scenarios 2026). [Plate 15]

20.3.6 Discussion of the LCS results

The London Carbon Scenarios work identifies the technologies, and the relative capacity of installations, that can best meet the carbon targets to 2026. It gives an indication of the likely costs of each strategy, and shows the most cost-effective route. The central element of the preferred scenario will involve selecting sites within London for large combined cycle gas plant and developing heat networks around that plant. Two combined cycle gas turbine (CCGT) CHP schemes are already planned: the Barking power station extension and the Tilfen Land scheme in Plumstead. While these schemes are still at the planning stage this at least gives an indication that a CCGT-based CHP approach is not only feasible but economically viable.

All of the scenarios would require persuading private householders to make changes to their homes, either through the installation of insulation, the integration of renewables, the use of micro-CHP systems or, particularly in the preferred scenario, connection to district heating.

However, the initial development of heat networks will most likely be in the area of new developments (and existing social housing) where it is easier to influence outcomes. Once these have a track record of providing reliable and affordable heat, these could be rolled out to existing stock.

The development of heat networks from new developments combined with the use of embedded renewables means that for the selected scenario to be realized, the planning system will be critical.

20.4 Use of the Planning System

The London Plan (2004) was published in February 2004, and the energy-related components became effective from the middle of that year. LSBU and ECSC carried out a review of the energy policy implementation (2006) between March and June 2006, based on data from the period May 2004 to January 2006. During this time there were 186 applications submitted to the mayor, but of these only 46 (mainly the latter submissions) were accompanied by comprehensive energy statements. The study was therefore based on a limited data set, although there were some significant trends arising from the analysis. The most significant results are in three areas:

1 The degree of success achieved by developers in meeting – and exceeding – the 10% renewables target.
2 The amount of energy efficiency savings achieved.
3 The rate at which the process matured.

20.4.1 The planning process

All planning applications for large or strategically significant new developments, together with high impact buildings and sites of particular interest, must be referred by the local borough authority to the mayor for approval. Under the London Plan these applications must show evidence of energy efficiency measures, the feasibility of CHP, and how the 10% requirement for renewables will be met. This evidence takes the form of a detailed energy statement that states the type of technologies to be used, the savings these will deliver, and the methodology by which the assessment has been carried out. A new activity is now required in the planning submission process to develop an energy statement showing that targets will be met on the future development. New skills sets have been needed in order to analyse the most appropriate technologies for a given type of development. To assist with this London Renewables (now part of the London Energy Partnership) and others have produced toolkits London Renewables (2004) designed for pre-feasibility assessments for the range of typical urban low and zero carbon technologies (LZCTs). This early energy assessment must not be confused with the later detailed design process that also requires an assessment of LZCTs (see below).

For each development the statement must show the baseline CO_2 emissions based on current building regulations, followed by the savings that will occur from additional energy efficiency measures (which include the use of CHP). Renewables must then contribute a further (minimum) 10% reduction from the improved situation (i.e. not 10% of the baseline emissions).

This change in the planning process has led to the engagement of engineering consultants earlier in the design process, which may be having a beneficial effect at detailed design, as there is a better knowledge of constraints that may be imposed by low energy requirements. This whole process has also meant that planners who receive the applications need a better understanding of energy delivery systems and their viability in different circumstances. These issues of raising awareness of energy issues for all players in the development process should have significant long-term benefits, but they have also created short-term conflicts as the system has bedded down. Some of these conflicts, and their resolutions, are discussed in the sections that follow.

20.4.2 Interpreting 10% renewables

There are several ways in which a renewable energy target may be defined. These are in terms of delivered energy, carbon savings, or installed capacity.

The last of these is easiest to specify and determine, but is probably least helpful. Installed capacity (in kW or MW) of an energy system does not reliably indicate the contribution (of energy or carbon savings) that the system will provide, and this is particularly true for intermittent renewables (solar and wind), but also as a result of demand load factors (i.e. demand divided by installed capacity) on plant such as heat pumps, biomass boilers or CHP. However, system costs are strongly related to installed capacity, so this information is important and can provide statistics that can be highly informative about technology take-up.

The requirements in the London Plan were initially written in a way that implied energy efficiency savings and renewables should be identified in terms of delivered energy (kWh, MWh or GWh, etc.). This is certainly a better way of identifying more realistic energy savings, but is more problematic than stating installed capacity as it requires an understanding of expected on-site energy demand (which varies over time), and how different technologies can impact on this. For early assessments where building details are unknown, this can at best only be based on a number of assumptions and (rather crude) models of demand forecasting. The various toolkits mentioned above use simple rules of thumb about energy demand in buildings, and the energy that can be delivered by various technologies. They necessarily have to assume that the systems will be designed and installed in such a way as to properly make use of this energy.

The use of delivered energy is, of course, more relevant than installed capacity when identifying renewable energy contributions, but it misses the important distinction between the types of energy being displaced. In general (and certainly in London), electricity from the grid has a higher associated carbon content than heat supplied by natural gas. One way to deal with this is to convert everything to primary energy by incorporating conversion and distribution losses. However, it is a simple step from primary energy to associated carbon or CO_2 emissions.

The UK has adopted CO_2 as the main measure of efficiency and LZCT contributions within the 2006 Building Regulations; (2006) this has become the industry standard. This is, after all, the key parameter that defines the environmental impact of energy. Energy statements submitted under the London Plan have therefore provided information in terms of CO_2 saved. Unfortunately this has rarely been accompanied by installed capacity, which would provide a more detailed picture of the mix of technologies actually being deployed.

20.4.3 Findings of the study

The review study (2006) looked at two key areas: the contribution of CO_2 savings from energy efficiency and renewables, and an analysis of the planning process.

20.4.3.1 CO_2 savings
Table 20.5 shows the overall savings from energy efficiency (including CHP) and renewable energy for the period May 2004 to January 2006. This shows savings from those developments that presented full energy statements, and an extrapolation to the whole 186 developments submitted during that period. The extrapolation is based on the assumption that similar trends are followed across all developments as those presenting with full energy statements. Around 14 of the energy statements included no detail about renewable energy contributions, and again similar assumptions were made to conduct that extrapolation.

Table 20.5. Summary of CO_2 savings from energy statement data and extrapolated to 186 developments. (Extrapolations based on 186/46 for energy efficiency and 186/32 for renewable energy.)

	Savings to Jan 2006 from energy statements		Savings to date – extrapolated for 186 developments	
	(tonnes CO_2/y)	(%)	(tonnes CO_2/y)	(%)
Energy efficiency	45 867	25%	185 500	25%
Renewable energy	4709	4.4%	27 400	4.9%
Total	50 576	27%	212 900	28%

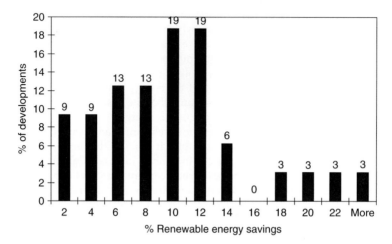

Fig. 20.5. Percentage renewable energy savings versus percentage frequency of projects.

The lack of energy statements does not mean that renewables were not included in the planning application, but that data were not available to include in the study. This is in fact more indicative of the development of the process, which will be discussed later.

It is important to note that the percentage savings from renewables are the average across all the developments, and will depend on the size of each development as well as the renewable energy contribution. So the average percentage attainment will differ significantly. This is largely due to the fact that the very first development considered under the London Plan was very large, but only achieved a relatively small percentage renewable energy contribution which skewed the data. As will be seen the renewable energy contribution improved significantly over time.

Figure 20.5 shows the frequency with which different percentage renewable energy attainments were reached. This shows that 37% of projects met or exceeded the 10% target, and the average attainment was 9.3%, close to the 10% requirement. Where accurate floor area data were available, this showed that more than half of the developments (57%) could achieve or exceed 4 kg CO_2 m^{-2} savings, with one development claiming 10.5 kg CO_2 m^{-2} (where UK offices typically operate in the range 36–160 kg CO_2 m^{-2}). Energy Consumption Guide (2003).

Figure 20.6 shows the numbers of installations for each renewable energy technology. Note that each site may use a mix of technologies, so the total installations exceed the

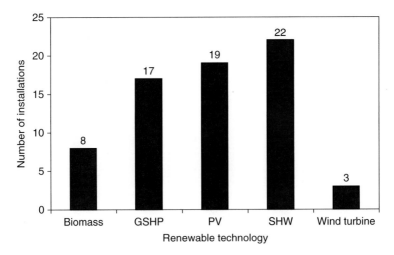

Fig. 20.6. Numbers of proposed installations by renewable technology.

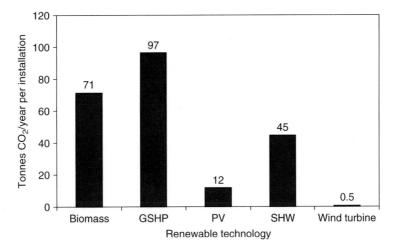

Fig. 20.7. CO_2 savings/year per installation.

number of developments. Solar hot water is a popular choice, probably on the grounds of its relative cost or perceived ease of installation, while photovoltaic (PV) installations are relatively common, but more likely on the grounds of their iconic status. The choice of technologies may be influenced by individual preferences on the part of developers, architects or engineers based on previous experience or perceptions of appropriateness.

When looked at from the point of view of carbon savings the picture is rather different. Figure 20.7 shows the carbon saving per installation. Ground source heat pumps (GSHP) and biomass show larger savings because of the size of the installations. (It is interesting also to note that with savings over 6 kg CO_2 m^{-2} GSHP or biomass is generally employed.) Likewise, PV installations are small due to cost. It is unfortunate that installed capacity was not available as these values could then have been normalized for a true comparison.

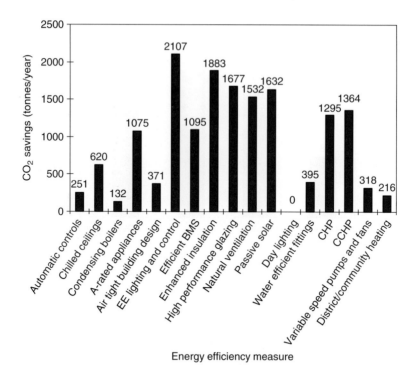

Fig. 20.8. CO_2 savings by energy efficiency measure.

Nevertheless, this shows that GSHP can make significant contributions to savings, as can biomass. There are technical and commercial issues with these technologies. GSHP may have long-term effects on local ground cooling (and therefore experience a deterioration in performance) unless these are reverse cycle to include summer cooling operation. This begs the question whether the use of GSHP might encourage the use of air conditioning where it might otherwise have been avoided. Biomass, on the other hand, provides a straightforward low carbon solution for heating. The problems in urban applications of biomass are fuel transportation and storage and, in a city such as London, the existence of a reliable supply chain.

By contrast, energy efficiency offers much greater savings. A quarter of all developments shows savings in the region 14–20 kg CO_2 m^{-2}, with a further 25% exceeding 20 kg CO_2 m^{-2} (one claiming a very large 83 kg CO_2 m^{-2}). The range of energy efficiency measures adopted, with their associated CO_2 savings, are shown in Fig. 20.8. The size of savings shown from energy efficiency directly impacts on the renewables contribution – the more the site emissions are reduced by energy efficiency, the smaller the renewables contribution. Since energy efficiency technologies are more established, it is cheaper to increase savings in this area. Interestingly, developers and their designers are coming to recognize this and are more focused on energy efficiency as a result of the renewables target. The difficulty is proving the savings are real, that they are realistic given the baseline scenario, and that they are fully implemented. With improvements to building regulations in 2006 the baseline building now has to comply with tighter requirements which already demand significant efficiency measures. Therefore finding additional savings will become increasingly

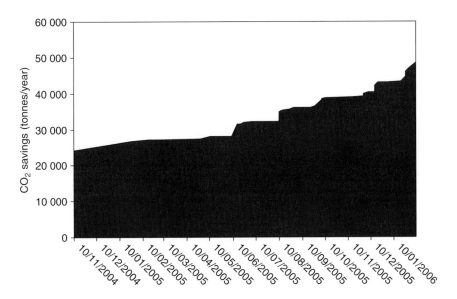

Fig. 20.9. Cumulative total CO_2 savings for all energy statements.

difficult if the technologies shown in Fig. 20.8 are already adopted. This highlights the need for transparency in the energy statement with respect to baselines, efficiency savings calculations, and choice of renewables.

20.4.3.2 Trends

An interesting aspect of London's planning policy is the time it took to take effect. The study was able to reveal some of the influencing factors in its implementation. Figure 20.9 shows the cumulative CO_2 savings for the developments submitted under the London Plan, and Fig. 20.10 shows the trend in these savings projected forwards for a further year. These figures are based solely on the 46 developments for which there are firm data. Table 20.6 shows these projected savings for energy efficiency, renewables and in total, but extrapolated to all 186 developments (which again assumes similar measures and trends occur across all developments).

The fact that the rate at which the savings occur increases over time is not surprising. It shows that for the initial project developers were unready or unwilling to fully comply with the requirements. Being unready is probably the most common explanation since developers did not have the expertise to conduct the necessary analysis, which meant establishing new ways of working, particularly with engineering consultants. In addition, the planners also had to learn about the implications of the new regime. They had to gain new technical knowledge in order to understand what was feasible and what was not, and be able to take a reasonably hard line on negotiations with developers. This inevitably took time to bed down.

This is perhaps most apparent in the uptake of renewables. Figures 20.11 and 20.12 show the cumulative savings and projections for renewables only. The most striking issue is that apart from the first large project (Stratford Olympic Village), almost no renewables were specified until the middle of 2005, almost a full year after implementation of the Plan. The

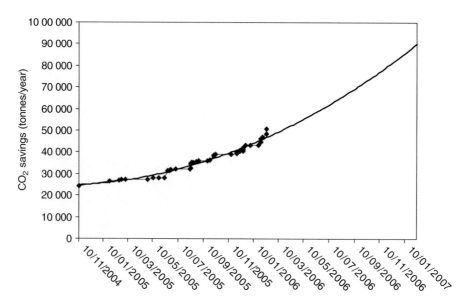

Fig. 20.10. Projected cumulative total CO_2 savings including Stratford Olympic Village.

Table 20.6. Approximate CO_2 savings in tonnes/year projected to January 2007 from actual data and extrapolated for 186 developments. (Note that energy efficiency extrapolations are based on 186/46, and renewable energy is based on 186/32.)

	From available data			Extrapolated for 186 developments		
	Linear	Mid-range	Max	Linear	Mid-range	Max
EE	60000	68000	75000	243000	275000	303000
RE	8000	13000	18000	47000	75000	105000
Total	68000	85000	93000	290000	350000	408000

cause of the sudden change was largely due to planning officers becoming more experienced in the negotiations and having the confidence to take a harder line. Once a firmer policy was adopted there was a rapid increase in renewable energy specification.

The maturing of the policy is also underlined by the pattern of energy statement submissions as shown in Fig. 20.13. The inclusion of an energy statement with a planning application may be seen as a proxy for developer engagement with the policy. Significant numbers of energy statements only started coming forward towards the end of 2004, with a further pick-up in 2005. The earliest energy statement pre-dated the London Plan, but this involved consultants who were already involved in the sustainability agenda for London. Much of the improvement may be attributed to the release of the London Renewable Toolkit (2004) in September 2004.

A further indication of improvement is the length of the planning application process. Figure 20.14 shows the length of each application (in days) against the application start date. There is a clear trend of reducing process time. There is little doubt that this was due to increased experience of the planning team, coupled with a better understanding of what was

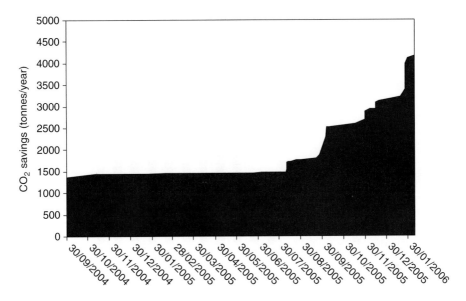

Fig. 20.11. Cumulative RE savings.

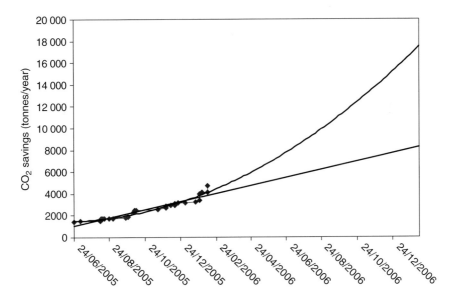

Fig. 20.12. Projected cumulative CO_2 savings from renewables.

required on the part of the developers. The developers needed to appreciate the need for pre-feasibility studies to be complete before submitting the application, while the planners needed to understand whether the best intentions of the London Plan were being met. There are clear advantages for developers, in terms of cost and avoided delays, to get the application right first time. This makes the planning policy a powerful mechanism for effecting change.

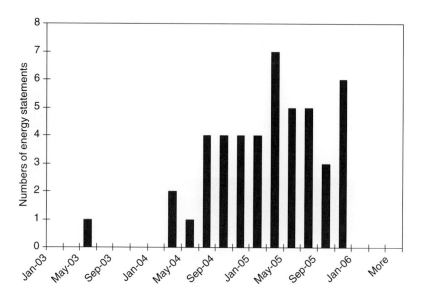

Fig. 20.13. Number of energy statements over time.

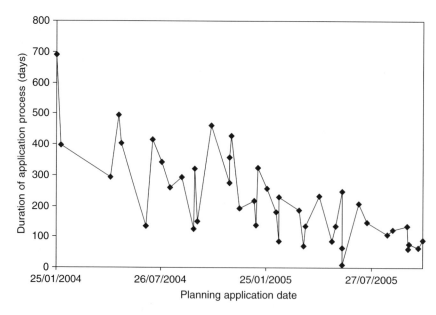

Fig. 20.14. Length of application process over time by start date.

20.4.3.3 *Issues*
Several issues arise from the implementation of the London Plan that need to be addressed if maximum impact from such a policy is to be achieved.

First, there is a need for good feasibility and assessment tools. Designing robust energy efficiency and energy generation systems is a complex engineering task, so early pre-feasibility assessments of carbon are necessarily approximations. It is essential to have standardized and validated tools that can perform these assessments. The London Energy Partnership commissioned the development of one such tool – the London Renewables Toolkit, (2004) which has been widely adopted in the submission of energy statements. However, the tool has proved to have limited capability, and the LEP has since commissioned a second generation tool, which has not been released at the time of writing.

Such tools must be usable by both developers (and their design consultants) and planners. These different players have different levels of technical understanding, and these tools need to address this fact. Some work is going on to tailor such tools for specific end users, with interfaces that cater for specific skills sets. Dunsdon *et al.* (2006)

Closely coupled to the use of tools is the need for training. Local authorities need to be properly resourced with good technical capability for dealing with energy issues. The LEP set up a program of training planners to use the London Renewables Toolkit. This program was highly successful in raising planners' awareness of energy systems, and how to evaluate submissions. Since there is a considerable turnover of planning officers in London boroughs, there is a need for regular ongoing training to ensure the knowledge base is maintained. In the longer term it will be necessary for university courses to incorporate this aspect as a core skill for planners.

Similar considerations are necessary for building control officers as well as developers, although in the case of developers, it is more usual that they will engage specialist design consultants with suitable expertise. These consultants carry out option appraisals and will develop/submit energy statements to the planning departments on behalf of their clients. Nevertheless, developers need to have in-house awareness of the needs of the planning system, and some knowledge of how to comply with these new requirements. Even the design community is climbing a steep learning curve in supporting their clients in this way and significant effort is under way to provide the necessary training.

The most challenging aspect of these pre-feasibility studies is the potential incompatibility between technologies. This has led to disagreements between developers and planners, with the occasional impasse that results in planning delays. Typical examples of such technology conflicts are the use of CHP with solar thermal systems. CHP will invariably provide heat demand, particularly in the summer when demand is low, but this is when solar thermal potential is high. A requirement for CHP before renewables (which in the light of the London Scenarios work is a sensible policy) may therefore drive the renewables component towards more expensive or unproven options (photovoltaics or urban wind). The alternative of using biomass CHP is viable, but has associated risk until fuel supply chains are fully established. There are other such conflicts, all of which require careful option appraisals and design to overcome these difficulties.

Another major issue arises in the way the baseline is established. Building regulations should provide the basis for the baseline energy consumption at planning stage. However, full building regulation compliance calculations come at a later stage in the design process. There may be a temptation to start with a more relaxed baseline, since the design will inevitably be less well developed for the planning submission. This requires the close attention of the planners, and again greater knowledge of an area normally associated with their colleagues in the building control department. In addition, as building regulations get tighter it becomes even more difficult to meet a percentage target. At some stage, when data become available, it may be necessary to move to targets expressed in kg CO_2/m^2 in order to provide clear guidance on what is expected from our building stock.

Finally, it should be stressed that theoretical savings calculated at the building planning stage are not yet proven. A building must have these technologies correctly installed,

commissioned and maintained if actual savings are to be realized. This will require building control officers to be fully conversant with renewable energy systems, which again
will generally mean new knowledge and skills within that profession. It would be a waste
of time and effort if carbon savings were just confined to a paper exercise and not realized and audited in practice: it is real carbon savings that count if the policy is to be truly
effective. Since developments approved under the London Plan are yet to be built in large
numbers this is an area for future concern, but one that needs to be planned for now.

Further work will be needed in future to corroborate energy statement claims with on-site
assessments to establish what has actually been saved. This may indicate shortcomings in
the original energy statement but may also indicate other problems such as poor installation,
incorrect commissioning and even deficiencies in the technologies themselves. Monitoring
of post-planning implementation will therefore be an important step in ensuring the success
of low carbon policies, and must feed back into the design tools to improve the process.

20.5 Conclusions

London has set itself clear policy objectives to reduce its carbon footprint, and move
towards a new paradigm in energy delivery. The aim is to increase embedded energy
generation substantially, while improving the energy efficiency of the building stock. In
order to meet these objectives studies have been conducted to assess what technologies
will be best deployed to meet carbon emissions reductions at reasonable cost. This helps
to inform policy and decision-makers about how to move towards this low carbon future.
More importantly, London has adopted mandatory requirements for new developments
to include new technologies. Without these measures it is very likely that aspirations for
deployment of CHP and renewables would not become a reality. If the free market is left
to decide, then high cost options simply will not be specified, irrespective of the recognition by society of the need for action.

The evidence so far has shown that the planning system is a good way to influence
energy efficiency of buildings. It gets developers and designers thinking about energy at
an early stage and provides a more strategic approach to site energy consumption. An
important effect of this policy is the early engagement of engineers in the design process. It
has long been debated about how multi-disciplinary design teams need to come together
earlier, and this is an encouraging trend in the UK construction industry.

The London Plan originally called for 10% contribution in carbon savings from renewables. After a slow start it has become apparent that this is reasonably straightforward to
achieve, and higher targets would be justified in future. There may be arguments for providing targets set against prescribed maximum emissions levels (in kg CO_2/m^2) for different building types, rather than the relativistic percentage reductions, which will become
harder to achieve in the future as regulations become tighter. There is a pressing need for
some work in this area to set realistic benchmarks and targets for typical building types.
The UK government has announced its intention to move towards zero carbon domestic
buildings by 2016, which will have a significant effect on the way targets are developed
over the next few years.

It has become clear that success in meeting percentage carbon reduction targets is heavily dependent on planners and developers understanding the requirements and setting up
clear mechanisms to negotiate the final design. What is important is the need for clear,
unambiguous targets to facilitate good proposals prior to first submission. The experience
in London is highly valuable in this respect, as the different parties have learned to work

within the new regime. However, there is a need to ensure training and education provision keeps pace with the skill improvements required to make it all work.

It is important that whatever regulatory instruments are used, the right technologies are deployed to achieve real carbon savings. The London Carbon Scenarios work clearly shows that CHP is a winning technology, and in particular utilizing large-scale community-based CHP can make substantial savings at reasonable cost. However, this should not be at the expense of other technologies, which may be zero carbon.

The hybrid scenario shows that, while not the most cost effective, it is realistic to expect a mix of technologies. Indeed, as we move beyond 2026 more zero carbon technologies will be needed to make deeper cuts in carbon emissions. Assisting the growth of the market (and therefore providing future cost reductions) is an important aspect of the energy future. By that time it may be possible that other technologies will have been developed or made cost effective – particularly hydrogen-based technologies – that can wean cities off their dependence on fossil fuels.

At present the London Plan is encouraging a mix of technologies to be specified, and not always the most cost-effective solution. So policy may indeed be driving in the direction of the (non-optimized) hybrid approach. However, as experience improves it may well prove that various winning technologies will start to be specified on cost advantage grounds. In this case other mechanisms may be required if it is seen necessary to maximize the diversity of energy delivery systems installed in the urban environment.

There are still many issues that need to be resolved if cities are to become truly low carbon emitters. These involve attitudes to investment in technology and infrastructure, engagement of the public and the need for an appropriately skilled workforce to deliver change. These are all complex subjects in their own right, and beyond the scope of this chapter. What is clear, however, is that cities need to make a start on reducing carbon emissions, and that involves developing policies, setting targets, and using regulatory powers to get things in motion. Once the benefits are demonstrated it should be easier to bring everyone along for the ride.

Acknowledgements

The authors would like to thanks the following people from the Greater London Authority and the London Energy Partnership for their advice and factual checking of the manuscript, and for their original support in the production of the reports upon which this is based: Andy Deacon, Suzanne LeMiere, Paula Kirk, Lucy Padfield and Tom Carpen.

References

Clarkson, R. and Deyes, K. (2002). Estimating the Social Cost of Carbon Emissions, *Government Economic Service Working Paper 140*, January. HM Treasury and Department of the Environment Food and Rural Affairs.

Dunsdon, A., Jones, P.G. and Day, A.R. (2006). Towards a computer based framework to support the low carbon building design process. *International Journal of Low Carbon Technologies*, 1(4), 355–371.

Energy Consumption Guide 19, 2003. Carbon Trust.

Green Light to Clean Power: The Mayor's Energy Strategy, February 2004. Greater London Authority.

London Carbon Scenarios to 2026, a report by SEA/RENUE for the London Energy Partnership, 2006. Greater London Authority.

London Renewables, integrating energy into development: toolkit for planners, developers and consultants, September 2004. Greater London Authority.

Meinshausen, M. (2006). What does a 2°C target mean for greenhouse gas concentrations? A brief analysis based on multi-gas emission pathways and several climate sensitivity uncertainty estimates, in *Avoiding Dangerous Climate Change* (eds) L.H.J. Schellnhuber *et al*. Cambridge University Press.

Review of impact of the energy policies in the London Plan on planning applications referred to the mayor, May 2006. Greater London Authority (Restricted circulation).

Royal Commission on Environmental Pollution's 22nd report, *Energy: the Changing Climate*, June 2000. RCEP.

The Building Regulations Part L 2000 (2006 edition), Conservation of Fuel and Power, April 2006. Department of Communities and Local Government.

The Energy Challenge, Energy Review, July 2006. Department of Trade and Industry.

The London Plan: Spatial Development Strategy for Greater London, February 2004. Greater London Authority.

The Low Carbon Buildings Programme, Department of Trade and Industry, retrieved 5 February 2007 from http://www.lowcarbonbuildings.org.uk

The Renewables Obligation Order 2006, Statutory Instrument No. 1004, Office of Public Sector Information, retrieved 5 February 2007 from http://www.opsi.gov.uk/SI/si2006/20061004.htm. (The Renewables Obligation for England and Wales compels electricity suppliers to supply a proportion of their electricity from renewable energy sources. For 2006 this proportion was 6.7%. Suppliers demonstrate compliance by purchasing Renewable Obligation Certificates (ROCs) from accredited renewable electricity generators. Each ROC relates to 1 MWh of renewable electricity. ROCs can be traded in conjunction with, or separately from, the electricity to which they are associated, which has established a market in tradable 'green' certificates. Suppliers who fail to meet the target have to pay a penalty, known as the buy-out. The buy-out price for 2006/07 was set at £33.24/MWh, and the buy-out fund that this creates is ultimately redistributed to suppliers in proportion to the number of ROCs they submitted in a given year. This has led to ROCs being traded at around £40/MWh – although this does fluctuate depending on the shortfall of renewable supply.)

The economics of climate change. *The Stern Review*, January 2007. HM Treasury.

Glossary of Terms

CCGT	Combined cycle gas turbine
CCL	Climate change levy
CHP	Combined heat and power
DHW	Domestic hot water
EEC	Energy efficiency commitment
EU ETS	European Union Emissions Trading Scheme
GLA	Greater London Authority
GSHP	Ground source heat pump
IRR	Internal rate of return
LCBP	Low Carbon Buildings Program
LCCA	London Climate Change Agency
LEC	Levy exemption certificate
LEP	London Energy Partnership
LZCT	Low and zero carbon technologies
NPV	Net present value
PV	Photovoltaic
RO	Renewables Obligation
ROC	Renewable Obligation Certificates
SELCHP	South East London CHP (energy from waste station)

Chapter 21
Urban Energy and Carbon Management in Leicester

PETER WEBBER AND PAUL FLEMING

21.1 Introduction

This chapter describes the approach of Leicester city in implementing measures to reduce greenhouse gas emissions. This includes monitoring its progress, and communicating this progress to decision-makers and to the general public, in an attempt to change their behaviour and move towards a low carbon economy. Leicester has a long history of implementing policies to reduce energy consumption. The city has set an ambitious target of reducing its energy use by 50% of the 1990 levels by the year 2025. An energy strategy (Leicester City Council 1994) was developed which gave details of energy demand and supply measures that should be implemented. A further target of meeting 20% of the city's energy requirements from renewable energy by 2020 was also agreed, and in 2003 a climate change strategy was produced that addressed adaptation, mitigation, evaluation and communication (Leicester Partnership 2003).

To address these ambitious targets, Leicester has undertaken many activities aimed at managing energy consumption and carbon emissions in different energy end-use sectors (domestic, non-domestic buildings and transport). It has been attempting to monitor the implementation of its energy and climate change strategies, for both city-wide energy consumption and emissions and its own corporate consumption and emissions. A number of different approaches to measuring this progress have been used including models of energy use and emissions in Leicester (Titheridge *et al.* 1996), and the International Council for Local Environmental Initiatives (ICLEI) Councils for Climate Protection software (Webber *et al.* 2001; Torrie Smith Associates 2000) to quantify energy consumption and monitor changes in greenhouse gas emissions. Since climate change is a global problem Leicester has been able to work internationally (with Energie-Cités and ICLEI) to learn and share experience. It was one of the earliest European Union cities to establish a European Commission funded Local Energy Agency (Leicester Energy Agency and Barnagel 1999). It has a comprehensive approach to energy efficiency and sustainable development.

The following sections review local carbon management policies in Leicester. Progress in achieving the ambitious targets is discussed and experiences with monitoring and evaluating energy consumption are reviewed, including developments in data availability. Finally the city's progress in engaging in a dialogue with the public is described.

21.2 Policy Background

21.2.1 *European Union and United Kingdom Policy*

The European Union (EU) has developed a number of policy measures and programmes that are relevant to energy and carbon management at the local level. For example, the White Paper on energy policy (European Commission 1995) and the White Paper on renewable energy (European Commission 1997) have outlined EU policy objectives. The EU has a target of reducing its greenhouse gas emissions by 8% of 1990 levels by 2008/12 under the Kyoto agreement and an EU Climate Change Programme has been established (e.g. European Commission 2006).

The UK has a Kyoto target of reducing its greenhouse gas emissions by 12.5% of 1990 levels by 2008/12 and a voluntary target of reducing carbon dioxide (CO_2) emissions by 20% from 1990 levels by 2010. The UK has produced a climate change strategy (Department of the Environment and Transport and the Regions 2000), which has been reviewed recently (DEFRA 2006a). The updated climate change programme highlights policy measures to reduce emissions in the energy supply, business, transport, domestic, agriculture, forestry and land management, and public and local government sectors. The 2003 Energy White Paper (DTI 2003) identified the ambitious aspirational target of reducing greenhouse gas emissions by 60% by 2050, which was subsequently confirmed in the government's 2006 energy review (DTI 2006) and 2007 Energy White Paper (DTI 2007).

In recent years there has been an increasing amount of relevant legislation (e.g. Home Energy Conservation Act 1995, Warm Homes and Energy Conservation Act 2000, Sustainable Energy Act 2003, Energy Act 2004, Climate Change and Sustainable Energy Act 2006 (The Stationery Office 2006), and the proposed Climate Change Bill). There has been an increase in emerging national policy relating to the management of carbon emissions, and there has been all party political support. Indeed, there has been a 'mainstreaming' of sustainable development and action to manage energy consumption and greenhouse gas emissions. There have been several recent policy initiatives and studies at the national level, with a recent example being the Stern Review (Stern 2006) which has highlighted the economic benefits of not delaying action on climate change.

The government has a number of support programmes for energy efficiency and renewable energy in various sectors, through the Energy Saving Trust, and programmes focused at the business sector through the Carbon Trust, including the public sector such as the Local Authority Carbon Management program. There has been growing national interest in the scope for reducing emissions through action at the local and regional level (Shackley *et al.* 2002). Best Value Performance Indicators for local government have existed which have been relevant to the management of energy and carbon dioxide emissions. As highlighted in the local government White Paper in 2006 the local authority performance framework is to be used to include action on climate change.

The Local Government Association provides support for local authority action on energy and carbon management. Examples include the 'Environmental good practice guide' in 1991 through the 'Energy for sustainable communities' in 1997 (LGA and LGMB 1998) and 'Leading the way' in 2004 (LGA/EST/EEPH 2005). Leicester helped develop these policies with the LGA and has been involved with the Beacon Councils for Sustainable Energy mentoring scheme where the seven Beacon Councils have provided support to other local authorities. At the regional level, Regional Assemblies and Regional Development Agencies have been established and in the East Midlands, a Regional Energy Strategy (EMRA 2004 and EMRA 2007) has been developed.

21.2.2 Leicester

Leicester, in the East Midlands region of the UK, covers an area of 7337 hectares and has a population of almost 300 000. It is the tenth largest city in the country and is a culturally diverse city with about one third of the population from an ethnic minority. Leicester was one of the first UK cities to become completely smoke free, following the introduction of the Clean Air Acts in the 1950s and 1960s (Russell 2007).

Leicester has a long history of energy-related work. It started in the 1970s when solar water heating was installed on buildings, continued in the 1980s when it was a lead city for combined heat and power and became Britain's first Environment City in 1990. It received honours at the United Nations conference on Environment and Development in 1992, was a European Sustainable City in 1996 and a UK Beacon Council for Sustainable Energy in 2005. The city has set an energy reduction target of 50% below 1990 levels by 2025. The City Council developed an Energy Action Plan in 1990, a home energy strategy in 1993 and has a target for 20% of Leicester's electricity to be obtained from renewable energy sources by 2020. The Leicester Environment City initiative published the city-wide energy strategy in 1994 (Leicester City Council 1994). This strategy adopts a comprehensive approach, addressing both energy supply, and energy demand in the domestic, services, industrial, and transport sectors. The effectiveness of the strategy has been monitored via the Leicester Energy Agency and Barnagel (1999). The Leicester Partnership (the local strategic partnership for Leicester) adopted a climate change strategy in 2003. This included addressing adaptation and mitigation.

Although action on local energy use and local responses to climate change has not been mandatory and has been a voluntary activity, Leicester has undertaken many activities aimed at managing energy consumption and carbon emissions in different sectors (domestic, non-domestic buildings, waste and transport).

21.3 Local Energy and Carbon Management Policies in Leicester

Leicester has adopted a number of strategies, which relate to energy and carbon management in the city. The two key ones are the Leicester Energy Strategy and the Leicester Climate Change Strategy.

Leicester Energy Strategy: Leicester's approach to local energy management, as described in the 1994 Leicester Energy Strategy (Leicester City Council 1994), has considered a range of actions on local energy/carbon management for Leicester to meet its target of reducing its CO_2 emissions to half the 1990 level by 2025. The strategy covered the following areas:

- Energy supply
- Energy demand measures
  - Domestic
  - Non-domestic
  - Transport
  - Advice
- Monitoring and implementation.

The key parts of the strategy were the ambitious target of a 50% reduction and the need to establish structures to implement energy savings on a large scale. The implementation of a city-wide combined heat and power scheme was seen as a key deliverable.

Leicester Climate Change Strategy: The City Council in partnership with other major organizations in the city, through the Leicester Strategic Partnership, developed the Climate Change Strategy (Leicester Partnership 2003). This built on the Leicester Energy Strategy, reviewing progress since the energy strategy and looking at wider greenhouse gas emissions management as well as the need to adapt to climate change. It considered carbon sequestration, and the role of awareness raising.

Key recommendations included the establishment of a city-wide combined heat and power scheme, the need to ensure that new buildings are built to standards that dramatically reduce greenhouse gas emissions and the need to engage with the public and decision-makers to change attitudes and behaviour.

21.3.1 Energy supply

The single biggest opportunity to reduce greenhouse gas emissions is through the delivery of a city-wide combined heat and power (CHP) scheme. In the 1980s Leicester was a UK 'lead city' for CHP and developed plans for various large-scale schemes. A company was formed to deliver this, draft contracts agreed to purchase the output from a North Sea gas field, build a combined cycle gas turbine electricity generating power station and to distribute the 'waste heat' through a city-wide heat distribution network. This was the first time in the UK that such draft agreements had been in place. However, at the time, the electricity and gas industries were being privatized and the uncertainties associated with a relatively small city-wide CHP scheme meant that the proposals never proceeded and the company was wound up. Since the 1980s Leicester has being continuing to try to implement city-wide CHP. Several other potential schemes were attempted, but, in the absence of a city-wide scheme, small-scale CHP plants have been installed in leisure centres and district heating schemes.

Leicester has also established a number of solar energy projects on its own buildings, demonstrating the application of renewable energy, and contributing to CO_2 emissions savings.

21.3.2 Energy demand measures

The main focus of the city's work has been in terms of reducing greenhouse gas emissions through reducing energy consumption; that is demand-side management.

21.3.2.1 Domestic

Leicester has a housing stock of about 116000 homes, with the largest proportion of these being solid wall, terraced houses (about 41%). In 1991, about 58% of homes were owner occupied, while 27% of homes were owned by the City Council. The number of homes the council has owned has decreased with ownership by housing associations and the private sector increasing. The City Council has been able to work directly to improve the energy efficiency of its own housing stock, for example with ongoing improvement programmes. In the 1990s it was able to insulate the homes of all the council housing stock with cavity walls (approximately 13000 homes). It pioneered the sale of low energy lamps to council tenants. It has replaced a range of council housing which had become substandard with new housing built to very high standards of energy efficiency.

Fig. 21.1. Leicester Energy Efficiency Centre.

It has used a range of financial incentives for residents to improve the energy efficiency of their homes. These have included grant schemes for improving private sector housing, discounts on energy efficiency equipment for residents through the Energy Sense scheme, and an interest-free loan scheme for home energy efficiency measures.

Energy surveys of homes have been undertaken and energy efficiency advice and information have been made available through the city's Energy Efficiency Advice Centre (EEAC) (Fig. 21.1), which is situated near the market in the city centre. Regular energy efficiency awareness raising activities are carried out for the general public and energy efficiency is promoted through the local media. Through partnership work with energy utilities (who have had energy efficiency obligations to meet through the Energy Efficiency Commitment) discounted energy efficiency measures have been made available, for example free low energy lightbulbs have been provided through the EEAC.

Solar energy demonstration projects have been undertaken by Leicester's environmental charity, Groundwork Leicester and Leicestershire (formerly Environ), who have successfully run the city's Eco House (the UK's first low energy show house) and delivered other innovative environmental projects in the city.

21.3.2.2 Non-domestic
Leicester has experienced industrial decline in recent decades but there remains a range of industrial and commercial businesses in the city, with major industries including textiles and clothing, printing and publishing.

Energy efficiency improvements to the non-domestic sector not only reduce greenhouse gas emissions but also result in energy cost savings and potentially improved businesses' competitiveness and productivity. The Leicester Energy Agency works with businesses providing advice and information, and signposting businesses to financial support for energy improvements. An energy efficiency roadshow has been used to promote energy efficiency to businesses and to provide energy advice and information. Staff training has been provided. Also local energy efficiency advice and information has been available to businesses through Groundwork/Environ. Regeneration funding (e.g. ERDF) has been used to provide free energy surveys, energy efficiency support and grants towards energy efficiency and clean technology improvements and to support businesses to work towards their own environmental management systems. Businesses have been made aware of

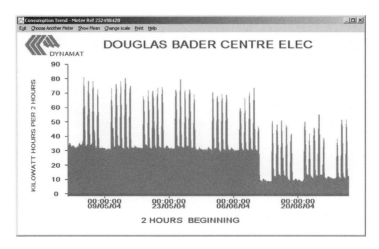

Fig. 21.2. Intelligent metering.

relevant other support for energy efficiency, e.g. from national schemes run by the Carbon Trust. A local energy and environment manager's group operated providing support for energy management in local businesses and a forum for the exchange of experiences.

Energy use in Leicester City Council's buildings and schools contributes an estimated 3–4% of CO_2 emissions in Leicester. The council has carried out energy management activities in its own buildings. For example, it regularly monitors energy consumption in its buildings and in recent years it has been undertaking a programme of installing automatic half hourly energy monitoring (intelligent metering – Fig. 21.2) in individual council buildings. This system records the electricity, gas and water consumption data every half hour and enables unusual consumption to be identified and energy saving opportunities to be investigated. Leicester is a leading local authority in this area, having connected over 200 buildings to the 'intelligent metering' system, and is unique in the UK in having all the main municipal buildings on the system.

Within its buildings the council has carried out energy surveys and staff training, and runs an interest-free revolving loan fund to support investments in energy saving projects, with the loan being repaid from the savings. Renewable energy demonstration projects have been set up on council buildings. A proportion of the council's electricity has been purchased from renewable energy or from combined heat and power. Some of the electricity generated in the council's CHP has been used by other buildings.

Leicester City Council was accredited to the Eco Management and Audit Scheme in 1999, which includes an annual report on progress towards EMAS targets, including energy reduction targets for the council's buildings and the uptake of renewable energy.

21.3.2.3 Transport
Transport is the fastest growing source of greenhouse gas emissions, and traditionally transport policy has not focused on reducing these emissions. Activities addressing emissions include land use planning to reduce the need for travel, activities to encourage the use of fuel efficient vehicles and the greater use of public and non-motorized transport.

Fig. 21.3. Bus travel Star Track information system.

Activities have been carried out to encourage behaviour change. Advice and information has been provided for the public, travel plans developed with local employers and schools, and infrastructure improvements to support public transport have been implemented. A road charging trial on one route in Leicester, alongside public transport provision, was trialled in the late 1990s. There is some further consideration being given to this with nearby cities. Park and ride provision has been made available.

Leicester City Council has arranged infrastructure improvements to support public transport, e.g. bus lanes, bus priority devices at traffic lights, and a real-time bus information system has been installed (Fig. 21.3). While the council does not operate the public transport in the city it is able to take a partnership approach and encourage bus operators, for example who can use lower emission vehicles and take steps to improve usage, e.g. with low access vehicles (Randall 2007). Leicester has an extensive network of cycle routes and a city centre bike park has been in place for a number of years.

There has been extensive air quality monitoring in Leicester. A unique set of data have been gathered, with a significant quantity of transport and air pollution data available. Action has been taken targeted at managing air quality in the city.

Leicester City Council has also established EMAS targets for its own vehicle fleet fuel use and staff business mileage. It has been trialling the use of alternative fuelled vehicles and a 5% biodiesel mix is used for its vehicle fleet. It has arranged discounts with bus companies for council staff, and cycle-related discounts for staff.

21.3.2.4 Waste
Greenhouse gas emissions from waste are relatively small compared to other sectors, with methane recovery in place at landfill sites. Leicester has had a recycling/composting target of 40% of household waste in the city. It has undertaken recycling initiatives in its own buildings, and about 27% of Leicester's waste is being recycled. Leicester takes an integrated approach to waste management, with various initiatives in place to reduce the amount of household waste going to landfill, increasing recycling. In recent years, a waste processing/recycling plant has come into operation to help Leicester meet its recycling/composting target. This separates recyclables from the domestic waste. Organic waste recovered from the plant is composted using anaerobic digestion. Also, kerbside and civic amenity site recycling of separate materials has been in place for a number of years.

21.3.2.5 Planning/development
Leicester has considered scope for managing energy and related emissions connected with planning, development and new construction. It has produced guidance for developers on energy efficiency and renewable energy in new developments. In the 1990s Leicester attempted to use the planning regulations to encourage greater energy efficiency in new development. However, the national planning inspectors did not allow this, saying that energy efficiency was not a planning concern. This view has now changed, and in 2006 targets for on-site renewable energy generation in new developments were set. Ten per cent of the energy requirements of new major developments must be met by renewable energy, rising by 1% a year (Leicester City Council 2007). The city has implemented the Leicester Better Buildings program (LCC 'Leicester Better Buildings' 2007) (Fig. 21.4) to support and encourage developers to build quality buildings, incorporating high energy and environmental standards.

21.3.2.6 Adaptation
The city recognizes that the climate is changing and that it needs to adapt to this change. For example, summers are anticipated to become warmer and drier, while winters are expected to be warmer and wetter with more heavy downpours. New development needs

Fig. 21.4. Leicester Better Buildings website.

to be able to adapt to these changes and to be designed accordingly; the City Council provides advice on this.

21.4 Local Modelling and Emissions Accounting

A key part of Leicester's approach is to monitor and evaluate progress in implementing the 1994 energy strategy and the 2003 climate change strategy. In order to monitor progress, data are collected at an individual building level, council-wide level and city-wide level. Leicester has taken a leading approach, being one of the first cities to measure its CO_2 emissions and the first city to install comprehensive intelligent metering in its own buildings (for electricity, gas, water and solar).

The quality of data gathered for many of the council's own buildings is very high. This includes electricity, gas, water and renewable energy data gathered from meter reading every half hour. The accuracy of data at the city-wide level has been more questionable. In developing the energy strategy in the 1980s and early 1990s energy data by postcode were available from the electricity and gas utilities. These data helped apply the DREAM (Titheridge *et al.* 1996) model used in the energy strategy. However, following privatization of the electricity and gas industries the postcode energy data became 'commercially sensitive' and was no longer available. The Local Government Association lobbied the DTI for several years to obtain these data again and in recent years the DTI (now BERR) has begun publishing regional statistics. The accuracy of these data has increased with subsequent years and is now used to help monitor the city's emissions.

21.4.1 Leicester City Council

Leicester monitors its progress towards a range of environmental management targets annually through the Eco Management and Audit Scheme (EMAS) (Leicester City Council 2006). This enables trends in a number of aspects relevant to local energy and carbon management to be monitored, including energy consumption in council buildings, the share of energy requirements met from renewable energy sources, vehicle fleet fuel consumption, and staff business mileage.

The City Council's Energy Management Team has monitored overall annual energy consumption in the council's buildings for several years. Buildings' energy consumption information and other information collected for EMAS purposes, with some additional information (e.g. on street lighting and staff commuting), has been used to estimate greenhouse gas emissions for the council.

Leicester City Council's 2005/06 greenhouse gas emissions have been estimated as 54 000 tonnes CO_2. This includes energy used in council buildings, fuel use by the council's vehicle fleet and staff business mileage and staff commuting, and energy use for street lighting. About 73% of 2005/06 greenhouse gas emissions were related to energy use in council buildings (about 40 000 tonnes), 15% were estimated from staff commuting (about 8000 tonnes), 10% were from the vehicle fleet (and waste vehicles) and 2% from staff business mileage.

For council operations as a whole, from 1990 to 2005/06 (Table 21.1) it is estimated there has been about a 25% reduction in total greenhouse gas emissions. The reduction in emissions associated with council buildings relates to different factors, for example purchasing some electricity for council buildings from renewable energy sources, a change from the use of coal to gas, the installation of energy efficiency measures (McKay 2000), the use of CHP (such as in community heating schemes), and the installation of solar energy projects. Also there has been a change in the shares of fuels used to generate electricity nationally, in

Table 21.1.   Leicester City Council estimated greenhouse gas
emissions 1990 and 2005/06.

	Annual greenhouse gas emissions CO_2 (thousand tonnes)	
	1990	2005/06
Buildings	47	39
Vehicle fleet (including staff business mileage), and staff commuting	14	15
Street lights	12	0
Waste	n/a	n/a
Total	74	54

particular a decrease in the use of coal and an increased use of gas, which has led to a reduction in carbon dioxide emissions per unit of electricity (Webber *et al.* 2001). Finally the electricity for street lighting has been sourced from renewable energy.

21.4.2   Community emissions

Monitoring and modelling of the city's emissions started in 1989. A number of different models and software tools, which have been available to help local authorities in quantifying energy use and greenhouse gas emissions in their communities, have been considered. One of the most widely used has been the Cities (or Councils) for Climate Protection (CCP) Greenhouse Gas Emissions Software, which was developed by Torrie Smith Associates (Canada) for the International Council for Local Environmental Initiatives (ICLEI) Climate Protection Campaign (Torrie Smith Associates 2000). This campaign started in 1993 and has grown to include over 800 participants worldwide. It provides support for local authorities interested in taking action on climate change. The software was developed to help local authorities to prepare inventories of greenhouse gas emissions, set targets, consider actions and monitor progress, for their corporate operations and for the wider community in their administrative area. This software was adapted for use in the UK as part of a government supported Councils for Climate Protection pilot project which began in 2000, involving the Improvement and Development Agency (IDeA) of local government, ICLEI, and 24 local authorities including Leicester. While local authorities have been progressing at different rates with action on greenhouse gas emissions (Allman *et al.* 2004), following the conclusion of the CCP pilot the approach has been taken on by the Carbon Trust's Local Authority Carbon Management programme which has been a spreadsheet tool to help quantify corporate emissions.

Accurate data were a key issue in the 1990s – and still are a key issue today. With the deregulation of the energy utilities and the introduction of competition into gas and electricity supply it became more difficult for many local authorities in the later 1990s, up to the time of the CCP pilot, to obtain actual local electricity and gas data. (Reasons for this included concerns about commercial confidentiality and perhaps the data not being readily available in a format to give information for the local area.) However, estimations or modelling could be used to determine consumption and emissions. Recently national statistical work has been carried out to obtain local energy (DTI 2007) and carbon dioxide

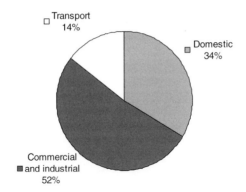

Fig. 21.5. Estimated Leicester city-wide greenhouse gas emissions 2004 (about 2 million tonnes CO_2).

emissions (DEFRA 2006b) figures, using actual electricity and gas data which has been made available from energy companies.

Accuracy and availability of data is a major issue. However, if you are a local authority and have money to spend on implementing energy efficiency and renewable energy measures, then you will want to spend all this money on implementation rather than on gathering data to evaluate your performance.

21.4.2.1 Leicester

The Dynamic Regional Energy Analysis Model (DREAM) was used to model energy use and emissions in Leicester for 1990 and 1995 (Leicester City Council 1994; Open University, 1997). In the early 1990s it was possible to collect local electricity and gas consumption data from regional electricity and gas companies to use the model to help to verify its outputs and establish a 1990 consumption baseline (Leicester City Council 1994). More recently, the ICLEI Councils for Climate Protection software was used to monitor greenhouse gas emissions. The predicted energy consumption outputs from the DREAM model were used as inputs for the CCP software to calculate greenhouse gas emissions (CO_2 equivalent for CO_2, methane and nitrous oxide) for 1990 and 1995. As DTI local energy data have been available in recent years these have been used as the data input. The DTI data have been revised over the years since they have been published, to improve their accuracy and resolve other anomalies. However, generally the DTI data have not been able to compare different years. They have been 'experimental data' and the method of collecting and interpreting the data has been changing; so it has not been possible to make comparisons between years.

Leicester's greenhouse gas emissions (2004) (Fig. 21.5) are estimated to be about 2 million tonnes CO_2, with energy use being about 28 000 TJ (terajoules) (about 7770 million kWh). About 34% of CO_2 emissions are from energy use in the domestic sector, about 52% from energy use in the commercial and industrial sectors, and about 14% from energy use in the transport sector.

In comparing local emissions figures for different years it has been necessary to consider the accuracy of the figures and take into account any differences in the methodology used to determine the figures. Greenhouse gas emissions figures for Leicester are given in Table 21.2. The transport figure for 1990 in the table has been adjusted using a national emissions trend to relate it to the 2004 figure.

Table 21.2.   Leicester greenhouse gas emissions estimates 1990 and
2004.

	Estimated annual greenhouse gas emissions CO_2 (thousand tonnes)	
	1990	2004
Domestic	630	670
Commercial and industrial	1380	1030
Transport	260	290
Total	2260	1980

Based on the data for Leicester in Table 21.2, it is estimated that there has been an over-all reduction of about 10% in greenhouse gas emissions from 1990 to 2004. This compares to national changes in emissions over this period of a 15% reduction in the six key green-house gas emissions (DEFRA 2006a), and a 6% reduction in CO_2 for which the key reason has been the reduction in power generation emissions (e.g. the switch from coal to gas).

While there is a view that Leicester has achieved an above average improvement in energy efficiency in housing the exact change in domestic emissions from 1990 and 2004 is not cer-tain from the table owing to differences between the data used. However, nationally there has been a decrease in CO_2 emissions of about 2% from 1990 to 2004. The effect on emissions of an increased demand for home heating, hot water and home electronic entertainment equipment has been offset especially by energy efficiency improvements and a change in the national electricity generating mix (DEFRA 2006a).

The exact overall reduction in CO_2 emissions for the services and industrial sectors in Leicester from 1990 to 2004 is not clear from the table (for the reason mentioned above). However, nationally CO_2 emissions have reduced by about 14% from 1990 to 2004 in the business and public sector. It is considered that this is due to changes in fuels used to gen-erate electricity, other fuel mix changes, and changes in 'energy intensity'(DEFRA 2006a).

Despite difficulties in obtaining a series of accurate local energy data previous work has been carried out to review Leicester's carbon management work in the domestic and busi-ness sectors (e.g. Bulkeley and Betsill 2003; and Fleming and Webber 2004). For example, as part of the work of the EC funded Leicester Energy Agency from 1996 to 1999, work was carried out to assess the energy and emissions savings from a wide range of energy effi-ciency activities for homes and businesses which had been taking place in the city. Savings were estimated from individual programs and compared to 1990 baseline consumption and future energy scenarios which were included in the Leicester energy strategy. The estimated energy-related CO_2 savings from energy efficiency programmes in the domestic and busi-ness sectors between early 1996 and early 1999 were about 23700 tonnes/year (Fleming and Webber 2004).

21.5 The City's Progress in Engaging in a Dialogue with the Public

A key element of the 2003 climate change strategy is engaging with the public and decision-makers in the city. This involves working in partnership with the key organizations to jointly address local energy and carbon management. The targets from the 2003 Climate Change Strategy are included in the Local Area Agreement. Leicester has undertaken activities to communicate climate change with the public to engage community groups, residents and businesses in considering sustainable energy approaches. Surveys have been undertaken to investigate attitudes related to sustainable energy (Colbourne *et al.* 1999), and in the mid-1990s an environmental awareness campaign focusing on different issues including energy was run (known as 'Turning the Tide'). More recently a climate change-related campaign called 'Keep Leicester Cool' was held and currently a 'Climate Change What's Your Plan' campaign, where individuals and organizations in the city are being encouraged to develop their own climate change action plans, is being run. This links with a current European Commission supported project 'Building Local Intelligent Energy Forums in Europe' (BELIEF) which involves a local forum related to a local sustainable energy action plan, and IMAGINE which is being coordinated by Energie-Cités – http://www.energie-cites.org/-IMAGINE,120-.

At the individual building level, training has been developed for building users (e.g. www.intelmeter.com) and pupils have been engaged in the design of their new school buildings as part of the 'Building Schools for the Future' initiative.

21.6 Lessons Learnt and Conclusions

Local energy and carbon management is very complex. For example, at the city-wide level, CO_2 emissions result from the combustion of different fuels to deliver a range of energy services for homes, industry, the services/commercial sector and transport. The actions of thousands of individual energy consumers have an influence on emissions. Work in this field is multi-professional, involving people from a range of organizations and jobs. It is clear that there is no single solution to successfully reducing emissions and both a range of technologies and also non-technical (e.g. behavioural) aspects are key to reducing emissions.

The ability to measure success with implementing local energy/carbon management projects is essential. Leicester has carried out work to monitor and evaluate its progress towards its energy and greenhouse gas emissions targets. In addition to providing CO_2 savings, many schemes have economic and social benefits. These benefits can be recognized and communicated to decision-makers and the public. The need for regular ongoing monitoring has been recognized. Implementing ambitious targets on reducing CO_2 emissions is not necessarily straightforward. It has not always been possible to establish exemplar low carbon buildings. Leicester has some very good examples of local energy/carbon management initiatives (e.g. combined heat and power in district heating, Leicester Better Buildings, and intelligent metering).

Leicester has taken a comprehensive approach to addressing climate change, reducing energy consumption, setting standards for new developments, and working to change attitudes and behaviour towards reducing greenhouse gas emissions. All this would not have been possible without the target set in 1994, the ambitious vision, strong leadership and a strong technical capability among City Council staff and staff in partner organizations.

The technology required to reduce carbon emissions already exists. Support is needed to implement this technology and to maintain progress with achieving carbon savings. The energy efficiency technology is often regarded as 'unexciting' and 'boring'. We need to show that 'being energy efficient' is enjoyable and that it is something that people and organizations aspire to.

Acknowledgements

The work of staff at Leicester City Council's Energy Efficiency Centre, De Montfort University's Institute of Energy and Sustainable Development and the City Council's Environment Team is gratefully acknowledged.

Also, acknowledgements are due to a number of people including: Prakash Patel, Alan Gledhill, Anna Dodd, Carol Brass, Mark Jeffcote, Alison Lea, Ann Branson, Nick Morris, Stewart Conway, Eddie Tyrer, Nick Hodges, Chris Randall, Humphrey Davis, Steve Weston and others at Leicester City Council.

Dr Simon Taylor, Dr Zoë Fleming, Iraklis Argyriou, Myira Khan, Institute of Energy and Sustainable Development, De Montfort University.

Michael Cooke, Local Strategic Partnership.

David Nicholls, Ben Dodd, Charlie Lewis, Groundworks Leicester and Leicestershire.

Don Lack, Faber Maunsell Ltd.

References

Allman, L., Fleming, P.D. and Wallace, A. (2004). The progress of English and Welsh local authorities in addressing climate change. *Local Environment*, **9**(3), 271–283.

Bulkeley, H. and Betsill, M. (2003). *Cities and Climate Change*. Routledge.

Colbourne, D., Lorenzoni, I., Powell, J. and Fleming, P. (1999). Identifying social attitudes to assist urban energy planning in Leicester. *International Journal of Sustainable Development and World Ecology*, **6**, 265–280.

DEFRA (2006a). *Climate Change. The UK Programme 2006*. The Stationery Office.

DEFRA (2006b). *Local and Regional CO_2 Emissions Estimates for 2004 for the UK*. DEFRA.

DTI (2003). *Energy White Paper. Our Energy Future-Creating a Low Carbon Economy*. The Stationery Office.

DTI (2006). *The Energy Challenge Energy Review Report 2006*. The Stationery Office.

DTI (2007). Regional Energy Consumption Statistics. Retrieved 23 February 2007 from http://www.dti.gov.uk/energy/statistics/regional/index.html

DTI (2007). Meeting the Energy Challenge. A White Paper on Energy. May 2007. The Stationary Office.

Department of the Environment, Transport and the Regions (2000). *Climate Change: the UK Programme*. DETR.

EMRA (2004). *The East Midlands Energy Challenge*. EMRA.

EMRA (2007). The East Midlands Energy Challenge. The Regional Energy Strategy (Part 2). A Framework for Action. EMRA.

European Commission (1995). *An Energy Policy for the European Union*. COM (95) 682 final.

European Commission (1997). *Energy for the future – renewable sources of energy*. White Paper COM(97) 599 final.

European Commission (2006). *The European Climate Change Programme*. EU Action against Climate Change.

Fleming, P.D. and Webber, P.H. (2004). Local and regional greenhouse gas management. *Energy Policy*, **32**, 761–771.

LCC 'Leicester Better Buildings'. Retrieved 23 February 2007 from http://www.leicester.gov.uk/your-council--services/ep/business--the-economy/betterbuildings

LGA and LGMB (1998). *Energy Services for Sustainable Communities. The Local Government Position*. Local Government Association.

LGA/EST/EEPH (2005). *Leading the Way: How Local Authorities can Meet the Challenge of Climate Change*. LGA.

Leicester City Council (1994). *The Leicester Energy Strategy*. Leicester Environment City.

Leicester City Council (2006). *Environmental statement April 2004–March 2005*. Leicester City Council.

Leicester City Council (2007). Leicester City local development framework Energy efficiency & renewable energy supplementary planning document, November 2005 (minor revisions March 2007).

Leicester Energy Agency and Barnagel (1999). *Leicester Energy Agency/Barnagel. Final Report to the European Commission.* Leicester Energy Agency.

Leicester Partnership (2003). *The Leicester Climate Change Strategy.* Leicester Partnership.

McKay, I. (2000). Energy Audit 1989–1999. Draft.

Open University (1997). *Modelling and Evaluation of Sustainable Energy Strategies for Cities: Final Project Report to the EPSRC.* Open University.

Randall, C. (2007). Personal communication.

Russell, A. (2007). Personal communication.

Shackley, S., Fleming, P. and Bulkeley, H. (2002). *Low Carbon Spaces.* Sustainable Development Commission.

Stern, N. (2006). The Stern Review: The Economics of Climate Change. http://www.hm-treasury.gov.uk/independent_reviews/stern_review_economics_climate_change/sternreview_index.cfm.

The Stationery Office (2006). *Climate Change and Sustainable Energy Act 2006.* The Stationery Office.

Titheridge, H., Boyle, G. and Fleming, P. (1996). Development and validation of a computer model for assessing energy demand and supply patterns in the urban environment. *Energy & Environment,* 7(1), 29–40.

Torrie Smith Associates (2000). *Greenhouse Gas Emissions Software – U.K. Edition. User's Guide. Release 4.6, December 2000.* Torrie Smith Associates.

Webber, P.H., Lack, D. and Fleming, P.D. (2001). Quantifying Greenhouse Gas Emissions In UK Local Authorities. *Proceedings of the First International Conference on Ecology and the City,* January–March 2001. Internet Conference.

Chapter 22
Reducing Carbon Emissions From Oxford City: Plans and Tools

RAJAT GUPTA

Department of Architecture, School of the Built Environment, Oxford Brookes University, Headington Campus, Gipsy Lane, Oxford OX3 0BP, United Kingdom

This chapter reviews two key initiatives undertaken in the historical city of Oxford (UK) to bring about reductions in energy-related CO_2 emissions on a city-wide scale. The author has been part of both the initiatives. A GIS-based DECoRuM® (domestic energy, carbon-counting and carbon-reduction model) model has been developed, demonstrated and validated by applying to existing dwellings in Oxford to estimate and map baseline energy use and CO_2 emissions on a house-by-house level, identify 'pollution' hotspots, predict the potential for reductions in CO_2 emissions and monitor reductions achieved as a result of deploying energy efficiency measures and renewable energy systems. The application of the DECoRuM® model to a case study in Oxford shows that CO_2 emission reductions above 60% are possible, at a cost of between £6 to £77 per tonne of CO_2 emissions saved, depending upon the package of measures used, and the scenario of capital costs (low or high) employed. The DECoRuM® project has led to the development of an action-oriented Oxford Climate Change Action Plan (OCCAP) which constructs an accurate CO_2 emissions inventory for Oxford city for a baseline year, establishes CO_2 reduction targets and proposes action for each of the energy-related sectors to meet those targets. The OCCAP is currently being implemented by Oxford City Council and provides a useful example for other cities in their endeavour for emission reductions.

22.1 Introduction

It is now increasingly recognized that actions at the local to regional scale are needed to deliver extensive carbon emission reductions, but to date most strategic thinking has focused on national mechanisms. There is great untapped potential for bottom-up-led carbon reduction (Shackley *et al.* 2002, p. 9). Area-based carbon emission reductions will require concerted action by all sectors of the community. In fact most of the measures required to reduce CO_2 emissions will be implemented at the city level, working with city councils since they have the ability to make communities more energy efficient by using their influence over the community and existing statutory powers, e.g. land use and planning decisions, infrastructure investment and municipal service delivery, etc. (McEvoy *et al.* 1998).

As a city, Oxford (located in the south of the United Kingdom) has had a strong commitment to energy efficiency and renewable energy technologies. For example, Oxford City Council has been procuring green electricity for many years, and owns a fleet of low emission electric and LPG fuelled vehicles. Most recently Oxford has been one of 24 local authorities to become part of the UK Councils for Climate Protection (CCP) Pilot Programme. Oxford is also home to numerous environmental organizations – academic, non-governmental and commercial. Many of these have an international reputation for their work on greenhouse gas accounting and/or climate change. These include the Oxford Institute for Sustainable Development at Oxford Brookes University, the Environmental Change Institute at the University of Oxford, Oxford Ecohouse, Best Foot Forward and the UK Climate Impacts Programme. Furthermore, in April 2006, Oxford played host to the 2nd International Solar Cities Conference (see www.solarcities.org.uk).

Within such a context, this chapter reviews two key initiatives undertaken in the historical city of Oxford to bring about reductions in energy-related CO_2 emissions on a city-wide scale. These include: the ongoing further development of a GIS-based DECoRuM® (domestic energy, carbon-counting and carbon-reduction model) model to count, cost and reduce domestic CO_2 emissions on an urban scale; and the Oxford Climate Change Action Plan which has been recently completed and underpins future work on carbon emission reductions from the city of Oxford.

22.2 Development and Demonstration of the DECoRuM® Model

DECoRuM® is a GIS-based domestic energy, carbon-counting and carbon-reduction model, which has been developed, demonstrated and validated by the author using Oxford as a case study (Gupta 2005a,b). Recently, the DECoRuM® model has also received the Royal Institute of British Architects' President's medal for outstanding research 2006. The DECoRuM® model estimates and maps baseline energy use and CO_2 emissions on a house-by-house level, identifies 'pollution' hotspots, predicts the potential for reductions in CO_2 emissions and monitors reductions achieved as a result of deploying energy efficiency measures and renewable energy systems on an urban scale. Also, it has an additional unique feature of assessing the cost/benefits of various CO_2 reduction measures and putting a financial cost to CO_2 emission reduction (£/tonne of CO_2 saved).

22.2.1 Developing DECoRuM®

The physically based annual version of Building Research Establishment's domestic energy model, BREDEM-12, combined with the government-approved Standard Assessment Procedure (SAP) 2001 home energy rating methodology, are the underlying energy models in DECoRuM®, which calculate annual energy use, fuel costs and CO_2 emissions resulting from space heating, water heating, cooking, lights and appliances, as well as a SAP energy rating (scale of 1 to 120), which is based on calculated annual energy cost for space and water heating for a dwelling (Anderson *et al.* 2002; BRE 2001). The net annual cost (NAC) method is chosen to assess the cost-effectiveness of deploying individual CO_2 reduction measures. Microsoft Excel dynamically linked to the Geographical Information System (GIS) software, MapInfo, are the operating platforms for DECoRuM®. The GIS software is the user interface.

Table 22.1.   List of categories used for data reduction in DECoRuM®.

No.	Category used for data reduction	Number of parameters	Percentage of parameters
1	Data common to all dwellings	50	52.7%
2	Data derived from built form	5	5.3%
3	Data derived from age	18	19.0%
4	Data to be collected for individual dwellings	22	23.0%
	TOTAL (BREDEM-12 calculation)	95	100%
5	Data collected for estimating the solar potential	4	

22.3 Estimating Baseline Energy Use and CO₂ Emissions

Although BREDEM-12 is a reputable, validated model, which performs calculations rapidly and at a level of detail appropriate to city planners, the quantity of input data (95 parameters) it requires is not easy to obtain in practice, owing to the high cost of detailed on-site surveys. This poses problems for energy modelling on an urban scale. In DECoRuM®, to overcome the problem of data collection, data reduction techniques have been developed to enable most of the dwelling-related data required by the underlying energy models to be supplied from traceable sources. Data reduction techniques classify the 95 input data parameters required by BREDEM-12 (linked to SAP model) into four categories, according to the source of data. Category 5 includes data to be collected for estimating the solar potential of dwellings (Gupta 2006a). The categories along with the number of dependent parameters are listed in Table 22.1.

The data for categories 1, 2 and 3 are derived from a range of secondary sources including traceable national statistics, UK Building Regulations, BRE reports, English House Condition Survey (EHCS) 2001, and locally relevant Home Energy Survey Forms, while data for categories 4 and 5 are obtained from primary sources, which include a GIS urban map and walk-by survey. As a result of the data reduction techniques, only *one fifth* of the data items required for a BREDEM-12 calculation need to be collected for individual dwellings in a case study, and only *one tenth* are to be collected by a walk-by survey. Results of energy use, CO_2 emissions, fuel costs and SAP energy ratings can be displayed in the form of thematic maps in MapInfo GIS, with an individual dwelling displayed as the basic unit of resolution. This helps to pinpoint *hot spots* of energy use and CO_2 emissions.

22.4 DECoRuM® CO₂ Reduction and Cost/Benefit Model

A wide range of literature was reviewed and the following measures were identified for reducing domestic CO_2 emissions on the demand and supply side of energy. For these measures to be incorporated in DECoRuM®, the baseline energy model is used to filter suitable dwellings for every CO_2 reduction measure by use of appropriate criteria. Subsequently, for the 'passed' dwellings, the measure is incorporated in their baseline energy models using

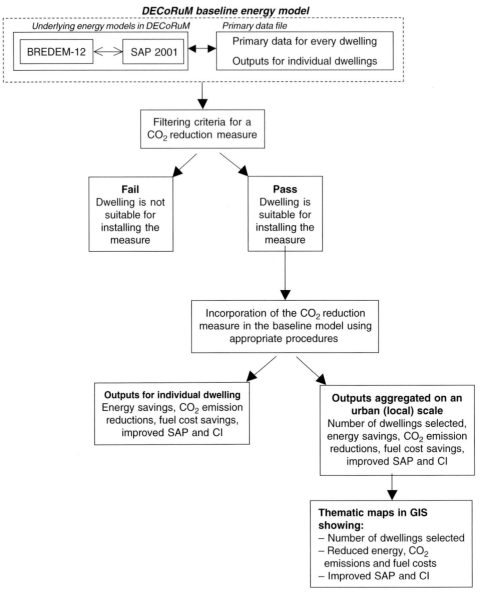

Fig. 22.1. Approach for assessing CO_2 reductions in DECoRuM®.

appropriate procedures to quantify the energy savings and reductions in CO_2 emissions that occur on an individual dwelling level, and also aggregated to an urban scale. This approach developed in DECoRuM® is presented in Fig. 22.1.

Depending upon the measure, the filtering criteria and procedures are derived from field trials, laboratory experiments, theoretical calculations, or from a realistic mixture of these (EST 2001, 2002; HEEBPp 2003; Shorrock *et al.* 2001). For instance, in case of the solar hot water systems, dwellings with roof orientation lying between $\pm 45°$ of south (Northern hemisphere), roof inclination lying between $0°$ and $60°$ and roof area >3.9 m^2 are selected. When insulation-related energy efficiency measures are run in DECoRuM$^®$, the procedure is to change the appropriate U-values and ventilation rate; a more detailed calculation is performed to quantify the potential for solar hot water yield of the selected dwellings (BRE 2004, p. 46). The results are displayed in GIS in the form of thematic maps.

The net annual cost (NAC) methodology is incorporated in DECoRuM$^®$ as the cost/benefit model to assess the cost effectiveness of individual CO$_2$ reduction measures. The cost effectiveness is assessed for high and low capital costs, with the expectation that in most circumstances, the real figure would lie somewhere between them. This cost/benefit model requires standard input parameters relating to typical capital costs per dwelling for installing a measure, lifetime of the measure, the cost and CO$_2$ emission factor of the fuel displaced, as well as case study-specific parameters. One of the key results of using the DECoRuM$^®$ cost/benefit model is that the *net annual cost per tonne of CO$_2$ emissions saved* can be derived, and used to assess the cost effectiveness of that measure. A measure is taken to be cost effective if the NAC is negative; the larger its absolute value, the more cost effective that measure is. The DECoRuM$^®$ cost/benefit model also gives the *cost for reducing a tonne of lifetime CO$_2$ emissions* by using one or a combination of measures.

22.4.1 Demonstration of the DECoRuM$^®$ model: application to a case study in Oxford

The DECoRuM$^®$ model was applied to a case study in Oxford covering approximately 318 dwellings, to demonstrate and validate its capabilities (Gupta 2005a). The case study dwellings contained all the built forms and age-bands present in UK housing, although in different proportions from those of the national stock. Data for individual dwellings were derived from the GIS urban map, local authority records and walk-by surveys. The DECoRuM$^®$ model calculated the baseline energy consumption and associated CO$_2$ emissions for every dwelling in the case study area (318 dwellings) and aggregated them together to give a total baseline energy consumption figure of 49 699 GJ/year and total CO$_2$ emissions of 3026 tonnes/year. A thematic map is created in MapInfo GIS, showing the estimate of baseline total annual CO$_2$ emissions from every dwelling (Fig. 22.2 [Plate 16]). Also, in the GIS map, through *hot links*, digital images of street-facing facades of dwellings can be presented to get an idea of their construction.

As expected, space heating consumed the most energy (61%), producing 55% of the total CO$_2$ emissions, and was responsible for 44% of the total running costs. The breakdown of baseline results according to the different built forms showed that mid-terraced dwellings in the case study had the highest average SAP rating at 47.9 (and a Carbon Index (CI) of 3.6) and the lowest average fuel cost of £657/year, while bungalows (open from all four sides) had the lowest SAP rating of 36.3 and the lowest CI of 2.7.

When the baseline figures of the case study were broken down by age group, as expected, dwellings built between 1996 to 2002 had the highest average SAP rating, 90.3, highest CI, 7.3, and the lowest running costs of £676/year. On the other hand, pre-1930 dwellings have the lowest average SAP rating and CI. Furthermore, SAP ratings increased as the dwellings became more recent (Fig. 22.3).

These results were extensively validated by comparisons with national- and city-level statistics, as well as with case study-specific databases. This instilled confidence in the predictions from DECoRuM$^®$, and built the case for extrapolation of findings from the case

Fig. 22.2. Representation of annual CO_2 emissions in the form of GIS thematic maps using DECoRuM®
for dwellings in north Oxford. This figure can be seen in the colour plate section as Plate 16.

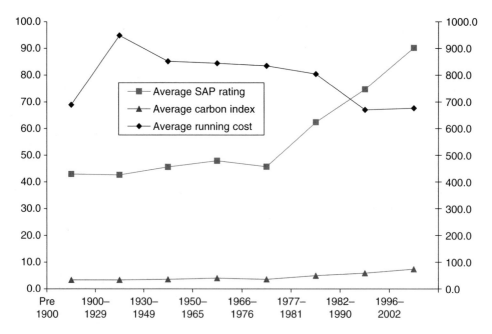

Fig. 22.3. Distribution of SAP rating, Carbon Index and annual running cost by age band in the
Oxford case study dwellings.

Fig. 22.4. Estimating the solar potential of dwellings in north Oxford using DECoRuM®. [Plate 17]

study to the national level. Each CO_2 reduction measure embedded in DECoRuM® was then applied individually in the case study, using appropriate filtering criteria to select the most suitable dwellings (see Fig. 22.4 [Plate 17] for solar potential of dwellings). The corresponding savings in total annual energy use, total annual CO_2 emissions, fuel costs, and improvement in SAP ratings and CI were calculated.

Results indicated that external solid wall insulation saved the most energy, CO_2 emissions and fuel costs in the case study dwellings. The cost/benefits of applying these measures were evaluated at 2001 fuel prices, and it was shown that measures which were the most cost effective did not necessarily save the most CO_2 emissions. Energy-efficient appliances, for instance, have considerable savings in expenditure only when installed collectively. However, hot water cylinder insulation, replacement with condensing boilers, roof insulation and cavity wall insulation appeared to be cost effective in any circumstance (low capital cost and high capital cost scenario), and this justifies the wide promotion they have received over the years. While domestic micro-CHP units appeared to be cost-effective in any circumstance, GSHP systems emerged as cost-effective only when government grants were included. SHW and solar PV systems were not cost effective at current prices. Although the simple payback period for a 1 kWp solar PV system reduced to 30 years from 95 years if government grants were included and if it was assumed that domestic PV systems qualified for selling Renewable Obligation Certificates (ROCs) to energy suppliers.

The overall potential for reducing CO_2 emissions from the case study dwellings was then predicted. Five alternative packages were developed, using a combination of energy efficiency measures, low carbon technologies, solar energy systems and green tariff electricity. The analyses showed that reductions in excess of 60% were possible from the case study dwellings at a cost of between £6/tonne of CO_2 and £77/tonne of CO_2 emissions saved. This cost depended upon the package of measures used, and the scenario of capital costs (low or high) employed. These figures compared favourably with the UK's estimated social cost of CO_2 emissions (£19/tCO_2). Cost-effective savings of 2044 tonnes of CO_2 emissions per year were estimated for the case study dwellings. This is equivalent to 67.5% reductions of CO_2 emissions over baseline if green tariff electricity is included and 57% if it is not. This indicated a total expenditure of between £0.4 million and £1 million as currently necessary to install all the cost-effective CO_2 emission reduction measures in all the case study dwellings.

The Oxford DECoRuM® model has been used in the Oxford Solar Initiative project to undertake targeted marketing to increase the uptake of the energy efficiency measures and solar energy installations. The model could be further used by the City Council in Oxford to develop and review the energy efficiency strategy and targets for improvement. In particular, the cost/benefit analysis of the measures and financial cost of CO_2 emission reduction is likely to be useful in supporting funding applications, or to make the case for reallocating internal resources.

The DECoRuM® model is presently funded as a 'proof of concept' to be further developed as a toolkit for carbon emission reduction planning for use by UK local authorities to report, monitor and improve the energy efficiency of both public and private housing, as required by the Home Energy Conservation Act. It is expected that a robust GIS-based toolkit would be available to UK local authorities, energy advisors, building surveyors and real estate professionals to assist them in counting, costing and reducing domestic carbon emissions. For further information on DECoRuM®, visit www.decorum-model.org.uk.

22.5 Oxford Climate Change Action Plan

The OSI and DECoRuM® projects have led to the development of an action-oriented Oxford Climate Change Action Plan (OCCAP) by the author (commissioned by the Oxford City Council). Although Oxford has had a strong commitment to energy efficiency and renewable energy technologies, so far no complete CO_2 emissions assessment of the city has been undertaken. The OCCAP therefore constructs an accurate CO_2 emissions inventory for the city of Oxford for a baseline year, establishes CO_2 emissions reduction targets, proposes action and measures for each of the energy-related sectors to meet those targets, and suggests ways of monitoring and verifying the reductions achieved (Gupta 2005b,c).

22.5.1 Assessing baseline CO_2 emissions in Oxford city

The most important part of any climate change action plan is to have an accurate CO_2 assessment for a baseline year. This not only enables one to identify the main energy-using sectors and set reduction targets, but it also provides a benchmark to measure the effectiveness of actions and programs adopted, and monitor the progress towards targets. The OCCAP constructs a disaggregated CO_2 emissions inventory for Oxford city by assessing the CO_2 emissions generated *directly*, in the jurisdiction of Oxford (such as a home heating system powered by natural gas, or a car or truck powered by petrol or diesel), and *indirectly* in case of generation of electricity that is used within Oxford city. There is no attempt made to detail the embodied energy of goods consumed or energy use undertaken by residents outside Oxford, for instance air travel or water-borne transport or emissions related to food. A combination of both top-down methods (national datasets) and bottom-up approaches (the use of localized data sets) are selected to calculate the major sources and levels of CO_2 emissions (by sector) in Oxford. While the top-down data sets provide an

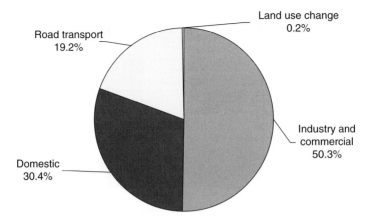

Fig. 22.5. Percentage breakdown of CO_2 emissions by sector in Oxford in 2003.

overall picture of the sectoral CO_2 emissions in Oxford, local data sets provide disaggregated figures for taking action.

22.6 Top-Down Approaches (National Data Sets)

The top-down data sets included are those published by various UK government departments: National Atmospheric Emissions Inventory (NAEI), Department of Trade and Industry (DTI) (DTI 2004, 2005) and Department of Environment, Food and Rural Affairs (DEFRA) (DEFRA 2005). The NAEI inventory differs fundamentally from DTI and DEFRA in its approach to collecting data, by focusing on the source of emissions (point, line or area source), while both DTI and DEFRA focus on the consumption of energy. DTI Energy Statistics, on the other hand, focus on gas and electricity consumption, and road transport fuels. According to DEFRA, CO_2 emissions from Oxford in 2003 are estimated to be 987 853 tCO_2, with domestic energy use and road transport responsible for almost half the total CO_2 emissions (Fig. 22.5).

The sectoral emissions in Oxford are indicative of the trends observed regionally and nationally, especially the domestic sector which in 2003 was responsible for around 30% of the total emissions at all levels. In fact the average annual gas consumption per household in Oxford was calculated as 21 150 kWh in 2003, almost 5% more than the average consumption of gas per household across Great Britain, while the average electricity consumption per household in Oxford was 1% less than the GB average in 2003 (Fig. 22.6). This results in higher overall annual CO_2 emissions from an Oxford dwelling (6478 kgCO_2) as compared to the GB typical of 6000 kgCO_2, clearly offering potential for reductions.

Furthermore, it is realized that the average annual gas and electricity consumption of, and CO_2 emissions from, the commercial and industrial sector in Oxford in 2003 were almost 20% more than corresponding figures for Great Britain. Obviously this sector provides another opportunity for taking action to reduce CO_2 emissions. Nevertheless the per capita annual CO_2 emissions of an Oxford resident (at 6.9 tCO_2 (DEFRA statistics) are much lower than that of the southeast of England (8.8 tCO_2) and UK (9.5 tCO_2) averages. However, in a global context they appear very high, almost three times the sustainable level of 2.5 tCO_2 per person (see Fig. 22.7).

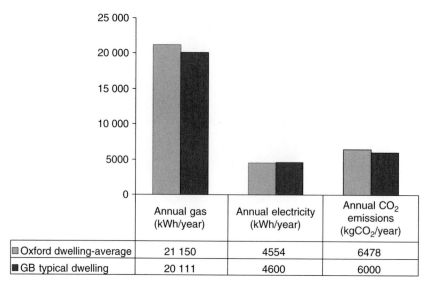

Fig. 22.6. Comparison of annual average gas, electricity consumption and related CO_2 emissions of an Oxford dwelling with GB typical.

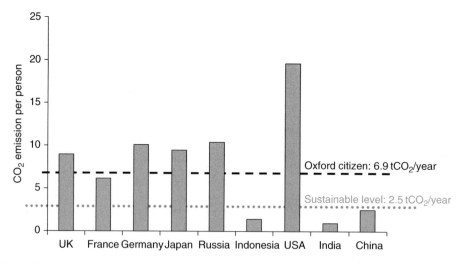

Fig. 22.7. Comparison of per capita annual CO_2 emission figures of an Oxford citizen with international figures.

22.7 Bottom-Up Approaches (Local Data sets)

Although the national data sets construct an overall inventory of CO_2 emissions from Oxford city by sector, these do not identify which particular areas within a sector could be targeted for future action and improvement. Such information is generally provided by bottom-up local data sets. For the domestic sector in Oxford, various organizations have quantified its energy consumption and CO_2 emissions for different years. While the Oxford House Condition Survey carried out in 1995 estimates the average SAP (energy rating) of Oxford

dwellings to be 43, Elmhurst energy surveys in 2001 calculate it as 44, while DECoRuM®
studies conducted in 2003 assess the average SAP as 45 for north Oxford dwellings (Gupta
2005b). Most recently, the private house condition survey undertaken in 2004 estimates the
average SAP as 54. It is realized that DECoRuM® is the only GIS-based domestic energy and
emissions model deployed in north Oxford which displays results on an individual dwelling level, pinpoints pollution hotspots, and predicts the potential for emissions reduction.
Its capabilities could be extended to evaluate energy consumption in, and CO_2 emission
reductions from, the entire Oxford housing stock, and the cost of a range of measures to
reduce both.

For the non-domestic sector, the only local data set available is the regular monitoring
of energy use and CO_2 emissions for only 16 high energy consuming council buildings.
These 16 monitored buildings are responsible for only 1% (4463 tCO_2/year) of Oxford's
total industrial and commercial CO_2 emissions, so it is important to extend the collection
of bottom-up energy data to include more non-domestic buildings. Even performance of
these 16 buildings could considerably be improved as most of them compare poorly with
both UK typical and good practice standards.

22.7.1 Setting city-wide CO_2 reduction targets

The CO_2 emissions inventory of Oxford has revealed that there is much scope for reducing CO_2 emissions from the various energy-related sectors in the city. The next step is
to establish targets for Oxford to reduce CO_2 emissions in the medium to long term, and
propose action to achieve these targets in each of the energy-related sectors. Worldwide,
many cities, as part of ICLEI (www.iclei.org), International Solar Cities Initiative (www.
solarcities.co.kr) and the European Solar Cities Initiative (www.eu-solarcities.org/), have
set targets for reducing future CO_2 emissions at levels consistent with stabilizing levels of
atmospheric CO_2. The UK target under the Kyoto Protocol is a 12.5% reduction based on
1990 levels. However, the UK has also set its own domestic target of a 20% reduction in
CO_2 emissions on 1990 levels by 2010. In addition, the UK government has stated in the
Energy White Paper published in 2003 to cut emissions of CO_2 by 60% by around 2050
(DTI 2003). To help achieve this, the government has set a target of producing 10% of UK
electricity from renewable sources by 2010. In line with such national goals (both legal and
aspirational) and the pressing need to stabilize atmospheric concentration of CO_2 emissions, the overall aim for OCCAP should be to ultimately achieve CO_2 emission reductions in Oxford above 60% by 2050 over the 2005 baseline emissions. This long-term target
should be accompanied by the following intermediate goals (objectives) of CO_2 reduction
(over 2005 baseline emissions) in the near to medium term:

- 15% reduction in CO_2 emissions by 2010
- 20% reduction in CO_2 emissions by 2015
- 30% reduction in CO_2 emissions by 2020
- 40% reduction in CO_2 emissions by 2030.

Such emission reduction targets will no doubt require concerted action by all sectors of the
community in Oxford city, which is described in the following section.

22.7.2 Identifying actions to achieve targets

It is clear that no one individual, group or institution can effect the changes necessary to
meet the challenging target of above 60% reductions in CO_2 emissions from Oxford city
by 2050. Instead a partnership approach is required across the city, with all energy-related

sectors fully engaged in taking action. It is therefore important to identify the key strategies (which have the greatest impact) that Oxford City Council could adopt in its effort to bring city-wide CO_2 emission reductions. These include:

- *Raising awareness and understanding of the impact of lifestyle on climate change:* Awareness is to be raised in the community about the effect of their lifestyle on climate change such that a minimum of 25% of the local community has knowledge of climate change and mitigation measures within the next five years. Workshops and open days are held regularly, and climate change action leaflets are also produced for the local community, which include information linking lifestyle and CO_2 emissions, as well information on measures to reduce emissions and available government grants.
- *Mapping CO_2 emissions in Oxford using GIS-based modelling:* The foundation of the OCCAP is based on developing a GIS-based carbon map of Oxford city of all the sectors to estimate baseline emissions, target pollution hotspots, predict resulting CO_2 savings and cost/benefits by deploying CO_2 reduction measures and finally to monitor emission reductions as measures are implemented. Such a robust and validated model exists for the domestic sector (DECoRuM®), and the first step should be to extend it to the entire Oxford housing stock to evaluate the baseline energy consumption, potential for CO_2 emission reductions, and the cost of a range of measures to reduce both. Such a carbon map of the existing Oxford dwellings would provide the opportunity to effectively target householders in less energy-efficient dwellings with promotional material. Moreover this would be a replicable and exportable methodology to enable other local councils to develop a carbon reductions planning capability.
- *Encouraging energy conservation and local generation of energy:* A recently published report by Sustainable Consumption Roundtable reveals that choosing to install micro-generation technology or living in a house where it has been installed can significantly shift awareness, attitudes and behaviour of occupants around energy and energy efficiency (SDC 2005). The mere presence of on-site micro-generation makes occupants more energy conscious much beyond the simple analysis of kilowatts produced and CO_2 emissions averted.
- *Setting and achieving targets for domestic energy efficiency:* To improve domestic energy efficiency in Oxford by 10% by 2010 and 20% by 2020 on 2005 levels. This would in turn also help to alleviate fuel poverty by 2010 in vulnerable households.
- *Energy audits and surveys of non-domestic buildings:* As a result of energy auditing, the aim should be to reduce CO_2 emissions from all council-owned non-domestic buildings by at least 25% on 2005 levels by 2010.
- *Improving energy efficiency standards of new developments:* It is necessary to ensure that by 2010 all new residential developments in Oxford are built to the *best practice* standard, which could be introduced for approximate capital costs of 6% above the datum for traditional masonry build, and 8% for timber frame constructions (EEBPH 2005).
- *Increasing the uptake of low carbon systems:* 10% of all houses in Oxford (in 2005) to have solar systems (solar hot water or solar PV) by the year 2010 and 25% of all houses to have solar systems by 2020. All new residential and non-residential developments (>1000 m^2) in Oxford to supply 10% of expected electricity consumption by on-site renewable energy systems.
- *Minimizing transport impact of new developments:* New developments should be located close to local travel destinations (such as the city centre) or where there is good access to public transport. It should be ensured that facilities in new developments are included to support the use of low/zero carbon fuelled vehicles.

Table 22.2. Objectives of a Solar City.

Renewable energy goals	Targets set for the future share of energy from renewable energy
CO_2 goals	Future CO_2 emissions targets set, usually on a city-wide or per capita basis
SHW	Incentives for solar hot water systems enacted
Solar PV	Policies and/or incentives for solar power enacted
Transport	Policies and/or urban planning approaches for sustainable transport enacted/being used
Buildings	Energy-efficient building codes and standards
Planning	Overall urban planning approaches with consideration for future energy consumption
Demonstration projects	Specific projects, subsidized by public funds or otherwise financed as one-time demonstrations

- *Transforming Oxford into a Solar (sustainable) City:* The objectives in Table 22.2 would need to be fulfilled for Oxford to become a leading Solar City in the UK, Europe and world-wide; most of these objectives have been addressed in the above-mentioned strategies.

22.7.3 Further work

Although this Climate Change Action Plan operates on a long-term perspective, it also includes targets and actions to be achieved in both the short and the medium term. The action plan therefore is a flexible document reviewable on an annual basis. Therefore it is essential that an in-house Climate Change Team within the Oxford City Council be set up with the responsibility of delivering the OCCAP. While climate change later this century will be determined by the emissions we allow now, climate change we expect in the next 30–40 years will be due to our past greenhouse gas emissions (Roaf *et al.* 2005). Further work needs to be done to develop a climate change adaptation strategy for Oxford city, bearing in mind that adaptation measures that lead to increasing greenhouse gas (CO_2) emissions must be avoided.

22.8 Conclusions: Future Work for Oxford City

Both the initiatives discussed above have led to the reduction of energy-related CO_2 emissions from Oxford city, to install carbon reduction measures in their homes, by providing a robust scientific model to measure, benchmark and verify the reductions in emissions achieved through the DECoRuM® model, and finally, by developing a strategic action plan to assess and reduce energy-related emissions from all sectors in the future. In particular, the actions identified in the OCCAP are being currently implemented by Oxford City Council. Two new posts for climate change officers have also been created in Oxford City Council to take forward the recommendations suggested in the OCCAP. Gupta (2006) has been appointed to perform post-occupancy evaluations (energy, environmental and occupant satisfaction) of 35 City Council-owned non-domestic buildings. The next step is to monitor and verify the reductions in CO_2 emissions achieved as a result of those actions. A GIS-based CO_2 model extended to the whole of Oxford would enable the City Council to track progress and assess the emission reductions achieved. Therefore, work on building the GIS CO_2 model (especially for the domestic sector) should be commissioned as soon as possible.

Undoubtedly, these experiences from Oxford provide a useful example for other cities in their endeavour for emission reductions. But importantly these two initiatives also make the case for all of us to do our bit as urban city dwellers, since the delivery of emissions reductions would require a partnership approach across a city, with all sectors including individuals fully engaged in taking forward an agreed action plan.

Acknowledgements

The author would like to express his thanks to Pilkington Energy Efficiency Trust, Royal Institute for Chartered Surveyors and Oxford Brookes University for co-funding the development of the DECoRuM® model; and to Oxford City Council for commissioning the Oxford Climate Change Action Plan.

References

Anderson, B.R. *et al.* (2002). *BREDEM-12 Model Description.* Building Research Establishment, Watford.

BRE (2001). *The Government's Standard Assessment Procedure for Energy Rating of Dwellings, 2001 edition.* Published on behalf of DEFRA by BRECSU, Building Research Establishment, Watford.

BRE (2004). *The Government's Standard Assessment Procedure for Energy Rating of Dwellings, Draft 2005 edition.* Published on behalf of DEFRA by BRECSU, Building Research Establishment, Watford.

DEFRA (2005). *Experimental Statistics on Carbon Dioxide Emissions at Local Authority and Regional Level.* Defra Statistics Summary. Department for Environment, Food and Rural Affairs. London.

DTI (2003). *Our Energy Future: Creating a Low Carbon Economy.* Department of Trade and Industry, London.

DTI (2004). *Energy Trends – December 2004.* HMSO, London.

DTI (2005). *Energy Trends – March 2005.* HMSO, London.

EEBPH (2005). *Best Practice in New Housing: A Practical Guide.* Energy Efficiency Best Practice in Housing, Watford.

EST (2001). *Easy Ways to Save Money by Insulating Your Home.* Energy Saving Trust, London.

EST (2002). *Is Your Home Behaving Badly?* Energy Saving Trust, London.

Gupta, R. (2005a). *Investigating the Potential for Local Carbon Dioxide Emission Reductions: Developing a GIS-Based Domestic Energy, Carbon-Counting and Carbon-Reduction Model.* Oxford Brookes University, Department of Architecture, Oxford.

Gupta, R. (2005b). Investigating the potential for local carbon dioxide emission reductions: developing a GIS-based domestic energy, carbon-counting and carbon-reduction model. Refereed Technical Paper, in *2005 Solar World Congress*, Orlando, Florida, USA.

Gupta, R. (2005c). *Oxford Climate Change Action Plan.* Submitted to Oxford City Council. Institute for Sustainable Development, Oxford.

Gupta, R. (2006a). *Applying CO_2 Reduction Strategies to Existing UK Dwellings using GIS-Based Modelling.* Sixty-second summary. Royal Institute of Chartered Surveyors. London.

Gupta, R. (2006b). Carbon emission reduction planning for cities: developing a climate change action plan for the city of Oxford, UK, in *Proceedings of Solar 2006 Congress*, Denver, USA.

HEEBPp (2003). *Domestic Energy Efficiency Primer.* Housing Energy Efficiency Best Practice Programme, Watford.

McEvoy, D., Gibbs, D.C. and Longhurst, J.W.S. (1998). Urban sustainability: problems facing the local approach to carbon-reduction strategies. *Environment and Planning C: Government and Policy*, **16**, 423–432.

Roaf, S., Crichton, D. and Nicol, F. (2005). *Adapting Buildings and Cities for Climate Change.* Architectural Press, Oxford.

SDC (2005). *Seeing the Light: The Impact of Micro-Generation on the Way We Use Energy.* Sustainable Consumption Roundtable, Sustainable Development Commission, London.

Shackley, S., Fleming, P. and Bulkeley, H. (2002). *Low Carbon Spaces Area-Based Carbon Emission Reduction: A Scoping Study.* A report to the Sustainable Development Commission prepared by the Tyndall Centre for Climate Change Research.

Shorrock, L.D. *et al.* (2001). *Carbon Emission Reductions from Energy Efficiency Improvements to the UK Housing Stock.* Building Research Establishment and Department for Environment, Food and Rural Affairs, Watford.

Chapter 23
Integrating Energy in Urban Planning in the Philippines and Vietnam

Independent Consultant, Energy and Sustainable Development

23.1 Introduction

One of the characteristics of developing countries is increasing urbanization. The Philippines and Vietnam are no exception. In fact, the urban population growth rates of the two countries in 2000–2005 are slightly higher than the average for all developing Asian countries during the same period and significantly higher than the world average (see Table 23.1). Expanding urban population (which to a very large extent is caused by or coincides with industrialization) is one of the factors driving a country's energy demand growth. Conversely, energy is a key force in the process of urbanization and has profound influence on the direction of urban development. In other words, the utilization of energy and the quality of the natural environment in the city cannot be divorced from one other.

All too often, energy issues are not taken into account or given priority in the planning of cities. Indeed, energy services are not singled out as among the critical issues in urban environmental governance (see, for example, UNESCAP (2005)) even if energy production, transformation, and use are recognized as the main contributors to air pollution and greenhouse gas (GHG) emissions. The traditional concept used in the formulation of the development plans of cities is that urban space is planned and designed to be livable. The human is said to be the focus of all city development planning: the overall concern is the satisfaction of the needs of the city population and the promotion of human dignity through constant improvement of people's welfare and health. Although adequate and reliable energy supply and clean and healthy environment contribute to man's well-being, energy issues are not fully addressed in this traditional concept of urban planning. This is ironic, given the severely deleterious impact on both health and well-being conventional urban energy systems can have on a city's population.

There are several reasons why many cities have not integrated energy in urban development planning. Among these are:[1]

---

[1]This is adapted from a similar project proposal written by Mr. Manuel Soriano, Regional Technical Advisor-Climate Change, UNDP/GEF.

Table 23.1. Urbanization trends in selected developing Asian countries.

Countries	Mid-year population (million)		Population growth rate (%)	Urban population as % of total population		Urban population growth rate (%)
	2000	2005	2000–2005	2000	2005	2000–2005
China	1267.4	1307.6	0.6	36.2	43.0	4.1
Cambodia	12.6	13.8	1.9	16.0	17.7	4.5
Indonesia	205.8	219.9	1.3	42.0	48.1	4.1
Lao PDR	5.2	5.6	1.4	19.3	21.6	3.8
Malaysia	23.5	26.1	2.2	62.0	63.0	2.5
Philippines	*76.9*	*85.2*	*2.1*	*58.5*	*62.7*	*3.5*
Thailand	62.2	64.8	0.8	19.0	32.5	12.2
Vietnam	*77.6*	*83.1*	*1.4*	*24.2*	*27.0*	*3.6*
Bangladesh	128.1	137.0	1.4	23.2	25.1	3.0
India	1019.0	1107.0	1.7	27.7	28.7	2.4
Nepal	22.6	25.3	2.3	13.4	15.8	5.7
Pakistan	139.8	154.0	2.0	33.0	34.0	2.6
Total Developing Asia	**3289.0**	**3495.0**	**1.2**	**33.9**	**37.7**	**3.4**
World	**6085.6**	**6464.8**	**1.2**	**46.7**	**48.7**	**2.1**

Source: ADB 2006.

- Energy suppliers control the energy market in cities. This is because energy supply in these cities is dependent on the market planning of the energy suppliers, which, except for electric utilities, are usually a private commercial concern.
- Only the energy (mainly electricity) requirements of the development programs and local government operational activities are covered in the urban development plans of many cities. This is because there is no perceived need for energy planning in cities, thus energy planning has not been a local government concern. Energy planning is a national government concern in many if not all developing countries.
- City planners and officials are not aware of the direct links between energy and urban development.
- Environmental concerns are addressed separately by cities. This is because energy–environment links to spatial and development planning is not well known among city planners and local officials.
- There is no information about the magnitude of energy usage and consumption trends in many cities as energy supply and consumption surveys are not conducted in these cities.
- There are neither regulations nor official policies on the practice of energy conservation in many cities. This problem is due to the generally low level of awareness about energy conservation among city officials, city planners, and the general public. The low level of awareness is partly due to the general lack of awareness campaigns on energy conservation in many cities.

Naga City in the Philippines and Can Tho City in Vietnam were 'host cities' to an EU-funded project that had aimed at preparing the integration of energy concerns in urban planning in ASEAN cities.[2] Naga City and Can Tho City represent a small city and large city respectively, operating under different political systems of government with balanced yet diverse economic structures. Both exhibit high economic growth yet different levels of economic development, and a large potential for integration of energy in urban development planning – albeit at different stages in their evolution to more sustainable forms of local governance. The present article chronicles the experience of the two cities in that project. Specifically, it describes a methodology for integrating energy in urban planning, highlights key conditions for its successful application, and presents some results.

23.2 Socioeconomic Background of Naga City and Can Tho City

23.2.1 Naga City

Naga City is located within the heartland of the province of Camarines Sur. It is 377 kilometres (km) south of Manila and 100 km north of Legazpi City, the capital of the Bicol region (or Region V). Occupying a total land area of only 84.48 square kilometres (sq. km) and with a population of only around 150 000, Naga City may be considered a small city by national and regional (Asian) standards. The city comprises 27 *barangays*[3] (or villages) and remains predominantly agricultural (accounting for 75% of total land use). A big portion of the population (more than 60%) is within the productive 15–64 years old bracket. The population of the city of Naga grows at a rate of 1.65% per year and is expected to double in 42 years. Like many cities in the Philippines, and probably in many parts of the globe, Naga City's population expands during school months and national holidays, as well as during working hours of the day.[4]

Naga City is the commercial, educational, as well as religious centre of the Bicol region. It is one of the fastest growing economies of the country with 6.5% annual growth rate. Naga captures 21% of the whole investments in the Bicol region and was once chosen as the Most Business Friendly City by the Philippines Chamber of Commerce and Industry. The city's commercial and industrial activities are largely related to agricultural production. Major crops produced by the agricultural sector include rice, sugarcane, coconuts, fruit trees, and corn, but their values of production have been on the decline. Backyard livestock and poultry production shares the biggest portion of total value of agricultural output. The average family income of Naga City's residents is 126% higher than the average in Bicol and 42% higher than the national average.

23.2.2 Can Tho City

Can Tho City is located at the centre of the Mekong river delta and 160 km south of Ho Chi Minh City, Vietnam's biggest city. Can Tho City, on the other hand, is the biggest city in

---

[2] The project was funded through the EU-ASEAN Energy Facility, which was executed from Jakarta, Indonesia by the ASEAN Centre for Energy (ACE). From 12 initially surveyed cities, the project selected three ASEAN cities – Naga City, Can Tho City, and Luang Prabang in Lao PDR – mainly because of their strong willingness to cooperate in the project and they could be good examples under different economic, social and political conditions. However, due to time and space constraints primarily, Luang Prabang is not included in this article. Second, the case of Naga City and Can Tho City may be viewed as examples from a small city and large city, respectively, with balanced economic structure. Luang Prabang is a medium-sized city that is heavily dependent on tourism.

[3] The *barangay* is the smallest administrative unit of government in the Philippines.

[4] Many residents from neighbouring municipalities or towns work or study in Naga City during the day.

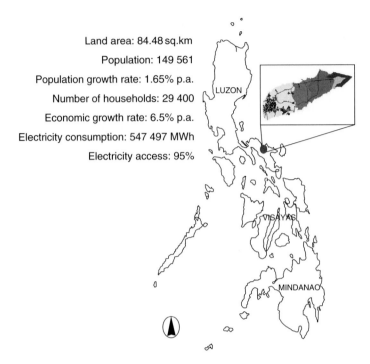

Land area: 84.48 sq.km

Population: 149 561

Population growth rate: 1.65% p.a.

Number of households: 29 400

Economic growth rate: 6.5% p.a.

Electricity consumption: 547 497 MWh

Electricity access: 95%

LUZON

VISAYAS

MINDANAO

Fig. 23.1. Location of and basic facts on Naga City, Philippines.

the Mekong delta region, occupying a total land area of 1390 sq. km and with a population of 1.127 million (see Fig. 23.2). The city is divided into four urban districts and four suburban districts. But more than 80% of Can Tho City's land area remains dedicated to agriculture. Like Naga, a big portion (69%) of the total population is within the 15–64 years old bracket.

But more than Naga City, Can Tho City's economy is growing at the rate of 14.95% per year. Most of this growth comes from commercial or trade activities, which in terms of number of establishments, are growing at 33.1% per year and account for 43.7% of local economic output. The number of industrial establishments is also growing significantly at 5.3% per year, and industrial output contributes 35.1% to local economy. But agriculture remains a very important sector. It accounts for the remaining 21.2% of local economic output and employs 37% of the labour force, even more than those employed in the industrial and commercial (trade) sectors combined.

23.3 Background on the Two Cities' Urban Planning Process

As expected, both Naga City and Can Tho City have dedicated departments responsible for urban planning and development. In Naga City, like in all other cities and municipalities in the Philippines, urban planning falls within the responsibility of the City Planning and Development Office (CPDO). In Can Tho City, urban planning and development is the responsibility of the Department of Planning and Investment (DPI), which is an extension of the country's Ministry of Planning and Investment at city level. Except for nature conservation, which was reportedly not covered by Can Tho City, both Naga City's CPDO

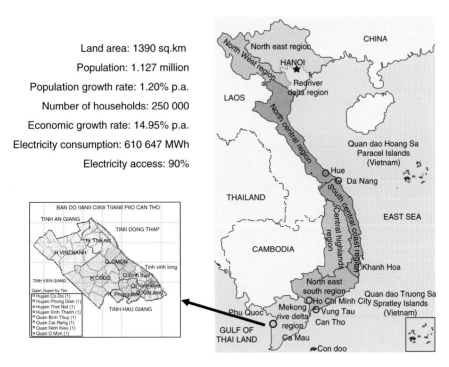

Land area: 1390 sq.km
Population: 1.127 million
Population growth rate: 1.20% p.a.
Number of households: 250 000
Economic growth rate: 14.95% p.a.
Electricity consumption: 610 647 MWh
Electricity access: 90%

Fig. 23.2. Location of and basic facts on Can Tho City, Vietnam.

and Can Tho City's DPI cover a wide range of urban planning sectors, including housing, water supply, roads, transport and traffic, waste management, land management, energy supply, energy demand management, and industrial zones. It is important to note that both cities claim to be dealing with energy supply and energy demand management in their urban planning.

However, urban planning is a multi-sectoral and multi-stakeholder process in Naga City, whereas Can Tho City reports that other agencies are not involved in the city's urban planning. Figure 23.3 shows the steps in urban planning in Naga Ctiy. Table 23.2 lists the agencies and other organizations representing each of the sectors covered by urban planning in Naga City. In fact, Naga City has received more than 140 citations from national and international bodies for its success in innovative and participatory governance. Box 23.1 summarizes the Naga City 'model'.

Nevertheless, while this model has given Naga City considerable success in urban planning, the two cities admit a number of weaknesses that tend to be possessed by cities in developing countries:[5]

- Data gaps – lack of accurate and up-to-date data banking.
- Lack of adequate technology (e.g. computers, training).
- Lack of communication/coordination among planning bodies.

---

[5] The project finds more or less these same weaknesses in all the 12 cities surveyed at the start of the project.

Fig. 23.3. Urban planning in Naga City.

Table 23.2. Other sectors and stakeholders involved in Naga City urban planning.

Other organization involved	Responsibilities and sector
• Housing and Land Use Regulatory Board (HLURB) • Housing and Urban Development Coordinator Council (HUDCC) • Urban Poor Affairs Office (UPAO) • Housing and Urban Development Board (HUBD)	• Housing
• Department of Public Works and Highways (DPWH) • City Engineer's Office (CEO)	• Roads
• Department of Environment and Natural Resources Office (DENR) • City Environment and Natural Resources Office (CENRO) • Solid Waste Management Board (SWMB) • LANCO (MRF)	• Waste management
• Protected Area Mgt Board (PAMB) • DENR • CENRO	• Nature conservation
• Metro Naga Chamber of Commerce and Industry (MNCCI) • Naga City Investment Board (NCIB) • Investment Promotions and Action Center (IPAC)	• Industrial zones
• Metro Naga Water District (MNWD)	• Water supply
• Phil. National Police (PNP) • Transport and Traffic Management Office (T&TMO)	• Transport and traffic
• Department of Agrarian Reform (DAR) • HLURB	• Land management
• Camarines Sur Electric Cooperative II (CASURECO II)	• Energy supply
• Camarines Sur Electric Cooperative II (CASURECO II)	• Energy demand and management

- Duplication of task/functions.
- Lack of monitoring of different projects and programs.
- Level of technical skills/knowledge of personnel involved in urban planning are not adequate.

Box 23.1.
The Naga City 'model'

Three elements form the foundation of Naga City governance model: (1) Progressive Development Perspective; (2) Functional Partnerships; and (3) People Participation.

The first element is the philosophy that anchors all development efforts and seeks to mainstream all sectors of society in accepting their role in local development. This is a function of leadership that the local administration must provide. This element is also associated with the concept of 'growth with equity' that has been embraced by Naga City as its core philosophy. 'Growth with equity' seeks to promote economic development and to sustain the implementation of pro-poor projects to ultimately build prosperity for the community at large.

The second element encompasses the vehicles that enable the city to tap community resources for priority undertakings, in the process multiplying its capacity and enabling it to overcome resource constraints that usually hamper government. The city's partnership mechanisms can be between the local government unit (LGU) and other community groups or government agencies, or between the LGU and individuals. They can be for growth programs or equity-building strategies. They may also be government initiated or private sector/community led.

The last element includes mechanisms that ensure inclusion of individuals and the community in government decision-making. They promote long-term sustainability by generating broad-based stakeholdership and community ownership over local undertakings. One such mechanism is the landmark local legislation 'Empowerment Ordinance of the City of Naga' that established the Naga City People's Council (NCPC). NCPC is composed of all accredited business, non-government, and people's organizations within Naga City. NCPC has representatives in all special bodies in the city and participates in all decision-making activities of the city. The latest mechanism to enhance citizen's involvement in governance is the *i*-Governance program of the city. *i* stands for information openness, interactive engagement, inclusive governance, and innovative management. *i*-Governance focuses on the individual member of the community. It has two basic tools: the Naga City website and the Naga Citizens Charter.

Source: Extracted from (Robredo and Jesse 2003).

23.4 Energy-Integrated Urban Planning (EIUP): Introduction to the EIUP Methodology[6]

One of the objectives of the project was to introduce a methodology for integrating energy strategies in the selected cities with high potential for replication in other ASEAN cities. In this regard, the project used a methodology that was labelled EIUP (energy-integrated urban planning). EIUP is a series of activities or steps that will be undertaken by the cities to help them define their short- and long-term strategies and action plans in solving

---

[6] The methodology was developed by Dr Gerhard Weihs of Centric Austria International (CAI), the lead partner for the EU-funded project involving the two cities. This section derives from the Final Technical Report of the project.

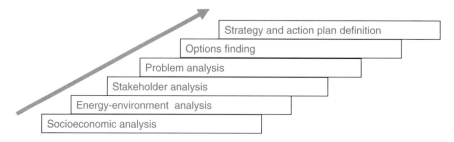

Fig. 23.4. The EIUP methodology.

energy problems that can be addressed at the local level. It comprises the activities as shown in Fig. 23.4.

The EIUP is guided by the following principles:

- EIUP should be based on a reliable database.
- EIUP should be in line with other policy and legal frameworks both at city and national levels.
- EIUP should be interrelated with other planning sectors.
- EIUP should involve all stakeholders and generate benefits for the whole citizenship.
- EIUP should contribute to sustainable development.
- EIUP should respect the local conditions (demands, constraints, resources).
- EIUP should identify feasible and justified improvements.
- EIUP should prepare the reliable ground to take concrete action.

The EIUP uses a concrete tool, the EIUP Working Report, which is a visible guide for undertaking the steps and activities of energy-integrated urban planning, monitoring the progress of each step, and documenting the results. The EIUP Working Report is structured into seven chapters that parallel the steps and activities in EIUP:

1. Socio-economic city profile
2. Energy and environment
3. Stakeholder analysis
4. Definition of problems and targets
5. Option finding
6. Action plan
7. Summary

Chapters 1 and 2 are meant to provide sufficient information to describe the existing situation in social, economic, environmental and energy-related aspects, from which the city starts its efforts of EIUP. Chapters 3 and 4 have an analytic character and identify the scope of required integration in the context of pressing problems. Chapters 5 and 6 identify and clarify the options in terms of their feasibility and implementation aspects. Chapter 6, in particular, provides clear recommendations to take action according to clear priorities and feasibility. Chapter 7 presents a summary that can be used for communicating the results of EIUP to decision-makers, stakeholders, and third parties.

The EIUP Working Report is a very comprehensive document and requires very comprehensive information, especially if the recommended actions mean justifying investment or raising capital. It is likely that not all information required by the EIUP Working Report is readily available in the cities. Data and information can be added step by step in the future, as they become available. Indeed, some data required by the EIUP Working Report may not be available at the city level, but can be obtained from national agencies or relevant institutions. For example, the data required to build the energy balance table may be available from various sources and institutions. In this case, it may be necessary to devise a way of collecting, organizing, and synthesizing those data. But it is important to get the necessary data before implementing any of the proposed actions.

Lastly, but very important, the EIUP Working Report should be prepared by a team, not by a single person, in the same way that EIUP is coordinated by a team of local technical experts. This team, acting as a technical working group, should work on data collection first; then answer main questions, discuss and rank individual city problems, identify priorities and targets, explore options, and draw strategy and a feasible action plan.

23.5 EIUP: Motivations and Conditions for Success

23.5.1 Host city motivations

Naga City expected EIUP to fulfil its promise – to provide good inputs to city development planning. At the same time, Naga City expected that EIUP would enhance local governance. With an international reputation of effective local governance, Naga City wanted nothing less from EIUP.

In the case of Can Tho, the EIUP was seen as providing opportunities for developing indigenous renewable energy sources. Indeed, Can Tho has huge potential for developing renewable energy resources because of its large agricultural base and favourable location that lends itself to solar energy development.

Both Naga City and Can Tho were conscious that the growing energy demand in their respective cities could cause deterioration in the local environment. The two cities recognized, for example, that their mounting solid waste problem is caused by increasing material and energy consumption. They understood that inefficient energy supply and use leads to deterioration of energy services and further worsens environmental conditions. They saw that the increasing energy demand in the transport sector is causing increase in local air pollution as well as carbon dioxide emissions more than any other sector. The two cities expected EIUP to provide solutions to their environmental problems.

23.5.2 Political commitment (to environment-friendly development)

This project would not have been made possible without the strong commitment from the highest political leadership of the two cities. The project sought to formalize this commitment through an agreement, a Memorandum of Understanding (MOU), between the host city and the international partners. The MOU, on the one hand, committed the international partners to conduct the planned activities of the project and, on the other hand, committed the local partners or host cities to actively participate in those activities. Through the MOU, the city and the international partners agree to successfully deliver the promised benefits of the project together.

To be sure, this strong political commitment emanates from the presence of effective and visible environment-friendly local governance in both cities. Naga City even declares this environment-friendly governance in its vision statement (see Box 23.2). But more importantly, the two cities practise it by dedicating an office that focuses on local environmental

---

Box 23.2.
Naga City vision statement

'An Maogmang Lugar'[7]
  We envision Naga
As home of God-loving, happy people

- Who are child-friendly, gender fair and globally competitive
- Rooted in positive moral and cultural values, and
- Belonging to strong, unified and mainstreamed sectors

As a progressive and peaceful community

- Considered a centre of people-focused education, development and governance
- With a clean, healthy and sustainable environment, and
- An economy that provides livelihood opportunities for all, ensuring equitable distribution of wealth and proper utilization of resources.

Source: Naga City Citizens Charter, second edition, August 2004, p. 15.

---

management, through the regular delivery of environmental services and actual execution of local environmental initiatives, and by continuously building their capacity in environmental management through active participation in international urban environment networks.

23.5.2.1 Dedicated office in urban environmental management

Within Naga City Hall is the 68-man strong Environment and Natural Resources Office (ENRO), which claims to be the largest department in the City Hall. On a regular basis, ENRO offers and has 'pledged'[8] to provide the following frontline services (City of Naga 2004):

- Receive, record and file communication and applications for environmental certificates.
- Prepare quarrying permits, permits to cut trees, and other environmental certificates.
- Sign and issue permits and certificates upon approval of the local chief executive (LCE), or the mayor.
- Attend to requests related to solid waste management, including: (1) responding to garbage collection complaints; (2) garbage collection schedule; and (3) request collection of garbage.
- Attend to requests related to watershed management, including: (1) request for grass cutting and trimming of trees; (2) assistance for tree planting activities; (3) inspect/ monitor quarrying activities; (4) proposed or actual construction of building.

---

[7] This literally means 'A Happy Place'.
[8] Each major department or office of the Naga City Hall has declared 'Performance Pledges', which is a public statement, embedded in the Naga City Citizens Charter, to perform their functions with 'the highest possible service performance'.

- Attend to requests related to Socio-Cultural Management Programs, including: (1) information education campaign for ENRO; (2) assist in the preparation of training modules and conduct of training desired; (3) link clients to entities and agencies concerned.
- Respond to garbage collection complaints.
- Clean city street gutters and pick-up points and maintain cleanliness daily.
- Collect garbage in pick-up points based on schedule and area coverage.

It should be noted that each local government unit in the Philippines has been mandated to be responsible for environmental management at their level since 1991 through the Local Government Code. Thus, it is expected that each would set up a similar office as Naga City's ENRO. However, garbage collections, by far the most important environment management-related function of a city or municipality, remain usually with the Engineering Office of many municipalities and cities, and one or few staff under an existing department (for example, the city planning and development office, which is present in all LGUs) maybe given the task of all other environment management-related functions.

In Can Tho, urban environmental management is handled by the Department of Science and Technology of Can Tho (DOST). Below is a list of the present responsibilities and undertakings of DOST:

- Evaluating and reporting to the Chairman of the People's Committee on concrete regulations, policies, and strategies regarding science, technology and the environment, in harmony with the local characteristics and not contravening the regulations and guidelines of the central and provincial government.
- Setting up and submitting to the Chairman of the People's Committee the decisions on tasks regarding science, technology, and environmental protection development programs in accordance with the social and economic development demands of the province.
- Providing guidelines for science organizations to formulate and develop plans to apply new technology to production units and residential developments; summarizing and submitting to the People's Committee for decision, supervision and the implementation of development programs and projects funded by the government budget.
- Managing the technology transfer activities; taking part in the government's appraisal of new technology for important investment projects in the province. Supervising and guiding the application of new technology into the development projects of the provincial economic production organizations.
- Formulating plans and local environmental protection methods in accordance with laws and regulations. Following upon and supervising the implementation of environmental protection measures. Presiding, appraising, and commenting on environmental impact assessment reports, monitoring investment projects belonging to the provincial authority; checking and supervising the environmental activities in the province.
- Monitoring Standard – Methodology – Quality for manufactured products and commodities in accordance with laws and regulations.
- Managing the activities regarding industrial ownership according to the industrial ownership protection decrees and other legal documents.
- Implementing the control function in executing regulations, guidelines and laws dealing with science, technology and the environment. When the inspection is required, submitting the results to the People's Committee for approval.

23.5.2.2. Local initiatives and accomplishments
Both ENRO of Naga City and DOST of Can Tho have led the actual implementation of a number of local environmental initiatives, beyond their regular functions, to further urban environmental management.

Naga City owns and operates the Balatas Material Recovery Facility (MRF) which processes 40–50 tons of garbage daily[9] and produces organic fertilizers in the form of soil enhancers. Through the sale of the fertilizer, the MRF earns for Naga USD168 000 per year (International Council for Local Environmental Initiatives (ICLEI) 2006), as well as secures the employment and livelihood of close to 155 families (Robredo and Jesse 2006). The City's Clean Air Initiative has succeeded in continuously improving local air quality. For example, the city reported that average TSP readings had further improved to 80.6 µg/NCM in 2006 (Robredo and Jesse 2007), from 110.65 µg/NCM in 2003, sustaining its good air quality rating. The city's water quality improvement program has improved water quality of the Naga River, maintaining its Class C grade, which means that it could be used for irrigation and navigation purposes. Naga City through the Isarog Garden Society, a local NGO, also maintains an Eco-Park.

Can Tho City has been chosen to become a model of Sustainable Urban Management by the Prime Minister of the People's Republic of Vietnam, who takes a strong and personal interest in the promotion of environmental issues. (In fact, he has directed the individual provincial governments to give priority to protection of the environment.) With assistance from Green City Denmark, Can Tho City has organized the Green Ambassadors that increase the public's awareness and encourage participation in the city's solid waste management program.

23.5.2.3. Active participation in international urban environment networks
The two cities' political commitment in environment-friendly local governance is also demonstrated by their active participation and membership in international urban environment networks that aim, among other goals, to increase the capacity of cities in urban environmental management.

Naga City is a prominent member of the International Council on Local Environment Initiatives (ICLEI) and a pilot city of ICLEI's Cities for Climate Protection Campaign (see Box 23.3). ICLEI provides the city technical support in the formulation, adoption, and monitoring of measures that would reduce GHG emissions. Through ICLEI, Naga also establishes links with other cities and organizations.

Can Tho City, on the other hand, is one of the few Asian member-countries of the Green City Network of Green City Denmark (GCD). Through its membership in this network, Can Tho City has received technical and financial assistance from GCD in undertaking solid waste management.

23.5.3 Organization of local EIUP teams

EIUP is designed to be carried out by a team, not by a single person. Another key therefore to the successful implementation of EIUP is the formation of a team of dedicated and competent local experts that will carry out the activities of the EIUP, including collecting and analysing data, liaising or coordinating with key stakeholders, identifying problems and

---

[9] But the facility can handle 60–70 tons of municipal solid wastes daily.

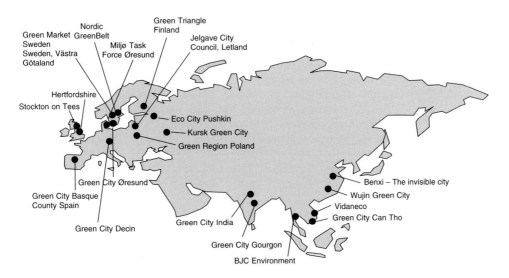

Fig. 23.5. Green City Network.

Table 23.3.   Composition of the local EIUP team in Naga City.

Institution and position within this institution	Major concerns and tasks within the EIUP team
City councillor, team leader	Legislate ordinance
Chief of City Environment Regulation Office (ENRO), asst. team leader	Executive implementation
Socio-Cultural Division Chief, City ENRO	Assist in planning and implementation
Planning officer IV, City Planning and Development Office (CPDO)	Assist in planning
Project officer I, CPDO	Planning
City engineer, City Engineer's Office (CEO)	Regulate project implementation
Engineer II, CEO	Assist in project designing
Illustrator I, City ENRO	Design projects
Program director, Naga City People's Council (NCPC)	Support planning development
NGO staff, NCPC	
Technical writer, City ENRO	Technical writing

priorities, and recommending options. This technical working group is headed by a leader that will report the progress and results of the EIUP to the local chief executive – Mayor of Naga City or Chairman of Can Tho People's Committee. In Naga, one of the city councillor's was appointed to lead the local EIUP team, but he was assisted by the Chief of ENRO, which is the unit responsible for the city's environmental management. This arrangement actually proceeded from the established practice of the city to share the executive functions with members of the legislative council, particularly on ad hoc committees. But the technical work of the Naga City EIUP team was still led by the Chief of ENRO, while the city councillor actually performed an advisory role. Representatives from the City Planning and Development Office, City Engineer's Office, and Naga City People's Council, which is an umbrella organization of all the active local NGOs in Naga City, completed the Naga City local EIUP team.

Similarly, the Vice-Chairman of the Can Tho People's Committee, which performs both executive and legislative functions, was part of the local EIUP team as an advisor, and the Chief of the DOST, which is responsible for the city's urban environmental management, was the designated team leader of the technical working group. Representatives from the Department of Planning and Investment, Department of Industry (which oversees the energy sector), Department of Natural Resources and Environment, and the Can Tho City People's Committee completed the local EIUP team for Can Tho. Representatives of other agencies and the energy companies were part of the local EIUP team on an ad hoc basis.

While it seems that the Can Tho EIUP team had more diverse representation from government agencies than that of Naga City, the formation of the local EIUP teams was dependent on the organizational structure of each city as shown earlier. What is important in the organization of the local EIUP team, as demonstrated in the experience of Naga City and Can Tho, is that key stakeholders are adequately represented in the team. This is essential for accomplishing the purpose of EIUP. In the Naga City EIUP team, the local legislators, the economic planning sectors, the infrastructure sector, and the NGOs were represented. In the Can Tho EIUP team, the economic planning sectors, industry and energy, environment, and infrastructure sectors were represented.

Table 23.4.   Composition of the local EIUP team in Can Tho.

Institution and position within this institution	Major concerns and tasks within the EIUP team
Vice-Chairman	Project supervisory role
People's Committee of Can Tho	
Director, Department of Science and Technology	Team leader
Vice Director, Department of Planning and Investment	Support information about related planning
	Considering and ratifying projects
Vice Director, Department of Natural Resources and Environment	Technical support
Chief of Technology Bureau, Department of Industry	Technical support
The main Expert of People's Committee Office	Administration
Chief of Environment Monitoring Station, Department of Natural Resources and Environment	Technical support
Director, Can Tho Power Company	Ad hoc basis
Director, Tay Nam Bo Petrol Company	Ad hoc basis
Manager of Technical Division, Department of Transportation and Public Works	Ad hoc basis

23.6 Application of EIUP: Key Results for Naga and Can Tho

23.6.1 Opportunities in energy and environment

Through the application of EIUP, Naga City saw the following opportunities:

- Potential for development of decentralized RES systems.
- National government supports power supply companies to deal with service improvement.
- Political support from the city government to energy efficiency and RES.
- National legislation supports development of local energy resources through various fiscal and non-fiscal incentives.
- Possibility to benefit from local energy resources – non-power and power related in the case of electricity.
- Improvement of quality of energy supply.
- High energy prices will promote fuel switching and implementation of new technologies.
- Requirement of national government to use biodiesel to replace a part of oil fuels in transport (1%) in the case of government agencies.
- Waste-to-energy project development and implementation.

Can Tho, on the other hand, became more convinced that urbanization guided by the principles of sustainable development will:

- Promote environmental protection and improvement in the government and communities;
- Improve transportation systems, architectural landscape, and manufacturing technologies; and

- Ensure support from the central government for improving the city's energy production and use.

23.6.2 Stakeholder analysis

Naga City and Can Tho had the common opinion that practically all the broad sectors of their local economies are the stakeholders of EIUP (see Table 23.5). Unsurprisingly, the common expectation or interest of these sectors was for EIUP to achieve an adequate, reliable, and low cost or reasonably priced and affordable energy supply, with minimal impacts on the environment. The stakeholder analysis also confirmed the potential of some sectors for developing and investing in renewable energy sources and energy efficiency. Equally importantly, the stakeholder analysis stressed the role of other sectors (particularly the general public and final consumers) in policy development, implementation and monitoring, as well as in increasing the social acceptability of energy projects (if the stakeholders are involved in the early stages of project development). At the same time, the stakeholder analysis identifies the lack of awareness and capacity on technologies, finance, and regulation as permeating most, if not all, sectors. These deficiencies could be addressed through the EIUP.

The stakeholder analysis included the identification of the responsible bodies for the different planning sectors and the related laws and policies and their target sectors in each of the two cities. These provided the institutional, policy, and legal context of EIUP that could support or limit the feasibility of its results or recommendations. For example, Naga City should comply with the Philippine Energy Plan that sets direction and targets related to energy supply and demand, and the Electric Power Industry Reform Act that outlines the new policies in the electricity sector. On the other hand, national environmental laws provide opportunities for Naga City in exploiting waste-to-energy potentials. In the case of Can Tho, however, the lack of clear directions on renewable energy development and energy efficiency and conservation at the state level tends to limit the identification and assessment of clean energy options at the level of local government.

23.6.3 Problem analysis

The SWOT analysis was necessary for identifying the real energy concerns of the two cities. The stakeholder analysis was necessary for identifying the sectors affected or affecting the problems and eventually engaging them to jointly rank and prioritize these problems and analyse their impacts through a consultation. Both cities went through the latter process without direct intervention from the international team of experts, except undergoing a small group workshop with selected members of the local EIUP team to become familiar with the steps in problem analysis.[10]

The stakeholder consultations in the two cities resulted in the identification of priority problems and their ranking as shown in Table 23.6.

Except for the availability of a reliable energy (electricity) supply and to some extent the use of renewable energy sources, notice the contrasting priorities of Naga City and Can Tho when it comes to energy issues. In particular, while Can Tho gave greater importance to having an energy policy than efficient energy end-use, Naga City gave more importance

---

[10]This workshop was conducted during the study tour in Europe when representatives from the three host cities and the EU and ASEAN partners exchanged experiences with stakeholders from selected EU cities and organizations.

Table 23.5. Stakeholder analysis of Naga City and Can Tho.

Stakeholders	Interests and expectations on EIUP	Potential	Deficiencies
Agriculture	Electricity supply at low cost	Large biomass resource	Lack of awareness on technologies and financial resources
Residential	Reliable and affordable energy supply; reduced environmental impacts of energy production and use		Lack of awareness on energy conservation and renewable energy resources
Commercial, trade, and industrial sectors	Adequate, reliable, and reasonable cost energy supply; opportunities for producing or selling electricity	Participation in EIUP development and monitoring; application of clean energy technologies; ability to invest in energy production and energy efficiency and savings	Lack of awareness on technologies, regulatory framework, monitoring and enforcement
Utilities and energy suppliers	Development of renewable energy resources; efficiency in electricity distribution and end-use; distributed generation	Support for energy efficiency programs; cooperation in electricity business; rehabilitation and extension of distribution network; big (and established) market on fossil fuel consumption	Few or inadequate expertise in developing renewable energy resources and energy efficiency programs; lack of information on fossil fuel end-use consumption
NGOs/POs	Reliable, accessible, and low priced energy supply; reduced environmental impact of energy use	Involvement in policy development, implementation, and monitoring; preparation of regulations; involvement in social preparation of projects	
Local government	Develop the tool for future improvement of energy supply, reduction of energy costs, improvement of the environment, improvement of standard of living, create climate for business development and employment	Develop capacity for energy planning; conduct awareness campaign for general public; facilitator of implementation of activities identified in the Action Plan; provider of co-financing of projects	
Other government agencies (public sector)	Monitor implementation of other or related government policies		Possible lack of coordination between the city government and government offices; lack of support to implementation

Table 23.6.   Problem storage (ABC Analysis) for Naga City and Can Tho.

Priority problems	Ranking
Can Tho	
1. Lack of energy development planning and policy	A
2. Shortage of energy	A
3. Inefficient energy end-use	B
4. Lack of know-how on sustainable waste-to-energy and other renewable energy technologies	B
Naga City	
1. Reliability of electricity supply	A
2. Lack of comprehensive policies and energy management	B
3. Insufficient use of alternative energy sources	B
4. Not all communities are connected to power supply	C
5. Insufficient use of alternative motor fuels	B
6. Insufficient extension of energy conservation and audits	A

Note on ranking:
A = problem with big pressure and urgent need for solution.
B = problem of great importance, which should be solved in the medium term.
C = important, but does not need urgent solution.

to extending its energy conservation efforts rather than addressing the lack of comprehensive energy policy. These differences in energy priorities between the two cities could be explained by the differences in the stage of their economic development. Although Can Tho is 16 times larger and seven times more populous than Naga City, the poorest segment of this Philippine city is more than five times richer than the equivalent portion of Can Tho City's residents. This also explains why the per capita electricity consumption of Naga City is higher than Can Tho, thus actually providing more opportunities for energy efficiency and conservation. Moreover, with a population of more than one million and lower electrification rate, there are actually more people in Can Tho without access to electricity than in Naga City, justifying the higher importance given by Can Tho to availability of energy supply compared to Naga City's concern on supply reliability. (It is estimated that about 112 000 Can Tho residents do not have access to electricity compared to Naga City's 7500 residents.)

In options finding, these problems are translated into concrete objectives. These objectives are then explored as to possible activities for achieving them, the obstacles to the realization of these activities, recommendations for overcoming the barriers, and concrete first steps.

23.6.4 Strategy and action plan

Options finding is also a bottom-up approach in defining the cities' EIUP strategy and formulating the final plan of actions in solving the problems identified earlier. The EIUP strategy is a statement of the city's energy policy, vision, mission, and goals. These are shown in Box 23.4 and Box 23.5 for Naga City and Can Tho, respectively.

The actions identified to address the priority problems mentioned and described earlier are also listed in Boxes 23.6 and 23.7.

Box 23.4.
EIUP strategy of Naga City

EIUP Policy:
To overcome identified barriers, the focus of the EUIP will be:

- Implementation of energy planning activities, establishing suitable policies, solutions and methods to attract ownership of the city by adding new energy sources (renewable energy, waste-to-energy solutions, energy efficiency, clean energy, etc. . . .) to the energy portfolio of the city.
- Development of waste-to-energy projects in order to generate energy from wastes, to produce compost from wastes processing, and to improve environmental conditions (land use and pollution).
- Promotion of a wider use of biofuels in transport with the aim of meeting the national strategy and reduce air pollution in Naga City.
- Promotion of energy efficiency and saving activities through expansion of energy audit and conservation campaign to other sectors in order to reduce energy consumption and production, as well as cost of energy.
- Improvement of reliability and quality of electricity supply in order to increase the quality of life of the population, to improve economic performance of industry and services and to increase the interest of entrepreneurs to locate their business in the City.
- Providing full access to electricity to the entire population, including those in remote areas, in order to increase quality of life and attract entrepreneurs to locate their business in these areas.

Motto:
Clean Energy for Quality Life.

Vision:
A vibrant and sustainably developed city with each citizen enjoying quality life supported by clean, affordable and efficiently used energy sources.

Goals:

- Have prepared EIUP and integrated energy plan in the City Medium-term Development Plan in one year.
- Have all municipal waste processed by recycling and by waste-to-energy conversion in case of non-recyclable waste in five years.
- Have at least 5% biofuels in the consumption of motor fuels in the year 2010.
- Have energy audit and conservation campaign expanded to all sectors in five years.
- Have reduced system losses below 10% and reached national standards in power supply quality in 5 years.
- Have 100% connection of households of Naga City in 5 years.

Note: This is an unedited version of the original statement, which is awaiting approval by the City's Executive and Legislative Committees.

Naga City also designed an 'EIUP logo' to increase the visibility and awareness of this new planning activity.

Box 23.5.
EIUP strategy of Can Tho

Eiup Policy:
To solve the problem of lack of energy for future development, the focus will be:

- Implementation of energy planning activities, establishing suitable policies, solutions and methods to attract ownership of the city by adding new energy sources (renewable energy, waste-to-energy solutions, energy efficiency, clean energy, etc. . . .) to the energy structure of city.
- Development of waste-to-energy projects in order to generate energy from wastes, to produce compost from wastes processing, and to improve environmental conditions (land use and pollution).
- Promotion of energy efficiency and saving activities in order to reduce energy consumption and production, as well as cost of energy.
- Development of renewable energy technology to solve the energy supply shortage and meet increasing energy demand caused by the rapid socio-economic development of the city.

Vision:

- Industrial and trade and service activities will have fast grow rate in the future. Moreover, living standards of people will also increase. Therefore, lack of energy, increasing waste, and deteriorating environment will result without suitable planning instruments. These negative impacts will also affect the potential growth rate of the city in the future. Suitable planning will help to limit these negative effects.

Goals:

- Having better coordination between various stakeholders and the city for the development through the establishment of an energy planning unit at city level, able to set-up an energy-environment data base and information system and to prepare city energy planning forecasts; having action plans for the improvement of the city energy systems to tackle future lack of energy and to propose solutions in relation with loses of energy and new sources of energy.
- The city will setup an energy plant for the generation of energy from wastes, production of compost from wastes processing as well as improvement of environmental conditions.
- Preparation and implementation of energy audits and surveys in Industry, commerce, households, agriculture – Identification and assessment of priority sectors and actions for improvement in energy efficiency and savings. Following assessment, preparation and implementation of specific measures for the various sectors of the economy (industry, handicraft, trade and services, households and agriculture).
- Increase additional energy production and increase awareness and participation of private sector in the development of RETs. Moreover, increase capacity of universities and research centers in the field of renewable technologies through RD&D, trainings, etc. . . . and improve environmental standard for production of energy.

Note: This is an unedited version of the original statement, which is currently pending approval by the Can Tho City People's Committee.

Box 23.6.
Action plan for Naga City

Approval of EIUP and inclusion into the City Development Plan

- Conduct stakeholders consultations
- Develop EIUP as a part of the City development plan
- Communicate to the general public and stakeholders to get feedback and support for implementation of activities.

Waste to energy use

- Development of the Feasibility Study for identified priority projects
- Search for potential investors/co-funding
- Priority project implementation.

Promotion of a wider use of biofuels in transport

- Identification of the volume of transport sector activities
- Prepare the plan for marketing of biodiesel and communicate it with the transport sector
- Demonstration of good practice of biodiesel use by cars of the city government
- Promulgation of a city ordinance on the use of biodiesel, public consultation on the ordinance.

Expansion of energy audit and conservation campaign to other sectors

- Identification of potential sectors to be involved in the energy efficiency and auditing campaign
- Issuing of ordinance for implementation of energy efficiency and auditing campaign
- Dissemination of results of activities done so far to demonstrate potential benefits energy efficiency and auditing campaign
- Provide support to other sectors in running energy audits and identification of energy conservation measures
- Presentation of results to show benefits.

Improvement of electricity supply

- Coordination with power supplier in activities to make sure that there is sufficient level of maintenance and monitoring; to monitor and act in minimizing illegal connections or pilferage, and decrease of system losses.

100% electricity supply

- Identification of households not yet connected
- Preparation of connection plan and its implantation
- Identification of financial sources for implementation of the plan.

Note: This is an unedited version of the original statement, which is currently pending approval by the city.

Box 23.7.
Action plan for Can Tho

Development of energy planning and policies at City level

- Creation of an energy planning unit at city level, will need a decision from the People's Committee; however, a realistic budget should be allocated to this unit to perform its activities and duties;
- Creation of an energy and environment Information System and Data Base (energy and environmental statistics and indicators) will require the collaboration and participation of many sectors of the city's economy;
- Prepare yearly energy balances and forecasts, will require that energy companies release information, and that surveys will be conducted by the energy planning unit to know better the energy end-uses in the city;
- Study and propose solutions for future lack of energy (waste-to-energy, renewables, etc. . . .) will also require the cooperation of universities and research centres to study and propose adequate technological solutions;
- Study and propose solutions for energy efficiency and savings measures (regulations, monitoring and enforcement) will require the cooperation of all sectors of the city, including universities and research centres.

Building of a waste treatment plant to produce energy

- An integrated plan for the development of energy in the city should be prepared and proposed to the city's government for action guidelines;
- Establishing pre-feasibility and feasibility studies to be submitted to the city's steering committee. These studies should specify location of plant, chosen technology, financial resources, partners and socio-economic impacts;
- To find partners, investors, international organizations paying attention to electricity production from waste in order to solve the problems (technology, finance, and training);
- Stop or go decision from the municipal administration.

Energy Saving

- Through preliminary integrated survey/audit on energy saving, submit the main results to the people's government and ask for action guidelines;
- Preparing an energy saving programme for the city for the next five years with recommendations up to 2015 to be submitted to the city's government for ratification and implementation;
- Implementing and executing the program with 2 main contents:
- Enhance knowledge and awareness on energy saving for community, government, organizations, businesses, etc. . . .;
- Building demonstration projects on energy saving in residential areas, business, services and industries;
- Establish a steering committee to liaise with communication agency, union, government, etc. . . .

Apply technological achievements to supplement new energy sources being suitable for the city's condition such as solar energy, biogas, biodiesel, etc. . . .

- Writing a report on the necessity and urgency for the city to develop and manufacture new energy technology systems;
- Establishing applied research programmes based on data to be collected, analysed;
- Study existing experiences and benchmarks in the field of renewable energy and organize capacity building;
- Establishing an organization to research and apply technology to the promotion of new energy technologies.

Note: This is an unedited version of the original statement, which is currently pending approval by the Can Tho City People's Committee.

23.7 Conclusions

Prior to participating in the EIUP project, Naga City and Can Tho had built considerable experience in urban environmental management and had some knowledge on energy projects and technologies. Naga City had a solid and international reputation on participative local governance, and both cities had embraced sustainable development principles, even if Can Tho had had limited experience in involving stakeholders. However, they had a vague idea of integrating energy concerns in urban development planning. To start with, reliable data about sectoral energy consumption and future demand were not available. Even national energy policies were seen more as restrictions than a framework under which local initiatives can be drawn. In general the former approaches of the cities in dealing with energy issues were restricted to single projects and a comprehensive framework was missing.

The EU-funded project presented in this chapter has equipped Naga City and Can Tho with the capacity and the tools to tackle energy as an integrated part of their development. This was achieved through the EIUP methodology, which had been used as an instrument for both capacity development and starting the process of integrating energy in urban planning, from building a reliable database to drawing the energy strategies and action plans through a participative process.

Intelligent decision-making builds on reliable and adequate information. So a necessary condition for a successful EIUP is a reliable socioeconomic and energy-environment database. The baseline information generated by this database provide knowledge and understanding of the constraints within which EIUP has to evolve and resources and opportunities through which EIUP goals can be achieved. Another necessary condition for a successful EIUP is knowledge and understanding of the sectors affected by and influencing energy policy. The project has built on the solid experience of Naga City and has built the capacity of Can Tho in identifying and engaging stakeholders.

But what this chapter highlighted as key conditions for a successful EIUP are a strong political commitment on sustainable development and a dedicated and competent team of local experts representing the diverse sectors of the local economy. The first of these conditions cannot be provided by any bilateral or multilateral technical assistance project.

At best, some technical assistance could aim at educating local government administrators and politicians to plant the seed of commitment to sustainable development. Fortunately, a number of local governments (cities and municipalities) have embraced such commitment. And Naga City and Can Tho are glaring examples in this regard. Then it will be easy to achieve the second condition through learning-by-doing and learning from the experiences of other cities. While the actual adherence to the strategies and implementation of the action plans remain to be seen, the two cities can very well share their newly acquired skills and expertise in integrating energy strategies in urban planning.[11]

Acknowledgements and Disclaimer

This article draws heavily from the work of a team of international consultants and the local EIUP teams set up in the two cities in the framework of the project 'Feasibility study for the implementation of energy integrated urban planning (EIUP) in ASEAN cities and of applicability of European approaches'. The team of international consultants included Dr Gerhard Weihs and Dr Miroslav Maly of Centric Austria International; Ms Susanne Hansen and Mr Ole Mynster Herold of Green City Denmark; Mr Xenarios Stefanos of University of Panteion (Greece); and Prof. Lefevre, Mr Lars Moller, and Mr Jessie L. Todoc of CEERD-FIHRD. The local EIUP team in Naga was led by Councillor William del Rosario, member of Naga City's Legislative Council, and Mr Oscar Orozco, Chief of the Environmental and Natural Resources Office of Naga City, while the local EIUP team in Can Tho was led by Mr Nguyen Minh Thong, Director of Department of Science and Technology of Can Tho City. The conduct of the project activities would not have been possible without the blessing of Mayor Jesse M. Robredo of Naga City and Mr Nguyen Thanh Son, Vice-Chairman of the Can Tho City People's Committee. The project is funded by the European Union through the EU-ASEAN Energy Facility program, which is executed by the ASEAN Centre for Energy, based in Jakarta.

The views in this article are solely of the author and should not be attributed to any of the above-mentioned persons or organizations. The accuracy of the information presented is solely his responsibility.

References

Aarhus County (Denmark) (2002). 'Green City Can Tho – Sustainable Urban Development', grant application form to the Asia Urbs Programme.
Asian Development Bank (ADB) (2006). *Key Indicators 2006: Measuring Policy Effectiveness in Health and Education*. ADB.
City of Naga (2004). *The Naga City Citizens Charter*. City of Naga.
International Council for Local Environmental Initiatives (ICLEI) (2006). *Cities for Climate Protection: ICLEI International Progress Report*. ICLEI.
Robredo, Jesse M. (2003). *Making Local Governance Work: The Naga City Model*. City of Naga.

---

[11]As indicated in the respective cities' action plans, one important result of this technical assistance project in the two cities is the possibility of developing energy projects from their municipal solid wastes. In fact, the project also taught and discussed with the two cities the possibility of putting up an anaerobic digestion facility to extract biogas and produce electricity at their existing dumpsites. The preliminary assessments have showed the feasibility of such projects, and the two cities are now considering the conduct of full feasibility studies.

Robredo, Jessie M. (2005). 2004 State of the City Report: Soaring to New Heights.

Robredo, Jesse M. (2006). 2005 State of the City Report: Doing More with Less.

Robredo, Jesse M. (2007). 2007 State of the City Report: Putting Our People First – Development that Matters.

UNESCAP (2005). *Urban Environmental Governance for Sustainable Development in Asia and the Pacific: A Regional Overview*. United Nations.

Chapter 24

Sustainable Energy Systems and the Urban Poor: Nigeria, Brazil, and the Philippines

JOY CLANCY, OLU MADUKA AND FERI LUMAMPAO

24.1 Introduction

Despite efforts at rural development to improve poor people's livelihoods,[1] urbanization in the south continues unabated. It has been estimated that half of the world's absolute poor would be living in towns and cities by 2000 and this will have risen to almost two thirds by 2025 (UNDP 1999). Rural poor people migrate to urban areas in search of perceived better livelihood opportunities, where they join other poor people with similar goals. This migration can be as part of a coping (temporary) strategy in response to a specific short-term disruption, such as drought, or an adaptive long-term strategy of permanent relocation.

The basis for urban livelihoods is complex; contexts are shifting and uncertain. As a consequence, members of low income households employ varied strategies, often living on credit and earning income through a variety of means, primarily in the informal sector. Evidence suggests that poor urban households' livelihood strategies often do not meet the most basic of needs, increasing the vulnerability[2] of those already marginalized (CARE 1999). It is tempting for policymakers to assume that the urban poor compared to the rural poor benefit from increased access to modern forms of energy (liquid fuels and electricity), contributing to better chances for moving out of poverty and reducing vulnerability by a decrease in household energy budgets or an increase in income (informal sector and small businesses become more profitable or new business opportunities are created) in which access to energy plays a significant role.

Energy is one of the most essential inputs into sustaining people's livelihoods. At the most basic level energy provides cooked food, boiled water and warmth. Lack of access to clean and affordable energy can be considered a core dimension of poverty. Energy use can also increase vulnerability. The health impacts from indoor air cooking pollution are well documented (Naeher *et al.* 2007) (see, for example, ENERGIA News 4-4). The illness of an income earner has been identified as the major significant negative impact on

---

[1] A livelihood can be defined as the capacity (ability and opportunities) to enjoy long, healthy lives in a manner of one's choosing in harmony with one's physical and social environment (Clancy *et al.* 2006).
[2] Vulnerability can be defined as a condition in which there is limited capacity to respond to adverse natural or economic events, such as droughts and price rises, and social obligations such as weddings and funerals (derived from (Ellis 2000)).

533

low income households (Amis 2002). Energy use can increase vulnerability in other ways. Homes in poor urban settlements are built of readily available, often scavenged, materials, such as wood and cardboard. These materials are also flammable and the physical assets of the poor can rapidly be destroyed by fire caused by sparks from a fire or a knocked over kerosene lamp (CARE 1999).

Rapid urbanization is associated with a rise in energy demand which puts stress on the environment through housing construction, transportation, economic activities and the generation of heat and electricity (Reddy 2000). However, urban areas do offer possibilities for increasing use of more environmentally friendly energy sources. The application of new energy efficient technologies may be more easily accelerated than in rural areas since the environment created by the agglomeration of social networks makes people, both in households and business, more acceptable to awareness raising campaigns. There are potentials for economies of scale due to the higher population densities reducing delivery costs, including those for energy services. Education opportunities are more readily available than in rural areas and people can learn about the opportunities energy services offer for their own use or as a business opportunity.

Issues of access to modern energy services by the urban poor should not be reduced to issues of purely technical supply (increasing quantity, quality and reliability), since energy access has a social dimension. The urban poor are not a homogeneous entity but are distinguished by multiple identities: marital status, age, ethnicity, and gender. Many single and married men, and increasingly young single women, migrate to urban areas leaving rural families behind. What strategies do they adopt to replace their disrupted social networks, which in rural areas enabled their access to resources? Amis (2002) states that particular groups, often organized on ethnic lines, tightly control the most profitable parts of the informal sector. Does this control also apply to access to energy sources?

Urban women face similar inequalities to their rural sisters: low capabilities, low rewards in the labour market, a lack of productive assets and resources compared to men as well as exclusion through social stigma and discrimination (Amis 2002). This lack of capacity to respond to changes in the social and economic environment puts women in a particularly vulnerable position. This situation is further exacerbated for poor women since they are over-represented among the chronically poor.[3] How do these inequalities manifest themselves in women gaining access to energy services? How are women able to meet their energy needs (individual and collective)? Are women able to mobilize new social networks?

Against this background, what roles can improved energy services[4] play in strengthening the livelihood strategies of the urban poor? Does increased income improve access to affordable modern energy services which give less indoor air pollution than traditional fuels, better cooked food, and more boiled water and hence contribute to better well-being? If the answer to these questions is positive the challenge is to provide urban households with clean affordable energy which requires a holistic understanding of the role of energy in sustainable urban livelihoods. This chapter aims to:

1. Summarize current knowledge on the livelihoods of the urban poor and energy;
2. Review current approaches to providing energy services for the urban poor;

---

[3] The chronically poor are those living in poverty for a considerable period of time.

[4] 'Improved energy services' is a relative and judgemental term. We have taken it to mean at the very least that the urban poor have better access to low cost, high quality energy services. It is possible that the urban poor might wish to extend the criteria.

3. Review the existing evidence of the impacts of commercialization and privatization in the energy sector on urban poor people's access to energy services and the sustainability of their livelihoods;
4. Map the impacts of fuel price increases on sustainable livelihoods of poor urban households; and
5. Review the options for energy services based on renewable energy sources.

The chapter is based on a review of literature (journals, newsletters, project reports and internet sites) related specifically to *urban* livelihoods and energy and empirical evidence from a study carried out in Nigeria, the Philippines and Brazil (Clancy *et al.* 2006).[5]

24.2 Energy Use Patterns in Poor Urban Households

24.2.1 Energy use

The two dominant end-uses for energy in urban households are cooking and lighting (although in certain locations there may be a demand for space heating or cooling). Urban households, irrespective of income, use a mix of fuels for both end-uses, although the fuel of preference does vary with household income. Lower income households rely on biomass fuels[6] (or coal in some countries such as China and South Africa), whereas higher income households will opt for electricity and LPG (Hosier and Kipondya 1993; Future Energy Solutions 2002). Kerosene is used both as a cooking and a lighting fuel, although electricity is the preferred option for lighting. Urban households often seem to retain a mixture of fuels, not only to safeguard against supply uncertainties, but to match cooking styles and time constraints. In Dar es Salaam, urban households with LPG and electricity tend to use these energy forms for breakfast and for making hot drinks in the evening when time saving is a particular advantage and to use charcoal for cooking other meals (Leach and Mearns 1988).

What can be said about the assumption that the urban poor have better access to modern energy carriers compared to the rural poor? Based on an analysis by ESMAP and UNDP of energy use data from official household living standards surveys in six countries,[7] held at different times between 1993 and 1999, it can be concluded that electricity access in urban areas is better than in rural areas even for the poorest households (UNDP/ESMAP 2003). However, there are variations. Nearly 100% of households in the first quintile (that is the poorest 20%) in Brazil and Vietnam were electrified, while in the comparable group in South Africa, Ghana and Nepal only around 30% of households were electrified. Some caution has to be exercised in any attempt at cross-comparison since the data were not collected under comparable conditions in terms of timing and the form of questions posed. Also the reliability of the data is questionable since some poor urban communities manage to get an electricity connection through a combination of illegal hook-ups and private generators (Schutyser 2003). The former are unlikely to be reported in official surveys. Also having an electricity connection does not mean that the electricity is actually used. A study

---

[5] The urban communities involved in the studies were Plataforma and Canabrava in Salvadore, Brazil; Ijaje and Amukoko in Lagos and Kwali and Karmo in Abuja, Nigeria; and Manila City and Marikina City in Metropolitan Manila, the Philippines.
[6] Wood, charcoal, agricultural residues and dung.
[7] Brazil, South Africa, Guatemala, Nepal, Nicaragua, Vietnam, Ghana and India.

in Arusha, Tanzania, found that poor households, especially those with women as the household head, had higher rates of disconnection than wealthier ones (Meikle 2004). What we can conclude is that in the majority of countries there are still a substantial number of urban households and enterprises without access to electricity.

There is a very different picture in terms of LPG use. In Brazil, nearly all urban households use LPG but in other countries it is mainly the high income households using LPG. The majority of the urban poor continue to use solid fuels or kerosene. Whether or not that puts those poor urban households using kerosene in a better position than poor rural ones is debatable. While kerosene is cleaner and easier to use than woodfuels,[8] many people complain that the taste kerosene imparts to food is unacceptable while that from woodfuels is actively sought by many consumers.

An interesting finding in the UNDP/ESMAP study is the use of dung and straw in urban households in Asia. Fifty per cent of urban households in the first quintile in Nepal were reported to use dung and/or straw with a similar percentage in Vietnam and 25% in the first quintile in India.

24.2.2 Access to energy carriers

The available evidence suggests that urban poor people do buy their fuel even though there are little quantitative data on the use of non-purchased biomass fuels. There are different mechanisms for accessing biomass. In the Philippines, some poor urban households still have rural plots of land which provide them with fuelwood and charcoal (Clancy *et al.* 2006). In some urban areas it is still possible for households to gather fuelwood. In a detailed household survey in three cities in Tanzania (Dar es Salaam, Mbeya and Shinyanga), it was found that low income households in Mbeya were able to find sufficient fuelwood from peri-urban wood lots to meet their needs and this significantly altered the household fuel purchasing profile compared to the other two urban centres. Leach and Mearns consider that 'free' wood, obtained from a variety of sources such as timber yards, discarded packaging and the urban hinterland, is an important energy source for low income households but one ignored by many energy planners (Leach and Mearns 1988). Time constraints and the availability of transport are limiting factors in access to these sources.

Poor people prefer to purchase fuels in patterns that match their incomes: small amounts on a daily basis. This purchasing pattern influences the types of fuel they use. Wood, charcoal and kerosene can be bought in small amounts on a daily basis. A consequence of this purchasing pattern is that they are paying a higher unit cost than for 'bulk' purchases.[9] Reddy has argued that lack of reliable income forces poor households to use much higher discount rates than the rich when making decisions about energy forms which means that they think solely in terms of first cost rather than life cycle costs (Reddy 2000). This tends to work against the purchase of energy efficient devices since they tend to have higher first costs, irrespective of whether or not these use market-based energy forms or traditional gathered fuels.

At the beginning of the 1990s, the World Bank carried out a global survey of 45 cities and 20 000 households. This study found that poor urban households spend a significant

---

[8] Woodfuels are used in this chapter to mean fuelwood and charcoal.
[9] However, it should not be forgotten that the provision of fuel in small quantities is often a service provided by other low income, informal sector suppliers (see section 24.3).

portion (15 to 22%) of their cash incomes on energy (Barnes 1995). This was attributed in part to the heat content of the fuels used and the conversion efficiency of the technologies influencing the amount of useful energy produced. The poorest 20% of households spend a higher proportion of their incomes than wealthier ones on lower quality fuels (primarily biomass and kerosene) both for cooking and lighting. (ESMAP 1999)

In the Tanzania household survey referred to above, it was found that woman-headed households use a higher average percentage of their income than man-headed households for purchasing energy (Hosier and Kipondya 1993). The authors attributed this difference to woman-headed households having lower incomes than man-headed households, rather than using more energy. The implication of such a finding is that woman-headed households will suffer more from rapid energy price rises than man-headed households.

24.2.3 Influences on urban energy use

Apart from the availability of biomass, what other factors influence the patterns of energy use by poor urban households? Urban size is found to influence a shift from traditional to modern fuels. A survey in Pakistan reported that 93% of households in towns of under 25 000 people depended on woodfuels for cooking and heating, 65 to 75% in medium sized cities and in the large cities (more than 250 000 inhabitants) the figure dropped to 25% (Government of Pakistan 1982) quoted in (Leach and Mearns 1988).

Hosier and Kipondya (1993) found that household energy use responds to price and availability, and that the patterns of use are distorted by subsidies. It is not only the availability of the fuel but also of appropriate appliances which determines the fuel used. Hosier and Kipondya considered that the greater number of households using kerosene in Dar es Salaam (the largest city and port) compared to the number in the other two cities could in part be attributed to the greater availability of kerosene stoves and lamps. Likewise, an increase in cooking with electricity which occurred during a specific three-year period could be linked to the removal of import duties on electric stoves. The unreliability of supply was considered a significant barrier to more households switching to LPG, and the cause of a number stopping using it. Instead, charcoal was used as the main cooking fuel because of its reliability. This provides an example of the private supplier satisfying customer needs while the public supplier fell short (despite the social equity aspect of the energy policy in operation in Tanzania at that time). Inefficiencies in public supply can also lead to the development of black markets which raise the price of fuel and limit access by those on low incomes (for example, kerosene in India). In Peru, competitive private distribution of fuels has increased the number of outlets and led to greater access by all households (Doig 1998).

However, is the type of fuel used by households merely a question of price and availability, essentially economic arguments, or are more complex issues involved? Hosier and Kipondya (1993) estimated that, at the time of their survey, electricity was the cheapest fuel for cooking, even when taking into account the cost of appliances and conversion efficiencies. Yet connected households, usually in the higher income groups, were not showing a significant switch to using a cleaner and safer fuel for cooking. It is possible to offer non-economic explanations for these findings – one relates to cooking practices and the other to intra-household decision-making. Long, slow simmering is required to cook two staples in Tanzania (ugali and beans) and without the use of specialized cookers (e.g. a slow cooker) the heat output of electric cookers is difficult to regulate and it takes time to learn the skill. One also needs a degree of confidence in supply reliability to commit one's basic food to being cooked using electricity. Cooks, usually women, may actively choose not to use an energy form they find impractical. For example, Leach and Mearns quote a World Bank study in Niger which found that despite cooking being cheaper with kerosene than wood, wood was still the preferred fuel (Leach and Mearns 1988). Three reasons were cited: (1) the power output of the kerosene stove was significantly lower than the traditional wood fire, and so cooking took longer; (2) the kerosene stove did not support the round-bottomed cooking pots used in the area which tended to overbalance during the frequent stirring necessary with staple local foods; and (3) the kerosene stoves were not robust. Kerosene stoves were used for rapid cooking and water boiling, while wood and charcoal were used with staples. A similar pattern with LPG and electricity was also found in Dar es Salaam, where these fuels were used when time was of the essence: at breakfast and for hot drinks in the evening. In Nigeria, poor urban households use kerosene rather than fuelwood for cooking, despite it being more expensive to cook the family meal with kerosene than wood. A possible explanation is that fuel choice here is governed by the nature of the housing in which the urban poor live: in cramped multiple occupancy dwellings where cooking is usually carried out in shared corridors. Under such circumstances, fuelwood is considered both impractical – where mobility of the stove is important – and a hazard.

A second alternative explanation for not switching to modern fuels for cooking rests on the fact that households have to make choices about expenditures. While economists tend to see households as a homogeneous entity making rational choices based only on price, social scientists using gender analysis consider this not to be the case. Eberhard and van Hornen (1995) considered that intra-power relations within the household play a significant role in decisions about household energy use. In households where there are adult men and women, the gendered division of labour generally allocates to women the responsibility for energy provision related to their spheres of influence in the household, in particular activities centred on the kitchen. However, when energy has to be purchased, men enter the decision-making process, for example men although not usually responsible of cooking will often decide on the stove technology if it is to be purchased (Tucker 1999). Men also make important decisions on other factors that influence cooking and kitchen comfort, for example material for kitchen walls and roofing (Dutta 1997). In some households, recreational equipment, such as TVs and radios, was bought before labour-saving equipment for domestic chores (Makan 1995).

The switch to electricity is hindered by high initial connection charges, high cost of wiring and high standing charges. Utilities are often reluctant to provide a service where there are doubts about the legal tenure of property and where the dwelling is not considered to be a permanent construction. Many low-income households fall into one or both of these categories. The consequence can be getting a connection illegally or through a third party, often a 'slum landlord or lady' who has a legal connection and charges exorbitant rates to tenants for their supply (Annecke and Endelli 2006).[10] Illegal connections are a problem for both utilities and households. For the utilities, illegal connections mean unpredictable loads, damage to the infrastructure and a loss of revenue. For example, in Bahia State, Brazil, it was estimated that, at the beginning of this century, around 11% of the electricity distributed was diverted to illegal connections (Andrade 2004). For the consumer, illegal connections are unreliable and dangerous with connections being made without the usual safety mechanisms.

---

[10] Not all electricity used through illegal connections is with the explicit compliance of the end-user. Research in Ghana found poor urban households were the victims of deception with unscrupulous fellow residents making illegal connections but collecting the payments on the pretence of making the payment to the utility (Bannister 2002).

Unfortunately many electricity utilities tend to see poor households as a problem not as potential clients and when faced with substantial loss of revenue through illegal connections generally resort to removing the connections instead of developing innovative ways to increase their client base. (See section 24.6 for further discussion on this point.)

However, there are examples of better practice. In Nigeria, the utility found that where meters are installed, payments are not made regularly so in response introduced slot meters in some areas. To its credit, the utility has not charged a higher tariff for slot meters, as is commonly the case (Friends of the Environment 2005). In Argentina, the utility Edenor has introduced a prepayment system using a voucher system with a printed number which has to be keyed into a pad on the meter – similar to pay-as-you-go mobile phones. The utility ensures easy access to buying the vouchers by using machines similar to automatic teller machines (ATMs) installed in convenience stores. Ninety per cent of respondents in a survey of 150 residents of the district in Buenos Aires where the meter trials took place reported that they appreciated the utilities' policy of agreeing to convert illegal connections into legal ones with the provision of a pre-paid meter (Annecke and Endelli 2006).

In the petroleum sector, improved availability of supplies seems to be linked to energy market liberalization. Kerosene appears to be widely available from a range of outlets (service stations, shops and dedicated kerosene pumps). A similar picture is also reported for LPG. In Kenya, for example, the distribution system has expanded and a variety of outlets sells and refills cylinders. The market has also responded to the purchasing patterns of low income households. Small cylinders are available and saving and loan schemes, which enable access to the cylinder and gas, are in operation (SPARKNET 2004). Kerosene is popular because it can be bought on a daily basis in small quantities enabling poor households to balance daily needs with daily income which gives kerosene the advantage over LPG. There are schemes which help low income households overcome the high initial costs but Sinha reports that in India the refill cost of the cylinder in itself is a barrier to sustained access and low income households resort to selling their connections once the first cylinder is empty (Sinha 2007, forthcoming).

LPG providers appear to have a different client-oriented service approach to the electricity utilities. In the Philippines suppliers deliver cylinders by truck and an order can easily be placed by telephone. The sector also appears to be responding to consumer demand, for example supplying different sizes of cylinder. However, it is not all good news. A negative fallout from the proliferation of LPG retail outlets in the Philippines has led to tampering with LPG cylinders by some retailers. However, the government has been quick to react to maintain consumer confidence that cylinders provide the right quantity and good quality of LPG. Routine and complaint-related inspections and investigation of LPG establishments and gasoline stations are currently being pursued to protect the public against illegal and unfair practices (Clancy *et al.* 2006). In Nigeria, urban poor households which use LPG purchase it from petrol stations or private dealers which can be very expensive. In many cases, a cylinder passes through up to three levels of middle men before it gets to the end-user, at each level there is a markup (Clancy *et al.* 2006). This type of informal sector energy sector service company (ESCO) exists in low income urban areas to fill a market gap that the main suppliers leave. The ESCOs can be very inventive in their mode of delivery. In Lima, Peru, shopping trolleys, bicycles, small motorbikes and pick-up trucks are reported as transport vehicles (Wakelin *et al.* 2003). The informal sector on the one hand is prepared to go to inaccessible areas (for example, where roads are of soft sand or up steep slopes); on the other hand, since they are unregulated, improper business practices can creep in. Wakelin *et al.* (2003) report of informal kerosene distributors in Lima putting a false bottom in the kerosene tins which they sell at what appears to be a lower price per litre.

24.2.4 Transitions in urban energy use

Bringing improvements to urban poor livelihoods through energy services is not only a question of increased access to energy but also the type of energy carrier they use. Are the urban poor making the transition to better quality, modern fuels? Such a transition results in positive outcomes for the household in terms of health gains (less indoor air pollution), time savings (from more convenient fuels) and potential cost savings for a particular activity (more efficient fuels).[11]

The available evidence shows that, since the 1980s, there have been some signs of fuel switching with poor urban households moving from woodfuels. A comprehensive survey in Hyderabad, India, on urban energy use found that a substantial shift had occurred in household energy use: over a 20-year period poor households have moved from wood to kerosene and LPG, while the middle classes had moved to LPG reducing their competition with the poor for kerosene (ESMAP 1999). In Addis Ababa, Ethiopia, wood accounted for 13% of the total energy used in 2000, compared with 70% in 1980 (Shanko and Rouse 2005). In both of these cities switching to modern fuels has occurred as a direct result of government policies introduced to reduce environmental impacts of clearing peri-urban forests to meet the demands of urban fuel supply. In Lima, Peru, almost 100% of the households have made the transition to LPG and kerosene for cooking. A significant factor in this switch was identified as a competitive private distribution network which increased the number of fuel distribution outlets thereby making LPG and kerosene more widely available (ITDG 1998). The UNDP/ESMAP study found only significant switching for the poorest households (the first decile) in Brazil and South Africa where in both cases around 75% of households had completely switched to modern energy carriers. In the other countries in the study there was little switching to LPG by poor households.

However, there are signs that these gains in terms of access to high quality fuel are also losses in terms of household spending power. The opening years of the twenty-first century have been accompanied by a significant increase in the traded price of oil on international markets. At the same time government policies for energy sector reform and energy market liberalization have often been matched by price increases. An example of the types of price rises can be seen in Table 24.1 which gives data for the petroleum sector in Nigeria where the government is pursuing a privatization policy. In Brazil, the cost of LPG cylinders has increased 550% in the last ten years which is three times the rate of inflation (Winrock International 2005). Electricity prices have also risen. For example, in the Philippines, the basic household tariff rose from 3.4329 pesos/kWh in June 2003 to 4.8970 pesos/kWh in 2006 (Clancy *et al.* 2006).

What are the consequences of these energy price rises? In their three country study, Clancy and her co-workers found that fuel switching to lower quality fuels depends on the availability of alternatives (Clancy *et al.* 2006). For example, in Salvador (Brazil) the supply of charcoal has virtually disappeared and environmental degradation has reduced fuelwood availability so people continue to use LPG but adopt other conservation measures. A study in Ghana and Indonesia reported a reduction in the number of cooked meals. In Ghana there has been a switch to cheaper fuels (wood and low quality charcoal) and in Indonesia a reduction in the use of kerosene and electricity as a result of energy price increases (Future Energy Solutions 2002). In the Philippines poor urban households are

---

[11] It is possible that a household's *total* energy payments will not go down. The household may decide to use the particular equipment more often (for example, cook more meals) or invest the savings in bringing a quality of life improvement, for example, buy a fan or TV.

Table 24.1. Prices per litre of petroleum products in Nigeria (1990–2004).

Products	1990	1991	1993	1994	1998	2000	2001	2002	2003	2004
Gasoline	0.51	0.6	3.25	11	20	22	42.50	32/34	40.23	42.80
Diesel	0.35	0.5	3.0	9	19	8	42.00	32	38/39	40.50
Kerosene	0.15	0.4	2.75	6	17	19	32.00	32	32/53	41.25
Fuel oil	0.30	0.5	2.75	9	12.40	230	230	230	275	275

Source: NNPC (Nigeria National Petroleum Corporation) (quoted in (Maduka 2004)) All prices are in naira $1=130 naira; exchange rate January 2004).

reported to be switching away from LPG and kerosene to using wood and charcoal to reduce energy costs. A possible explanation for this shift is that many poor urban households in the Philippines still have rural plots of land where they have access to fuelwood and charcoal (Clancy *et al.* 2006).

Higher electricity tariffs cause considerable resentment and hardship among low income households which can lead to increased levels of electricity theft.

24.2.5 Transitions brought about by urban life

Urban lifestyles differ in many respects to rural ones. One significant difference is that urban dwellers are more dependent than rural dwellers on third parties for the supply of essential services, including water and fuel.

Many urban residents purchase food in restaurants. In Thailand, a survey at the end of the 1990s found that 20% of households were eating most or all of their meals away from their homes or bring cooked food home from their workplace (ITDG 1998). Is this an alternative to purchasing cooking fuels, a time saving strategy, a sign of absent cooking facilities or an improved lifestyle associated with urban living? Paying for cooked food is not confined to higher income groups. The urban poor also eat outside the home often at informal road-side restaurants. In part this phenomenon is due to the significant number of men living without partners in urban areas. However, in the Philippines a small number of households reported switching to buying cooked food in response to the increase in energy prices,[12] while in Nigeria, the need for women to work long hours outside the household was cited as the reason for the switch (Clancy *et al.* 2006).

The use of informal food vendors has significant implications for the health of the urban poor. The vendors are often located along road sides so their food is not prepared under hygienic conditions. Since they are also operating often without licences they will choose sites not supported by essential services such as clean water and garbage disposal. The Department of Health in the Philippines reports that seven out of ten food handlers are infected with parasites (*Philippine Daily Inquirer*, 7 April 2006 quoted in (Lumampao *et al.* 2006)).

24.3 Energy, Urban Enterprises and Poverty

The urban poor depend largely on small-scale enterprises for generating income. Many of these enterprises are located in the informal sector. In Kenya, the informal sector in the

---

[12] 3.7% in a sample size of 1000 households in Metropolitan Manila.

second part of the 1990s employed around 2.2 million people – approximately 65% of the urban employed (Karakezi and Majoro 2002). Street food vendors, small-scale manufacturing and repair services are common types of businesses. The informal sector forms an important part of coping strategies particularly for women. UN statistics show that the informal sector is a larger source of employment for women than for men (cited in (BRIDGE 2001)).

The number of informal sector enterprises is on the increase. For example, in the Philippines, a large number of factories and small businesses closed due to the financial crises in Asia, but this has been accompanied by a five-fold increase in street food vendors over the past three years (Lumampao). These enterprises are often using process heat and since they operate in commercial markets they are vulnerable to shocks from energy price rises. The Intermediate Technology Development Group (ITDG), in Bangladesh, examined the role of energy in informal sector businesses, in particular the cooking activities of street food vendors. They found that any improvement in household energy would improve the livelihoods of street food vendors since the production of food for sale is a family-based activity and a large part of the food production takes place in the household (Tedd 2001). These vendors are also serving urban low income groups, possibly with their main meal of the day. Failure to store food at correct temperatures, or to sufficiently reheat food, can have a significant impact on the health of customers. Box 24.1 illustrates a project which aims to improve the hygiene standards of street food vendors by paying particular attention to energy efficiency.

Based on an extensive review of the literature, Meadows and her co-workers considered that the linkages between modern energy and micro-enterprises were:

1. Modern energy can, but does not necessarily, affect the emergence, development, productivity and efficiency of micro-enterprises. However, research in South Africa has shown how the introduction of electricity into urban low income areas can be associated with a diversification in the types of services available to poor people (see Table 24.2)
2. While the lack of access to modern energy is often characterized as a barrier to micro-enterprise development, removing this barrier (through, for example, area electrification) does not necessarily result in the desired changes. Rather, modern energy should be viewed as one of a suite of critical enabling factors that act individually and/or in concert to create a suitable environment in which micro-enterprises can operate.
3. The linkages between modern energy and micro-enterprises, and the effects of the former on the latter, can have a gender-specific dimension (see section 24.4) (Meadows *et al.* 2003).

Most of the literature reviewed by Meadows would appear to be linked to the effects of rural electrification on enterprises in rural areas. It is not clear, since there appears to be a lack of empirical evidence, as to whether small-scale urban enterprises have their own specific characteristics, challenges and better access to modern energy services than rural enterprises. Indeed it is difficult to provide specific data on informal sector energy use since many urban enterprises are located in the household or its backyard/compound and use household energy resources for their processes. Table 24.3 gives a list of household enterprises taken from official statistics in Botswana. The list also illustrates the range of small-scale enterprises found in urban areas.

Urban enterprises use both traditional and modern energy carriers. Electricity use is linked to size and enterprise status. The larger the enterprise the more likely it is that electricity, where available, will be used. Formal sector enterprises are more likely than

Box 24.1.
Energy efficiency improvements increase profits in informal sector enterprise

Josephine Cruz from Manila in the Philippines is not untypical of many women from low income households around the world. Circumstances propel them into the role of breadwinner of the family. In Josephine's case her husband has been unemployed since early 2006. So what to do to feed the family? For many women they turn to the skills they know best: those based on households chores such as food preparation and sewing.

In Josephine's case, she started selling fish balls and *kikiam* using the capital saved from her husband's income. This income generating activity is in the informal sector and the inputs have to be bought on a daily basis. From their savings they could afford the daily 400 pesos to buy the first inputs but all the hardware – the cart which forms production and the selling point, the traditional charcoal stoves and the utensils – had to be borrowed from relatives. The level of investment in equipment was beyond their reach and banks will not lend to informal sector entrepreneurs. With many competitors and little profit, Josephine began to lose heart until she tried selling potato chips. With the same amount of capital, she now realizes a daily profit of up to 300 pesos.

In January 2007, Josephine participated in a project funded by UNDP and run by the NGO Approtech Asia: 'Ambulant Food Vendors: Energy-Efficient Stoves and Hygienic, Healthy Food – A Pilot Project for the Urban Entrepreneurial Poor in the Philippines'. Through participating in this project Josephine has improved her business skills, such as record keeping and auditing, as well as gaining the very important Health Card from the City Health Office which has given her a clean bill of health from transmittable diseases such as tuberculosis. Such a card helps boost her business since customers know that they can enjoy her food free from the risk of infection. The two traditional charcoal stoves have been replaced by ones based on a design from Thailand and improved in Cambodia. These improved charcoal stoves are reported by users to show 50 to 70% fuel savings over the traditional charcoal stoves. In Josephine's case these fuel savings have enabled her to pay herself the minimum daily wage rate of 250 pesos per day.

So successful is Josephine's business that she was able after four months' saving from her income to acquire her own equipment. Her husband constructed a new cart and returned the borrowed one along with the utensils. Seven months after she joined the project, with the assistance from Approtech Asia for the down payment, she has been able to rent a more hygienic food stall near a large market. The annual rent is 6000 pesos although she is able to spread her payments to match her daily cash flow. The rent is paid in trances of 200 pesos per day. When the income of the eatery becomes stable, Josephine and her husband are planning to hire an employee to deliver packed meals ordered through cellular phone text messages by other stall owners and market vendors.

1USD = 42 pesos

Source: Approtech Asia

Table 24.2.   Changes in the commercial activities before and after
electrification in Elandskraal, South Africa.

Enterprises before electrification	Enterprises after electrification
Manufacturing	*Manufacturing*
Welding	Furniture
	Tailors
	Candle makers
	Brick makers
	Bakeries
Retailing	*Retailing*
Beer halls	Electrical goods
Wood and charcoal supply	Electrical hardware
General trading	Butcheries
Services	*Services*
Butcheries	Upholsterers
Garages	Shoe maker
Dry cleaning	Tailors
	Disco
	Panel beater
	Radio/TV repair
	Builders
	Hairdressers

Source: (Fauira 1994) quoted in (Karakezi *et al.* 2006).

Table 24.3.   Household enterprises in Botswana.

Activities	No. of enterprises	Activities	No. of enterprises
Property rentals	6791	Gathering/catching fish, etc.	284
Hawker/vendors	3864	Vehicle repair/panel beating	248
Brewing/selling of beer	2908	Haircutting/dressing	208
Making/selling of clothes	2274	Making/selling furniture	208
Selling crops, fruits, vegetables	620	Selling poultry/livestock	151
Cooking/selling of food	587	Selling milk/eggs, other livestock products	105
Taxi service	430	Making handicraft	62
Building/plumbing	391	Blacksmith/tinsmithing	29
General dealers	344		
Other repairs	286		

Source: (Central Statistics Office 1995) quoted in (Karauezi 2002).

informal sector ones to have access to electricity through their own connection. However, informal sector entrepreneurs can be very creative and many make use at night of community street lighting or security lighting to operate their business. Karekezi and Majoro report rather worryingly that in Zimbabwe, during the late 1990s, small and medium enterprises using the low voltage commercial tariff were being charged at a rate which resulted in them paying a higher tariff than large consumers and households (Karakezi and Majoro 2002). Not only does this practice raise issues of equity, but also concerns about the consequences that this can have, such as encouraging illegal connections or enterprises using the household tariff. The latter can act as a break on business expansion since household loads may not be sufficient to take more equipment or the type of equipment requiring a three phase supply. There are also concerns about the quality of supply. While market liberalization has seen a decrease in the number of power cuts, brownouts and power surges which damage equipment continue to be a problem. Given the linkage between household energy and enterprises, this equipment damage has a double impact on household income: the cost of replacement of broken equipment and the loss of earnings until the equipment is replaced. The uncertainty about electricity causes entrepreneurs to purchase stand-by generators and urban small enterprises often use diesel or petrol generators for electricity or mechanical power generation (ITDG 1998).

Many enterprises use standard household equipment for operating their businesses. For example, food processing enterprises use blenders, refrigerators, freezers, and stoves while hair salons use hair dryers, irons and electric razors. In Brazil, this has meant that enterprises have been able to benefit from packages designed to benefit the family, for example the LPG voucher which entitles them to a small free cylinder of LPG (Winrock International 2005).

Urban enterprises have been hit by energy price rises which are not only in the modern energy sector but have also hit the traditional woodfuels. In Nigeria, in the last few years, there has been a significant price increase in fuelwood because the cost of transporting the wood has gone up. A 10 kg bundle has gone up from 80 naira (approximately 35p) to 100 naira (approximately 44p) in 2005. Kerosene is suffering a double burden – the retail price reflects not only the increase in the wholesale price but also the transport costs. The cost of gasoline has risen from 42 naira/litre (approximately 17p/litre) in December 2004 to 65 naira/litre (approximately 26p/litre) in August 2005 (Friends of the Environment 2005).

Energy services are certainly important inputs in enterprise viability although they are not the only factors that influence viability, for example transport systems and communications are also important. Paying for energy services can use a significant portion of an enterprise's profits. A survey in Kenya reported that 60% of enterprises spent 10 to 50% of their profits on energy (Karakezi and Majoro 2002).

In their three country study, Clancy and her co-workers found that transport costs appeared as a major concern for all the respondents in their survey. The costs of transport for raw materials, goods and people were found to be rising significantly since urban areas rely on diesel- or petrol-based transport systems (Clancy *et al.* 2006). Informal sector enterprises are often serving local markets where customers are on low incomes so passing increased costs to the customer is not an option readily undertaken. Sometimes informal sector entrepreneurs have to transport their goods to other parts of the city, for example food vendors prepare food at home and take the bus to office or factory areas to sell their products. Certainly, entrepreneurs are reported as reducing the numbers of such journeys as a response to the increase in public transport costs (Clancy *et al.* 2006).

Energy services can also be provided by the poor. For example, fuelwood harvesting and charcoal production/distribution has become an important source of income for urban poor in Zimbabwe, Mozambique, Zambia and Malawi (Mika). The numbers involved can be quite significant. In Addis Ababa, in a single market day in 1984, 42000 suppliers were counted transporting traditional fuels into the city (Shanko and Rouse 2005). However, these numbers have shown a dramatic decline. By 2001, the number of suppliers had reached 3500. This decline has been attributed to the adoption of policies to promote fuel switching from wood to kerosene and electricity as a response to rapidly depleting forests. While the environmental benefits of such policies are not to be denied, the socioeconomic impacts are less certain. This is an under-researched area. Clearly in the case of Addis Ababa, a large number of people involved in supplying traditional fuels to the city have lost their livelihood. Some new jobs will have been created in the modern fuels sector. Shanko and Rouse estimate that around 2000 jobs have been created by small businesses manufacturing electric, kerosene and improved biomass stoves (Shanko and Rouse 2005). Kerosene is usually sold through existing petrol stations so the opportunities here are not likely to be at the level to match those displaced in the wood supply chain. Other parts of the energy sector require the sorts of levels of education and skills those working in the informal sector rarely display. Policymakers need to be conscious of the full consequences of their policies, in particular that well-meaning policies in one sector do not result in potentially more serious problems manifesting themselves elsewhere in society.

24.4 Gender, Energy and Urban Livelihoods

Whether the men and women in urban areas benefit equally from improving access to energy is not clear. A desk study for DFID found that the urban gender-energy-poverty nexus is under-researched (Clancy *et al.* 2002) and there is a lack of empirical data beyond health impacts (Barnett 2001).

Evidence would suggest that household energy in urban areas primarily remains a woman's responsibility. Based on evidence from urban livelihoods research in other sectors than energy (Beall and Kanji 1999), this responsibility can be extended to the provision of services for the community which in the case of energy services would include electricity connections. It is not unreasonable to assume that women will face increased stress as a consequence of responding to increases in energy prices. There is a worrying sign that households are economizing on cooking as a result of higher energy costs (22% of the 1000 households interviewed in a survey in Metropolitan Manila (Clancy *et al.* 2006)).

It would appear that there is a gender division in the types of enterprises owned and operated by men and women. Women's enterprises tend to be home based and use process heat. Where this is the case, women might benefit by access to clean modern fuels if they are substituting for the use of traditional biomass in confined spaces. There is evidence that women are benefiting from access to electricity which increases the range of services they are able to offer in a more commercial market that exists in urban areas. In the Philippines, many women's enterprises provide services such as clothes washing and ironing which are using electric appliances,[13] and in Brazil hairdressing and clothes making also benefit (Clancy *et al.* 2006). Women value electrical appliances for the reduction in drudgery they bring.

The study by ITDG in Bangladesh referred to above explored gender aspects in relation to income generation. They found that women are able to control the production process and hence keep the profits generated, which it was concluded (although no evidence was provided to support this statement) would lead to their empowerment and recognized the need to involve women in technology selection due to their key role in food preparation (Tedd 2001). However, since gender is culturally determined it would be wise not to generalize this finding. Meikle found in Tanzania, in terms of decision-making about energy, that:

- Men are most likely to make decisions about expenditure on household appliances.
- Women are most likely to decide on the type of cooking stove to be used unless the stove costs more than they are allowed to spend in which case they defer to their husbands (Meikle 2006).

---

[13] Focus group discussion.

Since many women's enterprises are home based, decisions about energy and appliances affect the viability of enterprises. As was stated in section 24.2 when energy has to be purchased, men enter the decision-making process, not only in relation to appliances but also in relation to other factors that influence working conditions in the kitchen. Clancy *et al.* (2006) found that there was a correlation between shared decision-making and national income level. In the Philippines and Brazil, both middle income countries in the United Nations Human Development Index (UNHDI), women were more likely to report joint decision-making between the man and woman whereas in Nigeria, a low income country in the UNHDI, it is still predominantly the man who makes the decisions about energy and appliances (226 out of 311 households responding).

There have been concerns expressed about access to electric lighting leading to longer working days, thus adding to women's burdens rather than reducing them. Disturbing data have been found in Nigeria about the existing length of the working day for women. Women working in pottery and fish smoking worked 12 hours a day, while those involved in cassava processing work 14 hours, six days a week (Friends of the Environment 2005). There is real cause for concern how little rest these women are getting especially since the work for their enterprises is in addition to household chores. Therefore, access to modern energy should be focused on reducing women's work day rather than increasing it. Women's businesses are often owner-managed; as a result there is often no one in the family to take over from the owner due to lack of skills or time. Women therefore work when they are sick which raises concerns not only in general for the burden it places on them but also in particular for those enterprises involved in food processing.

24.5 Energy Use, the Urban Poor and the Environment

The environmental impacts of the extraction, production and use of energy are well known. These impacts manifest themselves at the local level (for example, poor air quality indoors and outdoors), at the regional level (for example, acid rain) and at the global level (climate change). All income groups are affected to differing degrees by the regional and global environmental impacts, although the poor are in a much weaker position than the wealthy to respond to those impacts. At the local level, all urban dwellers are affected by air pollution caused by transport and industry but again the degree of exposure has a strong linkage to income levels. However, it is mainly the poor who bear the burden of indoor air pollution. Lack of space prevents a detailed review of all of these levels so this section focuses primarily on those issues which are particularly relevant for the urban poor: the localized impacts that have negative consequences for their health. Also reference is made to a particular part of the energy chain, where there are strong linkages between energy use by the urban poor and the environment: peri-urban charcoal production.

The urban poor are exposed to pollution as a consequence of energy use from three perspectives: (1) direct exposure to indoor air pollution from using poor quality fuels in confined spaces; (2) direct exposure in their places of work (there may be some overlap with the first condition due to location of many informal sector businesses in the household); and (3) indirect exposure due to inadequate urban planning.

In poor urban homes, biomass fuels and coal are typically burnt in open fires or poorly functioning stoves, often indoors due to a lack of space around the dwelling with inadequate ventilation for the smoke. Poor conversion efficiency leads to very high levels of pollution in the homes, with especially women and young children exposed on a daily basis. Smoke from these fuels contains many pollutants which together are known to be capable of irritating the airways and lungs, reducing the resistance to infection, and increasing the risk of cancer, particularly in women due to their greater time of exposure to the pollution (Bruce undated). Evidence has begun to emerge which suggests that indoor air pollution (IAP) in developing countries may also increase the risk of other important child health problems, such as low birth weight, perinatal mortality (stillbirths and deaths in the first week of life), asthma, and middle ear infection in children (Bruce *et al.* 2000).

Similar fears were also expressed about the environmental health impacts on women and children of the shift from fuelwood to kerosene in India which has been accompanied by a shift to cooking indoors, in poorly ventilated rooms (ESMAP 1999; Dasgupta 1999).

However, even if cooking is done outdoors with solid fuels, the level of exposure to pollutants is likely to be damaging to health, but probably not as great as for use indoors with the same fuel for the same time exposure. Other family members are more likely to benefit than the cook if the stove is located outdoors (Smith).

As was pointed out in section 24.3, the urban poor tend to find employment in the informal sector. Many of the activities are using process heat generated by unprocessed biofuels and residual oil products. There appears to be little research into the consequences of occupational exposure to local-level pollution in informal sector enterprises. Some level of effect could be estimated by extrapolating from findings related to kitchen indoor air pollution. An added hazard is created when entrepreneurs use discarded oil drums as make-shift combustion devices. Maduka (2006) describes the use of fuelwood in old oil drums for fish smoking in urban communities in Lagos, Nigeria, where women stand smoking fish for seven hours a day exposed not only to the wood smoke but also other pollutants of unknown toxicity from residual oil. Many informal retail outfits are located along the side of roads where the entrepreneur is exposed to exhaust fumes from transport for several hours a day.

Inadequate urban planning also leads to the poor carrying a greater environmental burden than people in higher income groups. Many poor urban settlements are also illegal which means that utilities are often not prepared to provide services.[14] Water quality which is strongly linked to health is a considerable concern for poor urban households. Households often have no internal piped water. In the four communities surveyed in Nigeria less than 20% of households obtained their water from a tap internal to the house or compound (Clancy *et al.* 2006). Where there are services they are subject to disruption due to electricity cuts. Supply disruptions can mean long periods waiting in line for the supply to be restored. Disruptions are also linked to quality problems. During these periods, those who can afford it buy bottled water.[15] Others will boil water, although this is both expensive in terms of energy and time consuming.

The location of poor settlements can lead residents to high levels of air pollution caused by urban transport and industry. A study of the environmental impacts of urban energy use found that the urban poor suffer disproportionately from the impacts of air pollution because they tend to live in areas with higher concentrations of roads and industrial areas which are residential areas higher income groups can avoid (Watkiss undated). However, where solid fuels are used for space heating, the indoor air pollution from this source can be of greater significance than from air traffic pollution for poor people's health.

The design of and the type of construction materials used in low income housing can lead to homes which are either too cold or too hot, leading to a demand for space heating or cooling which could be avoided by greater attention to the design of low cost dwellings.

---

[14] Although there are exceptions, for example in India, slum areas can be totally electrified (ITDG 1998).
[15] In 2002, water in Keularah Kramat, Jakarta, Indonesia, was reported as costing Rp700 per litre (approximately US$0.09 per litre), which was more expensive than kerosene (Rp600 per litre; approximately US$0.08 per litre) (Meikle and Bannister 2003).

Also the layout of informal settlements can make the physical provision of energy services (e.g. electricity lines or LPG cylinder delivery) difficult.

As was indicated in section 24.2, woodfuels continue to play an important part in the fuel mix in urban households. Charcoal production to serve urban markets is considered to contribute to peri-urban deforestation. A positive impact on deforestation in peri-urban and rural areas has been achieved in India by a shift from charcoal to kerosene (ESMAP 1999), although this might have a negative effect on peri-urban and rural livelihoods through loss of income from wood sales (see section 24.4). However, in Africa, Leach and Mearns (1988) considered that peri-urban deforestation is caused by land clearance for agriculture to produce crops for growing urban markets rather than specifically for charcoal production. Any charcoal is a by-product of this land clearance. There appear to be little empirical data on the environmental effects of a decline in urban charcoal use, for example in Senegal with the butane gas program introduced to reduce the pressure on natural forests (Watkiss undated). Has the cutting of wood declined in total as a consequence in the fall in urban demand or has this merely freed up more wood for rural household use?

Alternatively, has the switching back to wood and charcoal linked to energy price increases had any negative environmental impacts in peri-urban areas? The shortage of coal, high cost and intermittent availability of paraffin and electricity in Zimbabwe has seen fuelwood emerging as the cheapest and most easily accessible source of energy, which is thought to be contributing to massive deforestation around cities (Mika).

24.6 Energy Use Improving the Livelihoods of the Urban Poor

So far we have seen that the urban poor use a mix of energy sources to meet their household and income generating needs. There had been signs of a transition to modern fuels, particularly electricity for lighting and entertainment, across all countries and the use of LPG appears to be linked to the national economic status. However, in many places these transitions have been halted by liberalization of the energy markets and the rise in the price of crude oil in international markets. These changes have had the greatest impact on the poorest which means that women suffer the most hardship since women form the majority of people categorized as poor. Access to energy by the urban poor remains precarious both because incomes are volatile and location affects supply availability. In the last section, we also saw that the urban poor carry additional environmental burdens related to energy use that higher income groups can avoid through their purchasing power. These burdens are: indoor air pollution from cooking fuels (within unknown impacts from occupational exposure to emissions from fuel use) and ambient air pollution from vehicle exhausts and industrial activities. Any discussion about addressing the environmental impacts from fuel use generally considers renewable energy sources as a cleaner option to fossil fuels. However, renewable energy technologies, as small-scale decentralized energy systems, have not been so actively promoted in urban areas as they have in rural areas. In this section, we will review what the potential options for renewable energy systems in urban areas, why they have not been seen as a major contender for urban households (in general and not only for the poor) and what measures can be used to overcome any barriers.

24.6.1 *Which renewable energy systems and for which applications?*

There are two main questions which have to be answered in relation to improving the quality of life for the urban poor through renewable energy use: which urban applications can be met by renewable energy systems and at what level (decentralized or centralized)?

There are three general categories of energy needs in urban areas: household, productive and community (hospitals, clinics, schools, community centres, religious institutions, barracks, and prisons). Each category uses process heat, motive power (usually for transport) and electricity (batteries and grid). The renewable energy options which could meet those needs in urban areas are biomass and solar (we will assume that the urban poor would also benefit from large-scale electricity generation from renewable energy sources such as hydro and wind). Renewable energy technologies can be used at both the decentralized and centralized levels. First, we will look at the options for the decentralized level. Passive solar could be used in building design, while solar energy could be used for water heating and solar home systems could meet some electricity needs but not all including a very important end-use for poor people: clean, ironed clothes.[16] Electricity systems that displace batteries not only have environmental benefits from the lack of adequate disposal facilities for batteries but also could be a potentially cheaper way of powering radios and TVs.[17] Solar energy can also be used for cooking but not so easily for process heat in commercial applications where output size is a constraint. The urban poor continue to use woodfuels for cooking and space heating. Here the major challenge is to ensure that these fuels are produced on a sustainable basis as well as improving conversion efficiency and cleanliness of household and commercial equipment for cost and health improvements. The main advantages of solid fuels biomass cooking over solar cooking is that it does not require a change in cooking practice or the need for new utensils.

At the centralized level, anaerobic digestion for treating urban wastes and for communal latrines could provide a general environmental improvement (particularly since many low income households are located close to rubbish tips) as well as generating a useful gas for process heat at levels useful to industry as well as other users. The use of biogas could contribute to reducing air pollution produced by industry.

However, the main cause of air pollution is urban transport. Increased energy efficiency through better vehicle maintenance could make significant improvements. Fuel substitution for diesel with cleaner burning fuel such as compressed natural gas (CNG) can be effective in the short term as happened in New Delhi, India, where public transport has all been converted to run on CNG with immediate improvements in urban air quality. Biomass liquid fuels also offer medium- to long-term alternatives to fossil fuels and the Brazil fuel alcohol program is a well-known technical success story from the south. Vegetable oils as diesel substitute are technically feasible but still await the scaling up to the level of fuel alcohol in Brazil. Many countries in the south have the possibility of developing their own biofuels programs which hold the potential for reducing dependency on oil imports and an export commodity. However, there are environmental concerns about fertilizer and water use as well as the possibilities for the displacement of food production. Also we do not have any experience with a significant percentage of transport being run on biodiesel to test impact on urban air quality.

So if renewable energy technologies offer benefits to the urban poor why are they not being taken up in significant numbers? The next section will examine this question.

---

[16] In a major study for the World Bank, having clean, ironed clothes to wear were cited by poor people as one of the indicators of not being poor (Deepa *et al.* 2002).

[17] Barnett points out that although dry cell batteries are widely used in radios and tape players, they are a very expensive way of buying electricity (about US$400 per kWh). Survey data from Uganda in 1996 showed that rural households not connected to the electricity grid used dry cell batteries and were estimated to be spending about US$6 per household per month on them (Barnett 2000).

24.6.2 What are the barriers?

What motivates people to adopt an energy carrier? Understanding the motivations for adopting a new energy carrier and its associated technology are important since it provides an indicator of the appropriate mechanisms that need to be introduced to stimulate energy end-users to make the change. Unfortunately, often something gets in the way of the motivations to switch energy carriers. Identifying these barriers is also important for developing mechanisms for promoting switching energy carriers. The barriers to RETs in urban areas fall into two categories: those related to fuel switching in general and those that are specific to RETs. These barriers exist at the micro- (household/enterprise), meso- (institutional) and macro- (policy) levels. Although much of the debate about the uptake of RETs centres on cost there are also non-financial factors that influence switching between energy carriers. First, we will look at the financial factors as barriers.

A simple economic model for households and small enterprises would assume that energy carriers are economic goods. Modern energy carriers (electricity and gas) can be seen as superior goods and traditional energy carriers (such as wood and charcoal) can be seen as inferior goods. If this is the case then it can be expected that as a household's income increases, it will switch from relying on traditional fuels to modern fuels. By extension, higher income households will make greater use of modern fuels than low income households do. However, Leach and Mearns (1988) found such a model too simplistic. Based on survey work in Dar es Salaam, they concluded that energy carrier price was not the single determining factor in encouraging energy carrier switching, unless the price difference was 'very large'. They considered that there were two driving forces of energy carrier switching:

1. Access to dependable supplies of modern energy carriers in sufficient quantities; and
2. Sufficient income to invest in equipment to use modern energy carriers.

Additional empirical evidence supports Leach and Mearns' findings. The cost of an electricity connection, cost of wiring and high standing charges have hindered the switch to electricity. The costs of LPG cylinder and stove have hindered the switch to LPG (von Molthe *et al.* 2004). Availability of LPG cylinders has acted as another barrier (Hosier and Kipondya 1993) and continued unreliability of electricity encourages the use of stand-alone diesel generators (Schutyser 2003).

Significantly, Leach and Mearns (1988) identified having to buy new *equipment* for the new energy carrier, for example the need to buy a different set of cooking utensils when switching from charcoal to solar cookers, as an important barrier to switching. Also the nature of a stove can be a deterrent to switching energy carrier. In Niger, urban households were found to still prefer wood to kerosene despite the latter being cheaper because the available stove for kerosene was not suitable for the long slow cooking of staples (Leach and Mearns 1988). Kerosene was used for rapid cooking, particularly boiling water for tea.

We would argue that gender-based intra-household negotiations also play a significant role in switching and these are not taken into account in economic models. A number of researchers (see sections 24.2 and 24.5) have found evidence to show that men make the decisions about purchasing the energy equipment in the household. So it is men as well as women who must be motivated for switching energy carriers.

An often promoted argument from the renewable energy community is that RETs are not operating on a level playing field with fossil fuels, the latter often benefiting from subsidies.[18] How significant are subsidies in influencing energy carrier access? A study of energy services for the urban poor in East and Southern Africa (Mapako and Dube 2002) looked at the impact of subsidies and incentives on urban poor household energy use. Subsidies were found not to be decisive for the affordability of energy by the urban poor, but significantly the removal of subsidies would impact more on the poor than on the non-poor. Other factors such as upfront costs, proximity and availability of energy sources were found to be more decisive in creating barriers to access.

Reddy has argued that lack of reliable income forces poor households to use much higher discount rates than the rich when making decisions about energy forms which means that they think solely in terms of first cost rather than life cycle costs (Reddy 2000). This tends to work against the purchase of non-woody biomass renewable energy technologies since they tend to have higher first costs than fossil fuel and traditional energy carriers.

So what are the non-technical factors which act as a barrier to energy transitions? At the macro- and meso-levels, we find that legislation and the attitude of energy utilities also play a role in energy carrier access. Utilities are often reluctant to provide a service where there are doubts about the legal tenure of property and where the dwelling is not considered to be a permanent construction. Many low income households fall into one or both of these categories. Often energy service providers have a poor perception of low income consumers. When it comes to supplying electricity to poor districts, people in poor urban areas are seen as a problem rather than as (potential) customers. There appears to be a reluctance to *engage* with low income end-users to work towards solutions. Two examples from Brazil are given in Box 24.2 which are illustrative of this point.

Decisions (even well-meaning ones) in one sector can have negative outcomes for the urban poor and their access to energy. For example, in Cairo, buildings need an official certificate to prove they meet certain standards. This measure was introduced in response to buildings collapsing as a result of poor construction. However, a building standards certificate is also required by the utility before it will connect homes to the electricity supply. Many poor people regard this certificate as too expensive (UNDP 1999). Lack of tenure can also make it less likely that utilities will be prepared to offer services to informal settlements. Also lack of tenure means that people on low incomes will be prepared to invest in their dwellings if they have no long-term commitment to their houses and enterprises.

The poor construction of many informal settlements makes them prone to the theft of possessions, including items such as LPG cylinders.

All of the above barriers are not energy carrier specific. Do small-scale RETs have their own specific barriers in *urban* areas? The main dissemination mechanism to date for RETs has been through projects which have been aimed usually at rural areas. The market is not generally well developed. This may in part explain why wealthy households have not taken up small-scale renewable energy technologies in any significant numbers. (See, for example, South Africa where there is a suitable market in terms of numbers of wealthy households using electric water heater and sunshine levels are appropriate for solar water heaters, yet the switch to solar water heating has been insignificant (Sowazi 2005)).

---

[18] A particular argument against subsidies is often presented from an equity perspective, although this can be used to disguise an economic efficiency argument: blanket subsidies on fuels should be removed since middle income and better-off households are considered to reap a disproportionate share of the benefits (see, for example, (Barnes 1995)). However, this is not a universal truth. In urban Zambia, poor urban households have managed to capture the bulk of the kerosene subsidy (87%) (Kalumiana 2002).

Box 24.2.
Utilities fail to engage with low income urban households

In Canabrava (a low income district in Salvador, NE Brazil) a biogas plant has been installed to make use of a near-by rubbish dump. The plant is part of a pilot project producing approximately 75 kW/h of electricity which is enough to support around 100 households. However, the energy generated is used to light the park that was also built on the site rather than nearby households. It is estimated that the dump has the capacity to generate electricity to meet the needs of up to 50 000 families, at low cost, for about 18 to 20 years. Meanwhile the community feel resentful that they have not been consulted or benefited from a 'resource' that has for many years been seen as the cause of a number of problems for them (for example, health).

It is not unusual for people to have to pay for the installation of their own meter. In Brazil, the utility (COELBA) claims to run a monthly instalment plan to pay for meters; however, there appears to be little effort to publicize the scheme. In a survey of five hundred households in two low income districts in Salvador (NE Brazil) none of the respondents knew of the scheme.

Source: Winrock International, 2005.

There are practical considerations that make renewable energy systems less attractive in urban areas. For example, lack of space surrounding dwellings can mean no room for installing a household biogas digester, or shading from surrounding buildings makes solar water heaters or PV panels inoperable. On the other hand residents in high-rise buildings complain of reflections from PV panels installed on top of lower-level buildings. The question of theft or vandalism in locations of greater anonymity could also be a barrier. For solar cookers, unless there is a secure compound or flat roof where a cooker can be left there is a need for continuous supervision which reduces the cooker's convenience. Low income households do not have paid help to operate or supervise the cooker. For food processing enterprises, solar cookers do not come in sizes capable of dealing with larger quantities of food.

A pilot project in South Africa which tried to promote solar water heaters in low income households found that high initial costs and lack of awareness about the technology and its benefits were significant barriers to uptake of the heaters (ITDG 1998). The insubstantial nature of informal sector housing, as mentioned above, is also problematic in terms of whether or not the structure is strong enough to support solar technologies. Also the lack of security of tenure of informal sector housing makes people reluctant to invest in equipment that cannot be easily moved to another location.

What we find in urban households in terms of an energy transition is that there is no simple switch from one energy carrier to another to meet a certain need based on cost alone. At the household level more complex factors than a simple model of adopting clean efficient energy sources enabled by higher incomes are at play in what motivates the energy transition in households. Households retain the capacity to use a mixture of energy carriers for different needs and switch between carriers as circumstances dictate. Management decisions are made balancing preferences and habits with flexibility (influenced by access and availability) and time constraints. So it is not enough to focus on cost

alone. So what can be done to increase access to RETs? Some suggestions drawing on practical experience are described in the next section.

24.6.3 *How can barriers to RETs use by the urban poor be overcome?*

In the last section we examined the barriers to RETs take-up in low income urban areas.
What can be done to overcome these barriers? Some useful indicators can be found from
an analysis of how the transition from charcoal to LPG in urban areas of Senegal was managed. The key instruments identified were:

- Establishment of an effective and reliable supply system.
- Adoption of technology appropriate to local needs.
- Introduction and enforcement of regulations to discourage deforestation.
- Adoption of appropriate pricing and taxation policies.
- Provision of attractive incentives for distributors and consumers.
- Mounting effective information and awareness-raising campaigns (von Molthe *et al.*
  2004).

The basic barrier is one of availability not only of the conversion technologies and the
equipment to use with them (for example, switching from kerosene to electric lighting
needs totally different lamps) but also the skills to implement and promote the technology.
It is one thing to advocate the use of biogas with its obvious environmental benefits for
poor urban households but space consideration makes community systems a more likely
option. However, a community-scale biogas digester uses a different type of digester than
the one usually used in individual rural households and it requires different types of management structures. Consumers also need education in the use of piped gas. Such skills
need to be locally available. Local government bodies responsible for waste treatment
need to invest in developing these skills or they should allow the private sector to operate
the service in which case the legality of informal settlements needs to be addressed.

However, communities can also mobilize themselves. In the last section the situation in
Cairo was mentioned where a building standards certificate is also required by the utility
before it will connect homes to the electricity supply. In some areas, communities have
mobilized themselves and through credit associations have been able to extend infrastructure to their homes (UNDP 1999). Indeed, community involvement in projects has been
identified as a key factor of success by creating a sense of ownership and hence responsibility (ITDG 1998). In this regard the utilities have to change their attitude towards low
income consumers. Working with low income communities can become a win-win situation (see Box 24.3). Local governments can be important partners for enabling low income
households to gain access to houses designed with lower energy demands for space heating or cooling (see Box 24.4).

Local availability of supply is a key factor in the uptake of an energy carrier. This is
one of the reasons woodfuels remain popular in poor urban households. The woodfuel
market is dispersed throughout urban areas and the supply is regular. This is the type of
service a RESCO has to match. The informal sector has created a niche market supplying
low income settlements with energy carriers. However, the entrepreneurs who run these
mini-ESCOs would not have the capital to build a stock of RET equipment since they often
buy from suppliers as needed. Large conventional commercial banks would also not lend
them the capital. Larger RESCOs therefore need to develop innovative ways of working
with the entrepreneurs from mini-ESCOs to act as agents for RETs. In the informal sector

---

Box 24.3.
Utilities working with communities

In Salvador, NE Brazil, the utility (COELBA) ran energy conservation campaigns for low income households in 2001. The utility sent conservation teams to households to look for energy conservation opportunities. In most cases, households had very old refrigerators with old rubber seals around their doors that no longer were able to properly seal the doors. The company helped many customers replace these seals thereby benefiting the consumers with lower bills and helping the utility avoid investment in new generating plant.

A pilot program was started by the utility Edenor in 2003 in the suburb of Merlo, Buenos Aires, Argentina, to address the problem of illegal connections. These connections were not only directly affecting the utility in terms of lost revenue but also the network was being damaged and becoming dangerous through overloading. Four and half thousand households who had a history of non-payment and/or illegal connections were offered the chance of having their debt cancelled in return for agreeing to have a pre-payment meter installed. A survey of 150 consumers found 90% of respondents to be happy with the scheme much preferring to have legal connections. The meters also help households with tight budgets monitor their consumption and manage their expenditure on a daily basis.

Source: Winrock International 2005; Annecke and Endelli 2006.

Box 24.4.
Local government working with communities

Cape Town City Council in partnership with the non-profit developmental organization, SouthSouthNorth, has over a three year period developed the Kuyasa Low Income Urban Housing Energy Upgrade Project. This involves retrofitting 2300 low income houses with energy efficient lighting, insulated ceilings and solar water heaters. The project provides numerous additional sustainable development benefits such as improved health, access to energy services, and employment creation. Savings of up to 40% on electricity bills are reported due to the reduced need for artificial heating and cooling.

Source: http://www.reeep.org/

---

gender issues are never far removed. Women face additional resource constraints to men, in particular skills and access to finance. They may therefore need additional support to establish RESCOs (see Box 24.5).

NGOs can also play an important role in educating consumers about RETs. They can have direct contact with the urban poor and hence develop a good understanding of meeting their needs (see Box 24.6).

Box 24.5.
Women as energy entrepreneurs

In Addis Ababa, Ethiopia, in the mid 1990s, ILO supported a project to organize women fuelwood carriers, who lost their livelihoods as a result of government policies to stop deforestation, into an association to provide alternative employment opportunities. Over 100 women were offered training and technical support in alternative income generating schemes. In spite of serious resource constraints that hampered scaling up of its activities, membership nearly doubled within ten years. One of the activities has been to prepare a proposal for the women to be involved in sustainable management of existing fuelwood plantations around Addis Ababa. Without the empowerment of the association, it is unlikely that women would have been considered for such work.

Source: Shanko and Rouse 2005.

Box 24.6.
Low cost but effective water treatment using solar energy

If a low income urban area is fortunate enough to have piped tap water it is not always so fortunate in the quality of the water which can be poor due to the dilapidated infrastructure. Households who can afford it buy bottle water. Others are not so lucky and can spend 35 pesos daily on bottle water. In the Philippines, Approtech Asia has been promoting a simple low cost method for treating tap water to reduce pathogen levels. The method uses transparent plastic soft drink bottles of about 1 to 1.5 litres filled with water and left in direct sunlight for six hours after which the water is considered safe from common pathogens such as *E. coli* and *Salmonella* (although the chemical content is unaffected).

Source: Approtech Asia.

24.7 Conclusions

This chapter has reviewed the available literature which is dedicated to urban livelihoods and energy. The documentary and empirical evidence here is considerably thinner than for rural livelihoods and energy.

The patterns of urban fuel use are dynamic and complex. As in rural households, urban household energy provision is primarily the responsibility of women. Households use a mix of fuels and there are signs that even poor households use modern fuels. All income groups still use considerable amounts of woodfuels. All fuel types can be purchased, but this is not always the case. For urban households energy costs can form a significant part of household budgets, although the amount paid is in part influenced by the cash flow patterns of poor people (small amounts purchased frequently). In this context women-headed households are considered to be in a worse position than men-headed households.

The factors that influence switching to modern energy are extra- and intra-household. Where there are no supply availability problems, the most significant factor for low income households are the high 'entry costs', such as connection fees for electricity, LPG cylinder deposits, renewable energy equipment costs and conversion equipment prices. Extra-household factors include the size of the urban area influencing biomass fuel availability. Legal issues such as tenure of the property can affect the possibility for an electricity connection or LPG delivery. Questions of supply reliability prevent a complete transition from wood or charcoal to modern fuels, as does the availability of appropriate and affordable conversion equipment.

The intra-household factors can be divided into two: the preference for one energy form over another and gender issues. Energy preferences are complex. Taste imparted to particular foodstuffs by wood or charcoal is a common reason for not switching away from these fuels. However, there are other pragmatic reasons such as pots not fitting new stoves and the power output controllability of the stove. Although energy provision is the women's responsibility, decisions related to purchase are still made by men who determine household priorities. As a consequence, both men and women should be the target of any promotional campaigns to encourage households to switch energy sources.

In terms of energy services, the urban poor still appear to be underserved. Oil products (kerosene, LPG and natural gas) are fairing better than grid-based electricity. A possible explanation for this is that these fuels are easier to buy in batches, particularly kerosene which can be bought in small quantities to match cash flow patterns in poor households. This makes kerosene a direct competitor to wood and charcoal which again are available for purchase in small quantities. An additional benefit of kerosene, compared to the other modern fuels is that it has no 'upfront entry cost', with the exception of owning an appropriate stove. The poor buy electricity in batches, in the form of batteries, although the end-use is restricted mainly to entertainment and emergency lighting. Ironically batteries provide electricity at a much higher price per kWh than many grid tariffs. The pressure on governments to remove energy subsidies is likely to be detrimental to poor urban households, in particular gaining access to grid electricity. LPG suppliers do appear to be responding to the need for small cylinders to reduce the replacement cost.

Modern energy carriers are advocated for a number of reasons to displace traditional woodfuels. However, it appears that only in large cities or where there are enforced environmental policies to prevent deforestation are woodfuels being displaced by other energy carriers. For the urban dweller (not only those of low income) woodfuels continue to offer a number of advantages and consumers seem reluctant to totally give them up. It would be a mistake to see the role of RETs to displace woodfuels since woodfuels are in principle also renewable energy carriers. Unfortunately they are often produced in a non-sustainable way, often undermining the resource base of a community, nation or region. As was pointed out above, the wood fuel chain provides a significant number of jobs. It is also an indigenous resource (which can contribute to reducing the demand for fossil fuel imports) supplied in a manner appropriate to local needs (matching purchasing power, cash flow and existing equipment). So the challenge is not to displace woodfuels but to make their production more sustainable. The key here is to work with producers to develop a sustainable supply system to develop mechanisms that provide win-win situations. The rural informal woodfuel ESCO needs to be given a stake in sustainable production, for example by giving communities rights to sustainably manage forests intended for woodfuel and other forestry products.

But what of small informal sector entrepreneurs taking up the opportunities afforded under market sector reforms to establish ESCOs to serve the urban poor, such as those described in the study by Wakelin and his colleagues in urban Peru? Based on the literature, it is not possible to comment on the extent that informal ESCOs are being established. However, based on the conclusions of the study by Wakelin and his colleagues, it will require more than a created opportunity to be transformed into a reality of a service that matches poor people's needs. For example, start-up and working capital are often problematic for small entrepreneurs and hence form a barrier to establishing a business. This point is particularly pertinent to the promoters of RETs since these technologies are less well known by bankers. Also formal banks do not lend money to informal sector entrepreneurs. Women, due to their lack of convertible assets such as land, have particular barriers in accessing credit. While it is true that women have benefited from access to micro-credit, the levels of finance offered in such schemes are often at the level of a couple of dollars over very short periods of time and certainly not at the levels that would allow access to RETs (except improved household stoves). To enable women's access to RETs requires the establishment of financing organizations sensitive to the particular situation of women in terms of their access to and control over resources.

Market sector reforms are currently influencing the dynamic of urban energy availability. However, there is little micro-level data about the impact of these reforms particularly on the urban poor. In the evidence available the liberalization of energy markets does not appear, at least in the short term, to be promoting the urban fuel transition. In fact, it appears to be having the opposite effect due to the significant price increases which appear to accompany liberalization.

In the move to commercialization there has been some concern over the fate of subsidies particularly by those who regard this instrument as key to enabling access by the urban poor to modern energy. Indeed some authors consider that there would have been no fuel switching without subsidies, while others consider that they merely speed up a process that would have occurred as a 'natural' event of urbanization (Hosier and Kipondya 1993). If the latter is the case, the environmental benefits, for example, of reducing deforestation or improving urban air quality, have to be accounted for to justify the cost of subsidies. While there is little support for blanket subsidies, being generally regarded as not sustainable in the long term, there is support for the use of short-term, targeted (so-called 'smart') subsidies. There is no universal law for subsidies. Each case and the form of the subsidy have to be decided on its merits.

In managing the urban energy transition to more sustainable energy carriers, energy market reform alone is not enough to produce a positive outcome. Nor are the undoubted environmental, social and economic benefits of RETs. A number of other factors are considered to play a role in enabling a transition, for example, the cost of a stove or charcoal which requires good coordination between ministries to ensure coherence in policies – the Finance Ministry must be aware of the need to reduce the import tax on PV panels or the Forestry Department must enforce regulations on charcoal production. The energy transition does not take place in isolation within the energy sector. Factors in other sectors are also influential. For example, although there are improvements in the availability of LPG in main urban centres, the poor quality of roads can hinder the distribution of cylinders. However, informal sector suppliers can move in to fill the gap and supply low income households in ways that match their circumstances, particularly cash flow. The challenge for the formal sector companies is to build alliances with these informal sector ESCOs.

The literature reviewed in this chapter was written before the impacts of the 2004 oil price rises could have felt the consequences researched and published. However, there are data related to the effects of the price rises linked to commercialization in the energy sector. Not surprisingly the impacts on the urban poor have not been positive. There has been a reduction in quality of life with households adopting energy conservation measures, of the nature of 'not using equipment' rather than switching to more energy efficient devices. There appears to be some return to woodfuels but the extent and the environmental impact on peri-urban areas has yet to be quantified. For the electricity sector, the increase in tariffs has seen a rise in the number of illegal connections. It can only be surmised that the 2004 oil price rises have exacerbated the situation for poor urban households in terms of access to modern energy carriers derived from fossil fuels. At the same time, the rise in fossil fuel prices should open up opportunities for RETs. However, while RETs are gaining support for large-scale electricity generation in developing countries (Winrock International 2005), stand-alone systems do not appear to be adopted in significant numbers in poor urban areas, with the possible exception of improved charcoal stoves. Why? As we have stressed throughout this chapter, cost is only one aspect of the motivation for an energy transition. The lack of awareness of policymakers about the energy situation for poor urban households has to be at the root of the problem. As we stated in the introduction, the assumption is that all income groups in urban areas have access to modern energy services through the market. However, as we have shown here, for many poor households this is not the case. Should energy policy continue to be directed towards promoting access by these households to kerosene and LPG? The evidence suggests that this policy issue should be reflected on, since the fossil fuel transition at the moment appears to be under threat due to the price rises, which although subject to fluctuation are unlikely to return to the days of cheap oil. Use of fossil fuels does place a burden on the budgets of low income households and this is likely to increase (the World Bank data quoted earlier showed such households were spending around 20% of their household budget on energy and those percentages were at a time when oil prices were considerably lower than today). An alternative scenario is to promote a technological leapfrogging to RETs. However, governments have to be convinced that RETs are an option in urban areas just as much as in rural areas. The type of project described in Box 24.4 can provide convincing evidence that it is feasible, both to policymakers and the intended beneficiaries. It also shows that helping the poor urban dweller to a sustainable livelihood through affordable energy services requires a holistic approach and policy cohesion, for example before investing in an energy system a household needs security of tenure.

Acknowledgements

The authors would like to thank the UK Department for International Development who funded through their Knowledge and Research Programme (KaR) the study on which a substantial portion of this chapter is based. We would also like to acknowledge the contribution that Tanya Andrade, Adriana Alvarez and their colleagues from Winrock International, Brazil, made to this study.

References

Amis, P. (2002). Thinking about chronic urban poverty. CPRC Working Paper No. 12. Chronic Poverty Research Centre, University of Manchester.

Andrade, T. (2004). *Inception Report: Brazil.* Prepared for DFID KaR Project R8348 'Enabling urban poor livelihoods policy making: understanding the role of energy services'.

Annecke, W. and Endelli, M. (2006). Gender and prepayment electricity in Merlo, Argentina. *ENERGIA News*, **9**(1), 18–20.

Bannister, A. (2002). The Sustainable Urban Livelihoods Framework – a tool for looking at the links between energy and poverty. *Boiling Point*, **48**, 7–10.

Barnes, D. (1995). *Consequences of Energy Policies for the Urban Poor*. FPD Energy Note No. 7, The World Bank.

Barnett, A. (2000). Energy and the fight against poverty. Paper given as part of a series of Economic Research Seminars at the Institute of Social Studies, The Hague, 29 June 2000.

Barnett, A. (2001). Looking at household energy provision in a new way: the sustainable livelihoods approach. *Boiling Point*, **46**, 30–32.

Beall, J. and Kanji, N. (1999). Households, livelihoods and urban poverty. Urban Governance, Partnership and Poverty Theme Paper 3. Department of Social Policy and Administration, London School of Economics.

BRIDGE (2001). The feminisation of poverty. Briefing paper prepared for the Swedish International Development Cooperation Agency, BRIDGE Report No. 59, April.

Bruce, N. (undated). Public health and household energy – an introduction to the key issues. Theme paper for SPARKNET (Knowledge Network for Low-Income Households in East and Southern Africa). http://db.sparknet.info/goto.php/SparknetMainPage (accessed by author 10 July 2005).

Bruce, N., Perez-Padilla, R. and Alablak, R. (2000). Indoor air pollution in developing countries: a major environmental and public health challenge. Paper in *Bulletin of the World Health Organisation, WHO Bulletin 2000*, **78**, 1078–1092.

CARE (1999). *Household Livelihood Security in Urban Settlements*. CARE International UK Urban Briefing Notes. December (1).

Central Statistics Office (1995). *Household Income and Expenditure Survey: 1993/4*. Central Statistics Office, Ministry of Finance and Development Planning, Gabarone, Botswana.

Clancy, J.S., Alvarez, A., Maduka, O. and Lumampao, F.G. (2006). *Enabling Urban Poor Livelihoods Policy Making: Understanding the Role of Energy Services*. Synthesis report prepared for DFID Knowledge and Research Programme R8348.

Clancy J., Skutsch, M.M. and Batchelor, S. (2002). The gender-energy-poverty nexus: can we find the energy to address gender concerns in development? Position paper for DfID. Project CNTR998521.

Dasgupta, N. (1999). *Energy Efficiency and Poverty Alleviation*. DFID Energy Research Newsletter, Issue 8, May 1999. http://www.etsu.com/dfid-kar-energy/html/8_-_ee_pa.html (accessed by author 10 July 2005).

Deepa, N. with Patel, R., Schaft, K., Rademacher, A. and Koch-Schulte, S. (2002). *Voices of the Poor: Can Anyone Hear Us?* Oxford University Press.

Doig, A. (1998). *Energy Provision to the Urban Poor*. Report of the Household Energy Development Organisation Network IX Meeting. DFID Energy Research Newsletter, Issue 7, November 1998. http://www.etsu.com/dfid-kar-energy/html/7-urban_poor.html

Dutta, S. (1997). Role of women in rural energy programmes: issues, problems and opportunities. *ENERGIA News*, **4**, 11–14.

Eberhard, A. and van Horen, C. (1995). *Poverty and Power*. Pluto Press.

Ellis, F. (2000). *Rural Livelihoods and Diversity in Developing Countries*. Oxford University Press.

ESMAP (1999). *Household Energy Strategies for Urban India; the Case of Hyderabad*. Report 214/99, The World Bank.

Fakira, H. (1994). *Energy for Micro-Enterprises*. Energy Development Research Centre. University of Cape Town, Cap Town, South Africa.

Friends of the Environment (2005). *Country Study Report: Nigeria*. Prepared for DFID KaR Project R8348 'Enabling urban poor livelihoods policy making: understanding the role of energy services'.

Future Energy Solutions *et al.* (2002). *Energy, Poverty and Sustainable Urban Livelihoods*. DFID KaR R7661.

Government of Pakistan (1982). *Housing Census of Pakistan 1980: Summary Results*. Population Census Organisation, Government of Pakistan.

Hosier, R.H. and Kipondya, W. (1993). Urban household energy use in Tanzania. *Energy Policy*, May, 454–473.

IPIECA and UNEP (1995). *Technology Cooperation and Capacity Building: The Oil Industry Experience.*

ITDG (1998). *Energy Provision to the Urban Poor.* Issues Paper written for Department for International Development Knowledge and Research Contract No. R7182.

Kalumiana, O. (2002). Energy services for the urban poor: the Zambian perspective. *Proceedings of a National Policy Seminar: Zimbabwe's Policies on Urban Energy for the Poor.* Mapako, M. and Dube, I. (eds), AFREPREN Occasional Paper No. 20.

Karakezi, S. and Majoro, L. (2002). Improving modern energy services for Africa's urban poor. Energy Policy September, 1015–1028.

Leach, G. and Mearns, R. (1988). *Beyond the Woodfuel Crisis: People, Land and Trees in Africa.* Earthscan Publications Ltd.

Lumampao, F.G. Approtech Asia, personal communication.

Lumampao, F.G., R.S. Mataga, R.S. and Parado, B.C. (2006). Energising poor women entrepreneurs. *ENERGIA News*, **9**(2), 18–20.

Maduka, J.O. (2004). *Inception Report: Nigeria.* Prepared for DFID KaR Project R8348 'Enabling urban poor livelihoods policy making: understanding the role of energy services'.

Maduka, J.O. (2006). Smoke gets in their eyes: the women fish smokers of Lagos. *ENERGIA News*, **9**(1), 12–13.

Makan, A. (1995). Power for women and men: towards a gendered approach to domestic policy and planning in South Africa. *Third World Policy Review*, **17**(2)

Mapako, M. and Dube, I. (eds) (2002). *Proceedings of a National Policy Seminar: Zimbabwe's Policies on Urban Energy for the Poor.* AFREPREN Occasional Paper No. 20.

Meadows, K., Riley, C., Rao, G. and Harris, P. (2003). *Modern Energy: Impacts on Micro-enterprises.* Report of Literature Review for DFID KaR Project R8145.

Meikle, S. (2004). A study of the impact of energy use on poor women and girls' livelihoods in Arusha, Tanzania. *Urbanisation Issue*, **18**, 4–5. Department for International Development

Meikle, S. (2006). The impact of energy use on poor urban livelihoods in Arusha, Tanzania. *ENERGIA News*, **9**(1), 21–22.

Meikle, S. and Bannister, A. (2003). Energy, poverty and sustainable urban livelihoods. Working paper for Department for International Development Knowledge and Research. Contract No. R7661.

Mika, L. Practical Action Southern Africa, private communication.

Naeher, L.P., Brauer, M., Lipsett, M. *et al.* (2007). Woodsmoke health effects: a review. *Inhalation Toxicology*, **19**, 16–106.

Reddy, A.K.N. (2000). Energy and social issues, in UNPD *World Energy Assessment.*

Schutyser, P.J. (2003). *Small Diesel Generators in Rural Sudan: Social, Management and Technical Issues.* Unpublished Technical Internship Report, Department of Technology and Sustainable Development, University of Twente.

Shanko, M. and Rouse, J. (2005). The human and livelihoods cost of fuel-switching in Addis Ababa. *Boiling Point*, **51**, 31–33.

Sinha, S. (2007, forthcoming). Department of Technology and Sustainable Development, University of Twente, unpublished PhD research.

Smith, Kirk. University of Berkeley, private communication.

Sowazi, S. (2005). Exploring electricity conservation instruments for South Africa's mid to high income urban households: What are international best practices? Masters Thesis, 2005, University of Twente, The Netherlands.

SPARKNET 2004, Kenya Country Report Synthesis. http://db.sparknet.info/goto.php/ SparknetMainPage (last accessed by author 10 July 2005).

Tedd, L. (2001). Energy and street food vendors. *Boiling Point*, **47**, 10–12.

Tucker, M. (1999). Can solar cooking save forests?. *Ecological Economics*, **31**, 77–89.

UNDP (1999). *Rural and Urban Poverty: Similarities and Differences.* Retrieved from http://www.undp. org/sl

UNDP/ESMAP (2003). *Household Energy Use in Developing Countries: A Multi-Country Study.*

von Molthe, A., McKee, C. and Morgan, T. (2004). *Energy Subsidies: Lessons Learned in Assessing their Impact and Designing Policy Reforms*. UNEP/Greenleaf Publishing Ltd.

Wakelin O., Sohail Khan, M. Rob, A. and Sanchez, T. (2003). *Private Infrastructure Service Providers Learning from Experience*. Final Report prepared for DFID Knowledge and Research Programme R8177.

Watkiss P. (undated). *Urban Energy Use: Guidance on Reducing Environmental Impacts*. DFID KaR Project R7369. Project Report Summary. http://www.etsu.com/dfid-kar-energy/html/r7369.html

Winrock International (2005). *Country Report: Brazil*. Prepared for DFID KaR Project R8348 'Enabling urban poor livelihoods policy making: understanding the role of energy services'.

http://www.ren21.net/globalstatusreport/download/RE_GSR_2006_Update.pdf (accessed by the author 22 May 2007).

Chapter 25
Energy Planning in South African cities

MARK BORCHERS,[1] MEGAN EUSTON-BROWN[2] AND LEILA MAHOMED[3]

[1]Director Sustainable Energy Africa; [2]Sustainable Energy Africa; [3]Director Sustainable Energy Africa

25.1 Introduction

South Africa's largest 16 cities, comprising three quarters of national GDP, but occupying less than 3% of the land area, use approximately half of the country's energy (Sustainable Energy Africa 2007). These cities depend almost entirely on fossil fuels to meet their energy needs. Industrial development, based on the availability of cheap coal-fired electricity, and a still predominantly white, well-serviced suburban core contribute to global carbon emissions on a par with those of the industrialized countries of Europe and elsewhere.

This alongside extreme poverty: the municipality of King Sabata Dalindyebo, housing the ex-homeland 'capital' Umtata, has a per capita carbon emission of just over a ton per capita per year – equal to the African average and some 60% of their residents remain dependent on paraffin to meet their energy needs. This is reflected within the largely black peri-urban settlements on the margins of most South African cities (Sustainable Energy Africa 2007).

The low density and sprawling pattern of urban development within South African cities results in transport being the largest fuel consuming sector; this is emerging as *the* key energy issue facing South African cities. Indicative modelling based on energy data emerging from a recent energy measures process across a range of South African cities indicates that without intervention energy consumption in these cities is likely to double over the next 20 years (Sustainable Energy Africa 2007). There is both a great urgency to move towards sustainable energy systems and substantial challenges to be faced in such a transition.

South Africa emerged in 1994 with a highly centralized, supply-oriented energy sector, dominated by coal-fired electricity, originally established to service the Mineral-Energy complex at the heart of the South African economy. Energy choices were historically the realm of white male engineers and considered to be the business of national government. Local government had no energy capacity beyond that of electricity reticulation and distribution within the 'white' residential areas. The sale of electricity was (and remains to a large extent) a cornerstone of local government revenue.

The political transition in South Africa in 1994 opened up the way for a reshaping of the energy and development landscape. New legislation introduced demand-side and integrated energy planning approaches and in 1996 the new constitution gave local government

563

a new developmental mandate, specifically in the mandate to provide *equitable* and *sustainable* infrastructure and services to meet basic needs.

South Africa's cities, representing the centre of the nation's wealth, but also containing its most abject poverty, are critical players in the government's bid to tackle poverty and development. Energy is central to meeting basic socio-economic rights and the provision of electricity to households is a core task of local government in South Africa.

The Constitution of South Africa also establishes the right to a clean and safe environment for all. While energy is necessary for life and development, current dependence on fossil fuels means that energy is also highly destructive in terms of local air quality, health, disasters and global greenhouse gas emissions responsible for climate change. Given the substantial global concern and attention around climate change, activities involving the burning of fossil fuels are likely to come under increasing pressure. South African city economies are highly vulnerable to upward trends in the costs of carbon-related energy.

The Sustainable Energy for Environment and Development (SEED) program of Sustainable Energy Africa (SEA) was established within this environment. It recognized the driving role of *energy* in sustainable urban development and *people* as the key to transformation. SEED has worked to integrate sustainable energy approaches and practices into urban development in South Africa. This has been done through building the skills, knowledge and confidence of local government officials and development workers through training, mentoring and development of networks for learning exchange and facilitating local/national linkages.

Since 1998 SEED has generated a bottom-up movement of city energy transformation, which has taken root and is rapidly expanding in South Africa. This is being achieved through the application of city energy and climate change strategies, development of city targets, networking and training, institutional development and development of a broad city energy monitoring and data collection process. At the heart of the movement lies the development of a new cadre of local energy professionals working towards a new, sustainable urban energy landscape.

Transformation remains challenging. Addressing, simultaneously, local and global energy and environmental concerns is enormously demanding for local government. The coming decade, as South Africa moves beyond the 'visionary' post-democracy phase and into the realities of action and implementation, will reveal whether South African cities are able to make sustainability a reality.

25.2 South African Cities within the National Energy and Development Picture

Understanding the potential scope of, and constraints on, urban energy transformation requires a sense of the broad energy picture in South Africa and the role of cities within the national development framework.

25.2.1 South Africa's energy story

South Africa is well endowed with resources to meet its energy needs. Abundant coal has resulted in a heavy dependence on fossil fuels for energy. Large, brown coal resources supply three quarters of South Africa's energy, mostly in the form of coal-generated electricity and synthetic fuels, including petrol and paraffin. South African coal is cheap to mine and turn into electricity, but environmentally very damaging. South African coal is relatively low in sulphur but high in ash. This is significantly associated with respiratory health problems – and South African coal power stations are among the dirtiest in the world with high levels of particulate matter and carbon dioxide emissions. These environmental costs are disregarded in pricing power supply. Consequently South African electricity prices are among the cheapest in the world.

The Apartheid government's energy policy focused on energy supply security in the face of international sanctions. It established the Sasol oil-from-coal industry and the Mossgas off-shore gas utilization project to reduce dependence on imported oil. The nuclear industry was also considered to be very important, primarily for nuclear weapon capacity development (Ward 2002). These politically motivated projects were mostly poor economic investments, which ignored the energy needs of the majority of people and the need to establish a diverse, integrated energy economy. Compared to similar economies, such as Brazil, South Africa has an extraordinary dependence on a single energy source – electricity from coal power – for meeting most of its major needs.

Choices about energy use in South Africa, historically determined by white, male, engineers, failed to address equity or sustainability. The profession focused almost entirely on supply with little attention given to understanding energy poverty and the energy-service needs of people, or environmental issues. These were brought onto the agenda by NGOs, community representatives and research units of the universities in the early days of democracy.

Since 1994 the South African Department of Minerals and Energy (DME) has been developing new energy policy, with two primary objectives – to meet the basic needs of all people and to promote economic growth. The policy also attempts to address environmental issues more adequately. More recently, international attention to climate change has led to efforts to incorporate carbon dioxide emissions into policy and planning considerations.

Policy visions of national government are set against an entrenched energy intensive economic base and increasing electricity supply security concerns. Overcapacity of electricity in the 1980s–1990s meant that electricity intensive industries were actively encouraged in order to 'mop up' the excess supply. Energy consumption grew during this period, without concomitant economic growth (Fig. 25.1).

Electricity demand grew substantially during the 1990s. During this time government stalled on new capacity development in the hope that this would be forthcoming

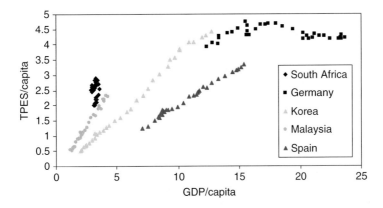

Fig. 25.1. Energy – economic trends 1971–2004. (*Source*: Dr Chris Cooper, University of Johannesburg using data sourced from the IEA).

from new, black, independent power producers, but without the prerequisite legal and support framework in place this did not take place. This has constrained electricity supply: South Africa has 40 GW of installed capacity and peak demand is currently around 35 GW. In the face of the looming electricity crisis the national Department of Minerals and Energy has put its consultative, integrated energy planning process on hold and forges ahead with bringing mothballed power stations on line and pursuing nuclear power plans.

25.2.2 Urban South Africa and the role of cities in a sustainable future

South Africa has some 17 primary cities that form the economic backbone of the country (Luus and Krugell 2005). These towns are concentrated around the mineral resources (predominantly gold and coal) in the north east and the ports that serve as economic gateways to the country. These cities house approximately 50% of the population. The six largest cities in South Africa (the 'metros') all have populations over a million – ranging from Nelson Mandela Bay to Metropolitan Municipality with a million residents to the Gauteng conurbation of nearly 8 million, consisting of Johannesburg, Ekurhuleni and Tshwane (StatsSA National Census 2001).

Below the primary-level cities South Africa is home to a tier of secondary towns, important as government service and industrial centres, or servicing rural areas grouped in this chapter as either 'industrial' or 'non-industrial' cities and towns. A further 170 small towns across South Africa, characterized by low populations (with average densities of less than 50 persons/square kilometre) and very small contributions to GDP, provide the third tier of 'urbanized' South Africa.

Despite the fact that the mining and heavy industry base of the South African economy means that substantial production and energy consumption in South Africa occurs outside of cities, cities still stand out as energy intensive nodes in the national fabric. The 'metro' cities account for 25% of national energy consumption; a grouping of 16 cities for which data has been gathered together account for 37% of total energy consumption (Sustainable Energy Africa 2007) – these include 14 of South Africa's 17 primary cities. In addition cities are substantial centres of economic power, with the metros alone producing more than half the nation's wealth. They are also sites of cultural and societal transformation. Cities represent both the centre of the nation's wealth, but also contain some of its most abject poverty (Parnell 2004). In these regards cities are critical drivers of national development.

To a great extent patterns of population growth and movement are still being established in post-apartheid South Africa. The rapid urban growth in cities that took place following the relaxation of the apartheid separate development laws has slowed down and current growth rates in city populations are, on average, in line with national population growth rates, at 2%. The Gauteng conurbation shows slightly higher than average growth rates, as do some of the industrial towns (uMhlatuzi, Saldanha Bay, Sedibeng) (South African Cities Network 2006).

South African cities exhibit a range of differences in terms of economic development, but all display similar levels of poverty and the existence of substantial levels of wealth disparity (Table 25.1). Although there has been growth in total South African GDP over the past decade, this masks an ongoing socio-economic crisis, namely that unemployment is extremely high and has been rising since 1993. Real wages have also been in decline: formal sector real wages fell by 0.5% over the period 1997–2003; informal sector real wages and self-employment incomes fell by 7.8% and 11.4% per annum, respectively (Kingdon and John Knight 2005).

Table 25.1. Population, economic production, unemployment and poverty levels in 15 South African cities, 2004.

	Population 2004		GGP			Unemployment rate*	Poverty level
	People	% SA	2004 million ZAR	% SA	Per capita	(% population)	(% Households with income <R1600 (US$228)/ month)
South Africa	46 586 607		1297		27 849		
City of Cape Town	3 069 404	6.6	144	11.1	46 913	29	39
City of Johannesburg	3 585 545	7.7	207	15.9	57 765	37	51
City of Tshwane	1 678 806	3.6	107	8.2	63 643	32	47
Ekurhuleni	2 761 253	5.9	92	7.1	33 270	40	56
eThekwini	3 269 641	7	118	9.1	36 170	43	57
Nelson Mandela	1 013 883	2.2	35	2.7	34 049	46	58
Metro total/ave	**15 378 532**	**33**	**703**	**54.2**	**45 688**	**38**	**51**
uMsunduzi	562 373	1.2	12	0.9	20 839	48	60
Saldanha Bay	79 315	0.2	3	0.2	37 020	21	39
Sedibeng	883 772	1.9	18	1.4	20 402	44	62
uMhlatuzi	360 002	0.8	14	1.1	37 956	41	61
Industrial total/ave	**1 885 462**	**4.1**	**46**	**3.6**	**24 583**	**39**	**56**
Buffalo City	702 671	1.5	15	1.1	20 906	53	69
King Sabata Dalindyebo	421 233	0.9	5	0.4	11 026	57	80
Mangaung	662 063	1.4	18	1.4	27 120	40	67
Potchefstroom	129 075	0.3	4	0.3	31 338	37	56
Sol Plaatje	196 846	0.4	7	0.5	34 474	42	56
Non industrial total/ave	**2 111 888**	**4.5**	**48**	**3.7**	**22 786**	**46**	**66**
Study cities total/ave	**19 375 882**	**41.6**	**7971**	**61.4**	**41 138**	**41**	**58**

Source: from data collated for the State of Energy in South African Cities 2006 Report (Sustainable Energy Africa 2007).

*These rates include 'discouraged workers', those unemployed who have not actively sought work over the past month, but would most likely like to be employed.

25.3 City Energy: Key Issues

25.3.1 Transport sector dominance in the city energy picture

South African city economies are all virtually 100% dependent on carbon-based energy sources (Fig. 25.2). Coal supplies electricity generation and industry, together accounting for 49% of the fuel consumption mix across the study cities. It is also used in the national manufacture of 30% of liquid fuels.

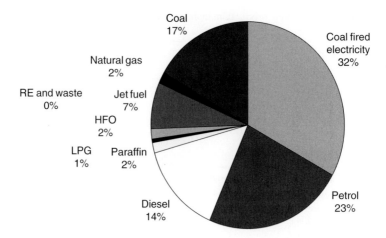

Fig. 25.2. Fuel mix across South African cities, 2004. (*Source*: Sustainable Energy Africa 2007): State of Energy in South African Cities, 2006 – Setting a baseline).

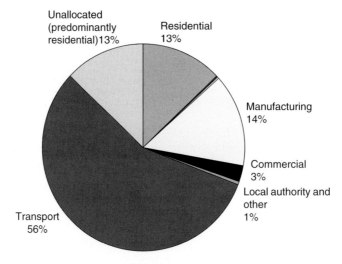

Fig. 25.3. South African metro energy consumption by sector, 2004. (*Source*: Sustainable Energy Africa 2007): State of Energy in South African Cities, 2006 – Setting a baseline).

Transport fuels comprise as much as half of all the energy used in South African cities. Within South Africa's metro cities the transport sector accounts for 56% of energy consumption (Fig. 25.3). The relative size of the transport sector in South African city energy consumption is, for example, in striking contrast to Asian megacities in which transport accounts for 5–10% of energy consumption (Institute for Global Environmental Strategies (IGES) 2004). The South African transport profile is influenced by a range of factors. Apartheid cities were built on the assumption that there was cheap, coal-based energy and low fuel costs for transporting poor people over the long distances required by racially divided urban spaces. The relatively mild climate in South Africa means that building and

Table 25.2.   Energy consumption per unit of GGP, 2004.

	KJ/2004 ZAR
South Africa	**1856**
City of Cape Town	900
City of Johannesburg	618
City of Tshwane	841
Ekurhuleni	1736
eThekwini	1030
Nelson Mandela MM	830
Average 6 metros	**938**
Msunduzi	2373
Saldanha Bay	9602
Sedibeng	4093
uMhlatuze	4087
Average industrial	**4009**
Buffalo City	1191
King Sabata	1462
Mangaung	992
Potchefstroom	1071
Sol Plaatje	895
Average non-industrial	**1084**

household energy costs are also relatively lower than many other countries that require substantial quantities of energy for thermal heating – distorting relative sector shares.

Transport modes, excluding freight transport, break down into non-motorized commuters (40%), public transport (30%) and private vehicle transport (30%) (Statistics South Africa (StatsSA) 2001). The emphasis on walking in South African cities is not due to ease of access by foot, but due to poverty. Census data indicates a strong correlation between income level and transport mode in South Africa. Although car ownership only becomes significant once households reach a R3000/month income level, the trend is towards a dramatic increase in car ownership in South Africa: currently standing at six million vehicles, car ownership is expected to increase to 15 million registered vehicles by 2012 (Kevin Nassiep 2007).

The institutional configuration of the public transport system across the country is fragmented and inefficient. This, alongside the failure to invest over the years in the maintenance and extension of the public transport infrastructure, has resulted in public transport that offers poor levels of service and imposes a high financial burden in operational subsidies on both state and households. From an energy, emissions and poverty alleviation perspective, enhanced mass public transport appears to be the most critical urban development need in South African cities.

25.3.2 Energy and urban economic development

The South African economy has one of the highest values for energy intensity (energy per unit of economic production) in the world. This is echoed within its city economies (Table 25.2): a function of sprawling cities, and an energy system characterized by highly inefficient coal-powered electricity generation and coal-fed liquid fuel plants. South African electricity is amongst the cheapest in the world, and the national economy has developed around exploiting this 'competitive advantage'.

In an era when trends in the costs of energy – both source costs of fuel and consequent costs of emissions related to energy use – exhibit increasing upward pressure (possibly at an exponential rate), cities with strong economic linkages to energy intensive activities face mounting challenges. Although currently an Annex 2 country within the Kyoto Protocol, as one of the world's highest carbon emitters, South Africa may well face carbon reduction targets in the next round of negotiations. This will increase the cost of local coal-generated energy. In addition, despite plentiful coal resources, South Africa is facing severe electricity generation capacity constraints and the costs of building new power stations are expected to substantially increase the price of electricity over the next few years. Oil prices, both to do with conflict in the Middle East and 'oil peak', also appear to be exhibiting an upward trend. Tertiary activities within South African metros are frequently upstream activities of the energy intensive industrial economy and are thus similarly vulnerable to rising trends in the price of energy.

Given the well-documented inequities within the South African economy, per capita energy consumption figures are to be interpreted with caution. Exploring the 'Pasternak curve' (Pasternak 2000) which identifies a strong correlation between human development (defined by UN Human Development Indices) and electricity consumption up to 4000 kWh/capita per annum, the following can be noted: electricity consumption within the South African metros, within about 20% above and below the 4000 kWh capita mark, implies that in underdeveloped areas the per capita consumption is well below that required to enable a minimum level of development; within non-industrial towns where the per capita electricity consumption level is less than half the 4000 kWh level this implies a severe constraint on human and economic development (Sustainable Energy Africa 2007).

Poverty and unemployment play out most dramatically at the local level: cities in South Africa face dwindling revenue bases as citizens reach the 'affordability limit' and are unable to meet service charges, among all the other manifestations of poverty that undermine local development – crime, social instability, disease, etc. Efforts by South African cities to proactively ensure their economically productive bases through diversifying energy usage and energy sources are in their infancy. This is critical to the future development of cities in South Africa.

25.3.3 Global carbon emissions

South African metros exhibit local per capita carbon emissions levels (on average 6.5 tons/capita) on a par with cities such as Berlin, Paris and London, even though the per capita GDP is far lower. Carbon dioxide emissions levels follow energy consumption patterns broadly, but are also affected by the local emissions structure. South African city reliance on coal-based electricity and energy intensive industry, with electricity prices among the cheapest in the world, are all factors that contribute to relatively high CO_2 emissions levels.

Poor rural municipalities in South Africa, in contrast, exhibit extremely low levels of carbon emissions – with that of King Sabata (1.4 ton/capita) equalling the African average of 1.1 ton/capita (Sustainable Energy Africa 2007). Yet the impacts of climate change, increasing disease incidence levels, water scarcity, extreme weather events and agricultural disruption, will be as, if not more, costly to poorer municipalities.

In terms of emissions structure, South African electricity production being 95% (brown) coal derived results in a larger volume of carbon emissions in those sectors reliant on electricity. Electricity in 16 South African cities is responsible for 66% of carbon emissions (Sustainable Energy Africa 2007). However, these calculations do not reflect the fact that

Table 25.3.   Carbon emissions in South African
study cities by end-use fuel type, 2004.

Fuel	CO_2 tons/yr	%
Electricity	102536369.4	66
Petrol	16611724.1	11
Diesel	10592249.9	7
Paraffin	1140868.8	1
LPG	415243.0	0
Jet fuel	5219851.1	3
Heavy furnace oil	1217702.0	1
Coal	15979685.9	10
Natural gas	2199180.2	1
RE	0	0
Total	**155912874.4**	**100**

a significant portion (some 30%) of South African liquid fuels is derived from coal. This would increase carbon emissions attributable to liquid fuels by a factor of between two and three for fuels manufactured in this manner. (Table 25.3)

25.3.4 Household energy transitions

A cornerstone of the new, democratic government has been an emphasis on facilitating physical access to grid electricity, through the National Electrification Program. Census data for 2001 indicates that while, on average, city households in South Africa are 85% electrified (using electric lighting as a proxy for electrification), lack of access to grid electricity persists in the poor, rural towns (and indeed peri-urban areas of larger cities). In addition, access to a grid connection is not sufficient to ensure a full transition to electricity use. Table 25.4, a composite of data across 15 cities, reveals the variegated nature of the urban energy transition.

The variegated nature of the urban household energy transition and persistent lack of physical access to grid electricity particularly within poor, rural and black areas means that policy targeting energy poverty needs to be well informed by local household energy conditions and flexible in meeting the needs of different urban household sectors. Affordability of energy is primarily targeted by national government through the Free Basic Electricity grant of 50 kWh/household per month. A concern here is that the subsidy fails to reach the poorest sectors of society, those without access to electricity. A fuel subsidy has been in place for paraffin through tax exemption, and gas price regulation is being explored. However, a broader suite of energy services beyond conventional fuel supply subsidies, including thermally efficient low cost housing construction support such as ceilings and plaster and hot water (solar water heating), would pursue poverty alleviation aims more productively, while addressing the environmental footprint and national capacity issues associated with growing electricity demand.

Mid and high income households in South African cities meet their energy needs almost exclusively through electricity. Electricity consumption figures within the cities of Cape Town and Sol Plaatje municipality (State of Energy, City of Cape Town 2003) indicate that higher income households consume disproportionably greater quantities of electricity than poorer households. More work is needed to determine the drivers of household energy consumption. However, initial data supports household energy transition theories

Table 25.4.   Energy sources for cooking, heating and lighting in
South African study cities, 2001.

Fuel source	Percentage households		
	Lighting	Cooking	Heating
Electricity	84	71	68
Paraffin	5	23	17
Gas	0.3	2	1
Coal		2	6
Wood		2	5
Candles	10		
Solar	0.1	0.2	0.2
Animal dung		0.5	0.3

Source: SA Stats, National Census 2001 analysed for 15 cities in the
State of Energy in South African Cities 2006 Report.

that point to an initial absorption of the increase in energy consumption accompanying
a shift from traditional to modern fuels through the greater efficiency of 'modern' fuels
(more useful energy may be obtained and thus households do not initially consumer a
greater mass). Beyond this there is a steady rise in electricity consumption that is largely
attributed to ongoing appliance purchase (Barnes *et al.* 2005).

25.4 Building Sustainable Energy Approaches in Urban Development in South Africa

25.4.1 *Planting the SEED*

In the heady days after the first democratic elections in South Africa there was great enthusiasm to take on the challenge of fundamentally changing South Africa. The activist Women's
Energy Group, a local energy NGO and a University of Cape Town-based energy research
unit, worked on the new Energy White Paper to ensure that poor households, gender and
environmental issues were considered in energy policy. This work included an intensive
programme of capacity building for community representatives from rural and urban areas,
particularly in preparation for the first National Energy Summit held in 1995. This summit
marked the first occasion in South Africa where community representatives were drawn
into high-level policy consultation. An explicitly demand-side led approach, and renewable
energy and energy efficiency were key features of the emerging national energy policy document. Not all the progressive content of the discussion document made it into the final white
paper, but it was a significant step forward.

During this process, confident and outspoken community representatives repeatedly
prioritized two issues: the need for access to appropriate energy information, and the need
for capacity to address energy needs at the local level. This laid the foundations for the
Sustainable Energy for Environment and Development (SEED) program, established
in 1998.

SEED is a national capacity-building program that aims to integrate sustainable energy
practices and approaches into urban development in South Africa (a rural SEED program ran
alongside the urban focus, but will not be explored here). SEED is run by Cape Town-based

NGO Sustainable Energy Africa, and funded through Danish foreign aid. People lie at the heart of the program: it is focused on those making energy service decisions by training predominantly young, black South Africans at the local level through local governments and relevant NGOs as 'sustainable energy advisors'. To work towards better choice being made in energy, SEED supports these 'advisors' in integrating sustainable energy work into their organizations.

Ten years ago there was little or no understanding of energy issues or capacity to make energy choices at the city or household level. In this context the 'sustainable' energy approaches of SEED emphasized a holistic approach to development. Energy provision, in this approach, begins with an understanding of the earth's energy system and how energy use affects this. SEED thus led a move away from supply-led modes of provision, towards a clear and explicit focus on the developmental needs of users (social, economic and environmental) as the driver of energy choices.

The sustainable development approach of SEED also looked to reducing dependence on fossil fuels, the introduction of cleaner fuels (and a more robust fuel mix), the increased use of energy efficiency and renewable energy, economic development based on efficient resource use and efficient public transport. 'Sustainability' extended into the conceptualization and structure of the program itself. SEA does not encourage dependency within the SEED partnership, but provides ongoing support that ensures that partners take full ownership of projects and processes. The partnership is adapted or terminated as the needs of the organization evolve.

As a partner organization within the program SEA offers sound technical knowledge in the area of energy provision and development. Staff is multi-skilled with backgrounds in engineering, environmental management, town planning, energy studies and economics. The organization, with staff members who have been engaged in energy work, policy development and political activism for years, has resources and skills that enable it to engage and manage issues and processes that emerge as the project develops.

While the strategic emphasis of the SEED program has changed over its nearly ten year lifespan, the central approach has remained constant. This is the emphasis on building capacity at the local level through the provision of networking support (building links and facilitating the exchange of local experience between local government, NGOs, provincial and national government and other stakeholders), training courses and mentoring, knowledge development and information provision and lobbying. SEED has worked closely with the major metro cities in South Africa, as well as two smaller cities and has a network of additional cities and towns with which it has been engaged through training, data collection, and information provision.

25.4.2 *Partnerships and poverty: SEED Phase One*

SEED's establishment began as part of the Household Energy Action Training (HEAT) strategy on capacity building around rural and urban energy provision that was developed for and funded by the Department of Minerals and Energy (DME). When SEED began in 1998 energy issues were mostly invisible for local government. The early focus was on energy in low income housing development, as this was the county's poverty alleviation focus at the national and local level at the time and the most strategic route 'in' for sustainable energy approaches. SEED's basic tenets were 'learn by doing' and 'communicate those lessons to decision-makers in parliament, the national and local government'.

A cornerstone of the SEED program has been the partnerships it forges. Phase One of SEED was implemented in three municipalities (metros), which were identified as having a strong drive to provide houses and electrification as part of their poverty-alleviation strategy. The program's key intervention was the placement of SEED advisors in the different partner organizations (municipalities and local, relevant civil society organizations). These young, black professionals were identified by SEA and its municipal and NGO counterpart. Importantly, the partners employed the advisors from the outset, with SEA providing a salary for a period of time in order to 'kick start' the initiative. Partners agreed to the ongoing employment of the advisors post the initial funded period. Advisors were mentored by a senior staff member within their organization, as well as by the SEA staff. They participate in all aspects of the program – training, support, networking, advice – offered by SEA.

As the program represented a paradigm shift in approaches to energy and service provision, resistance to SEED work has always been present. The cross-cutting nature of the work, requiring the cooperation of different departments and portfolios across cities, was also demanding. SEED supervisors, who were senior staff within the partner organizations, managed this with support and mentorship. This person's role as champion of the SEED work, providing leadership and advice to the advisor and project more broadly, emerged as critical in the success of the work.

The use of three metros in Phase One allowed for a cross-fertilization of ideas and strategies and provided useful lessons on working with partners who have different needs and resources. The three metros in this period were the City of Cape Town, with housing and development NGO Development Action Group the civil society partner; the Durban Metro Housing in partnership with the Built Environment Support Group; and Midrand Eco-City (which subsequently became part of the Johannesburg metro) and the Greenhouse Project in Johannesburg.

25.4.3 Innovative program implementation: city energy strategy development

Phase One of the SEED program (1998–2001) focused on poverty alleviation through promoting more efficient energy use within poor communities and access to better (safer, cleaner) energy sources. The flagship development program of the new government post 1994, the Reconstruction and Development Program (RDP), with its strong drive towards providing housing for low income earners, provided an excellent opportunity to implement SEED work around housing.

By 2001 energy issues were far less 'hidden' and increasingly becoming an important part of local government agendas. This was in part through the initiative of SEED, but also important was the Cities for Climate Protection Campaign of the Toronto-based International Council on Local Environmental Initiatives (ICLEI) that led city-to-city networking around decreasing city emissions. International processes, such as the World Summit on Sustainable Development (WSSD), also raised awareness of the importance of local governance in delivering sustainable development. Within South Africa the crisis of resource scarcity, notably water and electricity, was becoming increasingly apparent to local government.

While a number of energy projects now began to emerge within cities, the enduring success of SEED, and consequently the sustainable energy approach in urban development in South Africa, was its ability to see the need for an overarching 'grand plan' to draw together and direct various ad hoc energy projects. Without this, projects were spawned, had a brief lifespan and tended to fizzle out, with little overall impact on the system. South African cities are littered with 'pilot' projects that were entirely worthy, in which invaluable lessons were learnt, but which never managed to enter into mainstream service delivery.

In its second phase, with ongoing partnerships with City of Cape Town/DAG and the City of Joburg/Greenhouse Project and with the South African Local Government Association (SALGA), the national departments of Housing, Minerals and Energy, and Environment and Tourism, as well as new partnerships with the City of Tshwane, Sol Plaatjie Municipality and the Ekurhuleni Metro, SEED began to extend its work into integrated energy planning for whole cities.

SEED worked with the City of Cape Town to pioneer the first integrated city energy strategy in Africa. Cape Town had completed a State of Environment Report and was working towards the completion of a broad environmental management plan. The energy strategy work located itself alongside these initiatives and was driven by a SEED advisor, who'd been in place for three years, and an extremely dedicated, relatively high-level supervisor who championed the energy and environment cause. The Cape Town initiative was also boosted by a well-established energy research unit at the university and easy access to SEA (located in Cape Town).

The process began with a city-wide energy audit that informed the first State of Energy in Cape Town Report (2003). City goals and policies informed what information should be prioritized. In South Africa energy supply information is relatively easily available. Energy demand data, what users' energy service needs are, how they meet these through different energy sources and social, environmental and economic problems relating to energy, take considerably more effort to acquire. Cape Town's State of Energy Report provided the first insight into city energy issues within South Africa.

Turning this initial energy audit exercise into a city-wide strategy was the next step. To be successful this process had to be participatory, drawing in all areas and levels of city governance. Workshops were held across the city, with all departments, and the alignment of the strategy with broad city goals was emphasized and re-emphasized. From a political perspective it was extremely important to align energy strategy issues not just with the 'green' goals of the city, but also with areas such as city financial sustainability.

The strategic planning process was an effective means for all city players to rank priority issues and set goals. Strategic energy visions were linked to goals, which in turn had targets, measures and projects assigned to them. Cape Town's Energy Vision 2 is illustrated in (Table 25.5).

Cape Town became a flagship energy strategy. A SEED led South African City Energy Strategies Conference held in Cape Town in 2003 and ongoing SEED work that kept city energy work boiling through workshops, training events, meetings and local championship ensured the spread of the local integrated energy strategy approach to a number of other cities and towns in South Africa.

Four cities in South Africa – Cape Town, Tshwane, Ekurhuleni and Sol Plaatje – have completed detailed audit processes of energy consumption across their areas of jurisdiction and have (Sol Plaatje excluded) gone on to develop Energy and Climate Change Strategies. eThekwini (Durban), the City of Johannesburg, Nelson Mandela Metro (Port Elisabeth), Mangaung (Bloemfontein) are currently in the process of conducting similar detailed energy data exercises and strategy development processes.

25.4.4 A city energy measures and review process

Despite a growing body of city energy strategy work based on city-wide energy data and various sector-level urban energy studies (notably household level energy work), by 2005 there was still no comprehensive picture of the role of cities within the energy and emissions picture in South Africa, or a developed understanding of the urban energy drivers

Table 25.5.   City of Cape Town, Energy and Climate Change Strategy, Energy Vision 2.

City of Cape Town Energy Vision 2: A leading African city in meeting its energy needs in a sustainable way, and thus fulfilling its constitutional obligations and global responsibilities.

Goal	Targets	Measures	Projects
Increasing renewable and clean energy contribution to the energy supply mix, and reducing dependence on unsustainable sources of energy (starting with most financially viable options).	10% of all households to have solar water heaters by 2010. 10% of City-owned housing stock to have solar water heaters by 2010.	Measures include short terms (2 years) promotion of solar water heaters through installation on city owned housing stock, assessment market, putting codes and standards in place; Longer-term measures look at a SWH financing scheme and making SWH installation on all new homes mandatory.	Public building retrofits. Employee SWH installation financing project. SWH by-law development. (List not exhaustive.)

across the country. In a bid to consolidate the work of the preceding decade and work towards ensuring a good, demand-led database for sustainable energy policy development and implementation evaluation, a Cities Energy Review exercise was undertaken, under the auspices of the SEED program, across 15 cities in South Africa.

Analyses of energy and related emissions at the national scale have been conducted fairly extensively in a range of countries. At the city scale only a few such analyses are reported and methodological development for such local-scale studies is still under way. The City Energy Data Review process in South Africa involved the participation of an emerging network of city officials and related stakeholders, including national government and NGOs. The drive behind the review was to establish the role of city energy in the national energy and sustainability picture – and make the links between national and local policy and planning; to explore the drivers of urban energy use; to identify key local energy and environmental challenges; and develop a database to assist with energy planning and decision-making and against which to monitor and target urban energy development over time.

At the heart of the process was the participation of a core group of city energy professionals, NGOs, government and other stakeholders in the development of a set of energy measures for sustainable cities. These measures lined up with the energy and sustainable development mandates of South African cities. They also drew on energy indicators for sustainable development developed at the international level – notably those of the International Atomic Energy Agency (IAEA) (2005). The measures align with many international measures, enabling South African cities to begin to benchmark themselves against international developments. Complex indicator development was not achieved in this 'first cut' process. Distilling a coherent, measurable, defensible set of indicators from these will be an important part of taking the urban energy agenda forward.

Municipalities participating within an energy or sustainability city network formed the basis of the group of cities in the study. Additional municipalities were included in order to bring a better geographic, social and economic spread: coastal and inland towns, metro and smaller cities, industrial and more rural, ex-'homeland' towns. The study cities include a variety of income and development levels and institutional capacity.

As decisions relating to energy have traditionally been made at the national level, obtaining city scale energy data is extremely difficult and improving this area and level of regular, systematic data collection is an important challenge in urban energy development. Despite such challenges, the emerging picture, captured in the *State of Cities in South Africa 2006 – setting a baseline* report, provides an important baseline and platform from which to build the urban energy transition agenda.

25.5 Urban Energy Policy Implementation – and Challenges

25.5.1 Policy implementation

Great strides have been made in many South African cities in terms of energy policy development. Cities now face the far harder task of moving from policy to implementation. At times the barriers appear insurmountable and progress extremely constrained. However, some important experiences are emerging. There is also an increasing body of organizations and stakeholders beginning to coalesce around the issue of sustainable energy implementation within cities. To a great extent this is being fuelled by the electricity generation capacity issues being faced by the country, but also by a growing awareness of the role of cities in meeting national and international objectives.

The third and final Dandia-funded phase of SEED locates itself here. Its efforts are directed to support a core of six South African cities and towns in a substantial 'rollout' of energy interventions. The lessons and experiences that emerge in the process will be spread to a wider network of cities and towns. Key focus areas of implementation identified by cities include solar water heating and efficient building regulation. Cities are also being drawn into some pioneering renewable source electricity production. Transport is a key area to be tackled, but the fragmented nature of the institutional framework around the sector makes this difficult.

25.5.1.1 Solar water heater by-law development

Given South Africa's extremely favourable conditions for solar water heating, this application has been the focus of recent attention. The technology, although requiring an upfront capital input, is extremely cost effective. The City of Cape Town Energy and Climate Change Strategy establishes a target of 10% of all city households to have solar water heaters by 2010. The City of Cape Town is in the process of developing a local Solar Water Heater By-law towards meeting this target. The law is motivated in terms of energy saving (high on the city agenda given recent experiences of blackouts) and potential job creation through industry stimulation. Environmental benefits of carbon emissions reduction are captured as an additional benefit, but are perceived by city players as an insufficient driver of such a process.

The draft by-law requires all new buildings within the city, and all additions to existing buildings over a stipulated value which will require the use of hot water, to utilize solar water heating systems. The law exempts those buildings unsuited for such systems – for historical or structural reasons. Balancing inclusion with affordability was a substantial challenge for policy developers. Inclusion is an important goal to maximize impact, in particular in order to ensure that the growing electricity market of poorer to mid income city residents was also included. The current 'pegging' of the law is fairly inclusive in its range.

The city anticipates that a national solar water heating subsidy scheme may shortly alleviate the affordability concern.

Enacting a by-law is a first step. Creating an impact in terms of the intended objectives of the law, through participation by residents and enforcement of the law by the city, will be the real challenge. This requires levels of technical assurance as well as city inspection and enforcement capacity. SWH standards testing capacity has recently been established in the country, while implementation standards and qualifications are still being developed.

Cape Town forges ahead and anticipates many of the necessary conditions for successful implementation of the by-law will be created around their initiative. A number of other cities have indicated interest in the process and may well follow suite. SEED offers support to such processes, ensuring the spread of learning and experience into the broader city energy network through manuals, workshops and training events. Other innovative mechanisms for implementing solar water heating on a mass scale in cities will also be disseminated. Nelson Mandela Bay, for example, is pursuing a 'fee for service' approach that appears to offer an exciting opportunity around solar water heater penetration in cities. This approach utilizes the existing billing system and infrastructure of the city to provide a private entrepreneur with an easy and attractive set-up through which to offer hot water to households on a fee for service basis.

25.5.1.2 *Energy efficiency in buildings guidelines*

Much work has been done in South Africa around energy efficiency in buildings, yet the path towards putting any comprehensive programs in place (through regulation or guidelines) has been decidedly bumpy. In terms of the government low cost housing development program, numerous pilot housing projects contributed invaluable lessons towards a national set of draft environmentally efficient guidelines for implementation, produced in 1999. These, after a long period of dormancy, have been revived in the National Department of Housing's Framework for Environmentally Sound Housing, currently out for public comment.

The framework is sound, but no plan appears to be in place to ensure that the framework carries some weight in terms of the enforcement of recommendations. Slow progress appears to be due in part to government concern that any requirements would be too costly for them to deliver through their housing program, given the political pressure and demand for housing.

The national Department of Minerals and Energy's Energy Efficiency Strategy (2005) makes reference to establishing standards for energy efficiency in buildings within the National Building Standards. National energy efficiency standards for both commercial and residential buildings are under development. These will only provide standards and will not be mandatory. Government is exploring including such standards within the National Building Regulations, but may face lengthy law-making processes as this could involve changes to the National Building Act.

Some cities have begun work on local guidelines. Johannesburg has a Sustainable Housing Policy and Framework and in Cape Town a Green Building Guideline has been drafted, but has not yet been approved. Initiatives exist, but at this stage city-level implementation appears to be limited to promoting voluntary participation in energy efficiency building initiatives through guidelines and recommendations.

Within the commercial building sector a set of South African Energy and Demand Efficiency Standards (SAEDES) guidelines for new and existing commercial buildings have been developed and form the basis for the new set of national standards under development.

The City of Potchefstroom has shown interesting local initiative whereby they have made the SAEDES standards guidelines mandatory for all new municipal buildings. Where commercial building applications must go through an environmental impact assessment process, the city works to introduce the SAEDES standards as a condition for development.

25.5.1.3 *Renewable electricity generation and 'green' electricity purchase*
The contribution of renewable source electricity to city energy consumption is virtually non-existent in South Africa. The absence of feed-in tariffs (a regulatory framework for renewable energy, focussing on a feed-in tariff has recently got underway) for renewable source electricity, and low coal fired electricity prices make it almost impossible for renewable energy sources to compete. Critical for renewable generation projects is the existence of a Power Purchasing Agreement (PPA) through which a customer makes an upfront agreement to purchase the 'green' electricity at a premium. Although still in their infancy, local government is emerging as a key power purchaser, and thus driver, within such projects.

Tshwane has a PPA with the Tongaat-Hulett bagasse fired power plant. This 'green' power, though produced in Kwa-Zulu Natal, is bought and sold on by Tshwane Municipality, at a premium, to a multinational customer within its electricity distribution area. In this instance the process was initiated by the customer making the request to the municipality.

The City of Cape Town has a PPA with the Darling wind farm, South Africa's first commercial wind farm (looking at a final size of 13 MW from ten 1.3 MW turbines) being built outside Cape Town. The city will pay a premium for the purchase of the renewable electricity as part of its commitment to a renewable energy target of 20% of electricity from renewable sources by 2020. Without the assured customer (the city) the wind farm would be unable to go into production. The City of Cape Town, in terms of regulations set out by the National Energy Regulator of South Africa (NERSA), may not, however, dispel this additional cost by increasing local electricity tariffs. The 'green' electricity must be sold on at the premium charge to a willing customer(s). The city has been struggling to find customers willing to pay a premium for the 'green' electricity.

Nelson Mandela Metro has been actively pursuing renewable power options within the municipality. The initiative includes the production of wind-powered electricity, looking at a similar production capacity to that of the Darling wind farm, and solar water heating delivery, among other components. Strong mayoral leadership brought the project to life. In an interesting innovation, the service contract put out by the municipality stated that the energy services provided to the city through the initiative would need to come in at no additional cost to the municipality. This places substantial financial risk on the project developers. The development of the project is still under way and will provide interesting lessons for city renewable power development in South Africa.

25.5.1.4 *City transport*
The area of transport planning is vast and institutionally highly fragmented in South Africa. In addition, transport policy issues tend to overshadow energy issues within the sector. Cities, including Cape Town, Potchefstroom and the Gauteng cities, have initiated projects looking to encourage cycling as an alternative mode of transport. The model used has been the provision of bicycles to a target group, with cycle maintenance support programs looking to encourage local entrepreneurship in the area. Potchefstroom is in the process of launching a local cycle initiative that provides access to bikes, on a smart card style scheme, for cycle trips from the 'township' area into the CBD.

Sustainable Energy Africa, with funding from the British High Commission, has begun a two-year capacity building project along the lines of the SEED program with the City of Cape Town. The Tran:sit project, which employs a sustainable transport officer within the city transport department, aims to develop capacity in sustainable approaches to transport at the city level. Given the huge institutional barriers within transport, with functions variously aligned between local, provincial and national government, Tran:sit will initially focus on Transport Demand Management. The project will lead with TDM initiatives around city employee transport, including incentives for public transport use, encouraging high occupancy in vehicles through a 'matchmaking' scheme and providing a park and ride service. It is anticipated that the Tran:sit project will extend to other cities across the country and hopefully support increased awareness of transport as a key sustainability issue confronting local government.

South African cities are only just beginning to explore the range of mechanisms – from regulation to economic and institutional approaches, mandatory to voluntary – through which to give effect to policy. While some important developments are taking place, the common experience is one of struggling to get sustainable energy into overarching urban development policies and decisions of all city sectors. However, conditions appear to be ripe for city-level sustainable energy implementation, both externally – global and national pressures requiring new approaches to energy service delivery – and internally. SEED has meant that South African cities have laid solid foundations and developed capacity to take up the opportunities arising from global/national energy challenges.

25.5.2 The challenge of a developing country context

In the context of enormous wealth disparity and poverty, energy policies at the local level must address multiple issues – access, development, local and global environmental challenges. A policy focused on greenhouse gas mitigation is difficult to motivate within South African cities.

Energy and climate change strategies in South African cities require an integrated energy, environment and development approach, with basic services, local health issues and poverty at their heart. A challenge is to ensure that the energy plans become fully integrated in mainstream city planning processes – notably the Integrated Development Plan – which would ensure that projects receive budgets, resources, and commitments to outputs placed within specific line department 'scorecards'. This requires that local benefits are emphasized.

As with many developing countries, the challenge facing South African cities is that local, regional and global energy and environment issues are compressed, rather than phased: cities face enormous local and immediate environmental burdens such as air pollution, fuel-related fires, and energy poverty; at the same time global issues of climate change and resource scarcity cannot be ignored. In many instances the two can work together. However, tensions often exist, such as the need for the rapid establishment of houses versus ensuring sustainable urban planning and energy efficient housing; in such instances local, short-term issues trump longer-term poverty and sustainability approaches.

If the face of urgent poverty-related issues, notably disease, unemployment and crime, cities struggle to incorporate new, and therefore apparently 'riskier', approaches necessary for sustainability. Limited resources, along with the political time-frames for decision making – rather than the longer horizons required for sustainable development decisions – results in an emphasis on 'end of pipe' solutions rather than targeting major drivers of energy transition. For example, fuel subsidies are more readily adopted than targeting reducing the need for energy through efficient housing interventions, such as including ceilings within low cost housing developments.

The challenges are great and although relatively better endowed than many developing countries and with a strong local government sector, South Africa faces financial, regulatory and institutional constraints in the implementation of sustainable energy approaches. These complex and diverse issues must be tackled through mobilizing a finely tuned mix of different policy instruments suitable for local conditions.

25.5.3 Institutional development

A variety of institutional arrangements for sustainable energy management have emerged across cities. This includes a dedicated energy portfolio within the City of Cape Town, a Sustainable Energy Task Team coordinated by the Environmental Health Department of Tshwane and bringing together a multiple line of departments and an Electricity and Energy Business Unit established within the Nelson Mandela Municipal Metro.

Those cities furthest advanced in institutionalizing city energy management have in common the existence of supporting networks of NGOs and research institutions. This appears to be an important foundation for drawing in different levels of government as well as the extensive range of technical, financial and legislative expertise required to tackle integrated energy planning and implementation.

Institutional capacity varies between cities, but certain constraints to implementation exist within most local governments. Energy work in cities introduced through the SEED program consciously focused on building capacity outside of the traditional electricity departments. This was in order to counter the dominance of these departments and their very established, technical, supply-oriented approaches, but runs the risk of sustainable energy issues being relegated to 'green ghettos'. Ensuring that sustainable energy approaches are integrated into all sectors of local government planning and policy has been a challenge: environmental departments do not hold substantial political weight within local government and tend to be seen as 'elitist' and/or 'additional' to business as usual.

More broadly, energy-related responsibilities and powers are held variously at different levels of government and parastatals. No common platform is in place for achieving the necessary consensus for action around what has to be done and the role of each particular body. This also inhibits the cross-subsidization of energy projects – a core feature of implementing sustainable energy projects.

Local government is also restricted in its financial dealings. Council procurement procedures are necessarily restrictive and not designed to promote the flexibility needed for energy business and investment in energy projects. Local government is not designed to participate in financial markets, such as those developing around carbon and emissions trading, which are becoming an increasingly important part of the energy project financing landscape.

A key institutional constraint for sustainable energy development is that those currently in charge of energy – namely, local government and the national utility, Eskom – are focused on electricity supply and distribution and are dependent on income from electricity consumption. They are therefore poorly placed to engage with demand-side management and with diversification of supply sources and ownership.

Many of the cities internationally that have had success with implementing their sustainable energy plans have found it necessary to establish a dedicated energy vehicle to achieve their ends. Cape Town is in the process of exploring 'best fit' energy agency options. This will look to the creation of a more flexible, dedicated unit of energy managers better able to implement policy tools and draw in a range of stakeholders. In particular, such a unit would be better able to respond to opportunities and channel finance, such as that available through carbon trading.

25.5.4 *Issue of national/local cooperation*

While South Africa follows the decentralization trend, national direction remains a powerful and important component of local government. National commitments to energy for development, integrated resource planning and management of health and environmental impacts of energy generation, as articulated in the Energy White Paper of 1998, appear to have become diluted within current national energy priorities that look at achieving universal access to electricity, ensuring security of energy supply and the transformation of the energy industry through Black Economic Empowerment (as articulated by the South African National Energy Research Institute at the recent Networks of Expertise in Energy Technology workshop held in Johannesburg 2007).

South Africa is a signatory to the Kyoto Protocol, though as a non-Annex 1 country is only committed to a general carbon emissions reduction target. National government has also developed renewable energy and energy efficiency targets (DME 1998). Although much of what the country can do is at the local level, national commitments have not been devolved to cities.

A few guidelines and regulatory frameworks exist, but few action-oriented partnership activities between national and local government are in place. The role of national/local cooperation to date has centred on a minimal degree of research support, technical feasibility and promoting donor assisted projects in cities. Without active support, or mandatory obligations, it is difficult for local government to enforce local-level commitments. In the absence of such direction local government are forced to rely on building consensus – an enormously difficult project given vast differences in wealth and needs among city residents.

25.6 Conclusion

In the first ten years of democracy in South Africa the emphasis was on the aspirational – where do we go? Among many other new projects of democracy, this period saw the flourishing of new sustainable energy approaches in city development and their manifestation in visions, strategies and plans. The challenge now facing South Africa is how to move into the second phase, that of translating the visions and ideals into practical actions – what do we do?

Within the South African context of high global emissions alongside deepening poverty addressing both global and local energy and environmental concerns is critical, but highly challenging. A complex set of dynamic policy tools, tailored to meet specific local conditions, is required. This requires a sophisticated understanding of the drivers of urban energy transitions – a process under way, but with much work still to be done.

The challenge is enormous: without intervention city energy consumption is set to double over the next 15 years; energy poverty remains an enormous burden for a great number of urban residents; dependence on fossil fuels and energy intensive industrial development renders South African cities highly vulnerable to upwards trends in the costs of carbon energy – highly likely in the face of substantial and increasing concern and global attention around climate change.

A solid and substantial movement within cities towards sustainable energy management approaches has been established in South Africa. This has ensured the development of capacity in the heart of local government – a key locus of sustainability delivery. The ability of local government to move from sustainable energy policy to implementation, against the national tide of supply-side, coal- and nuclear-based power station development, will indeed be a real test of the strength of the democratic system of governance in South Africa and likely sustainability of our cities' future.

Acknowledgments

Megan Euston-Brown's time on this chapter and research development was funded by monies from Danida and the Wallace Global Fund. Thanks to Sarah Ward for input, review and comment. The views and opinions expressed are those of the authors alone. Our web site is www.sustainable.org.za.

References

As articulated by the South African National Energy Research Institute at the recent Networks of Expertise in Energy Technology workshop held in Johannesburg, 20–22 February 2007.

Barnes, D.F., Krutilla, K. and Hyde, W.F. (2005). The Urban Household Energy Transition – Social and Environmental Impacts in the Developing World, Resources for the Future, Washington, DC, USA, 30.

DME (Department of Minerals and Energy) (1998). White Paper on energy policy of South Africa. Pretoria, DME. This commits government to a target of 10000 GWh of national energy to come from renewable sources by 2013. The emphasis within the strategy is on the development of a local biofuels industry. Budgetary allocations towards renewable energy development, however, remain negligible. DME draft national Energy Efficiency Strategy (2004) provides specific targets for reduction in energy demand by 2014 within given demand sectors, with an overall target of 12% reduction in consumption and specific demand sectors reduction targets by the year 2014.

Institute for Global Environmental Strategies (IGES) (2004). Urban Environmental Management Project Report: urban energy use and greenhouse gas emissions in Asian mega-cities – policies for a sustainable future (draft). Kitakyushu, Japan, April, 37.

International energy indicators for sustainable development consulted were those of the International Atomic Energy Agency (2005). *Energy Indicators for Sustainable Development: Guidelines and Methodologies*, Vienna, Austria. Local indicator and measures referred to included Spalding-Fecher, R. (2002). Energy sustainability indicators for South Africa. Report prepared for the Sustainable Energy and Climate Change Partnership. Cape Town, Energy and Development Research Centre, University of Cape Town; Cape Town, Ekurhuleni and Tshwane and Sol Plaatje City State of Energy Reports, 2003, 2004, and 2005, respectively.

Kevin Nassiep, CEO South African National Energy Research Institute, personal communication. Networks of Expertise in Energy Technology Workshop, 22–22 February 2007.

Kingdon, G. and John Knight, J. (2005). Unemployment in South Africa 1995–2003. Causes, problems and policies. Presented at the Year Review of the South African Economy Conference, October 2005. Retrieved 20 February 2007 from http://www.csae.ox.ac/conferences/2006-EOI-RPI/papers/case/kindon.pdf.

Luus, M. and Krugell, W.F. (2005). Economic specialization and diversity of South Africa's cities. Paper submitted to the Biennial Conference of the Economic Society of South Africa, 10–11.

Parnell, S. (2004). Constructing a developmental nation – the challenge of including the poor in the post apartheid city. For DBSA, UNDP, HSRC conference Overcoming South Africa's Second Economy, October 2004, 15–16.

Pasternak, A.D. (2000). Global energy futures and human development: a framework for analysis. Study done in partnership with US Department of Energy and Lawrence Livermore National Laboratory, University of California.

South African Cities Network (2006). *State of the Cities Report 2006*. SACN, Gauteng.
State of Energy, City of Cape Town (2003) and State of Energy in Sol Plaatje (2005).
Statistics South Africa (StatsSA), National Census 2001.
StatsSA National Census 2001 and Development Bank South Africa 2006, Dataset: Total Population Projection at Local Authority Level, 1996–2006.
Sustainable Energy Africa (2007). State of Energy in South African Cities 2006 – Setting a baseline, 82.
Sustainable Energy Africa (2007). State of Energy in South African Cities 2006 – Setting a baseline, 80.
Sustainable Energy Africa (2007). State of Energy in South African Cities 2006 – Setting a baseline, 53.
Sustainable Energy Africa (2007). State of Energy in South African Cities 2006 – Setting a baseline, 118.
Sustainable Energy Africa (2007). State of Energy in South African Cities 2006 – Setting a baseline.
Sustainable Energy Africa (2007). State of Energy in South African Cities 2006 – Setting a baseline.
Sustainable Energy Africa (2007). State of Energy in South African Cities 2006 – Setting a baseline, 54.
Ward, S. (2002). *The Energy Book*. Sustainable Energy Africa. Over 90% of the country's energy budget was allocated to the Atomic Energy Corporation.

Chapter 26
Household Markets for Ethanol – Prospects for Ethiopia

ERIN LAURELL BOYD

School of Oriental and African Studies

26.1 Introduction

As for all nations, energy use in Africa is not without its problems. Two of the most ser-ious of these are a stifling dependency on imported fuels and a shortage of fuelwood, the primary fuel for the majority of African households. Compared to other nations, however, these energy-related problems place a disproportionate drain on the economy, environ-ment, and overall well-being of the population.

Fortunately, these difficulties are well known, and effort is being made to mitigate the asso-ciated negative impacts. However, there is a dynamic whose workings, given the absence of its full consideration in policy, will eventually override the intended outcomes of any interven-tions. This dynamic is urbanization, whose rate of growth is not only immense – sub-Saharan Africa has both the highest rates of population growth and urbanization in the world, with many other African nations not far behind – but whose pace is unlikely to slow in the near future given that the vast majority of the African population resides in rural areas.

The reasons for this counterbalance derive from the massive shift in energy use which will accompany both this rapid rural/urban migration and brisk population growth. To see how, consider Ethiopia, or rather Ethiopian household energy patterns associated with cooking, which make up all but a small percentage of household energy fuel use, between rural and urban inhabitants. In rural areas, nearly 85% of the population depends on fuel-wood as their primary fuel for cooking, with the next largest primary dependency ratio being 12.65% for crop residue; only 0.21% of the rural population depends on kerosene for their primary cooking fuel, while the numbers for electricity and LPG are 0.05 and 0.07%, respectively. Contrast this with the capital city of Addis Ababa, where 42% of residents depend on kerosene as their primary fuel, compared to just 6.5%, each, for LPG and elec-tricity; approximately one quarter of the population in Addis Ababa depends on fuelwood for their primary fuel, with 8% depending on crop residue and 4.5% depending on char-coal (Ethiopia Central Statistical Authority 2004). When all fuels used are considered, over 90% of the population in Addis Ababa depends on kerosene;[1] although no official statistics are available on all fuels used by households in rural areas, the percentage of households

---

[1] This figure is based on data collected as part of the Project Gaia pilot survey, described further later in this chapter.

who use kerosene is not likely to vary greatly from those who use it as their primary fuel given kerosene is not widely available in rural areas.

Clearly, then, as rural Ethiopian households migrate to urban centres, which they are doing at a rate of over 3.5% per year, the energy balance of the country will shift; overall, fuelwood use will decline as households switch to kerosene (World Bank Urbanization Rate). This is beneficial in that it mitigates the pressure on fuelwood, but detrimental in that dependency on petroleum imports increases; as dependency on petroleum imports increases, so, too, will expenditures of valuable foreign exchange, an outcome that is particularly damaging to the economy and a point well understood by governments. Yet, recognition of this damage seems to be quite short-sighted from a policy perspective, in both Ethiopia and Africa alike. Efforts focus entirely on reducing dependency in the transport sector at the expense of a sector that will become, over time, an ever more predominant component of imports; the current policy of choice is the blending of ethanol with transport fuels, itself also short-sighted, as will be discussed below.

Certainly, the use of ethanol in the transport sector is a worthy policy, its benefits extending well beyond those related to reductions in foreign exchange expenditures. To begin, ethanol's production is environmentally sustainable, with use also resulting in reduced environmental impacts from better resource management along with lower greenhouse gas emissions. In addition, as ethanol can be produced from a number of wasted or underused resources, such as molasses currently dumped into African rivers, the creation of a household market for ethanol will turn a no-value by-product into a high-value industrial product. This will spur productivity and increase employment in industry, agriculture, manufacturing, and the service sector; these increases, in turn, will translate into greater household wealth on a micro-economic level, reduced gender inequality, and income and wealth effects on a macro-economic level.

While the scope of benefits is immense the transport sector is not the only potential market, and, in the face of rapid urbanization, the transport sector may not be the best market either. Consider, once again, Addis Ababa. As with the many African urban centres whose fuel use patterns are not that different from those of the Ethiopian capital, Addis Ababa is in need of an energy policy which recognizes the realities of urban energy use and which does not assume electrification of an urban area to be a success (while the city is electrified, only 6.5% of the population depends on it as their primary cooking fuel, and those who do often continue to utilize kerosene as well as other fuels); these realities include relative reliance on 'dirty' intermediate fuels, lack of reliance by all but the wealthiest households on modern energy sources (e.g. electricity), greater familiarity with markets, particularly with regard to fuel purchases, and relatively developed infrastructure.

Coincidentally, ethanol is one of the cleanest of household fuels when burned in proper appliances and whose use by households fits into these realities quite well: ethanol is relatively low in cost, compared to fuels such as electricity and LPG, and similar in cost to kerosene; it can be distributed through the infrastructure and markets most common to the majority of households; and it has a high efficiency, low smoke/soot level, and high 'cleanliness' compared to those fuels most prevalent for cooking. Given these realities, would countries in Africa be better poised to nurture an urban household market for ethanol rather than a transport market, a trade-off which must be considered? Neither current nor potential short-term ethanol supply is large enough to meet the demand of both markets simultaneously. Undoubtedly, those benefits which accrue from the transport market also accrue in an urban household market.

Unfortunately, such a question is absent in discussions surrounding potential uses for ethanol. When the question does enter the debate, it is essentially dismissed. This stems, in part, from current foreign exchange expenditures on transport fuels being greater than those on kerosene, with substitution away from gasoline and diesel towards a domestically produced fuel seemingly offering greater savings. Likewise, there is perceived ease with which a transport market for ethanol could be implemented relative to a household market; a government may enact regulation to blend ethanol with gasoline, creating a market overnight, while a household market for ethanol may take time to build given cultural issues associated with cooking and time needed to create demand. As such, benefits in the transport market are seen as immediate, while those of the household market are seen as occurring further into the future. Finally, the question is often seen as irrelevant given most African countries consider electrification to be the policy of choice for the household energy sector; if anything has shown itself to be true, it is that access does not imply use, even in urban areas.

Yet, all of these arguments can themselves be dismissed; more importantly, the question not only deserves proper consideration, but is one whose answer is apt to be in the affirmative – African governments would, most likely, be better off nurturing a household market for ethanol rather than a transport market. These conclusions are based on the analysis conducted here within, which compares the benefits of an urban household market to those of an urban transport market for the case of Ethiopia; this is possible given data collected as part of a pilot study, conducted in 2005, which examined the potential for an ethanol-based household market in Addis Ababa, the capital city. As is clear from the pilot, urban households in Addis Ababa, who are accustomed to markets and dependent on kerosene, will readily and enthusiastically accept ethanol into their homes; as the situation is likely to be quite similar in other urban areas of Africa, the trade-off with regard to the timing of benefits for many African nations shortens, thus eliminating one of the main reasons cited for not pursuing an urban household market.

With regard to the remaining reasons offered for not considering an urban household market for ethanol, consider first foreign exchange savings; again, as calculated here, foreign exchange savings resulting from a household market are greater than those resulting from a transport market. This is true from the initial start date of each market; as household demand matures and urbanization takes hold, this difference will only widen. As for the belief that electricity is to be promoted as the energy of choice in urban areas, as discussed above, both the statistics for Addis Ababa, as well as those of other African urban centres whose patterns are similar, show electrification has simply not proven to be the hoped-for success.

Thus, there is no reason that the household market should continue to be dismissed. Furthermore, nations should ensure the costs and benefits of both alternatives are explored before proceeding; to do otherwise is of great disservice and detriment to the nation and its citizens alike. As the analysis here shows for Ethiopia, savings on a per household basis, aggregate savings to the economy, and reductions in greenhouse gas emissions are, in general, greater for the urban household market than for the transport market. When foreign exchange savings are also added to the list, the debate should, in the future, only concern how best to promote an urban household market within the Ethiopian capital to achieve maximum benefits from domestic production of ethanol.

With this being said, this chapter will set out to calculate the costs and benefits of an urban household versus urban transport market for Ethiopian ethanol, whose methodology and results should inform other African nations facing similar decisions. To begin, a discussion of the supply side, both with regard to Ethiopia and more generally, will be presented, including both current production levels as well as potential for expansion, the latter being important given current supply is unable to support simultaneous markets; also included here will be a discussion of some of the key economic and environmental factors related to ethanol production. Having outlined the supply side, the chapter will move to the demand side, where the costs and benefits of each market will be separately estimated for the case of Addis Ababa. Once this has been achieved, the outcomes of each market will be compared, allowing for conclusions to be drawn about which market provides the greater return on Ethiopia's ethanol supply; this will be the focus of the final section of this chapter. The conclusion will draw broader conclusions for Africa, not the least of which is to forego a hasty policy which favours a transport market in the absence of proper analysis which considers the potential of all possible markets, especially an urban household market.

26.2 Supply Side

This section will begin first by outlining current and potential ethanol supply in Ethiopia, followed by a discussion of the economic and environmental implications of this ethanol production.

26.2.1 Current and future ethanol supply in Ethiopia

Ethiopia currently has in reserve 8 million litres of ethanol, a product of the state-owned and managed sugar industry. Derived from the sector's accompanying waste by-product, molasses, only one of the three sugar factories – Finchaa, with the remaining two being Wonji-Shoa, and Metehara – currently has the capability to distil ethanol. This capability, however, is sufficient to process the molasses from the remaining two factories without any additional investment. As displayed in Table 26.1, this would equate to an additional 16.8 million litres, based on 2005/2006 molasses production at the remaining two sugar factories, and would result in a total potential ethanol supply of 25 million litres. Given the almost doubling of projected molasses production at Wonji-Shoa in 2007/2008, plus the projected increase of ethanol production by 4 million litres at Finchaa, also in 2007/2008, total ethanol production could reach 33.5 million litres should all molasses be distilled. (Ethiopian Sugar Factor Support Centre Share Company 2006)

Despite this potential with no extra investment, increasing production of ethanol is not an assumed reality; Metehara and Wonji-Shoa are waiting to explore the logistical realities given Finchaa has yet to find a market for its own supply. Originally, ethanol was intended to act as a gasoline substitute; oil companies, however, resisted this outcome. With storage at or near capacity given lack of a domestic market, molasses is currently being shipped, at a loss, to Djibouti for disposal (Tilimo and Kassa 2001).[2]

Notwithstanding the absence of a market, should Ethiopia wish to expand its ethanol production beyond its potential 33.5 million litres, further expansion is feasible. In (Ethiopian Sugar Factor Support Centre Share Company 2006), it is estimated that 388 000 hectares could be placed under immediate sugarcane cultivation, which would result in an additional 288.7 million litres of ethanol being supplied. Furthermore, based on an examination of rainfall levels, potential for irrigation, temperatures, agro-climatic conditions and soil types throughout Ethiopia, (Tilimo and Kassa 2001) estimate that there are 10 million hectares of land which could support the cultivation of sugar crops. Sensitive to the needs of local populations, though, (Tilimo and Kassa 2001) conceive that not all of this land could feasibly be converted. Despite being sparsely populated due to the presence of malaria and

---

[2]This loss is due to escalating costs, including those related to transport, tanker rentals and handling, relative to the export price secured.

Table 26.1. Current and potential ethanol production, '000 litres.

	Current	Potential		Total
	Finchaa	Wongi-Shoa	Metehara	
2005/06	8000	5454	11340	24794
2006/07	8000	6021	11340	25361
2007/08	12000	10152	11340	33492

Table 26.2. Employment and total population supported by sugar factories.

Factory	Permanent	Seasonal	Total	Total population supported
Finchaa	4500–5000	NA	4500–5000	25000
Metehara	2460	7540	10000	66000
Wonji-Shoa	2608	3744	6352	28584
Total	**9568–10068**	**11284**	**20852–21352**	**119584**

tryponasonmiases/tsetse fly, among other human and livestock diseases, the land is still used for shifting cultivation and for seasonal grazing by pastoralists when the prevalence of disease is absent or minimum. Likewise, population growth in these areas is occurring, which itself is creating competition for land as new populations seek to convert it for cultivation, limiting the pastoralists' historical right to graze freely. So as to not exacerbate this competition, (Tilimo and Kassa 2001) conclude that only 35 to 40% of the 10 million hectares could be converted to the cultivation of sugar crops; this, in combination with both cultivation of various crops, including sweet sorghum, a staple food, as well as implementation of particular cropping methods, would result in an additional 10.5 to 12 billion litres of ethanol per annum, bringing total production levels to 10.8 to 12.3 billion litres of ethanol per annum.

26.2.2 Economic considerations of ethanol production

26.2.2.1 Job creation

The production of ethanol in developing countries has the potential to provide a number of economic benefits to the country, not the least of which is jobs. Table 26.2 displays current employment levels, including both permanent and seasonal, for each of the three sugar factories, plus the total population supported by each factory, including families of employees. *In toto*, approximately 10000 permanent jobs are supported by the sugar industry in Ethiopia, with an additional 11284 seasonal jobs arising at times of harvest, etc. (Tilimo and Kassa 2001). Converting the common measure of employee days per hectare, the average across all three factories is 114; this falls at the high end of labour requirements for ethanol production in sub-Saharan Africa, estimated to be between 74 and 126 person days per hectare of sugarcane (Phillips 2002).

As will be seen in Section 26.4, land expansion to meet demand in the household market would result in an additional 2300 to 7000 additional permanent jobs and 5200 to 15800 seasonal jobs; to meet demand in a transport market would result in an additional 8544 permanent jobs and 19173 seasonal jobs. If both markets occur simultaneously, 10859 to 15555 permanent jobs will be created and 24368 to 34906 seasonal jobs will be created.[3] With a labour force estimated at 27583800 in 2000 (World Bank 2002), the increase in employment from simultaneous markets would be 0.04 to 0.06% for permanent employment only, and, if seasonal jobs are included, 0.09 to 0.13%; while not immense on a percentage basis, absolute levels are still significant.

Not only employees benefit, but so, too, do both their families as well those who provide vital support services, such as the selling of groceries, other goods, etc. Currently, just under 120000 people are economically reliant on the sugarcane industry (Tilimo and Kassa 2001); this results in approximately six individuals for every one employee. As direct employment expands with the growth in ethanol production, so, too, will economic opportunity for others.

26.2.2.2 Value created from a previously non-valued by-product

As mentioned above, ethanol in Ethiopia is produced from the otherwise waste by-product of molasses. Disposal has taken, and takes, numerous paths, all of which have an associated net cost, whether monetary or environmental; molasses has been spread on roads and, as discussed in Section 26.2.1, exported at a financial loss. Indeed, just a few years back, the primary means of elimination was the dumping of molasses into waterways: the major one being the Awash River flowing through large areas of federally protected habitat. Such dumping is quite damaging to both water quality as well as the surrounding habitat. A domestic market would create value not only from a non-valued by-product, but, rather, a by-product which currently has a negative value.

26.2.2.3 Impact on the trade balance

Ethiopia depends entirely on imports to meet its petroleum-based needs: diesel and gasoline for the transport market and kerosene for the household market. As expanded on in section 26.3, if a domestic market for ethanol were to arise in Ethiopia, imports of these refined fuels would unquestionably decline, and with them foreign exchange requirements to fund their purchase. The overall improvement to the balance of payments would undoubtedly be of benefit to the nation.

However, there is also a potential opportunity cost for a domestic market. World interest in the use of ethanol is growing rapidly, especially with regard to its use in the transport market. The largest consumer of gasoline, the United States, has taken a keen interest in promoting ethanol as a fuel substitute, with legislation encouraging its use likely in the near future. As many of the countries that consume large quantities of transport fuels are unlikely to domestically meet demand, imports will be necessary (Worldwatch Institute 2006). This offers great opportunity for those countries, such as Ethiopia, with an abundance of land and relatively low production costs; ethanol could be exported to lucrative markets in the developed world, bringing in greatly needed foreign exchange earnings. However, given the relatively high transport costs in the land-locked country of Ethiopia, not to mention the unlikely near-term development of an international market for ethanol free of tariffs and other trade barriers, the threat, in terms of lost export earnings, for Ethiopia of developing a domestic market is likely to be quite small in the near to medium term.

---

[3]These figures are based on the author's calculations undertaken for this survey; their derivation is described in section 26.4.

26.2.3 Environmental considerations of ethanol production

There is an ongoing debate about whether the overall life cycle of ethanol does more harm to the environment than good. One such issue, as part of this larger debate, concerns the net energy balance of ethanol, or whether the final fuel contains more useful energy than the energy required to produce it. While of great concern a decade ago, advances in technology have improved production efficiency such that ethanol now has a positive energy balance; based on an analysis of scholarly estimates of the energy balance of ethanol produced from sugarcane, current technology results in an overall energy balance of eight (Worldwatch Institute 2006). Furthermore, in a developing country such as Ethiopia, cultivation and processing are more heavily dependent on labour-intensive methods than capital-intensive methods, which helps to keep the energy balance relatively high on the positive side; while the labour to capital ratio may shift in the future, such a shift is unlikely to occur, given the low cost of labour and high unemployment, in the near to medium future.

Beyond issues surrounding the energy balance are those concerning the health of the ecosystem. Whenever there is the expansion of cultivation of a single crop, there is the potential to contribute to a reduction in biodiversity, the depletion of the soil's nutrient content, erosion, and the loss of habitat (Worldwatch Institute 2006). Fortunately, there are ways of mitigating these impacts, such as through diversification of monoculture ecosystems, and these should always be taken into consideration; likewise, care must be taken to properly dispose of toxic by-products and in the use of freshwater consumption in processing. With regard to Ethiopia, most of these issues are currently moot given production of ethanol from a by-product that otherwise is dumped into rivers; however, as cultivation expands, these issues will come to the forefront, and must be dealt with accordingly to ensure future production is not at the expense of the environment.

26.3 Demand Side

Having outlined the relevant issues for the supply side, the demand side will now be considered through two potential markets – an urban household market and an urban transport market. To estimate the potential impact of an urban household market, this chapter will make use of a pilot study,[4] conducted in 2004 to 2005 in Addis Ababa, undertaken to determine the feasibility of a household market for ethanol and accompanying ethanol stove. Consisting of 500 households, stratified by income, the pilot entailed two phases. The first was an initial baseline survey to capture household energy consumption patterns before introduction of the ethanol stove. Data gathered included, but was not limited to, fuels collected (e.g. who collects, quantity, frequency, etc.); fuels bought (e.g. frequency, quantity, etc.); cooking technology (e.g. type, frequency, location, mode of acquisition, identification of decision-maker, etc.); foods cooked on each technology; and behaviour patterns of cooking (e.g. cooking with or without a lid), supplemented by demographic data. Following the baseline survey, households took possession of the stove for a three month period, over which surveying took place every two weeks, with questions covering cooking technology employed, changes in cooking patterns, the reasons for these changes, perceptions of the improved cookstove, and willingness-to-pay. To estimate the potential impact of an urban transport market, estimates for Addis Ababa will be derived from country-wide and Addis Ababa specific data on car ownership and fuel consumption; the

---

[4]Called Project Gaia, the pilot was undertaken with support from the Shell Foundation and Dometic AB, the manufacturer of the stove.

choice of Addis Ababa as the defined market stems both from the Ethiopian government's aim to blend fuel for the Addis Ababa market only as well as current supply being unable to meet the full demand of this locale alone.

For each market, two scenarios will be considered. The first considers the level of benefits accruing in each market if supply were not to expand beyond distillation of molasses from the three sugar factories; this scenario reflects which market provides greater benefits for limited supply. The second considers the benefits accruing in the absence of a supply constraint, important to underscoring the point, pursued in Section 26.4, that, if feasible, supply should respond.

26.3.1 Urban household market

Households in Addis Ababa rely on a mix of fuels to meet their energy needs, with most households dependent on no less than two fuels. This portfolio approach to energy use is mirrored in the household fuel patterns of those in the pilot, shown in Table 26.3. The most prevalent fuel, across all income levels, is kerosene, with no less than 94% of households depending on it to meet their energy needs. The second most prevalent fuel is charcoal, with roughly equal usage across all income levels (67 to 74%). Thereafter, disparities between income levels begin to appear, with 58% of low income households using fuelwood, compared to just 32% of high income households; the pattern is reversed for kerosene, with 32% of upper income households using LPG, while no low income households, and just 4% of middle income households, do so.

Importantly, upon introducing ethanol into households, there was a clear shift in these energy patterns, as shown in Table 26.4. The greatest overall decline occurred for kerosene, with 88% of low income household decreasing their use by, on average, 83%; for middle and upper income homes, average reductions were 81% and 66%, respectively, with essentially similar percentages to those of low income households substituting kerosene for other fuels. Reductions in fuelwood were smallest, with just 37% and 36% of low and middle income households, respectively, reducing their usage; interestingly, 50% of upper income households reduced their fuelwood usage, the highest proportion of any income category, with the average reduction per household reducing its use being 50%. Reductions in charcoal range from 52% for upper income households to 66% for lower income households; the number of households substituting charcoal for other fuels is more varied, with 80% of middle income households reducing their usage, compared to just

Table 26.3. Fuel use prior to stove introduction, % of households.

	Low income	Middle income	Upper income
Kerosene	94%	98%	99%
Fuelwood	58%	40%	32%
Charcoal	69%	74%	67%
Eucalyptus leaves	15%	4%	1%
Electric	12%	18%	16%
Sawdust	14%	9%	8%
Agri-residue	5%	2%	0%
Dung	5%	4%	0%
Blt	5%	3%	1%
LPG	0%	4%	32%

Table 26.4. Substitution effects from introduction of ethanol.

Income	Per annum ethanol use (l)	Kerosene			Charcoal			Fuelwood		
		Households reducing baseline use	Reduction of those who reduced	Per annum absolute reduction (l)	Households reducing baseline use	Reduction of those who reduced	Per annum absolute reduction (l)	Households reducing baseline use	Reduction of those who reduced	Per annum absolute reduction (l)
Low	162	88%	83%	166	75%	66%	187	37%	44%	354
Middle	176	89%	81%	202	80%	60%	193	36%	66%	161
Upper	188	85%	66%	193	46%	52%	98	50%	50%	96

46% of upper income households. These reductions are replaced with ethanol usage of, on average, 162 litres per annum for low income households, 176 litres for middle income households, and 188 litres for upper income households.

Based on these data, projections for household demand can be constructed for two different scenarios: one where supply is constrained and one which traces a market from inception to full saturation. The first, or supply constrained scenario, considers the number of households which could be served given current molasses production and no expansion of land; such a scenario is important because it recognizes current conditions while placing upper bounds on the corresponding benefits if expansion were not to materialize. The second scenario estimates demand from the point of an ethanol stove and its fuel first being available for sale through to a fully mature market.

Before constructing either of these scenarios, it is important to understand what full market saturation would amount to, particularly for each income level. To do so, a number of factors must be considered. First, as the price of the stove is relatively more expensive than that for kerosene, fuelwood and charcoal, lower income households will be less likely to enter the market given the prohibitive upfront cost, although it should be noted that many low income households did express a willingness to participate in micro-credit schemes or other financing type options, if available. On the opposite end of the income spectrum, the ethanol-based stove is relatively less expensive than the price of an electric or LPG stove; as 32% of upper income homes use LPG, and 16% use electric stoves, it is unlikely that these will switch to an ethanol stove given the prior investment in a modern stove. However, it should be recognized that 99% of upper income households also use kerosene, so for those who are more reliant on kerosene than others in their income category, the tendency to switch will be greater. Finally, one last consideration: no less than 94% of households in any given income category use kerosene; as ethanol is likely to be kerosene's prime competitor, especially if the price of ethanol remains lower than that of kerosene, uptake, at a minimum, is likely to be considerable by both middle and upper income households.

Given these considerations, it is assumed that, upon reaching a fully mature market, uptake by low income households will be 30%, uptake by middle income households will be 85%, and uptake by upper income households will be 65%. As it will take time to reach these full saturation levels, it is assumed that such rates are not achieved until five years after the initial availability of the stoves in the market;[5] this also allows time for stove production to appropriately scale up.

Given these demand scenarios and market saturation levels, Table 26.5 displays the number of households which could be served, based on average per household ethanol demand, when supply is constrained, as well as the shortfall from demand if the market were fully mature. Note that demand is estimated for three separate prices of ethanol: 2.1 Ethiopian Birr (ETB) per litre (hereafter referred to as 'low'), 2.48 ETB/litre (hereafter referred to as 'medium'), and 3.3 ETB/litre (hereafter referred to as 'high'). The first is chosen as this is the price charged in the Project Gaia pilot, and the price which begets the demand data used here. The second and third prices are chosen as a means of capturing alternative potential initial investment costs and maintenance for the infrastructure and capital required to distribute the ethanol; this necessarily includes facilities to bottle the ethanol, or, as an alternative, distribute it through regular pumps at fuel stations, as well as transport to and from the points of sale. As an ordinary practice of business, the

---

[5]As other African stove programs, whose product was typically less expensive than the ethanol stove, have reached full saturation points in one to two years' time, five years' time for a more modern stove is quite reasonable.

final cost of a product should, necessarily, cover all costs; with a wholesale price of 1.65 ETB/litre in 2005, prices which include a margin of 50% and 100%, and hence retail prices of 2.48 ETB/litre and 3.3 ETB/litre, respectively, are included as a means of capturing and recovering any investment plus an appropriate profit. As the required investment to bottle and/or distribute through pumps is not immense, such margins seem adequate proxies.

Of course, with changes in price come changes in demand. However, as said above, ethanol's main competitor is kerosene; as the price of kerosene during the study was 2.43 ETB/litre, it is assumed that no loss of demand will occur with an ethanol price of 2.48 ETB/litre given both the high tendency in the study to prefer ethanol to kerosene and the near equality in prices. However, if the price of ethanol rises to 3.3 ETB/litre, demand is likely to fall considerably. To capture this drop, the quantity of ethanol for low income households is assumed to reduce to zero, with most of these households not being able to justify financially the high cost of the stove when the price of ethanol is considerably higher than the price of kerosene. For middle and upper income households, the number of households participating in the market is assumed to fall, as is demand for ethanol by those who participate; set at 15% for middle income households and 5% for upper income households, these reductions are chosen given these households are more able to absorb a higher price of ethanol, especially given their stated willingness to purchase a cleaner and more efficient fuel. Demand for ethanol by those households who do participate declines by 30% for middle income households, reflecting the large range of incomes and the ability of those at the lower end being much less able to absorb a price increase than those at the upper end, and by 5% for upper income households, again reflecting their ability to absorb a higher price for a fuel which is superior to kerosene. As demand for ethanol drops, households substitute back to alternative fuels at the corresponding rate of original substitution away from such fuels.

With regard to the ethanol prices used here, it is crucial that one see them as relative prices only, with the corresponding changes in ethanol demand depending entirely on the relative prices of all fuels used, especially kerosene; if the price of kerosene increases to 3.4 ETB/litre, it is likely that demand for ethanol will no longer decline, but rather will likely be similar to demand when the price of ethanol is 2.1 to 2.48 ETB/litre. As a basis of comparison, at the time of writing this study, the price of kerosene in Addis Ababa was over 4 ETB/litre.

Returning to Table 26.5, as indicated, when the price of ethanol is less than or approximately equal to the price of kerosene, roughly 143000 households could be served, based on 2005 and 2006 figures, when supply is constrained; compared to full market saturation levels, this is just 35% of full demand in the absence of a supply constraint. When molasses supply increases in 2007, the number of households able to be served increases to 191121, with now 45% of the full demand met. Compare this to a price of ethanol higher than

Table 26.5. Total number of households served with constrained supply.

Ethanol price		2005	2006	2007
Less than or equal to kerosene	# Households	141486	144722	191121
	Absolute shortfall	*260426*	*268686*	*234245*
	(% demand met)	35%	35%	45%
Higher than kerosene	# Households	164275	168032	221904
	Absolute shortfall	*151158*	*156423*	*111936*
	(% demand met)	52%	52%	66%

Table 26.6.   Demand in a maturing market (# households).

Ethanol price		2005	2006	2007	2008	2009
Less than or equal to kerosene	# Households	51965	126923	257381	384941	450118
	Ethanol demand ('000 litres)	*9141*	*22324*	*45223*	*67651*	*79065*
Higher than kerosene	# Households	39925	99151	200420	303696	353266
	Ethanol demand ('000 litres)	*5294*	*12984*	*26071*	*39233*	*45599*
		2010	2011	2012	2013	2014
Less than or equal to kerosene	# Households	462772	476008	489622	503626	518030
	Ethanol Demand ('000 litres)	*81287*	*83612*	*86004*	*88463*	*90994*
Higher than kerosene	# Households	363197	373585	384270	395261	406566
	Ethanol Demand ('000 litres)	*46881*	*48222*	*49601*	*51020*	*52479*

that of kerosene, i.e. a 'high' ethanol price; under this scenario, the number of households able to be served actually increases, as per household demand has decreased and supply remains unchanged. In 2005 and 2006, approximately 166000 households could be served, or 52% of full demand; in 2007, total households able to be served were 221904, or 66% of full demand.

Moving to the second scenario, Table 26.6 shows the number of households assumed to own a stove for each year of the ten year period over which the market matures, as well as the aggregate quantity of ethanol required to meet this demand.[6] As shown, in 2005, or in the 'assumed' year in which the stove and fuel become available on the market, total household demand for ethanol when the price is less than or roughly equal to that of kerosene – i.e. a low to medium ethanol price – is 9.1 million litres, well within current supply thresholds assuming all molasses is distilled; when the price of ethanol is significantly greater than that of kerosene, ethanol demand falls to 5.2 million litres, just under half of the current potential supply. The market, though, begins to take off thereafter and by the time full market saturation rates are reached in 2009, or year five of a maturing market, total demand is 79 and 45 million litres for a low/medium and high ethanol price, respectively. After ten years, these numbers, due to household growth, reach 91 and 52 million litres, respectively.

With estimates of demand specified as above, the remainder of this section will derive the corresponding benefits.

26.3.1.1 Household savings

As ethanol and the accompanying stove tested in the Project Gaia pilot are highly efficient, especially when compared to stoves for traditional and intermediate fuels, one of the greatest benefits of a household market is saved income. Despite the relatively high cost of the stove, the payback period for the average household is less than one year. This is exemplified in Table 26.7, which displays per household savings, both inclusive and exclusive of the price of the stove, assumed to be 450 ETB, for each income level. Middle income households are the greatest beneficiaries, in absolute terms, with annual savings of approximately 1150 ETB, inclusive of the cost of the stove, and 1600 ETB, exclusive of

---

[6]Note: these calculations also incorporate growth in the number of households in Addis Ababa, the rate being based on an extrapolation of recent trends.

Table 26.7. Per household savings, ETB.

Ethanol price	Income level	With stove	Without stove
<Kerosene	Low	1062	1512
	Middle	1198	1648
	Upper	634	1084
	Average	*965*	*1415*
~To kerosene	Low	1001	1451
	Middle	1132	1582
	Upper	563	1013
	Average	*899*	*1349*
>Kerosene	Low	0	0
	Middle	556	1006
	Upper	365	815
	Average	*157*	*607*

the cost of the stove, when the price of ethanol is approximately equal to or less than that of kerosene; net savings drop to 269 and 719 ETB, respectively, when the price of ethanol is relatively higher than the price of kerosene. While upper income households also see net savings inclusive or exclusive of the purchase price, low income households fare best given savings are not only near to those of middle income households, but, more importantly, by default, a greater proportion of income; although it is assumed that low income households will not participate in an ethanol market when ethanol's price is significantly higher than that of kerosene, when the two prices are roughly equal or the price of ethanol is lower than that of kerosene, low income households save, on average, approximately 1025 ETB, inclusive of the cost of the stove, and 1475 ETB, exclusive of the cost of the stove; this is approximately 36 and 51% of annual income, respectively, based on the corresponding income levels of low income households in the pilot. For middle income households, savings, as a percentage of annual income, are 22% of annual income, inclusive of the cost of the stove, and 15%, exclusive of the cost, when the price of ethanol is approximately equal to or less than that for kerosene, and, for a high price of ethanol, 5 and 9%, respectively. For upper income homes, whose annual expenditures are greater than 20 000 ETB and who make up just 12.45% of the total population of Addis Ababa, savings, when calculated based on 20 000 ETB, range from 2 to 5%, depending on the relative price of ethanol and whether the cost of the stove is included.

On an aggregate level, savings, when supply is constrained and excluding the cost of the stove, amount, on average, to 225 million ETB for a low ethanol price, 215 million ETB for a medium ethanol price, and 30 million ETB for a high ethanol price; this is 0.28, 0.27, 0.04% of 2005 GDP, respectively. Aggregate average savings inclusive of the stove are 153 million ETB, 143 million ETB, and 29 million ETB for a low, medium, and high price of ethanol, respectively; these savings now amount to 0.19, 0.18 and 0.04%, respectively, of 2005 GDP. When the market is able to mature, aggregate average per annum savings nearly triple for a low and medium price, while those for a high ethanol price jump quite dramatically – 287 million exclusive of the cost of the stove and 212 million ETB inclusive; savings now range from a low of 0.27% of 2005 GDP to a high of 0.74% of GDP, depending on the relative price of ethanol and whether the cost of the stove is included.

In addition to monetary savings from reduced energy expenditures, time is also saved. While there are no precise statistics on the number of hours saved by income, on average across all households, cooking time was reduced by 23%; these savings will mainly accrue to women, typically the primary cook. Less time spent cooking allows more time to be spent doing other, perhaps more productive, activities, such as pursuing education or employment.

26.3.1.2 Foreign exchange savings

Foreign exchange savings in an urban household market accrue from substitution away from kerosene, the entire supply of which is imported. As this price is unknown for Ethiopia, a proxy based on the wholesale prices of kerosene in the United States will be used (Energy Information Administration 2007). Clearly, the use of such prices is an imperfect indicator for Ethiopian import prices; the actual price paid by Ethiopia, while largely dependent on the world price of crude oil, will also depend on a diverse and immense set of factors – margins of the refiner, uncertainty, negotiating skills, politics, length of the contract, transport costs, etc. Yet, as crude oil prices are still the main driver of the price of kerosene, the use of US wholesale prices as a proxy will capture the underlying trend in the price of crude oil. In order to include influences other than those related to the price of crude oil, a range of foreign exchange savings will also be presented.

This being said, two separate 'proxy', or point, prices for kerosene are used: the average US wholesale price of kerosene in 2004 and the average US wholesale price of kerosene in 2005. The choice of these two prices has first to do with maintaining consistency with the data used throughout this study. Although 2005 is the reference year for much of the data, there is a likely lag between the time fuel is imported and the time it is distributed to the actual fuel pumps, especially as Ethiopia is characterized by poor transport links both into and within the country; to capture this lag, 2004 prices are used. However, as a check on the sensitivity of foreign exchange savings to prices – prices in 2005 were 43% higher than those in 2004–2005 prices are also used; further estimates are produced through both a 50% increase on the 2005 price and a 50% decrease on the 2004 price.

Caution is advised in considering the estimates. As the capital used to expand production of ethanol may be imported, these figures may be biased upward. However, as capital is a one-time deduction, while substitution away from kerosene is a recurring credit, the long-term gains from import substitution are likely to outweigh the short-term costs.

Table 26.8 presents estimates of savings in foreign exchange expenditures when supply is constrained, while Table 26.9 presents those for a maturing market. For constrained supply, the price of ethanol approximately less than or equal to the price of kerosene results in foreign exchange savings of 77 million ETB, or 8.8 million USD, based on the 2004 wholesale kerosene price and 2005 ethanol supply, and foreign exchange savings of 111 million ETB, or 12.5 million USD, based on the 2005 wholesale price; such savings would reduce total foreign exchange requirements in 2005 by 0.24 and 0.35%, respectively, with the total trade deficit reducing by 0.31 and 0.44%, respectively. When a range of 50% around these prices is estimated, foreign exchange savings based on 2005 ethanol supply for a low/medium price of ethanol range from a low of 38 million ETB, or 4.4 million USD, to a high of 167 million ETB, or 18.8 million USD; total foreign exchange expenditures are correspondingly reduced by 0.15 to 0.66%. When ethanol supply ramps up significantly in 2007, the foreign exchange savings range from 52 to 225 million ETB, or 5.9 to 25.4 million USD, with the 2004 and 2005 point proxies resulting in savings of 87 and 125 million ETB, or 11.8 and 16.9 million USD, respectively. Savings as a percentage of total foreign exchange expenditures now fall within a range of 0.16 to 0.70%, with the 2004 and 2005 point proxies resulting in a reduction in total foreign exchange expenditures of 0.33 and 0.47%, respectively; the reduction

Table 26.8. Foreign exchange savings in a constrained urban household market, ETB.

Ethanol price	Kerosene price		2005	2006	2007	Average
<Kerosene	2004	ETB	77 460 421	79 231 820	104 634 364	87 108 868
		USD	*8 752 590*	*8 952 748*	*11 823 092*	*9 842 810*
	2005	ETB	111 001 365	113 539 793	149 941 830	124 827 663
		USD	*12 542 527*	*12 829 355*	*16 942 580*	*14 104 821*
	50% increase	ETB	166 502 048	170 309 689	224 912 744	187 241 494
		USD	*18 813 791*	*19 244 033*	*25 413 869*	*21 157 231*
	50% decrease	ETB	38 730 210	39 615 910	52 317 182	43 554 434
		USD	*4 376 295*	*4 476 374*	*5 911 546*	*4 921 405*
~To kerosene	2004	ETB	77 460 421	79 231 820	104 634 364	87 108 868
		USD	*8 752 590*	*8 952 748*	*11 823 092*	*9 842 810*
	2005	ETB	111 001 365	113 539 793	149 941 830	124 827 663
		USD	*12 542 527*	*12 829 355*	*16 942 580*	*14 104 821*
	50% increase	ETB	166 502 048	170 309 689	224 912 744	187 241 494
		USD	*18 813 791*	*19 244 033*	*25 413 869*	*21 157 231*
	50% decrease	ETB	38 730 210	39 615 910	52 317 182	43 554 434
		USD	*4 376 295*	*4 476 374*	*5 911 546*	*4 921 405*
>Kerosene	2004	ETB	78 139 163	79 926 083	105 551 216	87 872 154
		USD	*8 829 284*	*9 031 196*	*11 926 691*	*9 929 057*
	2005	ETB	111 974 007	114 534 678	151 255 685	125 921 457
		USD	*12 652 430*	*12 941 771*	*17 091 038*	*14 228 413*
	50% increase	ETB	167 961 011	171 802 017	226 883 528	188 882 185
		USD	*18 978 645*	*19 412 657*	*25 636 557*	*21 342 620*
	50% decrease	ETB	39 069 581	39 963 042	52 775 608	43 936 077
		USD	*4 414 642*	*4 515 598*	*5 963 346*	*4 964 528*

in the overall trade deficit ranges from a low of 0.21% to a high of 0.90%, with the 2004 and 2005 point proxies resulting in a 0.42 and 0.60% reduction, respectively.

Interestingly, when supply is constrained, savings in foreign exchange expenditures actually increase ever so slightly when the price of ethanol is significantly higher than that of kerosene. As the higher price reduces per household ethanol consumption, more households are served in a supply constrained scenario. Furthermore, only middle and upper income households now participate in the market; even though these households are now using less ethanol per household, the absolute quantity is, on average, essentially the same absolute quantity as middle and low income households for a low/medium ethanol price. The balance between these two factors results in this slight increase.

The story is quite different for a maturing market. Whereas, when supply is constrained and the price of ethanol is less than or equal to that of kerosene, average foreign exchange savings over the three year period are 87 and 125 million ETB for the 2004 and 2005 proxy prices, respectively; average per annum foreign exchange savings over the ten year period for the maturing market jump almost three-fold to 216 and 310 million, respectively, or 24 and 35 million USD; the corresponding range, bounded by a 50% decrease in the 2004 proxy price and a 50% increase in the 2005 proxy price, of foreign exchange savings is 108 to 464 million ETB (12 to 52 million USD), or 0.34 to 1.44% of 2005 total foreign exchange expenditures. While foreign exchange savings start off relatively low – 30 and 43 million ETB for the 2004 and 2005 proxy prices, respectively, based on 2005 demand – the growth of the market more than makes up for low demand in the early years.

Table 26.9. Selected foreign exchange savings in a maturing urban household market, ETB.

Ethanol price	Kerosene price		2005	2014	Average
<Kerosene	2004	ETB	29 964 534	300 817 163	216 039 984
		USD	3 385 823	33 990 640	24 411 298
	2005	ETB	42 939 403	431 073 254	309 586 920
		USD	4 851 910	48 708 842	34 981 573
	50% increase	ETB	64 409 104	646 609 881	464 380 380
		USD	7 277 865	73 063 263	52 472 359
	50% decrease	ETB	14 982 267	150 408 581	108 019 992
		USD	1 692 912	16 995 320	12 205 649
~To kerosene	2004	ETB	29 964 534	300 817 163	216 039 984
		USD	3 385 823	33 990 640	24 411 298
	2005	ETB	42 939 403	431 073 254	309 586 920
		USD	4 851 910	48 708 842	34 981 573
	50% increase	ETB	64 409 104	646 609 881	464 380 380
		USD	7 277 865	73 063 263	52 472 359
	50% decrease	ETB	14 982 267	150 408 581	108 019 992
		USD	1 692 912	16 995 320	12 205 649
>Kerosene	2004	ETB	17 392 735	174 002 167	125 063 868
		USD	1 965 281	19 661 262	14 131 511
	2005	ETB	24 923 921	249 346 413	179 217 463
		USD	2 816 262	28 174 736	20 250 561
	50% increase	ETB	37 385 881	374 019 619	268 826 194
		USD	4 224 393	42 262 104	30 375 841
	50% decrease	ETB	8 696 368	87 001 083	62 531 934
		USD	982 640	9 830 631	7 065 755

Unlike when supply is constrained, foreign exchange savings decline in a maturing market when the price of ethanol is greater than that of kerosene; all who can participate in the market are now able to do so in any given year. On average, per annum foreign exchange savings are 125 and 179 million ETB, or 14 and 20 million USD, for the 2004 and 2005 proxy prices, respectively, with the corresponding 50% upwards and downwards range being 63 to 269 million ETB, or 7 to 30 million USD. The reduction in total foreign exchange expenditures now ranges from 0.19 to 0.84%, with the 2004 and 2005 point prices equating to a 0.39 and 0.56% reduction, respectively; the reduction in the overall trade balance ranges from 0.25 to 1.07%, with the 2004 and 2005 point prices generating a 0.50 and 0.71% reduction, respectively.

26.3.1.3 Improvements to health

Indoor air pollution, the primary cause of which is household fuel use, is estimated to cause 1.6 million deaths worldwide each year, 24% of which occur in Africa alone World Health Organization (2004). Beyond mortality, health impacts include, but are not limited to, coughing, wheezing, acute respiratory infection, chronic obstructive lung disease, adverse pregnancy outcomes in women and lung cancer. With these impacts come real costs to society, both with regard to the health care system (e.g. outpatient consultations, inpatient admissions) and those which accrue directly to the patient (e.g. health care, home treatment or traditional practitioner) (Hutton *et al.* 2006). In a developing country such as Ethiopia, where the expense of medical care and inadequacy of many hospitals is proportionally higher, such impacts are felt more severely by the population, with families incurring both financial harm, such as lost income due to an inability to work or the need to take care of ill family members, and personal sacrifices, such as foregoing school and lower nutrition; in aggregate, the economy as a whole suffers given lower productivity due to absences/sickness and lower overall education and skills levels, to name just a few.

Unfortunately, the degree of scientific evidence linking indoor air pollution to the health impacts listed above varies, as does the strength of evidence according to the population group (men, women, and children) (Hutton *et al.* 2006). As connecting the two is a difficult process made more complicated by the immense set of confounding factors, estimation here of the related benefits to health from a household market for ethanol is beyond the scope of this study. However, extensive air quality tests already conducted in Ethiopian homes show that World Health Organization (WHO) guidelines for carbon monoxide and particulates, two of the main indicators for the quality of air, can be achieved with the introduction of the ethanol-based stove; as levels prior to the introduction of the stove are several times the current WHO standards, one can only assume that, all else being equal, the introduction of ethanol into the home will result in health benefits, and thus associated declined in health-related costs.

26.3.1.4 Environmental benefits
With household use of kerosene, fuelwood, and charcoal comes the emission of the green-house gases carbon dioxide (CO_2), nitrous oxide (N_2O) and methane (CH_4), whose increasing concentration is, more and more, held by the scientific community to lead to global warming; with global warming comes changes in weather patterns (e.g. more frequent and severe storms, changes in rainfall patterns), the flooding of low lying areas, the spread of insect-borne diseases, and major disruptions to food supplies. Scientific consensus is fomenting around the belief that these changes in climate will occur unless greenhouse gas emissions are reduced. As Addis Ababa is highly dependent on these fuels, substitution away from these greenhouse gas emitting energy sources will help achieve this goal.

To estimate the reduction in greenhouse gas emissions from an urban household market for ethanol, and thus the benefits to the environment, emissions from CO_2 will be derived using the 2006 Intergovernmental Panel on Climate Change Guidelines for National Greenhouse Gas Inventories (Eggleston *et al.* 2006), including use of its emission factors. While nitrous oxide and methane are no doubt also important, they will not be estimated here given the high uncertainty in their calculation; this is due to their emission rates being highly dependent on the combustion technology used, which is not well known for the pilot study.

Calculated for each income level, aggregate results are shown for the constrained market, in Table 26.10, and, for the maturing market, in Table 26.11; also displayed in these tables are 95% confidence intervals for these estimates – i.e. there is a 95% probability that emissions reductions will fall somewhere within this range.[7] Given no variation in fuel reductions between a low and medium price for ethanol, the results for these two price levels are equal.

Beginning first with reductions in CO_2 emissions when supply is constrained, in 2005, expected reductions are approximately 190 000 metric tonnes when the ethanol price

---

[7]Confidence intervals were estimated by the IPCC assuming lognormal distributions, fitted to a data set, based on national inventory reports, IEA data and available national data.

Table 26.10. CO_2 reductions in a constrained urban household market, metric tons.

Ethanol price	Emissions factor	2005	2006	2007	Total	Average
	Default	189395	193726	255836	638957	212986
<Kerosene	Lower	170583	174484	230425	575491	191830
	Upper	212039	216888	286424	715350	238450
	Default	189395	193726	255836	638957	212986
~To kerosene	Lower	170583	174484	230425	575491	191830
	Upper	212039	216888	286424	715350	238450
	Default	158313	161934	213852	534099	178033
>Kerosene	Lower	144306	147606	194930	486843	162281
	Upper	175309	179318	236810	591438	197146

Table 26.11.   Selected CO_2 reductions in a maturing urban household market, metric tonnes.

Ethanol price	Emissions factor	2005	2014	Total	Average
	Default	71480	720589	5172694	517269
<Kerosene	Lower	64474	649798	4664653	466465
	Upper	79921	805862	5784673	578467
	Default	71480	720589	5172694	517269
~To kerosene	Lower	64474	649798	4664653	466465
	Upper	79921	805862	5784673	578467
	Default	39104	399432	2867634	286763
>Kerosene	Lower	35400	361122	2592779	259278
	Upper	43578	445653	3199262	319926

is roughly equal to or less than that of kerosene, with a 95% probability that emissions reductions will fall within a range of 170583 to 212039 metric tonnes; in 2007, expected reductions increase to approximately 256000 metric tonnes, with the 95% confidence interval being 230425 to 286424 metric tonnes. When the price of ethanol moves significantly above that of kerosene, per annum emissions decline; although the number of households able to be served in the market has increased, the participants (i.e. middle and upper income households) are now those whose overall fuel use prior to the introduction of ethanol was relatively cleaner, and whose substitution with regard to CO_2 reductions is less. Based on 2005 figures, CO_2 reductions are approximately 31000 metric tonnes lower – 158000 metric tonnes – or 84% of those levels when the price of ethanol is less than or equal to that of kerosene; the 95% confidence interval is 144306 to 175309. In 2007, CO_2 emissions reductions are now approximately 42000 metric tonnes less than those associated with low/medium ethanol price, with the expected value being 213852 metric tonnes and the 95% confidence interval being 194930 to 236810 metric tonnes.

For a maturing market, average per annum CO_2 reductions become much more significant, nearly tripling over the ten year period for a low/medium ethanol price, and doubling for a high ethanol price. In the final year of the ten year period, the expected reduction in CO_2 emissions for a price of ethanol approximately equal to or less than that of kerosene is

720 589 metric tonnes, or nearly three times 2007 levels in the supply-constrained scenario, with a 95% probability of falling between 649 798 and 805 862 metric tonnes; for a price of ethanol higher than that of kerosene, expected reductions in the tenth year are 399 432 metric tonnes, with a 95% probability of falling within a range of 361 122 to 445 653 metric tonnes. As the market has fully matured, such reductions will continue to be the norm.

In addition to reductions in greenhouse gas emissions comes reduced pressure on deforestation. While the use of fuelwood and charcoal by households is not the major cause of deforestation in Ethiopia, their use does, nonetheless, exacerbate the problem. Thus, while it will not result in a complete reversal of deforestation, the introduction of ethanol as an urban household fuel will alleviate some of this pressure. When supply is constrained, household use of fuelwood and charcoal will decline, on an annual basis, by approximately 25 000 and 32 000 metric tonnes, respectively, when the price of ethanol is roughly equal to or less than that of kerosene; when the price of ethanol rises above that of kerosene, annual fuelwood and charcoal usage decline by approximately 21 000 and 19 000 metric tonnes, respectively. For a maturing market, average annual reductions in charcoal for a low/medium ethanol price are now almost three-fold their levels when supply is constrained, rising to 68 000 metric tonnes; for fuelwood, a maturing market results in a doubling of average annual reductions – now 64 000 metric tonnes. When the price of ethanol is high, annual average reductions for charcoal and fuelwood reduce to 38 000 and 32 000 metric tonnes, respectively, although still nearly double per annum reductions when supply is constrained. While not begetting afforestation, substitution towards ethanol clearly will help alleviate pressure on deforestation within Ethiopia.

With the benefits in the household specified as above, the benefits of a transport market will now be quantified.

26.3.2 Urban transport market

To estimate the impact of an urban transport market, one must first have a picture of what vehicle ownership and fuel use in the capital city looks like, including the make, model, age, miles driven by each car, etc. Unfortunately, when it comes to such data, there is an immense dearth. As such, estimates for this market will be based more on assumption than those produced for the household market; these assumptions, however, will be tested through sensitivity analysis which, as will be seen in Section 26.4, still allows general conclusions to be drawn about the relative merits of each market.

That being said, in 2005, Addis Ababa was home to an estimated 100 033[8] vehicles, of which 30 507 were owned by households, and the remainder were owned by commercial enterprises (Ethiopia Central Statistical Authority 2004). To estimate the fuel type of these vehicles, country-wide ratios are applied to total ownership levels in Addis Ababa (Ministry of Transport and Communications 2006); based on these figures, 61%, or 61 020, of all vehicles in Addis Ababa are assumed to be gasoline powered, while the remaining 39 013 are assumed to be diesel powered.

With regard to fuel sales, estimates of consumption levels for both gasoline and diesel, as well as the corresponding levels of imports for each, are displayed in Table 26.12

---

[8]While household ownership of vehicles in Addis Ababa is known, that for commercial enterprises is not. To estimate commercial ownership, the ratio of households in Addis Ababa owning cars to total country-wide household ownership, also known, was used as a proxy for commercial ownership. This ratio – 60% – was then applied to country-wide ownership – 115 590 – to derive an estimate for Addis Ababa Ministry of Transport and Communications (2006). The known household figure and the estimated commercial figure were then added together to arrive at an overall figure.

Table 26.12.   Consumption and import of diesel and gasoline, litres.

Year	Consumption, country-wide		Consumption, Addis Ababa		Imports	
	Diesel	Gasoline	Diesel	Gasoline	Diesel	Gasoline
2005/06	848 138 824	220 180 000	510 147 501	132 436 193	1 104 007 059	202 650 000
2006/07	899 027 059	227 226 250	540 756 294	136 674 446	1 247 528 235	228 995 000
2007/08	952 968 235	234 042 500	573 201 403	140 774 356	1 409 707 059	258 763 750
2008/09	1 010 145 882	241 063 750	607 593 218	144 997 571	1 592 969 412	292 402 500

(Ethiopian Petroleum Enterprise MIS Unit 2006). These figures are constructed by applying the ratio of vehicles owned in Addis Ababa to country-wide ownership, or 60%, to the country-wide information; while clearly a simplistic assumption as actual consumption will be based on the relative driving distances, fuel efficiencies, the actual number of commercial vehicles located and purchasing fuel in Addis, among many other factors, such will have to be a second best calculation in the absence of more detailed data.

As with the household market, two scenarios will be explored. The first mirrors the supply constrained scenario constructed for the household market. The second, also similar to that of the household market, considers a ten year period over which supply is unconstrained.

With regard to calculating actual demand for ethanol in a transport market, ethanol is assumed to replace both diesel and gasoline on a one-to-one basis; fuel blending limits of 10% are imposed. While this is well short of the feasible limit of 25% for gasoline blending, at levels greater than 10%, liabilities, such as equipment corrosion and fuel knock, become a concern. Likewise, vehicle manufacturers will typically only warranty ethanol blends of 10% or less with gasoline in conventional spark-ignition engines (Worldwatch Institute 2006). Finally, beyond the 10% level, changes in the mechanics of the vehicle are required.

Returning to projected fuel demand, to estimate such when supply is unconstrained, trends for current consumption of gasoline and diesel are extrapolated out over a ten year time horizon, with the resulting quantity for each fuel then multiplied by 10% to arrive at the quantity of ethanol required to achieve maximum fuel blending levels.

With the assumptions thus described as above, it is possible to estimate potential demand in a transport market. Beginning with 'constrained supply', as shown in Table 26.13, if all ethanol is devoted to the transport market in Addis Ababa, fuel blending for diesel would reach just 4.9% based on 2005 diesel consumption levels; by 2007, this level increases to 5.8%, although still well short of the upper 10% feasibility limit. This is in contrast to gasoline, where, if all short-term ethanol supply is used as a one to one substitute for gasoline, blend levels could reach 18.7% in 2005. Given the imposed limit here of 10%, it would be feasible for all gasoline in Addis Ababa to be sold with a 10% ethanol blend, with the remaining used as a diesel substitute; this would imply diesel blend levels of 2.3% in 2005, and 3.4% in 2007.

As shown in Table 26.14, if supply of ethanol is not an issue – i.e. the market could meet demand for full blending of 10% for each fuel – total demand for ethanol would be 64 million litres in 2005, compared to 24.8 million litres when supply is constrained; of this, 51 million litres would go towards diesel blending, an increase of 39.5 million litres when supply is constrained, and 13 million which would be used for gasoline blending. In 2007, total ethanol demand would increase to 71 million litres, an increase of 37.5 million litres,

Table 26.13.   Maximum fuel blending levels given constrained supply in Addis Ababa.

	Ethanol devoted to either gasoline or diesel		10% ethanol devoted to gasoline, with remaining devoted to diesel	
	Diesel	Gasoline	Diesel	Gasoline
2005	4.9%	18.7%	2.3%	10.0%
2006	4.7%	18.6%	2.2%	10.0%
2007	5.8%	23.8%	3.4%	10.0%

Table 26.14.   Demand in a transport market when supply is unconstrained, '000 litres.

Year	2005	2006	2007	2008	2009
Demand	64258	67743	71398	75259	79349
Year	2010	2011	2012	2013	2014
Demand	83672	88240	93069	98173	103569

or 112%, on constrained supply of 33.5 million litres. By 2014, total demand for ethanol would be approximately 104 million litres of ethanol.[9]

With demand as specified when supply is both constrained and unconstrained, the benefits and costs are estimated as follows.

26.3.2.1 *Consumer savings*

Consumer savings accrue directly to households and commercial entities who own a vehicle. This is due to ethanol's 2005 wholesale cost of 1.65 ETB/litre being lower than either the import price of gasoline, estimated to be 3.78 ETB/litre in 2005, or of diesel, estimated to 3.94 ETB/litre in 2005; the estimation procedure of these import prices will be discussed in detail below. As a comparison, retail prices of diesel and gasoline were 5.44 and 7.77 ETB/litre, respectively, in 2005. It is worth noting that ethanol is likely to remain cheaper than either diesel or gasoline for some time, an outcome which requires the price for a barrel of crude oil to fall below approximately \$33. Savings will thus accrue as a lower priced commodity (i.e. ethanol) is blended with a higher priced commodity (i.e. gasoline and/or diesel), with the overall price of the blended commodity necessarily falling.

To estimate these savings, those ethanol prices used for estimation in the household market will also be used here. Not only does it make sense to use these prices so as to provide a basis of comparison when calculating the costs associated with an alternative market, but, as with the household market, multiple prices are necessary in the absence of data on the cost of building and maintaining the required infrastructure. In general, ethanol can be distributed through the same infrastructure used for petroleum products, although, in colder climates, dedicated pipelines may be required given ethanol's high affinity for water, which may cause it to separate from the petroleum products with which it is blended. Fortunately, while the climate of Addis Ababa provides no cause for concern, other problems requiring additional investments to be made into the existing infrastructure may

---

[9]To estimate consumption levels for gasoline and diesel over a ten year period, trends for current consumption of gasoline and diesel were extrapolated out to arrive at annual figures.

arise; for example, the relatively high solvency of ethanol may initially cause a release of deposits left by gasoline and diesel when introduced into tanks and facilities previously used for petroleum-based fuels only (Worldwatch Institute 2006).

Thus, as a means of capturing potential initial investment costs and maintenance for distribution infrastructure, margins employed for the household market are duplicated here; as the current infrastructure is likely suitable, such margins seem adequate proxies. Clearly, these estimated retail prices – i.e. recall that the 'high' price is 3.3 ETB/litre – are quite favourable when compared to the retail price of gasoline and petrol, although the 'high' retail price of ethanol is quite close to the corresponding imported wholesale prices; however, it is more appropriate to consider the retail prices for purposes of consumer savings because these are the prices which include the normal cost of distribution, maintenance and profit by the oil companies selling the petrol.

Given these prices, the blended per litre cost of gasoline and diesel are first calculated for the supply constrained scenario, the results of which are displayed in Table 26.15; this table also includes the corresponding per litre savings given blending; aggregate savings, along with average savings per vehicle, are displayed in Table 26.16.

Table 26.15. Savings for fuel blending in a transport market when supply is constrained, ETB.

		Ethanol < kerosene		Ethanol ~ kerosene		Ethanol > kerosene	
		Retail price	Per litre savings	Retail price	Per litre savings	Retail price	Per litre savings
2005	Diesel	5.36	0.08	5.37	0.07	5.39	0.05
	Gasoline	7.56	0.61	7.60	0.57	7.68	0.49
2006	Diesel	5.37	0.07	5.38	0.06	5.39	0.05
	Gasoline	7.20	0.57	7.24	0.53	7.32	0.45
2007	Diesel	5.33	0.11	5.34	0.10	5.37	0.07
	Gasoline	7.20	0.57	7.24	0.53	7.32	0.45

Table 26.16. Consumer savings in a transport market when supply is constrained, ETB.

		Average per vehicle savings, ETB			Aggregated total consumer savings, ETB		
		Ethanol < kerosene	Ethanol ~ kerosene	Ethanol > kerosene	Ethanol < kerosene	Ethanol ~ kerosene	Ethanol > kerosene
2005	Diesel powered	989	878	634	38 578 272	34 246 879	24 717 815
	Gasoline powered	1317	1236	1057	80 388 769	75 422 412	64 496 426
2006	Diesel powered	969	861	621	39 056 475	34 671 392	25 024 209
	Gasoline powered	1230	1148	969	77 494 411	72 369 119	61 093 477
2007	Diesel powered	1563	1388	1002	64 844 645	57 564 184	41 547 168
	Gasoline powered	1230	1149	970	79 819 060	74 540 021	62 926 137

On a per litre basis, consumer savings to owners of diesel-powered vehicles are not significant – 0.08 to 0.11 ETB/litre for a low ethanol price, dropping to 0.05 to 0.07 ETB/litre for a high ethanol price; this is mostly due to the low levels of fuel blending for diesel (i.e. 2.3 to 3.4%, depending on the year). Per litre savings for gasoline blending are better, with a low ethanol price yielding savings of approximately 0.60 ETB/litre, or about an 8% reduction off the retail price, and savings per litre of approximately 0.55 and 0.47 ETB for a medium and high ethanol price, respectively.

On a per vehicle basis, these sums become more impressive, with savings ranging from an annual low of 634 to 1002 ETB, depending on the year, per diesel-powered vehicle when ethanol prices are high, to an annual high of 989 to 1563 ETB per diesel-powered vehicle when ethanol prices are low; for those owning gasoline-powered vehicles, annual savings per vehicle range from 970 to 1057 ETB, depending on the year, when ethanol prices are high and 1230 to 1317 ETB when ethanol prices are low. On an aggregate level, overall annual savings from both diesel and gasoline blending range from a low of 90 to 105 million ETB, or 0.11 to 0.13% of 2005 GDP, to a high of 119 to 145 million ETB, or 0.15 to 0.18% of 2005 GDP, depending on whether the ethanol price is high or low, respectively.

As shown in Table 26.17, when supply is unconstrained, savings per litre from diesel blending nearly triple, increasing to 0.33 ETB/litre for a low ethanol price, 0.30 ETB/litre for a medium ethanol price, and 0.21 ETB/litre for a high ethanol price. On a per vehicle basis, annual savings range, on average over the ten year period,[10] from 3192 ETB when the price of ethanol is high, to a high of 4983 ETB when the price of ethanol is low. For gasoline, as 10% blending already occurs under the short-term scenario, there is no change in per litre savings; over the ten year period, average annual savings to gasoline-powered vehicles continue to be approximately 1320 ETB when ethanol prices are low and 972 ETB when ethanol prices are high. Savings to diesel vehicles are much higher at maximum fuel blending levels than those for gasoline blending given that, despite there being just over one and a half gasoline-powered vehicles for each diesel-powered vehicle, diesel-powered vehicles consume almost four times the quantity of fuel consumed by gasoline vehicles.

Of course, one glaring problem with these calculations is the assumption of a stable diesel and gasoline retail price over the ten year time frame, which is surely not realistic. To remedy this unlikely assumption, the retail price, before blending, of each fuel is varied upwards and downwards by 50%; this is done to estimate a range of savings which more appropriately captures the realm of potential savings over a ten year period given volatile and unstable prices for crude oil. For diesel, the resultant retail prices are 8.16 and 2.72 ETB/litre, for a 50% increase and decrease, respectively, in the original retail price of 5.44

Table 26.17. Savings for diesel-powered vehicles when supply is unconstrained.

Ethanol price	Retail price, ETB/L	Per litre savings, ETB	Average savings per vehicle, ETB	Aggregated savings, ETB
<Kerosene	5.11	0.33	4983	287 868 201
~Kerosene	5.14	0.30	4423	255 547 669
>Kerosene	5.23	0.21	3192	184 442 500

[10]These figures incorporate growth in the absolute number of vehicles in Addis Ababa over the ten year period, with the absolute number owned by commercial enterprises growing by 1% year on year and the number of vehicles owned by households both growing due to growth in the number of households residing in Addis Ababa, as detailed in the household market, and in the percentage of households owning vehicles.

ETB/litre, and for gasoline, 12.26 and 3.89 ETB/litre, respectively, based on the 2006/2007 original retail price of 7.77 ETB/litre. While consumption will undoubtedly respond to these changes in price, the response, even in developing countries, is often very minor; as no estimates exist for Ethiopia, it is assumed that demand will increase or decrease by 5% with a decrease or increase, respectively, in prices. Results for these sensitivities are shown in Table 26.18.

Beginning with sensitivities to changes in the diesel price, a 50% increase results in a blended price of 7.55, 7.59, and 7.67 ETB/litre, when the price of ethanol is low, medium, and high, respectively; on a per litre basis, corresponding savings are 0.61, 0.57, and 0.49 ETB, respectively. On a per vehicle basis, these per litre savings translate into annual average savings over the ten year period ranging from 6888 to 8588 ETB, depending on the price of ethanol. Compare this to a 50% decrease in the price of diesel. Savings per litre become negligible, and, in fact, turn negative when the price of ethanol is high; i.e. the 'high' price of ethanol (e.g. 3.3 ETB/litre) is now greater than that of diesel. The blended diesel price now ranges from 2.66 to 2.78 ETB/litre, with corresponding savings of 0.06 to −0.06/ETB/litre. Correspondingly, annual savings per vehicle average 971 ETB over the ten year period when the price of ethanol is low and, when the price of ethanol is high, consumers now pay more, or 908 ETB, per year. The range of estimates is thus immense – annual savings when the price of ethanol is low are almost one order of magnitude larger for a 50% increase than a 50% decrease – although owners of vehicles will always save annually unless a high price of ethanol comes about at the same time as a large drop in the price of diesel.

Similar results, excluding negative savings, occur for changes in the price of gasoline. A 50% retail price increase produces per litre savings of 0.90 to 1.02 ETB, depending on the price of ethanol, and a commensurate decrease produces per litre savings of 0.06 to 0.18 ETB, again, depending on the price of ethanol. On a per vehicle basis, an increase equates to an annual average saving of 1850 ETB when the price of ethanol is high, and an average annual saving of 2098 ETB when the price of ethanol is low; corresponding annual average per vehicle savings are 127 and 408 ETB, respectively, for a price decline. As the price of gasoline swings, then so do savings to owners of vehicles, with the range, although never negative, quite large.

Table 26.18.   Sensitivity of consumer savings to fuel retail prices in a transport market.

	Changes in retail price	Ethanol price	Retail price, ETB/L	Per litre savings, ETB	Average annual savings per vehicle, ETB	Aggregated annual savings, ETB
Diesel	50% increase	<Kerosene	7.55	0.61	8588	387 108 650
		~Kerosene	7.59	0.57	8057	363 153 907
		>Kerosene	7.67	0.49	6888	310 453 472
	50% decrease	<Kerosene	2.66	0.06	971	43 774 141
		~Kerosene	2.70	0.02	384	17 297 846
		>Kerosene	2.78	−0.06	−908	−40 950 003
Gasoline	50% increase	<Kerosene	11.24	1.02	2098	146 943 816
		~Kerosene	11.28	0.98	2021	141 517 531
		>Kerosene	11.36	0.90	1850	129 579 702
	50% decrease	<Kerosene	3.71	0.18	408	28 547 975
		~Kerosene	3.74	0.14	322	22 550 501
		>Kerosene	3.83	0.06	127	8 910 532

Unlike in the household market, it is difficult to estimate savings as a percentage of annual income given corresponding income figures for those owning vehicles are not available. However, with the starting price of a vehicle being approximately 100 000 ETB, per consumer savings are likely to be no more than 10% of income given the annual income required to own a vehicle in Addis Ababa, and even this is highly optimistic.

26.3.2.2 Foreign exchange savings

As Ethiopia imports its entire supply of gasoline and diesel, the use of ethanol as a substitute for one or both fuels would reduce imports by the same quantity, thereby saving valuable foreign exchange. As with the household market, to estimate the amount of foreign exchange that could be saved through fuel blending, proxy prices for the import prices are used. For gasoline, these prices are estimated using wholesale prices from three select international locations – the United States, Rotterdam, and Singapore (Energy Information Administration 2007). While the problems of using a 'portfolio' approach to estimating proxy prices for fuel imports described for the household market are also applicable here, this set of prices does a better job at capturing those regional variations unrelated to crude oil-although ranking in terms of which is most and least expensive vary the lowest price at any given time is approximately 70 to 90% of the highest price. For diesel, only United States wholesale prices are used (Energy Information Administration 2007). As with the household market, two separate 'proxy' prices are used: average prices in 2004 and average prices in 2005. For these prices, the 2005 gasoline price is 32% higher than that for 2004; for diesel, the corresponding difference is 43%.

Given these assumptions, estimated foreign exchange savings for the short-term market are displayed in Table 26.19. Using 2004 prices, foreign exchange savings accruing due to gasoline blending are approximately 37 million ETB, or 4.1 million USD, in 2005, increasing to 39 million ETB, or 4.4 million USD in 2007; for diesel, foreign exchange savings are 32 million ETB, or 3.6 million USD, in 2005, and 54 million ETB, or 6.1 million USD, in 2007. Total foreign exchange savings amount to 69 million ETB in 2005, or 7.7 million USD, increasing to 93 million ETB, or 10.5 million USD, in 2007.

In relation to overall foreign exchange requirements, fuel imports priced at the 2004 proxy price results in a reduction of 2005 import expenditures by 0.21%, based on 2005 blending levels, and by 0.29%, based on 2007 blending levels; the resultant decrease in the trade balance is 0.27 and 0.37%, respectively. As gasoline and diesel prices rise by 32 and 43%, respectively, between 2004 and 2005, foreign exchange savings accruing to blending with each fuel rise by an equivalent amount when 2005 prices are used, with total foreign exchange savings now rising by 37%; 2005 expenditures on imports would now be reduced by 0.29% based on 2005 fuel blending levels, and 0.40% based on 2007 levels, with the overall trade balance falling by 0.37 and 0.51%, respectively.

Finally, to reflect the volatility of the price of crude oil, consider a 50% increase on 2005 prices and a 50% decrease on 2004 prices, as was done for the household market. Indeed, the resultant range of potential foreign exchange savings is immense. Total foreign exchange savings, based on 2005 ethanol supply levels, range from a low of 34 million ETB to a high of 140 million ETB (3.9 to 15.9 million USD); this translates into a 0.11 to 0.44% reduction in 2005 expenditures on imports, or a 0.14 to 0.56% reduction in the trade balance. Based on 2007 supply, foreign exchange savings range from 46 to 192 million ETB (5.2 to 21.6 million USD), or a 0.14 to 0.60% reduction in expenditure on imports, and a 0.18 to 0.76% reduction in the trade balance. While the absolute range is large, the resultant impact on the trade balance is relatively less significant, although nonetheless still important.

Table 26.19. Foreign exchange savings in a transport market when supply is both constrained and unconstrained.

| | | | Foreign exchange savings | | | |
| | | | Supply constrained | | | Supply unconstrained |
			2005	2006	2007	10 year annual average
Diesel	2004 price	ETB	31879051	32274213	53584198	185586221
		USD	*3602153*	*3646804*	*6054712*	*20970194*
	2005 price	ETB	45452602	46016017	76399427	264605640
		USD	*5135887*	*5199550*	*8632704*	*29898942*
	50% decrease	ETB	15939525	16137107	26792099	92793111
		USD	*1801076*	*1823402*	*3027356*	*10485097*
	50% increase	ETB	68178903	69024026	114599141	396908460
		USD	*7703831*	*7799325*	*12949055*	*44848414*
Gasoline	2004 price	ETB	36684825	37858821	38994497	42191751
		USD	*4145178*	*4277833*	*4406158*	*4767430*
	2005 price	ETB	48471647	50022847	51523414	55747946
		USD	*5477022*	*5652299*	*5821855*	*6299203*
	50% decrease	ETB	18342413	18929411	19497248	21095876
		USD	*2072589*	*2138916*	*2203079*	*2383715*
	50% increase	ETB	72707470	75034271	77285121	83621918
		USD	*8215533*	*8478449*	*8732782*	*9448804*
Total	2004 price	ETB	68563876	70133034	92578694	227777972
		USD	7747331	7924637	10460869	25737624
	2005 price	ETB	93924249	96038865	127922841	320353586
		USD	*10612909*	*10851849*	*14454558*	*36198145*
	50% decrease	ETB	34281938	35066517	46289347	113888986
		USD	*3873665*	*3962318*	*5230435*	*12868812*
	50% increase	ETB	140886373	144058297	191884262	480530378
		USD	*15919364*	*16277774*	*21681838*	*54297218*

When supply is able to respond, annual average foreign exchange savings, also shown in Table 26.19, become much more considerable – more than double in 2007 when either 2004 or 2005 prices are used and, in ten years' times, reaching 286 million ETB when 2004 prices are used and 403 million ETB when 2005 prices are used (32.3 and 45.5 million USD, respectively).

While the savings potential is immense, it can also be quite volatile. When 2004 prices are decreased by 50%, annual average foreign exchange expenditures fall to a low of 113 million ETB, or 12.9 million USD; however, when the 2005 price is increased by 50%, average annual foreign exchange expenditures jump to 481 million ETB, or 54.3 million USD. With the latter being roughly four times that of the former, large swings, when aggregated, lead to large estimates of potential foreign exchange savings over the long term.

26.3.2.3 Improvements to health

Vehicle emissions are one of the key contributors to air pollution in urban areas the world over, with Addis Ababa being no different; passenger cars and commercial automobiles alike emit a long list of pollutants, some of which include carbon monoxide, carbon dioxide, particulate matter, methane, non-methane volatile organic compounds, sulphur dioxide, and oxides of nitrate. Evidence detailing the harm done to human health by this air pollution, particularly in urban settings, is on the rise (Krzyzanowski 2005). Long and growing, the list of ills includes not only the more widely known and understood respiratory ailments, such as reduced lung function and lung inflammation, but also those that impact the cardiovascular system. As with indoor air pollution, these ailments not only lead to increased medication and hospital admissions, but, more severely, a reduction in the life expectancy of the average population by a year or more. (Hutton *et al.* 2006)

Likewise, as with indoor air pollution, while recent assessments suggest these impacts are considerable, actual quantification poses quite a challenge. This is due, in part, to limited availability of information on the nature of the links between outdoor air pollution and health, and, in part, on the absence of actual data on exposure, both of which are further compounded by the complex and dynamic set of factors which determine this exposure (Cohen *et al.* 2004).

For all of the complexity and lack of data, even less is known about the impacts of interventions. While not all of the evidence is conclusive, fortunately, several studies do indicate that reducing pollution may reduce acute asthma attacks and the need for related medical care in children, while long-term decreases are shown to be correlated with gains in life expectancy, declines in bronchial hyperactivity, and a decrease in the average annual trend in deaths from both all causes and respiratory and cardiovascular disease. Unfortunately, for the few intervention studies that do exist, most are not specific to traffic-related air pollution (Krzyzanowski 2005).

Hence, as with the household market, in the absence of more concrete data and information, especially on interventions, any attempt to quantify benefits from a reduction in harmful emissions would be immensely difficult and certainly beyond the scope of this chapter. Here, then, the best that can be achieved is to acknowledge the likely benefit to health and assume its existence; to do more would be utterly circumspect.

26.3.2.4 *Environmental benefits*

The transport sector directly contributes to the production of the greenhouse gases carbon dioxide (CO_2), nitrous oxide (N_2O) and methane (CH_4), the impact of which has already been discussed above. As with ethanol's use in the household, ethanol blending with gasoline and diesel can act as a means of reducing such emissions.

To estimate the reduction in greenhouse gas emissions from fuel blending, and thus the benefits to the environment, emissions from CO_2 will be derived, as earlier, using the 2006 Intergovernmental Panel on Climate Change Guidelines for National Greenhouse Gas Inventories, including use of its emission factors (Eggleston *et al.* 2006). Again, while nitrous oxide and methane are no doubt also important, they will not be estimated here as emission rates are highly dependent on the combustion and emission control technology of the vehicle, and are highly uncertain in the absence of this information (Eggleston *et al.* 2006). Furthermore, unlike CO_2, emissions of CH_4 and N_2O are related to vehicle miles travelled rather than fuel consumption (United States Environmental Protection Agency 2005); while data on fuel consumption is known, miles travelled, not to mention the combustion and control technology of those vehicles, are unknown for Ethiopia.

As a word of caution, in engines that are poorly calibrated to run on ethanol, N_2O emissions can actually increase, as can emissions of volatile organic compounds. However, the next generation of biofuels, including Fischer-Tropsch diesel and di-methyl ether

Table 26.20. CO_2 reductions in a transport market when supply is both constrained and unconstrained, metric tons.

| | | | Avoided CO_2 emissions | | |
			Default	Lower	Upper
	2005/06	Diesel	30469	29853	30757
		Gasoline	29553	28785	31130
		Total	60022	58638	61888
Supply	2006/07	Diesel	30847	30223	31139
constrained		Gasoline	30498	29706	32127
		Total	61345	59929	63265
	2007/08	Diesel	51215	50178	51699
		Gasoline	31413	30597	33090
		Total	82628	80775	84789
	Annual	Diesel	182136	178449	183857
	average	Gasoline	34482	33586	36323
Supply		Total	216618	212035	220180
unconstrained	10 year	Diesel	1669695	1635896	1685468
	aggregate total	Gasoline	339889	331061	358036
		Total	2009584	1966956	2043504

(DME), can be tailored to meet certain emission specifications (Worldwatch Institute 2006). Likewise, it is worth noting that emissions of CH_4 and N_2O make up only approximately 5 to 6% of the GHG emissions from passenger vehicles, with CO_2 accounting for the remaining 94 to 95%. (Eggleston *et al.* (006)

Table 26.20 displays estimates of reductions in CO_2 for both constrained and unconstrained supply. When supply is constrained, CO_2 reductions from diesel blending in 2005 are estimated to be 30469 metric tonnes, with a 95% confidence interval of 29853 to 30757 metric tonnes. For gasoline, CO_2 reductions are estimated to be 29553 metric tonnes in 2005, with a 95% confidence interval of 28785 to 31130 metric tonnes. In sum, this equates to a total reduction in CO_2 emissions of 60022 metric tonnes. In 2007, total annual emissions reductions increase by approximately a third, reaching an estimated 82628 metric tonnes with a 95% probability of falling within a range of 80775 to 84789 metric tonnes.

If supply is unconstrained, approximately 2 million metric tonnes less of CO_2 will be emitted over a ten year period; with 95% probability, reductions in CO_2 emissions over the ten year period will fall by 1.97 to 2.04 million metric tonnes. Annually, this equates to roughly 216600 metric tonnes, with approximately 182100, or 84%, due to diesel blending, and the remaining 34500 due to gasoline blending.

26.4 Discussion/Results

Having outlined the benefits accruing to each market for ethanol, it is now possible to compare the two and determine which market provides a greater return to Ethiopia. As the overall question posed by this chapter concerns where to direct Ethiopia's limited supply, only those benefits in the constrained market will be considered. However, as shown throughout Section 26.3, with benefits in both markets increasing substantially when supply is unconstrained, a discussion of the required supply response to simultaneously serve demand in both markets should, and will, be considered. That being said, benefits accruing directly to consumers are first discussed, followed by a comparison of aggregate benefits.

26.4.1 Benefits to the consumer

Beginning first with savings accruing directly to the consumer, while average savings are quite similar when the price of ethanol is approximately less than or equal to that of kerosene, households save less, in absolute terms, per annum in the first year of participating in a household market compared to those consumers who own vehicles, and more in years thereafter; average per annum savings are approximately 1218 and 1108 ETB for gasoline- and diesel-powered vehicles, respectively, and 932, inclusive of the stove, and 1328 ETB, exclusive of the cost of the stove, for those participating in a household market. The story changes, though, when the price of ethanol is greater than that of kerosene. For those participating in a household market, average savings are now 157 ETB, inclusive of the cost of the stove, and 607 ETB, exclusive of the cost of the stove, compared to benefits of approximately 999 ETB for owners of gasoline-powered vehicles and 752 ETB for owners of diesel-powered vehicles; once the stove has been purchased, however, the average consumer for each market realizes benefits which are not immensely different from one another.

Yet, the use of averages and absolute values masks a glaring difference between these two markets – the recipient of the benefits and the savings as a percentage of their income. Indeed, in the transport market, it is only those in the higher reaches of the upper income bracket who benefit, as it is only at such levels that vehicle ownership is feasible, as well as commercial enterprises. Savings, as a percentage of annual income, are unlikely to be greater than 10%, and, given the income required to purchase a vehicle, likely fall closer to 5%; for a commercial enterprise, these savings are likely to be insignificant for most, excluding, of course, those whose costs and revenues are highly dependent on transport. Compare this to the household market, where, for a price of ethanol roughly equal to or less than the price of kerosene, not only do all income levels benefit but, more importantly, low income households gain the most. While savings, as a percentage of income, fall as income rises, benefits in the household market are essentially regressive, while those in the transport market are highly progressive, remaining out of reach by all but the wealthiest; while low income households do not participate in a household market when the price of ethanol is higher than that of kerosene, benefits still accrue to those who are relatively less well off, while those for the transport market simply translate to lower savings for the wealthy and businesses. For the third poorest country in the world, for a city whose unemployment rate is near 45%, and for an urban area whose residences have a 15% greater chance of being poor compared to those living elsewhere in the country (Ministry of Finance and Economic Development 2002), clearly such a difference not only matters, but is important for lifting households out of poverty.

Additional benefits accrue beyond monetary advantages. The household market makes better use of limited supply given such benefits as time savings due to shortened cooking periods and related gender impact. Likewise, although it is too difficult to say which market creates greater benefits for health in the absence of data, given that cooking in Addis Ababa typically occurs indoors, with those households who utilize the 'dirtiest' cooking fuels having little or no ventilation, one can but only surmise that the health benefits in an urban household market will outweigh those in an urban transport market, especially given the relative displacement of harmful emissions; however, such a conclusion is highly uncertain and should in no way be taken as fact.

26.4.2 Aggregate benefits

On an aggregate level, the overall balance of benefits tips towards the household market. Consider foreign exchange savings – those in the household market, no matter the wholesale import price of the fuel displaced nor the relative price of ethanol, are always higher than those in the transport market; in the household market, the range of savings is 43.5 to 188.8 million ETB, compared to the corresponding transport market range of 38.5 to 153.3 million ETB. For the actual point prices, foreign exchange savings are approximately 87.1 and 124.8 million ETB for the 2004 and 2005 prices, respectively; the corresponding savings in the transport market are 77 and 106 million ETB, respectively, not a great difference, but a difference nonetheless.

The results are more conclusive for CO_2 emissions, with reductions in the household market being, on average, three times those in the transport market when the price of ethanol is approximately equal to or less than that of kerosene; when the price of ethanol is greater than that of kerosene, the difference declines, although only to approximately two and half times. Likewise, the household market has the added benefit of reducing pressure on deforestation.

Finally, consider average per annum aggregate savings, the only set of benefits where the transport market produces savings greater than those in the household market, albeit in only one scenario. When the price of ethanol is less than or approximately equal to the price of kerosene, aggregate foreign exchange savings in the household market are, on average, approximately 220 million ETB, exclusive of the cost of the stove, and approximately 148 million ETB, inclusive of the cost of the stove; for the transport market, aggregate savings are approximately 121 million ETB. However, when the price of ethanol is higher than that of kerosene, savings in the transport market are significantly higher than those in the household market, indeed, over three times as high, with aggregate savings in the household market, both inclusive and exclusive of the cost of the stove, being approximately 30 million ETB, while those in the transport market are approximately 93 million ETB.

In toto, the household market is a better use of Ethiopia's limited ethanol supply. Not only are aggregate savings, in all but one scenario, greater in the household market than those in the transport market, but, per consumer savings in the household market, while relatively similar in absolute terms to those in the transport market, are substantially higher, as a percentage of income the lower the income level. Given such, one can only conclude that, in the relatively poverty-stricken city of Addis Ababa, which lies in a poverty-stricken country, the household market provides the greater return. Yet, both markets offer not insignificant benefits, especially when supply is unconstrained. What would it take, then, in terms of a supply response, to see one or both markets come to fruition? This will be the subject of the following, and final, section.

26.4.3 Phasing with a supply response

Given benefits for both markets when supply is unconstrained greatly exceed those when supply is constrained, perhaps the question asked at the end of the previous section is more important than the one considered by this chapter. Indeed, what would it take to serve full demand in both markets, and, furthermore, is such a supply response feasible?

To answer this question, consider, first, the household market. As displayed in Table 26.21, when the price of ethanol is less than or approximately equal to that of kerosene, current ethanol supply is enough to meet projected demand in the first two years of a maturing market. However, in the third year, demand becomes greater than supply

Table 26.21. Shortfall in demand and required land expansion to serve, household market.

		2005	2006	2007	2008	2009
Less than or equal to kerosene	Shortfall ('000 litres)	0	0	11731	34159	45573
	Incremental land required to meet (hectares)	0	0	3910	7476	3805
Higher than kerosene	Shortfall ('000 litres)	0	0	0	5741	12107
	Incremental land required to meet (hectares)	0	0	0	1914	2122
		2010	2011	2012	2013	2014
Less than or equal to kerosene	Shortfall ('000 litres)	47795	50120	52512	54971	57502
	Incremental land required to meet (hectares)	741	775	797	820	843
Higher than kerosene	Shortfall ('000 litres)	13389	14730	16109	17528	18987
	Incremental land required to meet (hectares)	427	447	460	473	486

by approximately 12 million litres; to meet this shortfall, approximately 3900 hectares of land would need to come under cultivation. Incremental land required to meet demand thereafter peaks the following year – 7476 hectares – and drops off thereafter; in 2010, or five years into the market, incremental land requirements stabilize at approximately 800 hectares per annum, the level of incremental land needed to meet household growth after full market saturation is achieved. A similar pattern for the price of ethanol greater than that of kerosene also emerges, although an extra year of lead time is gained in that demand does not exceed current supply until the fourth year of the maturing market. Peaking in 2008, or year four of the maturing market, incremental land required to meet demand in a household market stabilizes at around approximately 500 hectares per annum in 2010 and years thereafter.

Given that 388000 hectares is available for immediate expansion, meeting the incremental land expansion required to meet demand in a household market is not problematic, especially with regard to timing; with no less than two years before demand exceeds supply, sufficient lead time with regard to cultivation is almost guaranteed. The question then becomes whether there is sufficient lead time with regard to distillation capacity. Yet, even this does not pose a serious threat, both in terms of initial capacity and capacity thereafter, which, given growth in household demand, could grow at a relatively conservative rate. For example, in the first year that demand exceeds supply, the shortfall is approximately 6 to 12 million litres, depending on the price of ethanol; this shortfall could easily be met by just a handful of micro-distilleries, whose capacity ranges from 2500 to 40000 litres/ day. By 2014, or year 10 of a maturing market, the shortfall, in the absence of additional capacity, is 19 to 58 million litres, depending on the price of ethanol; yet, on the supply side, one to two medium-sized distilleries with a capacity of 48.6 million litres per annum could easily meet this demand, as could a single, large-scale distillery, whose capacity is 81 million litres per annum (Hodes *et al.* 2004).

Hence, not only would a household market fit nicely with any lead time required to expand cultivation and construct distillation capacity, but it would also allow for measured and steady expansion. With such a supply response comes the addition, when the price of ethanol is approximately less than or equal to that of kerosene, of 7011 permanent jobs and 15733 seasonal jobs and, when the price of ethanol is higher than that of kerosene, 2315 permanent and 5195 seasonal jobs.

With regard to the transport market, Table 26.22 displays the projected shortfall of supply over the ten year period given demand in each year, plus the level of incremental land expansion required to meet this shortfall. Given that full demand is immediate in the transport market, the shortfall between supply and demand – approximately 40 million litres – in the first year alone is quite large relative to current production levels; by 2014, or year ten of the market, the shortfall, in the absence of any expansion, is approximately 70 million litres. To meet this shortfall, land would need to expand by approximately 13 000 hectares prior to the first year of a transport market alone; while this requirement falls to zero in 2007 and 2008, given the increased production of molasses at both Wonji-Shoa and Finchaa in 2007, in 2009, expansion by 1400 hectares is required, with this increment steadily growing in the remaining years. Given the lead time required to plant and harvest the sugarcane, plus the time needed to build extra distillation capacity, the transport market would unlikely see fuel blending levels of 10% before three to four years' time. Full expansion over the ten year period would result in 8544 permanent jobs and 19 173 seasonal jobs. While job creation would be higher for the transport market, as will be seen in the paragraphs which follow, this point essentially becomes irrelevant.

Returning to the bigger picture, not only does a household market, in general, offer a greater return to Ethiopian society, but it better matches the ability of supply to respond. Yet, given suitable land in Ethiopia far exceeds the combined land requirements of both markets – 42 526 hectares when the price of ethanol is approximately less than or equal to the price of kerosene and 29 688 hectares when the price of ethanol is higher than that of kerosene – there is no reason why Ethiopia could not produce enough ethanol to eventually serve both markets.

To see how, assume that the household market is given first priority for all current ethanol production. Land would only need to expand by approximately 6000 hectares per annum, when the price of ethanol is approximately equal to or less than the price of kerosene, to meet full demand of a household market in every year of a maturing market, plus achieve fuel blending levels of 10% for both gasoline and diesel ten years after beginning

Table 26.22. Shortfall in demand and required land expansion to serve, transport market.

	2005	2006	2007	2008	2009
Shortfall ('000 litres)	39 464	42 382	37 906	41 767	45 857
Incremental land required to meet (hectares)	13 155	973	0	0	1158
	2010	2011	2012	2013	2014
Shortfall ('000 litres)	50 180	608	636	654	673
Incremental land required to meet (hectares)	1441	1523	1610	1701	1799

expansion; for a high ethanol price, this requirement drops to approximately 4700 hectares per annum. As distillation capacity could easily keep pace with this growth for either price of ethanol, there is no reason why both markets could not eventually be simultaneously met, with additional expansion occurring to meet demand for markets outside of Addis Ababa, should markets be encouraged there, or, should one be developed by the time surplus is available, an international market. Furthermore, as supply is able to respond to meet both markets, job creation will be the same no matter the market first served; by the time demand in both markets is being met, 10859 to 15733 permanent jobs and 24368 to 34906 seasonal jobs will be created, the exact number depending on the relative price of ethanol.

26.5 Conclusions

Ethiopia, like many nations in Africa, is currently at a crossroads with regard to the development of its ethanol industry. While the creation of a transport market for urban residents of the nation's capital, Addis Ababa, is taken on faith to be the demand of choice, particularly in the absence of suitable export opportunities, this is at the expense of an urban household market, whose return to the nation, in the presence of limited supply, is greater than that for a transport market, and, as discussed, whose benefits, by no means, do not accrue in the distant future. Rather, as shown by the pilot study, households across all income levels are enthusiastic about the use of ethanol in their homes. Likewise, urban centres, whose infrastructure is relatively developed, whose households are used to participating in markets for fuel, and whose residents depend on dirty and relatively expensive fuels, make development of a household market for ethanol much easier than is often supposed. Overall, the benefits, as clearly shown here, to the household, the economy and society in general outweigh those from an urban transport market in Addis Ababa, particularly in light of poverty alleviation aspects.

The choice, then, is easy, and other African countries should take heed; after all, the energy market dynamics in many of their own urban centres are not that different from Addis Ababa. At a minimum, African nations should understand that the trade-off is not as clear-cut as originally believed, although similar analysis should be pursued before committing to any one market. The results will likely be surprising, but reassuring as well, as not only are absolute benefits for an urban household market, given limited supply, likely to be higher than those in an urban transport market, but, from a developmental perspective, beneficial to overcoming poverty and encouraging economic growth.

Fortunately, such a choice is limited to the short term. As seen for Ethiopia, with not much expansion, demand in an urban household and an urban transport market could be simultaneously met in just ten years' time; certainly, this time could be shortened if so desired. Likewise, as expansion better matches a maturing household market, especially in the initial years, countries can progress at a balanced rate, and, in the not too distant future, be serving not just one market, but, at a minimum, two. The applicability of this argument is not limited to Ethiopia alone, a point other African nations would do quite well to remember.

Acknowledgements

The author wishes to thank Harry Stokes for motivating this chapter and for being such a strong mentor, willing sounding-board, and scholarly inspiration. The author would also like to thank Milkyas Debebe for his endless support and patience, both intellectually and technically, not only in preparing this chapter, but in all I do; never do I have an unanswered question about Ethiopia or ethanol, and never am I wanting for Ethiopian perspective and insight. Technically, this chapter would not have been possible without the amazing assistance of Firehiwot Mengesha, Wubshet Tadele, and Bilen Kassa.

Finally, the author would like to thank Ray Gronenthal for making sure that, at the end of the day, what I had written was rational, logical, and above all else, coherent.

References

Cohen, A. *et al.* (2004). Urban air pollution, in Comparative Quantification of Health Risks: Global and Regional Burden of Disease Attributable to Selected Major Risk Factors (eds A.D. Majid Ezzati, A.R. Lopez and J.L.M. Christopher), Vol. 1. World Health Organization.

Eggleston, S., Buendia, L., Miwa, K. *et al.* (eds) (2006). *2006 Intergovernmental Panel on Climate Change Guidelines for National Greenhouse Gas Inventories.* The Institute for Global Environmental Strategies.

Energy Information Administration, United States Department of Energy, 'Petroleum Price Data & Analysis' (2007). Retrieved on 10 January 2007 from http://tonto.eia.doe.gov/dnav/pet/pet_pri_top.asp

Ethiopia Central Statistical Authority (2004). *Ethiopia Welfare Monitoring Survey, 2004.* Ethiopia Central Statistical Agency.

Ethiopian PetroleumEnterprise MIS Unit (2006). *Statistics on Fuel Consumption.* Ethiopian Petroleum Enterprise MIS Unit.

Ethiopian Sugar Factor Support Centre Share Company (2006). *Data on Sugar & Ethanol Production.* Ethiopian Sugar Factor Support Centre Share Company.

Hodes, G., Utria, B. and Williams, A. (2004). Ethanol: re-examining a development opportunity for sub-Saharan Africa. World Bank's Energy and Poverty Thematic Group Working Paper.

Hutton, G., Rehfuess, E., Tediosi, F. and Weiss, S. (2006). *Evaluation of the Costs and Benefits of Household Energy Interventions at Global and Regional Levels.* World Health Organization and Swiss Tropical Institute.

Krzyzanowski, M. (2005). *Health Effects of Transport-Related Air Pollution: Summary for Policy-Makers.* World Health Organization.

Ministry of Finance and Economic Development, the Federal Democratic Republic of Ethiopia (2002). *Ethiopia: Sustainable Development and Poverty Reduction Program.* The Government of Ethiopia.

Ministry of Transport and Communications (June 2006). Strategic Plan and Research Department, *Annual Statistical Bulletin, 2004/2005.* Ministry of Transport and Communications.

Phillips, T.P. (2002). An agro-economic assessment of the potential to produce ethanol and millennium Gelfuel in Africa. World Bank's Energy and Poverty Thematic Group Working Paper.

Tilimo, S. and Kassa, M. (2001). *Ethiopian Millennium Gelfuel Initiative: Assessment of the Potential to Introduce the Local Production and Marketing of a 'Millennium Gelfuel' in Ethiopia as a Renewable Household Cooking Fuel.* World Bank.

United States Environmental Protection Agency (2005). *Emission Facts: Greenhouse Gas Emissions from a Typical Passenger Vehicle, EPA420-F-05-004.* Office of Transportation and Air Quality.

World Bank Urbanization Rate.

World Bank, Africa DataBase (2002).

World Health Organization (2004). *Indoor Air Pollution, Health and the Burden of Disease: Indoor Air Thematic Briefing 2.* Retrieved 14 June 2005 from http://www.who.int/indoorair/info/briefing2.pdf

Worldwatch Institute (2006). *Biofuels for Transportation: Global Potential and Implications for Sustainable Agriculture and Energy in the 21st Century.* Worldwatch Institute.

Chapter 27

Freedom from Fossil Fuel and Nuclear Power: The Scope for Local Solutions in the United States

TAM HUNT

Energy Program Director and Attorney, Community Environmental Council, Santa Barbara, California

27.1 Introduction

Santa Barbara County, located on the central coast of California, the most populous US state, has only about 400 000 inhabitants – but it looms large in proposing alternatives to the status quo. Santa Barbara County is popularly known as a weekend destination and wine growing area. Due to its great natural beauty and comfortable climate, the city of Santa Barbara promoted as the 'American Riviera', a title that elicits groans from some and pride in others. The area is also one of the most expensive real estate markets in the country, with a median home price on the southern coast of the county of about $1.2 million.

The Community Environmental Council ('CEC') is a non-profit organization based in Santa Barbara, established in 1970. In 2004, we shifted our course and began a new program to challenge the region's communities to wean themselves off fossil fuels and nuclear energy by 2033, through a program known as Fossil Free by '33 (www.fossilfreeby33.org).

Local energy solutions are being pursued throughout California and the US. However, CEC's efforts to hasten the renewable energy transition in Santa Barbara County are, as of early 2007, unmatched in their ambition. Our region is engaged in a level of planning detail also unmatched elsewhere. CEC completed a regional energy blueprint in early 2007. This detailed document spells out how the county can achieve the 'fossil-free' goal in a realistic and cost-effective manner. (Available at: www.fossilfreeby33.org).

CEC found in completing its regional energy blueprint that our county has sufficient potential for increased energy efficiency, conservation and renewable energy – both as electricity and as biofuels – to completely replace fossil fuel and nuclear power supply. More importantly, we found a high likelihood for achieving our goal cost effectively, promising significant savings in the long-term and greatly reducing the county's exposure to fossil fuel price volatility.

27.2 The Santa Barbara County Plan

The first step in local energy planning is assessing how much energy is consumed. In 2006, we used in Santa Barbara County about 184 million gallons of gasoline and 27 million gallons of diesel,[1] 525 000 gallons of aviation gasoline, 8.4 million gallons of jet fuel, 225 million therms of natural gas, and 3000 gigawatt hours (GWh) of electricity.[2] When we combine all of this consumed energy and convert to GWh as a common unit, we find that we used about 18 000 GWh in 2006.

We commissioned an energy forecast from the University of California, Santa Barbara's Economic Forecast Project. Our consultants project that our county's energy consumption will, under a business as usual scenario, rise to almost 21 000 GWh by 2030. We convert to GWh as a common unit because it is our primary goal to 'electrify' our transportation and building sector, substituting renewable electricity for petroleum and natural gas where possible.

As a guide to how much energy 21 000 GWh is, consider that we will, under a 'business as usual' projection, require the equivalent of *570 million* gallons of gasoline for our county's energy demand by 2030. Now imagine physically pushing your car 25 miles, the distance a normal car could travel on *one* gallon of gasoline. We will require the energy required for you to push your car 25 miles 570 million times over by 2030.

Fossil fuels are used today because they are a very concentrated form of energy. It will be a gargantuan challenge to find energy sources that can replace them.

Our plan has four primary elements:

- **Energy efficiency and conservation**. We identify ways to reduce energy use, particularly in the building sector and transportation sector, which constitute the lion's share of our current and projected energy demand (about 85%).
- **Hybrid cars and biofuels**. These technologies are currently available alternatives to petroleum. Hybrid car sales are booming. Biofuels like biodiesel and ethanol – preferably from cellulosic feedstocks (poplar trees, straw, and switchgrass, for example) grown in or near our region – can help us reduce petroleum demand immediately.
- **Renewable electricity**. We identify ways to produce large amounts of renewable electricity, such as wind, solar, biomass and ocean power, in or near our region. Having large amounts of renewable electricity substitutes for fossil fuel-based electricity and sets the stage for next generation vehicles that will use electricity as fuel.
- **Next generation vehicles**. We will transition to more efficient vehicles and vehicle fuels, such as plug-in hybrid vehicles, electric-only vehicles and potentially hydrogen vehicles, once they're available and affordable. These vehicles promise improved fuel efficiency while allowing electricity to become the primary fuel for transportation – something we call 'electrification' of transportation.

The one constant in the field of energy technology is change. We recognize the fast-changing nature of renewable energy technologies by allowing for additional chapters to be added over time. Creating our blueprint as a 'living document' – one that changes over time – will allow additional technologies to be 'plugged in' to the document as we explore additional options for our region.

The chapters of our blueprint are ordered based on each technology's 'realistic' potential for meeting our goal. Topics covered are, in order: general strategy for weaning our county off fossil fuels; reducing energy use in buildings; reducing energy use in transportation; next

---

[1] Gasoline use includes 5.7% ethanol in California in 2007. Diesel figures include a small amount of biodiesel used in the county.

[2] Sources: California Energy Commission, California Department of Transportation, Santa Barbara Air Pollution Control District.

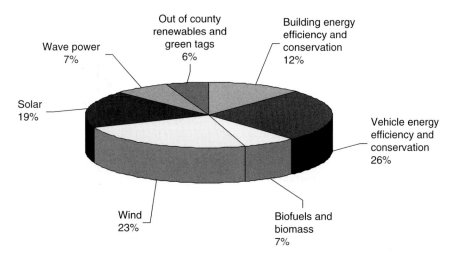

Fig. 27.1. CEC's plan to wean Santa Barbara County off fossil fuels.

generation vehicles; wind power; solar power; ocean power; and the economics of achieving our goals.

This order reflects our preference for conservation first, and then our assessment of the realistic potential of each technology in our county through 2030. We define realistic potential as *the technical potential of each technology tempered by an assessment of cost effectiveness and of the timeframe in which each technology can likely be developed in our region.*[3]

We considered geothermal[4] power and hydroelectric[5] power but decided not to include chapters on these technologies at this time due to limited potential *in our county* and finite funds for our project. Geothermal power, however, is experiencing a strong resurgence of interest due to new technologies that allow lower temperatures to be used to produce power as well as new surveys completed by the National Renewable Energy Laboratory and others indicating that deep geothermal potential is far more widespread than previously known.

---

[3] Typically, energy analysts use the following terms when discussing potential: 'Gross potential' indicates the full potential of a resource, such as the full amount of sunlight falling on a given area indicating the full potential for solar power; 'technical potential' is gross potential filtered by land use limitations (geographic or political) and technological limitations; 'economic potential' is technical potential filtered by cost-effectiveness considerations. Our term, 'realistic potential', is not commonly used but indicates a necessary additional consideration regarding what we can rationally expect in terms of future renewables development.

[4] Geothermal power is the largest source of renewable energy in California, with 5% of the total, under the state's definition of renewable energy (which excludes large hydroelectric facilities due to the negative environmental consequences of large dams).

[5] More than 20% of the state's electricity comes from large and small hydroelectric facilities, but we have limited potential in our region due to the lack of significant rivers that could be tapped. Also, there are many negative environmental consequences of dams, both large and small, such that we do not generally support new dam construction for power generation. There may, however, be significant potential for power generation in our region through retrofitting existing dams, though we have not yet studied this potential in detail.

We did not consider nuclear power – even though it is considered by some to be 'fossil free' – because of numerous problems associated with nuclear power. The traditional problems associated with nuclear power remain problems in 2007: waste storage and nuclear weapons proliferation. Federal solutions to these problems are progressing, but permanent waste storage facilities aren't expected to be completed for another ten years or so, if ever.

Additionally, though nuclear power is widely perceived to be an 'emissions free' technology, the only part of the fuel cycle that is emissions free is power production. Every other step requires significant amounts of energy – currently from fossil fuels for the most part – that result in greenhouse gas emissions equivalent to 40–100% of the emissions of a combined cycle natural gas plant, according to one team of researchers.[6] This is, admittedly, a minority view among energy analysts, but this team's conclusions hold up under analysis. Last, the cost of nuclear power has been mischaracterized repeatedly in the press and by industry advocates as 'cheap'. A recent analysis by Dan Kammen and his colleagues at the University of California, Berkeley, and Stanford University found that the levelized cost of nuclear electricity from 99 plants in the US (the large majority of plants ever built in the US) ranged from 3.2 cents per kilowatt hour to over 14 cents per kilowatt hour.[7] Industry advocates frequently cite costs for nuclear power in the lower end of this range, ignoring the many plants that produce power at far higher cost to consumers.

In identifying the best strategies for achieving our goal, we primarily considered two items: the *fossil fuel reduction potential* of each technology (including consideration of cost effectiveness) or strategy and the *potential for local influence* on energy use in a given sector of the economy. For example, as discussed below, local agencies have more influence over the building sector than over the transportation sector, so reducing energy in the building sector receives a higher priority.

27.2.1 Energy efficiency and conservation

The first element of our plan is energy efficiency and conservation because the starting point in any smart energy policy is to use energy wisely. With cost-effective energy efficiency and conservation measures, we believe our county can reduce energy demand in buildings 30%

California's Energy Mix

As an average, California's electricity is generated from the following sources:

Natural gas:	37.7%
Coal:	20.1%
Large hydro:	17.0%
Nuclear:	14.5%
Non-hydro renewables:	10.7%

Source: California Energy Commission's 2005 Net System Power Calculation.

---

[6] Storm van Leeuwen and Philip Smith, 'A Nuclear Power Primer', available online at: http://www.opendemocracy.net/globalization-climate_change_debate/2587.jsp
[7] Nathan Hultman, Dan Kammen and Jonathan Koomey, 'What history can teach us about the future costs of nuclear power', *Environmental Science & Technology*, April 1, 2007.

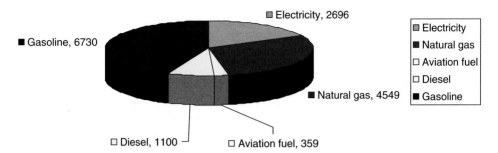

Fig. 27.2. Santa Barbara County energy use in 2005, in GWh.[8]

by 2030 and reduce energy demand in transportation 50% by 2030. These are ambitious figures, but we believe they are realistic due to the convergence of technological improvements; local, state and federal policies creating a more favourable environment for energy savings; and a growing and widespread concern about climate change. To a lesser degree, mainstream America is realizing that our oil and gas supplies are running out – and is thus looking for alternatives.

The distinction between energy efficiency and conservation is important. Energy efficiency allows us to do the same things we're doing today but with less energy. Conservation saves energy by ceasing wasteful activities. For example, today's refrigerators are larger than refrigerators sold in the 1970s, but they use 70% less energy due to greater energy efficiency. Turning off lights when rooms are unoccupied is an example of conservation.

California is a relatively energy efficient economy, due in large part to successful state policies promoting efficiency and conservation. Per capita electricity use, for example, has remained level since the early 1970s, while energy use per capita for the US as a whole has grown significantly. Per capita use in California for natural gas and petroleum has also remained level or dropped (see Fig. 27.3).

Considering the potential for local influence on energy use in our buildings, we found the most practical potential for fossil fuel reduction in this area. About 37% of the energy used in our county is to heat, cool or light buildings, compared to 48% nationally. Almost half of our energy is used in transportation, mostly by cars and trucks on our roadways, but because we can change building codes locally *we have a greater ability to influence energy use in buildings than on our roads.*[9] This is not to say that we cannot influence energy use on our roads at all. To the contrary, there are many ways we can reduce our petroleum demand through local action. But we (meaning all local entities) have more influence at the local level on building energy use than we do on transportation energy use.

Architecture 2030 is an organization dedicated to reducing energy consumption in buildings through its 'Architecture 2030 Challenge'. The goal of this movement is for new and renovated buildings to become carbon neutral by 2030. CEC is leading a coalition of architects, builders and local governments to implement the 2030 Challenge in our county and by doing so significantly decrease energy use in new buildings and renovated buildings. This is an area where local governments, architects and builders have significant control over energy use, so it is a natural focus area for CEC.

---

[8] Derived from California Energy Commission, Santa Barbara Air Pollution Control District and Department of Transportation figures.
[9] The remaining energy is used by industry, such as natural gas used for enhanced oil recovery.

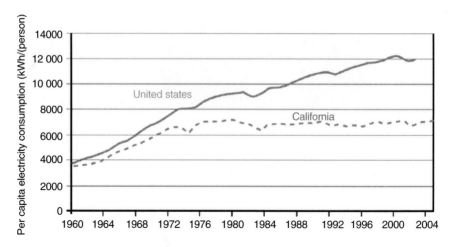

Fig. 27.3. Per capita energy use in California.

Transportation in Santa Barbara County represents about 47% of our total energy demand. A 50% reduction in energy consumption in the transportation sector will offset roughly 26% of our total energy demand by 2030. On the national level, oil consumption in the US fell after the introduction of Corporate Average Fuel Economy standards (CAFÉ) in 1975, when gasoline demand was 6.7 million barrels per day.[10] Gas demand didn't return to 1975 levels until 1985, aided by the 1979 oil crisis after the Iranian revolution. Unfortunately, CAFÉ standards have not been significantly strengthened since the early 1980s, so gasoline demand (and oil demand more generally) has continued to rise, slowing down only recently in light of historically high prices in 2005 and 2006.

Our county's residents have many options, however, for reducing energy use in the transportation sector. First and foremost, we can reduce petroleum demand by walking, biking, busing and taking the train. We can also carpool or 'car share', which is essentially car rental by the hour, an exciting new business model quickly spreading around the US and Europe. If normal driving is necessary, there are many highly efficient smaller cars available in the $15 000 range that can get up to 35 miles per gallon.

Public transit ridership, biking to work, and other alternatives to driving are increasing as gasoline prices rise and stay high. Small and more efficient car sales trends are also encouraging. There are a number of non-hybrid small cars, such as the Honda Fit, Toyota Yaris, Mazda 3, and Ford Focus, that can achieve 30 to 35 miles per gallon and only cost about $15 000.

Even though our 'business as usual' forecast shows petroleum demand rising through 2030, higher gas prices will lead to less demand as people find ways to drive less. There are also many potentially revolutionary technologies on the horizon that may significantly reduce petroleum demand, such as plug-in hybrid cars, discussed further below.

---

[10] Energy Information Administration, data available online at: http://tonto.eia.doe.gov/dnav/pet/hist/mgfupus1A.htm

27.2.2 Hybrid cars and biofuels

The theme of our second focus area, hybrid cars and biofuels,[11] is that these options are available today and can make a serious dent in our petroleum demand. There are 14 hybrid vehicle models available in early 2007 (the Toyota Prius and Honda Civic Hybrid being by far the most fuel efficient), and sales trends for these vehicles are very encouraging, with hybrid car sales up 59% in February 2007, compared to a year earlier.[12]

Hybrid vehicles cost more in up-front costs than their non-hybrid equivalents but can achieve much higher fuel efficiency than non-hybrid cars. Today's top seller, the Toyota Prius, averages about 45 miles per gallon in 'real world' driving. That's nearly twice the average fuel efficiency of all cars in the US. The Honda Civic Hybrid does almost as well. A recent study by Intellichoice, Inc.,[13] found that every hybrid vehicle on the market in 2007 in the US saved money even though they cost more up-front than their non-hybrid equivalents. Operating costs were reduced in terms of lower fuel costs, insurance, and depreciation – allowing the higher up-front cost to be recouped quickly.

Ethanol and biodiesel, both biofuels derived from plants, have significant potential to reduce our petroleum demand. We project that biofuels can supply about 12% of our projected transportation energy demand by 2030. Our plan is to utilize biofuels available in California today while also encouraging cultivation of switch grass, poplar trees, jatropha and other 'fuel crops' that can be used to create biofuels, in or near our county, in a more environmentally friendly way than corn-based ethanol or soy-based biodiesel (which are the predominant types available today). Many of these new fuel crops use less water than normal crops (except poplar trees, which are water intensive) and can actually build soil because their long roots pull nutrients up from below the surface. These crops have a much more favourable energy balance – and thus fewer greenhouse gas emissions – than biofuels produced from corn or soy beans. These options are described in detail in our regional energy blueprint.

In California in 2007, every gallon of gasoline included 5.7% ethanol, which, as an oxygenate, helps gas burn cleaner. California is the largest user of ethanol in the country. Also, approximately 400 000 'flex fuel vehicles', which can run on ethanol or gasoline, or any combination thereof, are currently on California's roads.

But even with so many flex fuel vehicles in California, there was still just one public ethanol fuel station in California in 2007. CEC is working to install four or more additional fuelling stations in our region in 2007, so fuel availability will not remain a problem for much longer. Once pump stations are installed, the cost of ethanol to consumers will remain as the only significant hurdle to greater ethanol use in our county. Unfortunately, at California's one public ethanol station, in San Diego, ethanol costs significantly more than regular gasoline. We expect the cost to drop once ethanol becomes more widely available and competitive forces work their magic. President Bush in early 2007 called for a radical increase in ethanol consumption in the US, so we expect to see the price of ethanol drop substantially as production is ramped up even more around the country.

---

[11] Our second focus area overlaps a little with our first category because hybrid cars use petroleum more efficiently, so can be considered a fuel efficiency technology.

[12] Hybridcars.com Market Dashboard.

[13] Study available at www.intellichoice.com

Biodiesel is already available in our county, at three stations.[14] Prices vary but are slightly more expensive than for regular diesel. Diesel cars are few and far between in California because of the historically higher emissions of certain pollutants from diesel engines. However, ultra low sulphur diesel (ULSD) recently became the only authorized petroleum-based diesel available in the US and many new diesel engine cars will soon be available in California due to the availability of this new and cleaner diesel fuel. Previously, California's strict air pollution laws severely limited diesel vehicle availability. With more diesel cars on the road, biodiesel use will likely become more widespread and we expect additional biodiesel stations to be installed in our region. Also, diesel cars are 20 to 30% more efficient than gasoline cars, so increased numbers of diesel cars will reduce our petroleum demand even if they don't run on biodiesel.

We have many acres of agricultural land in our county, much of it unused. We will, as a follow-up to our blueprint, analyse how much crop land would be required to grow up to 10% of our transportation fuel needs locally. We will also look at the cost effectiveness of doing so. If our results are favourable, we'll be reaching out to farmers and other landowners. With a new ethanol plant slated for construction in Santa Maria, we will have a facility that may be able to use locally grown feedstocks in the future. The developer, American Ethanol, Inc., plans to use Midwest corn for its first phase of operations but the company is currently examining the possibility of using cellulosic feedstocks like switch grass at a later date. Cellulosic feedstocks have a more favourable environmental footprint and can more substantially reduce greenhouse gas emissions when compared to corn grown in large-scale agricultural operations because of the large amount of fossil fuels corn cultivation requires.

Ethanol can also be made from agricultural waste and municipal solid waste. Making ethanol from these feedstocks allows us to solve some waste disposal problems at the same time as providing a sustainable source of transportation fuel. Accordingly, CEC will be examining the feasibility of using agricultural waste and municipal solid waste as feedstocks at the same time as we examine growing our own feedstocks locally.

27.2.3 *Renewable electricity*

Once we achieve the maximum energy savings through increased efficiency and conservation, hybrid cars and biofuels, there will still be much to do to wean our county off fossil fuels. We will need large amounts of renewable electricity to reach our goal. CEC found that wind power, solar power, biomass power and ocean power technologies have the most promise for our region to supply the rest of our power needs.

Electricity currently comprises only about one sixth of our county's energy demand, but we see electricity as the primary supply-side solution (versus demand-side solutions like energy efficiency and conservation) for getting off fossil fuels. CEC found that we can produce enough renewable electricity to meet all of our county's current demand for power *as well as* enough additional electricity to run our cars.

In other words, we will 'electrify' the transportation sector, and by doing so vastly increase our county's electricity demand. By 'electrification', we will substitute renewable electricity for petroleum on a relatively large scale – a truly momentous shift for the better if we succeed.

---

[14] For up-to-date information on biofuel stations around the country, visit the Department of Energy's Alternative Fuel Station Locator at http://afdcmap2.nrel.gov/locator/findpane.asp

This shift could happen if any of a number of new vehicle technologies becomes cost effective and widely available, discussed in section 27.2.4.

Wind power is cost competitive with fossil fuels today and, at over 11 000 megawatts (MW) of turbines installed around the United States in 2007 supplies the equivalent energy for about three million California-size homes. We project 1250 MW of wind power could be built in our county (onshore and offshore) by 2030, providing about 24% of our total energy demand in 2030.

'Concentrating solar power' has great promise because it can be backed up with a natural gas generator, as is the case with the 354 MW of facilities already installed near Kramer Junction, California. These facilities have operated at 108% of capacity, using solar power for 80% of that production and 20% natural gas. It is possible to achieve over 100% of capacity because these facilities were built as peak power facilities and the natural gas generators have been run at times above what was planned in order to generate additional revenue from selling more expensive peak electricity.[15] We project that over 1300 MW of new solar facilities of all types could be built in our county by 2030 – meeting 19% of future total energy demand.

Biomass and conversion technologies do not have as much potential in our region – about 1% of total energy demand by 2030 – because these technologies require less abundant fuel, generally agricultural waste and municipal solid waste. The advantages of these technologies are, primarily, that they help solve solid waste disposal problems and can also generate baseload electricity – which is generally not the case with wind power or most types of solar power.

Ocean power, while a much younger technology than wind or solar power, has tremendous potential in our region because we are a coastal county. A survey by the California Energy Commission found over 3500 megawatts of potential along our coastline – capable of generating far more electricity than our county of 400 000 needs. We project, however, that 500 megawatts or less will be built along our coast by 2030 – enough to meet about 7% of 2030 energy demand – due to environmental and technological issues limiting a full buildout of this resource. While no energy sources come with zero impacts to our environment, CEC believes that we can utilize our ocean energy resources with minimal disruption to the marine environment.

Promoting renewable energy in our region to the level we seek will probably require substantial help from local, state, and federal agencies. Under the state's current regulatory system, the utilities must achieve 20% renewable electricity by 2010 and 33% by 2020. The state's definition of 'renewable' excludes large hydroelectric plants, so we stand at about 11% renewables today for legally recognized renewable energy in our state. With electricity comprising only about one sixth of our county's energy use, 33% renewable electricity will, under a business as usual scenario, be only about 5% of our *total* energy demand in 2020.

In other words, the state's relatively aggressive renewable electricity goal will only take us 5% of the way toward our regional goal.

Accordingly, we need to find other ways to encourage the use and development of renewables in our county. Fortunately, a 2002 California law, known as 'Community Choice' (AB 117), allows local governments to build or buy as much as 100% of their total electricity demand. Essentially, Community Choice gives local governments control over the type of electricity they use, rather than our county's two private electric utilities, Southern California Edison and Pacific Gas & Electric, determining what type of electricity

---

[15] National Renewable Energy Laboratory report on file with author.

we use. Local agencies would still be subject to the 20% renewable electricity by 2010 requirement and other state laws. But local agencies implementing Community Choice would be able to move far beyond 20% if they chose. Community Choice is, as its name suggests, a way for communities to have more choice over the type of power they receive.

Community Choice may be the closest thing to a silver bullet we have for meeting our goal of weaning our region off fossil fuels and nuclear power.

Our goal of weaning our county off fossil fuels is equivalent to a 200 to 300% renewable electricity standard by 2033 – an order of magnitude more aggressive than the state's current goal. Going above 100% of our electricity demand may seem strange at first glance, but the additional electricity will be required to 'electrify' the transportation sector – a sector that we project will constitute fully half of our energy demand by 2020. Accordingly, Community Choice implementation by local agencies may be the only way to achieve our goals because it is unlikely under current market conditions that our goals will be met in the timeframe required.

New Jersey, Massachusetts and Ohio have their own versions of Community Choice. Essentially, any state or nation where electricity markets are at least partially deregulated – allowing private parties to build generation facilities – could enact some type of community choice legislation.

In sum, Community Choice is a powerful tool that should be considered by all jurisdictions – in areas where it is available – seeking more control over their power supply and more renewable power.

27.2.4 Next generation vehicles

Why do we need so much electricity from renewable sources? Again, our plan is to 'electrify' our transportation sector and vastly reduce our gasoline and diesel demand. There are three vehicle technologies that will allow electrification and together comprise the next generation of vehicles: plug-in hybrid cars, electric-only vehicles, and hydrogen vehicles.

Plug-in hybrids are like today's hybrids but with bigger batteries and a plug that allows the batteries to be charged from the grid. If the grid supplies only renewable power, the whole fuel cycle can be green. Plug-ins also can run on liquid fuels, however, providing the best of both worlds with a long travel range, while running on liquid fuels, and battery power for shorter trips. If the plug-in car can run on ethanol or biodiesel for their liquid fuel requirements, as is planned, the entire vehicle life cycle can be green.

Electric-only vehicles have been tried before – notably with GM's EV1, which was discontinued in 1999. However, a second round of electric-only vehicles is planned as battery technologies continue to improve. The Tesla Roadster, an electric only sports coupe that sells for $92 000, is available today. Tesla plans soon to release another model that will retail at about $60 000. These prices will likely come down radically if the vehicles are sold more widely. 'Neighbourhood electric vehicles', such as the GEM car, are available today for $10–$15 000. Electric-only vehicles must also charge from the grid or from a home's solar panels.

Hydrogen vehicles continue to improve, with BMW, GM and Toyota announcing plans to bring vehicles to market by 2010 or soon thereafter. Hydrogen vehicles use hydrogen either in a fuel cell – essentially a chemical battery that provides electricity – or in an internal combustion engine where it is burned much like gasoline is burned in today's cars.

Hydrogen does not occur by itself naturally – it has to be separated from other compounds such as water. This is why hydrogen is described as an 'energy carrier' and is not an energy source. The most environmentally friendly way to create hydrogen is using renewable electricity to split water into oxygen and hydrogen. In this case, renewable electricity from wind power or biomass power is the energy source and hydrogen becomes a way of using that electricity to move a car.

All three of these next generation technologies will require large amounts of electricity. If we can supply renewable electricity to our grid in large amounts, we will succeed in shifting our energy use from dirty petroleum and other fossil fuels to clean and renewable electricity and biofuels.

When we tally all these figures, we can obtain about 94% of our projected energy demand by 2030 from renewable energy sources. The rest can be supplied by out of county renewables and green tags, which represent the 'green' attributes of renewable energy outside of our county.

27.2.5 But what will it cost?

First and foremost, we recognize that we simply cannot afford *not* to shift rapidly from fossil fuels to energy efficiency, renewable electricity and biofuels. In other words, it will be too expensive for our county not to make the shifts we are suggesting. Our atmosphere and oceans cannot support more carbon dioxide and other greenhouse gases. Our current use of fossil fuels on a massive scale is clearly a very risky experiment with our planet when we consider that carbon dioxide levels are far higher today than they have been in at least 650 000 years.[16] The truth is, we don't know with much certainty how much temperatures will change under our currently sky high carbon dioxide levels. But do we really want to find out, given the risks?

Second, insulating our region from the vagaries of 'peak oil' and natural gas depletion makes very good economic sense, assuming alternatives are not that much more expensive. Again, no one knows with much certainty what the future holds for petroleum availability or gas prices, but we do know there is a substantial risk of extreme price shocks if any of a growing chorus of analysts is right about the risk of peak oil. Can we afford to ignore these risks?

A report from the well-respected consulting firm, SAIC, Inc., for the Department of Energy – known as the 'Hirsch Report' because of its primary author, Robert Hirsch – found that to mitigate the full impacts of peak oil, the US would have to start serious planning and transitioning activities 20 years ahead of the global peak. If a peak happens within the next ten years, we will evidently suffer some extreme disruptions to our economy because we have not, as a nation, begun serious planning for the transition.[17]

A recent comprehensive report from the UK's Sir Nicholas Stern, working on behalf of the UK Treasury, found that global climate change would probably cost the world's nations 5 to 20% of their gross domestic product each year over the coming decades. The good news: it will probably only cost about 1% of gross domestic product each year to mitigate the most serious consequences of climate change.[18]

---

[16] BBC News website, 'CO$_2$"highest for 650 000 years"', available at http://news.bbc.co.uk/1/hi/sci/tech/4467420.stm

[17] Hirsch, Robert *et al.*, 'Peaking of world oil production: impacts, mitigation and risk management', SAIC, Inc., February 2005, a summary of which is available online at http://www.acus.org/docs/051007-Hirsch_World_Oil_Production.pdf

[18] Stern Review Report, available at http://www.hm-treasury.gov.uk/independent_reviews/stern_review_economics_climate_change/stern_review_report.cfm

Closer to home, a number of reports for California state agencies have found meeting California's ambitious greenhouse gas reduction goals will in all likelihood *save money* by 2020. A report from UC Berkeley, using a sophisticated computer model for energy use and economics, found that reducing California's carbon dioxide emission back to 1990 levels by 2020 (the state's official goal) would probably result in $74 billion added to our economy.[19] Other reports have found similarly encouraging results. Specifically for our county energy blueprint, our consultant's rigorous analysis found that Santa Barbara county residents will in fact save substantially by switching to renewable energy. Due to projections from the UC Santa Barbara Economic Forecast Project that fossil fuel prices in our county and elsewhere will continue to trend upward, energy efficiency and renewable energy will save our county $418 million by 2020 and $1.5 billion by 2030. This is equivalent to annual savings of $830 per person in 2020 and $3015 per person in 2030.

Solutions available today to make the transition can be cheaper than the status quo – such as energy efficiency, wind power, certain types of solar power, and biomass power. And other solutions, such as concentrating solar power, plug-in hybrids, electric vehicles and deepwater offshore wind, are likely to become cost effective over the next decade. These technologies can jointly replace our current fossil fuel and nuclear energy supplies and save our county large sums of money in the process.

---

[19] David Roland-Holst, UC Berkeley, 'Economic growth and greenhouse gas mitigation in California' (August 2006), p. 3.

Chapter 28

Lagos, Nigeria: Sustainable Energy Technologies for an Emerging African Megacity

RICHARD INGWE,[1] EUGENE J. ANIAH[2] AND JUDITH OTU[3]

[1]*Centre for Research & Action on Developing Locales, Regions & Environment c/o LENF, Calabar, Nigeria;* [2]*Department of Geography and Regional Planning University of Calabar;* [3]*Department of Sociology, University of Calabar*

28.1 Introduction

The high rate of urbanization in Africa has not resulted in improved living standards (better paying jobs, infrastructure and services, clean and modern electricity, potable water and so forth). Increasing urbanization of the developing world has created a large mass of urban poor. The use of large quantities of fossil fuel to run the urbanizing world has led to dangerous and abrupt climate change, resulting in emissions of huge quantities of greenhouse and noxious gases that have created atmospheric obstruction to radiation from the earth's surface, and also depletion of the ozone layer on a global scale (Jaeger *et al.* 2004).

28.1.1 Sustainable energy: renewable energy and energy efficiency

Over the past half-century, renewable energy and energy efficiency rose from mere issues discussed at global meetings (e.g. the UN conference in Kenya in the 1960s) to the high ground of operational technologies that have been comprehensively and convincingly implemented in several areas and also demonstrated at the global conferences (e.g. in Bonn and China in 2004). Outcomes of the latter, especially the International Action Programme and Political Declaration, have been very successful in receiving the endorsement of governments, political and economic blocs, and international organizations (e.g. the G8, the World Bank, UN agencies, to name but a few). These profits might be attributed to the benefits (including decentralization, greater energy security, environmental friendliness, creation of jobs, and other advantages) derivable from the implementation of renewable energy and energy efficiency technologies (REEETs) and the way they are changing the energy characteristics of cities and regions employing them around the world. The impressive policy and programming outcomes of a few of these conferences include the creation of networks including the Renewable Energy Policy Network for

631

the 21st Century (REN21), the 2005 Gleneagles G8 meeting, and myriads of sub-sectoral professional societies and advocacy groups engaged in promoting rapid implementation of renewable energy and energy efficiency technologies. Simultaneously, other networks have emerged to promote general urban management and sustainable development programming (e.g. the megacity research teams and partnerships in Cologne, Germany; networks of built environmental professionals: architects, spatial-regional planners, building engineers and so forth) and promoters of urban energy transition in particular.

28.1.2 Sustainable energy as fitting for the third and subsequent development waves

The concept of the *third wave* succinctly captures inevitability of shifting urban energy supply from the conventional energy technologies including fossil fuel and nuclear energy towards renewable energy and energy efficiency technologies (Toffler 1990).

Similarly, the concept of Fordism and post-Fordism has been used to explain how cities have successfully accomplished the transition from centralized (or Fordist) approaches to managing social, political, economic and cultural including technological processes towards subnational approaches such as city or local government-based and regional (post-Fordist) approaches. Specifically, post-Fordist approaches towards analysing city transformation or reinvention of urban places by creating information and knowledge hubs have been recorded in Birmingham (Webster 2000), and in Eindhoven, Helsinki, Manchester, Marseilles and The Hague (Van den Berg 2000) and for stimulating sustainability in terms of job creation for the youth in three Finnish cities (Lampinen 2001).

Therefore, achievement of success in reducing the use of fossil fuels will depend to a large extent on how experiences gained from ICT implementation in European cities (and possibly beyond Europe) will be deployed to encourage urban development champions to transform urban consumption culture as well as urban managements to engender the transition from fossil fuel dependence towards alternatives such as renewable energy and energy efficiency (REEETs). There is a need also to recognize the impacts or successes that have been recorded from the use of urban marketing campaigns undertaken by some cities around the world as a means of attracting investors, knowledge workers and players capable of contributing towards the reinvention and repositioning of cities to develop their local economies and culture, and to create a virtuous cycle of prosperity. It is easy to argue that while ICT implementation by individual cities was propelled by the profit motive, the quest to reduce greenhouse gases (GHG) through implementation of REEETs may not be an economically attractive option. Therefore, REEETs implementation requires serious enforcement probably using existing frameworks (UN agencies and environmental protection agencies at national, regional or subnational and local scales).

28.1.3 Energy hunger and the resort to unsustainable nuclear power in Africa

Reliable sources estimated the population of sub-Saharan Africa in 2005 to be 732512000 (i.e. 11.4%) of the world population of 6453628000. Electricity consumption per capita (in kWh) in sub-Saharan Africa in 2001 ranged from 25 in Ethiopia, to 86 in Nigeria to 4546 in South Africa. The percentage of the population with access to electricity in 2000 in the region ranged from 4 in Uganda, 40 in Nigeria to 66 in South Africa. The total energy consumption – from all sources – in 1000 metric tonnes equivalent, toe, in 2001 ranged from 931 in Congo, 95444 in Nigeria to 107738 in South Africa (World Resources Institute 2005). Africa is believed to be notoriously deficient in power supply compared to elsewhere in the world. (Sub-Saharan) Africa suffers an enormous energy crisis: its large population, which forms a large proportion of the global population, consumes only 2.7% of the world's commercial energy, i.e. electricity consumption in industrialized countries is 150 times higher than in Africa. In Sahelian Africa the rural electrification level is less than 5%! Average per capita final commercial energy consumption is less than 300 kg of oil equivalent per inhabitant, compared with 7905 kg in North America and the world average of 1434 kg in 1996 (First-hand view of Africa's power crisis 2007). This is so irrespective of its possession of enormous energy resources: a consistently high level of solar insolation, large hydropower potential, in excess of 1100 TWh, 2% of world proven oil reserves, 6% of world proven gas reserves and 6% of world proven coal reserves and huge biomass resources, as well as Uranium deposits. Africa depends greatly on low quality, traditional fuel, i.e. fuelwood, and relies overly on imported commercial fuel, i.e. oil. These two issues can be summarized as an inaccessibility of Africans to their own energy resources (oil, hydroelectric potential), and an inaccessibility of populations within each country to all kinds of quality fuels. Togola, 2007; EDRC/ENDA 2002 (a); EDRC/ENDA 2002 (b). In Nigeria, a nation reputed to be the world's sixth largest oil (and leading gas) exporter, the energy crisis is even worse: the obsolete grid connects less than 45% of the population (about 150 million), rural areas have only 10% access to the grid, power sources (large hydro and thermal gas) are unreliable, while outages are frequent and prolonged. Nigeria presents a suitable framework for analysing and examining the electric energy situation in Africa. Therefore, a central question in this chapter is: How can urban regions reinvent themselves in a way so that sustainable energy can form the foundation for driving the local economy? More specifically, how can the energy transition be accomplished in developing urban regions?

28.2 Why Examine Energy Crisis in Lagos Megacity?

Lagos, Nigeria's prime city and Africa's megacity, provides an ideal place for addressing the latter question and to examine the existing conditions – as they constitute both potentials and problems as a basis for drawing experiences that are likely to be helpful in improving the existing energy crisis. About 80% of Nigeria's total population (projected to be over 150 million, in the 2006 census) are classified as poor. Electricity supply in Nigeria has remained varied over the past two decades and for a country with an estimated population of 150 million people, electricity generation from the Power Holding Company of Nigeria and existing independent power producing plants (IPPs) still hovers between 3200 and 4000 megawatts (MW). This is despite the spending of millions of dollars by the federal government in resuscitating ailing power plants and constructing of new ones. And while most homes go for days and sometimes weeks without energy supply for domestic use, local manufacturers, perhaps, have been the worse hit. For example, in a survey conducted by the Manufacturers' Association of Nigeria (MAN) in the first quarter of 2006, most of the industrial areas around the country suffered the effects of inadequate power supply. The power supply survey indicated an average of 14.5 hours of power outage per day as against 9.5 hours of supply. As a consequence, a major automobile tyre manufacturing multinational company moved out of the country early in 2007, citing inadequate business environmental conditions as the reason. Several other manufacturing companies are reported to be shutting down or relocating elsewhere.

28.2.1 Some aspects of Nigeria's power crisis

Nigeria's centralized national grid is the conventional energy technology that relies on a few geographically lopsided and unreliable large hydro and thermal gas generating stations. The grid's failure is due to connection problems: less than 45% of the mostly urbanized Nigerian population (over 71 million Nigerians) are excluded! These grid-connected urban residents get only 4000 MW of electricity, suffering frequent power shortages and prolonged power outages that last for about 12 hours a day to several days at a time. Over 50% of Nigeria's households (mostly urban residents) rely on unprocessed wood as cooking and heating fuel thereby exposing women and children to respiratory diseases leading to high mortality. The enormity of Nigeria's energy crisis made it an ideal case study for *Energetic Solutions* – a 2004 international conference in Nigeria, aimed at realizing renewable energy in developing nations.

The structure of electrical energy transmission in Nigeria comprises five regions with 4534 km of 330 kV lines and 19 major substations (The Punch 2007c). It is not clear how the above transmission structure serves power to the Lagos megacity. Apart from being Nigeria's prime city, one of Africa's commercial and cultural capitals and Nigeria's former political capital, Lagos has experienced one of the most rapid rates of urbanization and its population is predicted to rise to about 24.3 million by 2015. That demographic and physical transition will make Lagos' city-region the world's third most populous megacity! Within the context of economic stagnation and mismanagement, the demographic transition will compound Lagos' urban problems. Urban geographers have described the problems of Lagos and other Nigerian urban centres in the 1980s as revolving around: unlivability, unserviceability, unmanageability and unemployment (Ayeni 1978). These are problems that result from inadequate services and supplies including shortage of clean and modern energy, frequent social disorder and ethno-political conflicts, natural and human disasters, pollution, among others. This makes Nigerian urban centres ideal cases for advocating urban energy transition. Megacities generally present enormous problems but their problems assume more serious dimensions within the context of chronic economic decline and mismanagement, two areas that characterize Nigeria considering its history of prolonged military dictatorship. The recent innovative pioneering establishment of some of the independent power producing plants (IPPs) being developed in Nigeria by Lagos State Government – and compelled by its peculiar energy crisis which is more serious than the rest of Nigeria – presents opportunities for greater urban energy transition and a move worthy of emulation by the rest of urban Nigeria. Although it is easy to appreciate that the resolution of urban energy crisis is a *sine qua non* for achieving sustainable urban development, the great challenge is in programming towards realizing this laudable goal. This chapter explores the factors that are capable of engendering urban energy transition which can be replicated elsewhere. International policy aimed at implementing renewable energy and energy efficiency technologies, networks for promoting urban development and cognate factors are examined. Nigeria's huge population is spread across numerous cities, towns and rural regions thereby indicating the existence of the considerable unmet need for renewable energy and energy efficiency technologies and services in urban regions. The failure of the public grid operated by the Power Holding Company of Nigeria and the infancy of urban sustainable energy supply systems indicate that there are abundant resources for implementing renewable energy and energy efficiency in urban areas such as Lagos and others.

28.2.2 Unsustainable nuclear power offensive on Africa

Despite convincingly demonstrating that nuclear power technology worsens the current efforts to reduce carbon emissions (Matthes *et al.* 2006), there is an ongoing nuclear offensive on Africa with nuclear power investors and scientists claiming to promise to resolve the energy crisis on the continent with some African nations planning to establish nuclear power stations (Nigeria's nuclear power 'mix-up' 2007). The literature is replete with predictions of dire consequences of global warming and climate change for Africa. Some African nations such as South Africa have established nuclear power stations while many more including Nigeria, Algeria, Angola and others are planning to start their own nuclear power plants as a way of increasing power generation for their teeming population and socio-economic activities (Earth Life Africa (ELA) 2007).

28.3 Lagos: An African Megacity

The Lagos urban area is located in the rainforest area which is undergoing rapid deforestation in south west Nigeria. It covers a contiguous area of about 153 540 hectares of which about 22% is constituted by lagoons and with a coastline of about 180 kilometres (Bashorun and Olakulehin 2007). It comprises all local government areas in Lagos State and four local government areas in neighbouring Ogun State around the coastal parts of the ocean. It was formerly capital of Nigeria and currently commercial capital of most of Africa. Most of the megacity is administratively originally under the government of Lagos State – one of the 36 states forming the Federal Republic of Nigeria. Lagos State has been leading Nigerian states in confronting problems of development planning due to the huge burden of inadequacy of infrastructure and services and so forth. It has been considering and describing itself as a megacity for a long time without much effect, in terms of receiving the help it requires to deal with problems associated with its megacity status in a mismanaged economy like Nigeria's. In recognition of what it considers to be historical under-reporting of its population over the years and several national censuses as well as deprivations of its rights and privileges by the Federal Government of Nigeria, it was the only state – out of the 36 – to acquire the resources required for conducting a scientifically and technologically sound census in readiness for the 2006 national census. Dissatisfied with the National Population Commission's publication of its census calculated at 9 013 534, Lagos State recently quoted its population at about 17.55 million in local Nigerian newspapers and websites in early February 2007 (www.businessdayonline.com). Some factors that point to the fact that Lagos is a megacity are: the current huge and continuous drift of people into the megacity intending to reside there permanently, immense vehicular density, the huge volume of refuse generated, the national percentage of electricity consumed, the national percentage of fuel consumed, the number of houses and so forth. Additionally, international organizations have over the years and on the basis of objective criteria put the population of Lagos at between 15 and 20 million (Nigerian Tribune 2007). The UN projects the population of Lagos to be 23.2 million by 2015, making it the eighth largest urban area in the world, growing at a rate of 3.7% from 13.4 million in 2000 (UN World Urbanisation Prospects 2007). This disparity in population figures for Lagos betrays the way the demographic characteristics of Lagos are frequently ignored, misrepresented and underestimated by governments and international organizations thereby compounding the problems of infrastructural inadequacy of the city (including energy) development in the urban region. This attitude has led to the exclusion of Lagos from the list of emerging megacities receiving international research and planning attention. Under great pressure, the Lagos State Government grapples with this impediment and recently resorted to funding an alternate population census simul-taneously with the 2006 national census funded by the Nigerian Federal Government and international organizations (UNDP, the EU). However, while the literature acknowledges the considerable demographic dynamism exhibited by Lagos, there is considerable disparity in statements of the city's population and the Nigerian government-sponsored national census result for 2006.

Despite the fact that the megacity status of Lagos (including Lagos and Ogun states) has been tacitly supported by some agencies of the Federal Government of Nigeria, this urban status of the region is yet to be statutorily recognized and gazetted. Recently, a Presidential Committee, inaugurated in December 2005, submitted a bill for the establishment of the Lagos Megacity Development Authority (LMDA) to the national assembly for approval. The Committee proposes that the megacity be funded through a contribution by the following stakeholders: Federal Government of Nigeria 45%; Lagos State Government (40%); Ogun State (15%); and the capital market through the Urban Development Bank of Nigeria (The Punch 2007h). The purpose of the LMDA is to improve traffic flow, through the creation of 28 commercial activity centres each equipped with required infrastructure to attract and sustain investment and to decongest Lagos Island by drawing out traffic from it towards several new centres (The Punch 2007b).

28.4 The Dysfunctions of Lagos: The Monocentric Megacity

Dysfunctional urban structures (in terms of centralized transportation networks, construction machinery, industrial systems, manufacturing processes, intensive economic activity, labour markets) have resulted from historically intensive use of fossil fuels in urban centres (Droege (2004)). This dysfunctional structure is perhaps at its worst form in Lagos megacity, where Lagos Island constitutes the only centre of all social and economic activities for its several city sectors with large hinterland and huge human population. The notoriety of Lagos is well known for its suffocating vehicular and human traffic jams, inadequacy of facilities such as having only 0.4% of residents enjoying water closets at home. These have been deplored in various media including popular and academic literature and websites (Daily Sun 2007a).

28.4.1 Responses to the Lagos megacity dysfunction

Two responses that provide opportunities for introducing sustainable energy technologies in Lagos deserve mention. First, the Governor of Lagos State from 1999 to 2007, Mr Bola Tinubu, declared the twin highbrow sectors of the megacity (Victoria Island and Ikoyi), located in environs of Lagos Island, as 'model cities'. The declaration involves a plan to guide the physical development of the sectors for the next ten years; arrest their prolonged degradation and improve their aesthetics; and improve participation of private and public stakeholders in the urban governance and planning process. However, the plan fails to deal with the problem of greater infrastructural deterioration in the central business district (covering Balogun, Nnamdi Azikiwe, Frederick and Doherty streets), which have for long pushed the population to move into Victoria Island and Ikoyi to worsen existing pressure on infrastructure and services (The Punch 2007a). Second, a politician vying for governorship of Lagos State had earlier promised to create mini-satellite towns outside Lagos Island as a way of decongesting the Island, creating jobs, resolving near intractable traffic jams, and establishing new and more marketing service centres (Daily Sun 2007).

28.4.2 Dimensions of energy crisis in Lagos and Nigeria

Nigeria's energy crisis manifests itself in various dimensions: there is weakness in regional and local energy policy, institutions, structures, and attitudes for electrical energy supply generally and sustainable energy in particular. For example, like most of Nigeria's 36 states, Lagos State does not currently consider energy as a distinct sector that is significant enough to be organized as a separate institution of regional governance equipped with a distinctive policy making framework concentrating on energy supply or endowed with mechanisms for thorough examination and management of energy issues under a separate ministry of energy. For the past two years, the quantity of electricity generated and supplied to the 45% of Nigeria's population connected to the national grid, declined from a peak of about 3800 MW in August 2005 to 1200 MW on 7 March 2007. A number of excuses have been given by the Power Holding Company of Nigeria, which claims that the power shortage has resulted from damage to the gas pipeline at Escravos for the past 16 months. This damage has led to a drastic decline in power generation from the Egbin thermal gas generating station from about 1300 MW on 18 February 2006 to only 243 MW between 11 and 17 March 2007) – the lowest point in a decade; due to dwindling of the water volume in the Kainji dam, which only operates at full capacity in the rainy season – usually May. This has made the huge population of Lagos megacity to receive and share only 300 MW instead of the 1200 MW that it demands, resulting in a power outage that frequently lasts for about 22 to 23 hours a day or one to two hours of power supply (The Punch Again 2007).

28.5 Some Potentials for Realizing Urban Sustainable Energy in Lagos and Urban Nigeria

The enormity of the shortage of energy in the megacity summons its managers to urgently adopt decentralized sustainable energy systems, especially noting Albert Einstein's observation that: 'Problems could not be resolved by viewing them from the perspectives of those who created them', i.e. reliance on the current centralized energy system has/is bound to perpetuate the energy problems of individual cities (Quoted in the literature 2006). Therefore, Lagos megacity can promptly key into decentralized energy implementation including generation, distribution and transmission by exploiting opportunities presented by the recent unbundling of the Power Holding Company of Nigeria into 18 separate companies under three groups: 11 distribution companies; six generating companies; and one transmission company. While four out of 414 companies that have submitted expressions of interest (EoI) to participate in the transmission component of the power sector, it is not yet clear how many are interested in generation and distribution (The Punch 2007d). Urban development associations exist for the promotion of sustainable energy technologies: Victoria Island and Ikoyi Residents Association (VIIRA) and Victoria Island and Ikoyi Security and Environmental Trust (VIISET) have separately called for cooperation of stakeholders in the development of their jurisdictions and also endorsed government's efforts to offer a plan for creating model cities (The Punch 2007f).

28.5.1 The emergence and growth of energy networks in Nigeria

As described later, a number of IPPs' development firms have recently emerged to undertake projects in different parts of Nigeria. The unbundling of the state-owned enterprise formerly called the National Electric Power Authority (NEPA) into 18 separate companies including the Power Holding Company of Nigeria (PHCN) is aimed at creating a conducive environment for the emergence of energy service companies (ESCOs); however, it is currently not well known how many ESCOs have emerged. Fortunately, after the *Energetic Solutions* international conference on realizing renewable energies in developing nations hosted by *OneSky – The Canadian Institute for Sustainable Living* in Nigeria, a Council for Renewable Energy in Nigeria (CREN) was formed (www.onesky.ca/energetics).

Although it is currently caught up in the difficulty of its members to distinguish between the mission of an energy council and the interests of NGOs in seeking project funding as usual in Nigeria, there is hope that this problem will soon be overcome and will turn the CREN towards a more serious focus on promoting sustainable energy implementation. In 2005, a forum for promoting small-scale sustainable energy project development using the CDM as funding facilitation was undertaken by a Canadian NGO in Abuja. The good response of the Nigerian community of sustainable energy enthusiasts to the program indicates that similar programs, designed to focus on large-scale energy supply for urban and other regions, are likely to be well received. Enormous opportunities for implementing sustainable energy globally exist in the form of political endorsement by numerous nations after the Bonn *Renewables 2004* – easy gaining and acquisition of sustainable energy expertise, prospects of funding from the G8 endorsement of sustainable energy at the Gleneagles meeting and cognate resources resulting from international policy, and networks. How have these outside opportunities been exploited in Nigeria and what potentials are endemic to the nation and Lagos? We now turn to examine these conditions.

28.5.1.1 Development of independent power producing plants (IPPs) and creation of the Nigerian Electricity Regulatory Commission

As a way of increasing electricity supply, which has been historically low in Nigeria, some state governments and businesses recently undertook to develop independent power producing plants (IPPs) in some parts of the country. Some IPPs that have been under construction in Nigeria have been reviewed (Ingwe 2005). Recently, the Nigerian Electricity Regulatory Commission (NERC), which was created in 2005, issued 26 IPP licences to various power operators. The NERC claims it has evolved a plan to fast-track the issuance of additional IPP licences aimed at facilitating the addition of 1500 MW to the national power generating capacity annually in the coming years. The more recent IPPs include one being developed by a company incorporated in Nigeria. Sadly, most of the IPPs will generate electricity by burning petroleum and natural gas thereby saving associated gas flared during oil production in Nigeria's oil wells in the Niger Delta. Most of the IPPs are part of the program designed to stop notorious flaring of the gas in Nigeria. Since the advent of oil production in Nigeria nearly 50 years ago oil producing companies have flared vast quantities of the associated gas resulting in huge annual greenhouse gas emission to the extent that recently they were given an ultimatum to stop the obnoxious act by 2008. The IPPs face a number of problems: it is feared that it will take at least four years before they start generating electricity because of difficulties in financing; coping with compliance challenges; electricity pricing imbalances; and the fact that existing thermal gas-based power generating stations have been blaming shortfalls in power production and supply on problems they face in receiving gas supplies from oil and gas producing companies. Therefore, energy commentators and analysts have requested the government to intervene to make the cost of gathering and transmitting natural gas for domestic use to conform with the price at which gas is procured for electricity production (Financial Standard 2007).

28.5.1.2 Independent power producing plants for increasing power generation

After the inauguration of Nigeria's Fourth Republic in 1999, the Lagos State Government pioneered power generation based on IPP. Before providing details of these IPP power development projects, it is apposite to state that while they may not necessarily be sustainable, they certainly present opportunities for future initiation of sustainable energy generation and supply based on renewable energy. Apart from hosting the headquarters of the Manufacturers' Association of Nigeria (MAN), Lagos megacity currently manages a national- or regional-scale project that bears the name 'Lagos megacity project', which is capable of designing IPP projects. Most of the special districts and institutions, which are described as potentials for IPP development herein, are to be found in the Lagos megacity. Moreover, experience has shown that since returning to civil rule in 1999, the Lagos State Government has exhibited commendable resilience and insistence on dialogue and negotiation aimed at getting its rights and privileges from the Federal Government of Nigeria and to creatively undertake more IPPs to add to its existing ones. It has promoted and funded newly created local government councils against the rule of former President Obasanjo, who said that no state should create more local government councils to add to those that existed as at May 1999. Lagos State Government has tenaciously shown that it is responsible for earning huge amounts of value added tax to enrich the federal pool of financial resources yet has been denied and deprived of its funds as statutorily allocated and enshrined in the revenue sharing formula. What Lagos State needs to do further is to improve on such constructive programs such as renewable energy to earn incentives, and provide jobs for its teeming population.

28.5.1.3 *Manufacturers' Association of Nigeria (MAN)*

A national union of local manufacturers in Nigeria recently decided to collectively float two Independent Power Producing Plants (IPPs) as a means of improving competitiveness of its members in business within the hostile Nigerian business environment. MAN representatives signed an MoU with a company incorporated in Nigeria and affiliated to leading world renowned consultants in the specialized field of building and supply of power generation plants to construct the two IPPs aimed at producing high voltage overhead transmission lines, transformer substations and other infrastructure. The Infrastructure Committee of MAN is optimistic that the IPP project would ensure that the manufacturing sector has adequate and cost-effective power supply. This IPP development by MAN arises from the failure of the state electricity firm, Power Holding Company of Nigeria (PHCN), to meet the country's electricity demand. The IPPs would each be 1000 MW and will be sited in the northern and southern parts of the country. The IPPs are expected to be more beneficial compared to the manufacturers' usual resort to generating their electricity from fossil-fuel generating sets that burn diesel or petrol at high cost and which add to their gross manufacturing cost to make the business environment rather hostile. This has resulted in the high cost of engaging in all manufacturing businesses in Nigeria, which made the end-products of manufacturing far too expensive and not competitive when placed side by side with imported products. Additionally, inadequate and costly power supply also forecloses the prospect of export because it cannot withstand the price competition in the international market. Profitable manufacturing in Nigeria has been a tough challenge for members of MAN, most of whom have closed down or are contemplating doing so (Power failure: manufacturers opt for own independent power plants 2007).

28.5.1.4 *The Lagos IPP project*

The pioneering IPP in Nigeria was initiated by the slain former Minister of Mines, Power and Steel and later Attorney General of the FGN and Minister of Justice Chief Bola Ige in collaboration with Chief Bola Tinubu, the current Governor of Lagos State since May 1999. The woeful failure of the Nigeria State Owned Enterprise called the National Electric Power Authority (NEPA) mandated to generate and supply power compelled the creation of the project as a means of supplementing and complementing the shortfall in power supply by

NEPA. The Lagos State IPP was established to produce or derive the following benefits: generate employment for 2500 Nigerians; promote local industries, which have been disabled by inadequate power supply from NEPA; save gas being flared by injecting it into the Agbara Gas Turbine; cause investment of over $600 million in Lagos and Nigeria. Other benefits were to increase investor confidence through increased reliability of power supply and produce cheaper power (the proposed IPP aimed to generate power for NEPA at the cost of N8 and sell at N10.00 representing a Lagos State subsidy of about N3.00). Therefore, it was going to produce and sell power more cheaply compared to NEPA, which produced and sold power at higher costs. The first phase of the Lagos IPP was to cost $40m. The funding was expected from equity from Enron (which suffered a major trauma and collapsed in 2001). The funding for the second phase at Agbara was expected from the international capital market. The entire project was initially based on Power Purchases Agreement (PPA) between the Lagos State Government, Enron and the Federal Government of Nigeria (through the Guardian Guarantee and Project Agreement (GGPA)) while the Dispatch and Transmission Agreement (DTA) involves NEPA and Lagos State Government. NEPA was to perform the following functions: operate the IPP by receiving, transmitting and distributing the power generated via the directives of the Lagos State Government, and generate bills, collect and put revenue into an escrow account from which Enron could draw. The Federal Government of Nigeria facilitated the IPP project by guaranteeing it, and providing other support (Ingwe 2005).

28.5.1.5 The Omoku IPP project

Valued at N14 billion the Omoku IPP project is located in Omoku town in Rivers State; the project was planned to comprise six turbines imported from Italy to be powered by gas being released during the process of oil drilling. Although it was to be inaugurated in June 2004, only four out of the planned six turbines were installed as at May 2004. The delay in installing all the turbines was attributed to the unreliability of the contractor for the project – who is yet to deliver the remaining turbines (The Punch 2004). Initiated by the Rivers State Government, one of the 36 states comprising the Federal Republic of Nigeria, it was planned that the project would be extended to other parts of the state, which were poorly served by the national public electricity grid operated by the National Electric Power Authority (NEPA). It is apposite to state that Rivers State was the first out of Nigeria's 36 states to establish a local regional fully fledged Ministry of Power (Ingwe 2005).

28.5.2 Urban-based special districts (universities, tertiary educational institutions) as potentials for IPPs in Nigeria

There is every reason to hope that the move recently made by MAN to generate power through IPPs can be replicated by tertiary educational institutions and several other special districts (including military installations, barracks, railways, residential regions and so forth), which suffer prolonged outages from public grid dependence but have potentials to seek creative alternatives for independent power plant development. Since August 2007, Nigeria has established about 80 universities, even more tertiary educational institutions and a large number of military regions or districts that are capable of sustaining IPPs. The realization of IPPs by these special districts and institutions would depend on carefully targeted programs capable of showing how IPPs can be planned and implemented. Fortunately, a growing number of think-tanks and NGOs working towards promoting the renewable energy and energy efficiency technologies (REEETs) implementation in Nigeria and beyond have emerged and could be mobilized to undertake collaborations with the institutions that might show interest in IPPs.

28.6 Defects in Unsustainable IPPs being Developed: A Case for Transition to Sustainable Energy in Lagos and Urban Nigeria

While the development of IPPs indicates diversification of sources of power generation, some defects currently exist. Thermal gas power is unsustainable because gas is burnt to produce electricity thereby emitting greenhouse gases, which is unlike renewable energy technologies that are environmentally benign. In Nigeria, there have been frequent disruptions of gas supply to the thermal gas power stations due to frequent gas pipeline vandalization by militants of the Niger Delta region. These activists claim that they are angry about prolonged negligence of the acute poverty of the region which produces most of Nigeria's fossil fuels that earn a large share of the nation's wealth has been enjoyed by the elite who exclude the poor masses in the Niger Delta. Yet some of the gas stations involve the construction of rather long gas pipelines that must traverse large expanses of land thereby exposing them to the hazard of the angry protestors (The Punch 2007e). Contrary to propaganda which says that the gas power stations will rapidly produce power, most of them will take a good number of years before they start adding to the national grid – yet the nation needs much power to be generated and supplied rapidly. It is apposite to state that sustainable energies are not adversely affected by most of the foregoing problems plaguing the conventional power technologies. The aforementioned defects of the IPPs being developed easily indicate the need for the implementation of sustainable energy technologies in the country.

28.7 Recommendations

Lagos state needs to harness its renewable energy and energy efficiency potentials to generate and supply electricity and engender energy security in its urban region. Perhaps the first step is for the Lagos State Government, which is currently being hailed for the innovative efforts being made by its Governor Raji Fashola is to establish a separate ministry of energy and/or electricity with specific mandate of developing the power sector. For example, Lagos State Government ministry of energy can work with other littoral states and/or the Federal Government of Nigeria to establish offshore wind power farms available in its littoral territory in addition to others (biomass, concentrating solar power and so forth). To accomplish this, Lagos State can exploit its enormous sovereign power which it is entitled to within the Federal system of government in Nigeria to develop its own electrical energy generation and supply program that is separate from that of the centralized but ineffective national grid. Fortunately, Lagos State has gained experience from pioneering the development of independent power producing plants by a state government in the country. To rapidly introduce sustainable energy technologies, its managements must urgently work with existing programs such as the solar cities project to incorporate renewable energy strategies into its urban development programs while its redevelopment programs (model cities, or mini-cities programs) must be planned based on sustainable energy principles (Droege (2004)).

The Lagos megacity government needs to work with the Bureau for Public Enterprises to encourage renewable energy professionals and service providers to join the 205–414 entrepreneurs who have sent expressions of interests in the privatization of Nigeria's power sector. The Ministry of Power should expedite action aimed at completely privatizing Nigeria's power sector. This is justified because some people believe that the low or partial privatization of the power sector is responsible for the current poor response of entrepreneurs to the business opportunities available in the huge power sector. They argue further that full privatization of the power sector will generate the same level of rapid enthusiasm and growth that the global system of mobile telecommunications has brought to bear on the Nigerian economy since 2003 (The Punch 2007g and the *Calabar International Declaration* 2004).

Lagos needs education programs that are designed to increase the consciousness of its staff and residents of the opportunities and benefits of sustainable energy technologies and how they can be operationally implemented using project development strategies. Owing to the challenge of experiencing a rather indifferent private sector and governments in the face of worsening global warming and climate change, there is a need for ENGOs to create and strengthen local sustainable energy circuits as a means of pressurizing governments and the private sector to urgently address global warming.

The seemingly low consciousness of the seriousness of global warming and climate change in Nigeria warrants that ENGO organizes and implements events designed to raise the awareness of key players in the public and private sectors (civil society, faith-based organizations, professional lobbyists and so forth) of existing possibilities and a high degree of advancement in sustainable energy technologies (currently operational in several nations and cities but outside Lagos and Nigeria) to facilitate rapid implementation of the technologies to resolve the energy crisis in Lagos' and Nigeria's urban centres. It is recommended that there is a need to switch energy supply in Lagos from the moribund and failure-prone conventional energy technologies to the more sustainable energy supplies based on renewable energy and energy efficiency technologies.

28.7.1 Bio-fuels alternative for transportation

The megacity's huge vehicular population and notoriety for traffic jams makes a case for the production of bio-fuels outside the city for use in driving the city's huge transportation fleet. There are huge potentials for creating bio-fuels from several energy crops instead of from cassava as currently claimed by the Nigerian National Petroleum Corporation (NNPC). The Lagos State Government as well as energy institutions in the environs need to work with civil society and non-government organizations to implement sustainable energy. The Brazilian President recently stated that 'In the Africa-South American Summit in 2005, and in the two sessions of the Brazil-Africa Forum, we explored in depth the great potential of this alliance, which can be further strengthened and improved by biofuels'(The East African 2007). Several proposals are being made that deserve to be noted and capitalized upon by Lagos megacity management. For example, the leader of the German NGO's Forum on Environment & Development recently (in 2007) suggested that: 'What is necessary is not only more funds for renewables, but also the build-up of human capacity among energy decision-makers. Running decentralized energy systems requires different skills than the old-fashioned centralized systems that will never lift the two billion people who do not have access to modern energy services today out of energy poverty. This is where European development cooperation really has an important job to do. According to our partners the World Bank still enjoys a lot of authority as an advisory institution, yet its advice almost systematically underrates decentralized and renewable energy solutions'(German NGO's 2007).

Lagos megacity can draw credit facilities, through the issuance of bonds, from the international or global financial markets to develop its clean energy infrastructure. To achieve this, the megacity management needs to adopt a sound and transparent accounting framework to make it qualify for favourable rating by international financial institutions that assess the credit worthiness of cities. There is need for Lagos to lobby the authorities to review the unbundling of PHCN into only 18 companies: only one of which is currently to be responsible for energy transmission. It does appear that as a megacity, Lagos, and perhaps some of Nigeria's other urban regions, possesses enough peculiarities in terms of energy 'hunger' to make them deserve separate energy generation companies to work on power generation and transmission. Therefore, more than 18 companies are required to adequately deliver urban energy. The Lagos management should invite and work with local and international sustainable energy networks and institutions that are promoting urban energy transition and networks of communities in the quest to resolve the myriad of urban problems that are afflicting megacities in order to introduce innovations in sustainable energy supply.

References

Ayeni, M.O.A. (1978). Patterns, processes and problems of urban development (in Nigeria), in J.S. Oguntoyibo *et al. A Geography of Nigerian Development*. Heinemann; Mabogunje, A.L. (1968). *Urbanisation in Nigeria*. Arnold; Mabogunje, A.L. (1978). Towards an urban policy in Nigeria. *Nigerian Journal of Economic and Social Studies*, **16**(1), 85–97.

Bashorun, Y.O. and Olakulehin, J.O. (2007). The Lagos state fish farmers association. *LEISA*, **23**(1), 10–11, March.

Calabar International Declaration. (2004). Retrieved 7 January from www.onesky.ca/energetics

Daily Sun, 6 March 2007a, 47.

Daily Sun, 12 February 2007b.

Droege, P. (2004). Renewable Energy and the City. In: Cleveland, C. 2004. Encyclopedia of Energy. Amsterdam: Academic Press. (paper is now downloadable at: http://www.wcre.de/en/images/downloads/Renewable_Energy_and_City_Droege_june05.pdf)

Droege, P. (2004). *Op. cit.*

Earth Life Africa (ELA) (2007). *Sustainable Energy News* (Sustainable Energy and Climate Change Project (SECCP) under the ELA). Number 43, February.

EDRC/ENDA (2002)(a). Realising Africa's abundant energy for Africans: A picture of plenty energy in Africa.

EDRC/ENDA (2002)(b). Realising Africa's abundant energy for Africans: A picture of plenty energy in Africa. (Chapter on Oil and Gas).

Financial Standard, 6 March 2007, 1–2.

German NGO's Forum on Environment & Development/Juergen 2007.

http://allafrica.com/stories/200707310628.html. Africa: biofuels can allow all humanity to prosper, citing *The East African* (Nairobi), 31 July 2007, posted online 31 July 2007, retrieved 13 July 2007.

http://web.worldbank.org/WBSITE/EXTERNAL/COUNTRIES/AFRICAEXT/0,,contentMDK:21112588~menuPK:258649~pagePK:146736~piPK:226340~theSitePK:258644,00.html. First-hand view of Africa's power crisis. Retrieved 11 August 2007.

Ingwe, R. (2005). Final report of renewable energy research and development: Obudu Ranch Plateau case study (a report submitted to Onesky – the Canadian Institute for Sustainable Living). Retrieved 12 January 2007 from www.onesky.ca/energetics.

Jaeger, C., Hare, B. and Battaglini, A. (2004). What is dangerous climate change? A press statement at the 10th UN Climate Change conference, Buenos Aires (COP 10: 1-2, www.european-climate-forum.net).

Lampinen, P. (2001). How could information technology be a possibility for unemployed young people? in Antti Kasvio *et al.* (2001). *People, Cities and the New Information Economy* (materials from an International Conference, Helsinki, 14–15 December 2000, Palmenia Kustannus).

Matthes, F. Chr., Rosenkranz, G., Froggart, A. *et al.* (2006). *Nuclear Power: Myth and Reality (The Risks and Prospects of Nuclear Power)*. Saxonwold: Heinrich Boell Foundation.

Nigerian Tribune, 8 February 2007. Editorial Opinion: Census of controversies. UN World Urbanisation Prospects: the 1999 revision: key findings. Retrieved 27 February 2007 from http://www.un.org/esa/population/pubsarchive/urbanization/urbanization.pdf

Nigeria's nuclear power 'mix-up'. Retrieved 9 January 2007 from http://news.bbc.co.uk/2/hi/africa/3533225.stm; IAEA Media Line Staff 2007, IAEA Supports African Nuclear Efforts, 10 January; and recent personal communication with Citizens United for Renewable Energy and Sustainability Network, Germany.

Power failure: manufacturers opt for own independent power plants. Retrieved on 27 February 2007 from http://www.sunnewsonline.com/webpages/news/businessnews/2007/jan/22/business-22-01-2007-002.htm; uploaded Monday, 22 January 2007.

Quoted in the literature, e.g. Lohmann, L. (2006). Carbon trading (a critical conversation on climate change, privatization and power). *Development Dialogue* No. 48, Dag Hammaarksjoeld Foundation; Covey, S. (1989). *The 7 Habits of Very Effective People*. Simon and Schuster.

The East African, 30 July-5 August 2007, 17.

The Punch, 23 March 2007a, 9.

The Punch, 22 and 23 March 2007e.

The Punch, 22 March 2007f, 9.

The Punch, 22 March 2007, 22 and 23 March 2007g, 19.

The Punch, 26 March 2007h, 39.

The Punch, 9 April 2007b, 10.

The Punch, 2 April 2007c, 64.

The Punch, 2 April 2007d, 64.

The Punch Again. Egbin loses 600MW, 26 April 2005: 64; *Punch*, 14 March 2007, 56.

The Punch, 18 May 2004, 15.

Toffler, A. (1990). *The Third Wave*. William Morrow.

Togola, I. (2007). Sustainable energy future in Africa- a blueprint (draft of paper for the conference) "The European Union's Financing in the Energy Sector in Africa- which role for the European Investment Bank and Infrastructure Partnership Trust Fund", Berlin, October.

UN World Urbanisation Prospects: the 1999 revision: key findings. Retrieved 27 February 2007 from http://www.un.org/esa/population/pubsarchive/urbanization/urbanization.pdf (http://www.un.org/esa/population/)

Van den Berg, L. (2000). ICT as potential catalysts for sustainable urban development – experiences in Eindhoven, Helsinki, Manchester, Marseilles, and The Hague, in Antti Kasvio *et al.* (2001). *People, Cities and the New Information Economy* (materials from an International Conference, Helsinki, 14–15 December 2000, Palmenia Kustannus).

Webster, F. (2000). Reinventing place: Birmingham as an information city? in Antti Kasvio *et al.* (2001). *People, Cities and the New Information Economy* (materials from an International Conference, Helsinki, 14–15 December 2000, Palmenia Kustannus).

World Resources Institute, UNDP, UNEP, the World Bank (2005). World Resources 2005. World Resources Institute, UNDP, UNEP. The World Bank.

www.businessdayonline.com

www.onesky.ca/energetics

Table 28.1. Independent power producing plants (IPPs).

Generating station	Operator/ owner	Type/source of power	Planned/initial/ capacity: MW	Actual capacity power produced	Distance of gas pipeline required	Cost
Geregu (Kogi State)	NA	Gas turbine	414	137	500 m	
Omotoso (Ondo State)	NA	Gas turbine	355	NA	6.5 km	
Papalanto (Ogun State)		Gas turbine	355	NA	17 km	
Alaoji (Abia State)		Gas turbine	504		NA	
Gbarian (Bayelsa State)		Gas turbine	225		NA	
Ihovbor (Edo State)		Gas turbine	451		NA	
Sapele (Delta State)		Gas turbine	451		NA	
Egbema (Imo State)		Gas turbine	338		NA	
Calabar (Cross River State)		Gas turbine	561		NA	
Ibom (Akwa Ibom State)		Gas turbine	118		NA	
Omoku 2nd gas turbine (Rivers State: Ogba, Egbema/Andoni LGA)	Rivers State IPP	Gas turbine	230		NA	14B
2-phased A; Ijora Lagos State IPP (13/8/99)	NEPA Lagos State	Diesel	90 MW (3 barges of 30 MW each) i.e. 20% of Lagos + power		NA	$800
2nd phase Lagos IPP (20 year) Agbara (Lagos State)	NEPA	Gas turbine will use 290 km 24 in offshore natural gas from Nigeria Rivers Delta instead of flaring same	548		NA	Cost estimates: new thermal stations: US$359.70 m in Papalanto (Ogun State); US$360.68 m in Okitipupa (Ondo State); US$390 m in Apaokuta (Kogi State)
MAN I Southern Nigeria	NA		1000 MW	NA		NA
MAN II Northern Nigeria			1000 MW	NA		

Sources: The News, 24 January 2000, 18–22: www.sunnewsonline.com, 2007; *The Punch*, 22 March 2007, 22; 18 June 2007, 17; Allafrica.com, 28 July 2006, 3pp.; (Ingwe 2005).

Index